# Extracellular Vesicles

Qian Wang • Lei Zheng
Editors

# Extracellular Vesicles

## From Bench to Bedside

Science Press
Beijing

Springer

*Editors*
Qian Wang
Zhujiang Hospital
Southern Medical University
Guangzhou, China

Lei Zheng
Nanfang Hospital
Southern Medical University
Guangzhou, China

ISBN 978-981-99-8364-3 ISBN 978-981-99-8365-0 (eBook)
https://doi.org/10.1007/978-981-99-8365-0



This Springer imprint is published by the registered company Springer Nature Singapore Pte Ltd.
The registered company address is: 152 Beach Road, #21-01/04 Gateway East, Singapore 189721, Singapore

# Preface

In the early 1980s, researchers observed a fascinating phenomenon: cells releasing extracellular vesicles (EVs). Initially dismissed as mere "garbage bags" for disposing of metabolic waste, EVs have emerged as essential players in various biological activities. From antigen presentation to genetic material transfer and modification of cellular microenvironments, EVs serve as crucial messengers facilitating intercellular information transfer during biological processes. Studying EVs not only enhances our understanding of the fundamental principles of life but also profoundly influences the development of theories and technologies in fields like life sciences and medicine. The broad applications of EVs offer new avenues and hopes for disease diagnosis and treatment. Despite their tiny size, EVs harbor enormous potential. Over the past decade, the field of EV research has witnessed significant growth, yielding numerous findings and publications. However, the dispersed nature of the literature and the time-consuming process of gathering and organizing information present challenges in establishing a comprehensive and systematic knowledge system on EVs. Furthermore, the shortage of existing English monographs on EVs has created an urgent demand for such resources, enabling researchers to swiftly grasp the current status and methodologies of EV research.

As one of the world's largest communities, China has a substantial number of researchers actively studying extracellular vesicles in diverse contexts, leading to impressive research achievements. Led by the Chinese Society of Extracellular Vesicle (CSEV), we envisioned assembling a team to compile an English monograph focusing on extracellular vesicle research. The goal is to better connect the Chinese EV community with the global scientific community and to promote EV research.

This book comprises five parts with a total of 32 chapters. In the first part, Professor Wenrong Xu from Jiangsu University leads five chapters focusing on the biology and function of EVs. This part provides an overview of EVs, discussing their classification, biogenesis, mechanisms of action on target cells, and functions in physiological and pathological conditions. It serves as a foundation for subsequent sections that delve into EV research methods and clinical applications.

Professor Tong Wang from Jinan University leads the second part, which includes nine chapters dedicated to the technical methods of EV research. This comprehensive part covers methodologies used in EV research, including purification and characterization techniques, omics techniques, detection techniques, surface functionalization techniques, visualization, and tracking techniques.

Under the leadership of Professor Lei Zheng from Southern Medical University, the third part comprises eight chapters focusing on the use of EVs as disease biomarkers. This part introduces methods for screening EV-based biomarkers and discusses current EV biomarkers in major diseases.

Professor Yang Wang from the Sixth People's Hospital of Shanghai Jiao Tong University leads the fourth part, which includes seven chapters covering the utilization of EVs for disease therapy and prevention. This part explores the therapeutic potential and application value of EVs derived from different sources.

The last part, led by Professor Xiaomei Yan from Xiamen University, encompasses three chapters focusing on guidelines and resources for EV research. It offers an introduction to global guidelines, position papers, and online nucleic acid databases in the EV research field.

All the editors have extensive experience and have made significant contributions to research in this field, resulting in numerous outstanding findings. This book aims to provide a comprehensive and systematic overview of EV research, methodologies, clinical applications, and available resources. It caters to researchers and practitioners in this field, offering valuable insights and references.

As the publication of this book approaches, we express sincere gratitude to all members of the editorial board, editors, and original authors. Given the rapid advancement of EV research, the vast knowledge, and the limitations of the editors' expertise, there may be certain limitations to this book. We wholeheartedly welcome critiques and corrections from fellow experts and all readers who utilize this book. Your feedback will be instrumental in improving and refining the content.

Guangzhou, China Qian Wang
Guangzhou, China Lei Zheng

# Contents

# Editors and Contributors

## Associate Editors

**Tong Wang** The First Affiliated Hospital, Institute of Life and Health Engineering, Jinan University, Guangzhou, China

The First Hospital of Changsha, Changsha, China

**Yang Wang** Institute of Microsurgery on Extremities, Department of Orthopedic Surgery, Shanghai Sixth People's Hospital Affliated to Shanghai Jiao Tong University School of Medicine, Shanghai, China

**Wenrong Xu** Jiangsu Key Laboratory of Medical Science and Laboratory Medicine School of Medicine, Jiangsu University, Zhenjiang, China

**Xiaomei Yan** Department of Chemical Biology, College of Chemistry and Chemical Engineering, Xiamen University, Xiamen, Fujian, China

## Contributors

**Xiaochun Bai** State Key Laboratory of Organ Failure Research, Department of Cell Biology, School of Basic Medical Sciences, Southern Medical University, Guangzhou, China

**Tanxi Cai** Key Laboratory of Protein and Peptide Pharmaceuticals and Laboratory of Proteomics, Institute of Biophysics, Chinese Academy of Sciences, Beijing, China

University of Chinese Academy of Sciences, Beijing, China

**Zhijian Cai** School of Basic Medical Science, Zhejiang University, Hangzhou, Zhejiang, China

**Gang Chen** The State Key Laboratory Breeding Base of Basic Science of Stomatology (Hubei-MOST) and Key Laboratory of Oral Bio-medicine Ministry of Education, School and Hospital of Stomatology, Wuhan University, Wuhan, China

Department of Oral and Maxillofacial Surgery, School and Hospital of Stomatology, Wuhan University, Wuhan, China

Frontier Science Center for Immunology and Metabolism, Wuhan University, Wuhan, China

**Gao-Hong Chen** The State Key Laboratory Breeding Base of Basic Science of Stomatology (Hubei-MOST) and Key Laboratory of Oral Bio-medicine Ministry of Education, School and Hospital of Stomatology, Wuhan University, Wuhan, China

**Guofeng Cheng** Shanghai Tenth People's Hospital, Institute for Infectious Diseases and Vaccine Development, Tongji University School of Medicine, Shanghai, People's Republic of China

**Wei Cheng** The Center for Clinical Molecular Medical Detection, The First Affiliated Hospital of Chongqing Medical University, Chongqing, China

**Peng Chen** Department of Biomedical Engineering, The Key Laboratory for Biomedical Photonics of MOE at Wuhan National Laboratory for Optoelectronics-Hubei Bioinformatics & Molecular Imaging Key Laboratory, Systems Biology Theme, College of Life Science and Technology, Huazhong University of Science and Technology, Wuhan, China

**Xiulan Chen** Key Laboratory of Protein and Peptide Pharmaceuticals and Laboratory of Proteomics, Institute of Biophysics, Chinese Academy of Sciences, Beijing, China

University of Chinese Academy of Sciences, Beijing, China

**Yin-Xue Chen** The State Key Laboratory Breeding Base of Basic Science of Stomatology (Hubei-MOST) and Key Laboratory of Oral Bio-medicine Ministry of Education, School and Hospital of Stomatology, Wuhan University, Wuhan, China

**Yu Chen** Institute of Microsurgery on Extremities, Department of Orthopedic Surgery, Shanghai Sixth People's Hospital Affiliated to Shanghai Jiao Tong University School of Medicine, Shanghai, China

**Zhengsheng Chen** Institute of Microsurgery on Extremities, Department of Orthopedic Surgery, Shanghai Sixth People's Hospital Affiliated to Shanghai Jiao Tong University School of Medicine, Shanghai, China

**Zhuo-Kun Chen** The State Key Laboratory Breeding Base of Basic Science of Stomatology (Hubei-MOST) and Key Laboratory of Oral Bio-medicine Ministry of Education, School and Hospital of Stomatology, Wuhan University, Wuhan, China

**Zimei Chen** MOE Key Laboratory of Tumor Molecular Biology, and Key Laboratory of Functional Protein Research of Guangdong Higher Education Institutes, Institute of Life and Health Engineering, the First Affiliated Hospital, Jinan University, Guangzhou, Guangdong, China

**Binbin Cui** The Department of Mechanical and Aerospace Engineering, The Hong Kong University of Science and Technology, Hong Kong, China

**Zhifeng Deng** Department of Neurosurgery, Shanghai Sixth People's Hospital Affiliated to Shanghai Jiao Tong University School of Medicine, Shanghai, China

**Kaili Di** Department of Clinical Laboratory, The Affiliated Drum Tower Hospital of Nanjing University Medical School, Nanjing, China

**Shijia Ding** Key Laboratory of Clinical Laboratory Diagnostics (Ministry of Education), College of Laboratory Medicine, Chongqing Medical University, Chongqing, China

**Xu Duan** The State Key Laboratory Breeding Base of Basic Science of Stomatology (Hubei-MOST) and Key Laboratory of Oral Bio-medicine Ministry of Education, School and Hospital of Stomatology, Wuhan University, Wuhan, China

**Lutao Du** Department of Clinical Laboratory, The Second Hospital of Shandong University, Jinan, Shandong, China

**Boyue Fan** Department of Clinical Laboratory, The Affiliated Drum Tower Hospital of Nanjing University Medical School, Nanjing, China

**Weixiao Fan** Department of Clinical Microbiology, Greater Xi'an Area Medical Diagnostic Lab Center, Xi'an, Shaanxi, People's Republic of China

**Xingliang Fan** Otorhinolaryngology Hospital, The First Affiliated Hospital, Sun Yat-sen University, Guangzhou, China Institute of Biological and Food Engineering, Guangdong University of Education, Guangzhou, China

**Chuan Fang** Hebei Key Laboratory of Precise Diagnosis and Treatment of Glioma, Baoding, China

Department of Neurosurgery, Affiliated Hospital of Hebei University, Baoding, China

**Shu-Bin Fang** Otorhinolaryngology Hospital, The First Affiliated Hospital, Sun Yat-sen University, Guangzhou, China

**Houmei Feng** Department of Laboratory Medicine, Nanfang Hospital, Southern Medical University, Guangzhou, China

**Qing-Ling Fu** Otorhinolaryngology Hospital, The First Affiliated Hospital, Sun Yat-sen University, Guangzhou, China

China Key Laboratory for Stem Cells and Tissue Engineering, Ministry of Education, Sun Yat-sen University, Guangzhou, China

**Yang Gao** Transfusion Medicine Research Center, Institute of Blood Transfusion, Chinese Academy of Medical Sciences & Peking Union Medical College, Chengdu, Sichuan Province, People's Republic of China

**Mengyuan Geng** Department of Cell Biology, School of Medical Technology, Tianjin Medical University, Tianjin, China

**Yiqin Ge** Department of Laboratory Medicine, Shanghai General Hospital, Shanghai Jiao Tong University School, Shanghai, People's Republic of China

**Bikash R. Giri** Shanghai Tenth People's Hospital, Institute for Infectious Diseases and Vaccine Development, Tongji University School of Medicine, Shanghai, People's Republic of China

**Gang Han** Department of Cell Biology, School of Medical Technology, Tianjin Medical University, Tianjin, China

**Jin Hao** Sysmex Shanghai, Ltd., Shanghai, China

**Xiaoke Hao** Division of Life Sciences and Medicine, Northwest University, Xi'an, Shaanxi, People's Republic of China

Greater Xi'an Area Medical Diagnostic Lab Center, Xi'an, Shaanxi, People's Republic of China

**Wijnand Helfrich** Department of Surgery, Laboratory for Translational Surgical Oncology, University of Groningen, University Medical Center Groningen, Groningen, The Netherlands

**Jia Hu** Fudan University Shanghai Cancer Center and Institutes of Biomedical Sciences, Fudan University, Shanghai, China

**Xiumei Hu** Department of Laboratory Medicine, Nanfang Hospital, Southern Medical University, Guangzhou, People's Republic of China

**Yunyun Hu** Department of Chemical Biology, College of Chemistry and Chemical Engineering, Xiamen University, Xiamen, Fujian, China

**Jue Huang** The State Key Laboratory Breeding Base of Basic Science of Stomatology (Hubei-MOST) and Key Laboratory of Oral Bio-medicine Ministry of Education, School and Hospital of Stomatology, Wuhan University, Wuhan, China

**Mei Huang** Department of Laboratory Medicine, Nanfang Hospital, Southern Medical University, Guangzhou, People's Republic of China

**Rongrong Huang** Department of Clinical Laboratory, The Affiliated Drum Tower Hospital of Nanjing University Medical School, Nanjing, China

**Shenglin Huang** Fudan University Shanghai Cancer Center and Institutes of Biomedical Sciences, Fudan University, Shanghai, People's Republic of China

**Yiyao Huang** Department of Laboratory Medicine, Nanfang Hospital, Southern Medical University, Guangzhou, China

**Yuji Huang** Department of Laboratory Medicine, Shanghai General Hospital, Shanghai Jiao Tong University School, Shanghai, People's Republic of China

**Dehua Huo** Beckman Coulter Inc. Life Sciences Division, Indianapolis, IN, USA

**Jiajia Jiang** Aoyang Institute of Cancer, Affiliated Aoyang Hospital of Jiangsu University, Zhangjiagang, Jiangsu, China

**Yuanjun Jiang** State Key Laboratory of Organ Failure Research, Department of Cell Biology, School of Basic Medical Sciences, Southern Medical University, Guangzhou, China

**Cheng Ji** Zhenjiang Key Laboratory of High Technology Research on Exosomes Foundation and Transformation Application, Jiangsu Key Laboratory of Medical Science and Laboratory Medicine, Jiangsu University, Zhenjiang, Jiangsu, China

**Chunsheng Kang** Laboratory of Neuro-oncology, Tianjin Neurological Institute, Tianjin Medical University General Hospital, Tianjin, China

**Xiurong Ke** Department of Surgery, Laboratory for Translational Surgical Oncology, University of Groningen, University Medical Center Groningen, Groningen, The Netherlands

Shantou University Medical College, Shantou, Guangdong, China

**Mingqiang Lai** State Key Laboratory of Organ Failure Research, Department of Cell Biology, School of Basic Medical Sciences, Southern Medical University, Guangzhou, China

**Bo Li** Department of Laboratory Medicine, Nanfang Hospital, Southern Medical University, Guangzhou, China

**Li Li** Department of Laboratory Medicine, The Second Affiliated Hospital of Zhejiang University School of Medicine, Hangzhou, Zhejiang, People's Republic of China

**Ning Li** Department of Clinical Laboratory, The Second Hospital of Shandong University, Jinan, Shandong, China

**Qing Li** Institute of Microsurgery on Extremities, Department of Orthopedic Surgery, Shanghai Sixth People's Hospital Affiliated to Shanghai Jiao Tong University School of Medicine, Shanghai, China

**Shun Li** School of Life Science and Engineering, Foshan University, Foshan, People's Republic of China

**Xiao-lin Li** Department of Nephrology, South China Hospital, Health Science Center, Shenzhen University, Shenzhen, People's Republic of China

**Xinyu Li** Key Laboratory of Clinical Laboratory Diagnostics (Ministry of Education), College of Laboratory Medicine, Chongqing Medical University, Chongqing, China

**Yanning Li** Department of Laboratory Medicine, Shanghai General Hospital, Shanghai Jiao Tong University School, Shanghai, People's Republic of China

**Zhiyang Li** Department of Clinical Laboratory, The Affiliated Drum Tower Hospital of Nanjing University Medical School, Nanjing, China

**Qianxing Liang** Guangdong Engineering Research Center of Chinese Herbal Vesicles, The Third Affiliated Hospital of Guangzhou University of Chinese Medicine, Guangzhou, Guangdong, China

**Huanjin Liao** Department of Laboratory Medicine, Shanghai General Hospital, Shanghai Jiao Tong University School, Shanghai, People's Republic of China

**Huixian Lin** Department of Laboratory Medicine, Nanfang Hospital, Southern Medical University, Guangzhou, China

**Yusheng Lin** Institute of Precision Cancer Medicine and Pathology, Jinan University Medical College, Guangzhou, Guangdong, People's Republic of China

Department of Hematology, University Medical Center Groningen, University of Groningen, Groningen, The Netherlands

**Bi-Feng Liu** Department of Biomedical Engineering, The Key Laboratory for Biomedical Photonics of MOE at Wuhan National Laboratory for Optoelectronics-Hubei Bioinformatics & Molecular Imaging Key Laboratory, Systems Biology Theme, College of Life Science and Technology, Huazhong University of Science and Technology, Wuhan, China

**Caoyi Liu** Transfusion Medicine Research Center, Institute of Blood Transfusion, Chinese Academy of Medical Sciences & Peking Union Medical College, Chengdu, Sichuan Province, People's Republic of China

**Chao Liu** The Department of Mechanical and Aerospace Engineering, The Hong Kong University of Science and Technology, Hong Kong, China

**Chunchen Liu** Nanfang Hospital, Southern Medical University, Guangzhou, China

**Shouping Liu** Department of Laboratory Medicine, Nanfang Hospital, Southern Medical University, Guangzhou, China

**Tingjiao Liu** Department of Oral Pathology, Shanghai Stomatological Hospital and School of Stomatology, Fudan University, Shanghai, China

Key Laboratory of Craniomaxillofacial Development and Diseases, Fudan University, Shanghai, China

**Tong Liu** Department of Clinical Laboratory, The Second Hospital of Shandong University, Jinan, Shandong, China

**Xu Liu** Department of Hematology, Nanjing Drum Tower Hospital, Medical School of Nanjing University, Nanjing, Jiangsu Province, People's Republic of China

**Yu Liu** Transfusion Medicine Research Center, Institute of Blood Transfusion, Chinese Academy of Medical Sciences & Peking Union Medical College, Chengdu, Sichuan Province, People's Republic of China

**Tingting Luo** Department of Laboratory Medicine, Nanfang Hospital, Southern Medical University, Guangzhou, China

**Xianghang Luo** Xiangya Hospital of Central South University, Changsha, Hunan, China

**Lin-Li Lv** Institute of Nephrology, Zhongda Hospital, Southeast University School of Medicine, Nanjing, Jiangsu Province, China

**Wei Mu** School of Public Health, Shanghai JiaoTong University School of Medicine, Shanghai, China

**Xinyu Niu** Department of Neurosurgery, Shanghai Sixth People's Hospital Affiliated to Shanghai Jiao Tong University School of Medicine, Shanghai, China

**Zihao Ou** Department of Laboratory Medicine, Nanfang Hospital, Southern Medical University, Guangzhou, People's Republic of China

**Lu Peng** Beijing Institute of Heart, Lung and Blood Vessel Disease, Beijing Anzhen Hospital, Capital Medical University, Beijing, China

**Hui Qian** Zhenjiang Key Laboratory of High Technology Research on Exosomes Foundation and Transformation Application, Jiangsu Key Laboratory of Medical Science and Laboratory Medicine, Jiangsu University, Zhenjiang, Jiangsu, China

**An-Ran Shen** Institute of Nephrology, Zhongda Hospital, Southeast University School of Medicine, Nanjing, Jiangsu Province, China

**Hui Shi** Zhenjiang Key Laboratory of High Technology Research on Exosomes Foundation and Transformation Application, Jiangsu Key Laboratory of Medical Science and Laboratory Medicine, Jiangsu University, Zhenjiang, Jiangsu, China

**Xiubo Shi** Cancer Research Institute, School of Basic Medical Sciences, Southern Medical University, Guangzhou, People's Republic of China

**Bo Situ** Department of Laboratory Medicine, Nanfang Hospital, Southern Medical University, Guangzhou, People's Republic of China

**Fengtian Sun** Zhenjiang Key Laboratory of High Technology Research on Exosomes Foundation and Transformation Application, Jiangsu Key Laboratory of Medical Science and Laboratory Medicine, Jiangsu University, Zhenjiang, Jiangsu, China

**Gaoge Sun** School of Pharmaceutical Sciences, Key Laboratory of Bioorganic Phosphorous Chemistry and Chemical Biology (Ministry of Education), Department of Chemistry, Tsinghua University, Beijing, China

**Jiashu Sun** National Center for Nanoscience and Technology, University of Chinese Academy of Sciences, Beijing, China

**Yanli Tan** Hebei University, Baoding, China

**Tao-Tao Tang** Institute of Nephrology, Zhongda Hospital, Southeast University School of Medicine, Nanjing, Jiangsu Province, China

**Zhihua Tao** Department of Laboratory Medicine, The Second Affiliated Hospital of Zhejiang University School of Medicine, Hangzhou, Zhejiang, People's Republic of China

**Ye Tian** Department of Chemical Biology, College of Chemistry and Chemical Engineering, Xiamen University, Xiamen, Fujian, China

**Chuanxin Wang** Department of Clinical Laboratory, The Second Hospital of Shandong University, Jinan, Shandong, China

**Haobin Wang** The Third People's Hospital of Chengdu, Chengdu, Sichuan, China

**Jie Wang** Department of Biomedical Engineering, The Key Laboratory for Biomedical Photonics of MOE at Wuhan National Laboratory for Optoelectronics-Hubei Bioinformatics & Molecular Imaging Key Laboratory, Systems Biology Theme, College of Life Science and Technology, Huazhong University of Science and Technology, Wuhan, China

**Qian Wang** Laboratory Medicine Center, Zhujiang Hospital, Southern Medical University, Guangzhou, Guangdong, China

**Shuhong Wang** Institute of Precision Cancer Medicine and Pathology, Jinan University Medical College, Guangzhou, Guangdong, People's Republic of China

**Tong Wang** The First Affliated Hospital, Institute of Life and Health Engineering, Jinan University, Guangzhou, China

The First Hospital of Changsha, Changsha, China

**Xuchu Wang** Department of Laboratory Medicine, The Second Affiliated Hospital of Zhejiang University School of Medicine, Hangzhou, Zhejiang, People's Republic of China

**Yang Wang** Institute of Microsurgery on Extremities, Department of Orthopedic Surgery, Shanghai Sixth People's Hospital Affiliated to Shanghai Jiao Tong University School of Medicine, Shanghai, China

**Ying Wang** Cancer Research Institute, School of Basic Medical Sciences, Southern Medical University, Guangzhou, People's Republic of China

**Zhe Wang** The First Affiliated Hospital of Guangdong Pharmaceutical University, Guangzhou, Guangdong, China

**Bodeng Wu** Department of Laboratory Medicine, Nanfang Hospital, Southern Medical University, Guangzhou, China

**Peipei Wu** Jiangsu Key Laboratory of Medical Science and Laboratory Medicine School of Medicine, Jiangsu University, Zhenjiang, China

**Zhiwei Wu** Center for Public Health Research, Medical School, Nanjing University, Nanjing, People's Republic of China

State Key Lab of Analytical Chemistry for Life Science, Nanjing University, Nanjing, People's Republic of China

**Hou-Fu Xia** The State Key Laboratory Breeding Base of Basic Science of Stomatology (Hubei-MOST) and Key Laboratory of Oral Bio-medicine Ministry of Education, School and Hospital of Stomatology, Wuhan University, Wuhan, China

**Bo-Lin Xiao** The State Key Laboratory Breeding Base of Basic Science of Stomatology (Hubei-MOST) and Key Laboratory of Oral Bio-medicine Ministry of Education, School and Hospital of Stomatology, Wuhan University, Wuhan, China

**Hua Xiao** School of Life Sciences and Biotechnology, Shanghai Jiao Tong University, Shanghai, China

**Han Xie** Department of Biomedical Engineering, The Key Laboratory for Biomedical Photonics of MOE at Wuhan National Laboratory for Optoelectronics-Hubei Bioinformatics & Molecular Imaging Key Laboratory, Systems Biology Theme, College of Life Science and Technology, Huazhong University of Science and Technology, Wuhan, China

**Manlin Xie** Guangdong Engineering Research Center of Chinese Herbal Vesicles, The Third Affiliated Hospital of Guangzhou University of Chinese Medicine, Guangzhou, Guangdong, China

**Qi-Hui Xie** The State Key Laboratory Breeding Base of Basic Science of Stomatology (Hubei-MOST) and Key Laboratory of Oral Bio-medicine Ministry of Education, School and Hospital of Stomatology, Wuhan University, Wuhan, China

**Yujiao Xie** Department of Clinical Laboratory, The Second Hospital of Shandong University, Jinan, Shandong, China

**Lei Xin** Hebei University, Baoding, China

**Qi Xiu** Department of Laboratory Medicine, Nanfang Hospital, Southern Medical University, Guangzhou, China

**Jianglong Xu** Hebei Key Laboratory of Precise Diagnosis and Treatment of Glioma, Baoding, China

Department of Neurosurgery, Affiliated Hospital of Hebei University, Baoding, China

**Peipei Xu** Department of Hematology, Nanjing Drum Tower Hospital, Medical School of Nanjing University, Nanjing, Jiangsu Province, People's Republic of China

**Wenrong Xu** Jiangsu Key Laboratory of Medical Science and Laboratory Medicine School of Medicine, Jiangsu University, Zhenjiang, China

**Xiaomei Yan** Department of Chemical Biology, College of Chemistry and Chemical Engineering, Xiamen University, Xiamen, Fujian, China

**Fuquan Yang** Key Laboratory of Protein and Peptide Pharmaceuticals and Laboratory of Proteomics, Institute of Biophysics, Chinese Academy of Sciences, Beijing, China

University of Chinese Academy of Sciences, Beijing, China

**Xiangsheng Yang** Guangzhou Biotron Technology Co., Ltd., Guangzhou, Guangdong, China

**Shuhuai Yao** The Department of Mechanical and Aerospace Engineering, The Hong Kong University of Science and Technology, Hong Kong, China

**Teng Ye** Institute of Microsurgery on Extremities, Department of Orthopedic Surgery, Shanghai Sixth People's Hospital Affiliated to Shanghai Jiao Tong University School of Medicine, Shanghai, China

**HaiFang Yin** Department of Cell Biology, School of Medical Technology, Tianjin Medical University, Tianjin, China

**Hang Yin** School of Pharmaceutical Sciences, Key Laboratory of Bioorganic Phosphorous Chemistry and Chemical Biology (Ministry of Education), Department of Chemistry, Tsinghua University, Beijing, China

Beijing Advanced Innovation Center for Structural Biology, Tsinghua University, Beijing, China

Tsinghua-Peking Joint Center for Life Sciences, Tsinghua University, Beijing, China

**Siqi Yin** Zhenjiang Key Laboratory of High Technology Research on Exosomes Foundation and Transformation Application, Jiangsu Key Laboratory of Medical Science and Laboratory Medicine, Jiangsu University, Zhenjiang, Jiangsu, China

**Lijuan Yu** Department of Clinical Laboratory Medicine, Xijing Hospital, Fourth Military Medical University (Air Force Medical University), Xi'an, Shaanxi, People's Republic of China

**Zi-Li Yu** The State Key Laboratory Breeding Base of Basic Science of Stomatology (Hubei-MOST) and Key Laboratory of Oral Bio-medicine Ministry of Education, School and Hospital of Stomatology, Wuhan University, Wuhan, China

**Meng Yuan** Center for Public Health Research, Medical School, Nanjing University, Nanjing, People's Republic of China

**Shijing Yue** School of Medicine, Nankai University, Tianjin, China

**Gong Zhang** MOE Key Laboratory of Tumor Molecular Biology, and Key Laboratory of Functional Protein Research of Guangdong Higher Education Institutes, Institute of Life and Health Engineering, the First Affiliated Hospital, Jinan University, Guangzhou, Guangdong, China

**Guihua Zhang** Institute of Molecular Medicine, School of Medicine, Shanghai Jiao Tong University, Shanghai, China

**Hao Zhang** Institute of Precision Cancer Medicine and Pathology, Jinan University Medical College, Guangzhou, Guangdong, People's Republic of China

**He-Jing Zhang** The State Key Laboratory Breeding Base of Basic Science of Stomatology (Hubei-MOST) and Key Laboratory of Oral Bio-medicine Ministry of Education, School and Hospital of Stomatology, Wuhan University, Wuhan, China

**Huina Zhang** Beijing Institute of Heart, Lung and Blood Vessel Disease, Beijing Anzhen Hospital, Capital Medical University, Beijing, China

**Kaixiang Zhang** School of Pharmaceutical Sciences, Tsinghua University, Beijing, People's Republic of China

**Peng Zhang** Institute of Molecular Medicine, School of Medicine, Shanghai Jiao Tong University, Shanghai, China

**Rui Zhang** Center for Public Health Research, Medical School, Nanjing University, Nanjing, People's Republic of China

**Shujun Zhang** Department of Clinical Laboratory, The Second Hospital of Shandong University, Jinan, Shandong, China

**Ye Zhang** Department of Laboratory Medicine, Nanfang Hospital, Southern Medical University, Guangzhou, China

**Ying Zhang** School of Pharmaceutical Sciences, Tsinghua University, Beijing, People's Republic of China

**Kewei Zhao** Guangdong Engineering Research Center of Chinese Herbal Vesicles, The Third Affiliated Hospital of Guangzhou University of Chinese Medicine, Guangzhou, Guangdong, China

**Kun Zhao** The First Affiliated Hospital of Zhengzhou University, Zhengzhou, Henan, China

**Qing Zhao** Guangdong Engineering Research Center of Chinese Herbal Vesicles, The Third Affiliated Hospital of Guangzhou University of Chinese Medicine, Guangzhou, Guangdong, China

**Lei Zheng** Department of Laboratory Medicine, Nanfang Hospital, Southern Medical University, Guangzhou, People's Republic of China

Guangdong Engineering and Technology Research Center for Rapid Diagnostic Biosensors, Nanfang Hospital, Southern Medical University, Guangzhou, China

**Nali Zhu** Key Laboratory of Protein and Peptide Pharmaceuticals and Laboratory of Proteomics, Institute of Biophysics, Chinese Academy of Sciences, Beijing, China

Hangzhou Institute for Advanced Study, University of Chinese Academy of Sciences, Hangzhou, China

**He-qun Zou** Department of Nephrology, South China Hospital, Health Science Center, Shenzhen University, Shenzhen, People's Republic of China

**Wenchong Zou** State Key Laboratory of Organ Failure Research, Department of Cell Biology, School of Basic Medical Sciences, Southern Medical University, Guangzhou, China

**Yanfang Zou** Guangdong Engineering Research Center of Chinese Herbal Vesicles, The Third Affiliated Hospital of Guangzhou University of Chinese Medicine, Guangzhou, Guangdong, China

# Part I

# EVs Biology and Function

# Classification and Nomenclature of Extracellular Vesicles

# 1

Xiulan Chen and Fuquan Yang

## 1.1 The Discovery of Exosomes and Other Extracellular Vesicles (EVs)

The first description of exosomes dates back more than 40 years. In 1981, Trams et al. reported that cultures from various normal and neoplastic cell lines secreted vesicles with an average diameter of 500–1000 nm and often contained a second population of vesicles measuring approximately 40 nm in diameter. These secreted membrane vesicles carry 5′-nucleotidase activity and may mediate physiological processes. These secreted vesicles were referred to as "microvesicles" [1].

In 1983, Harding et al. [2] and Pan et al. [3] independently found that maturing blood reticulocytes can secrete small vesicles (~50 nm) into the extracellular space. The transferrin receptor [2, 3] and clathrin-uncoated ATPase/heat shock protein [4] have been identified in small vesicles. A few year later, the term "exosome" was adopted by Rose Johnstone as a label for small vesicles (30–100 nm) that are released during the differentiation of red blood cells as a consequence of the fusion of multivesicular endosomes or multivesicular bodies (MVBs) with the plasma membrane [5]. However, at that time, exosomes were only viewed as "garbage bags" and a mechanism to dispose unwanted components [6].

Over a decade after the discovery of exosomes in reticulocytes, B lymphocytes [7] and dendritic cells [8] were found to release vesicles of endosomal origin that participated in immune responses. These studies sparked a greater interest in

X. Chen · F. Yang (✉)
Key Laboratory of Protein and Peptide Pharmaceuticals and Laboratory of Proteomics, Institute of Biophysics, Chinese Academy of Sciences, Beijing, China

University of Chinese Academy of Sciences, Beijing, China
e-mail: fqyang@ibp.ac.cn



Q. Wang, L. Zheng (eds.), *Extracellular Vesicles*,
https://doi.org/10.1007/978-981-99-8365-0_1

exosomes because they suggested that exosomes may be important mediators of intracellular communication.

Subsequently, cells from different organisms, including all eukaryotes (i.e., amoebae, *Caenorhabditis elegans*, parasites, and mammals) and prokaryotes, have been shown to release vesicles into the extracellular space [6]. In fact, cells from all three domains of life (Archaea, Bacteria, and Eukarya) produce EVs [9]. EVs have also been isolated from various body fluids, including blood, urine, saliva, breast milk, amniotic fluid, ascites, cerebrospinal fluid, and semen [6].

Importantly, EVs play key roles not only in the maintenance of normal physiological processes, such as cell death and differentiation, blood coagulation, and tissue repair, but also in the pathogenesis of several diseases [10]. The therapeutic potential of EVs has also been demonstrated [11].

## 1.2 Classification and Nomenclature of EVs in Recent Decades

The classification of EVs can be based on their size, cellular origin, biological function, and/or biogenesis pathways. Over the past few decades, researchers have established names for dozens of different types of EVs. Based on cell and tissue origin and/or specific functions, EVs are classified as ectosomes (vesicles secreted by neutrophils or monocytes), microparticles (vesicles from platelets in blood or endothelial cells), tolerosomes (vesicles purified from serum of antigen-fed mice), prostasomes (prostrate-derived vesicles), cardiosomes (vesicles secreted by cardiomyocytes), vexosomes (vesicles linked with adeno-associated virus vectors), dexosomes (dendritic cell-derived exosomes), matrix vesicles (vesicles in bone, cartilage, and atherosclerotic plaques), oncosomes (vesicles derived from tumor cells), epididimosomes (vesicles from epididymal fluid), argosomes (basolateral membrane-derived vesicles), and synaptic vesicles (vesicles from neurons) [10, 12–14]. Although such names can be useful within a specialized field, they do not provide clues for classification with regard to the type of vesicles involved. More generic terms, such as "exosome" and "microvesicle," show broader utility.

Based on biogenesis, EVs have been classified into three main subgroups: exosomes, microvesicles (also called shedding vesicles or shedding microvesicles), and apoptotic bodies (also called apoptotic vesicles or apoptotic blebs) [10, 12]. Exosomes are vesicles measuring 40–120 nm in diameter that form within the endosomal pathway and are released upon fusion of MVBs with the plasma membrane [10]. By contrast, microvesicles are vesicles measuring 50–1000 nm in diameter that are derived from outward budding of plasma membrane [10], and apoptotic bodies are vesicles measuring 500–2000 nmin diameter that are released as blebs of cells undergoing apoptosis [10] (Table 1.1).

Specific protein markers have been identified in various EVs. For example, exosomes contain specific protein markers of the endosomal pathway, such as tetraspanins (CD9, CD63, and CD81), TSG101, and ALG-2-interacting protein X (ALIX), whereas microvesicles contain markers of plasma membrane, such as integrins,

**Table 1.1** Classification of EVs based on biogenesis

| | Characteristics | | |
|---|---|---|---|
| Vesicle types | Size (nm) | Origin | Main protein markers |
| Exosomes | 40–120 | Endosomes | Tetraspanins (CD63, CD9, CD81), TSG101, ALIX |
| Microvesicles (shedding vesicles/shedding microvesicles) | 50–1000 | Plasma membrane; outward budding of plasma membrane | Integrins, selectins, CD40 ligand |
| Apoptotic bodies (apoptotic vesicles/apoptotic blebs) | 500–2000 | Cell surface; outward blebbing of apoptotic cell membrane | Histones |

selectins, and CD40 ligand [10, 15, 16]. Similarly, apoptotic bodies selectively contain histones (Table 1.1). However, EV markers are not exclusively specific, and the same markers can also be present in other types of vesicles. Since there are no universal molecular markers to distinguish different subtypes of EVs, such as exosomes and microvesicles, the nomenclature and classification of EVs are very complex, challenging, and controversial and are still major points of debate within the research community.

## 1.3 Suggestions and Consensus on the Classification and Nomenclature of EVs from the International Society for Extracellular Vesicles (ISEV)

Determining the appropriate classification and nomenclature of EVs is challenging because EVs are heterogeneous and exhibit different sizes, densities, and intracellular sources; furthermore, there are no universal molecular markers or methods that can discriminate different subtypes of EVs. This rapid expansion of interest in EVs has also highlighted some problems related to the classification and nomenclature of EVs.

In 2012, at the first meeting of the ISEV in Gothenburg, Sweden, there was a session on vesicle nomenclature. Although a broad agreement on vesicle nomenclature was not reached in this session, researchers participating the session reached a consensus on two points: (1) the ISEV should strive to protect the scientific independence of its members on this issue and (2) Stephen J. Gould and Graca Raposo should articulate some of the relevant points of concern in *Journal of Extracellular Vesicles* [17].

In the face of the conflicting definitions of EVs, as discussed above, Gould and Raposo offered four suggestions for authors, reviewers, and editors as follows: (1) the authors should state the use of terms clearly, choose terms based on precedent and logical arguments, and apply them consistently throughout the study; (2) authors should clearly state their methods for obtaining and storing biological fluids, isolating vesicles, and selecting the specific terms used for the EVs; (3) reviewers and editors should respect authors' scientific freedom in their choice of vesicle nomenclature; and (4) authors should be encouraged to use the term "extracellular vesicle" (EV) as a generic term for all secreted vesicles [17].

Later, the ISEV published two detailed position papers and listed recommendations on EV isolation [18] and EV RNA analysis [19]. In 2014, board members of the ISEV published the "minimal experimental requirements for definition of extracellular vesicles and their functions" [20] and provided minimal information for studies on EVs (MISEV 2014), which covered EV separation/isolation, characterization, and functional studies.

In 2018, the ISEV updated the MISEV recommendations for EV studies, which presented recommendations in six areas: EV nomenclature, sample collection and preprocessing, EV separation and concentration, EV characterization, functional studies, and reporting requirements/exceptions. Regarding nomenclature, "extracellular vesicle" (EV) is the preferred generic term for "particles naturally released from the cell that are delimited by a lipid bilayer and cannot replicate, i.e., do not contain a functional nucleus." However, a consensus on specific markers of EV subtypes, such as endosome-origin "exosomes" and plasma membrane-derived "ectosomes" (microparticles/microvesicles), has not emerged. Accordingly, the ISEV suggested that unless authors can establish specific markers of subcellular origin that are reliable within their experimental system(s), the following operational terms for EV subtypes should be considered: (a) on the basis of physical characteristics of EVs, such as size, e.g., "small EVs" (<100 nm or <200 nm) and "medium/large EVs" (>200 nm), or density (low, middle, high, with each range defined); (b) on the basis of biochemical composition (CD63+/CD81+EVs, Annexin A5-stained EVs, etc.); and (c) on the basis of descriptions of conditions or cell of origin (podocyte EVs, hypoxic EVs, large oncosomes, apoptotic bodies). Other terms, such as extracellular particle, may be more appropriate if confirmation of EV identity cannot be achieved, according to the MISEV2018 publication [21].

Although "extracellular vesicle" (EV) is the expert consensus for the general term used to refer to small EVs, exosomes, medium/large EVs, and microvesicles, differences in properties, such as size, density, and protein composition, are considered as insufficient factors for clearly distinguishing among different EV subtypes. However, considering their different origins, "exosomes" seem to be different from microvesicles, and a major challenge for future research will be establishing methods that will allow the discrimination between MVB/exosomes and microvesicles [22]. With the development of new technologies for the analysis of EVs, this problem can be solved.

**Acknowledgments** This work was supported by grants from the National Key R&D Program of China (Grant Nos. 2018YFA0507103and 2018YFA0507801). We would like to thank Editage (www.editage.cn) for English language editing.

## References


1. Trams EG, Lauter CJ, Salem N Jr, Heine U. Exfoliation of membrane ecto-enzymes in the form of micro-vesicles. Biochim Biophys Acta. 1981;645(1):63–70.
2. Harding C, Heuser J, Stahl P. Receptor-mediated endocytosis of transferrin and recycling of the transferrin receptor in rat reticulocytes. J Cell Biol. 1983;97(2):329–39.
3. Pan BT, Johnstone RM. Fate of the transferrin receptor during maturation of sheep reticulocytes in vitro: selective externalization of the receptor. Cell. 1983;33(3):967–78.
4. Davis JQ, Dansereau D, Johnstone RM, Bennett V. Selective externalization of an ATP-binding protein structurally related to the clathrin-uncoating ATPase/heat shock protein in vesicles containing terminal transferrin receptors during reticulocyte maturation. J Biol Chem. 1986;261(33):15368–71.
5. Johnstone RM, Adam M, Hammond JR, Orr L, Turbide C. Vesicle formation during reticulocyte maturation. Association of plasma membrane activities with released vesicles (exosomes). J Biol Chem. 1987;262(19):9412–20.
6. Colombo M, Raposo G, Théry C. Biogenesis, secretion, and intercellular interactions of exosomes and other extracellular vesicles. Annu Rev Cell Dev Biol. 2014;30:255–89.
7. Raposo G, Nijman HW, Stoorvogel W, et al. B lymphocytes secrete antigen-presenting vesicles. J Exp Med. 1996;183(3):1161–72.
8. Zitvogel L, Regnault A, Lozier A, et al. Eradication of established murine tumors using a novel cell-free vaccine: dendritic cell-derived exosomes. Nat Med. 1998;4(5):594–600.
9. Gill S, Catchpole R, Forterre P. Extracellular membrane vesicles in the three domains of life and beyond. FEMS Microbiol Rev. 2019;43(3):273–303.
10. Andaloussi SEL, Mäger I, Breakefield XO, Wood MJ. Extracellular vesicles: biology and emerging therapeutic opportunities. Nat Rev Drug Discov. 2013;12(5):347–57.
11. Braicu C, Tomuleasa C, Monroig P, et al. Exosomes as divine messengers: are they the Hermes of modern molecular oncology? Cell Death Differ. 2015;22(1):34–45.
12. van der Pol E, Boing AN, Harrison P, Sturk A, Nieuwland R. Classification, functions, and clinical relevance of extracellular vesicles. Pharmacol Rev. 2012;64(3):676–705.
13. Akers JC, Gonda D, Kim R, Carter BS, Chen CC. Biogenesis of extracellular vesicles (EV): exosomes, microvesicles, retrovirus-like vesicles, and apoptotic bodies. J Neuro-Oncol. 2013;113(1):1–11.
14. Pocsfalvi G, Stanly C, Vilasi A, et al. Mass spectrometry of extracellular vesicles. Mass Spectrom Rev. 2016;35(1):3–21.
15. Shao H, Im H, Castro CM, et al. New technologies for analysis of extracellular vesicles. Chem Rev. 2018;118(4):1917–50.
16. Théry C, Ostrowski M, Segura E. Membrane vesicles as conveyors of immune responses. Nat Rev Immunol. 2009;9(8):581–93.
17. Gould SJ, Raposo G. As we wait: coping with an imperfect nomenclature for extracellular vesicles. J Extracell Vesicles. 2013;2.
18. Witwer KW, Buzás EI, Bemis LT, et al. Standardization of sample collection, isolation and analysis methods in extracellular vesicle research. J Extracell Vesicles. 2013;2.
19. Hill AF, Pegtel DM, Lambertz U, et al. ISEV position paper: extracellular vesicle RNA analysis and bioinformatics. J Extracell Vesicles. 2013;2.
20. Lötvall J, Hill AF, Hochberg F, et al. Minimal experimental requirements for definition of extracellular vesicles and their functions: a position statement from the International Society for Extracellular Vesicles. J Extracell Vesicles. 2014;3:26913.
21. Théry C, Witwer KW, Aikawa E, et al. Minimal information for studies of extracellular vesicles 2018 (MISEV2018): a position statement of the International Society for Extracellular Vesicles and update of the MISEV2014 guidelines. J Extracell Vesicles. 2018;7(1):1535750.
22. Bazzan E, Tinè M, Casara A, et al. Critical review of the evolution of extracellular vesicles' knowledge: from 1946 to today. Int J Mol Sci. 2021;22(12):6417.

# Biogenesis and Functions of Extracellular Vesicles

2

Wenchong Zou, Kaixiang Zhang, Mingqiang Lai, Yuanjun Jiang, Ying Zhang, and Xiaochun Bai

## 2.1 Extracellular Vesicle (EV) Biogenesis

EVs are vesicular structures that are released into the environment [1]. EVs contain various biomolecules, including proteins, nucleic acids, lipids, and metabolites. They exchange materials between cells and are involved in multiple physiological and pathological processes [2, 3]. EVs can be roughly divided into plasma membrane-derived vesicles and endosome-derived exosomes, according to their secretory mode. However, the differences in biogenesis and function between exosomes and other EVs are not clear [4, 5], and our understanding of these differences is constantly being updated as new information becomes available [6]. Thus, the biological processes of EVs, such as their biogenesis, secretion, and intercellular interactions, require further investigation. In this chapter, we focus on the biogenesis, secretion, and related mechanisms of EVs (Fig. 2.1).

Wenchong Zou and Kaixiang Zhang contributed equally to this work.

W. Zou · M. Lai · Y. Jiang · X. Bai (✉)
State Key Laboratory of Organ Failure Research, Department of Cell Biology, School of Basic Medical Sciences, Southern Medical University, Guangzhou, China
e-mail: baixc15@smu.edu.cn

K. Zhang · Y. Zhang (✉)
School of Pharmaceutical Sciences, Tsinghua University, Beijing, People's Republic of China
e-mail: z_ying@mail.tsinghua.edu.cn



Q. Wang, L. Zheng (eds.), *Extracellular Vesicles*,
https://doi.org/10.1007/978-981-99-8365-0_2

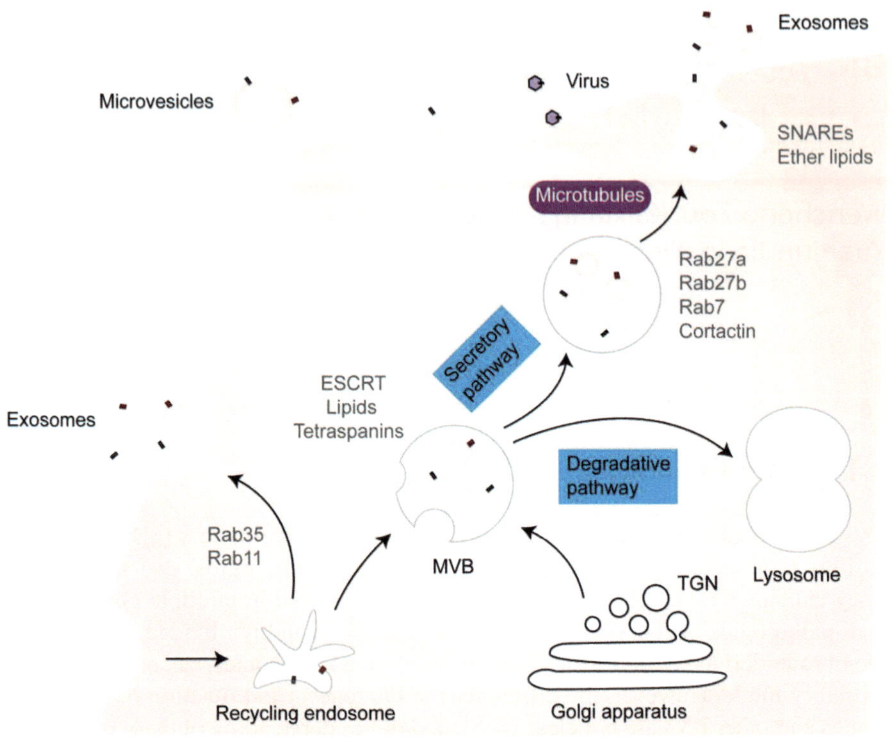


**Fig. 2.1** Biogenesis and secretion of extracellular vesicles (EVs). Multiple intracellular trafficking mechanisms are involved in the biogenesis and secretion of EVs. (1) The generation of exosomes can be divided into three steps: (a) exosome biogenesis, (b) multivesicular body (MVB) transfer, and (c) fusion of MVBs with the plasma membrane. (2) Microvesicles arise from outward budding and fission of the plasma membrane directly

### 2.1.1 Mechanisms of Exosome Biogenesis

Exosomes are EVs formed by the endosomal system and have a diameter of 40–200 nm [7]. The plasma membrane buds inward and forms endocytic bodies and multiple endocytic bodies. These two bodies fuse to form early endosomes. Trans-Golgi network-budding vesicles also fuse with endosomes [8]. After inward budding of the endosomal membrane and further modification, intraluminal vesicles (ILVs) give rise to multivesicular bodies (MVBs) [9]. MVB formation is mediated by at least two distinct pathways and involves sorting of various molecules into ILVs (Fig. 2.2). MVBs can not only degrade the cargo of lysosomes but also fuse with the plasma membrane to release ILVs into the extracellular space [10, 11].

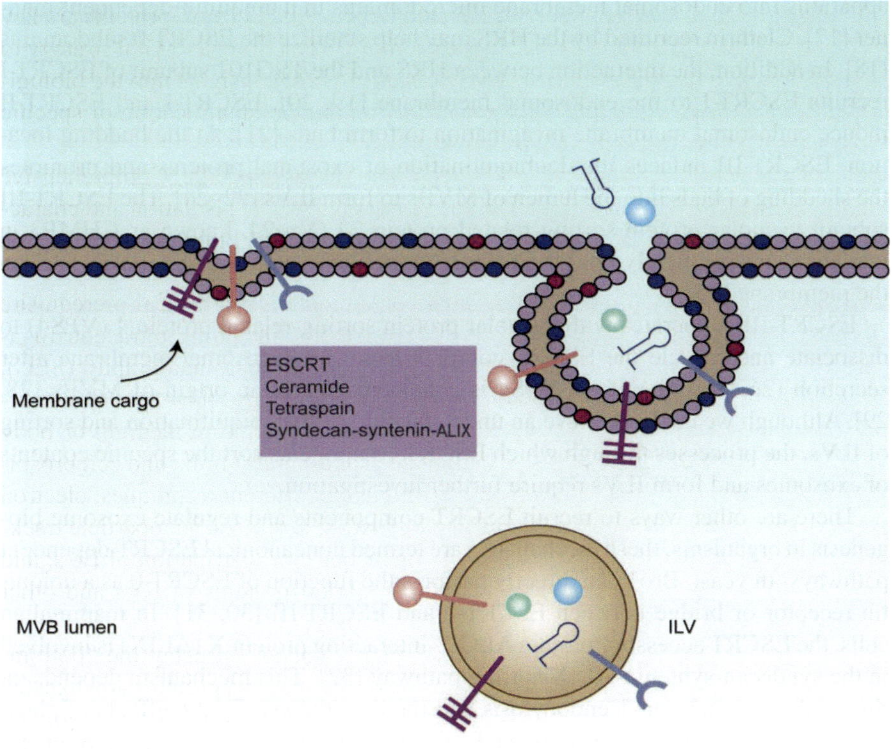


**Fig. 2.2** Biogenesis mechanism of MVBs. Multiple sorting machineries that generate intraluminal vesicles (ILVs) are illustrated. Several molecules, including ESCRTs, syndecan-syntenin-ALIX, ceramide, and tetraspanins, are involved in the biogenesis of ILVs in MVBs

#### 2.1.1.1 Endosomal Sorting Complex Required for Transport (ESCRT) Machinery

Exosome formation consists of a series of well-regulated processes that involve specific sorting mechanisms. In the early 2000s, the discovery of the ESCRT initiated exploration of the biogenesis of MVBs and ILVs [12]. Subsequently, ESCRT-classified proteins in endosomes and lysosomes have been shown to degrade MVBs [13, 14].

The ESCRT consists of a family of continuous complex proteins on the endosomal membrane (ESCRT-0, ESCRT-I, ESCRT-II, and ESCRT-III) and related accessory proteins. The ESCRT machinery is conserved in both yeast and mammals [15].

In the classical ESCRT-dependent pathway, phosphatidylinositol-3-phosphate (PtdIns3P) recruits the ESCRT-0 complex to early endosomes through the hepatocyte growth factor-regulated tyrosine kinase substrate (HRS) subunit [16]. Another subunit of ESCRT-0 is the signal transducing adaptor molecule (STAM), which has a ubiquitin-binding domain. ESCRT-0 aggregates exosome cargo from the Golgi apparatus into endosomal membrane microdomains in a ubiquitin-dependent manner [17]. Clathrin recruited by the HRS may help stabilize the ESCRT-0 subdomains [18]. In addition, the interaction between HRS and the TSG101 subunit of ESCRT-I recruits ESCRT-I to the endosomal membrane [19, 20]. ESCRT-I and ESCRT-II induce endosomal membrane invagination to form buds [21]. At the budding location, ESCRT-III induces the deubiquitination of exosomal proteins and promotes the shedding of buds into the lumen of MVBs to form ILVs [22–24]. The ESCRT-III subunit vacuolar protein sorting-related protein 24 (Vps24, known as CHMP3 in humans) can specifically bind to PtdIns3P, which may help stabilize ESCRT-III on the membrane [25].

ESCRT-III cooperates with vacuolar protein sorting-related protein 4 (VPS4) to dissociate and recycle the ESCRT complex from the endosomal membrane after secretion [26, 27]. Therefore, VPS4 is considered the basic origin of MVBs [28, 29]. Although we currently have an understanding of the ubiquitination and sorting of ILVs, the processes through which ESCRT components sort the specific contents of exosomes and form ILVs require further investigation.

There are other ways to recruit ESCRT components and regulate exosome biogenesis in organisms; these mechanisms are termed noncanonical ESCRT-dependent pathways. In yeast, Bro1 can directly perform the function of ESCRT-0 as a ubiquitin receptor or bridge between ESCRT-0 and ESCRT-III [30, 31]. In mammalian cells, the ESCRT accessory protein ALG-2-interacting protein X (ALIX) is involved in the syndecan-syntenin-ALIX sorting pathway [32]. This mechanism depends on Src-mediated syndecan-1 endocytosis and requires the synergistic effects of phospholipase D2 (PLD2) and GTPase ADP-ribosylation factor 6 (ARF6) [33]. Syntenin is a multivalent soluble protein that binds to ALIX through three LYPX L motifs and induces the inward budding of endosomal membranes. ALIX binds to protease-activated receptor-1(PAR1) and recruits ESCRT-III [34]. ALIX also directly recruits ESCRT-III to late endosomes by combining with high-density lipid lysobisphosphatidic acid (LBPA) in the inner membrane and promotes inward budding of the MVB-limiting membrane and formation of ILVs [35, 36]. Notably, noncanonical ESCRT-dependent pathways also require the participation of ESCRT-III and VPS24 [37].

Medium-throughput RNA interference with more than 23 components and related proteins of ESCRT indicated the various roles of this family in exosome biogenesis. The partial inactivation or overexpression of ESCRT components can affect the composition and release of exosomes. In HeLa and HEK293 cells, the expression of CD63 and CD81 in secreted exosomes is reduced after the inhibition of HRS and STAM1 [38, 39]. Knockout of HRS in head and neck squamous cancer cells can reduce exosome secretion [40], and depletion of TSG101, a subunit of ESCRT-I, leads to a decrease in the release of exosomes from tumor cells, such as HeLa, MCF-7, and immortalized RPE1 epithelial cells [41]. Interestingly, silencing of ESCRT-II-related genes significantly affects the release of tumor cell exosomes. In MCF-7 cells, ESCRT-III subunit *VPS32* (also known as *CHMP4*) gene silencing suppresses exosome release [38].

In the syndecan-syntenin-ALIX pathway, syntenin interacts with the tetraspanin protein CD63 to regulate the sorting of CD63 into ILVs and release of CD63-containing EVs. In addition, overexpression of syntenin induces exosome release in the ALIX-dependent pathway (containing CD63 and MHC II) [42]. In rat reticulocytes, ALIX can compete with heat shock homology 71 kDa protein (HSC70) to bind to the cytoplasmic domain of the transferrin receptor (TfR), which can promote sorting of TfR on ILVs and increase exosome release [43]. In HeLa cells, suppression of ALIX expression by silencing the programmed cell death 6 interacting protein gene can increase the release of MHC II-containing exosomes, but does not alter total EVs [44]. ESCRT machinery is involved in various cellular processes. When using suppression of ESCRT expression as a tool to inhibit exosome secretion, researchers should also consider that other cell cytokinesis and membrane repair functions may be altered in addition to MVB formation [13, 45].

#### 2.1.1.2 ESCRT-Independent Mechanisms

Silencing all key subunits of the ESCRT complex cannot prevent the formation of MVBs characterized by ILVs loaded with CD63 [46]. The formation of exosomes also proceeds in an ESCRT-independent manner, involving lipids, tetraspanins, or heat shock proteins.

The lipids in exosomes not only play important roles in maintaining the shape of exosomes but also participate in their biogenesis. In oligodendrocytes, neutral sphingomyelinase-2 (nSMase-2) hydrolyzes sphingomyelin phosphocholine to form ceramide [47]. Ceramide directly promotes MVB membrane budding through its conical shape [48]. Moreover, ceramide can also be metabolized to sphingosine 1-phosphate, which activates the G-protein-coupled sphingosine phosphate receptor on the MVB membrane and sorts the cellular contents of ILVs [49]. Inhibition of nSMase with GW4869 reduces the release of proteolipid protein (PLP)-bearing exosomes. In GW4869-treated HEK293 and tumor cells, the expression of classic exosomal marker proteins (e.g., CD63, CD81, and TSG101) is also decreased. However, inhibition of ceramide formation does not affect exosomal secretion in PC-3 cells [50–52]. This discrepancy may be related to the features of different cell types. In various types of cancer cells, the use of neutral sphingomyelinase to inhibit gene expression reduces the budding of vesicles in the lumen of MVBs and decreases exosome release. By contrast, C6 ceramide stimulation increases the release of exosomes [53]. Ceramide also regulates autophagy, which may indirectly affect MVB homeostasis.

Different pathways regulate the release of exosomes either individually or in parallel. Specific knockout of neutral nSMase inhibits the release of exosomes in the syndecan-syntenin-ALIX pathway [54–56]. The interaction between LBPA and ALIX in the inner membrane of MVBs promotes the formation of inward buds and ILVs [35]. Additionally, PLD2 can hydrolyze phosphatidylcholine to phosphatidic acid in an ESCRT-independent manner [57]. The membrane subdomains of exosomes are enriched in cholesterol and glycosphingolipids, which play important roles in signaling and sorting [58]. Drug treatment or gene mutations have been shown to induce cholesterol accumulation in MVBs, which increases the release of vesicles containing ALIX, CD63, and cholesterol in a Flotillin-2 dependent manner [59, 60].

Tetraspanins, a family of transmembrane proteins, regulate exosome release in an ESCRT-independent manner [61]. In human melanoma cells, CD63 anchors the pre-melanosome protein to apolipoprotein E and sorts into ILVs independent of ubiquitination or ceramide [62, 63]. Overexpression of CD9 or CD82 in HEK293 cells induces the secretion of β-catenin by exosomes [64, 65]. Recent studies showed that CD81 has a conical structure in the inner cavity of cholesterol-containing membranes. The aggregation of several cone-shaped tetraspanin proteins induces inward budding of microdomains and targets a series of ligands to ILVs [64].

Studies have shown that the expression of cell surface tetraspanin-TSPAN8 may alter the mRNA contents and protein compositions of exosomes in rat pancreatic adenocarcinoma cells, but does not affect total release [66]. Additionally, members of the tetraspanin family simultaneously regulate the transport of intracellular integrins and other components to MVBs, indicating that impairment of tetraspanin function may affect different steps of exosome biogenesis.

The small integral membrane protein of the lysosome/late endosome (SIMPLE, also known as lipopolysaccharide-induced tumor necrosis factor-α factor) also plays important roles in exosome release. Both TSG101 and Nedd4 (an E3-ubiquitin ligase that promotes lysosomal degradation) contain a SIMPLE-binding domain, which can bind to SIMPLE and promote exosome release [67]. SIMPLE binds to Nedd4 to help target cytoplasmic proteins (such as the tumor suppressor phosphatase and tensin homolog). Exosome secretion is then increased following transfection of COS cells with SIMPLE, and mutation of SIMPLE interferes with MVB formation. In a mouse model of SIMPLE formation disorder, exosome release is reduced [68].

The sorting of cytoplasmic contents into ILVs is a result of multiple sorting mechanisms. Phosphoglycerate kinase 1, glyceraldehyde 3-phosphate dehydrogenase, and various GTPases enter the ILVs via cooperation between the ESCRT machinery and transmembrane proteins [69]. Other mechanisms also help to select soluble and membrane-bound contents in exosomes; for example, molecular chaperones help to maintain the normal structure and function of the protein. Heat shock 70 kDa protein (HSP70) and HSC70 bind the soluble cytoplasmic protein containing the KFERQ sequence to target exosomes [43, 70]. These proteins are present in most exosomes derived from different cell types.

In addition, the affinity of some membranous cargo for lipid domains and lipid rafts can be directly involved in ILV generation via different pathways [60]. In addition to proteins and lipids, exosomes contain mRNAs and microRNAs (miRNAs). RNA-binding proteins have RNA-binding domains that enrich specific miRNAs in exosomes [71, 72]. The binding protein heterogeneous ribonucleoprotein A2B1 (hnRNPA2B1) is sulfonylated, and this modification is essential for the recognition and binding of the 3′ end of miRNAs [73]. KRAS also plays a role in miRNA sorting. Compared with wild-type cells, exosomes of mutant KRAS colorectal cancer cells have a unique miRNA profile [74]. In human hepatocytes, loading miRNAs on exosomes requires the specific binding of ALIX to the miRNA processing protein

Argonaute2 [75]. There may also be other mechanisms that target mRNA and miRNAs in exosomes.

Exosome biogenesis is complex and involves many different mechanisms. For example, exosome biogenesis is associated with exosome contents and cell type and is affected by many other signals and pathological stimuli [76]. The current discussion on the biological mechanism of exosomes inside cells is not exhaustive; however, it is clear that most cells contain various subpopulations of MVBs. Ultrastructural observations have revealed that the size and appearance of ILVs in MVB subpopulations are quite different [9], with different lipid and protein compositions. Cholesterol-positive and cholesterol-negative MVBs coexist in a single cell. Most cholesterol-containing MVBs tend to fuse with the plasma membrane, whereas lysophosphatidic acid-containing MVBs may be degraded [77]. In this case, different mechanisms contribute to MVB formation simultaneously or sequentially, allowing various cargoes to be classified at different stages of MVB maturity, and these mechanisms may be targeted using different approaches [78]. As the main regulator of exosome composition, the endosomal sorting mechanism appears to be essential for its functional properties. When exploring exosome generation and its biological significance, factors that affect other sorting mechanisms and their kinetics should also be considered.

MVBs fuse with lysosomes to degrade cargo or transport to the plasma membrane to release exosomes. However, little is known about the molecular mechanisms and cell states that regulate this balance. Gradual acidification along the endocytic pathway is necessary for degradation and recycling of intracellular components [79]. pH may also be a determinant of the function of MVBs as a secretory or degradation compartment [80]. The degradation or secretion of MVBs may depend on the cell state, and exosomes may play important roles in protecting cells from intracellular stress. Several studies have shown that exosome release is increased under conditions of cell irradiation-induced senescence or hypoxia [76]. However, it is still unclear why cells respond to external pressures by releasing more exosomes. This mechanism may be another process through which cellular waste is eliminated. Both autophagosomes and exosomes are derived from MVBs. In individuals with age-related degenerative diseases, lysosomal function defects can promote exosomal secretion and eliminate amyloid-like proteins, intracellular misfolded proteins, and aggregated proteins [81, 82]. Inhibition of phosphoinositide kinase generates phosphatidylinositol-3,5-bisphosphate, which results in reduced autophagic degradation and increased exosome release. This may be due to interference with organelle transport or fusion, leading to increased exosome release [83].

### 2.1.2 Biogenesis of Plasma Membrane-Derived Vesicles

Cells can release several different types of vesicles. Exosomes are released after MVBs fuse with the plasma membrane, whereas other types of EVs bud directly from the plasma membrane [84].

Microvesicles have diameters ranging from 100 to 1000 nm and are released from the plasma membrane (Fig. 2.3). The membrane composition of microvesicles is largely similar to that of the plasma membrane [85]. Endocytosis and recirculation of contents in the cell affect the production site of microvesicles. ARF6 selectively recruits proteins targeted to microvesicles, including β1 integrin receptors, MHC class I molecules, membrane-type 1-matrix metalloproteinase 1, and vesicle-associated membrane protein (VAMP) 3 [86, 87]. Targeted microvesicles, lipids, and other membrane-related cargoes anchor to plasma membrane lipids and are located at the sprouting position of microvesicles owing to their affinity for lipid rafts [60, 88]. It is unclear how nucleic acids in microvesicles target the cell membrane surface. Some studies in the field of cancer cell biology have suggested that conserved RNA sequences in the 3′-untranslated region of mRNA may target microvesicles; however, additional work is required to elucidate the specific details of these mechanisms [89, 90].

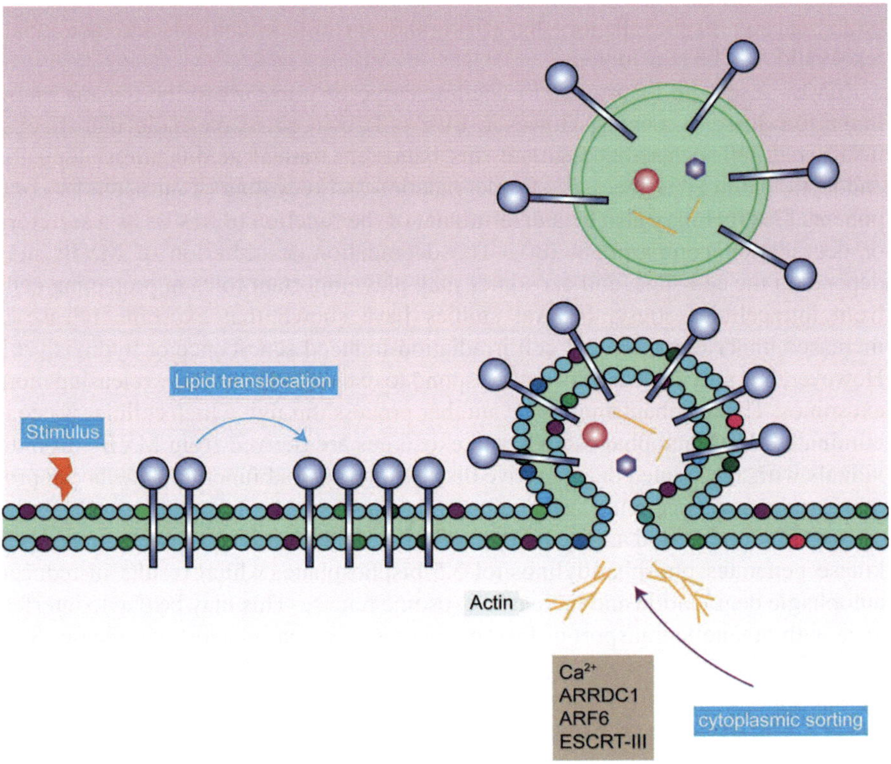


**Fig. 2.3** Formation and release of microvesicles. Microvesicles are generated at the plasma membrane in a constitutive manner or upon stimulation. The plasma membrane bilayer has an asymmetric distribution of phospholipids. Redistribution of lipids occurs and gives rise to secreted microvesicles after stimulation

Lipid components of the plasma membrane rearrange during the formation of microvesicles and in response to changes in $Ca^{2+}$levels and protein composition [91]. The influx of $Ca^{2+}$ ions into the cytoplasm activates phospholipid-modulating enzymes, which destroy the asymmetry of the phospholipid membrane, redistribute phospholipids in the cell membrane double layer, reshape the cytoskeleton, and promote budding of the cell membrane [92]. Sphingomyelinase is another lipid metabolism enzyme that is involved in vesicle budding. Acid sphingomyelinase produced by lysosomes sprouts from the plasma membrane of astrocytes and releases large vesicles after being triggered by P2X7 receptors [49, 93]. Cholesterol levels are relatively high in microvesicles, and cholesterol domains contribute to the formation of microvesicles in neutrophils [94]. The biogenesis of microvesicles also requires the cytoskeleton and its regulators. The RHO family of small GTPases and RHO-related protein kinases are important regulators of actin dynamics that induce microvesicles in different tumor cells [95].

Because most literature on microvesicles focuses on their potential functions rather than their origin, the molecular mechanisms of microvesicle formation remain unclear. Studies have shown that the formation of EVs derived from the plasma membrane of T lymphocytes is associated with the ESCRT-related proteins TSG101 and VPS4 [96]. Arrestin domain-containing protein 1 (ARRDC1) recruits TSG101 and VPS4 to the cell membrane [97]. Moreover, TSG101 interacts with the PXAP sequence present in the gag protein of retroviruses to mediate the formation of roughly uniform microvesicles with a diameter of approximately 50 nm from gag or ARRDC1 budding at the plasma membrane [98]. Although microvesicles bud from the plasma membrane, their formation involves TSG101 and VPS4 in the ESCRT. Vesicle budding is also thought to function in the repair of plasma membrane damage [96].

Apoptotic bodies are vesicular bodies with diameters of 500–2000 nm, and their morphology is more heterogeneous than that of other cell-derived vesicles [99, 100]. Apoptotic nuclear chromatin condenses and undergoes fragmentation to form clumps of chromatin of varying sizes. The cell encapsulates the cytoplasm, organelles, and nuclear fragments into different vesicles through sprouting, blistering, and shedding to form apoptotic bodies [101]. Binding of annexin A5, thrombospondin, and complement protein C3b acts as the main protein marker of apoptotic bodies [100].

There are many similarities between enveloped viral budding and small EV biogenesis. Viruses use transport mechanisms in host cells during their assembly and germination. Human immunodeficiency virus (HIV)-1 germination is independent of the ARF6 GTPase pathway [102] but involves many Rab proteins, including Rab9, Rab7a, Rab27a, Rab14, and Rab11 [103]. Soluble N-ethylmaleimide-sensitive factor attachment protein receptor (SNARE) also participates in HIV budding from the plasma membrane. The recruitment of the HIV-1 Gag protein at the budding site is necessary for final assembly [104]. This process also involves the formation of filamentous actin structures that disappear after viral budding [105]. The lipid components of HIV-1 virions and small EVs contain high proportions of cholesterol, sphingomyelin, and saturated lipids [106]. The HIV-1 membrane is also rich in certain tetraspanin proteins, such as CD63 [107] and CD81 [108] in small EVs. This similarity makes isolation of virosomes and EVs from virus-infected cells challenging. Because many aspects of these processes are still largely unknown, it is necessary to identify and characterize individual-specific molecules in EVs or virus production.

## 2.2 EV Secretion

The release of exosomes requires the transport of MVBs to the plasma membrane and the fusion of MVBs with the plasma membrane. ILVs (exosomes) are released into the extracellular microenvironment. Once plasma membrane-derived EVs are formed, they are released. Different transport mechanisms may involve various processes [109].

### 2.2.1 Rab GTPases

The small GTPase proteins of the Rab family regulate the molecular switches of different steps of intracellular vesicle transport, such as budding, transfer in the cell lumen, docking between vesicles and target chambers, and membrane fusion [110, 111]. The first Rab GTPase identified was Rab11 [112]. Rab11 acts mainly on early endosomes and increases the possibility of EV release from earlier endosomes. In human leukemia cells, the expression of Rab11 is related to HSC70 and TfR contained in exosomes [113]. Rab35 is primarily located in the plasma membrane. Rab35 inhibition reduces the release of PLP-containing exosomes into neuronal cells and primary oligodendrocytes, suggesting that knockdown of Rab35 may weaken the docking effect of MVBs and the plasma membrane [114].

During MVB docking, Rab27A rearranges the submembrane actin cytoskeleton of MVBs [115]. This protein is conserved in vesicular secretion mechanisms, and Rab27B and its effector protein SLAC2-B regulate MVB translocation to the plasma membrane. Knockdown of Rab27b results in MVB accumulation around the nucleus, and damage to Rab27A or Rab27B alters the morphology and docking state of MVBs with the plasma membrane [111]. In many tumor cell lines (including murine melanoma cells, breast cancer cells, and human squamous cell carcinoma cells), reduction of Rab27A and Rab27B expression can greatly reduce the secretion of CD63- and CD81-bearing exosomes and MHC class II molecules [116]. In response to nutrient or growth factor conditions, the mechanistic target of rapamycin complex 1 (mTORC1) regulates the release of exosomes in a Rab27A-dependent manner. Constitutive activation of mTORC1 inhibits exosome release from murine embryonic fibroblasts. Correspondingly, inhibition of mTORC1 increases the release of exosomes; however, additional studies are required to elucidate the underlying mechanisms [117].

Rab7 and related proteins promote the reverse transport-mediated degradation of MVBs that target lysosomes. Inhibition of Rab7 also leads to the release of small EVs, and its dual effects may depend on the ubiquitination status of Rab7 [118,

119]. The regulation of EV secretion by different Rab proteins may depend on cell type. In HeLa cells, depletion of Rab27A and Rab27B reduces EV secretion, whereas secretion of lethal factor EVs requires Rab11 and Rab35, but not Rab27, in RPE cells [41].

Small GTPases from other families, such as the Ral/Rho/Rac/cdc family, may also play roles in exosome release [120]. Inhibition of a small GTPase of the Ral family causes MVBs to accumulate near the plasma membrane, and exosome release is reduced by approximately 50% [121]. Notably, small GTPase proteins are not constitutively expressed in all cell types, suggesting that each cell type may have its own mechanisms for secreting exosomes [122].

### 2.2.2 Cytoskeleton

The transport of MVBs to the plasma membrane requires interaction with actin and the microtubule cytoskeleton. The actin-binding protein cortactin is involved in MVB transport and docking with the plasma membrane. Its knockdown and overexpression have been shown to decrease and increase exosome release, respectively [116].

The release of microvesicles also requires changes in actin polymerization under the plasma membrane. Activation of small GTP-binding proteins, such as ARF6 and ARF1, leads to the phosphorylation of myosin and ATP-dependent contraction of actin, which promotes vesicle budding [86, 123]. $Ca^{2+}$-dependent proteolysis degrades the membrane-bound cytoskeleton, remodels the cell membrane, and increases microvesicle release [124].

### 2.2.3 SNAREs

After MVBs fuse with the plasma membrane, exosomes are released into the extracellular environment. Several energy barriers must be overcome in this process. Protein-lipid interactions can decrease these energy barriers and promote fusion [6]. SNARE contributes to the fusion of membrane lipid bilayers [125]. The SNARE protein VAMP7 is essential for the release of exosomes from K562 human leukemia cells. Overexpression of the N-terminal domain of VAMP7 inhibits the formation of SNARE complexes, disperses MVBs in cells, and reduces exosome release [126]. Additionally, the release of exosomes also requires the R-SNARE protein YKT6. Knockout of YKT6 in HEK293 cells suppresses the release of exosomes containing TSG101 and VPS26/35. YKT6 is also necessary for the secretion of Wnt-bearing exosomes [127]. In HeLa cells, the plasma membrane-associated SNARE protein synaptosome-associated protein 23 spontaneously mediates MVB-plasma membrane fusion after histamine treatment and promote exosome release [128]. SNARE is mainly involved in the fusion of vesicles and the plasma membrane and may be involved in the secretion of secretory granules but does not interfere with upstream transport.

The release of EVs is closely related to the physiological and pathological states of cells [129]. Cancer, viral infections, diabetes, cardiovascular diseases, chronic inflammatory diseases, and neurodegenerative diseases are associated with large increases in EV numbers. Therefore, EVs may serve as specific biomarkers for these diseases [130], and further studies are needed to explore the biogenesis and release mechanisms of EVs in order to elucidate EV-related biological processes, such as generation and secretion, and to promote the application of EVs in the clinical setting.

## 2.3 EV Function

EVs participate in noncanonical pathways for the secretion of proteins, nucleic acids, and lipids. This confers EVs with diverse biological functions, many of which have been identified in recent years. The first report of EVs was published in 1982, although the term "EVs" was not formally used at that time [131]. In this original report, prostasomes were found to interact with cells and promote the motility of sperm cells. Subsequently, highly diverse biological functions have been attributed to EVs. These functions can be categorized into three major parts: autonomous regulators of biogenesis from released cells, components and remodelers of the extracellular environment and extracellular matrix, and transporters of molecules and signals for recipient cells (Fig. 2.4).

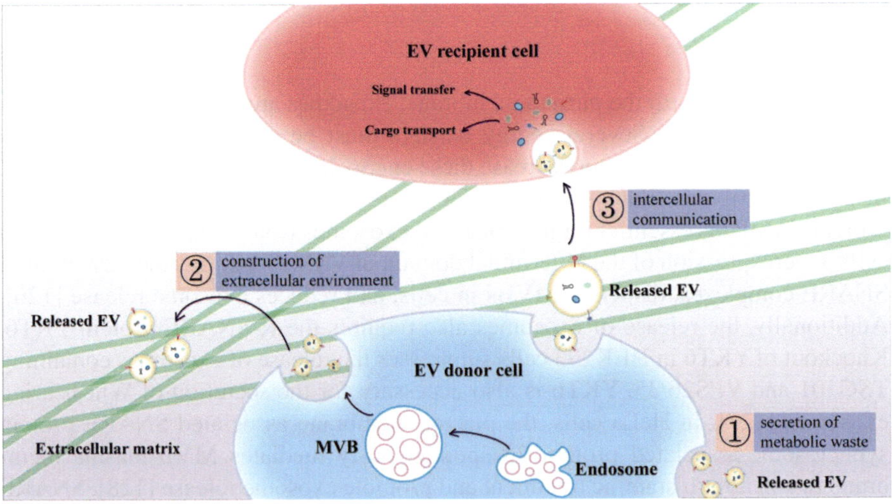


**Fig. 2.4** Functions of extracellular vesicles (EVs). EVs are released from cells and function via three important mechanisms: (1) secretion of metabolic waste, (2) construction of the extracellular environment, and (3) intercellular communication. *MVB* multivesicular body

### 2.3.1 Essential Roles of EVs in Fundamental Cellular Functions

#### 2.3.1.1 Clearance of Metabolic Waste

When EVs were first found in the 1980s [132, 133], they were thought to reflect a novel pathway through which cells could secrete unneeded compounds [134]. However, we now know that the cargo of EVs includes proteins, nucleic acids, lipids, and other biomacromolecules and small molecules. Indeed, EVs are an elegant mechanism for cells to deal with faulty or harmful molecules. This process maintains the stability of the cytoplasmic environment and prevents unwanted molecules from performing their functions. EVs can selectively eliminate misfolded or useless proteins by transferring them into the extracellular environment. One ubiquitous example of protein quality control is fertilization. Once fertilization occurs, the sperm receptor Juno is rapidly shed from egg cells by EVs [135]. Another classical example of protein elimination by EVs is the recycling of transferrin receptors in reticulocytes [133, 136]. The discovery of this process has led to the identification and verification of EVs. In addition to proteins, nucleic acids are also important cargo for EVs. The released EV RNAs are usually accompanied by genetic information transfer and modification of the biological functions of recipient cells. RNA release also contributes to the depletion of unstable or incorrect RNAs. Modified or unshielded RNAs are eliminated from the cytoplasm using EVs, and the increased levels of endogenous RNA after cellular stimulation induced by pathological changes may result in exposure of RNAs via separation from their normal binding proteins. These abnormal RNAs are transported into EVs and drive the inflammatory response [137]. The release of unshielded RNAs by EVs is strongly related to tumor growth and metastasis [138]. EVs can also decrease RNA levels in the cytoplasm and facilitate RNA degradation. Redundant RNAs are eliminated by processing into EVs and removed from cells to prevent protein overexpression [139]. EVs play important roles in the release of cellular waste, which can further relieve stress related to cellular metabolism. The released compounds are then taken up by other cells as nutrients.

#### 2.3.1.2 Balance of the Intracellular Environment

Cells autonomously generate signaling compounds that modulate cellular activities. The delicate balance of these signal molecules synergistically maintains the steady state of the intracellular environment. However, when stimulation occurs, balance may be disturbed. EVs provide a channel to reduce or remove these excessive signals, which may have already been detected and may then become redundant or even harmful. This process is observed in modulation of the cell cycle and immunity. For example, mixed lineage kinase domain-like pseudokinase (MLKL) modulates the cell cycle and triggers necroptosis once activated. The cellular membrane is fractured, which initiates cell death. Thus, excessive or unregulated MLKL signaling is harmful to cells. Notably, MLKL can bind to ESCRT proteins and is eventually excluded from cells by EVs [140]. Thus, EVs remove harmful compounds and maintain the balance of the intracellular environment. Cells release several stress proteins and initiate a downstream signaling pathway that affects cell growth and the cell cycle in response to stimulation or damage. Stress proteins are designated for degradation after the self-regulation of cells. In addition to being degraded by lysosomes or proteasomes, these proteins can also be secreted by EVs to reduce their copy numbers in the cellular plasma [141]. EVs contain many stress proteins, including HSP70, HSP60, and high mobility group box 1 [142–144]. In doing so, EVs also help cells degrade stress proteins and maintain the balance of cellular plasma. The released proteins were further taken up by recipient cells. Therefore, EVs act as damage-associated molecular patterns and initiate signaling pathways associated with the immune reaction [145].

#### 2.3.1.3 Generation of Cell Polarity

The proteins sorted into EVs can influence cell polarity or migration when taken up by cells. The polarity of cells is associated with their migration capability. The function of EVs in cell polarity during migration varies among different types of vesicles. EV release leads to the transfer of protein cargo into the cells, and the direction of EV motion in the cell generates the polarity of protein sorting and defines the site of uropod formation [88]. The protein cargo of EVs not only determines the location of the uropod but also increases the formation of uropods and further enhances the migration ability of the cell. In addition, uropods also function to modulate cell morphology. Therefore, EVs in these cells contribute to cargo transfer, affecting the morphology of recipient cells [146]. This exosome biogenesis pathway establishes a method for the polarized sorting of proteins and lipids to specific types of polar cells. EVs can also regulate cell migration via mechanisms other than cargo transfer. The discovery of migrasomes has shown that EVs can affect the direction and route of cell migration [147]. Migrated cells can release vesicles carrying signal molecules and affect recipient cells to maintain a polarized but ordered system. This type of polarization is involved in many physiological processes, including organ development and cell morphogenesis [148].

### 2.3.2 Roles of EVs in the Construction and Regulation of the Extracellular Environment

#### 2.3.2.1 EVs as Materials for the Extracellular Matrix

Given that EVs carry various molecular cargoes, including proteins and lipids, they can directly participate in extracellular matrix construction. These extracellular cargoes can either regulate the composition of the surrounding environment or act as matrix components, highlighting the general functions of EVs in the construction of many extracellular structures. For example, EVs can mediate calcification. Released vesicles from cardiovascular tissues contain specific mineralization-related compounds that inhibit or promote cargo under different conditions [149], including both normal and pathological conditions [150]. EVs can also contribute to osteogenesis as a seed for crystal formation during calcification. By contrast, imbalanced cell conditions may result in the unregulated release of calcification-regulated EVs and cause cardiovascular diseases, such as arteriosclerosis [151, 152].

#### 2.3.2.2 Protection Against Pathogens and Injury

EVs can protect cells from pathogen invasion and the environment from injury. EVs act as a barrier in response to pathogens or injury and show high immunoactivity to protect cells against infection [153]. For example, EVs in the peripheral blood show cardioprotective effects. Furthermore, HSP70 can activate Toll-like receptor (TLR) 4 signaling and induce cardioprotective signals at local sites of injury [154, 155]. Recent studies have also shown that EVs can behave as independent functional units to protect neighboring cells. After being secreted from donor cells, EVs can function in the absence of donor or recipient cells. For example, EVs can protect cells from bacterial toxins. Cell release EVs are enriched with disintegrin and metalloprotease 10 (ADAM10) after stimulation by pathogens. Extracellular ADAM10 leads to the neutralization of bacterial toxins and protection of cells [156]. Thus, these studies have shown that EVs can act as functional units in addition to their well-recognized roles as shuttles between cells. Several studies have shown that EVs can function independently in the extracellular environment. Accordingly, EVs seem to act as organelles rather than just simple structures.

### 2.3.3 EV Functions in Recipient Cells

Research on EVs in recent years indicates that EVs affect the transmission between cells and mediate signaling and molecular transfer. EV cargoes include lipids, proteins, nucleic acids, and other signaling molecules, and EVs can transfer multiple cargoes simultaneously and produce a downstream signal response [37]. Additionally, EVs regulate many important signal pathways, representing an elegant mechanisms for intercellular communication. Research on signal transmission is important not only for biological mechanisms but also for medical applications.

#### 2.3.3.1 Proteins as Major EV Cargo and Signaling Molecules

Proteins are one of the most important types of EV cargo that act on several pathways directly in recipient cells. In a recent study, EVs were shown to mediate intercellular communication via exosomal programmed cell death ligand 1 (PD-L1) excreted from tumors, affecting the immune system. Tumor cells express high levels of PD-L1, which is secreted by EVs, and PD-L1 remains in the extracellular environment until it is absorbed by recipient cells [157]. Extracellular PD-L1 can suppress T-cell activation and protect tumor cells from being eliminated [158]. Accordingly, tumor-derived EVs can affect the signaling pathways of recipient T cells, thereby mediating tumor escape from immune system surveillance. Moreover, the clearance of extracellular PD-L1 can inhibit tumor growth [159]. These studies have provided insights into cancer immune evasion and treatment methods.

Many other studies have emphasized the importance of EVs in delivering signal proteins. Epidermal growth factor receptor, a cancer cell-derived protein and active hepatocyte growth factor that promotes the growth and metastasis of tumors, has been shown to be delivered by EVs [160]. A similar phenomenon was observed for the transfer of HSP72 between cells via tumor-derived EVs [161]. Tumor cells can also use EVs to affect the growth rates of surrounding cells in the microenvironment. Similarly, EVs are important for intercellular regulation in the immune system. EVs from dendritic cells and macrophages contain several pattern recognition receptors, including TLRs, such as TLR4 and TLR7 [162, 163]. Active Cdc42 can also be carried by EVs to mediate communication among immune cells, functioning to propagate immune signals by enhancing EV endocytosis [164]. In future studies, additional related intercellular communication pathways should also be investigated.

#### 2.3.3.2 EVs as Carriers of Functional RNAs into Recipient Cells

RNAs are also common cargo used by EVs to transmit signals. In 2007, EVs were shown to carry miRNAs and mRNAs, which exert specific functions in recipient cells [165]. EVs can carry several miRNAs via interactions of miRNAs with RNA-binding proteins, including argonaute-2, hnRNPA2B1, and trinucleotide repeat containing 6 protein [73, 75, 166]. Additionally, miRNAs can be transmitted between immune cells via EVs, enabling the regulation of gene expression in the recipient cells. T cells release miRNAs through EVs into antigen-presenting cells (APCs), such as dendritic cells [56]. Regulation by miRNAs can modulate the immune response of APCs. For example, in recipient cells, this type of indirect miRNA transfer can affect cell growth, both in tumor cells and normal cells [167, 168]. Cancer cell-derived EVs contain cancer-promoting components, including miRNAs, and mediate the growth and biological functions of surrounding cells through gene regulation [169]. Extracellular miRNAs can bind to native RNAs in recipient cells to block their intrinsic functions and may have important roles in addition to their functions in signal transfer. Viruses can also send their viral miRNAs by infected cells to other recipient cells [170]. This may be a subordinate mechanism of viral infections. Importantly, extracellular miRNAs may have critical applications in the clinical setting. Indeed, owing to their excellent cytocompatibility, EVs could be used as novel carriers to deliver drugs, including miRNAs [171]. miRNAs are also suitable biomarkers for disease diagnosis, including the diagnosis of cancer and cardiovascular diseases. Therefore, analysis of miRNAs in EVs obtained by liquid biopsy may have applications as an effective diagnostic tool. Many studies have verified that using extracellular miRNAs as biomarkers for disease classification and diagnosis is possible and shows great potential [172–174].

Notably, EVs can also contain large numbers of mRNAs. However, compared with extracellular miRNAs, research on this topic is still preliminary. Unlike miRNAs, the mechanisms used to sort mRNAs into EVs have not yet been identified. mRNAs carried by EVs also regulate gene expression in recipient cells [175], and EV mRNAs can be taken up by recipient cells and used to express the encoded genes [176]. Although extracellular mRNAs may have other functions, owing to the limitations of currently available analytical techniques, research on the biological functions of mRNAs and recipient cells is still limited. EVs may also carry mRNAs for delivery to target cells in medical applications. With breakthroughs in this field, more extensive research may be conducted.

In addition to miRNAs and mRNAs, EVs also carry other types of nucleic acids for therapeutic use. For example, single guide RNA (sgRNA) in the clustered regularly interspersed short palindromic repeats/cas9 system can also be carried as cargo in EVs. After being released from the cells by EVs, the sgRNA can guide cas9 to knockout specific genes in recipient cells [177, 178]. Moreover, the use of EVs to deliver small interfering RNA (siRNA) has been demonstrated in a series of successful experiments [179]. EV-delivered siRNA induces gene silencing, which makes it an efficient tool for precisely regulating cell survival and death. Indeed, siRNA-loaded EVs have been used for gene regulation both in vitro and in vivo. The delivery of these functional RNAs shows that EVs have many potential therapeutic applications.

## 2.4 Conclusion

In this chapter, we discussed the various intracellular and intercellular functions of EVs, including their roles in various processes ranging from compound elimination to signal transmission. Although the specific functions of EVs among all cellular biological processes are still difficult to define, the importance of EVs in our living system is becoming clear.

Extensive research has been conducted to assess the functions of EVs. Mounting evidence indicates that EVs are responsible for several biological processes involved in intracellular and intercellular communication. Although current research has only shed light on a small portion of all the complex functions of EVs, EVs appear to be a promising tool for many medical applications. Therefore, the utilization of EVs is based on our understanding of their genesis and biological functions. However, future research efforts will require more reliable technical methods and standards. Precise and accurate methods for the separation and characterization of EVs will further expand our knowledge of these small but interesting vesicles.

**Acknowledgments** We acknowledge funding from the National Key R&D Program of China (grant no. 2021YFE0109300) and National Natural Science Foundation of China (grant nos. 81991510 and 21977061).

## References

1. Cocozza F, Grisard E, Martin-Jaular L, et al. SnapShot: extracellular vesicles. Cell. 2020;182:262.
2. Mathieu M, Martin-Jaular L, Lavieu G, et al. Specificities of secretion and uptake of exosomes and other extracellular vesicles for cell-to-cell communication. Nat Cell Biol. 2019;21:9–17.
3. Teng F, Fussenegger M. Shedding light on extracellular vesicle biogenesis and bioengineering. Adv Sci (Weinh). 2020;8:2003505.
4. van Niel G, D'Angelo G, Raposo G. Shedding light on the cell biology of extracellular vesicles. Nat Rev Mol Cell Biol. 2018;19:213–28.
5. Tkach M, Thery C. Communication by extracellular vesicles: where we are and where we need to go. Cell. 2016;164:1226–32.
6. Hessvik NP, Llorente A. Current knowledge on exosome biogenesis and release. Cell Mol Life Sci. 2018;75:193–208.

7. Shao H, Im H, Castro CM, et al. New technologies for analysis of extracellular vesicles. Chem Rev. 2018;118:1917–50.
8. Chen J, Chen ZJ. PtdIns4P on dispersed trans-Golgi network mediates NLRP3 inflammasome activation. Nature. 2018;564:71–6.
9. Colombo M, Raposo G, Thery C. Biogenesis, secretion, and intercellular interactions of exosomes and other extracellular vesicles. Annu Rev Cell Dev Biol. 2014;30:255–89.
10. Kowal J, Tkach M, Thery C. Biogenesis and secretion of exosomes. Curr Opin Cell Biol. 2014;29:116–25.
11. Dreyer F, Baur A. Biogenesis and functions of exosomes and extracellular vesicles. Methods Mol Biol. 2016;1448:201–16.
12. Katzmann DJ, Babst M, Emr SD. Ubiquitin-dependent sorting into the multivesicular body pathway requires the function of a conserved endosomal protein sorting complex, ESCRT-I. Cell. 2001;106:145–55.
13. Henne WM, Stenmark H, Emr SD. Molecular mechanisms of the membrane sculpting ESCRT pathway. Cold Spring Harb Perspect Biol. 2013;5.
14. Huotari J, Helenius A. Endosome maturation. EMBO J. 2011;30:3481–500.
15. Williams RL, Urbe S. The emerging shape of the ESCRT machinery. Nat Rev Mol Cell Biol. 2007;8:355–68.
16. Rusten TE, Vaccari T, Stenmark H. Shaping development with ESCRTs. Nat Cell Biol. 2011;14:38–45.
17. Raiborg C, Bremnes B, Mehlum A, et al. FYVE and coiled-coil domains determine the specific localisation of Hrs to early endosomes. J Cell Sci. 2001;114:2255–63.
18. Raiborg C, Bache KG, Gillooly DJ, et al. Hrs sorts ubiquitinated proteins into clathrin-coated microdomains of early endosomes. Nat Cell Biol. 2002;4:394–8.
19. Lu Q, Hope LW, Brasch M, et al. TSG101 interaction with HRS mediates endosomal trafficking and receptor down-regulation. Proc Natl Acad Sci U S A. 2003;100:7626–31.
20. Bache KG, Brech A, Mehlum A, et al. Hrs regulates multivesicular body formation via ESCRT recruitment to endosomes. J Cell Biol. 2003;162:435–42.
21. Wollert T, Hurley JH. Molecular mechanism of multivesicular body biogenesis by ESCRT complexes. Nature. 2010;464:864–9.
22. Babst M, Katzmann DJ, Snyder WB, et al. Endosome-associated complex, ESCRT-II, recruits transport machinery for protein sorting at the multivesicular body. Dev Cell. 2002;3:283–9.
23. Babst M, Katzmann DJ, Estepa-Sabal EJ, et al. Escrt-III: an endosome-associated heterooligomeric protein complex required for MVB sorting. Dev Cell. 2002;3:271–82.
24. Wollert T, Wunder C, Lippincott-Schwartz J, et al. Membrane scission by the ESCRT-III complex. Nature. 2009;458:172–7.
25. Huber ST, Mostafavi S, Mortensen SA, et al. Structure and assembly of ESCRT-III helical Vps24 filaments. Sci Adv. 2020;6:a4897.
26. Hurley JH, Hanson PI. Membrane budding and scission by the ESCRT machinery: it's all in the neck. Nat Rev Mol Cell Biol. 2010;11:556–66.
27. Saksena S, Wahlman J, Teis D, et al. Functional reconstitution of ESCRT-III assembly and disassembly. Cell. 2009;136:97–109.
28. Lata S, Schoehn G, Jain A, et al. Helical structures of ESCRT-III are disassembled by VPS4. Science. 2008;321:1354–7.
29. Adell MA, Vogel GF, Pakdel M, et al. Coordinated binding of Vps4 to ESCRT-III drives membrane neck constriction during MVB vesicle formation. J Cell Biol. 2014;205:33–49.
30. Pashkova N, Gakhar L, Winistorfer SC, et al. The yeast Alix homolog Bro1 functions as a ubiquitin receptor for protein sorting into multivesicular endosomes. Dev Cell. 2013;25:520–33.
31. Wemmer M, Azmi I, West M, et al. Bro1 binding to Snf7 regulates ESCRT-III membrane scission activity in yeast. J Cell Biol. 2011;192:295–306.
32. Baietti MF, Zhang Z, Mortier E, et al. Syndecan-syntenin-ALIX regulates the biogenesis of exosomes. Nat Cell Biol. 2012;14:677–85.
33. Imjeti NS, Menck K, Egea-Jimenez AL, et al. Syntenin mediates SRC function in exosomal cell-to-cell communication. Proc Natl Acad Sci U S A. 2017;114:12495–500.

34. Dores MR, Chen B, Lin H, et al. ALIX binds a YPX(3)L motif of the GPCR PAR1 and mediates ubiquitin-independent ESCRT-III/MVB sorting. J Cell Biol. 2012;197:407–19.
35. Matsuo H, Chevallier J, Mayran N, et al. Role of LBPA and Alix in multivesicular liposome formation and endosome organization. Science. 2004;303:531–4.
36. Larios J, Mercier V, Roux A, et al. ALIX- and ESCRT-III-dependent sorting of tetraspanins to exosomes. J Cell Biol. 2020;219.
37. Pegtel DM, Gould SJ. Exosomes. Annu Rev Biochem. 2019;88:487–514.
38. Colombo M, Moita C, van Niel G, et al. Analysis of ESCRT functions in exosome biogenesis, composition and secretion highlights the heterogeneity of extracellular vesicles. J Cell Sci. 2013;126:5553–65.
39. Gross JC, Chaudhary V, Bartscherer K, et al. Active Wnt proteins are secreted on exosomes. Nat Cell Biol. 2012;14:1036–45.
40. Hoshino D, Kirkbride KC, Costello K, et al. Exosome secretion is enhanced by invadopodia and drives invasive behavior. Cell Rep. 2013;5:1159–68.
41. Abrami L, Brandi L, Moayeri M, et al. Hijacking multivesicular bodies enables long-term and exosome-mediated long-distance action of anthrax toxin. Cell Rep. 2013;5:986–96.
42. Bissig C, Gruenberg J. ALIX and the multivesicular endosome: ALIX in wonderland. Trends Cell Biol. 2014;24:19–25.
43. Geminard C, De Gassart A, Blanc L, et al. Degradation of AP2 during reticulocyte maturation enhances binding of hsc70 and Alix to a common site on TFR for sorting into exosomes. Traffic. 2004;5:181–93.
44. Raiborg C, Stenmark H. The ESCRT machinery in endosomal sorting of ubiquitylated membrane proteins. Nature. 2009;458:445–52.
45. Jimenez AJ, Maiuri P, Lafaurie-Janvore J, et al. ESCRT machinery is required for plasma membrane repair. Science. 2014;343:1247136.
46. Stuffers S, Sem WC, Stenmark H, et al. Multivesicular endosome biogenesis in the absence of ESCRTs. Traffic. 2009;10:925–37.
47. Trajkovic K, Hsu C, Chiantia S, et al. Ceramide triggers budding of exosome vesicles into multivesicular endosomes. Science. 2008;319:1244–7.
48. Scarlatti F, Bauvy C, Ventruti A, et al. Ceramide-mediated macroautophagy involves inhibition of protein kinase B and up-regulation of beclin 1. J Biol Chem. 2004;279:18384–91.
49. Bianco F, Perrotta C, Novellino L, et al. Acid sphingomyelinase activity triggers microparticle release from glial cells. EMBO J. 2009;28:1043–54.
50. Shamseddine AA, Airola MV, Hannun YA. Roles and regulation of neutral sphingomyelinase-2 in cellular and pathological processes. Adv Biol Regul. 2015;57:24–41.
51. Menck K, Sonmezer C, Worst TS, et al. Neutral sphingomyelinases control extracellular vesicles budding from the plasma membrane. J Extracell Vesicles. 2017;6:1378056.
52. Vuckovic S, Vandyke K, Rickards DA, et al. The cationic small molecule GW4869 is cytotoxic to high phosphatidylserine-expressing myeloma cells. Br J Haematol. 2017;177:423–40.
53. Cheng Q, Li X, Wang Y, et al. The ceramide pathway is involved in the survival, apoptosis and exosome functions of human multiple myeloma cells in vitro. Acta Pharmacol Sin. 2018;39:561–8.
54. Dreux M, Garaigorta U, Boyd B, et al. Short-range exosomal transfer of viral RNA from infected cells to plasmacytoid dendritic cells triggers innate immunity. Cell Host Microbe. 2012;12:558–70.
55. Kosaka N, Iguchi H, Yoshioka Y, et al. Secretory mechanisms and intercellular transfer of microRNAs in living cells. J Biol Chem. 2010;285:17442–52.
56. Mittelbrunn M, Gutierrez-Vazquez C, Villarroya-Beltri C, et al. Unidirectional transfer of microRNA-loaded exosomes from T cells to antigen-presenting cells. Nat Commun. 2011;2:282.
57. Laulagnier K, Grand D, Dujardin A, et al. PLD2 is enriched on exosomes and its activity is correlated to the release of exosomes. FEBS Lett. 2004;572:11–4.
58. Skotland T, Sandvig K, Llorente A. Lipids in exosomes: current knowledge and the way forward. Prog Lipid Res. 2017;66:30–41.

59. Strauss K, Goebel C, Runz H, et al. Exosome secretion ameliorates lysosomal storage of cholesterol in Niemann-pick type C disease. J Biol Chem. 2010;285:26279–88.
60. de Gassart A, Geminard C, Fevrier B, et al. Lipid raft-associated protein sorting in exosomes. Blood. 2003;102:4336–44.
61. Escola JM, Kleijmeer MJ, Stoorvogel W, et al. Selective enrichment of tetraspan proteins on the internal vesicles of multivesicular endosomes and on exosomes secreted by human B-lymphocytes. J Biol Chem. 1998;273:20121–7.
62. Rochin L, Hurbain I, Serneels L, et al. BACE2 processes PMEL to form the melanosome amyloid matrix in pigment cells. Proc Natl Acad Sci U S A. 2013;110:10658–63.
63. van Niel G, Charrin S, Simoes S, et al. The tetraspanin CD63 regulates ESCRT-independent and -dependent endosomal sorting during melanogenesis. Dev Cell. 2011;21:708–21.
64. Perez-Hernandez D, Gutierrez-Vazquez C, Jorge I, et al. The intracellular interactome of tetraspanin-enriched microdomains reveals their function as sorting machineries toward exosomes. J Biol Chem. 2013;288:11649–61.
65. Chairoungdua A, Smith DL, Pochard P, et al. Exosome release of beta-catenin: a novel mechanism that antagonizes Wnt signaling. J Cell Biol. 2010;190:1079–91.
66. Nazarenko I, Rana S, Baumann A, et al. Cell surface tetraspanin Tspan8 contributes to molecular pathways of exosome-induced endothelial cell activation. Cancer Res. 2010;70:1668–78.
67. Zhu H, Guariglia S, Yu RY, et al. Mutation of SIMPLE in Charcot-Marie-tooth 1C alters production of exosomes. Mol Biol Cell. 2013;24(1619–37):S1–3.
68. Putz U, Howitt J, Doan A, et al. The tumor suppressor PTEN is exported in exosomes and has phosphatase activity in recipient cells. Sci Signal. 2012;5:a70.
69. Sahu R, Kaushik S, Clement CC, et al. Microautophagy of cytosolic proteins by late endosomes. Dev Cell. 2011;20:131–9.
70. Thery C, Boussac M, Veron P, et al. Proteomic analysis of dendritic cell-derived exosomes: a secreted subcellular compartment distinct from apoptotic vesicles. J Immunol. 2001;166:7309–18.
71. Qu L, Ding J, Chen C, et al. Exosome-transmitted lncARSR promotes sunitinib resistance in renal cancer by acting as a competing endogenous RNA. Cancer Cell. 2016;29:653–68.
72. Santangelo L, Giurato G, Cicchini C, et al. The RNA-binding protein SYNCRIP is a component of the hepatocyte exosomal machinery controlling microRNA sorting. Cell Rep. 2016;17:799–808.
73. Villarroya-Beltri C, Gutierrez-Vazquez C, Sanchez-Cabo F, et al. Sumoylated hnRNPA2B1 controls the sorting of miRNAs into exosomes through binding to specific motifs. Nat Commun. 2013;4:2980.
74. Cha DJ, Franklin JL, Dou Y, et al. KRAS-dependent sorting of miRNA to exosomes. elife. 2015;4:e7197.
75. Arroyo JD, Chevillet JR, Kroh EM, et al. Argonaute2 complexes carry a population of circulating microRNAs independent of vesicles in human plasma. Proc Natl Acad Sci U S A. 2011;108:5003–8.
76. Xu R, Rai A, Chen M, et al. Extracellular vesicles in cancer—implications for future improvements in cancer care. Nat Rev Clin Oncol. 2018;15:617–38.
77. White IJ, Bailey LM, Aghakhani MR, et al. EGF stimulates annexin 1-dependent inward vesiculation in a multivesicular endosome subpopulation. EMBO J. 2006;25:1–12.
78. Kobayashi T, Startchev K, Whitney AJ, et al. Localization of lysobisphosphatidic acid-rich membrane domains in late endosomes. Biol Chem. 2001;382:483–5.
79. Scott CC, Gruenberg J. Ion flux and the function of endosomes and lysosomes: pH is just the start: the flux of ions across endosomal membranes influences endosome function not only through regulation of the luminal pH. BioEssays. 2011;33:103–11.
80. Palokangas H, Ying M, Vaananen K, et al. Retrograde transport from the pre-Golgi intermediate compartment and the Golgi complex is affected by the vacuolar H+-ATPase inhibitor bafilomycin A1. Mol Biol Cell. 1998;9:3561–78.
81. Eitan E, Suire C, Zhang S, et al. Impact of lysosome status on extracellular vesicle content and release. Ageing Res Rev. 2016;32:65–74.

82. Villarroya-Beltri C, Baixauli F, Mittelbrunn M, et al. ISGylation controls exosome secretion by promoting lysosomal degradation of MVB proteins. Nat Commun. 2016;7:13588.
83. Hessvik NP, Overbye A, Brech A, et al. PIKfyve inhibition increases exosome release and induces secretory autophagy. Cell Mol Life Sci. 2016;73:4717–37.
84. Huang-Doran I, Zhang CY, Vidal-Puig A. Extracellular vesicles: novel mediators of cell communication in metabolic disease. Trends Endocrinol Metab. 2017;28:3–18.
85. Raposo G, Stoorvogel W. Extracellular vesicles: exosomes, microvesicles, and friends. J Cell Biol. 2013;200:373–83.
86. D'Souza-Schorey C, Chavrier P. ARF proteins: roles in membrane traffic and beyond. Nat Rev Mol Cell Biol. 2006;7:347–58.
87. Muralidharan-Chari V, Clancy J, Plou C, et al. ARF6-regulated shedding of tumor cell-derived plasma membrane microvesicles. Curr Biol. 2009;19:1875–85.
88. Shen B, Fang Y, Wu N, et al. Biogenesis of the posterior pole is mediated by the exosome/microvesicle protein-sorting pathway. J Biol Chem. 2011;286:44162–76.
89. Bolukbasi MF, Mizrak A, Ozdener GB, et al. miR-1289 and "Zipcode"-like sequence enrich mRNAs in microvesicles. Mol Ther Nucleic Acids. 2012;1:e10.
90. Gibbings DJ, Ciaudo C, Erhardt M, et al. Multivesicular bodies associate with components of miRNA effector complexes and modulate miRNA activity. Nat Cell Biol. 2009;11:1143–9.
91. Minciacchi VR, Freeman MR, Di Vizio D. Extracellular vesicles in cancer: exosomes, microvesicles and the emerging role of large oncosomes. Semin Cell Dev Biol. 2015;40:41–51.
92. Piccin A, Murphy WG, Smith OP. Circulating microparticles: pathophysiology and clinical implications. Blood Rev. 2007;21:157–71.
93. Thomas LM, Salter RD. Activation of macrophages by P2X7-induced microvesicles from myeloid cells is mediated by phospholipids and is partially dependent on TLR4. J Immunol. 2010;185:3740–9.
94. Del CI, Shrimpton CN, Thiagarajan P, et al. Tissue-factor-bearing microvesicles arise from lipid rafts and fuse with activated platelets to initiate coagulation. Blood. 2005;106:1604–11.
95. Li B, Antonyak MA, Zhang J, et al. RhoA triggers a specific signaling pathway that generates transforming microvesicles in cancer cells. Oncogene. 2012;31:4740–9.
96. Nabhan JF, Hu R, Oh RS, et al. Formation and release of arrestin domain-containing protein 1-mediated microvesicles (ARMMs) at plasma membrane by recruitment of TSG101 protein. Proc Natl Acad Sci U S A. 2012;109:4146–51.
97. Maas S, Breakefield XO, Weaver AM. Extracellular vesicles: unique intercellular delivery vehicles. Trends Cell Biol. 2017;27:172–88.
98. Garrus JE, von Schwedler UK, Pornillos OW, et al. Tsg101 and the vacuolar protein sorting pathway are essential for HIV-1 budding. Cell. 2001;107:55–65.
99. Gurunathan S, Kang MH, Jeyaraj M, et al. Review of the isolation, characterization, biological function, and multifarious therapeutic approaches of exosomes. Cell. 2019;8:307.
100. Akers JC, Gonda D, Kim R, et al. Biogenesis of extracellular vesicles (EV): exosomes, microvesicles, retrovirus-like vesicles, and apoptotic bodies. J Neuro-Oncol. 2013;113:1–11.
101. Hauser P, Wang S, Didenko VV. Apoptotic bodies: selective detection in extracellular vesicles. Methods Mol Biol. 2017;1554:193–200.
102. Ghossoub R, Lembo F, Rubio A, et al. Syntenin-ALIX exosome biogenesis and budding into multivesicular bodies are controlled by ARF6 and PLD2. Nat Commun. 2014;5:3477.
103. Spearman P. Viral interactions with host cell Rab GTPases. Small GTPases. 2018;9:192–201.
104. Reales E, Mora-Lopez F, Rivas V, et al. Identification of soluble N-ethylmaleimide-sensitive factor attachment protein receptor exocytotic machinery in human plasma cells: SNAP-23 is essential for antibody secretion. J Immunol. 2005;175:6686–93.
105. Gladnikoff M, Shimoni E, Gov NS, et al. Retroviral assembly and budding occur through an actin-driven mechanism. Biophys J. 2009;97:2419–28.
106. Brugger B, Glass B, Haberkant P, et al. The HIV lipidome: a raft with an unusual composition. Proc Natl Acad Sci U S A. 2006;103:2641–6.
107. Orentas RJ, Hildreth JE. Association of host cell surface adhesion receptors and other membrane proteins with HIV and SIV. AIDS Res Hum Retrovir. 1993;9:1157–65.

108. Grigorov B, Attuil-Audenis V, Perugi F, et al. A role for CD81 on the late steps of HIV-1 replication in a chronically infected T cell line. Retrovirology. 2009;6:28.
109. Stahl PD, Raposo G. Exosomes and extracellular vesicles: the path forward. Essays Biochem. 2018;62:119–24.
110. Mittelbrunn M, Vicente MM, Sanchez-Madrid F. Organizing polarized delivery of exosomes at synapses. Traffic. 2015;16:327–37.
111. Stenmark H. Rab GTPases as coordinators of vesicle traffic. Nat Rev Mol Cell Biol. 2009;10:513–25.
112. Koles K, Nunnari J, Korkut C, et al. Mechanism of evenness interrupted (Evi)-exosome release at synaptic boutons. J Biol Chem. 2012;287:16820–34.
113. Savina A, Vidal M, Colombo MI. The exosome pathway in K562 cells is regulated by Rab11. J Cell Sci. 2002;115:2505–15.
114. Hsu C, Morohashi Y, Yoshimura S, et al. Regulation of exosome secretion by Rab35 and its GTPase-activating proteins TBC1D10A-C. J Cell Biol. 2010;189:223–32.
115. Sinha S, Hoshino D, Hong NH, et al. Cortactin promotes exosome secretion by controlling branched actin dynamics. J Cell Biol. 2016;214:197–213.
116. Ostrowski M, Carmo NB, Krumeich S, et al. Rab27a and Rab27b control different steps of the exosome secretion pathway. Nat Cell Biol. 2010;12(19–30):1–13.
117. Zou W, Lai M, Zhang Y, et al. Exosome release is regulated by mTORC1. Adv Sci (Weinh). 2019;6:1801313.
118. Rocha N, Kuijl C, van der Kant R, et al. Cholesterol sensor ORP1L contacts the ER protein VAP to control Rab7-RILP-p150 glued and late endosome positioning. J Cell Biol. 2009;185:1209–25.
119. Song P, Trajkovic K, Tsunemi T, et al. Parkin modulates endosomal organization and function of the endo-lysosomal pathway. J Neurosci. 2016;36:2425–37.
120. Sedgwick AE, D'Souza-Schorey C. The biology of extracellular microvesicles. Traffic. 2018;19:319–27.
121. Hyenne V, Apaydin A, Rodriguez D, et al. RAL-1 controls multivesicular body biogenesis and exosome secretion. J Cell Biol. 2015;211:27–37.
122. Antonyak MA, Wilson KF, Cerione RA. R(h)oads to microvesicles. Small GTPases. 2012;3:219–24.
123. Schlienger S, Campbell S, Claing A. ARF1 regulates the rho/MLC pathway to control EGF-dependent breast cancer cell invasion. Mol Biol Cell. 2014;25:17–29.
124. Savina A, Fader CM, Damiani MT, et al. Rab11 promotes docking and fusion of multivesicular bodies in a calcium-dependent manner. Traffic. 2005;6:131–43.
125. Pfeffer SR. Unsolved mysteries in membrane traffic. Annu Rev Biochem. 2007;76:629–45.
126. Fader CM, Sanchez DG, Mestre MB, et al. TI-VAMP/VAMP7 and VAMP3/cellubrevin: two v-SNARE proteins involved in specific steps of the autophagy/multivesicular body pathways. Biochim Biophys Acta. 2009;1793:1901–16.
127. Ruiz-Martinez M, Navarro A, Marrades RM, et al. YKT6 expression, exosome release, and survival in non-small cell lung cancer. Oncotarget. 2016;7:51515–24.
128. Verweij FJ, Bebelman MP, Jimenez CR, et al. Quantifying exosome secretion from single cells reveals a modulatory role for GPCR signaling. J Cell Biol. 2018;217:1129–42.
129. Mashouri L, Yousefi H, Aref AR, et al. Exosomes: composition, biogenesis, and mechanisms in cancer metastasis and drug resistance. Mol Cancer. 2019;18:75.
130. Ohno S, Ishikawa A, Kuroda M. Roles of exosomes and microvesicles in disease pathogenesis. Adv Drug Deliv Rev. 2013;65:398–401.
131. Stegmayr B, Ronquist G. Promotive effect on human sperm progressive motility by prostasomes. Urol Res. 1982;10:253–7.
132. Pan BT, Johnstone RM. Fate of the transferrin receptor during maturation of sheep reticulocytes in vitro: selective externalization of the receptor. Cell. 1983;33:967–78.
133. Harding C, Heuser J, Stahl P. Receptor-mediated endocytosis of transferrin and recycling of the transferrin receptor in rat reticulocytes. J Cell Biol. 1983;97:329–39.

134. Johnstone RM, Adam M, Hammond JR, et al. Vesicle formation during reticulocyte maturation. Association of plasma membraneactivities with released vesicles (exosomes). J Biol Chem. 1987;262:9412–20.
135. Bianchi E, Doe B, Goulding D, et al. Juno is the egg Izumo receptor and is essential for mammalian fertilization. Nature. 2014;508:483–7.
136. Harding CV, Heuser JE, Stahl PD. Exosomes: looking back three decades and into the future. J Cell Biol. 2013;200:367–71.
137. Nabet BY, Qiu Y, Shabason JE, et al. Exosome RNA unshielding couples stromal activation to pattern recognition receptor signaling in cancer. Cell. 2017;170:352–66.
138. Matei I, Kim HS, Lyden D. Unshielding exosomal RNA unleashes tumor growth and metastasis. Cell. 2017;170:223–5.
139. Phinney DG, Di Giuseppe M, Njah J, et al. Mesenchymal stem cells use extracellular vesicles to outsource mitophagy and shuttle microRNAs. Nat Commun. 2015;6:8472.
140. Yoon S, Kovalenko A, Bogdanov K, et al. MLKL, the protein that mediates necroptosis, also regulates endosomal trafficking and extracellular vesicle generation. Immunity. 2017;47:51–65.
141. Collett GP, Redman CW, Sargent IL, et al. Endoplasmic reticulum stress stimulates the release of extracellular vesicles carrying danger-associated molecular pattern (DAMP) molecules. Oncotarget. 2018;9:6707–17.
142. Lancaster GI, Febbraio MA. Exosome-dependent trafficking of HSP70: a novel secretory pathway for cellular stress proteins. J Biol Chem. 2005;280:23349–55.
143. Gupta S, Knowlton AA. HSP60 trafficking in adult cardiac myocytes: role of the exosomal pathway. Am J Physiol Heart Circ Physiol. 2007;292:H3052–6.
144. Eppensteiner J, Davis RP, Barbas AS, et al. Immunothrombotic activity of damage-associated molecular patterns and extracellular vesicles in secondary organ failure induced by trauma and sterile insults. Front Immunol. 2018;9:190.
145. Park SJ, Kim JM, Kim J, et al. Molecular mechanisms of biogenesis of apoptotic exosome-like vesicles and their roles as damage-associated molecular patterns. Proc Natl Acad Sci U S A. 2018;115:E11721–30.
146. Dustin ML, Long EO. Cytotoxic immunological synapses. Immunol Rev. 2010;235:24–34.
147. Ma L, Li Y, Peng J, et al. Discovery of the migrasome, an organelle mediating release of cytoplasmic contents during cell migration. Cell Res. 2015;25:24–38.
148. Jiang D, Jiang Z, Lu D, et al. Migrasomes provide regional cues for organ morphogenesis during zebrafish gastrulation. Nat Cell Biol. 2019;21:966–77.
149. Bakhshian NA, Hutcheson JD, Aikawa E. Extracellular vesicles as mediators of cardiovascular calcification. Front Cardiovasc Med. 2017;4:78.
150. Yang W, Zou B, Hou Y, et al. Extracellular vesicles in vascular calcification. Clin Chim Acta. 2019;499:118–22.
151. Loyer X, Vion AC, Tedgui A, et al. Microvesicles as cell-cell messengers in cardiovascular diseases. Circ Res. 2014;114:345–53.
152. New SE, Goettsch C, Aikawa M, et al. Macrophage-derived matrix vesicles: an alternative novel mechanism for microcalcification in atherosclerotic plaques. Circ Res. 2013;113:72–7.
153. Olaya-Abril A, Prados-Rosales R, McConnell MJ, et al. Characterization of protective extracellular membrane-derived vesicles produced by *Streptococcus pneumoniae*. J Proteome. 2014;106:46–60.
154. Vicencio JM, Yellon DM, Sivaraman V, et al. Plasma exosomes protect the myocardium from ischemia-reperfusion injury. J Am Coll Cardiol. 2015;65:1525–36.
155. Emanueli C, Shearn AI, Angelini GD, et al. Exosomes and exosomal miRNAs in cardiovascular protection and repair. Vasc Pharmacol. 2015;71:24–30.
156. Keller MD, Ching KL, Liang FX, et al. Decoy exosomes provide protection against bacterial toxins. Nature. 2020;579:260–4.
157. Chen G, Huang AC, Zhang W, et al. Exosomal PD-L1 contributes to immunosuppression and is associated with anti-PD-1 response. Nature. 2018;560:382–6.

158. Ricklefs FL, Alayo Q, Krenzlin H, et al. Immune evasion mediated by PD-L1 on glioblastoma-derived extracellular vesicles. Sci Adv. 2018;4:r2766.
159. Poggio M, Hu T, Pai CC, et al. Suppression of exosomal PD-L1 induces systemic anti-tumor immunity and memory. Cell. 2019;177:414–27.
160. Zhang H, Deng T, Liu R, et al. Exosome-delivered EGFR regulates liver microenvironment to promote gastric cancer liver metastasis. Nat Commun. 2017;8:15016.
161. Chalmin F, Ladoire S, Mignot G, et al. Membrane-associated Hsp72 from tumor-derived exosomes mediates STAT3-dependent immunosuppressive function of mouse and human myeloid-derived suppressor cells. J Clin Invest. 2010;120:457–71.
162. Zhou M, Chen J, Zhou L, et al. Pancreatic cancer derived exosomes regulate the expression of TLR4 in dendritic cells via miR-203. Cell Immunol. 2014;292:65–9.
163. Majer O, Liu B, Kreuk L, et al. UNC93B1 recruits syntenin-1 to dampen TLR7 signalling and prevent autoimmunity. Nature. 2019;575:366–70.
164. Zhang Y, Jin X, Liang J, et al. Extracellular vesicles derived from ODN-stimulated macrophages transfer and activate Cdc42 in recipient cells and thereby increase cellular permissiveness to EV uptake. Sci Adv. 2019;5:v1564.
165. Valadi H, Ekstrom K, Bossios A, et al. Exosome-mediated transfer of mRNAs and microRNAs is a novel mechanism of genetic exchange between cells. Nat Cell Biol. 2007;9:654–9.
166. Chen CY, Zheng D, Xia Z, et al. Ago-TNRC6 triggers microRNA-mediated decay by promoting two deadenylation steps. Nat Struct Mol Biol. 2009;16:1160–6.
167. Katakowski M, Buller B, Zheng X, et al. Exosomes from marrow stromal cells expressing miR-146b inhibit glioma growth. Cancer Lett. 2013;335:201–4.
168. Ohno S, Takanashi M, Sudo K, et al. Systemically injected exosomes targeted to EGFR deliver antitumor microRNA to breast cancer cells. Mol Ther. 2013;21:185–91.
169. Hannafon BN, Ding WQ. Intercellular communication by exosome-derived microRNAs in cancer. Int J Mol Sci. 2013;14:14240–69.
170. Pegtel DM, Cosmopoulos K, Thorley-Lawson DA, et al. Functional delivery of viral miRNAs via exosomes. Proc Natl Acad Sci U S A. 2010;107:6328–33.
171. Ha D, Yang N, Nadithe V. Exosomes as therapeutic drug carriers and delivery vehicles across biological membranes: current perspectives and future challenges. Acta Pharm Sin B. 2016;6:287–96.
172. Li Y, Zheng Q, Bao C, et al. Circular RNA is enriched and stable in exosomes: a promising biomarker for cancer diagnosis. Cell Res. 2015;25:981–4.
173. Rabinowits G, Gercel-Taylor C, Day JM, et al. Exosomal microRNA: a diagnostic marker for lung cancer. Clin Lung Cancer. 2009;10:42–6.
174. Etheridge A, Lee I, Hood L, et al. Extracellular microRNA: a new source of biomarkers. Mutat Res. 2011;717:85–90.
175. Baj-Krzyworzeka M, Szatanek R, Weglarczyk K, et al. Tumour-derived microvesicles carry several surface determinants and mRNA of tumour cells and transfer some of these determinants to monocytes. Cancer Immunol Immunother. 2006;55:808–18.
176. Tomasoni S, Longaretti L, Rota C, et al. Transfer of growth factor receptor mRNA via exosomes unravels the regenerative effect of mesenchymal stem cells. Stem Cells Dev. 2013;22:772–80.
177. Gee P, Lung M, Okuzaki Y, et al. Extracellular nanovesicles for packaging of CRISPR-Cas9 protein and sgRNA to induce therapeutic exon skipping. Nat Commun. 2020;11:1334.
178. Kim SM, Yang Y, Oh SJ, et al. Cancer-derived exosomes as a delivery platform of CRISPR/Cas9 confer cancer cell tropism-dependent targeting. J Control Release. 2017;266:8–16.
179. Alvarez-Erviti L, Seow Y, Yin H, et al. Delivery of siRNA to the mouse brain by systemic injection of targeted exosomes. Nat Biotechnol. 2011;29:341–5.

# The Mechanisms of Interaction Between Extracellular Vesicles and Target Cells

3

Lutao Du, Tong Liu, Yujiao Xie, Shujun Zhang, Ning Li, and Chuanxin Wang

## 3.1 Introduction

Extracellular vesicles (EVs) are enclosed structures that contain various biomolecules released by cells and have been shown to act as a major form of communication between cells [1]. When EVs are docked onto the surface of a recipient cell, they can trigger a functional response in the recipient cell by directly transmitting signals from the cell surface and/or transferring their contents into the cell. EV recognition, binding, cellular uptake, and intracellular transport are critical but extremely complex processes (Fig. 3.1) [2].

EVs play important roles in many physiological processes, such as pregnancy, stem cell differentiation, inflammation, and blood coagulation [3, 4]. Studies have shown that embryo implantation is one of the earliest and most vital stages of pregnancy, and EVs play key roles in the process of blastocyst stage embryo attachment to the maternal endometrium and migration to the tissues. EVs produced by embryonic stem cells (ESCs) can recognize and bind to the surfaces of surrounding trophoblasts to promote trophoblast migration. These effects are mediated by the interaction between integrin/laminin receptors expressed along the trophoblast plasma membrane and fibronectin extracellular matrix (ECM) proteins distributed on the surfaces of ESC-derived EVs [5].

In addition to their roles in normal biology, EVs also participate in the occurrence and development of diseases, such as viral infections, metabolic and cardiovascular diseases, and tumors [6]. In the field of oncology, some studies have shown that EVs are involved in various processes related to cancer progression, including inflammation, angiogenesis, migration and invasion, proliferation, and

L. Du · T. Liu · Y. Xie · S. Zhang · N. Li · C. Wang (✉)
Department of Clinical Laboratory, The Second Hospital of Shandong University, Jinan, Shandong, China
e-mail: cxwang@sdu.edu.cn





Q. Wang, L. Zheng (eds.), *Extracellular Vesicles*,
https://doi.org/10.1007/978-981-99-8365-0_3

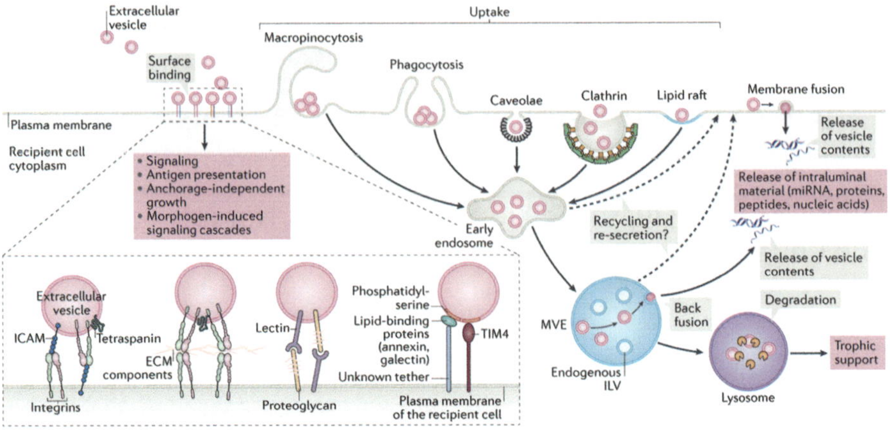


**Fig. 3.1** The process through which extracellular vesicles target recipient cells [2]. EVs recognize and bind to the surfaces of recipient cells based on the cell type. This binding initiates intracellular signaling pathways, such as antigen presentation. EVs may also be internalized by multiple pathways, including staying on the surface or fusing with the membrane to reach multivesicular endosomes and recycling their contents to fuel the metabolism of the recipient cell [2]. Reprinted with permission from Springer Nature

immunosuppression (Fig. 3.2a) [2]. Moreover, EVs derived from tumor cells can act on normal cells adjacent to the tumor. For example, EVs derived from a highly aggressive breast cancer cell line were found to promote the proliferation, migration, and invasion of immortalized fibroblasts and normal breast epithelial cells. This effect depends on the crosslinked form of fibronectin on the surface of EVs derived from tumor cells, which stimulates the expression of integrins on receptor fibroblasts and epithelial cells, activates signal transduction, promotes cell growth under unanchored conditions, and helps normal cells develop some characteristics of cancer cells [7]. Although research in the field of EVs is still in its infancy, many studies have assessed the production, content delivery, and biological functions of EVs. Further in-depth studies of the interaction mechanisms between EVs and target cells will provide insights into new approaches for the treatment and management of various diseases.

## 3.2 Mechanisms Through Which EVs Recognize and Bind to the Surfaces of Recipient Cells

EVs derived from different sources are biased toward interacting with certain cell types [8]. For example, EVs derived from oligodendrocyte precursor cells or mature oligodendrocytes can only be taken up by microglia and not by astrocytes, neurons, or other oligodendrocytes [9]. Therefore, the interaction between receptor cells and EVs may depend on the nature of the recipient cells. Interestingly, epithelial cells and astrocytes in the nervous system cannot take up EVs from transformed cells

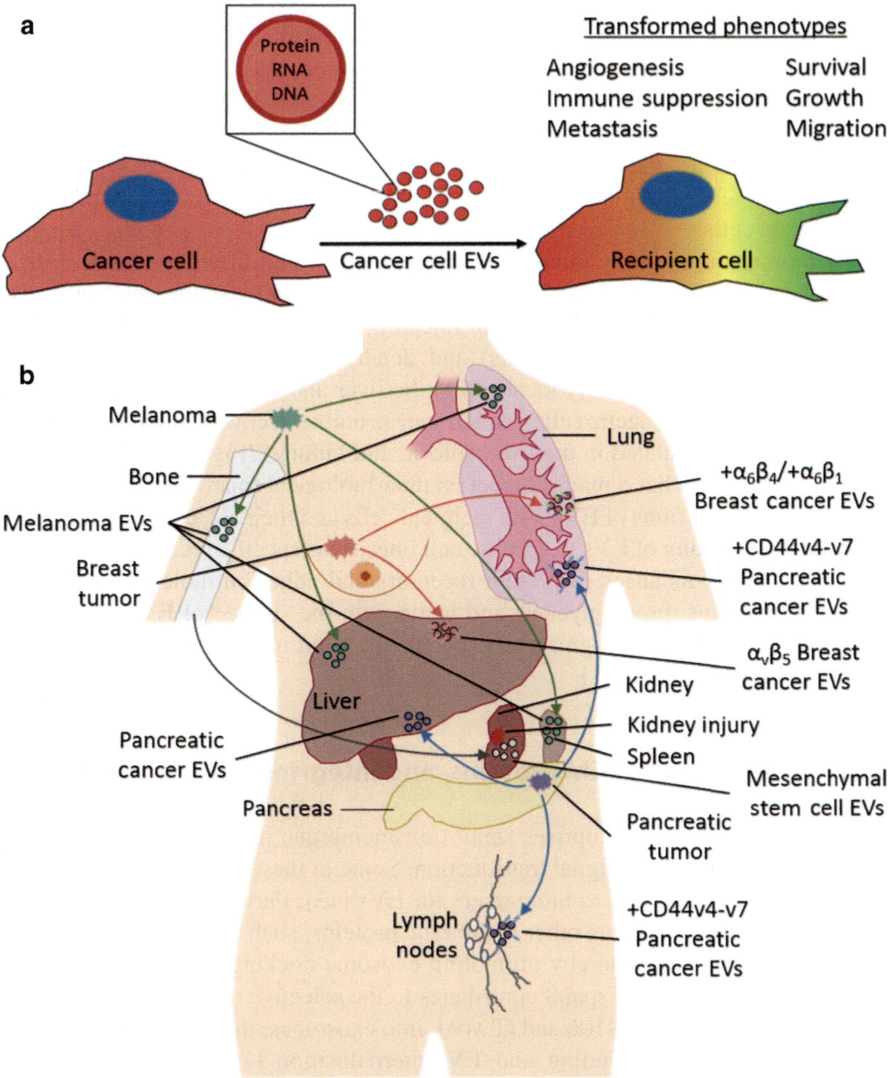


**Fig. 3.2** EV sources and interactions. (**a**) EVs can be produced by tumor cells; carry nucleic acids, proteins, and other substances; and be transmitted to other recipient cells; EVs derived from different tumor cells can specifically accumulate in EVs derived from certain organs. EVs can be produced by tumor cells; carry nucleic acids, proteins, and other substances; and be transmitted to other recipient cells. (**b**) EVs derived from different tumor cells can specifically accumulate in certain organs [12]. Reprinted with permission from Elsevier

[10]; however, after transformation by overexpression of oncogenes *Ras* or *c-Src* in the same cell type, these cells are transformed into epithelial cells and astrocytes that take up EVs. Thus, the induction of cell transformation may lead to the transformation of some cell types, making them more susceptible to EV uptake. EVs can

affect the functions of various cell types, and the mechanisms of recognition and binding between EVs and receptor cells vary substantially among different cell types. For example, when microglia-secreted EVs dock on astrocytes, they remain stationary and are not ingested; however, when they attach to the surface of other microglia cells, they move towards their nuclei on the cell surface before being taken up. Therefore, the mechanisms that regulate the targeted cells of EVs and the fate of EVs after attachment to the cell surface can determine their biological effects on recipient cells.

By monitoring the accumulation of specific EVs from different organs in vivo, we found that preferential interactions between EVs and specific cell types also occurred in animals (Fig. 3.2b). When injecting EVs derived from human embryonic kidney cells (HEK-293T cells) and dendritic cells into mouse blood, we observed that EVs are mainly localized to the liver and spleen, whereas EVs from human mesenchymal stem cells (which can promote recovery after acute tissue damage) are accumulated in the liver, spleen, and kidneys (Fig. 3.2b). Specific targeting of these EVs has a major impact on their biological functions. Therefore, the preferential interactions of EVs with recipient cells and their selective accumulation indicate the targeting of EVs to specific cell lines. This specific recognition depends mostly on the abundances of protein receptors and adhesion molecules (e.g., tetraspanin, integrins, proteoglycans, and lectins) on the surface of EVs (Fig. 3.3a), which enables EVs to recognize and specifically bind to cells expressing specific receptors on the surface [11].

### 3.2.1 Tetraspanin, ECM Proteins, and Integrins

The tetraspanin family comprises small transmembrane proteins that mediate cell adhesion, migration, and signal transduction. Some of these proteins, such as CD63 and CD81, are often used as biomarkers for EVs [13]. Furthermore, exosomal tetraspanin selectively recruits other membrane proteins, such as integrins, and forms tetraspanin complexes, thereby promoting exosome docking and uptake by target cells [14]. For example, Tspan8 contributes to the selective recruitment of proteins and mRNA, including CD106 and CD49d, into exosomes; these proteins are implicated in exosome-EV binding and EV internalization [15]. In addition, ECM protein-integrin interactions also play critical roles in the binding of EVs to receptor cells (Fig. 3.3a). Treating cells with RGD peptides (peptides that block fibronectin-integrin interactions) to inhibit fibronectin binding on the surface of EVs derived from tumor cells or activate integrin $\alpha 5\beta 1$ on receptor fibroblasts can inhibit the anchorage-dependent growth of fibroblasts. Similarly, treatment of trophoblasts with RGD and YIGSR peptides can attenuate the effects of EVs from embryonic stem cells in promoting the migration of trophoblasts; this mechanism involves RGD and YIGSR peptides blocking the binding of cell integrins with fibronectin and laminin from EVs. In addition, the recognition and binding of receptor cells to EVs also depends on the ECM proteins and their EV receptors, such as integrin $\beta 1$, $\alpha v$, $\beta 3$, $\alpha L$, and intercellular adhesion molecule 1 (ICAM-1).

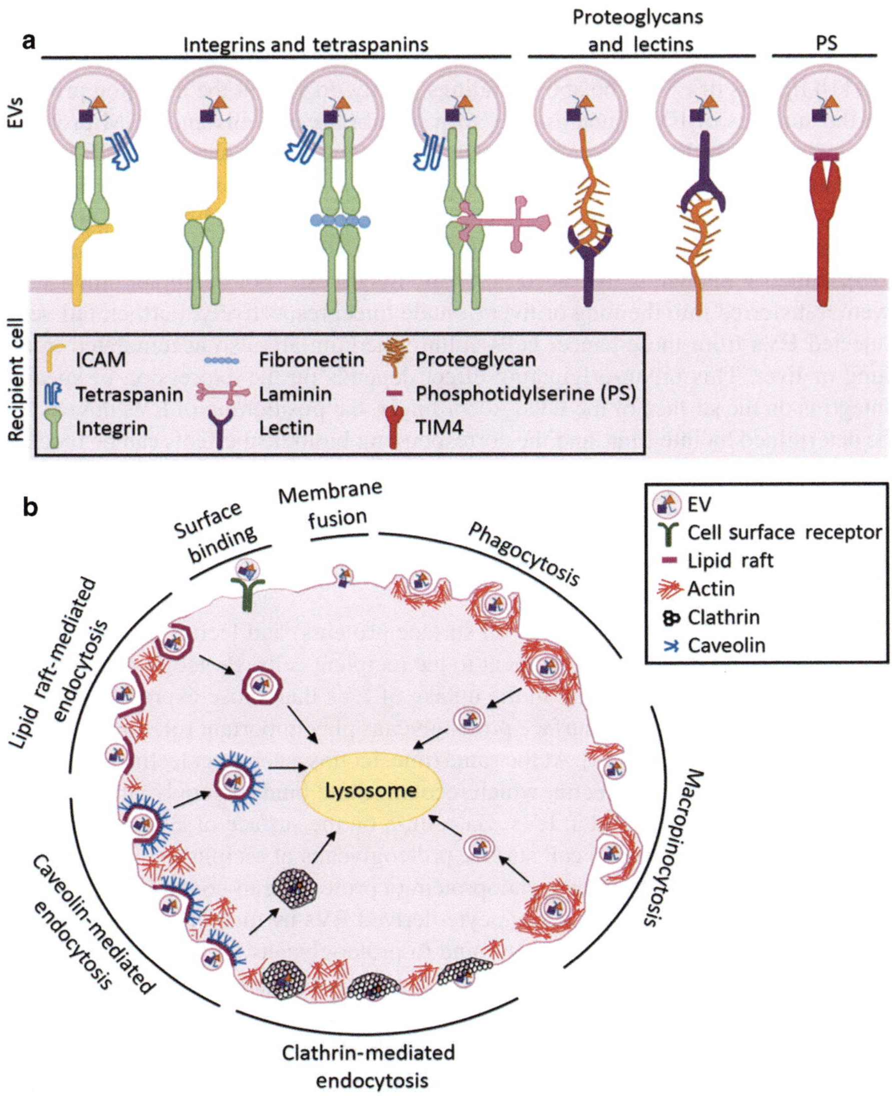


**Fig. 3.3** Schematic diagram of EVs binding to recipient cells. (**a**) EVs bind to the surface of receptor cells through various lipids and adhesion proteins (including tetrameric proteins, integrins, ECM proteins, and proteoglycans). (**b**) EVs bind to the cell surface; membrane fusion, phagocytosis, micropinocytosis, and endocytosis are mediated by clathrin and caveolin; and lipid rafts interact with the recipient cell and are internalized by the recipient cell [12]. Reprinted with permission from Elsevier

In in vivo experiments, integrins on the surface of recipient cells also play important roles in the targeting of EVs to specific recipient cells. EVs expressing ICAM-1 on the membrane surface can bind to T cells and lymph node dendritic cells that express integrin β2. Similarly, EVs derived from pancreatic ductal adenocarcinoma preferentially combine with F4/80$^+$ and CD11b$^+$ Kupffer cells in the liver (Fig. 3.3b) and promote liver metastasis of pancreatic cancer cells through the ICAM and CD11b ligands of EVs. The above findings jointly prove that the adhesion proteins on the surfaces of EVs, such as four-span membrane proteins and ECM proteins, can interact with the corresponding receptors (integrins) on recipient cells to promote the recognition and binding of EVs to recipient cells. However, some EVs can also regulate the recognition of certain cell types because they contain integrins. Moreover, certain types of cancer cells preferentially colonize at secondary sites [16], a theory known as the "seed and soil" hypothesis. For example, tumor cells were transferred into the lung or liver of nude mice, respectively. Further, tail vein-injected EVs from these cancer cells culture medium are also accumulated in the lung or liver. This organ-originating effect depends on the expression of specific integrins on the surface of the EVs. Accordingly, the positioning of EVs in vivo can be determined by integrins, and the corresponding biological effects can be blocked by obstructing integrins targeted by EVs.

### 3.2.2 Proteoglycans and Lectins

Proteoglycans (glycosyl-modified cell surface proteins) and lectins are enriched in EVs and can enhance their attachment to the recipient cells. Proteoglycan-deficient recipient cells are less efficient in the uptake of EVs than those expressing proteoglycans, indicating that cell surface proteoglycans play important roles in the uptake of EVs by recipient cells [17]. At the same time, lectins, such as galectin-1, galectin-3, and galectin-5 and E-selectin, which recognize and bind proteoglycans or glycolipids, have also been found in EVs. Galectin-5 on the surface of EVs mediates the uptake of EVs by binding to cell surface proteoglycans of recipient cells, and treatment with EVs harboring excess sialoprotein (a proteoglycan-containing galectin-5 ligand) reduces the uptake of reticulocyte-derived EVs by macrophages, indicating that galectin-5 in recipient cells may bind to proteoglycans on EVs. In addition to promoting the interaction between EVs and recipient cells in vitro, proteoglycans (CD44v4-v7) in EVs and cell surface lectins (CD169) affect the distribution of EVs in vivo. EVs derived from primary B cells are localized on the surface of CD169-positive cells, whereas the distribution of EVs in CD169-knockout mice is altered, indicating that EV proteoglycans and cell surface lectins can target EVs to specific recipient cells [18]. In conclusion, integrins help target EVs to specific organs in the body, and four-span membrane proteins, proteoglycans, and lectins can participate in the targeting of EVs and the recognition mechanism of recipient cells. In this chapter, we explain the mechanisms through which regulatory receptor cells recognize and bind to EVs. A better understanding of the mechanisms through which EVs target recipient cells will improve our knowledge of the functions of EVs in various fields of biology and facilitate the development of the clinical applications of EVs.

## 3.3 Uptake Mechanisms in Receptor Cells After Binding EVs

When EVs are recognized and bound to the cell surface, they function in two ways. First, the proteins on EVs bind to and activate the receptor expressed on the recipient cell, but the recipient cell does not take up EVs (Fig. 3.3b). Second, EVs can directly fuse to the plasma membrane or be endocytosed. The contents are transferred to recipient cells, thereby playing important biological roles (Fig. 3.3b). EV-mediated signal transduction from the cell surface and the transfer of EV contents into the cell can trigger the biological functions of recipient cells. In this chapter, we focus on the biological mechanisms of EV uptake by recipient cells. The most extensively studied mechanism of EV uptake is endocytosis, which is a biological process involving phagocytic particles or biomolecules. The detection of double-membrane EVs in cells after internalization and the colocalization of EVs with various endosomal and lysosomal markers have demonstrated that recipient cells take up EVs through an active endocytosis process, rather than passive membrane fusion [19].

### 3.3.1 Clathrin-Dependent Endocytosis, Phagocytosis, and Macropinocytosis

The uptake of EVs can be suppressed by disrupting the actin dynamic system of recipient cells, suggesting that the uptake of EVs requires endocytosis mechanisms involving cytoskeletal remodeling, including clathrin-dependent endocytosis, phagocytosis, and macropinocytosis (Fig. 3.3b). When the accessory protein recruited to the plasma membrane induces cell membrane bending, clathrin-dependent endocytosis begins, and the clathrin-coated membrane invades, producing a clathrin-coated sunken lattice. EVs are released from the plasma membrane by membrane separation in the presence of dynein [20]. The uptake of EVs is significantly reduced when clathrin expression in cells is low or when recipient cells are treated with a clathrin inhibitor (chlorpromazine), indicating that EV uptake depends on clathrin-mediated endocytosis. Phagocytosis and actin-dependent and receptor-mediated uptake of large cellular particles are also involved in EV uptake. The process begins when cells recognize EVs and rearrange their cytoskeleton to form a cup-shaped extension around the EVs. Subsequently, the tips of the cup extensions fuse with each other, the membrane ruptures to produce endosomes, and the uptake process ends (Fig. 3.3b). EVs can colocalize with the phagocytosis marker lysosome-associated membrane protein 1 but not with the clathrin-coated depression marker transferrin. Moreover, several proteins known to participate in phagocytosis, such as T-cell immunoglobulin mucin domain, are required for EV uptake. Giant pinocytosis is also involved in EV endocytosis and can be combined with other endocytosis mechanisms. Giant pinocytosis occurs when the cytoskeleton rearranges the membrane and its folds, extending to the laminar pseudopod, folding back on itself, and fusing with the plasma membrane (Fig. 3.3b). Studies have shown that by disrupting dynein, $Na^+/H^+$ exchange, or Rac1 function, macropinocytosis can be inhibited, thereby reducing the uptake of EVs by receptor cells and enhancing the micropinocytosis signaling pathway. The degree of factor 1α activation can also affect EV uptake, indicating that EVs can be taken up through macropinocytosis.

### 3.3.2 Lipid Carriers and Microcapsule Proteins

EVs can also be taken up in different regions of the plasma membrane via a process called lipid carrier transport, which is required for the formation of cave-like invaginations (small cave-like invaginations in the plasma membrane; Fig. 3.3b). EVs can colocalize with lipid carrier markers on recipient cells and disrupt lipid carriers by depleting cellular cholesterol, thereby reducing the cellular uptake of EVs; this mechanism highlights the involvement of lipid carriers in EV uptake. In addition, microencapsulin-1 knockout significantly suppresses the uptake of EVs by recipient cells, suggesting that microencapsulin-1 is also involved in the uptake of EVs. Although the destruction of lipid carriers reduces EV uptake, EVs do not fully colocalize with microencapsulation protein-1 in recipient cells. Thus, EV endocytosis may occur through a lipid carrier-dependent pathway, and the involvement of microencapsulation protein-1 is organ- and cell-specific. In summary, the mechanisms mediating EV uptake are extremely diverse and depend mainly on cell type. EVs can trigger various biological effects by activating surface receptors or signaling pathways in recipient cells through the delivery of contents such as transcripts, signaling proteins, and transcription factors [21, 22].

## 3.4 Effects of EV Contents on Recipient Cells

The third and final step of EV uptake is delivery of the contents to recipient cells. However, the end processes of EV uptake (e.g., delivery of contents, EV degradation, and re-secretion) have not been studied, hampering further functional studies of EV-mediated message transmission. To date, most researchers have concluded that EVs enter cells via cytosis or cytokinesis, deliver EV contents, and induce subsequent biological effects [7, 23]. Therefore, endosomes are a putative location for the delivery of EV contents, and acid pH-induced membrane fusion may mediate content release and function. Lipids, such as cholesterol and phosphatidylserine, also play important roles in membrane fusion. For example, annexin V binds phosphatidylserine exposed on certain EV surfaces and prevents membrane fusion between monocyte-derived EVs and activated platelets. However, many EVs do not have phosphatidylserine on their surfaces, suggesting that other molecules may be involved. Topological analysis of EVs has shown that EV fusion with the endosomal membrane requires a specific mechanism different from the intracellular vesicle fusion process involving SNARE proteins (Fig. 3.4). Second, the mechanisms through which EVs escape from endosomes are passively regulated by the release of partially degraded substances from ruptured endosomes or lysosomes. In addition, when the nucleus or endoplasmic reticulum binds to EV-containing

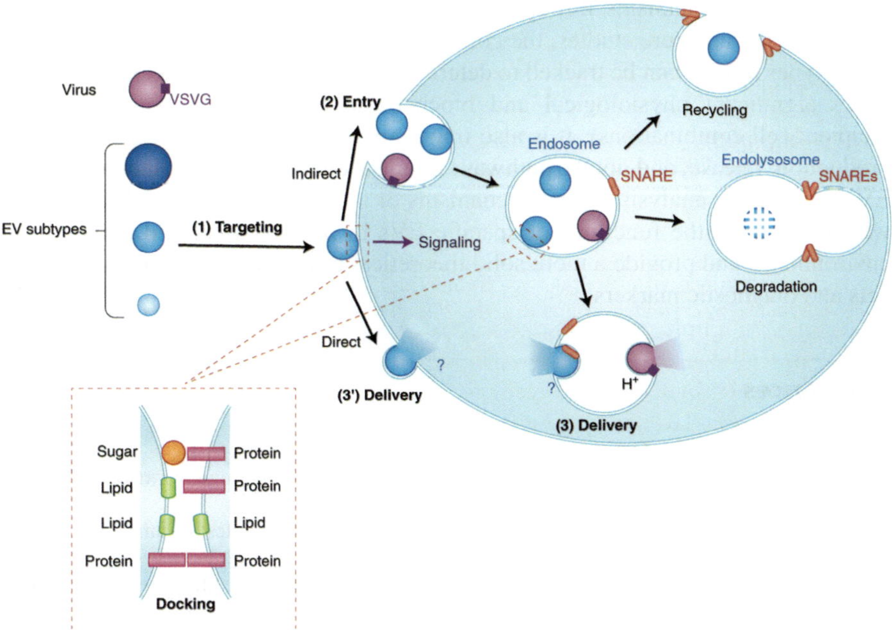


**Fig. 3.4** The process of intracellular vesicle fusion. (1) EVs or specific EV subtypes targeting recipient cells. Docking may occur through specific molecular interactions, including membrane exposure of proteins or lipids (inset) or through nonspecific macropinocytosis or micropinocytosis. (2) The EV subtype enters recipient cells. (3) EV transports within cells. Internalized EVs release their contents through an indirect mechanism of endosome escape, fuse with the plasma membrane, and release the cargo directly into the cytosol [11]

endosomes, the substances contained in the EVs can be directly transferred to the nucleus or endoplasmic reticulum to play important biological roles.

In summary, EV recognition, binding, and uptake underlie the precise functions of EVs. Despite the growing interest in this field, our understanding of the cellular and molecular mechanisms that regulate the biological roles of EVs remains limited. A key limitation to the accurate characterization of EV function is the isolation and characterization of specific subsets of EVs, as the methods we currently use lead to confounding of EVs with different subcellular origins [24]. For example, although many articles use the term "exosomes" to refer to separation from larger EVs by means of hypercentrifugation, ultrafiltration, or precipitators, they are likely to refer to a mixture of exosomes and nonexosomes (small EVs). Therefore, unless their origin from multivesicular bodies is clear, the best general term is "small EVs." Currently, our understanding of EV physiology, diversity, uptake, and content delivery is still too limited to reach precise conclusions regarding how EVs interact with recipient cells and the mechanisms through which they modify recipient cells [25]. To further advance the field of EVs, a comprehensive approach involving molecular, cellular, and functional characterization is needed to differentiate as many EV subtypes as possible. These approaches are essential for identifying the specific molecules or mechanisms through which certain subtypes of EVs interact with receptor cells. In future studies, the common soluble contents present in many different types of EVs can be tracked to determine whether information exchange produces significant physiological and biochemical effects (including donor and recipient cell combinations). It is also important to consider the effects of various production, release, and uptake pathways on intracellular functions, independent of EVs. Systematic analysis of the mechanisms of interaction between EVs and cells will help clarify the functions of specific EVs in physiological and pathological environments and provide a more solid theoretical basis for EVs as therapeutic targets and diagnostic markers.

## References


1. Pegtel DM, Gould SJ. Exosomes. Annu Rev Biochem. 2019;88:487–514.
2. van Niel G, D'Angelo G, Raposo G. Shedding light on the cell biology of extracellular vesicles. Nat Rev Mol Cell Biol. 2018;19(4):213–28.
3. Machtinger R, Laurent LC, Baccarelli AA. Extracellular vesicles: roles in gamete maturation, fertilization and embryo implantation. Hum Reprod Update. 2016;22(2):182–93.
4. Robbins PD, Morelli AE. Regulation of immune responses by extracellular vesicles. Nat Rev Immunol. 2014;14(3):195–208.
5. Hur YH, Feng S, Wilson KF, et al. Embryonic stem cell-derived extracellular vesicles maintain ESC stemness by activating FAK. Dev Cell. 2021;56(3):277–91.e6.
6. Kalluri R, LeBleu VS. The biology, function, and biomedical applications of exosomes. Science. 2020;367(6478).
7. Colombo M, Raposo G, Thery C. Biogenesis, secretion, and intercellular interactions of exosomes and other extracellular vesicles. Annu Rev Cell Dev Biol. 2014;30:255–89.
8. Ratajczak J, Wysoczynski M, Hayek F, et al. Membrane-derived microvesicles: important and underappreciated mediators of cell-to-cell communication. Leukemia. 2006;20(9):1487–95.
9. Fitzner D, Schnaars M, van Rossum D, et al. Selective transfer of exosomes from oligodendrocytes to microglia by macropinocytosis. J Cell Sci. 2011;124(Pt 3):447–58.
10. Lee TH, Chennakrishnaiah S, Meehan B, et al. Barriers to horizontal cell transformation by extracellular vesicles containing oncogenic H-ras. Oncotarget. 2016;7(32):51991–2002.
11. Mathieu M, Martin-Jaular L, Lavieu G, et al. Specificities of secretion and uptake of exosomes and other extracellular vesicles for cell-to-cell communication. Nat Cell Biol. 2019;21(1):9–17.
12. French KC, Antonyak MA, Cerione RA. Extracellular vesicle docking at the cellular port: extracellular vesicle binding and uptake. Semin Cell Dev Biol. 2017;67:48–55.
13. Yang D, Zhang W, Zhang H, et al. Progress, opportunity, and perspective on exosome isolation—efforts for efficient exosome-based theranostics. Theranostics. 2020;10(8):3684–707.
14. Hu W, Liu C, Bi ZY, et al. Comprehensive landscape of extracellular vesicle-derived RNAs in cancer initiation, progression, metastasis and cancer immunology. Mol Cancer. 2020;19(1):102.
15. Nazarenko I, Rana S, Baumann A, et al. Cell surface tetraspanin Tspan8 contributes to molecular pathways of exosome-induced endothelial cell activation. Cancer Res. 2010;70(4):1668–78.
16. Liu Y, Cao X. Characteristics and significance of the pre-metastatic niche. Cancer Cell. 2016;30(5):668–81.
17. Cerezo-Magana M, Bang-Rudenstam A, Belting M. The pleiotropic role of proteoglycans in extracellular vesicle mediated communication in the tumor microenvironment. Semin Cancer Biol. 2020;62:99–107.
18. Saunderson SC, Dunn AC, Crocker PR, et al. CD169 mediates the capture of exosomes in spleen and lymph node. Blood. 2014;123(2):208–16.

19. Mulcahy LA, Pink RC, Carter DR. Routes and mechanisms of extracellular vesicle uptake. J Extracell Vesicles. 2014;3.
20. Costa Verdera H, Gitz-Francois JJ, Schiffelers RM, et al. Cellular uptake of extracellular vesicles is mediated by clathrin-independent endocytosis and macropinocytosis. J Control Release. 2017;266:100–8.
21. Ridder K, Sevko A, Heide J, et al. Extracellular vesicle-mediated transfer of functional RNA in the tumor microenvironment. Onco Targets Ther. 2015;4(6):e1008371.
22. Squadrito ML, Baer C, Burdet F, et al. Endogenous RNAs modulate microRNA sorting to exosomes and transfer to acceptor cells. Cell Rep. 2014;8(5):1432–46.
23. Schorey JS, Cheng Y, Singh PP, et al. Exosomes and other extracellular vesicles in host-pathogen interactions. EMBO Rep. 2015;16(1):24–43.
24. van der Pol E, Boing AN, Harrison P, et al. Classification, functions, and clinical relevance of extracellular vesicles. Pharmacol Rev. 2012;64(3):676–705.
25. Raposo G, Stoorvogel W. Extracellular vesicles: exosomes, microvesicles, and friends. J Cell Biol. 2013;200(4):373–83.

# Biological Functions of Extracellular Vesicles in Various Systems

# 4

Peipei Wu, Jiajia Jiang, and Wenrong Xu

## 4.1 Components of Extracellular Vesicles (EVs)

EVs are rich in various biologically active substances, mainly proteins, nucleotides, and lipids. In recent years, the application of technologies such as mass spectrometry and high-throughput sequencing has ushered in major advances in protein, nucleotide, and lipidomic analysis technologies, enabling large-scale screening and identification of EV contents. The results of these and many other studies on mammalian EVs are summarized in the Vesiclepedia database (http://microvesicles.org), which also includes continuously updated data from other types of EVs. The Vesiclepedia database released the latest version of the statistical results of EVs from a variety of cells and various extracellular vesicle components, including 41 species, with a total of 1254 research results; 349,988 proteins, 27,646 mRNAs, 10,520 microRNAs (miRNAs), and 639 types of lipids were identified. Another database includes studies on nonmammalian EVs of all sizes (http://evpedia.info). These two databases contain data on proteins, nucleic acids, and lipids as well as details of the purification processes used.

Peipei Wu and Jiajia Jiang contributed equally to this work.

P. Wu · W. Xu (✉)
Jiangsu Key Laboratory of Medical Science and Laboratory Medicine School of Medicine, Jiangsu University, Zhenjiang, Jiangsu, China
e-mail: icls@ujs.edu.cn

J. Jiang
Aoyang Institute of Cancer, Affiliated Aoyang Hospital of Jiangsu University, Zhangjiagang, Jiangsu, China



Q. Wang, L. Zheng (eds.), *Extracellular Vesicles*,
https://doi.org/10.1007/978-981-99-8365-0_4

Proteins, important components of EVs, are rich in variety and complex in structure (Fig. 4.1). EV proteins are mainly divided into two types. The first type is ubiquitous proteins that participate in the formation of their own structures, including cytoskeletal components, such as tubulin, actin, and microfilament binding protein; membrane fusion and transport-related protcins, such as annexins, ALG-2-interacting protein X, and tumor susceptibility 101 protein, and the four-pass transmembrane protein superfamily, including CD9, CD63, and CD81; and heat shock proteins (HSPs), such as HSP70 and HSP90. The second type is related to the source of cells; for example, EVs derived from antigen-presenting cells are rich in major histocompatibility complex (MHC)-I, MHC-II, CD80, and CD86, whereas platelet-derived EVs contain hemophilia factor and integrin CD41a, and tumor cell-derived EVs exhibit overexpression of some proteins, including Fas ligand (FasL), transforming growth factor (TGF)-β, and other components [1].

The nucleotides in EVs are approximately 200-bp fragments and can be translated into functional proteins that affect the biological functions of recipient cells.

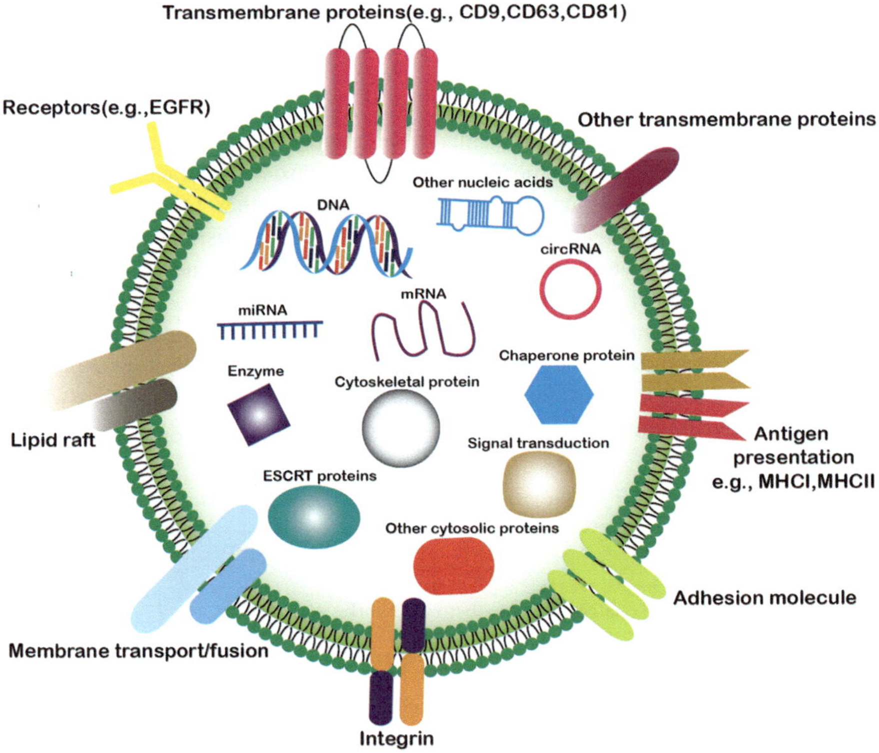


**Fig. 4.1** The main contents of EVs consist of nucleic acids, proteins, lipids and metabolites. EV proteins can be divided into two categories: non-specific proteins and specific proteins; EV nucleic acids mainly include genomic DNA, mitochondrial DNA, mRNA, microRNA, lncRNA, and circular RNA; EVs are also rich in lipids such as cholesterol, polyglycerol, phospholipids, diglycerides, and, phosphatidyl ethanolamine.

EVs contain many types of nucleic acids, including genomic DNA [2], mitochondrial DNA [3], and RNAs (e.g., mRNAs, miRNAs, long noncoding RNAs, and circular RNAs) [4, 5]. As a class of extremely important biological macromolecules, RNAs play crucial roles in regulating gene expression and functions in target cells. These molecules can regulate the translation and posttranslational modification of mRNAs and proteins through various regulatory mechanisms, thereby affecting cellular biological functions and disease progression. Accordingly, RNAs play important roles in normal physiology, pathology, and biological evolution.

EVs are also enriched in cholesterol, phospholipids, phosphatidylethanolamine, and other lipids, and their lipid compositions are closely related to their cell of origin. Similarly, in addition to participating in the biosynthesis and uptake of EVs, these lipids are also important biologically active molecules involved in various biological processes, such as immune surveillance, tumor microenvironment modification, and inflammation regulation. Therefore, lipids in EVs may also be used as biomarkers for disease diagnosis and treatment [6].

## 4.2 Biological Functions of EVs

The earliest research on the biological functions of EVs was related to the relationship between exosomes and reticulocyte maturation. Reticulocyte exosomes selectively remove transferrin receptors, membrane-related enzymes, and proteins to promote the maturation of reticulocytes [7]. To date, EVs have been shown to be released by almost all mammalian cells and to be widely distributed in various bodily fluids, including the saliva, urine, and blood. EVs act as a medium for cell communication and modify the biological behaviors of neighboring or distant cells through paracrine or endocrine methods.

### 4.2.1 The Normal Physiology of EVs

The physiological processes regulated by EVs are as follows: (1) antigen presentation, (2) immune response, (3) neuronal communication, (4) organ development, and (5) reproduction (Fig. 4.2). EVs can activate the immune system, induce macrophages to release pro-inflammatory cytokines [8] and tumor necrosis factor [9], enhance the activity of natural killer (NK) cells [10], promote dendritic cell (DC) activity [11], mediate the delivery of MHC peptide molecules [12], and alter antigen presentation [13]. EVs also participate in the mutual signal transfer among sensory and motor neurons, interneurons, and glial cells [14]. EVs are also involved in multiple stages of reproductive development, including gamete maturation, fertilization, embryo implantation, and even successful pregnancy. EVs rely on immune communication between the mother and fetus. At the same time, they can affect fetal health by regulating local and/or systemic immune responses. For example, EVs mediate the communication between epithelium and mesenchyme and therefore play key roles in organ development. EVs secreted by keratinocytes can also

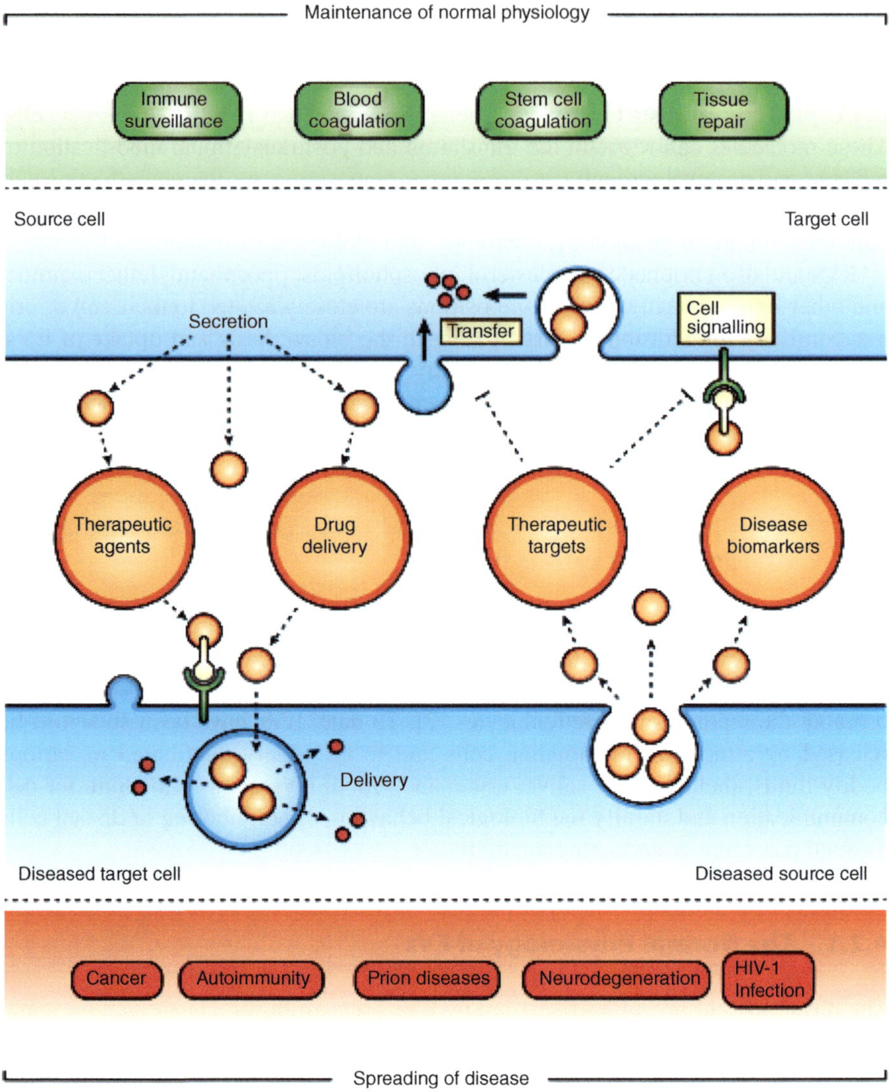


**Fig. 4.2** Biological functions of EVs [22]

enhance melanin synthesis, suggesting that EVs have important roles in human skin pigmentation [16–19]. In addition, EVs modulate the aging of the nervous system, and miRNAs in EVs derived from hypothalamic neural stem/progenitor cells can delay aging [15]. The effects of EVs on cell proliferation cannot be neglected. Indeed, hepatocyte-derived EVs can transfer sphingosine kinase 2 to target cells, thereby facilitating hepatocyte proliferation and liver regeneration [20]. Additionally, CrkI-containing EVs derived from apoptotic cells can induce the proliferation of neighboring cells, thereby contributing to restoration of homeostasis [21].

## 4.2.2 The Relationship Between EVs and Disease

### 4.2.2.1 EVs and Tumor Occurrence, Development, Diagnosis, and Treatment

As a novel cell-to-cell communication mechanism, EVs play important roles in the occurrence and development of tumors. They can transfer biologically active molecules to recipient cells, leading to the exchange of genetic information and reprogramming of the recipient cells.

Accumulating evidence has shown that tumor cells can release excessive amounts of EVs containing oncogenic proteins and nucleotide components and deliver them to target cells, which can induce normal cells to become cancerous, promote neovascularization, improve tumor cell migration and invasion, and enhance drug resistance. In addition, EVs can transfer tumor cell-derived components to cells in the tumor microenvironment (including immune cells and stromal cells) for various interactions; assist tumors in evading immune surveillance; promote the formation of tumor niches; and affect tumor occurrence, progression, metastasis, and resistance. Different types of cells in various disease states can secrete EVs with specific contents, including different proteins, nucleic acids, lipids, and other components; these components vary with the type of tissue and cell of origin and can therefore reflect the pathological state of the derived cells. Given that EVs have cell-type specificity and stability and can be obtained from body fluids, their various components can act as effective biomarkers for early disease diagnosis, treatment, and prognosis. The use of EVs as biomarkers has a broad significance for individualized disease diagnosis and monitoring.

### 4.2.2.2 EVs and Immunosuppressive Effects

EVs can regulate the development, differentiation, and maturation of tissues and organs and thereby exert substantial effects on pathological processes, including cardiovascular diseases, neurodegenerative diseases, and viral infection and transmission. EVs can inhibit inflammation and exert immunosuppressive effects by reducing NK cell activity or DC differentiation, inducing FasL-mediated T-cell apoptosis, inhibiting T-cell proliferation, and delivering EVs harboring interleukin-10 [22]. In addition, tumor-derived EVs expressing HSP72 on their surface can promote antigen nonspecific immunosuppression through myeloid-derived suppressor cells. Breast cancer cells produce more EVs under hypoxic conditions than normoxic conditions, and these EVs can inhibit T-cell proliferation through the TGF-β pathway [23]. Recent studies have shown that cancer EVs can provide various signals to immune cells, overcome immune system monitoring, and induce the immune system to specifically ignore or tolerate cancer cells. This tolerogenic function of cancer cell-derived EVs may be useful in the context of tumor immune evasion [24]. Therefore, EVs may have dual conflicting roles in immune regulation.

### 4.2.2.3 EVs and Pathogenic Infections

In addition to toxic proteins, EVs may transmit various pathogens and pathogenic factors, which can be incorporated into EVs derived from virus-infected cells.

EV-mediated factor delivery can affect the immune response of infected recipient cells. For example, EVs can transfer viral miRNAs to recipient cells to inhibit subcellular gene sites, which also participate in the N-terminal modification of the prion protein (PrP) and selectively deliver different PrP sugar forms to neuronal cells. In addition, human immunodeficiency virus-1 achieves cell entry through the EV-mediated transfer of chemokine receptor 5 to recipient cells. These findings indicate that viruses (particularly retroviruses) and EVs produced in virus-infected cells may be closely related [25–28].

#### 4.2.2.4 EVs and Regenerative Medicine

Many studies have shown that EVs, particularly stem cell-derived EVs, can play important roles in the treatment of many diseases and may have applications in tissue regeneration and repair and tumor-targeted therapy. Increasing evidence has confirmed that the therapeutic activity of EVs is mediated by adult stem cells through paracrine action, which has given rise to a new concept of "cell-free" stem cell therapy. EVs derived from mesenchymal stem cells (MSCs) are good alternatives to therapeutic stem cells. Bone marrow-, umbilical cord-, and adipose tissue-derived MSCs could act as pivotal components of parental cell paracrine function; EVs have been shown to play key roles in tissue regeneration and repair and in damage to the heart, liver, kidneys, and skin [29]. EVs derived from human umbilical cord MSCs can reverse insulin resistance in peripheral tissues, reduce islet β-cell damage, and decrease blood glucose levels in a rat model of type 2 diabetes. These studies further highlighted the important roles of these EVs in disease treatment [30]. EVs derived from other cells, including macrophages and DCs, are also expected to become important therapeutic tools for biological drug delivery and tissue regeneration and repair.

#### 4.2.2.5 Research on Other Applications of EVs

As natural carriers, EVs can be isolated from various cells. EVs from different sources possess specific markers and functions. Current research has found that EVs can not only be developed into vaccines for disease prevention and treatment but can also be modified into good drug or gene delivery tools to optimize the existing treatment efficiency. In view of the role of EVs in promoting pathological progress, researchers have proposed three promising strategies to thoroughly develop EVs for the treatment of diseases: (1) inhibiting the assembly, release, and uptake of EVs; (2) changing harmful components; and (3) blocking the spread of EVs and eliminating EVs from the circulatory system. Emerging theories and strategies provide new hope for the treatment of many diseases [22].

## 4.3 Conclusion

In this chapter, we reviewed the key components of recently identified EVs and focus on the important roles of EVs in the physiological and pathological states of multiorgan systems.

EVs play essential roles in the maintenance of normal physiological stability. Tissues and organs of the human body maintain normal functioning by secreting EVs to regulate the functions of their own and neighboring tissue cells. However, more studies have focused on the pivotal roles of EVs in different pathological processes. During the occurrence and development of various tumor diseases, a large number of EVs are released, creating favorable conditions for tumor proliferation, metastasis, drug resistance, and immune escape. In addition, when other diseases occur, such as cardiovascular diseases, the number and contents of EVs also change significantly. These EVs, which carry specific genetic information, also affect the course of the disease in different ways. With advances in approaches to study EVs, the functions of EVs in multiple systems have been discovered. Regulating the functions of EVs under physiological or pathological conditions to better maintain body homeostasis or for the treatment of diseases is particularly urgent.

**Acknowledgments** We obtained funding from the National Natural Science Foundation of China (grant no. 81971757) and Jiangsu Province's Major Project in Research and Development (grant no. BE2020680).

## References

1. He C, Zheng S, Luo Y, Wang B. Exosome theranostics: biology and translational medicine. Theranostics. 2018;8(1):237–55.
2. Thakur BK, Zhang H, Becker A, et al. Double-stranded DNA in exosomes: a novel biomarker in cancer detection. Cell Res. 2014;24:766–9.
3. Sansone P, Savini C, Kurelac I, et al. Packaging and transfer of mitochondrial DNA via exosomes regulate escape from dormancy in hormonal therapy-resistant breast cancer. Proc Natl Acad Sci U S A. 2017;114:E9066–75.
4. Colombo M, Raposo G, Théry C. Biogenesis, secretion, and intercellular interactions of exosomes and other extracellular vesicles. Annu Rev Cell Dev Biol. 2014;30:255–89.
5. Li S, Li Y, Chen B, et al. exoRBase: a database of circRNA, lncRNA and mRNA in human blood exosomes. Nucleic Acids Res. 2017;46(D1):D106–12.
6. Record M, Silventepoirot S, Poirot M, et al. Extracellular vesicles: lipids as key components of their biogenesis and functions. J Lipid Res. 2018;59(8):1316–24.
7. Johnstone RM, Adam M, Hammond JR, Orr L, Turbide C. Vesicle formation during reticulocyte maturation. Association of plasma membrane activities with released vesicles (exosomes). J Biol Chem. 1987;262:9412–20.
8. Bitencourt TA, Rezende CP, Quaresemin NR, et al. Extracellular vesicles from the dermatophyte trichophyton interdigitale modulate macrophage and keratinocyte functions. Front Immunol. 2018;9:2343.
9. Multhoff G, Pockley AG, Schmid TE, Schilling D. The role of heat shock protein 70 (Hsp70) in radiation-induced immunomodulation. Cancer Lett. 2015;368(2):179–14.
10. Skokos D, Botros HG, Demeure C, et al. Mast cell-derived exosomes induce phenotypic and functional maturation of dendritic cells and elicit specific immune responses in vivo. J Immunol. 2003;170:3037–45.
11. Pang XL, Wang ZG, Liu L, et al. Immature dendritic cells derived exosomes promotes immune tolerance by regulating T cell differentiation in renal transplantation. Aging (Albany NY). 2019;11(20):8911–24.
12. Olejarz W, Dominiak A, Żołnierzak A, Kubiak-Tomaszewska G, Lorenc T. Tumor-derived exosomes in immunosuppression and immunotherapy. J Immunol Res. 2020;2020:6272498.

13. Budnik V, Ruiz-Cañada C, Wendler F. Extracellular vesicles round off communication in the nervous system. Nat Rev Neurosci. 2016;17:160–72.
14. Zhang G, Li J, Purkayastha S, et al. Hypothalamic programming of systemic ageing involving IKK-beta, NF-kappaB and GnRH. Nature. 2013;497:211–6.
15. Zhang Y, Kim MS, Jia B, et al. Hypothalamic stem cells control ageing speed partly through exosomal miRNAs. Nature. 2017;548:52–7.
16. Yanez-Mo M, Siljander PR, Andreu Z, et al. Biological properties of extracellular vesicles and their physiological functions. J Extracell Vesicles. 2015;4:27066.
17. Arck PC, Hecher K. Fetomaternal immune cross-talk and its consequences for maternal and offspring's health. Nat Med. 2013;19:548–56.
18. Jiang N, Xiang L, He L, et al. Exosomes mediate epithelium-mesenchyme crosstalk in organ development. ACS Nano. 2017;11:7736–46.
19. Cicero AL, Delevoye C, Gilles-Marsens F, et al. Exosomes released by keratinocytes modulate melanocyte pigmentation. Nat Commun. 2015;6:7506.
20. Nojima H, Freeman CM, Schuster RM, et al. Hepatocyte exosomes mediate liver repair and regeneration via sphingosine-1-phosphate. J Hepatol. 2016;64:60–8.
21. Gupta KH, Goldufsky JW, Wood SJ, et al. Apoptosis and compensatory proliferation signaling are coupled by CrkI-containing microvesicles. Dev Cell. 2017;41:674–84.
22. Andaloussi SEL, Mäger I, Breakefield XO, Wood MJA. Extracellular vesicles: biology and emerging therapeutic opportunities. Nat Rev Drug Discov. 2013;12:347–57.
23. Rong L, Li R, Li S, Luo R. Immunosuppression of breast cancer cells mediated by transforming growth factor-β in exosomes from cancer cells. Oncol Lett. 2015;11:500–4.
24. Yin Z, Yu M, Ma T, et al. Mechanisms underlying low-clinical responses to PD-1/PD-L1 blocking antibodies in immunotherapy of cancer: a key role of exosomal PD-L1. J Immunother Cancer. 2021;9(1):e001698.
25. Wu J, Yang J, Ding J, et al. Exosomes in virus-associated cancer. Cancer Lett. 2018;438:44–51.
26. Dreux M, Garaigorta U, Boyd B, et al. Short-range exosomal transfer of viral RNA from infected cells to plasmacytoid dendritic cells triggers innate immunity. Cell Host Microbe. 2012;12:558–70.
27. Yoshikawa FSY, Teixeira FME, Sato MN, Oliveira LMDS. Delivery of microRNAs by extracellular vesicles in viral infections: could the news be packaged? Cell. 2019;8(6):611.
28. Mack M, Kleinschmidt A, Brühl H, et al. Transfer of the chemokine receptor CCR5 between cells by membrane-derived microparticles: a mechanism for cellular human immunodeficiency virus 1 infection. Nat Med. 2000;6:769–75.
29. Wu P, Zhang B, Shi H, Qian H, Xu W. MSC-exosome: a novel cell-free therapy for cutaneous regeneration. Cytotherapy. 2018;20(3):291–301.
30. Sun Y, Shi H, Yin S, et al. Human mesenchymal stem cell derived exosomes alleviate type 2 diabetes mellitus by reversing peripheral insulin resistance and relieving β-cell destruction. ACS Nano. 2018;12(8):7613–28.

# The Roles and Mechanisms of EVs in Disease Processes

# 5

Tingjiao Liu

## 5.1 Introduction

Extracellular vesicles (EVs) mediate cell-to-cell communication and play critical roles in the development and progression of many diseases. EVs can transfer bioactive molecules to recipient cells, leading to genetic information exchange and reprogramming of the recipient cells. Increasing evidence shows that cancer cells transmit oncogenic proteins and nucleotide components to target cells via EVs to induce the carcinogenesis of normal cells, promote angiogenesis, enhance the migration and invasion abilities of tumor cells, and increase drug resistance. EVs can also transfer the products of tumor cells to other cells, including stromal, endothelial, and immune cells, in the primary tumor microenvironment, resulting in enhanced invasion and metastasis, immune tolerance, and drug resistance. Furthermore, EVs participate in the development of diabetes and neurodegenerative diseases as well as the spread of viral infections. This chapter summarizes the roles and mechanisms of EVs in the development and progression of tumors, diabetes, neurodegenerative diseases, and infectious diseases.

T. Liu (✉)
Department of Oral Pathology, Shanghai Stomatological Hospital and School of Stomatology, Fudan University, Shanghai, China

Shanghai Key Laboratory of Craniomaxillofacial Development and Diseases, Fudan University, Shanghai, China
e-mail: tingjiao_liu@fudan.edu.cn



Q. Wang, L. Zheng (eds.), *Extracellular Vesicles*,
https://doi.org/10.1007/978-981-99-8365-0_5

## 5.2 Roles of EVs in Cancer

EVs not only mediate communication between tumor cells and nontumor cells but also participate in information sharing between tumor cells, thereby regulating the tumor cell proliferation and invasion, angiogenesis, and immunosuppression. In particular, EVs have a protective effect on their contents and can therefore accurately and efficiently deliver molecules from primary tumor cells to distant organs, mediate information exchange between the primary tumor and future metastatic organs, and promote tumor metastasis.

### 5.2.1 EVs Mediate Molecular Exchange Between Tumor Cells

Tumors show intratumoral heterogeneity in genotype and phenotype, reflecting the high complexity and diversity of cancer evolution. EVs from tumor cells can transfer oncogenic molecules from one tumor cell subgroup to another. Al-Nedawi et al. [1] showed that glioma cells shared epidermal growth factor receptor variant III (EGFRvIII) by intercellular transfer of large EVs (microvesicles). Large EVs containing EGFRvIII are released from EGFRvIII-expressing glioma cells and merge with the plasma membranes of cancer cells lacking EGFRvIII. Consequently, the transferred EGFRvIII activated the mitogen-activated protein kinase (MAPK) and Akt signaling pathways, altered the expression of EGFRvIII-regulated genes (e.g., *VEGF*, *Bcl-xL*, and *p27*), and increased anchorage-independent growth capacity. Mutations in the gene encoding the Ser/Thrkinase BRAF are found in 40–60% of patients with melanoma. Cesi et al. [2] reported that the activation of a novel truncated form of anaplastic lymphoma kinase (ALK) is associated with acquired drug resistance in patients with melanoma. ALK knockdown or inhibition enables resistant cancer cells to become sensitive to BRAF inhibition. EVs containing truncated ALK are vehicles for transferring drug resistance between tumor cells.

### 5.2.2 EVs Promote the Epithelial-Mesenchymal Transition (EMT)

The EMT is a biological process through which epithelial cells lose their epithelial phenotype and gain a mesenchymal phenotype. This process is characterized by the loss of tight junctions and cell polarity among epithelial cells, which results in increased migration, invasion, and extracellular matrix (ECM) degradation. The EMT is an important biological mechanism underlying the migration and invasion of malignant cancer cells derived from epithelial cells. EVs have been shown to have important roles in this process.

Mesenchymal cells promote the EMT in tumor cells. Luga et al. [3] demonstrated that EVs secreted by fibroblasts or carcinoma-associated fibroblasts (CAFs) promote the EMT in breast cancer cells in vitro and the metastasis of BCCs in vivo. Additional studies have suggested that the activation of Wnt-planar cell polarity signaling is an important molecular mechanism through which EVs promote BCC migration and metastasis, and CD81-positive EVs play critical roles in this process. Donnarumma et al. [4] found that EVs secreted by CAFs can present three different microRNAs (miRNAs; *miR-21*, *miR-378e*, and *miR-143*) to BCCs to induce the EMT. Furthermore, Li et al. [5] found that EVs secreted by ovarian CAFs expressed high levels of transforming growth factor (TGF)-β compared with normal omentum fibroblasts. These EVs promote the EMT in ovarian cancer cells by activating the SMAD signaling pathway. Zhao et al. [6] demonstrated that EVs derived from human umbilical mesenchymal stem cells can promote the EMT in A549 lung cancer cells and that silencing of TGF-β1 in these cells reversed the EMT-promoting effect.

Highly metastatic tumor cells secrete EVs, which can induce the EMT in tumor cells with low metastatic ability, thereby increasing their invasion. Chen et al. [7] demonstrated that EVs derived from highly metastatic tumor cells could communicate with weakly metastatic liver cells, inducing the EMT and activating MAPK/extracellular signal-regulated kinase signaling, thereby increasing their migration and invasion. Rahman et al. [8] reported that EVs secreted by highly metastatic lung cancer cells induce the EMT in human bronchial epithelial cells and enhance their migration, invasion, and proliferation. Additionally, Schillaci et al. [9] reported that EVs released by highly metastatic colon cancer cells may influence the morphology of less aggressive colon cancer cells and enhance their motility and migration.

Hypoxia can induce the EMT in tumor cells. Ramteke et al. [10] showed that human prostate cancer cells exposed to hypoxic conditions released EVs with smaller average sizes and higher levels of CD63, CD81, heat-shock protein (HSP) 70, HSP90, and annexin II compared with cells exposed to normoxic conditions. EVs secreted by hypoxic prostate cancer cells induce the EMT in prostate cancer cells under normoxic conditions. Further studies demonstrated that prostate cancer cells secrete EVs with higher metalloproteinase activity and increased levels of TGF-β2, tumor necrosis factor-1α, interleukin (IL)-6, tumor susceptibility 101, Akt, integrin-linked kinase-1, and β-catenin under hypoxic conditions. Aga et al. [11] demonstrated that hypoxia-inducible factor-1α (HIF-1α) was detectable in EVs derived from nasopharyngeal carcinoma (NPC) and that latent membrane protein (LMP1) significantly increased the levels of HIF-1α in EVs. LMP1 is a key protein in Epstein-Barr virus (EBV)-induced NPC and plays a key role in the malignant progression of NPC. HIF-1α in EVs induces the EMT in EBV-negative cells and enhances their migration and invasion.

### 5.2.3 EVs in Angiogenesis

Compared with normal tissues, tumors usually exhibit high microvessel density, which is closely related to active angiogenesis. Angiogenesis refers to the formation of new blood vessels that sprout from preexisting vessels. Angiogenesis involves ECM degradation, vascular endothelial cell (EC) proliferation and migration, and lumen formation. EVs secreted by tumor or stromal cells contain pro-angiogenic cytokines or miRNAs and promote angiogenesis.

Tumor cell EVs can transfer pro-angiogenic factors to ECs and induce angiogenesis. Millimaggi et al. [12] reported that EVs released by epithelial ovarian cancer cells are rich in CD147, which can induce the expression of matrix metalloproteinases (MMPs) and tube formation in human umbilical vein ECs in vitro. Tang et al. [13] found that soluble E-cadherin (an 80-kDa soluble form) was highly expressed in the malignant ascites of patients with ovarian cancer as a potent inducer of angiogenesis. Further studies demonstrated that soluble E-cadherin was abundantly released in the form of EVs and transferred into ECs, resulting in activation of the β-catenin and nuclear factor-κB signaling pathways. Al-nedawi et al. [14] reported that EGFR-positive EVs secreted by human squamous cell carcinoma cells can be internalized by ECs, thereby activating the vascular endothelial growth factor receptor (VEGFR) 2 signaling pathway, which is one of the most important signaling pathways regulating angiogenesis. Liu et al. [15] reported that EVs from the cerebrospinal fluid of patients with glioblastoma, rather than plasma EVs, could promote EC migration, proliferation, and angiogenesis through the β-catenin/Akt pathway.

EVs carry nucleic acids, particularly miRNAs, which play important roles in promoting angiogenesis. Hsu et al. [16] reported that EVs released by lung cancer cells under hypoxic conditions can induce angiogenesis and increase permeability. Further studies have revealed that elevated levels of *miR-23a* in EVs released from hypoxic lung cancer cells may be a target for lung cancer treatment. Umezu et al. [17] found that EVs secreted by multiple myeloma cells under hypoxic conditions contained *miR-135b*, which could be internalized by ECs and enhance endothelial tube formation.

Li et al. [18] reported that EVs derived from CAFs were the main regulators of angiogenesis. VEGF interacts with heparan sulfate proteoglycans on the surface of CAF EVs and induces angiogenesis in a bevacizumab-resistant manner. EVVEGF can be dissociated by heparinase and neutralized by bevacizumab.

### 5.2.4 EVs Contribute to Immunosuppression

It was initially hypothesized that EVs may contribute to immune regulation because EVs secreted by immune cells carry MHC I, MHC II, and T-cell costimulating molecules. Over the past decade, studies have shown that EVs secreted by cells other than immune cells (such as epithelial and tumor cells) also carry MHC I molecules and/or tumor antigens that regulate the immune system. Most EVs secreted by tumor cells show immunosuppressive effects, thus promoting tumor progression and metastasis; however, there is also evidence that tumor EVs can both promote tumor progression and exert antitumor effects. This complex relationship between EVs and the immune system may be related to the cell origin and EV cargo.

Chen et al. [19] reported that metastatic melanomas release EVs with programmed cell death ligand 1 (PD-L1) on their surface. Interferon-γ (IFN-γ) stimulation increases the amount of PD-L1 on EV surfaces and suppresses the functions of $CD8^+$ T cells. Circulating EV-associated PD-L1 levels are positively correlated with

IFN-γ levels and vary during anti-programmed cell death-1 therapy in patients with metastatic melanoma.

Tumor EVs can directly stimulate immune cells or deliver antigens to dendritic cells (DCs), which present them to T cells and induce primary cytotoxic reactions. Chen et al. [20] found that EVs from heat-shocked mouse B lymphoma cells contained more molecules involved in immunogenicity than control EVs secreted by the same cells. Furthermore, heat-shocked EVs induced both phenotypic and functional maturation of DCs more efficiently and activated T-cell responses more potently. In addition, heat-shocked EVs induced enhanced antitumor immune responses in prophylaxis and therapeutic in vivo lymphoma models compared with control EVs from the same cells. They further demonstrated that $CD8^+$ T cells were the predominant T-cell subset responsible for the antitumor effects of heat-shock EVs.

Tumor-associated macrophages (TAMs) are important immune cells in the tumor microenvironment and have been shown to promote the formation of an immunosuppressive tumor microenvironment. Wen et al. [21] isolated tumor cell-released autophagosomes from multiple murine tumor cell lines and biofluids of patients with cancer. They found that these autophagosomes converted macrophages into an immunosuppressive M2-like phenotype with PD-L1 and IL-10 expression. M2-like macrophages inhibit the proliferation of both $CD4^+$ and $CD8^+$ T cells in vitro.

Cooks et al. [22] found that human *TP53* mutant (mutp53) cancer cells reprogram macrophages to a tumor-supportive and anti-inflammatory state. Colon cancer cells harboring oncogenic mutp53 predominantly secrete *miR-1246*-rich EVs. Macrophages internalize these EVs and transform into a cancer-promoting state in an *miR-1246*-dependent manner. These mutp53-reprogammed TAMs formed an immunosuppressive microenvironment with increased TGF-β activity.

Regulatory T cells (Tregs) and helper T cells 17 (Th17), two subsets of $CD4^+$T cells, are important tumor immune cells. The Treg/Th17 ratio is associated with the histological grade of ovarian cancer and is an independent prognostic factor for predicting the overall survival of patients. The immune microenvironment of epithelial ovarian cancer (EOC) includes TAMs, Tregs, and Th17 cells. Zhou et al. [23] reported that EVs secreted by TAMs delivered miRNAs that induced a Treg/Th17 ratio imbalance in EOC. Microarray analysis of ECs revealed that *miR-29a-3p* and *miR-21-5p* were enriched in TAM EVs. These two miRNA mimics directly suppress signal transducer and activator of transcription 3 and regulate the Treg/Th17 ratio when transfected into $CD4^+$ T cells. These findings suggest that EVs mediate the communication between TAMs and T cells to form an immunosuppressive microenvironment.

DC-derived EVs (DC-EVs) can stimulate and improve antigen-specific T-cell responses. Li et al. [24] reported that DC-EVs effectively stimulate naïve T-cell proliferation and induce T-cell activation to become antigen-specific cytotoxic T lymphocytes. The antitumor immune response inhibits the progression of hepatocellular carcinoma. In addition, pretreatment of DC precursors with DC-EVs effectively triggers the MHC I-restricted cytotoxic T-lymphocyte response and allows DCs to make full use of minor antigen peptides. Consequently, specific immune responses against hepatocellular carcinoma are maximally activated. Thus, DC-EVs are promising candidates for mature DC-derived cancer vaccines or natural antitumor adjuvants.

### 5.2.5 EVs Mediate Premetastatic Niche (PMN) Formation

The PMN is a supportive tissue microenvironment in future metastatic organs reconstructed by tumor-derived factors to facilitate tumor metastasis. The PMN is characterized by vascular leakage, ECM remodeling, and bone marrow-derived cell (BMDC) recruitment [25]. EVs carry molecules such as oncoproteins, mRNAs, miRNAs, small RNAs, and DNA fragments from donor to recipient cells, initiating profound phenotypic changes in the PMN [26]. Lyden et al. [27] pioneered research on the PMN in 2012. They found that EVs from highly metastatic melanomas were involved in PMN formation through the receptor tyrosine kinase MET, which mediates the provasculogenic effect and BMDC recruitment. They also confirmed that RAB27A, a regulator of membrane trafficking and exosome formation, is highly expressed in melanoma cells. In 2015, Lyden et al. [28] demonstrated that EVs could facilitate ECM remodeling in the PMN during the liver metastasis of pancreatic cancer. EVs can colonize hepatic Kupffer cells and release macrophage migration inhibitory factor to activate the TGF-β pathway. CAFs contribute to the initiation and progression of various malignancies. Kong et al. [29] demonstrated that CAF-EVs participate in ECM remodeling of the PMN in the lung by activating the TGF-β signaling pathway in salivary adenoid cystic carcinoma (SACC). They found that CAF-EVs mainly target lung fibroblasts via integrin α2β1 and promote tumor-permissive activities. They also recommended that plasma EVs may be promising biomarkers for predicting SACC metastasis at an early stage. An integrated strategy targeting both tumor and stromal cells is necessary to prevent SACC metastasis. The increase in vascular permeability caused by tumor-derived EVs has also attracted the attention of researchers. Zeng et al. suggested that colorectal cancer-derived EVs contain *miR-25-3p* and upregulate the expression of VEGFR2, zona occludens-1, occludin, and Claudin-5 in ECs by targeting Kruppel-like factor (KLF) 2 and KLF4, consequently increasing vascular permeability and lung or liver metastasis. Moreover, circulating EV-associated *miR-25-3p* levels are significantly higher in patients with metastasis than in those without. These results suggest that EV-derived *miR-25-3p* may be a biomarker for the diagnosis of colorectal cancer [30]. Cancer EVs can also form the PMN by regulating their metabolism. Fong et al. [31] showed that *miR-122* carried by breast cancer EVs is involved in reprogrammed glucose metabolism in different premetastatic organs. EV-derived *miR-122* mainly suppresses glucose uptake by nontumor cells in the PMN and increases the glucose required for BCC proliferation and metastasis [31]. The colonization of tumor-derived EVs is organotropic. In 2015, researchers from Weill Cornell Medical found that exosomal integrins α6β4 and α6β1 on the membrane of tumor EVs were associated with lung and liver metastasis [32].

Many recent studies have established that tumor-derived EVs can generate suitable microenvironments in secondary organs prior to the dissemination of metastases. However, Plebanek et al. [33] showed that EVs from poorly metastatic melanoma cells could potently inhibit PMN formation in the lungs. They found that "nonmetastatic" EVs could simulate the expansion of Ly6C$^{low}$ patrolling monocytes that eliminate cancer cells at the PMN via the recruitment of natural killer cells and by macrophage-dependent killing of melanoma cells. This event is a membrane-bound pigment epithelium-derived factor in EVs. Importantly, serum from patients with "nonmetastatic" melanoma also suppresses lung metastasis when injected into nude mice. These findings suggest that nonmetastatic tumor-derived EVs can trigger immune responses to eliminate cancer cells in the PMN [33].

The above studies indicate that EVs from different types of tumors or different subsets of the same tumor exhibit variations. They carry various cargo molecules and facilitate or inhibit tumor progression.

## 5.3 Roles of EVs in Diabetes

Diabetes is typically characterized by elevated blood glucose levels, which can lead to serious complications, most frequently those occurring in the cardiovascular system. Diabetes can be divided into two types; type I diabetes is an autoimmune disease, in which the immune cells destroy the islets of the pancreas and kill islet cells, leading to the failure of insulin secretion, whereas type II diabetes is characterized by insulin secretion or dysfunction and is often associated with obesity and aging.

A recent study has shown that adipose tissue constitutes a major source of circulating exosomal miRNAs, and these miRNAs can regulate gene expression in distant tissues (e.g., the liver) to regulate metabolism, exhibiting functions similar to those of adipokines [34]. Huang-Doran et al. [35] reported that adipocyte-derived EVs can enter the liver and regulate their metabolism, which can induce inflammation and fibrosis and initiate nonalcoholic hepatitis. Nonalcoholic hepatitis is an important predictor of type II diabetes in middle-aged individuals. EV-derived *miR-122* and *miR-192* are found in the blood circulation of animals with nonalcoholic fatty liver disease. This suggests that type II diabetes is associated with EV-related miRNAs [36].

Insulin resistance is an important cause of type II diabetes. Ying et al. [37] found that adipose tissue macrophages (ATMs) in obese mice can secrete EV-containing miRNAs that are absorbed and ingested by insulin target cells in vitro and in vivo, leading to insulin resistance and glucose intolerance at both the cellular and systemic levels. Further studies have shown that *miR-155* is highly expressed in the EVs of ATMs in obese individuals. *MiR-155*-knockout mice show insulin sensitivity and glucose tolerance phenotypes, and *miR-155* can inhibit insulin signaling and glucose tolerance by suppressing the expression of its target gene peroxisome proliferator-activated receptor γ. In conclusion, miRNA-containing EVs produced by ATMs affect islet target cells through paracrine or endocrine pathways and regulate insulin sensitivity and glucose homeostasis in vivo.

The development of cardiovascular complications in diabetes is commonly caused by endothelial dysfunction. Zhang et al. [38] revealed that EVs from diabetic mice are internalized by aortic endothelial cells, thereby severely impairing endothelial function in nondiabetic mice. These researchers found that EV-derived RNAs, rather than EV proteins, were the main factors for EV-induced endothelial dysfunction. Proteomic analysis showed that EV arginase 1 levels were significantly upregulated in the serum of diabetic mice, and silencing or overexpression assays confirmed the essential roles of arginase 1 in EV-induced endothelial dysfunction in diabetic serum. Therefore, EVs deliver arginase 1 to endothelial cells, representing a cellular mechanism during the development of diabetic endothelial dysfunction. Kamalden et al. [39] reported that increased*miR-15a* levels in the plasma of patients with diabetes are correlated with disease severity. *miR-15a* plays important roles in insulin production in pancreatic β cells. Pancreatic β cell-secreted EVs are an important source of circulating *miR-15a*, which induces retinal injury via circulation in the bloodstream. Diabetic retinopathy is a complication of diabetes and the leading cause of vision loss. Huang et al. [40] showed that $IgG^+$ EVs are increased in the plasma of patients with diabetes. EVs can activate the classical complement system and induce vascular injury in retinopathy. Moreover, researchers have found that a lack of IgG in exosomes results in a reduction in retinal vascular damage in diabetic mice. Therefore, complement activation by IgG-laden plasma EVs can contribute to the development of diabetes in patients. Overproliferation of vascular smooth muscle cells is the main cause of cardiovascular complications in diabetes. Togliatto et al. [41] found that $CD31^+$ EVs from vascular endothelial cells were enriched in membrane-bound platelet-derived growth factor-BB, which can account for the resistance of vascular smooth muscle cells to apoptosis.

Type 1 diabetes is an autoimmune disease initiated by the death of pancreatic islet cells. Guay et al. [42] found that mice and human T lymphocytes release EVs containing *miR-142-3p*, *miR-142-5p*, and *miR-155*, which can be taken up by β cells and induce apoptosis. Inactivation of these miRNAs can prevent apoptosis and diabetes in mice. Moreover, T-lymphocyte EVs trigger apoptosis and the expression of the C-C motif chemokine ligand (CCL) 2, *CCL7*, and C-X-C motif chemokine ligand (CXCL) 10 genes, which are involved in the recruitment of immune cells and exacerbate the death of β cells.

## 5.4 Roles and Mechanisms of EVs in Neurodegenerative Diseases

Neurocytes, including neurons, oligodendrocytes, astrocytes, and microglia, can release EVs. Studies have found that proteins related to Alzheimer's disease (AD) and Parkinson's disease (PD) are released by cells in the form of EVs, suggesting that EVs play important roles in the progression and transmission of these neurodegenerative diseases.

### 5.4.1 Roles and Mechanisms of EVs in AD

The cause of AD is not known. Accumulation of β-amyloid (Aβ) peptides in senile plaques is a hallmark of disease progression. The Aβ peptide is formed by amyloidogenic cleavage of amyloid precursor protein (APP) by β- and γ-secretases. Researchers have found that β-cleavage of APP occurs in early endosomes, followed by translocation of Aβ into multivesicular bodies [43]. Cells then secrete Aβ in the form of EVs into the extracellular space. ALG-2-interacting protein X, a biomarker of EVs, accumulates in the brain plaques of patients with AD, suggesting a role for EVs in the pathogenesis of AD. Recent studies have revealed the toxicity of EVs in AD progression. Inhibition of neurogenic EV secretion can reduce the formation of amyloid plaques and improve cognitive function in AD model mice [44, 45]. The gradual deterioration of cognitive function in AD is paralleled by the hierarchical progression of Aβ and tau brain pathology. Although the underlying mechanism is not fully understood, recent studies have indicated that small EVs from the brains of patients with AD contain high levels of Aβ and can transfer such toxic species to recipient neurons cultured in vitro. Moreover, blocking the formation, secretion, or uptake of EVs can reduce both the spread of oligomers and the associated toxicity [46]. These findings imply that EVs are centrally involved in AD and could serve as targets for the development of new diagnostic and therapeutic principles. However, there are different views regarding the roles of EVs in AD. Yuyama et al. [47] reported that EVs from neuroblastoma N2a cells could bind to soluble Aβ, promote the fibrosis of Aβ on its surface, and then be cleared by microglial cells. A recent study also showed that EVs from primary neuron cultures could capture Aβ, thereby reducing the accumulation of Aβ in the brains of APP transgenic mice [48]. Therefore, infusion of neuronal EVs into the brains of APP transgenic mice decreases Aβ and amyloid deposition. In addition, An et al. [49] demonstrated that EVs derived from N2a cells or human cerebrospinal fluid can abrogate the synaptic plasticity-disrupting activity of both synthetic and AD brain-derived Aβ. Mechanistically, this protective effect may be due to fibrosis of Aβ on the surface of EVs, which reduces the toxic effects of soluble Aβ in the brain. However, recent studies have demonstrated that microglia cultured in vitro release EVs, promote the formation of soluble toxic Aβ by EVs containing Aβ, and greatly enhance the neurotoxic effects of Aβ. These results suggest that EVs from different subgroups may have different roles in AD progression. Accordingly, EVs act as a "double-edged sword"; its role in AD progression depends on its cell origin, relative abundance, and production conditions.

Another important pathological feature of AD is neurofibrillary tangles in neurocytes, which contain a highly phosphorylated protein called tau. The regulation of tau phosphorylation varies with development. There are more tau phosphorylation sites in the immature brain than in the mature brain. There are 6–8 phosphorylation sites in the former, but only 2–3 phosphorylation sites in the latter, indicating that the Tau protein is selectively dephosphorylated during brain maturation. Tau protein is considered a powerful cocatalyst for microtubule protein polymerization in vitro, and phosphorylation is an important mechanism for negatively regulating Tau protein binding to microtubules. Saman et al. [50] first reported that Tau is secreted by cells in the form of EVs. Other researchers have reported that EVs released from primary cultured neurons or N2a cells contain different tau constructs, which are transmitted from one neuron to another through synaptic connectivity [51]. The fact that EVs released by astrocytes and microglia contain Tau or phosphorylated Tau suggests that the Tau protein can be transmitted by the EV synaptic pathway in addition to trans-synaptic transmission.

### 5.4.2 Roles and Mechanisms of EVs in PD

PD is a neurodegenerative disease with insidious onset and slow progression. The first symptom is usually tremor or clumsiness in one limb, which later also affects the opposite limb. Genetic and biochemical data indicate that increased expression of α-synuclein is sufficient to cause neurodegeneration in PD. Researchers have found that cell-produced α-synuclein is secreted via EVs and that α-synuclein secretion amplifies and propagates PD-related pathology [52]. Although α-synuclein in EVs is only a portion of the total amount of synuclein, α-synuclein has been shown to play important roles in PD dissemination and to accelerate the accumulation of free α-synuclein [53, 54]. In addition, decreased lysosomal activity promotes the formation of EVs [55], and autophagy-lysosomal pathway dysfunction is a marker of sporadic and hereditary PD. Therefore, EVs may be the most important form of α-synuclein dissemination.

## 5.5 Roles and Mechanisms of EVs in Viral Infectious Diseases

EVs play important roles in viral infection and transmission. EVs derived from virus-infected cells can deliver pathogens and elements to other cells. EVs are involved in the spread of human immunodeficiency virus (HIV-1) between different cells in the patient's body. C-C motif chemokine receptor 5 (CCR5) is the principal coreceptor for macrophage-tropic HIV-1. Mack et al. [56] reported Chinese hamster ovary cells, and peripheral blood mononuclear cells can release $CCR5^+$ EVs, which can be transferred to $CCR5^-$ cells, rendering these recipient cells $CCR5^+$. CCR5 delivered by EVs can function as a coreceptor, enabling the original $CCR5^-$ monocytes to be infected with HIV-1. Therefore, EV-mediated CCR5 cell-to-cell transmission may lead to HIV-1 infection in tissues without endogenous CCR5 expression.

Pegtel et al. [57] found that EBV-encoded miRNAs are secreted by EBV-infected B cells through EVs and internalized by monocyte-derived DCs to inhibit the expression of EBV target genes, including *CXCL11/ITAC*. In addition, the authors also found that EBV-miRNA was present in infected B cells and in uninfected non-B cells in the peripheral blood of patients, suggestive of miRNA transfer.

In addition to toxic proteins, EVs may also transmit diverse pathogens and pathogenic factors, which can be incorporated into EVs derived from virus-infected cells. EV-mediated delivery of various factors can trigger an innate immune response in the recipient infected cells. For example, EVs can transfer viral miRNAs to the recipient cells. Meckes et al. [58] reported that EVs secreted by NPC cells harboring latent EBV contain EBV LMP1and virus-encoded miRNAs. These EVs transfer both signaling molecules and viral miRNAs to neighboring cells to manipulate the tumor microenvironment and influence tumor growth. Dreux et al. [59] reported that hepatitis C virus RNA-containing EVs mediate the cell-to-cell transfer of immunostimulatory viral RNA from infected cells to cocultured plasmacytoid DCs and trigger the production of type 1 IFN.

## 5.6 Conclusions and Perspectives

In summary, EVs released by donor cells transfer their cargoes, including proteins, lipids, and regulatory RNAs, to recipient cells. EV-mediated intercellular communication is critical for the progression of several diseases. However, the processes directing EV entry into recipient cells remain poorly understood, and EVs may induce signaling at the cell surface. Further studies are required to elucidate the mechanisms through which EVs interact with the recipient cells. The critical functions of EVs in disease progression make them suitable therapeutic targets. Approaches for inhibiting EV release and blocking EV uptake are currently being explored. However, challenges remain in the development of compounds that target pathogenic EVs but not healthy EVs. Another important aspect of EV function is the interaction between EVs and the ECM. Studies have shown that EVs may bind to the ECM and directly modulate the ECM. Membrane type 1 MMP can be secreted in EVs and degrade type 1 collagen and gelatin [60]. Activated polymorphonuclear leukocytes secrete EVs with high levels of neutrophil elastase and degrade collagen I [61]. Overall, these studies of EV functions in disease progression have established EVs as potential therapeutic targets.

## References

1. Al-Nedawi K, Meehan B, Micallef J, et al. Intercellular transfer of the oncogenic receptor EGFRvIII by microvesicles derived from tumour cells. Nat Cell Biol. 2008;10(5):619–24.
2. Cesi G, Philippidou D, Kozar I, et al. A new ALK isoform transported by extracellular vesicles confers drug resistance to melanoma cells. Mol Cancer. 2018;17(1):145.
3. Luga V, Zhang L, Viloria-Petit AM, et al. Exosomes mediate stromal mobilization of autocrine Wnt-PCP signaling in breast cancer cell migration. Cell. 2012;151(7):1542–56.
4. Donnarumma E, Fiore D, Nappa M, et al. Cancer-associated fibroblasts release exosomal microRNAs that dictate an aggressive phenotype in breast cancer. Oncotarget. 2017;8(12):19592–608.
5. Li W, Zhang X, Wang J, et al. TGFbeta1 in fibroblasts-derived exosomes promotes epithelial-mesenchymal transition of ovarian cancer cells. Oncotarget. 2017;8(56):96035–47.

6. Zhao X, Wu X, Qian M, et al. Knockdown of TGF-beta1 expression in human umbilical cord mesenchymal stem cells reverts their exosome-mediated EMT promoting effect on lung cancer cells. Cancer Lett. 2018;428:34–44.
7. Chen L, Guo P, He Y, et al. HCC-derived exosomes elicit HCC progression and recurrence by epithelial-mesenchymal transition through MAPK/ERK signalling pathway. Cell Death Dis. 2018;9(5):513.
8. Rahman MA, Barger JF, Lovat F, et al. Lung cancer exosomes as drivers of epithelial mesenchymal transition. Oncotarget. 2016;7(34):54852–66.
9. Schillaci O, Fontana S, Monteleone F, et al. Exosomes from metastatic cancer cells transfer amoeboid phenotype to non-metastatic cells and increase endothelial permeability: their emerging role in tumor heterogeneity. Sci Rep. 2017;7(1):4711.
10. Ramteke A, Ting H, Agarwal C, et al. Exosomes secreted under hypoxia enhance invasiveness and stemness of prostate cancer cells by targeting adherens junction molecules. Mol Carcinog. 2015;54(7):554–65.
11. Aga M, Bentz GL, Raffa S, et al. Exosomal HIF1alpha supports invasive potential of nasopharyngeal carcinoma-associated LMP1-positive exosomes. Oncogene. 2014;33(37):4613–22.
12. Millimaggi D, Mari M, D'ascenzo S, et al. Tumor vesicle-associated CD147 modulates the angiogenic capability of endothelial cells. Neoplasia. 2007;9(4):349–57.
13. Tang MKS, Yue PYK, Ip PP, et al. Soluble E-cadherin promotes tumor angiogenesis and localizes to exosome surface. Nat Commun. 2018;9(1):2270.
14. Al-Nedawi K, Meehan B, Kerbel RS, et al. Endothelial expression of autocrine VEGF upon the uptake of tumor-derived microvesicles containing oncogenic EGFR. Proc Natl Acad Sci U S A. 2009;106(10):3794–9.
15. Liu S, Sun J, Lan Q. Glioblastoma microvesicles promote endothelial cell proliferation through Akt/beta-catenin pathway. Int J Clin Exp Pathol. 2014;7(8):4857–66.
16. Hsu YL, Hung JY, Chang WA, et al. Hypoxic lung cancer-secreted exosomal miR-23a increased angiogenesis and vascular permeability by targeting prolyl hydroxylase and tight junction protein ZO-1. Oncogene. 2017;36(34):4929–42.
17. Umezu T, Tadokoro H, Azuma K, et al. Exosomal miR-135b shed from hypoxic multiple myeloma cells enhances angiogenesis by targeting factor-inhibiting HIF-1. Blood. 2014;124(25):3748–57.
18. Li J, Liu X, Zang S, et al. Small extracellular vesicle-bound vascular endothelial growth factor secreted by carcinoma-associated fibroblasts promotes angiogenesis in a bevacizumab-resistant manner. Cancer Lett. 2020;492:71–83.
19. Chen G, Huang AC, Zhang W, et al. Exosomal PD-L1 contributes to immunosuppression and is associated with anti-PD-1 response. Nature. 2018;560(7718):382–6.
20. Chen W, Wang J, Shao C, et al. Efficient induction of antitumor T cell immunity by exosomes derived from heat-shocked lymphoma cells. Eur J Immunol. 2006;36(6):1598–607.
21. Wen ZF, Liu H, Gao R, et al. Tumor cell-released autophagosomes (TRAPs) promote immunosuppression through induction of M2-like macrophages with increased expression of PD-L1. J Immunother Cancer. 2018;6(1):151.
22. Cooks T, Pateras IS, Jenkins LM, et al. Mutant p53 cancers reprogram macrophages to tumor supporting macrophages via exosomal miR-1246. Nat Commun. 2018;9(1):771.
23. Zhou J, Li X, Wu X, et al. Exosomes released from tumor-associated macrophages transfer miRNAs that induce a Treg/Th17 cell imbalance in epithelial ovarian cancer. Cancer Immunol Res. 2018;6(12):1578–92.
24. Li J, Huang S, Zhou Z, et al. Exosomes derived from rAAV/AFP-transfected dendritic cells elicit specific T cell-mediated immune responses against hepatocellular carcinoma. Cancer Manag Res. 2018;10:4945–57.
25. Liu Y, Cao X. Characteristics and significance of the pre-metastatic niche. Cancer Cell. 2016;30(5):668–81.
26. Xu R, Rai A, Chen M, et al. Extracellular vesicles in cancer—implications for future improvements in cancer care. Nat Rev Clin Oncol. 2018;15(10):617–38.

27. Peinado H, Aleckovic M, Lavotshkin S, et al. Melanoma exosomes educate bone marrow progenitor cells toward a pro-metastatic phenotype through MET. Nat Med. 2012;18(6):883–91.
28. Costa-Silva B, Aiello NM, Ocean AJ, et al. Pancreatic cancer exosomes initiate pre-metastatic niche formation in the liver. Nat Cell Biol. 2015;17(6):816–26.
29. Kong J, Tian H, Zhang F, et al. Extracellular vesicles of carcinoma-associated fibroblasts creates a pre-metastatic niche in the lung through activating fibroblasts. Mol Cancer. 2019;18(1):175.
30. Zeng Z, Li Y, Pan Y, et al. Cancer-derived exosomal miR-25-3p promotes pre-metastatic niche formation by inducing vascular permeability and angiogenesis. Nat Commun. 2018;9(1):5395.
31. Fong MY, Zhou W, Liu L, et al. Breast-cancer-secreted miR-122 reprograms glucose metabolism in premetastatic niche to promote metastasis. Nat Cell Biol. 2015;17(2):183–94.
32. Hoshino A, Costa-Silva B, Shen TL, et al. Tumour exosome integrins determine organotropic metastasis. Nature. 2015;527(7578):329–35.
33. Plebanek MP, Angeloni NL, Vinokour E, et al. Pre-metastatic cancer exosomes induce immune surveillance by patrolling monocytes at the metastatic niche. Nat Commun. 2017;8(1):1319.
34. Thomou T, Mori MA, Dreyfuss JM, et al. Adipose-derived circulating miRNAs regulate gene expression in other tissues. Nature. 2017;542(7642):450–5.
35. Huang-Doran I, Zhang CY, Vidal-Puig A. Extracellular vesicles: novel mediators of cell communication in metabolic disease. Trends Endocrinol Metab. 2017;28(1):3–18.
36. Povero D, Eguchi A, Li H, et al. Circulating extracellular vesicles with specific proteome and liver microRNAs are potential biomarkers for liver injury in experimental fatty liver disease. PLoS One. 2014;9(12):e113651.
37. Ying W, Riopel M, Bandyopadhyay G, et al. Adipose tissue macrophage-derived exosomal miRNAs can modulate in vivo and in vitro insulin sensitivity. Cell. 2017;171(2):372–84.e12.
38. Zhang H, Liu J, Qu D, et al. Serum exosomes mediate delivery of arginase 1 as a novel mechanism for endothelial dysfunction in diabetes. Proc Natl Acad Sci U S A. 2018;115(29):E6927–36.
39. Kamalden TA, Macgregor-Das AM, Kannan SM, et al. ExosomalmicroRNA-15a transfer from the pancreas augments diabetic complications by inducing oxidative stress. Antioxid Redox Signal. 2017;27(13):913–30.
40. Huang C, Fisher KP, Hammer SS, et al. Plasma exosomes contribute to microvascular damage in diabetic retinopathy by activating the classical complement pathway. Diabetes. 2018;67(8):1639–49.
41. Togliatto G, Dentelli P, Rosso A, et al. PDGF-BB carried by endothelial cell-derived extracellular vesicles reduces vascular smooth muscle cell apoptosis in diabetes. Diabetes. 2018;67(4):704–16.
42. Guay C, Kruit JK, Rome S, et al. Lymphocyte-derived exosomal microRNAs promote pancreatic beta cell death and may contribute to type 1 diabetes development. Cell Metab. 2019;29(2):348–61.e6.
43. Rajendran L, Honsho M, Zahn TR, et al. Alzheimer's disease beta-amyloid peptides are released in association with exosomes. Proc Natl Acad Sci U S A. 2006;103(30):11172–7.
44. Dinkins MB, Dasgupta S, Wang G, et al. Exosome reduction in vivo is associated with lower amyloid plaque load in the 5XFAD mouse model of Alzheimer's disease. Neurobiol Aging. 2014;35(8):1792–800.
45. Dinkins MB, Enasko J, Hernandez C, et al. Neutral sphingomyelinase-2 deficiency ameliorates Alzheimer's disease pathology and improves cognition in the 5XFAD mouse. J Neurosci. 2016;36(33):8653–67.
46. Sardar Sinha M, Ansell-Schultz A, Civitelli L, et al. Alzheimer's disease pathology propagation by exosomes containing toxic amyloid-beta oligomers. Acta Neuropathol. 2018;136(1):41–56.
47. Yuyama K, Sun H, Sakai S, et al. Decreased amyloid-beta pathologies by intracerebral loading of glycosphingolipid-enriched exosomes in Alzheimer model mice. J Biol Chem. 2014;289(35):24488–98.
48. Yuyama K, Sun H, Usuki S, et al. A potential function for neuronal exosomes: sequestering intracerebral amyloid-beta peptide. FEBS Lett. 2015;589(1):84–8.
49. An K, Klyubin I, Kim Y, et al. Exosomes neutralize synaptic-plasticity-disrupting activity of Abeta assemblies in vivo. Mol Brain. 2013;6:47.

50. Saman S, Kim W, Raya M, et al. Exosome-associated tau is secreted in tauopathy models and is selectively phosphorylated in cerebrospinal fluid in early Alzheimer disease. J Biol Chem. 2012;287(6):3842–9.
51. Wang Y, Balaji V, Kaniyappan S, et al. The release and trans-synaptic transmission of tau via exosomes. Mol Neurodegener. 2017;12(1):5.
52. Emmanouilidou E, Melachroinou K, Roumeliotis T, et al. Cell-produced alpha-synuclein is secreted in a calcium-dependent manner by exosomes and impacts neuronal survival. J Neurosci. 2010;30(20):6838–51.
53. Danzer KM, Kranich LR, Ruf WP, et al. Exosomal cell-to-cell transmission of alpha synuclein oligomers. Mol Neurodegener. 2012;7:42.
54. Grey M, Dunning CJ, Gaspar R, et al. Acceleration of alpha-synuclein aggregation by exosomes. J Biol Chem. 2015;290(5):2969–82.
55. Alvarez-Erviti L, Seow Y, Schapira AH, et al. Lysosomal dysfunction increases exosome-mediated alpha-synuclein release and transmission. Neurobiol Dis. 2011;42(3):360–7.
56. Mack M, Kleinschmidt A, Bruhl H, et al. Transfer of the chemokine receptor CCR5 between cells by membrane-derived microparticles: a mechanism for cellular human immunodeficiency virus 1 infection. Nat Med. 2000;6(7):769–75.
57. Pegtel DM, Cosmopoulos K, Thorley-Lawson DA, et al. Functional delivery of viral miRNAs via exosomes. Proc Natl Acad Sci U S A. 2010;107(14):6328–33.
58. Meckes DG Jr, Shair KH, Marquitz AR, et al. Human tumor virus utilizes exosomes for intercellular communication. Proc Natl Acad Sci U S A. 2010;107(47):20370–5.
59. Dreux M, Garaigorta U, Boyd B, et al. Short-range exosomal transfer of viral RNA from infected cells to plasmacytoid dendritic cells triggers innate immunity. Cell Host Microbe. 2012;12(4):558–70.
60. Hakulinen J, Sankkila L, Sugiyama N, et al. Secretion of active membrane type 1 matrix metalloproteinase (MMP-14) into extracellular space in microvesicular exosomes. J Cell Biochem. 2008;105(5):1211–8.
61. Genschmer KR, Russell DW, Lal C, et al. Activated PMN exosomes: pathogenic entities causing matrix destruction and disease in the lung. Cell. 2019;176(1–2):113–26.e15.

# Part II

# Technologies and Methods for EVs Research

# Methods for the Isolation of Extracellular Vesicles

# 6

Zimei Chen, Dehua Huo, Xiangsheng Yang, Zhijian Cai, Hua Xiao, and Tong Wang

## 6.1 Introduction

Isolation of purified extracellular vesicles (EVs) is a bottleneck in EV-associated biological studies and clinical applications. For numerous sample types, the size and physicochemical properties of EVs are quite similar to co-isolating contaminants. In 2018, International Society of EVs published a constructive guideline, the Minimal Information for Studies of Extracellular Vesicles (MISEV2018) [1]. According to MISEV2018, current situation is that this is no specific marker to denote all EV types. As such, it is hard to give definitive purity performance of

Z. Chen
MOE Key Laboratory of Tumor Molecular Biology, and Key Laboratory of Functional Protein Research of Guangdong Higher Education Institutes, Institute of Life and Health Engineering, the First Affiliated Hospital, Jinan University, Guangzhou, Guangdong, China

D. Huo
Beckman Coulter Inc. Life Sciences Division, Indianapolis, IN, USA

X. Yang
Guangzhou Biotron Technology Co., Ltd., Guangzhou, Guangdong, China

Z. Cai
School of Basic Medical Science, Zhejiang University, Hangzhou, Zhejiang, China

H. Xiao
School of Life Sciences and Biotechnology, Shanghai Jiao Tong University, Shanghai, China

T. Wang (✉)
The First Affiliated Hospital, Institute of Life and Health Engineering, Jinan University, Guangzhou, China

The First Hospital of Changsha, Changsha, China
e-mail: tongwang@email.jnu.edu.cn



Q. Wang, L. Zheng (eds.), *Extracellular Vesicles*,
https://doi.org/10.1007/978-981-99-8365-0_6

currently available EV isolation methods. When choosing methods, investigators have to face the balance between purity and yield as well as ease-of-operation. As we have discussed in reference [2], extremely high purity is desired when probing and claiming contents that specifically exist in EVs. However, if a certain method can acquire EVs that have enough power to solve clinical questions, it should be deemed sufficient enrichment of EVs [2]. In such a context, EV purity will not be the top priority for clinical translations. Therefore, understanding principles and pros and cons of EV isolation methods is an important topic in the field of EV research.

In this section, the following well-characterized methods will be introduced and discussed: differential and density gradient ultracentrifugation (UC), immunoaffinity capture, precipitation, tangential flow ultrafiltration (TFU), size exclusion chromatography (SEC), and microfluidic chip separation.

According to a worldwide survey, differential UC is the most widely used method for EV isolation and is regarded as the "gold standard" [3]. Using this method, cells and debris are first removed using low-speed centrifugation, and EVs can be directly pelleted by UC. In addition, density gradient UC can be used to obtain EVs with higher purity [4, 5]. The immunoaffinity capture largely employs imunomagnetic beads to capture EVs [6], and precipitation methods use reagents or polymers such as polyethylene glycol (PEG) to precipitate EVs [7]. TFU intercepts EVs with tangential membranes [8], while SEC retains smaller particles inside porous resin beads and allows EVs to elute at higher speed [9]. Each of the above-mentioned EV separation methods has its own advantages and disadvantages, while we will discuss them in detail in the following sections.

## 6.2 Differential Ultracentrifugation and Density Gradient Ultracentrifugation

As reviewed by Mathieu et al., EVs are mixed with gigantic amounts of non-EV particles with similar physiochemical properties in sample types of cell culture medium, blood, urine, milk, and tears [10]. These particles include lipoproteins, subcellular organelles, cellular debris, and viral particles. For example, very-low-density-lipoprotein (VLDL) and chylomicrons have overlapped particle sizes with EVs, while high-density-lipoprotein (HDL) has a very similar density to EVs (Fig. 6.1).

### 6.2.1 Basic Theory of Ultracentrifugation

The fundamental principle of UC can be explained by the sedimentation velocity and equilibrium equation:

$$v = \frac{d^2(\sigma - \rho)}{18\eta}\omega^2 r \qquad \text{(Equation 1)}$$

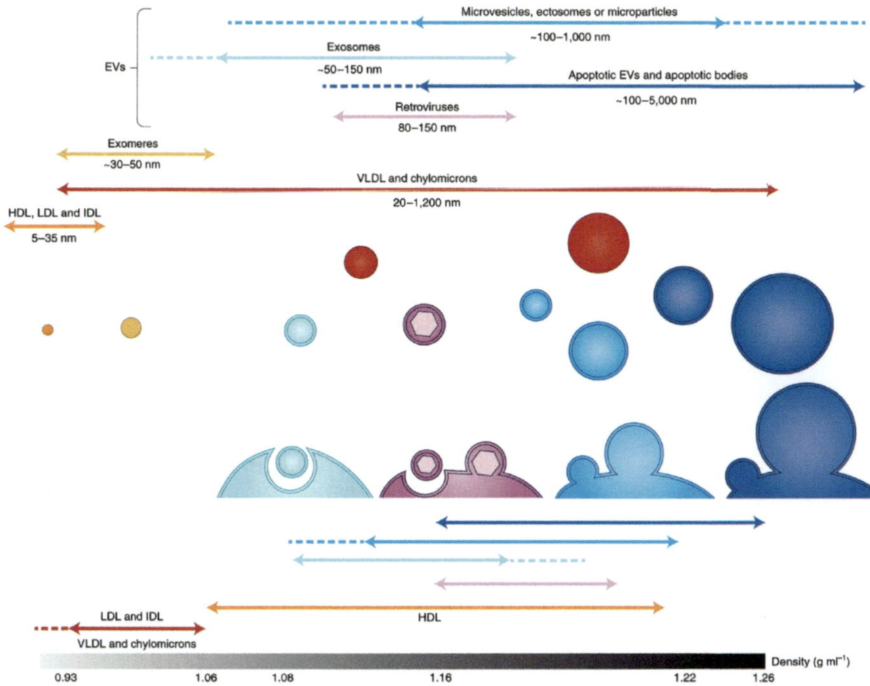


**Fig. 6.1** Common distributions of the sizes and densities of different EV subtypes and biomolecules [10]. *HDL* high-density lipoprotein; *LDL* low-density lipoprotein; *VLDL* very-low-density lipoprotein; *IDL* Intermediate-density lipoprotein

Where $v$ is the velocity or sedimentation rate of the particle, $d$ is the particle diameter, $\sigma$ is the density of the particles, $\rho$ is the density of liquid, η is the viscosity of liquid, $\omega$ and r are the angular velocity and radian of the rotor, respectively. Here, the centrifugal force (G) can be computed with $\omega 2r$.

According to Equation 1, when two or more particles have significant difference in diameter their centrifugal sedimentation velocities are different. Particles with greater diameter are pelleted more quickly, whereas small particles require a greater centrifugal force or more centrifugation time for sedimentation. For example, particles with a diameter of approximately 100 nm can be pelleted after centrifugation for 1–3 h at 100,000 × *g*, while smaller particles are still in the supernatant.

## 6.2.2 Differential Ultracentrifugation

Differential UC refers to a procedure of using different centrifugation speeds to sequentially remove unwanted impurities, while pelleting EVs in the final step (Fig. 6.2). As summarized by Thery et al., cells and cell debris (and/or large organelles) can by removed by centrifugations at 2,000 ×g and 10,000 ×g, respectively. EVs can

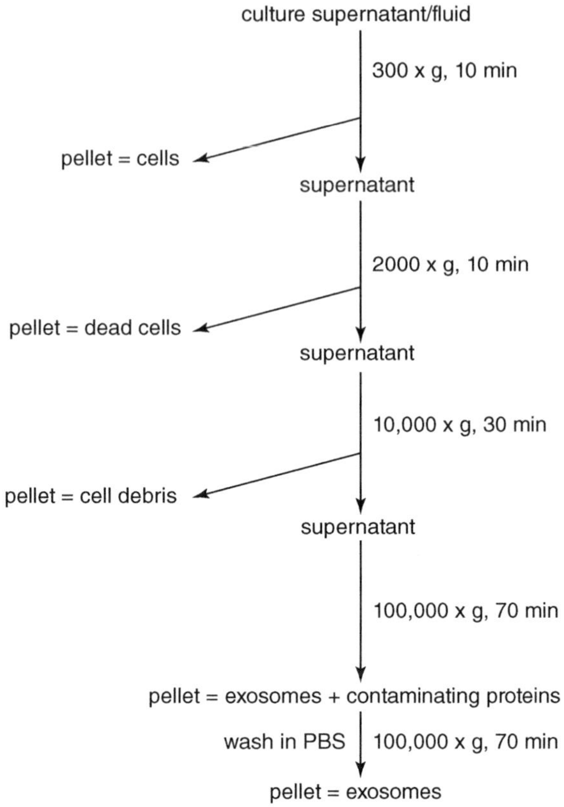


**Fig. 6.2** Typical experimental process of using differential ultracentrifugation to isolate and concentrate EVs [11]. Reprinted with permission from John Wiley

then be harvested and/or further washed by UC at 100,000 ×g [11]. Nonetheless, such a protocol is not always applicable to all sample types; further optimization and adjustment on centrifugation speeds and durations are highly suggested according to specific sample characteristics.

Per Equation 1, differential UC has an intrinsic problem, namely the sedimentation velocity is dependent on particle size (d). As such, particles such as VLDL, chylomicrons and proteins aggregates will automatically be co-isolated with EVs when using differential UC.

### 6.2.3 Density Gradient Ultracentrifugation

To cope with the above-mentioned coherent problem of differential UC, the liquid density ($\rho$) in Equation 1 can be operated. It is known that the overall average density of EVs is approximately 1.13–1.19 g/mL, while they encapsulate nucleic acids (common density: 1.4–1.7 g/mL) and proteins (common density: 1.2–1.4 g/mL). According to Equation 1, if the liquid density equals to EV density ($\sigma = \rho$), then EVs will not be sedimented in liquid (v = 0). Under this condition, the size (d) and

| Gradient Layer | Estimated Density (g/mL) | Volume (mL) | % Iodixanol (0.25M Sucrose; pH 7.5) |
|---|---|---|---|
| 1 | 1.223 | 3 | 40 |
| 2 | 1.127 | 3 | 20 |
| 3 | 1.079 | 3 | 10 |
| 4 | 1.054 | 2 | 5 |

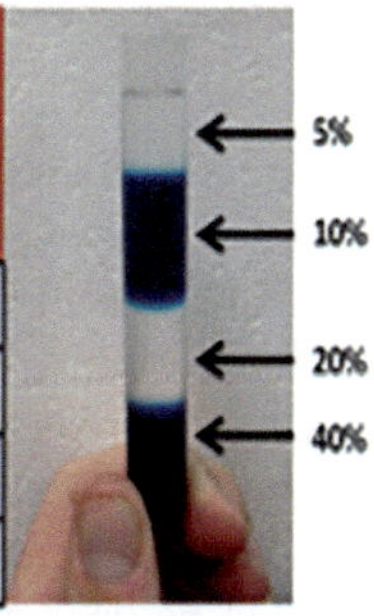


**Fig. 6.3** Typical density gradient medium preparation for the purification of EVs by isopycnic density gradient ultracentrifugation

centrifugal force ($\omega 2r$) will have no effect on EV migration in liquid, thus efficiently avoiding the co-isolation of most non-EV particles with similar sizes to EVs. For example, OptiPrep iodixanol solutions with different density distributions could be used to generate density gradients, while EVs can be retained in Layer 2 (Fig. 6.3) [11]. It should be noted that HDLs have very similar density to EVs, which makes HDL a primary co-isolating contaminant in density-gradient UC.

## 6.3 Immunoaffinity-Based Isolation of EVs

Immunoaffinity isolation allows for the capture of EVs based on the expression of specific membrane proteins, including CD9, epithelial cell adhesion molecule, CD63, CD81, and annexin [12]. Such proteins must be expressed on the surface of EVs; otherwise, antibodies cannot bind to these proteins. With this method, antibodies can be conjugated to magnetic beads or plates [e.g., enzyme-linked immunosorbent assay (ELISA) or microfluidic devices] [13]. Immunoaffinity isolation is featured by simple operation procedures, intact EV morphology, and highly specific and pure EV yields. However, the isolated EVs are specific subsets, not all types of EVs; therefore, the extraction efficiency and yield are low. In addition, EVs isolated by immunoaffinity are susceptible to variations in pH and salt concentrations [14].

### 6.3.1 Magnetic Bead Immune Separation

In magnetic bead immune separation, a biotinylated antibody against certain antigen of interest is attached to the surface of the streptavidin-coated magnetic beads. Incubation of preconcentrated EV samples with these magnetic beads enables specific pull-down of EVs expressing the target antigen. Some beads precoated with a general EV surface marker (e.g., CD81, CD9, and CD63) are commercially available. This method provides better efficiency and can handle large sample volumes, and the EVs captured on the beads can be well used for downstream analyses [13].

### 6.3.2 ELISA

ELISA is an immunoaffinity technique for isolating EVs from samples [15]. Specific antibodies against antigens of interest can be immobilized on the surface of a microplate. After pretreatment with UC or ultrafiltration, the EV sample is added into the well containing the immobilized antibodies, and EVs expressing specific antigens can be captured onto the plate due to the antibody-antigen interaction. The uncaptured EVs and sample contents can be washed away, and the captured EVs are then detected using another antibody containing an absorbent tag (Fig. 6.4) [16]. In the EV field, ELISA can be used not only to isolate EVs, but also to quantify antigens when standards are employed to create a calibration curve. However, microplate-based ELISA has a maximum volume limitation within the wells of a typical 96-well microplate.

### 6.3.3 Immunoaffinity-Based Microfluidic Devices

Immunoaffinity-based microfluidic systems are based on the interaction between proteins expressed on the surface of EVs and an antibody against the protein that is immobilized on a microfluidic chip (Fig. 6.5). There are two major types of such systems: 1) microchannel substrate surface modification-based immunoaffinity

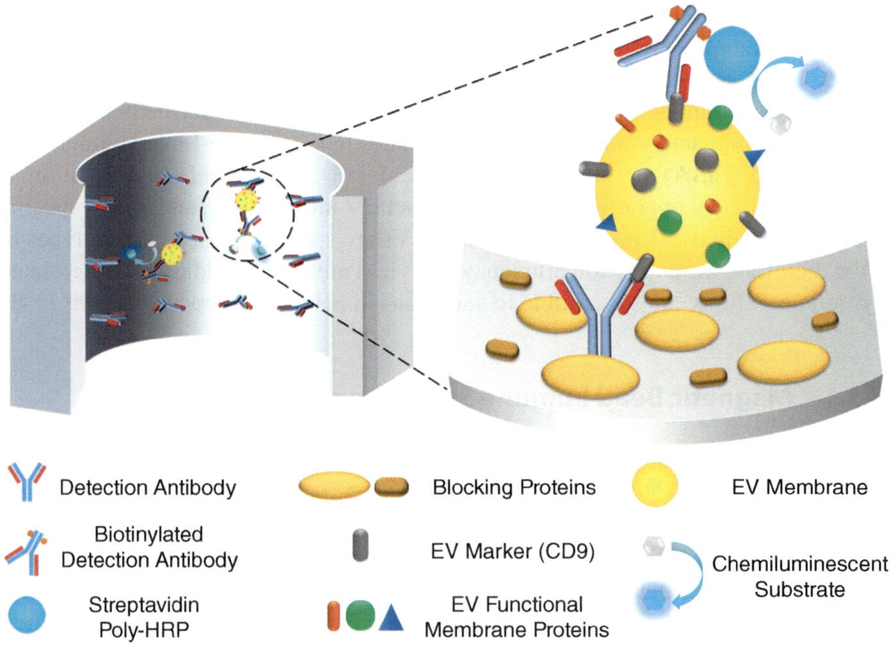


**Fig. 6.4** Illustration of microchip ELISA for detection/quantification of EVs [13]

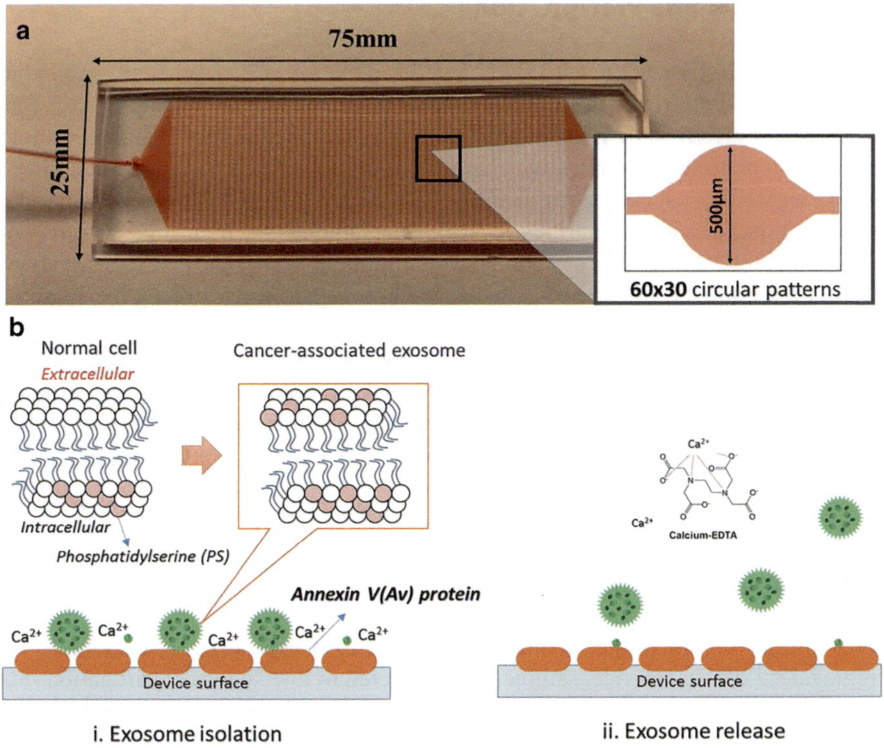


**Fig. 6.5** The microfluidic device design and working principle [18]. Reprinted with permission from WIELY VCH

isolation, and 2) manipulation of antibody-conjugated micrometer-sized beads. For the first system, microfluidic devices are modified with different types of specific antibodies to form functionalized microchannel surfaces for the immunocapture of EVs. For the second system, immunomagnetic beads can capture EVs and form a tight aggregate that can be separated in the downstream microfluidic chamber using magnetic force. Compared with the microchannel substrate surface method, the immunobead method requires additional procedures of premixing and incubation of immunobeads with samples; however, this approach has higher capture efficiency and analytical sensitivity owing to the larger surface area [17].

## 6.4 Polymer Precipitation

EVs can be precipitated by numerous methods based on charge neutralization, protein aggregation or polymer capture. EV precipitation methods have evolved rapidly; they largely include protamine precipitation, sodium acetate precipitation, organic solvent precipitation, and hydrophilic polymer precipitation.

### 6.4.1 Protamine Precipitation

It is known that EVs are negatively charged as their surfaces have highly abundant phosphatidylserine (PS). In the nature state, EVs tend not to aggregate with each other due to electrostatic repulsion. While protamine is a positively charged molecule, it can neutralize the negative charge of EVs and cause EV aggregation [19].

Practically, low-speed centrifugation (1500–3000 × g) needs to be performed to remove larger particle impurities in samples. Samples are then mixed with protamine buffer (0.25 mg/mL protamine) at a volume ratio of 4:1, and incubate overnight. Precipitated EVs can be harvested by centrifugation at 1500 × g for 30 min at room temperature.

This method is simple to operate and allows simultaneous enrichment of large volumes or multiple samples. Isolated EVs have a complete membrane structure and are relatively cost-efficient. The disadvantage is that the separation process is time-consuming. In addition, numerous negatively charged non-EV particles such as LDL can also be co-isolated.

### 6.4.2 Sodium Acetate Precipitation

Very similar to protamine, sodium acetate can also neutralize PS and cause EV aggregates [20]. In practice, samples are subjected to low-speed centrifugation (500 × g, 30 min; 12000 × g, 30 min), and treated with 0.1 volume of sodium acetate solution (1.0 M, pH 4.75). The mixture is then put in ice bath for 30–60 min, followed by incubation at 37°C for 5 min. EVs can then be pelleted by centrifugation at 5000 × g for 10 min. HEPES buffer can be used to wash off sodium acetate. This method shares similar pros and cons to the protamine precipitation.

### 6.4.3 Organic Solvent (PROSPR) Precipitation

The general principle of this method is to use acetone to precipitate proteins, while keeping hydrophobic EVs in the supernatant [21]. Typically, four-time volume of precooled acetone (–20°C) is used to treat the samples, and EVs can be harvested by centrifugation at 3000 × g for 1 min. This method is easy to operate and has high EV yields. However, acetone can destroy the lipid membrane structure of EVs, thereby affecting the subsequent functional studies if applicable.

### 6.4.4 Hydrophilic Polymer Precipitation

The most common polymer used in this method is 6,000–20,000 Da PEG. PEG is a negatively charged and hydrophilic volume-excluding polymer. At certain concentrations, PEG can form mesh structures to allow EV capture and precipitation. Typically, we need to remove cell debris from the sample with sequential low-speed and high-speed centrifugations. Samples are then mixed with PEG solution to reach a final concentration of 5%–8%, followed by incubation at 4°C. Incubation time could range from 60 min to overnight for different sample types or based on individual laboratory protocols. EVs will be aggregated and can be harvested through low-speed centrifugation. There are numerous PEG precipitation kits that are commercially available. They include ExoQuick (System Biosciences, USA), total exosome isolation reagent (Invitrogen, USA), ExoPrep (Estonia), Exosome Isolation Kit (Canada), miRCURY Exosome Isolation Kit (Denmark), EVast-S/EVast-TCU (Guangzhou Biotron Technology Co., Ltd., China).

The PEG precipitation method is simple to operate and does not rely on sophistic equipment; it can process large-volume samples and achieve high-throughput operations. However, it is easily contaminated by lipoproteins or even viruses, resulting in low EV purity. In addition, PEG is hard to be removed from the EV isolates. With this regard, Rider et al. suggested that the 8% PEG precipitation followed by UC could significantly improve EV purity [22]. In addition, Chang et al. also recommended that the use of PEG-coated Fe3O4 nanoparticles could effectively reduce the non-EV protein contaminations [23].

## 6.5 SEC

### 6.5.1 Basic Theory of SEC

SEC, also known as gel chromatography, is a chromatography technology that separates molecules in solution according to their sizes or molecular weights [24].

In SEC, the sample flows through the chromatographic column due to the driving force of the mobile phase. SEC columns are generally packed with porous resins. Bigger particles cannot enter the pores of these resins and move faster (Fig. 6.6). However, smaller molecules or particles can enter the pores of resins, resulting in their slower elution than big particles. Accordingly, particles can be fractionated and collected based on their sizes [25].

According to the different mobile phases used, SEC can be divided into gel filtration and gel permeation chromatography. Gel filtration chromatography uses an aqueous solution as the mobile phase and transports the sample through a chromatographic column for separation. This approach is different from gel permeation chromatography, which uses an organic solvent as the eluent buffer. SEC columns are typically packed with porous resins, including dextran polymer (Sephadex), agarose (Sepharose), and polyacrylamide (Sephacryl or BioGel P).

### 6.5.2 Advantages and Disadvantages of SEC for EV Isolation

SEC has several advantages in the separation and preparation of EVs. First, SEC can completely separate large particles from small particles, which is compatible with various water-based elution solutions [27]. Second, SEC is performed at low pressure

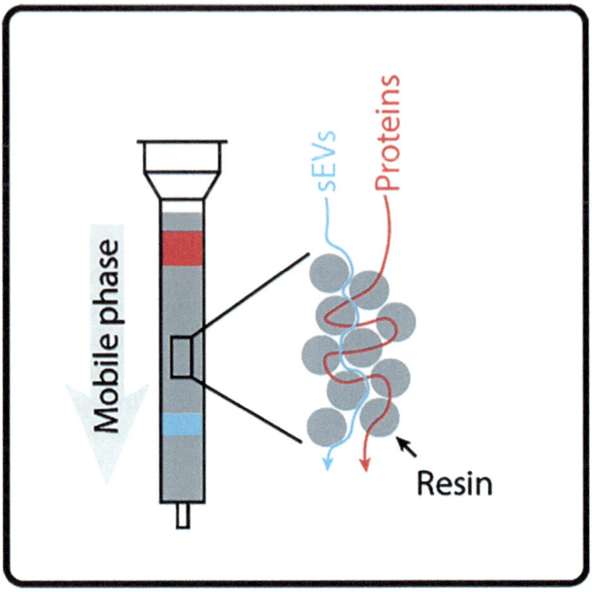


**Fig. 6.6** Schematic diagram of size-exclusion chromatography [26]. Reprinted with permission from Elsevier

and can be driven by gravity that has minimal effect on the structure and integrity of EVs. Third, a high ionic strength buffer can be used to further remove nonspecific contaminations. Fourth, SEC can be easily scaled up. For example, increasing the length of the chromatographic column can improve the purity of the separation, and increasing the column diameter can increase its capacity. Finally, SEC separates EVs with relatively stable retention time, resulting in good reproducibility. Theoretically, because EVs do not interact with the stationary phase, SEC does not cause sample loss, and the recovery rate is high. Because of these advantages, numerous commercial SEC columns are currently available for the preparation of EVs [28].

Although SEC can effectively separate most of the interfering particles in biofluids, it also suffers from co-isolation of contaminants of EVs, such as lipoproteins and immunoglobulins [29].

The amount of residual contaminants in EV separation depends on the stationary phase of the SEC column. In the above-mentioned methods, agarose resin columns are frequently used for serum EV isolation. It can separate particles smaller than 75 nm from larger particles. However, at this pore size threshold, the EV fractions will still contain large amounts of soluble proteins, HDL and LDL. Therefore, to obtain EVs with higher purity, it is recommended to combine SEC with other methods for the separation [29]. It was also proposed to combine the 75 nm CL-6B sepharose resin and increased bed volume to concentrate EVs in a single 8 mL PBS wash, which was called dichotomic SEC (dSEC) [26]. Guo et al. showed that although lipoprotein contamination was still one of its intrinsic problem, dSEC can remove remarkably more albumin than UC [26]. In addition, EVs separated by SEC are severely diluted and need to be concentrated prior to subsequent experiments. Furthermore, although the low pressure of the SEC helps to maintain the integrity and activity of the EVs, it also increases the operation time of the experiment.

## 6.6 TFU

To use ultrafiltration for EV isolation, it is very important to avoid concentration polarization, which means concentrated particles can jam filtration membranes when more and more particles are retained on membrane. As compared with the dead-end filtration, TFU has better tolerance to concentration polarization. For example, Busatto et al. showed that EV isolation from 200 mL cell culture supernatant could be completed within 1 h using TFU [30]. The principle of TFU is to allow the liquid flow to be perpendicular to the centrifugation force, which enables the retention of EVs on the membrane while efficiently reducing concentration polarization [31]. Various filter membranes are available based on different molecular weight cut-offs [32, 33]. For example, Xu et al. used filters with pore sizes of 0.1, 0.22, 0.45, and 0.65 μm to sequentially isolate EVs from the culture supernatants of LIM1863 human colon cancer cells [34]. They found that EVs of different sizes were of distinct biochemical characteristics and functions.

Although TFU is compatible with cell culture supernatants, it is also characterized by severe co-isolation of protein contaminants, particularly when using complex body fluids, such as serum/plasma, cerebral spinal fluid, and urine (see Sect. 6.2). Thus, combining TFU with density gradient UC could be used to separate therapeutic levels of EVs in clinical trials of cancer immunotherapy [35].

TFU is a convenient method for EV separation that does not necessitate the introduction of other solutions or interfering reagents. TFU also does not require special equipment and has the benefits of low cost, good experimental reproducibility, and high enrichment. However, TFU is also a size-based EV isolation method, it will intercept all extracellular particles, including EVs, lipoproteins and protein aggregates. To analyze nucleic acids or specific proteins carried by EVs isolated from serum/plasma, contaminants of lipoproteins may be tolerated. For example, as a clinically approved kit, the SupBio EV/exosome separation kit (Guangzhou, China) can quickly enrich extracellular particles (EPs) from human serum/plasma for subsequent analyses in laboratory medicine. The operation time required for a single sample is approximately 30 min with a regular benchtop centrifuge, and the procedure can be easily performed by under-trained staff in clinical settings. In addition, it is known that numerous EPs can carry nucleic acids. Thus, this TFU method can be used in the evaluation of clinical significance of EP-RNA or EP-DNA for specific diseases.

## 6.7 Summary

In summary, methods for EV isolation and purification are essential for providing first-line evidence for EV research. These methods have many advantages and disadvantages, as detailed in Table 6.1. EV purity is variable, and 100% purity cannot be achieved by any method alone or in combination. Therefore, we propose several factors that should be considered prior to the decision to adopt certain methods. They are (1) how much co-isolation contaminants can be tolerated; (2) how

**Table 6.1** Comparison of the advantages and disadvantages of EV extraction methods

| Extraction method | Separation principle | Advantages | Disadvantages |
|---|---|---|---|
| SEC | To use porous resins to retain small particles and molecule, while EVs are eluted faster | Obtaining high-purity EVs, preserving the integrity and biological activity of EVs, high EV yield | Contaminants with similar size to EVs cannot be sufficiently removed. Low throughput |
| Ultracentrifugation | To isolate EVs based on size and/or density in ultracentrifugation | Low reagent cost, low contamination risk | Major instruments required, long processing time, low yield, and possible interference of EVs per integrity and bioactivity |
| Ultrafiltration | Size based interception | Fast EV isolation, no major instruments needed, large sample volume compatible | Low EV purity, clogging filter membrane for processing complex sample types |
| EV precipitation | Alteration of EV charges, or capturing EVs by polymer | Easy-to-use, no major instruments required, high-throughput | Sever co-isolating contaminants for complex sample types, long processing durations |
| Immune affinity capture | Antibody-antigen interaction | High EV purity, and suitable for the separation of different EVs subtypes | High reagent costs, low EV yield. Cancer heterogeneity may affect immune recognition, and antigen epitopes may be blocked. The specificity of this method depends on the specificity of antibodies |
| Microfluidics | To use specific devices to capture EVs based on immune affinity, size, density, etc. | High speed, low cost, simple operation, and compatible with automation | The separation efficiency of this method depends on the design of the device and the separation principle adopted. The selectivity and specificity need to be verified and confirmed, and it is suitable for the extraction of a small amount of samples |

important is ease-of-use for conducting large-scale clinical or basic research; and (3) how critical is the EV yield for the subsequent analysis. Indeed, sufficient enrichment of EVs based on well-standardized methods could bypass the relentless effort for achieving purest EV isolates at least in laboratory medicine, when the diagnostic/prognostic power is already significant [2]. In addition, numerous emerging methods for EV isolation and characterization, including nano-flow cytometry, biosensors, and microfluidics, are expected to lead to major advancements in the field. These methods will be introduced and discussed in other chapters in this book.

## References


1. Thery C, Witwer KW, Aikawa E, Alcaraz MJ, Anderson JD, Andriantsitohaina R, et al. Minimal information for studies of extracellular vesicles 2018 (MISEV2018): a position statement of the International Society for Extracellular Vesicles and update of the MISEV2014 guidelines. J Extracell Vesicles. 2018;7(1):1535750.
2. Zhang L, Ma W, Gan X, Chen W, Guo J, Cui Y, et al. Adequate enrichment of extracellular vesicles in laboratory medicine. Interdiscip Med. 2023;e20220003.
3. Lobb RJ, Becker M, Wen SW, Wong CS, Wiegmans AP, Leimgruber A, et al. Optimized exosome isolation protocol for cell culture supernatant and human plasma. J Extracell Vesicles. 2015;4:27031.
4. Li P, Kaslan M, Lee SH, Yao J, Gao Z. Progress in exosome isolation techniques. Theranostics. 2017;7(3):789–804.
5. Miranda KC, Bond DT, Levin JZ, Adiconis X, Sivachenko A, Russ C, et al. Massively parallel sequencing of human urinary exosome/microvesicle RNA reveals a predominance of non-coding RNA. PLoS One. 2014;9(5):e96094.
6. Zarovni N, Corrado A, Guazzi P, Zocco D, Lari E, Radano G, et al. Integrated isolation and quantitative analysis of exosome shuttled proteins and nucleic acids using immunocapture approaches. Methods. 2015;87:46–58.
7. Zeringer E, Barta T, Li M, Vlassov AV. Strategies for isolation of exosomes. Cold Spring Harb Protoc. 2015;2015(4):319–23.
8. Cheruvanky A, Zhou H, Pisitkun T, Kopp JB, Knepper MA, Yuen PS, et al. Rapid isolation of urinary exosomal biomarkers using a nanomembrane ultrafiltration concentrator. Am J Physiol Renal Physiol. 2007;292(5):F1657–61.
9. Boing AN, van der Pol E, Grootemaat AE, Coumans FA, Sturk A, Nieuwland R. Single-step isolation of extracellular vesicles by size-exclusion chromatography. J Extracell Vesicles. 2014;3.
10. Mathieu M, Martin-Jaular L, Lavieu G, Thery C. Specificities of secretion and uptake of exosomes and other extracellular vesicles for cell-to-cell communication. Nat Cell Biol. 2019;21(1):9–17.
11. Thery C, Amigorena S, Raposo G, Clayton A. Isolation and characterization of exosomes from cell culture supernatants and biological fluids. Curr Protoc Cell Biol. 2006;Chapter 3:Unit 3.22.
12. Jiang Z, Liu G, Li J. Recent progress on the isolation and detection methods of exosomes. Chem Asian J. 2020;15(23):3973–82.
13. Lane RE, Korbie D, Trau M, Hill MM. Purification protocols for extracellular vesicles. Methods Mol Biol (Clifton, NJ). 2017;1660:111–30.
14. Fitzgerald J, Leonard P, Darcy E, Sharma S, O'Kennedy R. Immunoaffinity chromatography: concepts and applications. Methods Mol Biol (Clifton, NJ). 2017;1485:27–51.
15. Doyle LM, Wang MZ. Overview of extracellular vesicles, their origin, composition, purpose, and methods for exosome isolation and analysis. Cell. 2019;8(7).
16. Hosseini S, Vázquez-Villegas P, Rito-Palomares M, Martinez-Chapa SO. General overviews on applications of ELISA: enzyme-linked immunosorbent assay (ELISA); 2018.
17. Hao N, Zhang JXJ. Microfluidic screening of circulating tumor biomarkers toward liquid biopsy. Sep Purif Rev. 2017;47(1):19–48.
18. Kang YT, Purcell E, Palacios-Rolston C, Lo TW, Ramnath N, Jolly S, et al. Isolation and profiling of circulating tumor-associated exosomes using extracellular vesicular lipid-protein binding affinity based microfluidic device. Small. 2019;15(47):e1903600.
19. Deregibus MC, Figliolini F, D'Antico S, Manzini PM, Pasquino C, De Lena M, et al. Charge-based precipitation of extracellular vesicles. Int J Mol Med. 2016;38(5):1359–66.
20. Brownlee Z, Lynn KD, Thorpe PE, Schroit AJ. A novel "salting-out" procedure for the isolation of tumor-derived exosomes. J Immunol Methods. 2014;407:120–6.
21. Gallart-Palau X, Serra A, Sze SK. Enrichment of extracellular vesicles from tissues of the central nervous system by PROSPR. Mol Neurodegener. 2016;11(1):41.

22. Rider MA, Hurwitz SN, Meckes DG Jr. ExtraPEG: a polyethylene glycol-based method for enrichment of extracellular vesicles. Sci Rep. 2016;6:23978.
23. Chang M, Chang YJ, Chao PY, Yu Q. Exosome purification based on PEG-coated Fe3O4 nanoparticles. PLoS One. 2018;13(6):e0199438.
24. Burgess RR. A brief practical review of size exclusion chromatography: rules of thumb, limitations, and troubleshooting. Protein Expr Purif. 2018;150:81–5.
25. Brooks DE, Haynes CA, Hritcu D, Steels BM, Muller W. Size exclusion chromatography does not require pores. Proc Natl Acad Sci U S A. 2000;97(13):7064–7.
26. Guo J, Wu C, Lin X, Zhou J, Zhang J, Zheng W, et al. Establishment of a simplified dichotomic size-exclusion chromatography for isolating extracellular vesicles toward clinical applications. J Extracell Vesicles. 2021;10(11):e12145.
27. Wang Y, Teraoka I, Hansen FY, Peters GH, Hassager O. A theoretical study of the separation principle in size exclusion chromatography. Macromolecules. 2010;43(3):1651–9.
28. Welton JL, Webber JP, Botos LA, Jones M, Clayton A. Ready-made chromatography columns for extracellular vesicle isolation from plasma. J Extracell Vesicles. 2015;4:27269.
29. Karimi N, Cvjetkovic A, Jang SC, Crescitelli R, Hosseinpour Feizi MA, Nieuwland R, et al. Detailed analysis of the plasma extracellular vesicle proteome after separation from lipoproteins. Cell Mol Life Sci. 2018;75(15):2873–86.
30. Busatto S, Vilanilam G, Ticer T, Lin WL, Dickson DW, Shapiro S, et al. Tangential flow filtration for highly efficient concentration of extracellular vesicles from large volumes of fluid. Cell. 2018;7(12).
31. Musumeci T, Leonardi A, Bonaccorso A, Pignatello R, Puglisi G. Tangential flow filtration technique: an overview on nanomedicine applications. Pharm Nanotechnol. 2018;6(1):48–60.
32. Eppler A, Weigandt M, Schulze S, Hanefeld A, Bunjes H. Comparison of different protein concentration techniques within preformulation development. Int J Pharm. 2011;421(1):120–9.
33. Bhattacharjee S, Bhattacharjee C, Datta S. Studies on the fraction of beta-lactoglobulin from casein whey using ultrafiltration and ion-exchange membrane chromatography. J Membr Sci. 2006;275:141–50.
34. Xu R, Greening DW, Rai A, Ji H, Simpson RJ. Highly-purified exosomes and shed microvesicles isolated from the human colon cancer cell line LIM1863 by sequential centrifugal ultrafiltration are biochemically and functionally distinct. Methods. 2015;87:11–25.
35. Escudier B, Dorval T, Chaput N, Andre F, Caby MP, Novault S, et al. Vaccination of metastatic melanoma patients with autologous dendritic cell (DC) derived-exosomes: results of thefirst phase I clinical trial. J Transl Med. 2005;3(1):10.

# Technologies Used for Analysis of Extracellular Vesicle-Omics

7

Xiulan Chen, Nali Zhu, Tanxi Cai, and Fuquan Yang

## 7.1 Introduction

Extracellular vesicles (EVs) are naturally secreted membrane vesicles that contain many cellular components, such as proteins, lipids, RNAs (mRNAs, microRNAs [miRNAs], and long noncoding RNAs [lncRNAs]), and DNA, making them key players in cell-cell communication. EVs have been reported to play active roles in numerous biological processes, such as antigen presentation, regulation of immune responses, and the transfer of molecules between cells. EVs are also involved in the pathological processes of various diseases, such as tumor growth and metastasis, inflammatory diseases, cardiovascular diseases, and neurodegenerative disorders.

To elucidate the roles of EVs in physiology and pathology, different technologies, including some "-omics" technologies, such as transcriptomics [1], proteomics [2, 3], lipidomics [4], and metabolomics [5], as well as single-vesicle technologies, such as Raman tweezer microspectroscopy and fluorescence microscopy [6], have been applied to characterize the molecular compositions of EVs from different species, tissues, and cells. Chromatography and mass spectrometry (MS)-based -omics technologies have become essential tools for identifying the protein (proteome),

X. Chen · T. Cai · F. Yang (✉)
Key Laboratory of Protein and Peptide Pharmaceuticals and Laboratory of Proteomics, Institute of Biophysics, Chinese Academy of Sciences, Beijing, China

University of Chinese Academy of Sciences, Beijing, China
e-mail: fqyang@ibp.ac.cn

N. Zhu
Key Laboratory of Protein and Peptide Pharmaceuticals and Laboratory of Proteomics, Institute of Biophysics, Chinese Academy of Sciences, Beijing, China

Hangzhou Institute for Advanced Study, University of Chinese Academy of Sciences, Hangzhou, China



Q. Wang, L. Zheng (eds.), *Extracellular Vesicles*,
https://doi.org/10.1007/978-981-99-8365-0_7

lipid (lipidome), and metabolite (metabolome) components contained in EVs. Here, we provide an overview of the key technologies used to analyze the proteomes and lipidomes of EVs.

## 7.2 Proteomic Analysis of EVs

Upon being released from cells, EVs circulate in the extracellular space or travel long distances to deliver complex signals into target cells. Because proteins are key components of EVs, MS-based proteomic profiling is emerging as a vital strategy to analyze the protein composition of EVs [3], thus providing insights into the biological roles of EVs in health and disease and facilitating the identification of specific EV biomarkers for diagnosis and prognosis.

The protein compositions of EVs are cell- and disease-specific. Comprehensive analyses of the EV proteome involve three steps (Fig. 7.1): (1) isolation and characterization of EVs, (2) MS-based proteomic analysis, and (3) bioinformatic analysis of EV proteomic data.

Efficient isolation of EVs with minimal contamination is a critical prerequisite for proteomic analysis. Various EV isolation methods, including ultracentrifugation, density gradient ultracentrifugation, size exclusion chromatography (SEC), immunoaffinity, polymer-based precipitation, and asymmetric flow field-flow fractionation [7, 8], have been used to purify EVs from cell culture medium or body fluids, including blood, saliva, urine, malignant ascites, breast milk, and cerebrospinal fluid [9, 10]. EVs can then be characterized using different techniques; electron microscopy is used to visualize the morphology and size of EVs, nanoparticle tracking analysis is applied to measure the heterogeneous size distribution of EVs, and western blotting is employed for detection of marker proteins of EVs and other organelles to verify the purity of isolated EV samples [11].

MS-based proteomic analysis of EVs is usually conducted using the "bottom-up" model (also called "shotgun proteomics"). In this strategy, proteins are digested into peptides by proteases, and the peptides are analyzed by liquid chromatography (LC)-MS/MS. Before MS analysis, sample separation is needed to reduce the complexity of the samples in MS and thus increase the number of proteins identified. Sample separation is performed as follows: (1) sodium dodecyl sulfate-polyacrylamide gel electrophoresis (SDS-PAGE)-based separation of proteins prior to in-gel digestion of protein bands and (2) in-solution digestion of proteins into peptides. LC-based techniques, such as high-pH reversed-phase chromatography or hydrophilic interaction LC (HILIC), are optional methods for fractionating peptide mixtures into several fractions. LC-MS/MS has been used to identify the peptides in EV samples. MS analysis of EV proteomes can be performed in data-dependent acquisition (DDA) mode [12, 13] or data-independent acquisition (DIA) mode, such as Sequential Window Acquisition of all Theoretical Mass Spectra (SWATH-MS) [14] and $MS^E$ [15]. After MS analysis, a database search is performed to obtain protein identification results.

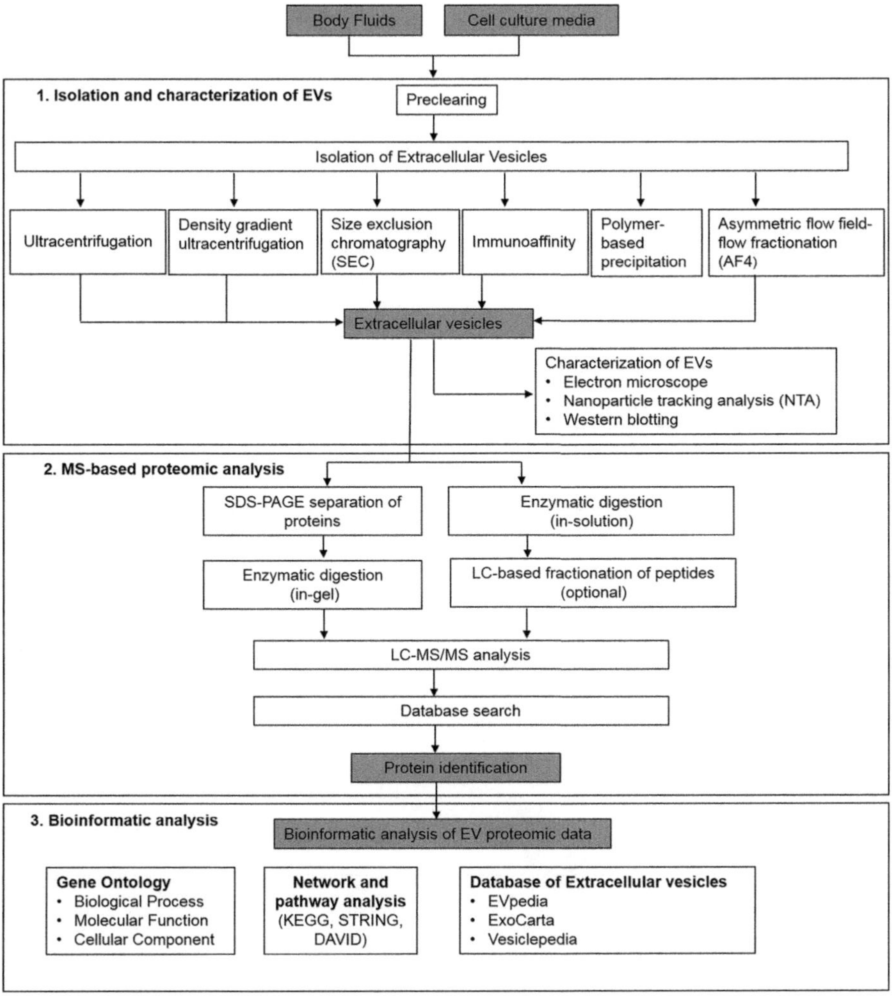


**Fig. 7.1** Approaches for comprehensive analysis of the EV proteome. In the first step for analyzing the EV proteome, EVs are isolated from body fluids or cell culture medium using different techniques, such as ultracentrifugation, density gradient ultracentrifugation, size exclusion chromatography (SEC), immunoaffinity, precipitation, or asymmetric flow field-flow fractionation (AF4). Purified EV samples are then characterized using various approaches, such as electron microscopy, nanoparticle tracking analysis (NTA), or western blotting. In the second step, LC-MS/MS analysis is utilized to identify proteins in EVs using either in-gel or in-solution digestion of proteins into peptides. Proteins in EVs are then identified using a database search. In the final step, the EV proteomic data are annotated using bioinformatic approaches, such as Gene Ontology and network and pathway analyses. Proteomic data can be classified using an EV database

With the protein list identified from EVs, several bioinformatics resources and tools have been applied to obtain insights into the functional roles of EV proteins. For example, Gene Ontology analysis is utilized to identify the enrichment of particular groups of biological processes, molecular functions, or cellular compartments in the proteins identified in EVs samples. Protein-protein interaction network and pathway analyses have been used to identify functional modules or pathways involving EV proteins. Moreover, three specialized web-based databases for EVs, i.e., ExoCarta [16], EVpedia [17], and Vesiclepedia [18], have been used for the systematic classification of EV data.

In the last two decades, the rapid development of MS-based proteomic technologies, coupled with improved purification techniques for EVs, has allowed in-depth proteomic analysis of EVs. Thousands of proteins have been identified in EVs derived from cell culture medium or body fluids [12, 13]. Additionally, several signatures of EV protein compositions have been discovered, with an increasing number of proteins being identified in EVs from different sources.

EVs contain two groups of proteins. The first group is a set of cytosolic and membrane proteins that are commonly identified in different studies and are conserved among different origins. Hurwitz et al. conducted a high-throughput study to analyze EVs isolated from 60 cell lines from the National Cancer Institute and identified 6071 proteins, among which 213 proteins were common to all isolated EVs. This result indicated that a group of proteins is conserved in EVs from different sources [19]. This common set of proteins includes cytoskeletal proteins (e.g., tubulins, actins, ezrin, and moesin), metabolic enzymes (e.g., glyceraldehyde 3-phosphate dehydrogenase [GAPDH], phosphoglycerate kinase 1, pyruvate kinase M [PKM] 2, enolase 1 [ENO1]), ribosomal proteins, annexins, heat shock proteins (HSPs; e.g.,

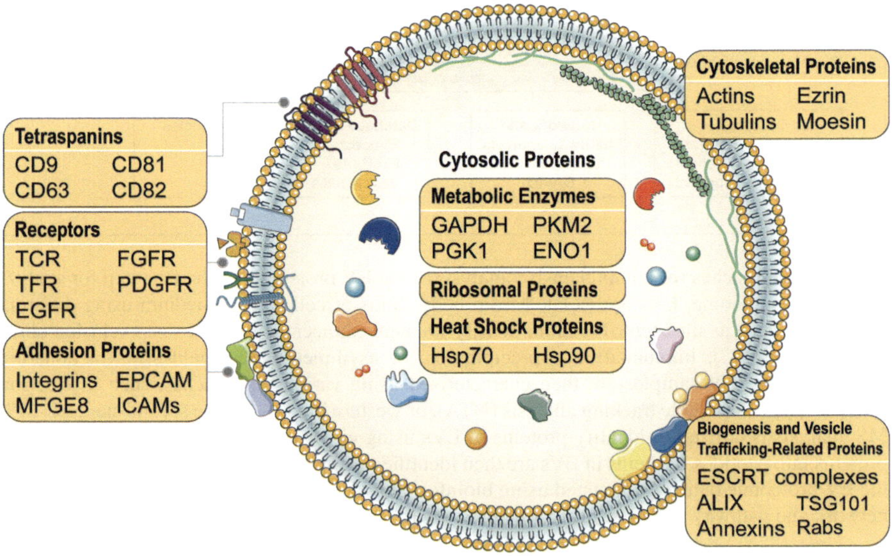


**Fig. 7.2** The protein compositions of EVs. EVs are enriched with cytoskeletal proteins, cytosolic proteins (such as metabolic enzymes, ribosomal proteins, and heat shock proteins), and membrane proteins (such as tetraspanins, receptors, adhesion proteins, and biogenesis and vesicle trafficking-related proteins)

HSP70 and HSP90), and tetraspanins (e.g., CD9, CD63, and CD81) [3] (Fig. 7.2). However, some cytosolic proteins (e.g., tubulins, HSP90) and glycolytic enzymes (e.g., GAPDH, PKM, and ENO1) were not found to be components of EVs in another study, which employed a high-resolution density gradient to separate small EVs from nonvesicular materials and used direct immunoaffinity capture technology to isolate other types of small EVs [20].

The second group of proteins are cell-specific proteins that account for cell-specific functions. Some cell-specific proteins are membrane proteins, including receptors and cell adhesion proteins, such as T-cell receptor, transferrin receptor, epidermal growth factor receptor, fibroblast growth factor receptor, platelet-derived growth factor receptor alpha, integrins, epithelial cell adhesion molecule, intercellular adhesion molecule, and milk fat globule-EGF factor 8. Because these proteins may participate in different pathophysiological processes, they have great potential for use as biomarkers. Melo et al. compared the compositions of exosomes derived from cancer cells and noncancer cells and found that 48 proteins were exclusively present in cancer exosomes [21]. Among these proteins, glypican-1 (GPC1) is a membrane-anchored protein overexpressed in breast and pancreatic cancers. GPC1 in serum exosomes has been validated as a potential diagnostic and screening biomarker to detect early stage pancreatic cancer [21]. These results indicate that EVs derived from different sources tend to retain unique signatures and carry cell-specific information to mediate disease progression.

Second, EVs contain a specific subpopulation of proteins rather than randomly selected molecules from the cells of origin. Proteins in EVs are mainly derived from the plasma membrane, intracellular vesicles, and the cytosol. The protein compositions of EVs are distinct from that of other subcellular organelles and compartments. Choi et al. analyzed the data for EVs from three cell types (SW480 cancer cells, B cells, and platelets) using the functional annotation tool DAVID [22] and compared the data for EVs with other subcellular organelles or compartments [2]. They found that proteins from EVs were enriched with vesicle structure- and vesicle biogenesis-related biological processes, such as protein localization, protein transport, vesicle-mediated transport, protein targeting and folding, regulation of protein modification processes, membrane organization, regulation of organelle organization, cytoskeleton organization, and small GTPase-mediated signal transduction. The enrichment of specific biological processes in EVs is different from that of intracellular organelles, such as the endoplasmic reticulum, Golgi apparatus, and lysosomes [2]. EVs are compartments that differ from other subcellular organelles.

Additionally, the composition of EVs is associated not only with cell types and physiological or pathological conditions in which EVs are secreted but also with EV subtypes. Jimenez et al. compared the protein compositions of small and large EVs purified from a colon cancer cell line using an isobaric tags for relative and absolute quantitation (iTRAQ)-based quantitative proteomic strategy. They discovered that proteins enriched in small EVs were associated with cell-cell junctions, cell-matrix adhesion, and exosome biogenesis machinery, whereas proteins enriched in large EVs were relevant to metabolism, ribosomes, RNA biogenesis, and processing [23]. However, it is very challenging to purify different subtypes of EVs, given the overlaps in the physical properties of different subtypes of EVs [24]. Some proteins enriched in EVs have been used as standard marker proteins to verify their purity. For example, tetraspanins (CD9, CD63, and CD81), HSPs (HSP70 and HSP90), ALG-2 interacting protein X (ALIX), and tumor susceptibility gene 101 protein (TSG101) are normally considered as "specific" markers of exosomes [25]. However, these proteins have also been identified in microvesicles and apoptotic bodies [3]. In another study, annexin A1 was identified as a specific marker for microvesicles shed from the plasma membrane [20]. Recently, Hoshino et al. conducted large-scale and comprehensive proteomic profiling of EVs and EV particles (EVPs) from 426 normal and cancer-associated samples, including cell lines, tissues, and body fluids, and identified 13 ubiquitous pan-EVP markers in human cell line-derived samples. However, several standard exosome markers, including CD63, TSG101, ALIX, and flotilins, are not well-represented in human plasma [26]. This result indicates that more efforts are needed to identify EV markers in human body fluids in the future.

Although thousands of proteins have been identified in EVs, there are many challenges associated with analysis of EV proteomes. First, it is difficult to isolate "pure" EVs without contamination. Cells secrete heterogeneous subpopulations of EVs of different sizes and distinct protein compositions, and the mixture of EV subpopulations makes isolation more challenging. Using a combination of different vesicle isolation techniques or application of a density gradient ultracentrifugation step can yield "purer" EVs [20]. Second, contamination with highly abundant nonvesicular proteins, such as uromodulin in urine exosomes, serum abundant proteins, or protein aggregates, is another challenge in EV proteomics. These highly abundant proteins can hinder the identification of low-abundance proteins in the EVs. Third, it is difficult to distinguish the genuine components of EVs from contaminated ones. More than two proteomic datasets are needed to catalog genuine proteins in EVs. Clark et al. defined the "true" exosomal cargo proteins by comparing the abundances of proteins derived from three fractions of differential ultracentrifugation, namely, the 10,000 × *g* cellular debris pellet, a 100,000 × *g* crude exosome pellet, and OptiPrep-enriched exosome pellet, using a tandem mass tag (TMT)-based quantitative proteomic approach [27]. Fourth, it takes a long time to prepare EV samples for proteomic analysis, which may hinder high-throughput analysis of samples from body fluids. The sensitivity and throughput of this analytical method should be enhanced for EV-based clinical use.

## 7.3 Quantitative Proteomic Analysis of EVs

Identification of proteins in EVs alone does not lead to a complete understanding of the roles of EVs because the proteomes of EVs are not only dependent on cell type but are also influenced by cell conditions. A more insightful understanding of the roles of EVs in health and disease can be provided by quantitative proteomic analysis, which compares the relative abundances of EV proteins under different conditions.

Quantitative proteomic analysis can be conducted using either gel- or LC-MS-based methods. The latter includes label-free and stable isotopic labeling methods.

### 7.3.1 Gel-Based Quantitative Proteomic Strategies in EV Research

Two-dimensional electrophoresis (2-DE) coupled with MS was the first technique used to compare proteomes under different conditions. In traditional 2-DE, proteins are separated according to their isoelectric points (pIs) in the first dimension by isoelectric focusing and their molecular mass in the second dimension by SDS-PAGE. After separation, gels are visualized by staining, and gel images are aligned using software to identify spots representing differentially expressed proteins. Protein spots are then cut out, digested, and identified using MS [28]. Shai et al. found that more microparticles were generated in platelets by shear stress than by thrombin stimulation. They then compared the proteomic profiles of microparticles secreted by these two platelet stimuli with 2-DE. Twenty-six proteins were differentially expressed between the two groups. Most of these proteins were shown to be related to cell assembly, organization, and cell morphology [29]. However, the inherent limitations of 2-DE have hindered its application in EV research. These limitations include low quantitative reproducibility, low dynamic range, and difficulties in detecting proteins with extreme molecular weights or pIs as well as hydrophobic membrane proteins. In particular, EVs contain high levels of membrane proteins, which comprise of an estimated 28% of biofluid-derived EVs and 34% of cell culture medium-derived EVs [30].

Difference gel electrophoresis (DIGE) improves the quantitative reproducibility of 2-DE by introducing fluorescent cyanine dyes into proteins. With DIGE technology, two to three samples can be labeled with different dyes, mixed together, and separated using the same two-dimensional (2D) gel; thus, the variability of the spot pattern is reduced, and the quantitative reproducibility is increased [31]. DIGE technology has been applied to investigate the immune regulatory roles of EVs in brain tumors [32] and alterations in B-cell exosome proteins upon γ-herpes virus infection [33]. However, DIGE technology is rarely applied in the EV field, likely because 2-DE technology is biased toward analyzing hydrophobic membrane proteins, as described above.

### 7.3.2 LC-MS/MS-Based Quantitative Proteomic Strategies in EV Research

With improvements in sensitivity, speed, mass accuracy, and resolution, LC-MS/MS has become a mainstream strategy for quantitative proteomics. Based on whether stable isotopes are introduced during sample preparation, LC-MS/MS-based quantitative proteomic strategies can be divided into label-free and stable isotopic labeling methods.

Label-free methods can be used for quantification without adding any labels to proteins or peptides, but by measuring the ion intensity/peak area in chromatography or counting the number of spectra identifying peptides from a given protein (spectral counting). Compared with stable isotopic labeling methods, label-free methods have several advantages. First, label-free methods are simpler and cheaper because they do not require special sample processing or expensive stable isotopic reagents. Second, there is no limit to the number of experimental conditions that can be compared in label-free methods because samples are processed separately, whereas only 2–18 experimental conditions can be compared directly in stable isotopic labeling methods. Third, label-free methods provide a higher dynamic range of quantification than stable isotope labeling methods [34].

Spectral counting is a commonly used label-free quantitative method in the field of EVs [35]. Choi et al. compared the proteomes of EVs isolated from primary and metastatic colorectal cancer cell lines using a spectral counting-based label-free quantitative method. They found that EVs from primary cancer cells were enriched with proteins that play roles in cell adhesion, whereas proteins identified in EVs from metastatic cancer cells were associated with cancer progression. These cancer progression-associated EV proteins could be used as potential biomarkers for cancer metastasis [36].

Another label-free quantification approach is based on the measurement of ion intensity or peak area in chromatography. Hoshino et al. profiled the proteome of exosomes derived from several tumor models using an ion intensity-based method and found that the expression pattern of integrins correlated with different metastatic potentials. For example, exosomal integrins $\alpha_6\beta_4$ and $\alpha_6\beta_1$ are associated with lung metastasis, whereas exosomal integrin $\alpha_v\beta_5$ is linked to liver metastasis. These results proved the "seed-and-soil" hypothesis of cancer metastasis and provided new insights into the mechanisms of organ-specific metastasis of cancers [37]. In another study, EVs were isolated from highly and poorly metastatic lung cancer cells, and LC-MS/MS analysis revealed that small EVs from cancer cells promoted cancer metastasis through the hepatocyte growth factor/c-Met pathway [38]. Because there is no limit to the number of experimental samples to be analyzed, intensity-based label-free quantitative proteomic strategies may have much potential for applications in the analysis of EVs isolated from body fluids, such as plasma [39], cerebrospinal fluid [40], urine [41], and saliva [42].

Although label-free quantitative methods are simple, they are highly dependent on instrument reproducibility. Moreover, peptides display different mass spectrometric responses owing to ion suppression effects when analyzing peptide mixtures. To solve these problems, stable isotope labels are introduced into proteins or peptides through metabolic or chemical methods to allow the relative quantification of proteins from different samples in the same MS analysis. Popular stable isotope labeling methods in EV research include stable isotope labeling with amino acids in cell culture (SILAC), iTRAQ, and TMT.

The SILAC-based quantitative proteomic workflow compares the protein expression under two (2-plex) and three (3-plex) experimental conditions. In SILAC, one population of cells is cultured in cell culture medium supplemented with "light" forms of lysine and arginine, whereas other populations of cells are cultured in cell culture medium supplemented with "medium" or "heavy" forms of lysine and arginine. After SILAC labeling is completed, proteins with different labels are extracted, mixed, and digested into peptides for MS analysis. Clark et al. compared the abundances of exosomal proteins between an immortalized normal bronchial epithelial cell line and two non-small cell lung cancer (NSCLC) cell lines using a 3-plex SILAC quantitative strategy. In total, 721 exosomal proteins were quantified in the exosomes of the three cell lines. Proteins associated with cell invasion, angiogenesis, and cell proliferation are enriched in NSCLC exosomes, indicating that these proteins play roles in lung carcinogenesis [43]. In addition, SILAC technology has been applied to compare proteome changes in exosomes upon human immunodeficiency virus-1 infection in lymphocytic H9 cells [44] and hepatitis B virus infection in Huh-7 cells [45].

Because SILAC is metabolically labeled, it can only be used to analyze EVs isolated from cell culture medium. The chemical labeling methods iTRAQ and TMT are more flexible because they can be used for EVs isolated from both cell culture medium and body fluids. iTRAQ and TMT are multiplex isobaric labeling techniques in which up to 18 experimental samples can be analyzed in a single analytical experiment [46]. Using isobaric labeling methods, peptides from several samples were labeled with different isotopic variants of mass tags, followed by peptide combination and LC-MS/MS analysis. Upon LC-MS/MS analysis, these isobaric tags had identical masses and chemical properties, allowing heavy and light isotopologues to coelute in MS1. The relative intensities of reporter ions generated upon peptide fragmentation in MS2 were used to obtain quantitative information on the labeled peptides/proteins in different samples. iTRAQ- and TMT-based quantitative proteomic strategies have been applied to compare the proteomes of EVs in response to different stimuli [47, 48] or from different disease states [49]. Cypryk et al. compared the proteomes of EVs released from human monocyte-derived macrophages upon β-glucan stimulation using an iTRAQ-based quantitative proteomic strategy and discovered that integrins and their cytoplasmic cargo proteins were highly abundant proteins in EVs released upon β-glucan stimulation [47].

Although studies using the quantitative proteomic strategies mentioned above have provided some insights into the biological roles of EVs, the consistency and reproducibility of data obtained from DDA are insufficient. DIA, a new data acquisition method, allows the fragmentation and analysis of all peptides across the MS window of a specific mass range in an unbiased model; therefore, the consistency of protein identification and reproducibility of protein quantification are improved [50]. This improvement offers DIA-based techniques with great advantages for identifying or validating biomarkers in EVs. Zheng et al. applied a TMT-based quantitative proteomic strategy to screen different proteomic signatures in EVs isolated from the plasma of patients with colorectal cancer (CRC) at different stages. DIA-MS was applied to validate these signatures in multiple cohorts. Compared with TMT experiments, DIA-MS has several advantages, such as no limitation to the number of samples to be analyzed and reduced operation costs and time [51].

DIA-MS analysis is expected to offer a high-throughput and reliable tool for screening biomarkers in clinical samples in the future [52].

With the development of different quantitative proteomic technologies, the contribution of LC-MS/MS-based quantitative proteomics to EV biology will increase in the near future, thereby improving our understanding of the biological roles of EVs and the discovery of new biomarkers.

## 7.4 Post-translational Modifications (PTMs) and EVs

In addition to identifying and quantifying EV proteins, another area of proteomic research on EVs involves the analysis of protein PTMs.

PTMs are covalent processing events that change the properties of proteins by introducing a small group to amino acid residues or by proteolytic cleavage of proteins. PTMs can affect the stability, activity, and subcellular localization of proteins. Many types of protein PTMs, such as phosphorylation, glycosylation, ubiquitination, sumoylation, palmitoylation, oxidation, and citrullination, have been identified in EVs [53–56]. Almost all proteins reported in EVs can be modified by different PTMs at certain stages [57]. In the past decade, the analysis of protein PTMs in EVs has been driven by their relevance in EV biology, such as in the regulation of biogenesis and release of EVs, cargo selection, uptake of EVs, and intracellular signaling.

### 7.4.1 Phosphorylation and EVs

Protein phosphorylation, the most commonly studied PTM, plays important roles in many cellular processes, such as cellular signaling, metabolism, division, differentiation, growth, apoptosis, membrane transport, and immunity. Phosphorylation is a reversible modification catalyzed by protein kinases and reversed by protein phosphatases.

Phosphorylation is essential in EV biology. Phosphorylation influences the biogenesis and release of EVs. The release of EVs from tumor cells into the surrounding environment depends on the ARF6-regulated and extracellular signal-regulated kinase-mediated pathways, which lead to the phosphorylation of myosin light chain (MLC). Inhibition of ARF6 activation is accompanied by protein kinase C-mediated phosphorylation of MLC, which blocks the release of microvesicles [58]. Furthermore, the release of EVs by mast cells is a phosphorylation-dependent process [59], and tyrosine kinase inhibitors reduce the release of exosomes from leukemia cells [60]. Second, EVs participate in cellular communication by transferring kinases or phosphatases between the cells. EVs have been reported to carry active kinases [61] and phosphophatases [62], which can be transferred to recipient cells and exert different functions through phosphorylation events.

Although phosphorylation plays important roles in EV biology, there are many challenges associated with analyzing phosphorylation in EVs. First, the abundance of phosphopeptides is very low. It is reported that only 1–2% of the entire protein amount is present in phosphorylated form [63]. The low abundance of phosphopeptides results in fewer identifications because only abundant peptides are fragmented, and low-abundance peptides are not selected for MS/MS in normal DDA experiments. Second, phosphopeptides display lower ionization efficiency compared with nonphosphorylated peptides, which results in lower signal intensities of phosphopeptides in MS analysis. Third, phosphorylation modifications are labile during MS analysis [64]. To increase the detection of low-abundance phosphopeptides in MS analysis, it is necessary to efficiently enrich phosphoproteins/phosphopeptides prior to MS analysis.

Phosphorylation can be enriched at the protein and peptide levels. Immunoprecipitation using antibodies is an efficient method for the enrichment of phosphoproteins. However, the specificity of antibodies against phosphoserine and phosphothreonine is low, and only highly selective phosphotyrosine-specific antibodies can be utilized for the enrichment of tyrosine-phosphorylated proteins. Enrichment for phosphoproteomic analysis is commonly performed at the peptide level. Immobilized metal ion affinity chromatography (IMAC) and titanium dioxide ($TiO_2$) chromatography are widely used to enrich phosphopeptides. IMAC enriches phosphopeptides based on the high-affinity coordination of negatively charged phosphate groups to immobilized metal ions ($Fe^{3+}$, $Al^{3+}$, $Ga^{3+}$, or $Co^{2+}$), whereas $TiO_2$ selectively enriches phosphopeptides because organic phosphates are effectively absorbed by $TiO_2$ under acidic conditions and desorbed under alkaline conditions. These two techniques have different selectivities for different pools of phosphopeptides; IMAC has a stronger selectivity for multi-phosphopeptides, whereas $TiO_2$ has a higher affinity for mono-phosphopeptides. The sequential elution from IMAC (SIMAC) method [65] and $TiO_2$-SIMAC-HILIC method [66] combines the use of IMAC and $TiO_2$ to improve phosphopeptide coverage from complex biological samples.

Several studies have investigated the phosphorylation of EV proteins in human urinary exosomes [67] and microvesicles from cytokine-stimulated β cells [68] using LC-MS/MS. However, the phosphoproteomic coverages reported in these two studies were low. Only a few large-scale phosphoproteomic studies have been performed on EVs derived from cells [69] or body fluids [70, 71]. Guo et al. performed a high-coverage phosphoproteome analysis of exosomes purified from SW620 CRC cells and identified 313 phosphoproteins with 1091 phosphosites, among which 202 proteins were novel. An in-depth analysis of the exosome phosphoproteome revealed that exosomes have a remarkably high level of tyrosine-phosphorylated proteins (6.4% in exosomes versus 0.6% in cells) and that tyrosine-phosphorylated proteins are functionally relevant to ephrin signaling pathway-directed cytoskeleton remodeling [69]. This result implies that tyrosine phosphorylation may influence specific exosome functions. Chen et al. identified 9643 unique phosphopeptides in EVs isolated from plasma, and the expression of 144 phosphoproteins was found to be significantly higher in patients diagnosed with breast cancer than in healthy controls using label-free quantitative proteomics. These proteins may be used as biomarkers for cancer screening and monitoring [70]. In the future, large-scale phosphoproteomic analysis of EVs from different cell types or body fluids will provide more information on the roles of phosphorylation in cell communication and the roles of EVs in physiology and disease.

### 7.4.2 Glycosylation and EVs

Protein glycosylation, an important PTM, is involved in various cell functions, such as cell adhesion, migration, cell-cell communication, and signal transduction. Aberrant glycosylation has been observed in many diseases, such as cancer, neurodegenerative diseases, and cardiovascular diseases [72]. Sensitive detection of disease-associated changes in glycosylation would help to identify biomarkers for diagnosis and prognosis. Protein glycosylation is one of the most abundant protein modifications, and over half of all proteins are glycosylated [73]. N-linked and O-linked glycosylation are the most widely studied types of protein glycosylation.

Glycosylation plays an essential role in EV biology [74]. First, glycosylation is involved in EV biogenesis. Clark et al. investigated the impact of the increased expression of α-(1,6)-fucosyltransferase (FUT8), a protein involved in the protein glycosylation pathway, on prostate cancer-derived EVs. They found that overexpression of FUT8 reduced the number of vesicles secreted by prostate cancer cells and increased the abundance of proteins associated with cell motility and prostate cancer metastasis. This result provides direct evidence that glycosylation plays a role in EV biogenesis [75]. Second, glycosylation influences protein sorting in EVs. Introducing a glycosylation site onto a protein can protect targeting ligands displayed on the surface of exosomes from degradation and lead to increased recruitment of glycoproteins into exosomes [76]. In another study, N-linked glycosylation was found to mediate protein sorting into exosomes, and inhibition of N-linked glycan processing with glycan inhibitors reduced the recruitment of proteins into exosomes [77]. These results suggest that glycosylation is involved in protein sorting and influences EV protein profiles. Third, the glycans of EVs are important for their uptake by recipient cells. Williams et al. reported that disruption of the native glycosylation of EVs by treatment with the glycosidase PNGase F or neuraminidase alters the uptake of EVs in a panel of different cell lines [78]. Removal of sialic acid modification by neuraminidase leads to increased EV uptake by RAW 264.7 osteoclast precursor cells [79]. A similar effect of sialic acid removal was observed in SKOV3 ovarian cancer cells [70].

EVs display a specific glycoconjugate composition and characteristic glycosignature. Batista et al. compared the glycosylation patterns of microvesicles from different biological sources, including different human cell lines (T cells, melanoma, and colon cancer), and breast milk, using lectin array technology. They observed that specific glycosignatures with enrichment in high mannose, ploylactosamine, α-2,6 sialic acid, and complex N-linked glycans and exclusion of terminal blood group A and B antigens were conserved in microvesicles from different biological sources [80]. Moreover, Escrevente et al. found that the glycosylation profiles of exosomes were distinct from those of the microsomal fraction or plasma membrane in SKOV3 ovarian carcinoma cells by lectin blotting. Complex N-glycans of the di-, tri-, and tetra-antennary type with or without proximal fucose as well as high-mannose glycans were detected in exosomes derived from SKOV3 cells by normal-phase high-performance LC (HPLC) and matrix-assisted laser desorption ionization time-of-flight MS (MALDI-TOF-MS) [81]. These results indicate that glycans and glycosylation affect the biogenesis, sorting, and uptake of EVs.

Although glycosylation plays an important role in EV biology, characterization of the EV glycoproteome is still a great challenge for several reasons. First, glycoproteins are present in low abundance compared with other proteins. Second, many proteins contain several glycosylation sites, each of which can be occupied by different glycan structures. This complex microheterogeneity of glycosylation hampers the analysis of glycopeptides with MS. Third, the ionization efficiency of glycopeptides is lower than that of the unmodified peptides. Therefore, selective enrichment and isolation of glycoproteins or glycopeptides from complex biological samples are essential for in-depth glycoproteomic analysis [73]. During the past decade, several enrichment methods, including lectin affinity, hydrazide chemistry, boronic acid, HILIC, and electrostatic repulsion hydrophilic interaction chromatography, have been applied in glycoproteomic research [73].

Several glycoproteomic studies have been conducted to analyze urinary EVs. Saraswa et al. enriched N-glycopeptides from urine exosomes by SNA affinity chromatography or SEC. They identified 126 N-glycopeptides from 51 glycosylation sites belonging to 37 glycoproteins. This result indicates that microheterogeneity in glycosylation sites exists in exosome proteins [82]. Huo et al. separated EV proteins and soluble proteins in urine using stepwise ultrafiltration and enriched N-glycopeptides from the two parts using HILIC. Using this strategy, they identified 1210 and 1256 N-glycopeptides belonging to 586 and 619 N-glycoproteins in EVs and soluble proteins, respectively. Distinct patterns of the N-glycoproteome have been identified for the two subproteomes in urine, and approximately 50% of N-glycoproteins were only identified in either EVs or soluble proteins [83]. Recently, Brown et al. conducted a large-scale characterization of the glycoproteome of human urinary exosomes using multilectin affinity chromatography and MS; they identified 378 glycoproteins and 3565 glycopeptides, corresponding to 3144 unique glycosylation events [84]. In another study, quantitative glycoproteomic analysis was conducted in EVs isolated from the plasma of patients with breast cancer and healthy individuals. In total, 1453 unique N-glycopeptides, corresponding to 556 glycoproteins, were identified. Among these glycoproteins, the abundances of 20 proteins were significantly higher in patients with breast cancer [85]. These results provide insights into the functional roles of EVs. However, except for glycoproteome analyses of EVs from body fluids, there are few large-scale glycoproteomic analyses of EVs from cell culture medium, possibly because large numbers of cells are needed to collect sufficient EVs for glycoproteomic analysis. In the future, more studies of the glycoproteome in EVs derived from cell culture medium will provide comprehensive information about glycosylation in EVs. In addition, more attention should be paid to EV isolation approaches when analyzing the EV glycoproteome because different isolation approaches result in EV populations with different glycosylation profiles [86, 87].

### 7.4.3 Other Modifications and EVs

In addition to phosphorylation and glycosylation, several other PTMs, such as ubiquitination, sumoylation, palmitoylation, myristoylation, citrullination, and oxidation, have been identified in EVs.

Ubiquitination is an important PTM that regulates protein activity by providing signals for degradation. As a major controller of specific loading of RNA species or proteins into EVs [56], ubiquitination is also necessary for sorting proteins to lysosomes for degradation by the endosomal sorting complex for transport machinery [88]. In addition, ubiquitination is a mechanism for trafficking soluble mycobacterial proteins within the phagocytic/endocytic network to exosomes [89].

Ubiquitinated proteins are found in EVs isolated from cell lines [90] and body fluids [91]. Large-scale identification of ubiquitination modifications in EVs with MS is challenging because of their low abundance. Therefore, enrichment of ubiquitinated proteins/peptides is necessary for their identification. Burke et al. identified 50 ubiquitinated proteins in exosomes purified from myeloid-derived suppressor cells (MDSCs) by immunoprecipitation with anti-ubiquitin antibodies at the protein and peptide levels. The number of identified ubiquitinated proteins was 10% of the number of proteins identified in MDSC-derived exosomes, and 35 of these proteins had not been previously reported to be ubiquitinated [92], indicating that enrichment enabled the identification of low-abundance ubiquitinated proteins in exosomes. Huebner et al. performed a large-scale analysis of ubiquitination sites in human urinary exosome proteins. The ubiquitinated peptides were immunoaffinity-purified using an anti-diglycine remnant (K-ε-GG) antibody. They identified more than 600 ubiquitinated proteins, which accounted for 15% of the proteins identified in human urinary exosomes. This result supports the model that deubiquitination of protein cargo is not essential during the process of protein cargo incorporation into multivesicular bodies (MVBs) [93].

In addition to ubiquitination, modifications with ubiquitin-like modifiers, such as SUMO, ISG15, NEDDs, andUBL3, are involved in sorting specific proteins or miRNAs into EVs or in controlling exosome secretion. Indeed, the loading of EXOmotif-containing miRNAs into exosomes is controlled by sumoylated heterogeneous nuclear ribonucleoprotein A2/B1 [94]. The sorting of α-synuclein into EVs is regulated by sumoylation [95]. Additionally, protein sorting into small EVs is also regulated by UBL3 modifications [96]. In addition to ubiquitination, ISGylation tags proteins for degradation. ISGylation has been reported to regulate exosome secretion by triggering MVB colocalization with lysosomes and promoting MVB protein degradation [97]. These studies suggest that ubiquitin-like modifications play important roles in the functions of EVs, although most of these PTM studies were small scale, and only a limited number of PTM sites have been reported. In the future, it will be necessary to develop methods for large-scale identification of these PTMs on EV proteins.

Palmitoylation, a reversible lipid PTM, modulates protein activity, stability, and multiprotein complex formation. Palmitoylation regulates the function of ALIX in skeletal muscle cells by influencing its interaction with CD9, a tetraspanin protein. Palmitoylation is important for maintaining the proper membrane structure organization of small EVs in skeletal muscle cells [98]. Recently, a comprehensive palmitoyl-proteomic analysis was conducted using a novel approach called low-background acyl-biotinyl exchange. Combined with label-free proteomics, this method enabled the quantification of palmitoyl-protein signatures for large EVs and small EVs derived from prostate cancer cells. This study suggested that palmitoylation maybe involved in the sorting and trafficking of proteins to EVs [99].

Many other modifications, such as myristoylation [100], citrullination [54], and oxidation [54], have been found to exist in EVs. However, the roles of other PTMs remain largely unexplored.

Although these PTMs play important roles in the biogenesis, release, and uptake of EVs, as well as protein and miRNA sorting into EVs, the exact mechanisms are not fully understood. Given the complexity of PTMs, we are still trying to understand the impact of PTMs in EVs on physiological and pathological processes. More studies are needed to assess whether these PTMs in EVs can serve as biomarkers or drug targets.

## 7.5 LncRNA-Encoded Microproteins/Peptides in EVs

LncRNAs are a family of noncoding RNAs exceeding 200 nucleotides in length. These molecules were initially considered as "junk RNAs" because they possess no apparent open reading frames (ORFs) for proteins greater than 100 amino acids in length. Recently, advancements in bioinformatics, transcriptomics, and proteomics technologies have led to the discovery of numerous new peptides encoded by small ORFs (smORFs) embedded in lncRNA transcripts (hereafter referred as microproteins) [101–103]. Accumulating evidence indicates that these microproteins are involved in diverse regulatory roles in many vital processes, such as $Ca^{2+}$ homeostasis, metabolism, development, stress signaling, and DNA repair [104–107]. Moreover, numerous cancer-related lncRNAs and their encoded functional microproteins have been shown to influence tumorigenesis, invasion, and metastasis [103, 108–110].

Nonetheless, although a limited number of microproteins have been functionally characterized, the mechanisms underlying their roles in cancer remain unclear. Many issues, such as possible modes of action, have not yet been explored. Of particular interest is whether microproteins participate in intercellular communication between tumor cells and stromal cells in either the local or distant microenvironment, considering that numerous lncRNAs have been found to be selectively sorted into EVs to regulate cancer onset and progression through various mechanisms.

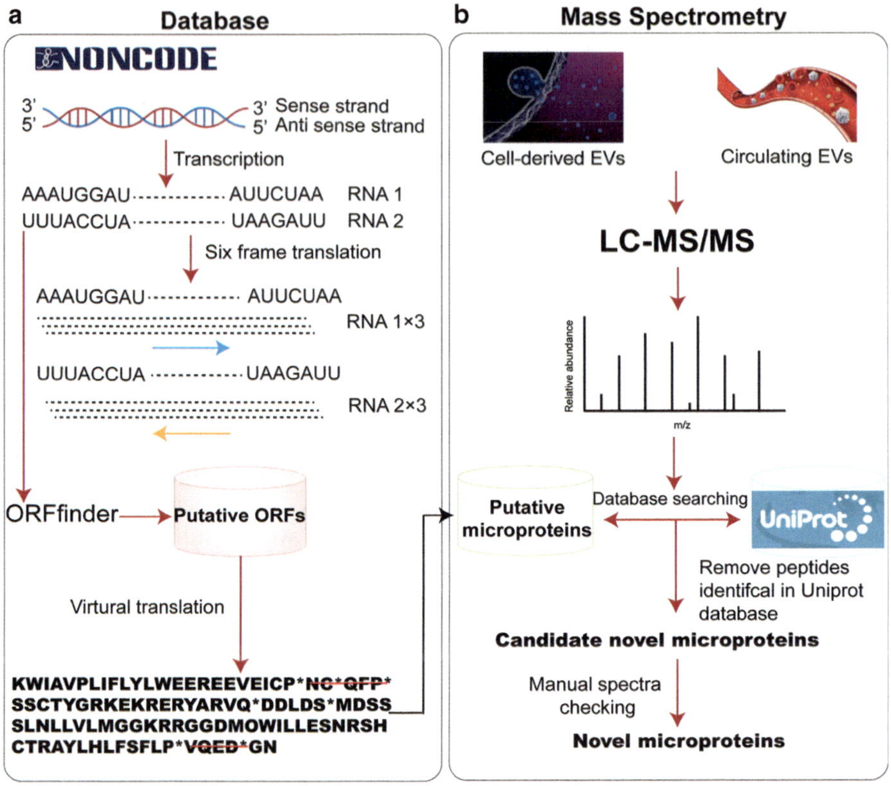


**Fig. 7.3** Schematic illustration of the workflow for MS-based discovery of microproteins encoded by lncRNAs. (**a**) Construction of the putative microprotein database. The human lncRNA transcripts deposited in noncode database were screened using oRF finder and six-frame translation to identify all possible smORFs and then theoretically translated into putative microproteins. All microproteins were collected in the human microprotein database. (**b**) MS-based identification of microproteins in cell culture-derived and circulating EVs

Recently, our group employed multiple technologies, including an improved microprotein analytical workflow and stringent criteria (Fig. 7.3), to provide the first experimental evidence of abundant microproteins in cell culture-derived EVs. This result indicates that microproteins are a new form of cargo in EVs and suggests that EV-mediated transfer of microproteins may represent a novel mechanism for intercellular communication [111]. The improved discovery rate of microproteins in EVs can be partially attributed to the newly constructed in-house microprotein reference database. By combining the six-frame translation mode and ORF finder, the database completely collected putative smORFs from the NONCODE database, an interactive database containing the most complete collection and annotation of lncRNAs. This database enabled us to mine substantially more microproteins from cell lines and tissues with stringent identification criteria.

Intriguingly, when examining human plasma, 48, 11, and 3 microproteins were identified from purified EVs, whole plasma, and EV-free plasma, respectively, suggesting that circulating microproteins were primarily enriched in EVs. Most importantly, some preliminary data showed that the expression profiles of EV microproteins in patients with glioma diverged from those of healthy donors. These results suggested that circulating microproteins in EVs may have potential diagnostic applications for identifying patients with glioma.

## 7.6 Lipidomic Technology for EVs

Lipids, a type of active molecule that is carried by EVs, not only participate in cell-to-cell communication but are also closely related to the biogenesis of EVs. Recent studies have shown that various lipids, such as sterols, phosphatidic acids, and sphingolipids, are involved in the production of EVs and play key roles in their functional applications [112–114]. In-depth research on the qualitative and quantitative analysis of the EV lipidome has been essential for the characterization of EVs and probing their biogenesis mechanisms.

With the development of analytical techniques, such as chromatography and MS, and the emergence of new data processing software, lipidomics has advanced rapidly in both basic and biomedical sciences. EV lipidomic research has a general workflow, as follows: (1) determination of scientific questions; (2) lipidomics analysis, including sample preparation, separation and identification, and data analysis; (3) bioinformatic analysis of lipidomic results to obtain biological information, such as lipid compound-related metabolic pathways; and (4) biological and/or clinical validation.

The lipidomic analysis of EVs from different biological sources requires robust separation techniques and detection methods. Currently, the main analytical techniques used in this field are chromatography-MS and nuclear magnetic resonance (NMR). Due to the limitations of NMR with regard to sample size and sensitivity, chromatography-MS technology is more commonly used in the lipidomic-based analysis of EVs, which is the focus of this chapter.

According to different research purposes, lipidomic analysis can be divided into nontargeted and targeted lipidomics. Nontargeted lipidomics is mainly used for the analysis and identification of all lipid compounds and their metabolites in biological samples, such as EVs. Furthermore, this approach is mainly applied in studies focusing on overall changes in lipid molecules, phenotype identification, or marker screening. Nontargeted lipidomics aims to analyze and identify as many lipid molecules as possible, in which fast and high-resolution MS instruments, such as quadrupole-TOF MS (Q-TOF MS) and quadrupole-orbitrap MS (Q-Orbitrap MS), are required [115–119]. Haraszti et al. used high-resolution MS (triple TOF MS 5600) to conduct nontargeted lipidomic research on exosomes and microvesicles in different cell lines. They found that exosomes were more enriched with glycolipids and free fatty acids, whereas microvesicles were more enriched with ceramide and sphingomyelin [119].

Targeted lipidomics is mainly used for qualitative and quantitative analyses of several specific target lipids. This requires high selectivity of the analytical methods. Therefore, triple quadrupole tandem MS (QQQ MS) with strong quantitative ability or high-resolution MS with precise determinism ability is used, with instruments such as Q-TOF MS, Q-Orbitrap MS, and Fourier transform MS [120, 121]. Khayrullin et al. performed targeted lipidomics with HPLC-QQQ-MS to analyze ceramides in human and primate serum EVs. They found that the long-chain ceramide C24:1 in EVs from older women was significantly increased compared with that in younger women and that this long-chain ceramide can induce the aging of bone marrow mesenchymal stem cells [121].

### 7.6.1 Sample Preparation

The first step in EV lipidomic analysis is to extract all lipids from the EV bilayer. Currently, lipid extraction methods include traditional liquid-liquid extraction methods and some automated sample extraction techniques. The latter includes solid-phase, microwave-assisted, and supercritical fluid extraction. Although these techniques can shorten the lipid extraction time and improve the convenience of experiments, they are not universally applicable to all lipid types. Therefore, the traditional liquid-liquid extraction method is preferred by numerous groups for the extraction of all lipid compounds.

Traditional liquid-liquid extraction methods use chloroform-methanol-water solvents (Folch method [chloroform:methanol:water = 8:4:3, v/v] [122], the Bligh-Dyer method [chloroform:methanol:water = 2:2:1.8, v/v] [123]) and methyl tert-butyl ether (MTBE)-methanol-water solvents (Matyash method [124]). These methods use chloroform-methanol-water solvents and have very high extraction efficiency and good reproducibility, enabling the extraction of approximately 98% of all lipid molecules. The Folch method and the Bligh-Dyer method have been widely used for a variety of samples, such as plasma, urine, tissues, cells, and plants. The Matyash method is a lipid extraction method developed in recent years that uses MTBE instead of chloroform. Because the density of MTBE is lower than that of water, when an organic-water two-phase system is formed, the MTBE organic phase containing the lipid components is in the upper layer. Compared with the Folch method and the Bligh-Dyer method, the MTBE method facilitates the collection of lipid components from the organic phase, allows multiple extractions, and is not easily contaminated by impurities in the water phase; therefore, this approach has been increasingly used in the extraction of lipids from various tissue samples and EVs [116, 117].

To improve the efficiency of lipid extraction, the three lipid extraction methods have been further optimized by changing the solvent ratio and adding inorganic salts or antioxidants. Haraszti et al. used a chloroform-methanol-LiCl aqueous solution (2:2:1, v/v) system to extract lipids and compared the lipidomes of exosomes and microvesicles from three different cell lines [119]. Their results showed that the types and concentrations of lipids extracted using the three lipid extraction methods were partially different. Therefore, it is necessary to carry out preliminary experiments to compare the coverages and extraction efficiencies of the three lipid extraction methods for specific types of lipids in lipidomics research of EVs, particularly for targeted lipidomic research.

### 7.6.2 Separation and Identification of Lipid Components

The total lipids extracted from EVs must be analyzed using chromatography-MS. Because lipids extracted from EVs are of many different types, ranging from nonpolar to polar, it is necessary to separate different lipids for accurate identification by subsequent MS analysis. In the early days, thin-layer chromatography-MS (TLC-MS) and gas chromatography-MS (GC-MS) were widely used. However, TLC is now rarely used because of its low separation efficiency. Moreover, GC-MS requires derivatization of lipid compounds, and the analysis steps are complicated and time-consuming. Thus, it is rarely used for nontargeted lipidomic research, but it is mainly used for the analysis of specific lipids, such as fatty acids and cholesterol in EVs. With the rapid development of ultra-high-pressure LC, high-resolution MS, and tandem MS technology, LC-MS has been widely used in lipidomic research. LC-MS has excellent qualitative and quantitative analytical capabilities and can be used to separate and analyze various lipid compounds without derivatization.

Currently, the most widely used LC-MS technologies in the analysis of EV lipidomics are reversed-phase chromatography-MS (RPLC-MS) technology, whereas normal phase chromatography-MS (NPLC-MS) technology and HILIC-MS technology are available but rarely used. When NPLC-MS and HILIC-MS were used to separate whole lipids, the lipid molecules eluted in the order from hydrophobic to hydrophilic, that is, from weak to strong polarity. This separation method of grouping by polarity greatly simplifies the qualitative analysis of MS. However, because nonpolar lipid components are very weakly retained in NPLC, some useful information on lipid metabolism can be lost. Thus, NPLC and HILIC have been gradually replaced by RPLC-MS technology in the study of nontargeted lipidomics. However, these approaches are still used for the separation and collection of certain types of specific lipid components in targeted lipidomics research. For example, normal-phase silica gel columns have been used to separate the lipid components of EVs, including sterols, glycolipids, and phospholipids, followed by MS-based identification [125].

In general, RPLC-MS is the most widely used technique in EV lipidomic research. With RPLC-MS, the lipid components are mainly separated according to the length of the fatty acyl chain and degree of unsaturation. For lipids with the same degree of unsaturation, those with shorter fatty acyl chains are eluted more rapidly; similarly, for lipids with the same acyl chain length, those with greater unsaturation are eluted earlier. This method can efficiently retain most lipid components, including polar and nonpolar components, which greatly enriches EV lipidomics research. However, because lipid molecules are separated according to polarity from strong to weak, different types of lipids are co-eluted with each other, which may cause difficulties in subsequent MS analysis and identification. To solve this problem, 2D-LC MS technology has been rapidly developed in EV lipidomics research. For example, the combination of HILIC and RPLC can greatly increase the identification of lipid species in nontargeted lipidomics analysis. Here, HILIC is used as the first-dimension column to separate lipids into the same type according to the polarity of the lipid molecules, and the second-dimensional column RPLC further separates the same "type" lipids into different species according to the hydrophobicity of the molecules (fatty acid chains). This method greatly increases the peak capacity of lipid samples and is a preferred approach for nontargeted lipidomic research.

In addition, the "shotgun" lipidomic method, which directly analyzes lipids using MS, has also been applied to research on EVs lipidomics [115, 126]. Jae et al. conducted an in-depth analysis of the phospholipid composition in tumor EVs by applying the "shotgun" lipidomics technology. They found that the phospholipid profile of tumor EVs was significantly different from that of normal cells and that 67 different lipids could be used to provide assistance for subsequent drug treatment [126]. "Shotgun" lipidomics is characterized by a fast analysis speed and strong identification ability; However, this approach also has the following limitations: (1) since all lipid molecules are ionized in the ion source, the ionization of low-abundance and difficult-to-ionize lipids is significantly inhibited, preventing some lipids from being detected; (2) lipid isomers cannot be accurately identified because they do not show chromatographic separation; and (3) high-resolution or even ultra-high-resolution MS is required to ensure the accuracy of mass spectrometric identification, resulting in significant increases of equipment costs.

### 7.6.3 Data Processing

Lipidomic analysis ultimately requires the association of lipid compounds obtained from MS analysis with metabolic pathways or networks to obtain biological information. According to different scientific questions, MS data must be searched in the lipid library based on their mass spectra and other information to obtain total or specific lipid structure information. In addition, bioinformatics and system informatics methods are required to integrate specific EV lipids into metabolic pathways. Currently, many software- and lipid-related websites are available. The commonly used data processing software consists of two types: one is free and open software, such as SECD and LIMSAI, which can analyze MS and MS/MS data in both positive and negative ion modes; the second type is commercial software, such as LipidView (SCIEX), MassHunter (Agilent), and LipidSearch (Thermo Scientific). Widely used lipid-related websites include LMSD (http://www.lipidmaps.org), LipidBank (http://lipidbank.jp), CyberlipidCenter (http://www.cyberlipid.org), Lipid Library (http://www.lipidlibrary.co.uk/inex.html), and Kyoto Encyclopedia of Genes and Genomes (https://www.genome.jp/kegg). In general, there is no consensus software with complete functions for lipidomic analyses. For example, most software programs cannot determine the positions of the fatty acid bonds. Researchers often need to optimize and combine commercial and open software according to their own instruments and analysis purposes. Sometimes, they need to independently develop their own software. There is still a lot of work to do in this area.

Compared with the proteomic analysis of EVs, the lipidomics of EVs is still in its infancy. Therefore, comprehensive studies of the lipidomics of EVs derived from various cell types and body fluids using LC-MS/MS would be helpful for elucidating the roles of lipids in vesicle biogenesis and their biological functions.

## 7.7 Conclusion

In summary, various sets of -omics data can provide different information regarding the components of EVs. In future studies, multi-omics analyses of EV samples (combinations of different -omics data) will provide more information regarding the cargo of EVs as well as the roles of EVs in health and disease. These data are expected to validate the potential clinical applications of EVs in disease monitoring or diagnosis.

**Acknowledgments** This work was supported by grants from the National Key R&D Program of China (Grant Nos. 2018YFA0507103and 2018YFA0507801). We would like to thank Editage (www.editage.cn) for English language editing.

## References

1. Turchinovich A, Drapkina O, Tonevitsky A. Transcriptome of extracellular vesicles: state-of-the-art. Front Immunol. 2019;10:202.
2. Choi DS, Kim DK, Kim YK, Gho YS. Proteomics of extracellular vesicles: exosomes and ectosomes. Mass Spectrom Rev. 2015;34(4):474–90.
3. Pocsfalvi G, Stanly C, Vilasi A, et al. Mass spectrometry of extracellular vesicles. Mass Spectrom Rev. 2016;35(1):3–21.
4. Subra C, Laulagnier K, Perret B, Record M. Exosome lipidomics unravels lipid sorting at the level of multivesicular bodies. Biochimie. 2007;89(2):205–12.
5. Guan F, Xiang X, Xie Y, et al. Simultaneous metabolomics and proteomics analysis of plasma-derived extracellular vesicles. Anal Methods. 2021;13(16):1930–8.
6. Bordanaba-Florit G, Royo F, Kruglik SG, Falcón-Pérez JM. Using single-vesicle technologies to unravel the heterogeneity of extracellular vesicles. Nat Protoc. 2021;16:3163.
7. Wu B, Chen X, Wang J, et al. Separation and characterization of extracellular vesicles from human plasma by asymmetrical flow field-flow fractionation. Anal Chim Acta. 2020;1127:234–45.
8. Coumans FAW, Brisson AR, Buzas EI, et al. Methodological guidelines to study extracellular vesicles. Circ Res. 2017;120(10):1632–48.
9. Simpson RJ, Lim JWE, Moritz RL, Mathivanan S. Exosomes: proteomic insights and diagnostic potential. Expert Rev Proteomics. 2009;6(3):267–83.
10. Cocozza F, Grisard E, Martin-Jaular L, Mathieu M, Thery C. SnapShot: extracellular vesicles. Cell. 2020;182(1):262.

11. Gurunathan S, Kang MH, Jeyaraj M, Qasim M, Kim JH. Review of the isolation, characterization, biological function, and multifarious therapeutic approaches of exosomes. Cell. 2019;8(4):307.
12. Li J, Chen X, Yi J, et al. Identification and characterization of 293T cell-derived exosomes by profiling the protein, mRNA and MicroRNA components. PLoS One. 2016;11(9):e0163043.
13. Li D, Zhang W, Chen X, et al. Proteomic profiling of MIN6 cell-derived exosomes. J Proteome. 2020;224:103841.
14. Chutipongtanate S, Greis KD. Multiplex biomarker screening assay for urinary extracellular vesicles study: a targeted label-free proteomic approach. Sci Rep. 2018;8(1):15039.
15. Moon PG, Lee JE, You S, et al. Proteomic analysis of urinary exosomes from patients of early IgA nephropathy and thin basement membrane nephropathy. Proteomics. 2011;11(12):2459–75.
16. Keerthikumar S, Chisanga D, Ariyaratne D, et al. ExoCarta: a web-based compendium of Exosomal cargo. J Mol Biol. 2016;428(4):688–92.
17. Kim DK, Lee J, Kim SR, et al. EVpedia: a community web portal for extracellular vesicles research. Bioinformatics. 2015;31(6):933–9.
18. Pathan M, Fonseka P, Chitti SV, et al. Vesiclepedia 2019: a compendium of RNA, proteins, lipids and metabolites in extracellular vesicles. Nucleic Acids Res. 2019;47(D1):D516–D9.
19. Hurwitz SN, Rider MA, Bundy JL, et al. Proteomic profiling of NCI-60 extracellular vesicles uncovers common protein cargo and cancer type-specific biomarkers. Oncotarget. 2016;7(52):86999–7015.
20. Jeppesen DK, Fenix AM, Franklin JL, et al. Reassessment of exosome composition. Cell. 2019;177(2):428–45. e18.
21. Melo SA, Luecke LB, Kahlert C, et al. Glypican-1 identifies cancer exosomes and detects early pancreatic cancer. Nature. 2015;523(7559):177–82.
22. Huang da W, Sherman BT, Lempicki RA. Systematic and integrative analysis of large gene lists using DAVID bioinformatics resources. Nat Protoc. 2009;4(1):44–57.
23. Jimenez L, Yu H, McKenzie AJ, et al. Quantitative proteomic analysis of small and large extracellular vesicles (EVs) reveals enrichment of adhesion proteins in small EVs. J Proteome Res. 2019;18(3):947–59.
24. Kreimer S, Belov AM, Ghiran I, et al. Mass-spectrometry-based molecular characterization of extracellular vesicles: lipidomics and proteomics. J Proteome Res. 2015;14(6):2367–84.
25. Rosa-Fernandes L, Rocha VB, Carregari VC, Urbani A, Palmisano G. A perspective on extracellular vesicles proteomics. Front Chem. 2017;5:102.
26. Hoshino A, Kim HS, Bojmar L, et al. Extracellular vesicle and particle biomarkers define multiple human cancers. Cell. 182(4):1044-+. 2020;182:1044.
27. Clark DJ, Fondrie WE, Liao Z, et al. Redefining the breast cancer exosome proteome by tandem mass tag quantitative proteomics and multivariate cluster analysis. Anal Chem. 2015;87(20):10462–9.
28. Görg A, Weiss W, Dunn MJ. Current two-dimensional electrophoresis technology for proteomics. Proteomics. 2004;4(12):3665–85.
29. Shai E, Rosa I, Parguina AF, et al. Comparative analysis of platelet-derived microparticles reveals differences in their amount and proteome depending on the platelet stimulus. J Proteome. 2012;76:287–96.
30. Raimondo F, Morosi L, Chinello C, Magni F, Pitto M. Advances in membranous vesicle and exosome proteomics improving biological understanding and biomarker discovery. Proteomics. 2011;11(4):709–20.
31. Tannu NS, Hemby SE. Two-dimensional fluorescence difference gel electrophoresis for comparative proteomics profiling. Nat Protoc. 2006;1(4):1732–42.
32. Graner MW, Alzate O, Dechkovskaia AM, et al. Proteomic and immunologic analyses of brain tumor exosomes. FASEB J. 2009;23(5):1541–57.
33. Meckes DG Jr, Gunawardena HP, Dekroon RM, et al. Modulation of B-cell exosome proteins by gamma herpesvirus infection. Proc Natl Acad Sci U S A. 2013;110(31):E2925–33.

34. Bantscheff M, Schirle M, Sweetman G, Rick J, Kuster B. Quantitative mass spectrometry in proteomics: a critical review. Anal Bioanal Chem. 2007;389(4):1017–31.
35. Harshman SW, Canella A, Ciarlariello PD, et al. Proteomic characterization of circulating extracellular vesicles identifies novel serum myeloma associated markers. J Proteome. 2016;136:89–98.
36. Choi DS, Choi DY, Hong BS, et al. Quantitative proteomics of extracellular vesicles derived from human primary and metastatic colorectal cancer cells. J Extracell Vesicles. 2012;1.
37. Hoshino A, Costa-Silva B, Shen TL, et al. Tumour exosome integrins determine organotropic metastasis. Nature. 2015;527(7578):329–35.
38. Qiao Z, Zhang Y, Ge M, et al. Cancer cell derived small extracellular vesicles contribute to recipient cell metastasis through promoting HGF/c-met pathway. Mol Cell Proteomics. 2019;18(8):1619–29.
39. Cheow ES, Cheng WC, Lee CN, et al. Plasma-derived extracellular vesicles contain predictive biomarkers and potential therapeutic targets for myocardial ischemic (MI) injury. Mol Cell Proteomics. 2016;15(8):2628–40.
40. Muraoka S, Jedrychowski MP, Yanamandra K, et al. Proteomic profiling of extracellular vesicles derived from cerebrospinal fluid of Alzheimer's disease patients: a pilot study. Cell. 2020;9(9).
41. Jung HY, Lee CH, Choi JY, et al. Potential urinary extracellular vesicle protein biomarkers of chronic active antibody-mediated rejection in kidney transplant recipients. J Chromatogr B Analyt Technol Biomed Life Sci. 2020;1138:121958.
42. Sun Y, Huo C, Qiao Z, et al. Comparative proteomic analysis of exosomes and microvesicles in human saliva for lung cancer. J Proteome Res. 2018;17(3):1101–7.
43. Clark DJ, Fondrie WE, Yang A, Mao L. Triple SILAC quantitative proteomic analysis reveals differential abundance of cell signaling proteins between normal and lung cancer-derived exosomes. J Proteome. 2016;133:161–9.
44. Li M, Aliotta JM, Asara JM, et al. Quantitative proteomic analysis of exosomes from HIV-1-infected lymphocytic cells. Proteomics. 2012;12(13):2203–11.
45. Zhao X, Wu Y, Duan J, et al. Quantitative proteomic analysis of exosome protein content changes induced by hepatitis B virus in Huh-7 cells using SILAC labeling and LC-MS/MS. J Proteome Res. 2014;13(12):5391–402.
46. Li J, Van Vranken JG, Pontano Vaites L, et al. TMTpro reagents: a set of isobaric labeling mass tags enables simultaneous proteome-wide measurements across 16 samples. Nat Methods. 2020;17(4):399–404.
47. Cypryk W, Ohman T, Eskelinen EL, Matikainen S, Nyman TA. Quantitative proteomics of extracellular vesicles released from human monocyte-derived macrophages upon β-glucan stimulation. J Proteome Res. 2014;13(5):2468–77.
48. de Jong OG, Verhaar MC, Chen Y, et al. Cellular stress conditions are reflected in the protein and RNA content of endothelial cell-derived exosomes. J Extracell Vesicles. 2012;1.
49. Lin Z, Gu Y, Zhou R, et al. Serum Exosomal proteins F9 and TSP-1 as potential diagnostic biomarkers for newly diagnosed epilepsy. Front Neurosci. 2020;14:737.
50. Ludwig C, Gillet L, Rosenberger G, et al. Data-independent acquisition-based SWATH-MS for quantitative proteomics: a tutorial. Mol Syst Biol. 2018;14(8):e8126.
51. Zheng X, Xu K, Zhou B, et al. A circulating extracellular vesicles-based novel screening tool for colorectal cancer revealed by shotgun and data-independent acquisition mass spectrometry. J Extracell Vesicles. 2020;9(1):1750202.
52. Hallal S, Azimi A, Wei H, et al. A comprehensive proteomic SWATH-MS workflow for profiling blood extracellular vesicles: a new avenue for glioma tumour surveillance. Int J Mol Sci. 2020;21(13).
53. Moreno-Gonzalo O, Villarroya-Beltri C, Sanchez-Madrid F. Post-translational modifications of exosomal proteins. Front Immunol. 2014;5:383.
54. Szabo-Taylor K, Ryan B, Osteikoetxea X, et al. Oxidative and other posttranslational modifications in extracellular vesicle biology. Semin Cell Dev Biol. 2015;40:8–16.

55. Klement E, Medzihradszky KF. Extracellular protein phosphorylation, the neglected side of the modification. Mol Cell Proteomics. 2017;16(1):1–7.
56. Moreno-Gonzalo O, Fernandez-Delgado I, Sanchez-Madrid F. Post-translational add-ons mark the path in exosomal protein sorting. Cell Mol Life Sci. 2018;75(1):1–19.
57. Zhang Y, Wu X, Andy TW. Characterization and applications of extracellular vesicle proteome with post-translational modifications. Trends Anal Chem. 2018;107:21–30.
58. Muralidharan-Chari V, Clancy J, Plou C, et al. ARF6-regulated shedding of tumor cell-derived plasma membrane microvesicles. Curr Biol. 2009;19(22):1875–85.
59. Masi G, Mercati D, Vannuccini E, et al. p66Shc regulates vesicle-mediated secretion in mast cells by affecting F-actin dynamics. J Leukoc Biol. 2014;95(2):285–92.
60. Mineo M, Garfield SH, Taverna S, et al. Exosomes released by K562 chronic myeloid leukemia cells promote angiogenesis in a Src-dependent fashion. Angiogenesis. 2012;15(1):33–45.
61. Rigogliuso S, Donati C, Cassarà D, et al. An active form of sphingosine kinase-1 is released in the extracellular medium as component of membrane vesicles shed by two human tumor cell lines. J Oncol. 2010;2010:509329.
62. Putz U, Howitt J, Doan A, et al. The tumor suppressor PTEN is exported in exosomes and has phosphatase activity in recipient cells. Sci Signal. 2012;5(243):ra70.
63. Reinders J, Sickmann A. State-of-the-art in phosphoproteomics. Proteomics. 2005;5(16):4052–61.
64. Engholm-Keller K, Larsen MR. Technologies and challenges in large-scale phosphoproteomics. Proteomics. 2013;13(6):910–31.
65. Thingholm TE, Jensen ON, Robinson PJ, Larsen MR. SIMAC (sequential elution from IMAC), a phosphoproteomics strategy for the rapid separation of monophosphorylated from multiply phosphorylated peptides. Mol Cell Proteomics. 2008;7(4):661–71.
66. Engholm-Keller K, Birck P, Storling J, et al. TiSH—a robust and sensitive global phosphoproteomics strategy employing a combination of TiO2, SIMAC, and HILIC. J Proteomics. 2012;75(18):5749–61.
67. Gonzales PA, Pisitkun T, Hoffert JD, et al. Large-scale proteomics and phosphoproteomics of urinary exosomes. J Am Soc Nephrol. 2009;20(2):363–79.
68. Palmisano G, Jensen SS, Le Bihan MC, et al. Characterization of membrane-shed microvesicles from cytokine-stimulated beta-cells using proteomics strategies. Mol Cell Proteomics. 2012;11(8):230–43.
69. Guo J, Cui Y, Yan Z, et al. Phosphoproteome characterization of human colorectal cancer SW620 cell-derived exosomes and new Phosphosite discovery for C-HPP. J Proteome Res. 2016;15(11):4060–72.
70. Chen IH, Xue L, Hsu CC, et al. Phosphoproteins in extracellular vesicles as candidate markers for breast cancer. Proc Natl Acad Sci U S A. 2017;114(12):3175–80.
71. Wu X, Li L, Iliuk A, Tao WA. Highly efficient phosphoproteome capture and analysis from urinary extracellular vesicles. J Proteome Res. 2018;17(9):3308–16.
72. Tian Y, Zhang H. Glycoproteomics and clinical applications. Proteomics Clin Appl. 2010;4(2):124–32.
73. Yu A, Zhao J, Peng W, et al. Advances in mass spectrometry-based glycoproteomics. Electrophoresis. 2018;39(24):3104–22.
74. Costa J. Glycoconjugates from extracellular vesicles: structures, functions and emerging potential as cancer biomarkers. Biochim Biophys Acta Rev Cancer. 2017;1868(1):157–66.
75. Clark DJ, Schnaubelt M, Hoti N, et al. Impact of increased FUT8 expression on the extracellular vesicle proteome in prostate cancer cells. J Proteome Res. 2020;19(6):2195–205.
76. Hung ME, Leonard JN. Stabilization of exosome-targeting peptides via engineered glycosylation. J Biol Chem. 2015;290(13):8166–72.
77. Liang Y, Eng WS, Colquhoun DR, et al. Complex N-linked glycans serve as a determinant for exosome/microvesicle cargo recruitment. J Biol Chem. 2014;289(47):32526–37.
78. Williams C, Pazos R, Royo F, et al. Assessing the role of surface glycans of extracellular vesicles on cellular uptake. Sci Rep. 2019;9(1):11920.

79. Inder KL, Ruelcke JE, Petelin L, et al. Cavin-1/PTRF alters prostate cancer cell-derived extracellular vesicle content and internalization to attenuate extracellular vesicle-mediated osteoclastogenesis and osteoblast proliferation. J Extracell Vesicles. 2014;3.
80. Batista BS, Eng WS, Pilobello KT, Hendricks-Munoz KD, Mahal LK. Identification of a conserved glycan signature for microvesicles. J Proteome Res. 2011;10(10):4624–33.
81. Escrevente C, Grammel N, Kandzia S, et al. Sialoglycoproteins and N-glycans from secreted exosomes of ovarian carcinoma cells. PLoS One. 2013;8(10):e78631.
82. Saraswat M, Joenvaara S, Musante L, et al. N-linked (N-) Glycoproteomics of urinary exosomes. Mol Cell Proteomics. 2015;14(8):2298.
83. Huo B, Chen M, Chen J, et al. A sequential separation strategy for facile isolation and comprehensive analysis of human urine N-glycoproteome. Anal Bioanal Chem. 2018;410(28):7305–12.
84. Brown CJ, Gaunitz S, Wang Z, et al. Glycoproteomic analysis of human urinary exosomes. Anal Chem. 2020;92(21):14357–65.
85. Chen IH, Aguilar HA, Paez Paez JS, et al. Analytical pipeline for discovery and verification of glycoproteins from plasma-derived extracellular vesicles as breast cancer biomarkers. Anal Chem. 2018;90(10):6307–13.
86. Gerlach JQ, Maguire CM, Krüger A, et al. Urinary nanovesicles captured by lectins or antibodies demonstrate variations in size and surface glycosylation profile. Nanomedicine (London, England). 2017;12(11):1217–29.
87. Freitas D, Balmaña M, Poças J, et al. Different isolation approaches lead to diverse glycosylated extracellular vesicle populations. J Extracell Vesicles. 2019;8(1):1621131.
88. Raiborg C, Stenmark H. The ESCRT machinery in endosomal sorting of ubiquitylated membrane proteins. Nature. 2009;458(7237):445–52.
89. Smith VL, Jackson L, Schorey JS. Ubiquitination as a mechanism to transport soluble mycobacterial and eukaryotic proteins to exosomes. J Immunol. 2015;195(6):2722–30.
90. Buschow SI, Liefhebber JM, Wubbolts R, Stoorvogel W. Exosomes contain ubiquitinated proteins. Blood Cells Mol Dis. 2005;35(3):398–403.
91. Bastos-Amador P, Royo F, Gonzalez E, et al. Proteomic analysis of microvesicles from plasma of healthy donors reveals high individual variability. J Proteome. 2012;75(12):3574–84.
92. Burke MC, Oei MS, Edwards NJ, Ostrand-Rosenberg S, Fenselau C. Ubiquitinated proteins in exosomes secreted by myeloid-derived suppressor cells. J Proteome Res. 2014;13(12):5965–72.
93. Huebner AR, Cheng L, Somparn P, et al. Deubiquitylation of protein cargo is not an essential step in exosome formation. Mol Cell Proteomics. 2016;15(5):1556–71.
94. Villarroya-Beltri C, Gutiérrez-Vázquez C, Sánchez-Cabo F, et al. Sumoylated hnRNPA2B1 controls the sorting of miRNAs into exosomes through binding to specific motifs. Nat Commun. 2013;4:2980.
95. Kunadt M, Eckermann K, Stuendl A, et al. Extracellular vesicle sorting of α-Synuclein is regulated by sumoylation. Acta Neuropathol. 2015;129(5):695–713.
96. Ageta H, Ageta-Ishihara N, Hitachi K, et al. UBL3 modification influences protein sorting to small extracellular vesicles. Nat Commun. 2018;9(1):3936.
97. Villarroya-Beltri C, Baixauli F, Mittelbrunn M, et al. ISGylation controls exosome secretion by promoting lysosomal degradation of MVB proteins. Nat Commun. 2016;7:13588.
98. Romancino DP, Buffa V, Caruso S, et al. Palmitoylation is a post-translational modification of Alix regulating the membrane organization of exosome-like small extracellular vesicles. Biochim Biophys Acta Gen Subjects. 2018;1862(12):2879–87.
99. Mariscal J, Vagner T, Kim M, et al. Comprehensive palmitoyl-proteomic analysis identifies distinct protein signatures for large and small cancer-derived extracellular vesicles. J Extracell Vesicles. 2020;9(1):1764192.
100. Shen B, Wu N, Yang JM, Gould SJ. Protein targeting to exosomes/microvesicles by plasma membrane anchors. J Biol Chem. 2011;286(16):14383–95.

101. Hanyu-Nakamura K, Sonobe-Nojima H, Tanigawa A, Lasko P, Nakamura A. Drosophila Pgc protein inhibits P-TEFb recruitment to chromatin in primordial germ cells. Nature. 2008;451(7179):730–3.
102. Nelson BR, Makarewich CA, Anderson DM, et al. A peptide encoded by a transcript annotated as long noncoding RNA enhances SERCA activity in muscle. Science. 2016;351(6270):271–5.
103. Matsumoto A, Pasut A, Matsumoto M, et al. mTORC1 and muscle regeneration are regulated by the LINC00961-encoded SPAR polypeptide. Nature. 2017;541(7636):228–32.
104. Jackson R, Kroehling L, Khitun A, et al. The translation of non-canonical open reading frames controls mucosal immunity. Nature. 2018;564(7736):434–8.
105. Huang JZ, Chen M, Chen D, et al. A peptide encoded by a putative lncRNA HOXB-AS3 suppresses colon cancer growth. Mol Cell. 2017;68(1):171–84.e6.
106. Makarewich CA, Baskin KK, Munir AZ, et al. MOXI is a mitochondrial micropeptide that enhances fatty acid β-oxidation. Cell Rep. 2018;23(13):3701–9.
107. Pircher A, Bakowska-Zywicka K, Schneider L, Zywicki M, Polacek N. An mRNA-derived noncoding RNA targets and regulates the ribosome. Mol Cell. 2014;54(1):147–55.
108. Jiang B, Liu J, Zhang YH, et al. Long noncoding RNA LINC00961 inhibits cell invasion and metastasis in human non-small cell lung cancer. Biomed Pharmacother. 2018;97:1311–8.
109. D'Lima NG, Ma J, Winkler L, et al. A human microprotein that interacts with the mRNA decapping complex. Nat Chem Biol. 2017;13(2):174–80.
110. Yang L, Tang Y, He Y, et al. High expression of LINC01420 indicates an unfavorable prognosis and modulates cell migration and invasion in nasopharyngeal carcinoma. J Cancer. 2017;8(1):97–103.
111. Cai T, Zhang Q, Wu B, et al. LncRNA-encoded microproteins: a new form of cargo in cell culture-derived and circulating extracellular vesicles. J Extracell Vesicles. 2021;10(9):e12123.
112. Klemm RW, Ejsing CS, Surma MA, et al. Segregation of sphingolipids and sterols during formation of secretory vesicles at the trans-Golgi network. J Cell Biol. 2009;185(4):601–12.
113. Egea-Jimenez AL, Zimmermann P. Phospholipase D and phosphatidic acid in the biogenesis and cargo loading of extracellular vesicles. J Lipid Res. 2018;59(9):1554–60.
114. Verderio C, Gabrielli M, Giussani P. Role of sphingolipids in the biogenesis and biological activity of extracellular vesicles. J Lipid Res. 2018;59(8):1325–40.
115. Dang VD, Jella KK, Ragheb RRT, Denslow ND, Alli AA. Lipidomic and proteomic analysis of exosomes from mouse cortical collecting duct cells. FASEB J. 2017;31(12):5399–408.
116. Durcin M, Fleury A, Taillebois E, et al. Characterisation of adipocyte-derived extracellular vesicle subtypes identifies distinct protein and lipid signatures for large and small extracellular vesicles. J Extracell Vesicles. 2017;6(1):1305677.
117. Yang JS, Lee JC, Byeon SK, Rha KH, Moon MH. Size dependent Lipidomic analysis of urinary exosomes from patients with prostate cancer by flow field-flow fractionation and Nanoflow liquid chromatography-tandem mass spectrometry. Anal Chem. 2017;89(4):2488–96.
118. Skotland T, Ekroos K, Kauhanen D, et al. Molecular lipid species in urinary exosomes as potential prostate cancer biomarkers. Eur J Cancer. 2017;70:122–32.
119. Haraszti RA, Didiot MC, Sapp E, et al. High-resolution proteomic and lipidomic analysis of exosomes and microvesicles from different cell sources. J Extracell Vesicles. 2016;5:32570.
120. Brzozowski JS, Jankowski H, Bond DR, et al. Lipidomic profiling of extracellular vesicles derived from prostate and prostate cancer cell lines. Lipids Health Dis. 2018;17(1):211.
121. Khayrullin A, Krishnan P, Martinez-Nater L, et al. Very long-chain C24:1 ceramide is increased in serum extracellular vesicles with aging and can induce senescence in bone-derived mesenchymal stem cells. Cell. 2019;8(1).
122. Folch J, Lees M, Sloane Stanley GH. A simple method for the isolation and purification of total lipides from animal tissues. J Biol Chem. 1957;226(1):497–509.
123. Bligh EG, Dyer WJ. A rapid method of total lipid extraction and purification. Can J Biochem Physiol. 1959;37(8):911–7.
124. Matyash V, Liebisch G, Kurzchalia TV, Shevchenko A, Schwudke D. Lipid extraction by methyl-tert-butyl ether for high-throughput lipidomics. J Lipid Res. 2008;49(5):1137–46.

125. Vallejo MC, Nakayasu ES, Longo LV, et al. Lipidomic analysis of extracellular vesicles from the pathogenic phase of *Paracoccidioides brasiliensis*. PLoS One. 2012;7(6):e39463.
126. Jung JH, Lee MY, Choi DY, et al. Phospholipids of tumor extracellular vesicles stratify gefitinib-resistant nonsmall cell lung cancer cells from gefitinib-sensitive cells. Proteomics. 2015;15(4):824–35.

# Flow Cytometry for Single Extracellular Vesicle Analysis

8

Ye Tian and Xiaomei Yan

## 8.1 Introduction

The analysis of extracellular vesicles (EVs) remains highly challenging owing to the inherently complex and vast heterogeneity of EVs in terms of size, composition, and origin [1–5]. With more functional implications proposed for EVs [6–8], the heterogeneity of EVs is increasingly recognized and has recently been highlighted in the position statement of the International Society for Extracellular Vesicles (ISEV) on EVs—Minimal Information for Studies of Extracellular Vesicles 2018 [9]. Thus, techniques that enable high-throughput multiparameter analysis of single EVs within the full size range (40–1000 nm) are urgently needed to elucidate the diversity of EVs and the roles of specific EV subtypes in the development of certain diseases [10]. Among the numerous single-particle techniques, flow cytometry (FCM), a powerful and well-established technique for the quantitative multiparameter analysis of single cells or micrometer-sized entities in a high-throughput manner, has gained popularity in the analysis of single EVs [11]. Although FCM has also been used for the characterization of protein markers of EVs by capturing EVs on the surface of micrometer-sized beads and then staining with fluorescent antibodies [12–14], this approach lacks single-vesicle profiling capabilities and results in the masking of different vesicular subsets. In this chapter, we focus on direct FCM analysis of single EVs. The challenges faced by conventional FCM and the development and application of ultrasensitive nano-FCM (nFCM) are also discussed.

Y. Tian · X. Yan (✉)
Department of Chemical Biology, College of Chemistry and Chemical Engineering, Xiamen University, Xiamen, Fujian, China
e-mail: xmyan@xmu.edu.cn



Q. Wang, L. Zheng (eds.), *Extracellular Vesicles*,
https://doi.org/10.1007/978-981-99-8365-0_8

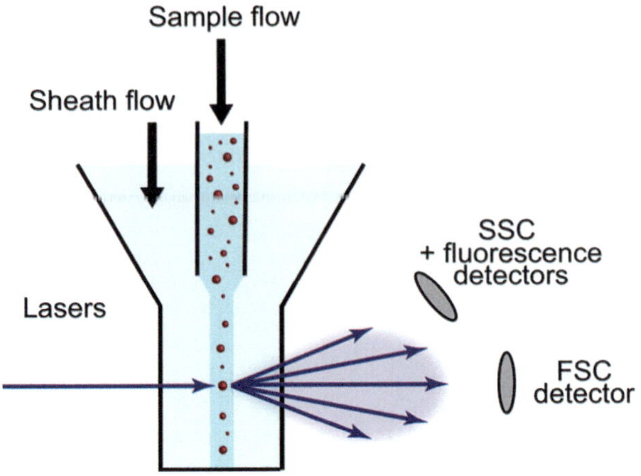


**Fig. 8.1** Schematic diagram of a typical flow cytometer, which utilizes a sheath flow to hydrodynamically focus the sample flow. The forward scatter (FSC) light and side scatter (SSC) light are collected by lenses. The fluorescence of particles is collected by the SSC lens and spectrally filtered to differentiate fluorescent signals from SSC. (*Reliable measurements of extracellular vesicles by clinical flow cytometry.* Reference [15]. *John Wiley & Sons, 2021*)

## 8.2 Principles of FCM and Challenges with Single EV Analysis

As shown in Fig. 8.1, a typical flow cytometer detects forward scatter (FSC) light, side scatter (SSC) light, and the fluorescence of single particles as they transit individually through a highly focused laser beam [15]. The particle size, granularity, concentration, and specific molecular components of samples can be profiled at the single-particle level using FCM analysis. However, most conventional flow cytometers are designed for the analysis of eukaryotic cells or micron-sized particles, rather than nanoscale EVs. Indeed, cells are on the order of 10 μm in diameter, whereas EVs are on the order of 100 nm in diameter, although both cells and EVs are spherical. This 100-fold difference in diameter results in a 10,000-fold difference in surface area and a 1,000,000-fold difference in volume between cells and EVs. Assuming a similar surface density of receptors and antigens, EVs will bear ~10,000 times fewer surface molecules than their parent cells. Moreover, if the concentrations of cytoplasmic contents are similar in cells and EVs, the cytoplasmic contents of an EV will be approximately 1/1,000,000th that of the parent cell [16]. Therefore, the differences in physical parameters between cells and EVs render both light scatter and fluorescence detection of EVs by FCM challenging.

For particles smaller than the wavelength of the illuminating laser, the light scatter decreases exponentially (as the fourth–sixth power of the radius). Therefore, the scattered light intensity of single EVs is 8–12 orders of magnitude smaller than that of cells. For particles ranging from smaller than 100 nm to submicron size, the Mie theory provides a general solution for scattering calculations. In addition to the particle size, the scattering intensity is also dependent on the refractive index (RI) of the particles and the optical configuration of the flow cytometer [17]. Using polystyrene beads and a Mie-theory-based light scattering model to establish the scatter-diameter relationship for EVs, van der Pol et al. evaluated 46 flow cytometers from 33 participants to investigate the size range of detectable EVs [18]. The RI of the EVs was assumed to be constant under an excitation wavelength of 488 nm, with a value of approximately 1.40, as previously reported [19, 20], although it was also highly heterogeneous [21, 22]. For most conventional flow cytometers, the detection limit for light scattering is 300–500 nm for polystyrene beads, with an RI of 1.61. According to the Mie theory, the scatter intensity of polystyrene beads with a diameter of 400 nm corresponds to that of EVs with a diameter of 700–900 nm on FSC and 1200–1900 nm on SSC, depending on the configuration of the FCM used. They also found that 22 of the 46 participating flow cytometers could measure EVs in the size range of 600–1200 nm, and only a few highly sensitive instruments (including Apogee A50-Micro, BD Influx, and BD LSR II) enabled the detection of EVs measuring 300–600 nm in diameter (Fig. 8.2).

Although EVs exhibit a broad size range of approximately 40–1000 nm, most are smaller than 300 nm (Fig.8.3) [23, 24]. For example, by examining more than 1000 EVs with cryo-electron microscopy (cryo-TEM), researchers found that EVs isolated by differential ultracentrifugation (UC) from the conditioned medium of HCT-15 human colorectal cancer cells exhibited a wide size distribution ranging from 30 to 600 nm, and approximately 65% and 86% of the vesicles fell within the size ranges of 30–100 and 30–200 nm, respectively (Fig. 8.3a). These data implies that, for light-scatter-based EV detection, the majority of EVs fall outside the detection limits of most conventional flow cytometers, and only the largest and brightest particles (the "tip of the iceberg") can be detected and measured.

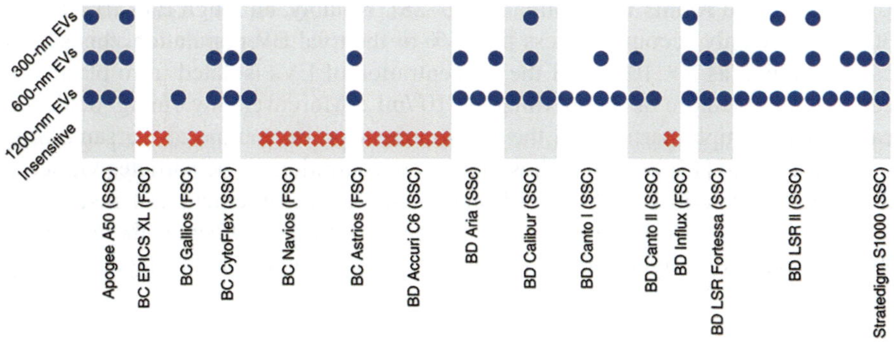


**Fig. 8.2** The sensitivity of flow cytometers for EV detection. Blue markers indicate that the flow cytometer could detect signals from vesicles measuring 1200, 600, and/or 300 nm at above the threshold level. Red crosses indicate that the threshold of these flow cytometers could not be determined because their scatter was too insensitive to detect 400-nm fluorescent polystyrene beads. Whether the model was applied to forward scatter (FSC) or side scatter (SSC) is shown in parentheses after the flow cytometer name. (*Standardization of extracellular vesicle measurements by flow cytometry through vesicle diameter approximation.* Reprinted with permission from *John Wiley & Sons*)

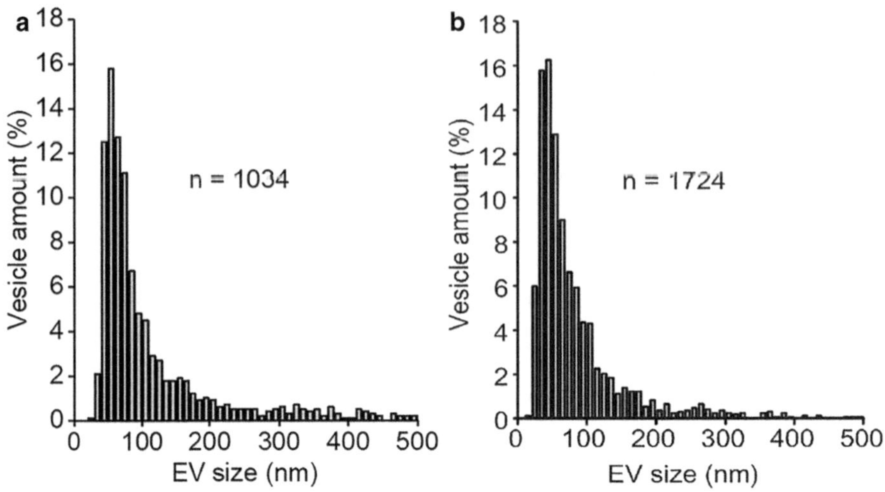


**Fig. 8.3** Size distributions of EVs derived from HCT-15 cells (**a**) and HMC-1 human mast cells (**b**) as determined by cryo-electron microscopy. (*Protein profiling and sizing of extracellular vesicles from colorectal cancer patients* via *flow cytometry*. Reference [24]. *American Chemical Society, 2018; Exosomes purified from a single cell type have diverse morphology*. Reprinted with permission from *Taylor & Francis*)

## 8.3 Analysis of Single EVs by Conventional FCM

To adapt conventional FCM for single EV analysis, experiments have mainly been performed using cytometers with small particle detection capability (e.g., Becton Dickinson Influx, Apogee A50-Micro, Beckman Coulter MoFlo Astrios and CytoFLEX, and Amnis ImageStream) [25–28]. Notably, although EVs larger than 300 nm may only account for less than 5% of the total EV population, this subset can be as high as $5 \times 10^7$/mL if the concentration of EVs isolated from plasma or cell culture medium is approximately $10^9$/mL. Moreover, swarming detection caused by multiple particles in the measurement probe volume at the same time must be avoided [15, 29, 30]. Adjusting the concentration of the particles via serial dilution is a practical method, and single-particle measurement can result in a decrease in the event detection rate in response to dilution, whereas the measured fluorescence and/or light scatter intensities of those events do not change.

### 8.3.1 Analysis of Single EVs upon Labeling with a Generic Fluorescent Probe

For FCM measurements, the fluorescence background is usually lower than the light scatter background, and a higher signal-to-noise (S/N) ratio is obtained by using fluorescence triggering as compared with scattering triggering [31]. Because there are no generic protein markers for all EVs, a more general approach is to use a fluorescent probe to discriminate EVs from other nonvesicular particles and trigger the detection of EVs. Among them, lipophilic membrane dyes, such as Paul Karl Horan (PKH) and dialkylcarbocyanine probes (DiI, DiO, DiD, etc.), and luminal dyes, such as carboxyfluorescein succinimidyl ester (CFSE) and calcein-acetoxymethyl ester, are commonly used [32]. The latter two are membrane permeable and nonfluorescent, requiring hydrolysis to be converted into a membrane-impermeable fluorescent analog. Using an optimized configuration of the BD Influx flow cytometer, upon PKH67 labeling, van der Vlist et al. demonstrated that the vast majority of vesicles produced by dendritic cells could be detected, allowing the quantitative and multiparameter analysis of fluorescent vesicles of ~100 nm [33]. Using the voltage-sensing dye di-8-ANEPPS and a custom-constructed high-sensitivity FCM, Stoner et al. reported successful enumeration and size estimation of individual EVs along with the identification of phenotypic subsets of EVs (Fig. 8.4) [34, 35]. Through double labeling with protein (CFSE) and lipid (FM) dyes, Pospichalova et al. developed a simple protocol to distinguish EVs

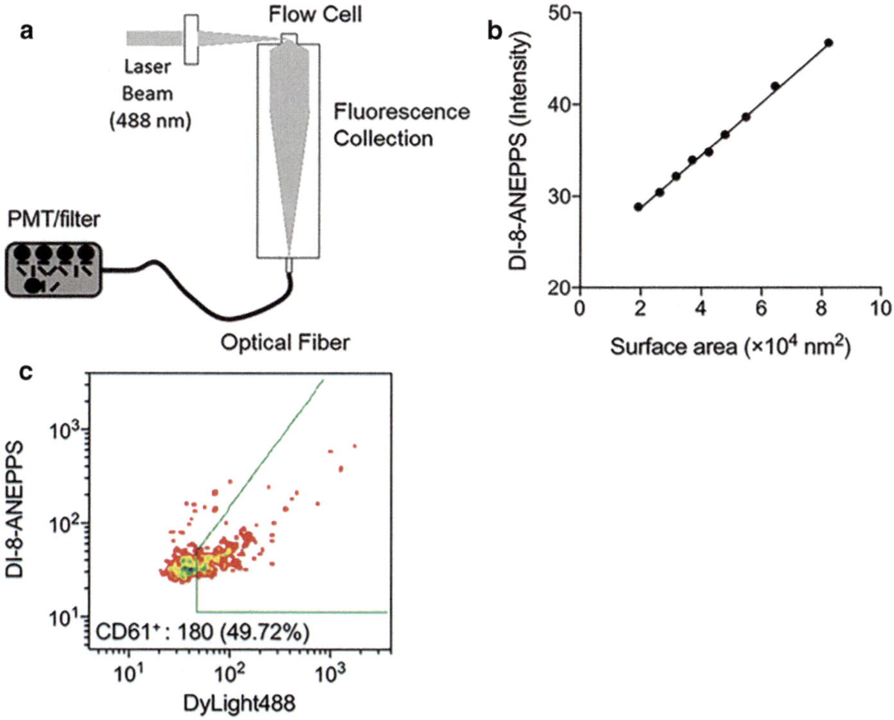


**Fig. 8.4** A custom high-sensitivity flow cytometer for EV analysis. (**a**) A high-sensitivity flow cytometer was constructed for EV analysis. (**b**) The fluorescent intensity from liposomes, labeled with di-8-ANEPPS, was calibrated for analysis of vesicle diameter. (**c**) EVs from platelet-rich plasma samples were labeled with di-8-ANEPPS for size measurement and anti-CD61 antibodies conjugated with DyLight488 for detection of platelet-derived EVs. (*New technologies for analysis of extracellular vesicles*. Reprinted with permission from *American Chemical Society*)

from common contaminants of EV preparations, such as protein aggregates or micelles formed by unbound lipophilic dyes, on an Apogee A50 microflow cytometer [36]. Although fluorescent labeling improves EV detection by FCM, the fluorescence signals of small EVs remain too dim to be detected by conventional FCM. Furthermore, several pitfalls need to be considered when working with lipid-anchored fluorophores, such as concomitant labeling of non-EV contaminants, including lipoproteins, and the aggregation of these amphiphilic dyes to form EV-like particles [37].

### 8.3.2 Analysis of EV Proteins

EVs exert biological functions through their luminal or surface proteins, and protein profiling of EVs is indispensable for understanding EV functions, distinguishing disease-related subsets for cancer diagnosis, predicting metastatic propensity, and determining organ sites for future metastasis [38, 39]. Using a multicolor CytoFLEX flow cytometer, structured illumination (SIM), and atomic force microscopy (AFM), Choi et al. demonstrated a highly heterogeneous distribution of EV components, including oncoproteins (e.g., epidermal growth factor receptor), clotting factors (tissue factor), prometastatic integrins (e.g., integrin subunit alpha [ITGA] 6, ITGA4), tetraspanins (e.g., CD63), and DNA [40]. Compared with SIM and AFM, FCM has the advantages of being able to quantify changes across the entire secretome. By analyzing EVs expressing prostate-specific antigens from peripheral blood via FCM, Logozzi et al. found that the level of EVs expressing both CD81 and prostate-specific antigen in patients with prostate cancer was higher than that in healthy donors, providing a noninvasive biomarker for the diagnosis of prostate cancer [41]. Using a retroviral envelope glycoprotein as a selection marker, Tang et al. discriminated retroviruses from EVs that overlap in size, buoyant densities, and refractive indices and share many cell-derived surface markers using a BD LSRFortessa flow cytometer [42]. In addition, Shen et al. employed DNA aptamer probes to bind to the surface markers of EVs, thereby triggering the generation of a DNA nanostructure via a hybridization chain reaction to enable the detection of single EVs in a conventional FACSCanto flow cytometer [43].

Imaging FCM (IFCM) with a time delay integration system and charge-coupled device (CCD) cameras combines FCM with imaging technology to enable high-throughput rapid analysis of individual EVs, where CCD cameras can quickly capture images of detected particles [44–47]. Using AMNIS ImageStream, Ricklefs et al. found that EVs with double-positive tetraspanin expression ($CD63^+/CD81^+$) were enriched in cancer cell lines and patient plasma samples [47]. Moreover, Görgens et al. performed a multiparameter analysis of single EVs based on fluorescent biological reference materials and quantified EV subpopulations in various clinical specimens via IFCM [46]. In addition to the increasingly recognized pathophysiological roles of EVs and their potential utility as diagnostic and therapeutic tools, protein profiling of EVs plays pivotal roles in the development of EV-based liquid biopsies and therapeutics. Thus, EV protein analysis using FCM has progressed from studies of the basic biology of EVs to clinical applications.

## 8.4 Development of nFCM for Single EV Analysis

To address the limitation of conventional FCM for nanosized particles in terms of sensitivity, researchers have been working on the construction of highly sensitive flow cytometers. In 1979, using a custom-built device, the Shapiro group was able to distinguish bacteriophage T2 (~100 nm) from a reovirus top component by scattering light, whereas the reovirus top component was discernible from background noise [48]. In 2004, Steen discriminated 74-nm polystyrene beads and virions with sizes of approximately 100 nm from noise by collecting scattered light over an angle of 16–70° on a modified flow cytometer [49]. Recently, Arkesteijn et al. optimized FSC detection of a BD Influx flow cytometer by adjusting the size of the obscuration bar, allowing a robust separation of 100-nm polystyrene beads from background signals within the FSC channel [50]. However, the sensitive detection of EVs as small as 40 nm requires a significant increase in sensitivity.

For spherical particles much smaller than the excitation wavelength (the case for the majority population of EVs), the scattering intensity scales off with a sixth-power dependence of particle size based on Rayleigh scattering theory, as shown in Eq. (8.1):

$$\sigma_{\text{scatter}} = \frac{2\pi^5 d^6 n_{\text{med}}^4}{3\lambda^4} \left| \frac{m^2 - 1}{m^2 + 2} \right|^2 \tag{8.1}$$

where $\sigma_{\text{scatt}}$ is the scattering cross-section, $d$ is the particle diameter, $\lambda$ is the wavelength of the incident light, $n_{\text{med}}$ is the RI of the medium surrounding the particle, and $m$ is the ratio of the RIs of the particle ($n_{\text{particle}} = n_{\text{rel}} + in_{\text{im}}$) to the medium ($n_{\text{med}}$). Thus, pushing the detection limit of conventional flow cytometers from 200 nm polystyrene beads to 40 or 25 nm corresponds to approximate fold increases of 15,625 or 262,144 in the S/N ratio, respectively. In addition to particle size, the RI also significantly affects the scattering intensity of the nanoparticles. For example, 60-nm polystyrene beads are approximately fourfold brighter than silica nanoparticles of the same size, and silica nanoparticles are approximately 3.5-fold brighter than EVs of the same size. Employing strategies for single-molecule fluorescence detection in a sheathed flow, Yan et al. developed nFCM, which enables light scattering detection of single gold nanoparticles, silica nanoparticles, and viruses as small as 7, 24, and 27 nm in diameter [51, 52]. For fluorescence detection, single R-phycoerythrin molecules were detected at an S/N ratio of 17. The limit of detection for organic dyes was determined to be three Alexa Fluor 532 molecules [51, 53]. This significantly enhanced sensitivity is based on the combined effects of (1) a reduced probe volume (~25 fL) for background reduction, (2) an extended dwell time (~1 ms) of the nanoparticles in the laser beam for enhanced photon generation, (3) a single-photon counting avalanche photodiode (APD) detector for enhanced quantum efficiency of photon detection, and (4) excellent optics to restrict the field of view of the APD detector solely to the sample stream for very good background rejection. Based on advanced nFCM, this group developed a series of methods for the analysis of nanoscale synthetic or naturally occurring biological nanoparticles, such as bacteria, mitochondria, viruses, liposomes, and quantum dots [54–56].

For EV analysis, light scatter-based detection of individual EVs as small as 40 nm in diameter was achieved, and the analysis rate was as high as 10,000 particles/min [24]. By using a mixture of five different sizes of monodisperse silica nanoparticles as a reference standard and with the aid of RI mismatch correction based on the Mie theory, a statistically representative particle size distribution of EVs can be acquired in just minutes with resolution and accuracy comparable to those of cryo-TEM (Fig. 8.5). In particular, subpopulations of EVs expressing CD9, CD63, and/or CD81 and the copy number distribution of each protein were measured by immunofluorescent staining. Notably, a significant difference in the concentrations of CD147-positive EVs in platelet-free plasma (PFP) was observed between patients with colorectal cancer and healthy donors. nFCM provides an advanced platform for insight into heterogeneous EV populations, which is highly desirable for deciphering the biology of EVs and promoting the development of EV-based liquid biopsy and therapeutics.

Additionally, this laboratory-built nFCM was employed to assess the quality and efficiency of EVs isolated from the plasma [57]. The performance of five widely used commercial isolation kits (ExoQuick, System Biosciences; Total Exosome Isolation [TEI] kits, Invitrogen; qEV, Izon; ultrafiltration [UF], Millipore; and

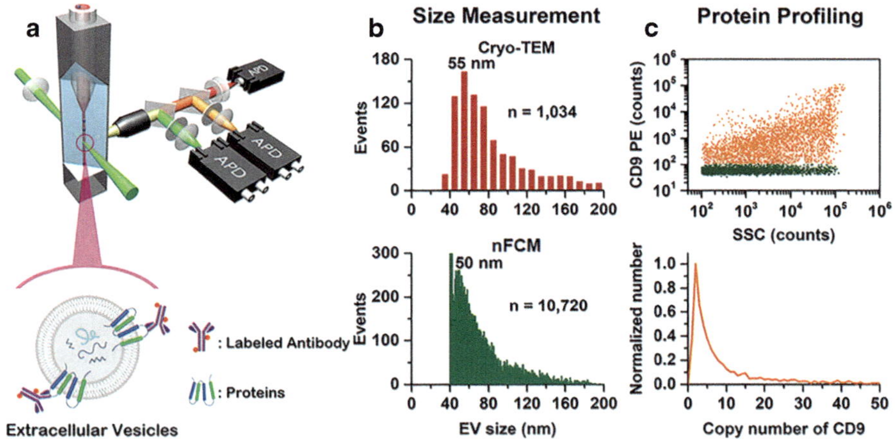


**Fig. 8.5** nFCM analysis of single EVs. (**a**) Schematic diagram of the laboratory-built nFCM for the simultaneous detection of side scatter (SSC) and two-color fluorescence of single EVs. (**b**) Histogram of particle size for EVs isolated from HCT-15 colorectal cancer cells by nFCM and cryo-TEM. (**c**) Analysis and identification of EVs expressing CD9 with immunofluorescent staining and the distribution profiles of the protein copy numbers of CD9 on single EVs. (*Protein profiling and sizing of extracellular vesicles from colorectal cancer patients* via *flow cytometry.* Reprinted with permission from *American Chemical Society*)

exoEasy, Qiagen) was examined and compared with traditional differential UC. The results showed that the particle concentrations of EV preparations from PFP by kits were two to four orders of magnitude higher than those of EV preparation by UC, although the purity was much lower. Moreover, the particle size distribution profiles of the EV preparations obtained using kits were quite narrow and closely resembled those of PFP, whereas the EV preparations obtained using UC showed a broader size distribution at relatively large particle sizes (Fig. 8.6). Additionally, the comparable particle counts between EV preparations from PFP and vesicle-depleted PFP confirmed again that a large quantity of nonvesicular contaminants could be isolated using these commercial isolation kits. Moreover, subpopulations of EVs expressing CD9, CD63, CD81, CD235a, CD45, CD41a, CD144, and phosphatidylserine in EV preparations from PFP were also assessed and quantified.

The technology of nFCM has been commercialized by NanoFCM Inc. (Xiamen, China), and Flow NanoAnalyzer instruments have become a powerful tool for EV researchers worldwide [58–70]. Wang et al. at the Mayo Clinic found that the number and size distributions of AQP4-positive EVs from cultured rat astrocytes were selectively affected by the frequency of electrical stimulation, demonstrating the potential of the frequency of electrical stimulation in treating particular neurological diseases [62]. Jiang et al. demonstrated that approximately 85.1% of pX protein (a structural protein of hepatitis A virus [HAV])-loaded EVs were CD63 positive by dual immunofluorescent staining, strongly supporting the importance of pX in HAV production and its potential for loading large protein complexes into exosome-like EVs for therapeutic purposes [67]. nFCM also provides an advanced platform for assessing the quality of EV-based drug delivery systems. Lv et al. fused genetically

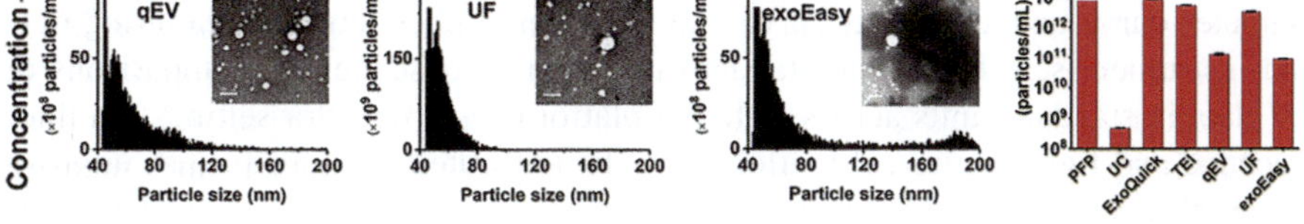


**Fig. 8.6** nFCM measurement of the particle size (diameter) distributions and concentrations of EV preparations from PFP using different isolation methods. (**a**) Particle size distribution histogram of PFP and a typical TEM micrograph (scale bar: 100 nm). (**b**–**g**) Particle size distribution histograms of EV preparations from PFP obtained using UC, ExoQuick, TEI, qEV, UF, and exoEasy. (**h**) Particle concentrations measured for PFP and its EV preparations. The error bars represent standard deviation of three experiments for each isolation method. (*Quality and efficiency assessment of six extracellular vesicle isolation methods by nano-flow cytometry.* Reprinted with permission from *Taylor & Francis*)

engineered CD47-expressing exosomes with thermosensitive liposomes to overcome delivery limitations and enhance the effects of hyperthermic intraperitoneal chemotherapy therapy [71]. The expression of CD47 in genetically engineered exosomes and the hybrid efficiency of engineered exosomes with drug-loaded liposomes were quantified and assessed using nFCM. On September 15, 2020, Codiak Biosciences Inc., a clinical-stage company pioneering the development of exosome-based therapeutics as a new class of medicines, announced the initiation of subject dosing in its phase 1 clinical trial of its novel exosome therapeutic candidate, exoIL-12 [72]. In the patent they filed, the particle-by-particle analysis of individual exosomes with Flow NanoAnalyzer indicated that almost all exosomes (~97%) expressed the proprietary scaffold protein PTGRFN, which provided an advanced platform for the quality control analysis of engineered exosomes [73]. Additionally, Kim et al. characterized the size and nucleic acid content of EVs secreted from two placental mesenchymal stem/stromal cell lines [60]. Lv et al. evaluated the encapsulation efficiency of nanovesicle-virus composites and demonstrated the successful preparation of oncolytic adenovirus-encapsulated bioengineered cell membrane nanovesicles [61]. Zhang et al. observed that the fluorescence intensity notably increased by ~20% in EVs derived from fluorescein isothiocyanate-oligodeoxynucleotide (unmethylated CpG dinucleotide-containing DNA)-treated cells [63].

## 8.5 Standardized Reporting of EV FCM Experiments

The application of FCM in the analysis of single EVs is facing many challenges, in part owing to wide variations in instrument configuration and sample preparation, lack of reference materials, incomplete descriptions of methods, and ambiguities in data interpretation. Standardized reporting of EV FCM experiments is necessary to normalize EV measurements between instruments and laboratories. A flow cytometer postacquisition analysis software program ($FCM_{PASS}$, https://nano.ccr.cancer.gov/software/) built upon Mie theory was developed to enable the approximation of flow cytometer geometric parameters by analyzing beads of known diameter and RI or by inputting the collection angle if known. The software can then create a scatter-diameter curve and scatter-RI curve that helps researchers convert scattering data and instrument sensitivity into standard units [11]. Subsequently, comparisons of EV data in standard units across different platforms and detector settings via fluorescence and light-scatter calibration were demonstrated [74]. To further improve the reliability and reproducibility of EV FCM measurements, a group of EV FCM researchers active in the ISEV, International Society for Advancement of Cytometry, and/or International Society for Thrombosis and Hemostasis developed a standard framework for reporting EV FCM methods and results, named "minimum information for the reporting of an EV FCM experiment" (MIFlowCyt-EV) [75]. Regarding the urgent need for developing reference materials [76], Varga et al. synthesized hollow organosilica beads to obtain scattering light comparable to that of EVs of similar diameter [77]. In addition to polystyrene and silica beads, biological reference materials, such as liposomes, viruses, and vesicles produced from erythrocytes, have been proposed as candidates to further match the RI of EVs [78]. In particular, trackable recombinant EVs have been generated as biological reference materials and have been shown to be stable and to exhibit the physical and biochemical traits of EVs [79]. With the availability of ultrasensitive nFCM and improved reference materials, the adoption of standardized reporting of EV FCM studies will greatly enhance our ability to quantitatively compare results from different laboratories and thereby support EV research.

## References

1. Tkach M, Thery C. Communication by extracellular vesicles: where we are and where we need to go. Cell. 2016;164(6):1226–32.
2. van Niel G, D'Angelo G, Raposo G. Shedding light on the cell biology of extracellular vesicles. Nat Rev Mol Cell Biol. 2018;19(4):213–28.
3. Mathieu M, Martin-Jaular L, Lavieu G, et al. Specificities of secretion and uptake of exosomes and other extracellular vesicles for cell-to-cell communication. Nat Cell Biol. 2019;21(1):9–17.
4. Pegtel DM, Gould SJ. Exosomes. Annu Rev Biochem. 2019;88:487–514.
5. Kalluri R, LeBleu VS. The biology, function, and biomedical applications of exosomes. Science. 2020;367(6478):eaau6977.
6. Kanada M, Bachmann MH, Hardy JW, et al. Differential fates of biomolecules delivered to target cells via extracellular vesicles. Proc Natl Acad Sci U S A. 2015;112(12):E1433–42.
7. Kowal J, Arras G, Colombo M, et al. Proteomic comparison defines novel markers to characterize heterogeneous populations of extracellular vesicle subtypes. Proc Natl Acad Sci U S A. 2016;113(8):E968–77.
8. Zhang H, Freitas D, Kim HS, et al. Identification of distinct nanoparticles and subsets of extracellular vesicles by asymmetric flow field-flow fractionation. Nat Cell Biol. 2018;20(3):332–43.
9. Thery C, Witwer KW, Aikawa E, et al. Minimal information for studies of extracellular vesicles 2018 (MISEV2018): a position statement of the International Society for Extracellular Vesicles and update of the MISEV2014 guidelines. J Extracell Vesicles. 2018;7(1):1535750.
10. Welsh JA, Holloway JA, Wilkinson JS, et al. Extracellular vesicle flow cytometry analysis and standardization. Front Cell Dev Biol. 2017;5:78.
11. Welsh JA, Horak P, Wilkinson JS, et al. FCMPASS software aids extracellular vesicle light scatter standardization. Cytom Part A. 2020;97(6):569–81.
12. Wan S, Zhang LQ, Wang S, et al. Molecular recognition-based DNA nanoassemblies on the surfaces of nanosized exosomes. J Am Chem Soc. 2017;139(15):5289–92.
13. Yang KS, Lin HY, Curley C, et al. Bead-based extracellular vesicle analysis using flow cytometry. Adv Biosyst. 2020;4(12):2000203.
14. Melo SA, Luecke LB, Kahlert C, et al. Glypican-1 identifies cancer exosomes and detects early pancreatic cancer. Nature. 2015;523(7559):177–82.
15. Kuiper M, van de Nes A, Nieuwland R, et al. Reliable measurements of extracellular vesicles by clinical flow cytometry. Am J Reprod Immunol. 2021;85:e13350.
16. Nolan JP, Jones JC. Detection of platelet vesicles by flow cytometry. Platelets. 2017;28(3):256–62.
17. Zhang W, Tian Y, Hu X, et al. Light-scattering sizing of single submicron particles by high-sensitivity flow cytometry. Anal Chem. 2018;90(21):12768–75.
18. van der Pol E, Sturk A, van Leeuwen T, et al. Standardization of extracellular vesicle measurements by flow cytometry through vesicle diameter approximation. J Thromb Haemost. 2018;16(6):1236–45.

19. Konokhova AI, Yurkin MA, Moskalensky AE, et al. Light-scattering flow cytometry for identification and characterization of blood microparticles. J Biomed Opt. 2012;17(5):057006.
20. van der Pol E, Coumans FA, Grootemaat AE, et al. Particle size distribution of exosomes and microvesicles determined by transmission electron microscopy, flow cytometry, nanoparticle tracking analysis, and resistive pulse sensing. J Thromb Haemost. 2014;12(7):1182–92.
21. Gardiner C, Shaw M, Hole P, et al. Measurement of refractive index by nanoparticle tracking analysis reveals heterogeneity in extracellular vesicles. J Extracell Vesicles. 2014;3(1):25361.
22. van der Pol E, Coumans FA, Sturk A, et al. Refractive index determination of nanoparticles in suspension using nanoparticle tracking analysis. Nano Lett. 2014;14(11):6195–201.
23. Zabeo D, Cvjetkovic A, Lasser C, et al. Exosomes purified from a single cell type have diverse morphology. J Extracell Vesicles. 2017;6(1):1329476.
24. Tian Y, Ma L, Gong MF, et al. Protein profiling and sizing of extracellular vesicles from colorectal cancer patients via flow cytometry. ACS Nano. 2018;12(1):671–80.
25. Nolte-'t Hoen ENM, van der Vlist EJ, Aalberts M, et al. Quantitative and qualitative flow cytometric analysis of nanosized cell-derived membrane vesicles. Nanomed-Nanotechnol. 2012;8(5):712–20.
26. Kibria G, Ramos EK, Lee KE, et al. A rapid, automated surface protein profiling of single circulating exosomes in human blood. Sci Rep. 2016;6:36502.
27. Morales-Kastresana A, Telford B, Musich TA, et al. Labeling extracellular vesicles for nanoscale flow cytometry. Sci Rep. 2017;7(1):1878.
28. Morales-Kastresana A, Musich TA, Welsh JA, et al. High-fidelity detection and sorting of nanoscale vesicles in viral disease and cancer. J Extracell Vesicles. 2019;8(1):1597603.
29. Van der Pol E, Van Gemert MJC, Sturk A, et al. Single vs. swarm detection of microparticles and exosomes by flow cytometry. J Thromb Haemost. 2012;10(5):919–30.
30. Kormelink TG, Arkesteijn GJA, Nauwelaers FA, et al. Prerequisites for the analysis and sorting of extracellular vesicle subpopulations by high-resolution flow cytometry. Cytom Part A. 2016;89a(2):135–47.
31. Arraud N, Gounou C, Turpin D, et al. Fluorescence triggering: a general strategy for enumerating and phenotyping extracellular vesicles by flow cytometry. Cytom Part A. 2016;89(2):184–95.
32. de Rond L, van der Pol E, Hau CM, et al. Comparison of generic fluorescent markers for detection of extracellular vesicles by flow cytometry. Clin Chem. 2018;64(4):680–9.
33. van der Vlist EJ, Nolte-'t Hoen ENM, Stoorvogel W, et al. Fluorescent labeling of nanosized vesicles released by cells and subsequent quantitative and qualitative analysis by high-resolution flow cytometry. Nat Protoc. 2012;7(7):1311–26.
34. Stoner SA, Duggan E, Condello D, et al. High sensitivity flow cytometry of membrane vesicles. Cytom Part A. 2016;89a(2):196–206.
35. Shao H, Im H, Castro CM, et al. New technologies for analysis of extracellular vesicles. Chem Rev. 2018;118(4):1917–50.
36. Pospichalova V, Svoboda J, Dave Z, et al. Simplified protocol for flow cytometry analysis of fluorescently labeled exosomes and microvesicles using dedicated flow cytometer. J Extracell Vesicles. 2015;4(1):25530.
37. Simonsen JB. Pitfalls associated with lipophilic fluorophore staining of extracellular vesicles for uptake studies. J Extracell Vesicles. 2019;8(1):1582237.
38. Xu R, Rai A, Chen M, et al. Extracellular vesicles in cancer—implications for future improvements in cancer care. Nat Rev Clin Oncol. 2018;15(10):617–38.
39. Hoshino A, Costa-Silva B, Shen TL, et al. Tumour exosome integrins determine organotropic metastasis. Nature. 2015;527(7578):329–35.
40. Choi D, Montermini L, Jeong H, et al. Mapping subpopulations of cancer cell-derived extracellular vesicles and particles by nano-flow cytometry. ACS Nano. 2019;13(9):10499–511.
41. Logozzi M, Angelini DF, Iessi E, et al. Increased PSA expression on prostate cancer exosomes in in vitro condition and in cancer patients. Cancer Lett. 2017;403:318–29.
42. Tang VA, Renner TM, Fritzsche AK, et al. Single-particle discrimination of retroviruses from extracellular vesicles by nanoscale flow cytometry. Sci Rep. 2017;7(1):17769.

43. Shen W, Guo K, Adkins GB, et al. A single extracellular vesicle (EV) flow cytometry approach to reveal EV heterogeneity. Angew Chem Int Ed Engl. 2018;57(48):15675–80.
44. Lannigan J, Erdbruegger U. Imaging flow cytometry for the characterization of extracellular vesicles. Methods. 2017;112:55–67.
45. Ofir-Birin Y, Abou Karam P, Rudik A, et al. Monitoring extracellular vesicle cargo active uptake by imaging flow cytometry. Front Immunol. 2018;9:1011.
46. Gorgens A, Bremer M, Ferrer-Tur R, et al. Optimisation of imaging flow cytometry for the analysis of single extracellular vesicles by using fluorescence-tagged vesicles as biological reference material. J Extracell Vesicles. 2019;8(1):1587567.
47. Ricklefs FL, Maire CL, Reimer R, et al. Imaging flow cytometry facilitates multiparametric characterization of extracellular vesicles in malignant brain tumours. J Extracell Vesicles. 2019;8(1):1588555.
48. Hercher M, Mueller W, Shapiro HM. Detection and discrimination of individual viruses by flow cytometry. J Histochem Cytochem. 1979;27(1):350–2.
49. Steen HB. Flow cytometer for measurement of the light scattering of viral and other submicroscopic particles. Cytom Part A. 2004;57a(2):94–9.
50. Arkesteijn GJA, Lozano-Andres E, Libregts SFWM, et al. Improved flow cytometric light scatter detection of submicron-sized particles by reduction of optical background signals. Cytom Part A. 2020;97(6):610–9.
51. Zhu SB, Ma L, Wang S, et al. Light-scattering detection below the level of single fluorescent molecules for high-resolution characterization of functional nanoparticles. ACS Nano. 2014;8(10):10998–1006.
52. Ma L, Zhu SB, Tian Y, et al. Label-free analysis of single viruses with a resolution comparable to that of electron microscopy and the throughput of flow cytometry. Angew Chem Int Edit. 2016;55(35):10239–43.
53. Yang LL, Zhu SB, Hang W, et al. Development of an ultrasensitive dual-channel flow cytometer for the individual analysis of nanosized particles and biomolecules. Anal Chem. 2009;81(7):2555–63.
54. Chen CX, Zhu SB, Wang S, et al. Multiparameter quantification of liposomal nanomedicines at the single-particle level by high-sensitivity flow cytometry. ACS Appl Mater Inter. 2017;9(16):13913–9.
55. Wang S, Li LH, Jin SH, et al. Rapid and quantitative measurement of single quantum dots in a sheath flow cuvette. Anal Chem. 2017;89(18):9857–63.
56. Lian H, He SB, Chen CX, et al. Flow cytometric analysis of nanoscale biological particles and organelles. Annu Rev Anal Chem. 2019;12:389–409.
57. Tian Y, Gong MF, Hu YY, et al. Quality and efficiency assessment of six extracellular vesicle isolation methods by nano-flow cytometry. J Extracell Vesicles. 2020;9(1):1697028.
58. Dong C, Huang L, Melo SA, et al. Multiple antibodies identify glypican-1 associated with exosomes from pancreatic cancer cells and serum from patients with pancreatic cancer. bioRxiv. 2018. https://doi.org/10.1101/145706.
59. Xu HY, Liao C, Zuo P, et al. Magnetic-based microfluidic device for on-chip isolation and detection of tumor-derived exosomes. Anal Chem. 2018;90(22):13451–8.
60. Kim SY, Joglekar MV, Hardikar AA, et al. Placenta stem/stromal cell-derived extracellular vesicles for potential use in lung repair. Proteomics. 2019;19(17):e1800166.
61. Lv P, Liu X, Chen XM, et al. Genetically engineered cell membrane nanovesicles for oncolytic adenovirus delivery: a versatile platform for cancer virotherapy. Nano Lett. 2019;19(5):2993–3001.
62. Wang Y, Melvin R, Bemis LT, et al. Programmable modulation for extracellular vesicles. bioRxiv. 2019. https://doi.org/10.1101/566448.
63. Zhang Y, Jin X, Liang JQ, et al. Extracellular vesicles derived from ODN-stimulated macrophages transfer and activate Cdc42 in recipient cells and thereby increase cellular permissiveness to EV uptake. Sci Adv. 2019;5(7):eaav1564.

64. Zhu QW, Ling XZ, Yang YL, et al. Embryonic stem cells-derived exosomes endowed with targeting properties as chemotherapeutics delivery vehicles for glioblastoma therapy. Adv Sci. 2019;6(6):1801899.
65. Chen X, Lin W, Huang Z, et al. Radiation-induced small extracellular vesicles as "carriages" promote tumor antigens release and trigger antitumor immunity. Int J Radiat Oncol. 2020;108(3):E561.
66. Hu GW, Xia YG, Zhang JT, et al. ESC-sEVs rejuvenate senescent hippocampal NSCs by activating lysosomes to improve cognitive dysfunction in vascular dementia. Adv Sci. 2020;7(10):1903330.
67. Jiang W, Ma PJ, Deng LB, et al. Hepatitis a virus structural protein pX interacts with ALIX and promotes the secretion of virions and foreign proteins through exosome-like vesicles. J Extracell Vesicles. 2020;9(1):1716513.
68. Ling XZ, Zhang GW, Xia YG, et al. Exosomes from human urine-derived stem cells enhanced neurogenesis via miR-26a/HDAC6 axis after ischaemic stroke. J Cell Mol Med. 2020;24(1):640–54.
69. Mallick ER, Arab T, Huang Y, et al. Characterization of extracellular vesicles and artificial nanoparticles with four orthogonal single-particle analysis platforms. bioRxiv. 2020. https://doi.org/10.1101/566448.
70. Yu S, Li Y, Liao Z, et al. Plasma extracellular vesicle long RNA profiling identifies a diagnostic signature for the detection of pancreatic ductal adenocarcinoma. Gut. 2020;69(3):540–50.
71. Lv QJ, Cheng LL, Lu Y, et al. Thermosensitive exosome-liposome hybrid nanoparticle-mediated chemoimmunotherapy for improved treatment of metastatic peritoneal cancer. Adv Sci. 2020;7(18):2000515.
72. https://ir.codiakbio.com/news-releases/news-release-details/codiak-initiates-subject-dosing-phase-1-clinical-trial-exoil.
73. http://patft.uspto.gov/netacgi/nph-Parser?Sect1=PTO2&Sect2=HITOFF&p=1&u=%2Fnetahtml%2FPTO%2Fsearch-bool.html&r=5&f=G&l=50&co1=AND&d=PTXT&s2=Codiak&OS=Codiak&RS=Codiak.
74. Welsh JA, Jones JC, Tang VA. Fluorescence and light scatter calibration allow comparisons of small particle data in standard units across different flow cytometry platforms and detector settings. Cytom Part A. 2020;97(6):592–601.
75. Welsh JA, Van Der Pol E, Arkesteijn GJA, et al. MIFlowCyt-EV: a framework for standardized reporting of extracellular vesicle flow cytometry experiments. J Extracell Vesicles. 2020;9(1):1713526.
76. Welsh JA, van der Pol E, Bettin BA, et al. Towards defining reference materials for measuring extracellular vesicle refractive index, epitope abundance, size and concentration. J Extracell Vesicles. 2020;9(1):1816641.
77. Varga Z, Van Der Pol E, Palmai M, et al. Hollow organosilica beads as reference particles for optical detection of extracellular vesicles. J Thromb Haemost. 2018;16(8):1646–55.
78. Valkonen S, van der Pol E, Boing A, et al. Biological reference materials for extracellular vesicle studies. Eur J Pharm Sci. 2017;98:4–16.
79. Geeurickx E, Tulkens J, Dhondt B, et al. The generation and use of recombinant extracellular vesicles as biological reference material. Nat Commun. 2019;10(1):3288.

# Microfluidics-Based Technologies for Extracellular Vesicle Research

# 9

Binbin Cui, Chao Liu, Guihua Zhang, Chunchen Liu, Fuquan Yang, Jin Hao, Peng Zhang, Shuhuai Yao, and Jiashu Sun

## 9.1 Microfluidics-Based Strategies for Extracellular Vesicle (EV) Isolation

Because EVs are generally mixed with other biofluid components, EVs can be isolated and concentrated from clinical samples prior to downstream molecular detection or therapeutic treatment. Compared with commonly used EV isolation approaches, such as ultracentrifugation and polymer-based precipitation, microfluidic methods provide fast and efficient EV isolation with high yield and purity. Most

---

Binbin Cui, Chao Liu and Guihua Zhang contributed equally with all other contributors.

B. Cui · C. Liu · S. Yao (✉)
The Department of Mechanical and Aerospace Engineering, The Hong Kong University of Science and Technology, Hong Kong, China
e-mail: meshyao@ust.hk

C. Liu · J. Sun (✉)
National Center for Nanoscience and Technology, University of Chinese Academy of Sciences, Beijing, China
e-mail: sunjs@nanoctr.cn

G. Zhang · P. Zhang (✉)
Institute of Molecular Medicine, School of Medicine, Shanghai Jiao Tong University, Shanghai, China
e-mail: pengzhang016@sjtu.edu.cn

F. Yang
Laboratory of Proteomics, Institute of Biophysics, Chinese Academy of Sciences, Beijing, China

J. Hao
Sysmex Shanghai, Ltd., Shanghai, China



Q. Wang, L. Zheng (eds.), *Extracellular Vesicles*,
https://doi.org/10.1007/978-981-99-8365-0_9

importantly, they enable integrated EV isolation and detection using a small volume of fluid samples. In the past decade, significant progress has been made in the separation of EVs using various microfluidic devices. These platforms are mainly based on three types of isolation strategies: immunoaffinity-based isolation, size-based sorting, and dynamic separation, as well as combinations of these approaches.

### 9.1.1 Immunoaffinity-Based EV Isolation

The most commonly used microfluidic approach for EV separation is the immunocapture of specific EVs using immobilized antibodies in microfluidic chips. EV membranes contain a variety of protein markers, including CD63, CD9, and CD81. Such protein markers can serve as binding targets to achieve specific capture of EVs through the modification of microchannel surfaces with protein-specific capture probes (such as antibodies or aptamers). In microfluidic devices, patterned microstructures, such as pillar or herringbone array structures, can enhance the interactions between EVs and chip interfaces. Chen et al. [1] reported the first immunoaffinity-based microfluidic exosome isolation device, which employed anti-CD63 IgG-coated microfluidic channel surfaces that selectively captured exosomes from 400 μL serum samples collected from patients with non-small cell lung cancer (NSCLC). Kanwar et al. [2] also developed an "ExoChip" platform with circular chambers interconnected by narrow channels to increase the retention time for exosome capture. Lo et al. [3] developed an "OncoBean" chip using a radial flow microfluidic device with antibody-modified bean-shape microposts to achieve high-throughput EV isolation (Fig. 9.1a). Another immunoaffinity-based microfluidic exosome isolation device was prepared using Y-shaped micropillars coated with nanostructured graphene oxide (GO)/polydopamine (PDA; Fig. 9.1b) [4]. The GO/PDA nanoporous interfaces provided numerous reactive sites for antibody immobilization to enhance the efficiency and specificity of immunocapture of exosomes.

Compared with microfluidic chips harboring immobilized antibodies on the channel surfaces, immunomagnetic beads endowed higher EV capture efficiency owing to the larger surface area and simplified sample preparation process. For example, Zhao et al. [5] designed an ExoSearch Chip consisting of a Y-shaped ejector for simultaneous loading of beads and exosomes and a serpentine mixer for dynamic capture of exosomes. Theoretically, the ExoSearch chip has a nearly unlimited capture capacity, which could facilitate high-throughput applications using large volumes of biological fluid samples (Fig. 9.1c). In order to further analyze exosome-coated microbeads at the single-bead level, Tayebi et al. [6] developed a microfluidic device composed of trapping arrays to capture individual microbeads. Within the trapping structures, molecular analysis could be achieved by on-chip exosome purification and nucleic acid extraction (Fig. 9.1d).

Immunoaffinity-based EV isolation methods can only capture EVs containing specific EV markers, making this strategy feasible when the marker proteins (antigens) are present on the surface of EVs with high proportions. Owing to the

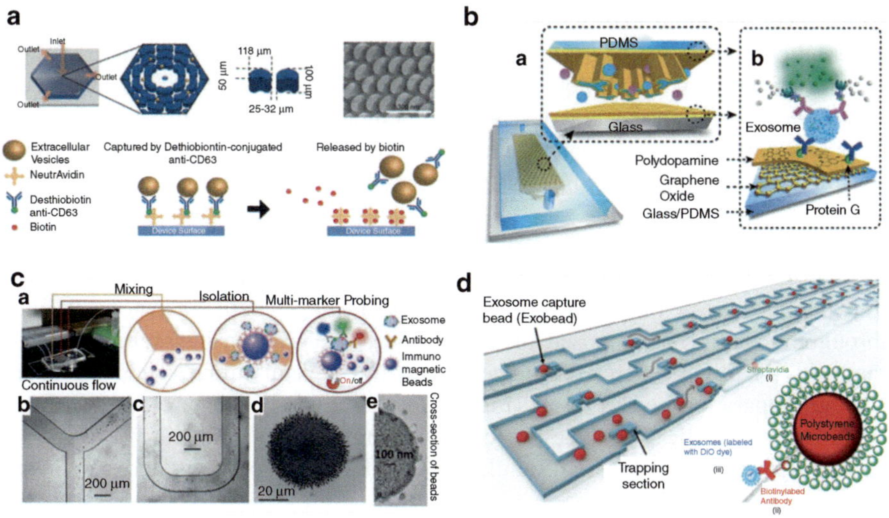


**Fig. 9.1** (**a**) Schematic of EV isolation on an OncoBean Chip. Reproduced with permission [3]. *(Copyright 2020, The Royal Society of Chemistry)*. (**b**) Schematic of the nanostructure-coated microfluidic chip. Reproduced with permission [4]. *(Copyright 2016, The Royal Society of Chemistry)*. (**c**) Workflow of the ExoSearch chip for continuous separation and detection of circulating exosomes. Reproduced with permission [5]. *(Copyright 2016, The Royal Society of Chemistry)*. (**d**) Schematic diagram of the trapping microchannels for exosome captured immunobeads. Reproduced with permission [6]. *(Copyright 2020, ACS American Chemical Society)*

molecular heterogeneity of EVs, the immunoaffinity-based capture performance of EVs varies substantially. Therefore, the selection of surface targets on EVs is crucial for EV separation.

### 9.1.2 Size-Based EV Filtration and Trapping

EVs are membrane-enclosed vesicles derived from mammalian cells. The widely accepted size range of EVs is 20–1000 nm. Size-based EV separation strategies have been frequently investigated because they provide label-free separation, ensure size uniformity, and minimize sample bias. Size-based methods usually incorporate nanofilters, nanoporous membranes, or nanoarrays positioned along microchannel cross-sections or physically linked to microchannel substrate surfaces to filter or trap EVs.

#### 9.1.2.1 Membrane-Based Filtration

Membrane-based filtration depends on the intrinsic size difference between EVs and contaminant vesicles. For example, Woo et al. [7] developed a tabletop-sized centrifugal microfluidic system integrated with two nanofilters to separate EVs of 20–600 nm. The recovery rates from cell-culture supernatants or urine from patients with cancer are more than 95% when evaluated within 20 min. Membranes with different pore sizes can be assembled into a microfluidic device in tandem to separate EVs from complex biological samples (Fig. 9.2a). Kim et al. [8] proposed a microfluidic filtration device consisting of membranes of three different size fractions (e.g., 200, 100, and 30 nm; Fig. 9.2b). Such technology simplifies the operation of separating EVs and improves the separation purity of EVs.

### 9.1.2.2 Nanowire-Based Trapping

Another EV trapping approach employs elaborately synthesized nanowires that can trap EVs using an interlaced structure. Yeh et al. [9] built an emergent high-throughput platform composed of three-dimensional carbon nanotube arrays for rapid capture of EVs excreted from neurons within several minutes and without any sample preparation requirements (Fig. 9.2c). Additionally, Wang et al. [10] fabricated a microfluidic prototype device consisting of porous silicon nanowires on micropillars to trap EVs and simultaneously filter out proteins and cell debris (Fig. 9.2d).

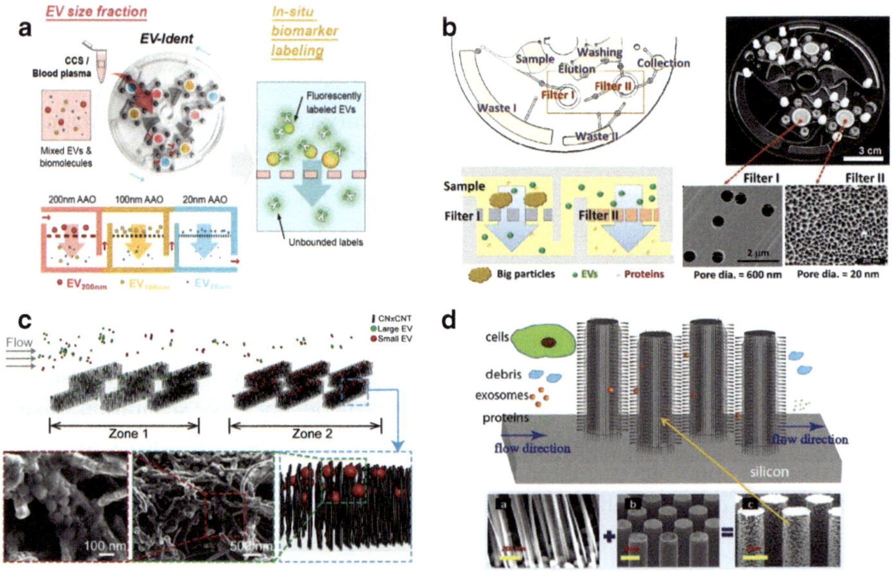


**Fig. 9.2** (**a**) Schematic of EV-Ident and EV isolation. Reproduced with permission [7]. *(Copyright 2020, ACS American Chemical Society)*. (**b**) Schematic of the device for the EV isolation microfluidic chip. Reproduced with permission [8]. *(Copyright 2017, ACS American Chemical Society)*. (**c**) Rapid size-based isolation of EVs using three-dimensional carbon nanotube arrays. Reproduced with permission [9]. *(Copyright 2020, ACS American Chemical Society)*. (**d**) Schematic of the ciliated micropillar array for exosome capture. Reproduced with permission [10]. *(Copyright 2013, The Royal Society of Chemistry)*

### 9.1.3 Dynamic EV Isolation

#### 9.1.3.1 Active Dynamic EV Isolation

Active dynamic EV isolation techniques exploit external forces between EVs and the surrounding flow medium to concentrate exosomes according to their mechanical (size, density, compressibility) and dielectric properties. This strategy is contactless and label-free, which facilitates the easy recovery of intact EVs and increases the separation throughput of EVs. To date, several separation methods have been developed employing sound, electric field and centrifugal force without complicated procedures.

Wu et al. [11] designed a microfluidic blood sorting platform that integrates microfluidic channels and interdigital transducers to separate not only red blood cells, white blood cells, and platelets from the blood but also EVs. The separation and extraction of nanoscale particles, such as apoptotic bodies and cellular microvesicles in cell-free plasma, is essential for the development of biomedical research related to biological particles (Fig. 9.3a). Furthermore, Lee et al. [12] used sound waves to subdivide EVs in a contactless manner. This separation uses standing ultrasonic waves to apply different acoustic interaction forces to the vesicles according to their size and density. The device consists of a pair of interleaved transducer electrodes to generate standing wave surface acoustic waves across the flow channel (Fig. 9.3b). The particles in the sound zone are subjected to radiation pressure and migrate to the pressure node. Since the acoustic interaction force is proportional to the volume of the vesicle, the larger vesicles move faster. In addition, the filter size can be electronically controlled on the spot to achieve flexible selection according to size requirements. The system has the ability to separate EVs quickly and with high throughput, thereby exhibiting the capacity to act as a multifunctional preparation tool for the early stages of EV analysis.

Another external force applied in microfluidic devices for dynamic isolation is an electric field, in which electrokinetic forces can be controlled by the setup of the device. Dielectrophoresis (DEP) is a versatile technique that leverages variations in DEP force (as a function of species size) to manipulate and discriminate different polarized particles in fluid suspensions exposed to nonuniform electric fields. Heller's group [13] designed an alternating current electrokinetic microarray chip to quickly separate EVs from human blood plasma (Fig. 9.1c). Only 30–50 μL plasma is required, and the EVs can be separated within 30 min and be used for downstream analysis. More recently, the same authors reported a rapid pancreatic cancer diagnostic approach involving the capture and analysis of EVs from raw biofluids (such as whole blood, plasma, or serum; Fig. 9.3d) [14]. Although the platform is powerful, the electrodes are gradually covered by EVs, and this method is not suitable for long-term continuous inflow EV separation, which results in a decrease in separation efficiency.

#### 9.1.3.2 Passive Dynamic EV Isolation

Microfluidics enables precise flow control in well-designed channels and can be used to adjust the trajectory of particles in the flow. Therefore, EVs can be

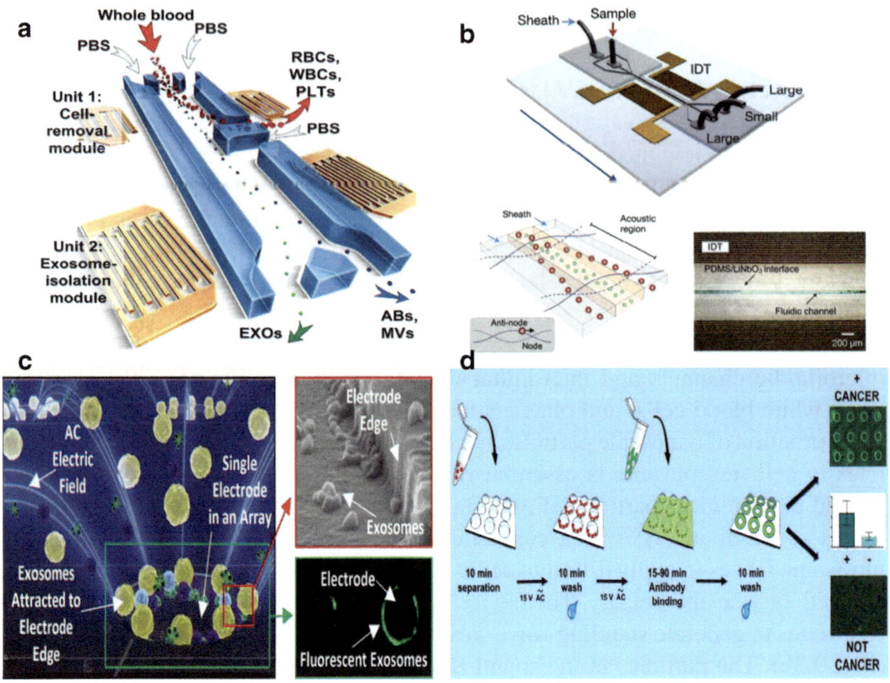


**Fig. 9.3** (**a**) Acoustofluidic module for separation of EVs from whole blood. Reproduced with permission [11]. *(Copyright 2017, National Academy of Sciences)*. (**b**) An acoustic nanofilter was designed to separate EVs (≤1 μm) through in-flow size fractionation. Particles in an acoustic field experience radiation forces and migrate toward the pressure nodes. Reproduced with permission [12]. *(Copyright 2015, American Chemical Society)*. (**c**) Abridged general views of the ACE microelectrode chip showing a collection of exosomes. Reproduced with permission [13]. *(Copyright 2017, ACS American Chemical Society)*. (**d**) Schematic diagram illustrating the AC electrokinetic (ACE) direct immunoassay procedure. [Reproduced with permission [14]. *(Copyright 2020, ACS American Chemical Society)*]

manipulated in a passive dynamic isolation model without any externally applied field, which only relies on physical constraints based on hydrodynamic properties. The trajectory of the particles is determined by the flow dynamics. Passive dynamic EV isolation technology is mainly divided into three categories: field flow fractionation (FFF), deterministic lateral displacement (DLD), and viscoelastic flow sorting.

FFF is another separation technique that applies a force field perpendicular to the sample flow to achieve separation of particles with different sizes and molecular weights. In recent years, an asymmetric flow field flow fractionation technology (AF4) has been applied to the separation of EVs [15]. AF4 has a permeable plate at the boundary of the channel. When the sample flows in the channel, due to the laminar flow, the fluid velocity at the boundary is slower than the velocity at the center; thus, a parabolic velocity profile is formed (Fig. 9.4a). When a vertical force field is applied, the analyte in the sample is pushed toward the boundary. Brownian motion produces a counteracting motion, causing smaller particles to tend to equilibrium

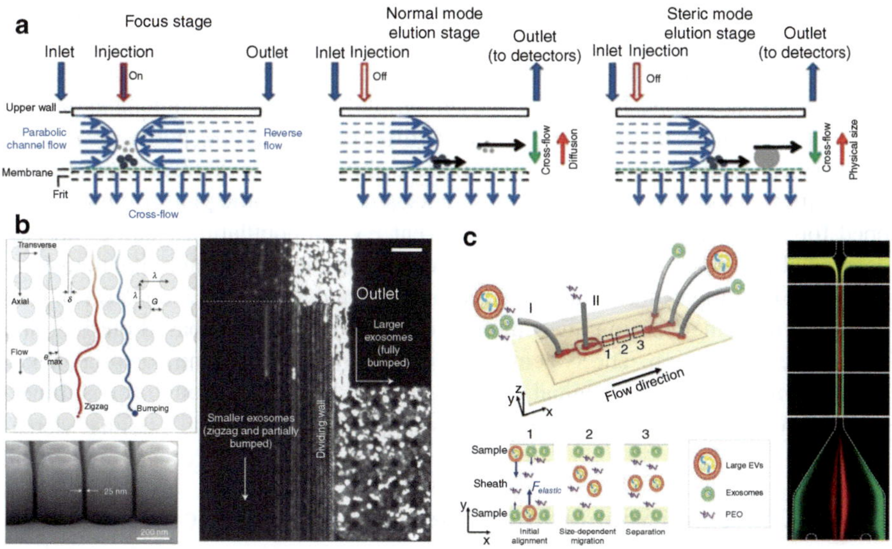


**Fig. 9.4** (**a**) Schematic of asymmetric-flow field-flow fractionation technology for exomere and small EV separation. Reproduced with permission [15]. *(Copyright 2019, Springer Nature)*. (**b**) Nano-DLD for EV separation using a nanopillar array. Reproduced with permission [16]. *(Copyright 2016, Springer Nature)*. (**c**) Viscoelastic separation of exosomes and microvesicles in a microfluidic device. [Reproduced with permission [17]. *(Copyright 2017, ACS American Chemical Society)*]

positions farther from the boundary. This type of separation has a wide range and is suitable for various eluents.

DLD is a passive microfluidic separation technique introduced by Austin and coworkers in 2004 [18]. Recently, Wunsch et al. [16] invented a nanoscale DLD array to sort, separate, and enrich EVs (Fig. 9.4b). At the micrometer scale, particles smaller than the critical diameter (Dc) will move laterally along the array in a collision mode, whereas particles larger than the Dc will move along the laminar flow in a zigzag pattern. The authors fabricated nanoscale DLD arrays with silica pillars with uniform gap sizes (ranging from 25 to 235 nm) to enable vesicles to be separated. During the operation, the larger vesicles are replaced by the nano-DLD array to the right side of the channel and collected in the side channel, whereas the smaller vesicles will flow from the array to the large channel in a zigzag pattern. The authors further applied the DLD array to divide fluorescently labeled human urine-derived EVs into two parts: complete collision and zigzag partial collision. Most EVs exhibit a partial collision mode with respect to the displacement distribution of exosomes, consistent with their size distribution (60–70 nm). This strategy demonstrated the size-based displacement of EVs, thus revealing the potential applications of microarrays to sort and quantify nanoscale biocolloids.

Viscoelastic microfluidics manipulates particle migration by inducing a size-dependent elastic lift force in a viscoelastic medium, which has been shown to enable precise separation of EV subpopulations. Liu et al. [17] proposed a method mainly based on the viscosity of polyethylene oxide fluid to successfully separate exosomes (Fig. 9.4c). This method, which combines the advantages of microfluidics and hydrodynamics, pushes EVs to the center line of the microchannel and demonstrates that vesicles of different sizes have different rates; 96% purity and 93.6% recovery of small EVs (sEVs, <200 nm) were obtained at a volumetric throughput of 200 μL/h. Moreover, an interfacial viscoelastic microfluidic platform was developed for size-selective separation of different EV subpopulations. Sun et al. [19] developed a λ-DNA and aptamer-mediated method that allows simultaneous size-selective separation of EVs and surface protein analysis of individual EVs. They also used machine learning algorithms to evaluate EV features based on their size and marker expression. The systems described above highlight the potential of viscoelastic microfluidics in clinical practice for liquid biopsy, although it is still necessary to validate these systems in clinical samples.

### 9.1.4 Combined Separation Strategies for EV Isolation

To further improve the separation efficiency of EVs, many efforts have been made to create a single device that combines two or more of the abovementioned separation strategies. For example, Vaidyanathan et al. [20] developed a microfluid chip modified with antibodies (e.g., anti-human epidermal growth factor receptor 2, anti-CD9, and/or anti-prostate-specific antigen [PSA]) and combined with a tunable alternating current electrohydrodynamic field for high-specificity capture of EVs (Fig. 9.5a). Moreover, Cho et al. [21] proposed a strategy combining size-based filtration and an electrical dynamic field to capture EVs onto a membrane with a 30-nm pore size (Fig. 9.5b). This method yielded an EV recovery rate of up to 65% (from RNA-level estimates) and a protein removal rate of up to 83.6% in approximately 30 min for EV isolation from mouse plasma. Therefore, combining multiple strategies to generate an EV isolation approach is an effective way to improve the isolation efficiency and purity of EVs.

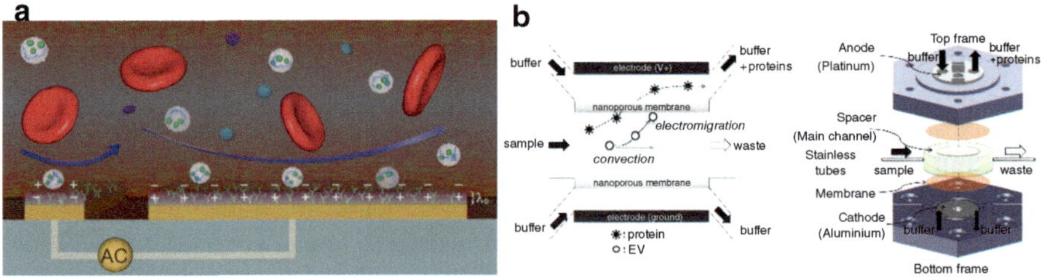


**Fig. 9.5** (**a**) Enhanced immunocapture of EVs via alternating current electrohydrodynamic field nanoshearing using on-chip electrodes to separate EVs. Reproduced with permission [20]. *(Copyright 2014, ACS American Chemical Society)*. (**b**) Schematic of electrophoretic migration isolation processes. [Reproduced with permission [21]. *(Copyright 2016, Elsevier)*]

Microfluidic chips possess the advantages of miniaturization, automation, and integration. The introduction of separation and various analysis techniques into microfluidic chips can provide a high-throughput, high-openness, and high-stability sample-to-result platform for the capture and analysis of EVs. Such a platform is expected to provide important insights into future medical research approaches based on EVs and body fluid biopsy technologies.

## 9.2 Microfluidic Protein Profiling of EVs

The heterogeneity of EVs is reflected in their morphological diversity, size variations, and cargo diversity, including a range of surface markers (proteins, carbohydrate antigens), lipids, free proteins, and nucleic acids carried inside the vesicles. The proteomic profiling of EVs may facilitate elucidation of the biogenesis and sorting mechanisms of cargo loading through analysis of conserved sets of exosomal proteins. Furthermore, this may enable the discovery of novel biomarkers for various types of diseases and provide guidance for the delivery of nanomedicines. For analysis of EV protein markers, conventional methods, such as western blotting, enzyme-linked immunosorbent assay (ELISA), and mass spectrometry, require large volumes of sample for input. Microfluidics coupled with other detection methods, such as fluorescence detection, nuclear magnetic resonance, plasmonic sensing, electrochemical sensing, and photon counting, may represent an integrated approach to studying EV subpopulations and analyzing proteins at the single EV level, thereby enabling profiling of the surface proteins of single EVs.

As a technology platform, microfluidic technology, also known as lab-on-a-chip, can be cross-compatible with various disciplines, such as functional materials [4], acoustic [11], optics [11], and electrochemistry [11]. Microfluidic technology has also been used in various studies of the isolation, sorting, and analysis of EVs. Compared with traditional analytical methods, microfluidic technology can reduce sample consumption, lower reagent costs, and achieve system miniaturization, integration, automation, and intelligence; therefore, this approach has become a research hotspot in the life science field [22]. In this section, we will illustrate state-of-the-art applications in microfluidics technology on EV analysis, with a focus on the advancement of traditional continuous microfluidic technology and discrete microfluidics in the analysis of individual EVs.

### 9.2.1 Bulk EV Protein Analysis with Microfluidics

#### 9.2.1.1 Integrated Immunoisolation and Protein Analysis

EVs are highly heterogeneous in terms of the size, cell origin, and biogenesis pathway, similar to nonvesicular particles of the same density; these features are major obstacles to delineating the compositional and functional properties of different EV subpopulations. To overcome the interference from unwanted vesicular contents and protein particles, He et al. devised a microfluidic platform to integrate the multistep analysis of circulating EVs directly from the blood plasma based on immunoaffinity capture with magnetic beads, including immunoisolation, EV lysis and protein capture, and intravesicular protein analysis (Fig. 9.6a) [5]. This approach enables selective isolation of certain subpopulations of EVs and subsequent quantitative analysis of surface and intravesicular biomarkers directly from as little as 30 μL blood plasma. However, owing to the application of ELISA, only one type of protein marker can be analyzed in each assay. Thus, Zhao et al. developed a microfluidic technology, termed ExoSearch, to isolate specific subpopulations of EVs using immunomagnetic immobilization in combination with multiplexed sandwich immunofluorescence assays; this approach enabled the integration of EV isolation and multiplexed protein detection from exosomes using blood plasma on a single microfluidic chip (Fig. 9.6a). Briefly, the magnetic beads conjugated with capture antibodies (e.g., anti-CD9 antibodies) and blood plasma from a patient with ovarian cancer were pumped into the Y-shaped microfluidic channel from two separate inlets and mixed thoroughly using a segment of the serpentine channel to maximize the capture efficiency. When arriving at the chamber, all of the magnetic beads, including the bounded EVs, were aggregated in situ using a magnet placed at the bottom of the structure. Finally, a panel of probing antibodies with fluorescent labels of different colors was used to quantify the abundances of different protein markers. Employing this approach, the authors performed a comparative analysis of a panel of three surface markers (CA-125, epithelial cell adhesion molecule, and CD24) on EVs isolated from patients with ovarian cancer and healthy individuals. Their findings validated the overexpression of different proteins in patients with ovarian cancer.

#### 9.2.1.2 Differentiated Immunostaining-Based Characterization of EVs (DICE)

To avoid biased enrichment and analysis related to the immunoaffinity-based capture of EVs, Kang et al. adopted a new strategy to impartially capture the EVs harvested from ultracentrifugation based on the specific interaction between biotin and streptavidin following the biotinylation of EVs [24]. Biotinylated EVs were first captured with a microfluidic chip functionalized with a NeutrAvidin binding surface, followed by multiplexed immunostaining for the profiling and identification of EVs. The device possessed a quadratic design with each quadrat immune-stained for a unique marker, thereby allowing multiplexed detection of EVs from one sample (Fig. 9.6c).

#### 9.2.1.3 μNMR

Magnetic labeling can boost the imaging contrast of background noise when detected using NMR because no ferromagnetic signal is detected for most natural biomolecules in the blood. However, the application of NMR to EV detection is a major engineering challenge because EVs are 1–2 orders of magnitude smaller than tumor cells. To overcome this limitation, Shao et al. developed a microfluidic system and analytical technology specifically for microvesicle detection and profiling in which magnetic nanoparticles (MNPs with a core diameter of 7 nm) were

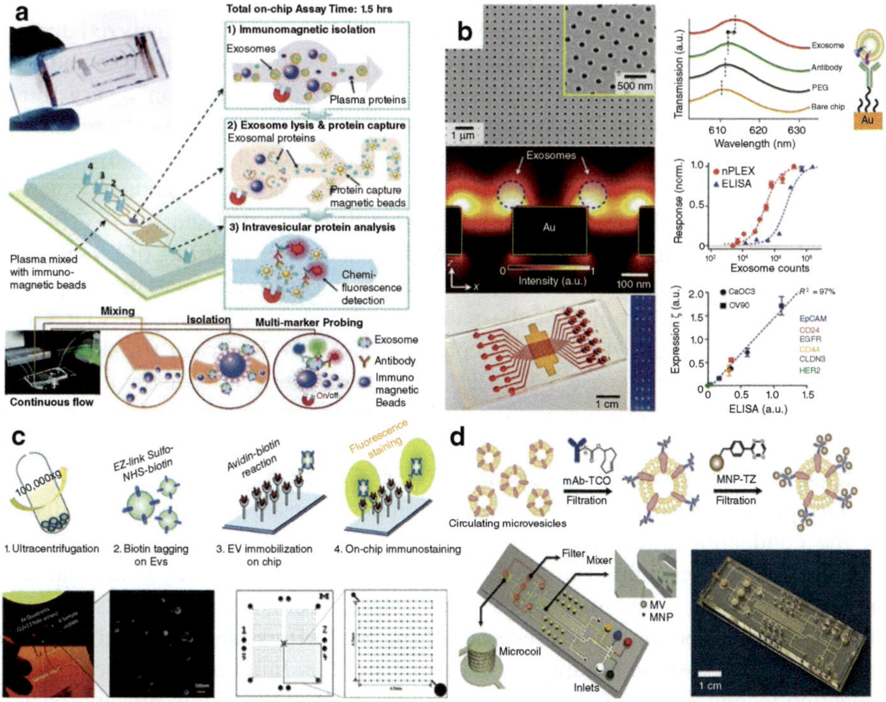


**Fig. 9.6** Microfluidic detection of proteins from bulk EVs. (**a**) Integrated immunoisolation and protein analysis. Exosomes immobilized on the magnetic beads are aggregated in the first chamber, followed by lysis of the exosomes. The exosomal lysate is co-incubated with magnetic beads conjugated with antibodies targeting the proteins of interest. Finally, ELISA is adopted to detect exosomal marker proteins. Reproduced with permission [5]. *(Copyright 2016, The Royal Society of Chemistry)*. (**b**) Nanoplasmonic exosome (nPLEX) sensor. A thin gold film (200 nm in depth) patterned with periodic nanoholes is deposited on a glass substrate as the bottom surface of the multichannel microfluidic device. The periodicity of nanoholes is adjustable to maximize the sensitivity of exosome detection. The gold surface is prefunctionalized using a layer of polyethylene glycol (PEG), and antibody conjugation and specific exosome binding are monitored by transmission spectral shifts. Comparative analysis indicates that this approach has $10^4$- and $10^2$-fold better sensitivity than western blotting and ELISA. Reproduced with permission [23]. *(Copyright 2017, Nature Publishing Group)*. (**c**) Schematic of DICE. Exosomes isolated with ultracentrifugation are biotinylated and immobilized by the avidin molecules coated on the bottom surface of the microfluidic chips. Immunostaining enables the detection of certain surface markers on the exosomes. Quadratic design enables the detection of four types of marker proteins simultaneously using chip. Reproduced with permission [24]. *(Copyright 2019, The Royal Society of Chemistry)*. (**d**) The micronuclear magnetic resonance (μNMR) system, including nanoparticle labeling of the exosomes and chip design. Using a two-step bio-orthogonal click chemistry, EVs labeled with magnetic nanoparticles (MNPs with a core diameter of 7 nm) can be analyzed directly on-chip with a miniaturized μNMR system for measuring protein abundances. [Reproduced with permission [25]. *(Copyright 2012, Nature Publishing Group)*]

chemically attached to antibodies that were bound to the EVs through two-step bio-orthogonal click chemistry [25]. The MNP-labeled EVs were then analyzed directly on-chip with a miniaturized μNMR system for detection of protein abundance. Because the size of the modification molecule was negligible, only very minor changes in the sizes of the MNPs and antibodies occurred, enabling easy separation between the labeled EVs and unbound antibodies or MNPs through passive filtering in the microfluidic chip (Fig. 9.6d). Compared with traditional methods, such as western blotting and ELISA, μNMR shows significantly better sensitivity ($10^4$- and $10^3$-fold higher sensitivity, respectively). Additional protein profiling of EVs derived from glioblastoma multiforme (GBM) cell lines confirmed that this approach reflected the protein profiles of the parental cells. A combination of four GBM markers (epidermal growth factor receptor [EGFR], EGFR variant III, podoplanin, and isocitrate dehydrogenase 1 R132H mutant) can be used to distinguish cancer EVs from EVs derived from normal cells.

#### 9.2.1.4 Nanoplasmonic Exosome (nPLEX) Sensing

Surface plasmon resonance (SPR) is a phenomenon caused by collective oscillation of the conduction electrons on a metal-dielectric interface when the surface is exposed to incident light. The local change in the refractive index is indicative of molecular binding to the metal surface. Therefore, in contrast to methods based on fluorescence or chemiluminescence detection, SPR offers label-free and real-time monitoring of exosome binding on the metal surface. Im et al. developed a novel detection technology for EV proteins, designated nPLEX sensing; this approach is based on the optical transmission of SPR through periodic nanoholes patterned on a thin gold film instead of conventional total internal reflection [23]. By exploiting the tunable periodicity of nanoholes, the electromagnetic field can be adjusted to overlap with the size of exosomes, thereby maximizing detection sensitivity. The change in the refractive index caused by exosome binding to antibodies conjugated on the gold nanohole surface leads to a spectral red shift or an intensity change proportional to the amount of exosome binding; this enables the nPLEX sensor to be used for quantitative analyses. Comparative analyses indicate that this approach has $10^4$- and $10^2$-fold better sensitivity than western blotting and ELISA (Fig. 9.6b). Embedded on the bottom of multiparallel microchannels, the nPLEX sensor array can be used to analyze multiple samples independently within 30 min, suggesting potential clinical applications in rapid, label-free, multiplexed protein analysis.

#### 9.2.1.5 ExoCounter for Quantification of the Number of Specific sEVs

ExoCounter has been proposed as a device that can quantify the number of specific sEVs in various samples without isolation procedures [26]. ExoCounter detects sEVs via a method in which a sandwich immunoassay is combined with optical disc technology (Fig. 9.7a). sEVs are captured on the surface of an optical disk having a diameter of 80 mm, and FG beads [27] are then bound to the captured exosomes as a label for laser spot-based detection. The disc surface and FG beads are coated or conjugated with antibodies against proteins present on the sEV surface to identify the specific sEVs (Fig. 9.7b, c). The surface of the optical disc provides a

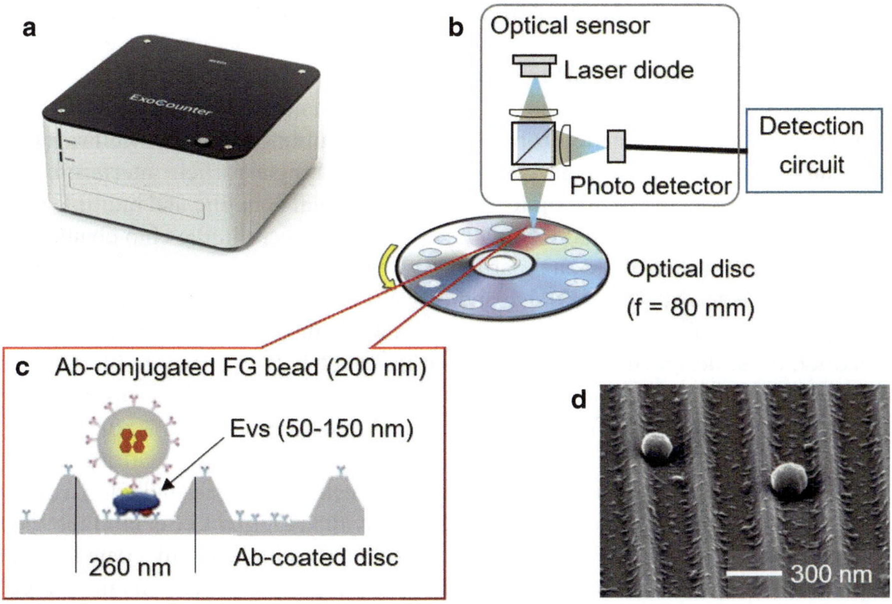


**Fig. 9.7** Schematic overview of the ExoCounter. (**a**) Photograph of the ExoCounter instrument. (**b**) sEV and FG bead complexes are detected by an optical sensor composed of a laser diode and a photodetector and then counted via a detection circuit. (**c**) Enlarged illustration of the disc surface. Each sEV is isolated in the groove of an optical disc coated with an antibody and labeled with an antibody-conjugated single FG bead. (**d**) A scanning electron microgram of the disc surface on which sEV and FG bead complexes are formed

nano-groove whose width (260 nm) is larger than that of sEVs (50–150 nm) and is slightly larger than the diameter of the FG beads (200 nm), thereby allowing only one FG bead to occupy the space in the nano-groove. Complexes of single sEVs and single FG beads are formed on the disc surface during incubation. The number of sEVs is measured by detecting the sEV/FG bead complexes with a dedicated optical disc drive at a speed of up to 200,000 sEVs/s. The optical disc drive is equipped with an optical sensor consisting of a laser diode with wavelength of 405 nm, an objective lens with numerical aperture of 0.85, and a photo diode. Figure 9.7d shows a scanning electron microscopic image of the disc surface on which sEV and FG bead complexes are formed. ExoCounter can simultaneously confirm the presence of specific proteins on sEVs and number of specific sEVs in body fluids and cell culture medium, which contains various contaminant particles, without the requirement for isolation procedures. Quantification of specific sEVs can not only accelerate basic research on sEVs but also act as a key technology for the implementation of sEV-related technologies in the clinical setting.

## 9.2.2 Single EV Protein Analysis with Microfluidics

### 9.2.2.1 Continuous Microfluidic Methods

Microfluidic devices facilitate the confinement and manipulation of liquid samples in parallel, offering tremendous opportunities for coupling with advanced imaging tools and analysis of surface markers on single EVs. Single-particle interferometric reflectance imaging sensor technology allows for digital counting and multiple phenotypic analyses of individual EVs captured within a microfluidic chip channel by enhancing the contrast of the nanoparticle scattering signal, thereby enabling highly accurate characterization of EV size and concentration. This approach to single EV multiple phenotypic analysis is conducive to the diagnosis of clinical diseases [28]. Friedrich et al. designed a nanofluidic chip to detect EVs in a sample through the parallel nanochannels in the chip and then successively analyzed the results using fluorescence microscopy. The sample loading volume was less than 20 μL, and the sensitivity for EV detection was as low as 170 fM [29]. Recently, a platform known as single EV analysis, based on photon-counting techniques, was applied in a study of multiplexed protein biomarker profiling of individual EVs by Lee et al. (Fig. 9.8a) [30]. In this approach, the EVs are first fixed in a microfluidic cavity coated with a capture antibody, and careful on-chip immunostaining and complex imaging procedures are then performed to differentiate single vesicles from protein complexes or other clusters with an intensity cutoff of 102 counts. This microfluidic chip can identify the heterogeneity of EV protein expression at the single-vesicle level, enabling detection of 11 different protein markers on one vesicle; further increases in the number of protein markers detected in one assay may also be possible [30]. Beekman et al. reported a multimodal analysis platform for individual EVs derived from the characterization of LnCAP prostate cancer cells using atomic force microscopy, scanning electron microscopy, and Raman spectroscopy on one microfluidic chip (Fig. 9.8c) [32]. This multiplatform combination strategy can be used to obtain comprehensive information on individual EVs, with the ability to distinguish EVs from contamination, and can reveal the size distribution and chemical fingerprint of EVs at the single EV level.

### 9.2.2.2 Discrete Microfluidic Methods

Compared with traditional continuous microfluidic technology, droplet- and well-based microfluidic technology utilizes the interaction between shear and surface tension to divide a continuous fluid into discrete femtoliter-sized droplets or a predesigned microwell array on a microfluidic chip to form separate liquid reaction compartments [36]. The number of target molecules in each droplet or well is not more than one; therefore, the interference of the background signal can be effectively avoided, and numerous single-molecule assays can be performed in parallel, e.g., digital polymerase chain reaction (PCR) [11] and digital ELISA [11] for the digital detection and tracing of protein molecules in plasma samples. In contrast to the above approaches, another discrete microfluidic platform equipped with thousands of nanopillars in parallel, called localized SPR imaging, enables single exosome multiplex detection, as reported by Raghu et al. (Fig. 9.8b). Because this

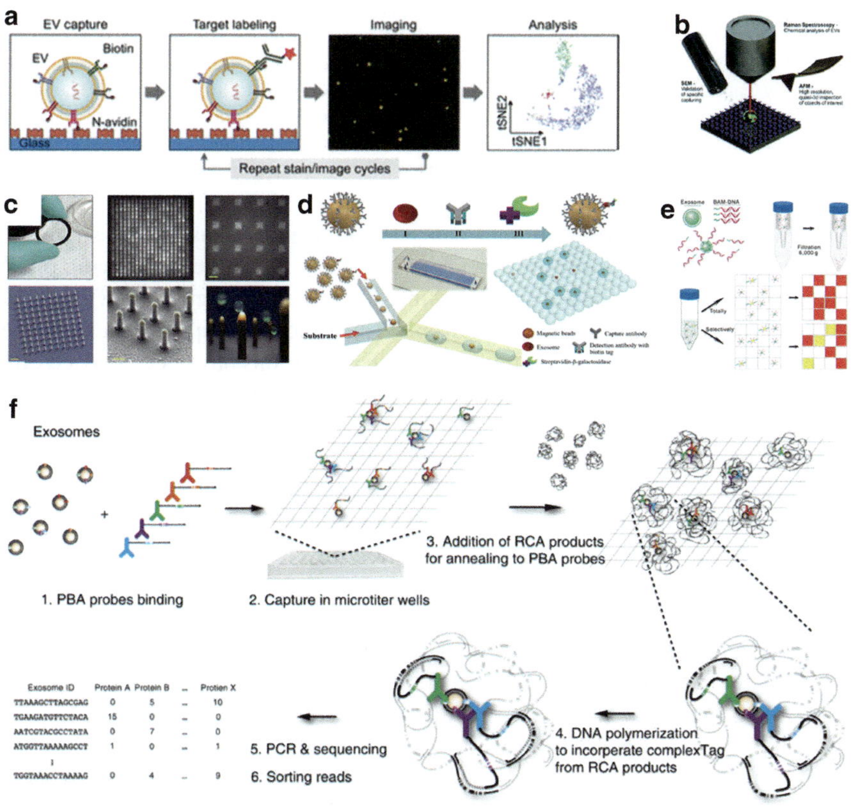


**Fig. 9.8** Single EV analysis using microfluidic chips. (**a**) Biotinylated EVs are first captured on microfluidic cavities with neutravidin-coated surfaces, followed by staining with fluorescent antibodies and imaging by microscopy. Subsequently, fluorophores are quenched, and the staining process is repeated for a different set of antibody-labeled fluorescent tags. Thus, multiplexed detection of single EV surface proteins is realized. Reproduced with permission [30]. *(Copyright 2018, American Chemical Society).* (**b**) In a multimodal analysis platform, target EVs are selectively captured on antibody-functionalized stainless-steel substrates, where Raman spectroscopy, scanning electron microscopy, and atomic force microscopy can be performed separately. Reproduced with permission [31]. *(Copyright 2019, The Royal Society of Chemistry).* (**c**) a shows a schematic diagram of single EVs captured using size matched nanopillars; b shows nanopillars under false-colored scanning electron microscopy, and c and d represent localized surface plasmon resonance imaging of 4 × 4 and 20 × 20 arrays, respectively. Reproduced with permission [32]. *(Copyright 2018, PLOS).* (**d**) Single-level EV quantification using a droplet digital ExoELISA. With a dual Poisson distribution through a two-step process involving magnetic bead immunocapture and droplet-based compartmentalization, individual EVs are encapsulated within water-in-oil droplets, which are subjected to single amplification and readout. Reproduced with permission [33]. *(Copyright 2018, American Chemical Society).* (**e**) Selective and quantitative analysis of EVs by digital detection integrated with nucleic acid-based amplification in a microwell-based microfluidic chip. DNA-anchored EVs are distributed into a sufficient number of microwells, resulting in one or fewer EVs per chamber based on the Poisson distribution. After DNA single amplification, target EVs can be digitally detected through the positive microwells with fluorescence signals.

(continued)

Reproduced with permission [34]. *(Copyright 2018, American Chemical Society)*. (**f**) High-throughput approach to simultaneous profiling of 38 surface proteins for their presence on individual EVs. EVs are first reacted with a proximity-dependent barcoding assay probe and then individually captured in microtiter wells by immobilized cholera toxin subunit B. The surface protein composition of individual EVs is evaluated using next-generation sequencing analysis of rolling circle amplification and polymerase chain reaction products. Reproduced with permission [35]. *(Copyright 2019, Nature Publishing Group)*

approach utilizes a nanosensor that can be engineered to match the size of the detected exosome, single exosomes can be analyzed. Dynamic monitoring of EVs secreted from live cells may also be possible [31]. In the detection of EVs in body fluids, microfluidic methods based on droplets and microwells provide a new detection platform for the study of individual EVs. Explorations have been carried out to assess the application of individual EV analysis using droplet microfluidic technology. For example, Liu et al. reported an innovative method for digital detection of exosomes by combining droplet microfluidic technology with exosome ELISA (droplet digital ExoELISA, Fig. 9.8d) [33]. Magnetic beads serve as a medium for capture and separation of the target EVs. A detection antibody tagged with an enzymatic reporter further recognizes the antigen on the captured exosomes, forming a single enzyme-linked immunocomplex on the bead. Using a droplet microfluidic platform, single EV complexes tagged with an enzymatic reporter can be encapsulated in droplets with a fluorescein-di-β-D-galactopyranoside (FDG) substrate to produce a fluorescence signal for detection. Ultra-high sensitivity (as low as 10 EVs/μL) and high specificity were demonstrated in the quantification of EVs harboring glypican-1 (GPC-1) from various EV subpopulations using serum samples from patients with breast cancer. Thus, droplet-based ExoELISA enabled recognition of particular proteins present on single EVs. As an improvement to the multiplexing capability, Ko et al. devised an immune-droplet digital PCR amplification method that utilizes multiplexed PCR in droplets to amplify target DNA molecules specifically conjugated on different types of antibodies [37]. Tian et al. reported a microwell-based microfluidics approach for digitally detecting EVs via surface-anchored DNA nucleic acid sequence-initiated DNA self-assembly signal amplification technology (Fig. 9.8e) [34]. In this approach, an external biocompatible anchor molecule conjugated with DNA oligonucleotides is anchored in the lipid bilayer membrane of EVs via surface self-assembly for total EV analysis. Then, specific antibody-DNA conjugates are applied to label selective EVs among the total EVs. The DNA-anchored EVs are distributed into microchip chambers with one or fewer EVs per chamber. The signal from the DNA on the EVs is then amplified using a rapid isothermal nucleic acid detection assay [34]. Another well-based approach, designated the proximity-dependent barcoding assay (PBA), was developed for profiling surface proteins on individual EVs [35]. In PBA, the EVs are first captured by cholera toxin subunit B at the bottom of the microwell, and antibody-DNA conjugates are then used to specifically bind to the proteins on the membrane of EVs. Each DNA sequence contains a unique molecular identifier sequence for distinguishing protein molecules on the EVs, and identification of the surface protein composition on individual EVs is finally performed by sequencing the PCR products (Fig. 9.8f).

## 9.3 Microfluidic Nucleic Acid Profiling of EVs

Nucleic acid cargo (e.g., mRNAs, microRNAs [miRNAs], and long noncoding RNAs [lncRNAs]) of EVs can alter recipient cell gene expression and function and serve as critical regulators of multiple biological processes in various diseases, such as cancer [11]. EV-associated RNAs are protected from RNase activity by the lipid bilayer, and their expression can reflect the physiopathological state of parent cells, rendering EVs attractive sources of RNA disease biomarkers [38, 39]. Although quantitative reverse transcription-PCR (qRT-PCR) remains the gold standard for RNA profiling, its application to EV-associated RNA analysis generally involves EV lysis, RNA extraction, cDNA conversion, and enzymatic amplification, which are laborious, sample-consuming, and prone to contamination from nonvesicular RNAs [11]. Microfluidics provides a promising avenue for detecting EV-associated RNAs by offering a multitude of advantages, including high-level integration, small sample consumption, and feasible implementation of diverse sensing strategies [11]. As such, various microfluidic sensors have been developed for rapid, sensitive detection of EV-associated RNAs, which can be categorized into two types: ex situ RNA detection from EV lysates and in situ detection directly from EVs.

### 9.3.1 Ex Situ Detection of EV RNAs Using Microfluidic Platforms

By using diverse signal amplification and detection strategies, microfluidic platforms have been applied for the detection of EV-associated RNAs with high sensitivity and specificity. Integrated microfluidic platforms with streamlined procedures have also been developed to realize stand-alone, single-chip analysis or digital analysis of EV-associated RNAs.

#### 9.3.1.1 Surface-Enhanced Raman Scattering (SERS) Sensing

MiRNAs are noncoding RNA species (~22 nt) that posttranscriptionally regulate gene expression through binding to complementary sequences of mRNAs. MiRNAs can be transported by EVs for intercellular communication, and EV-associated miRNAs have been recognized as important disease biomarkers [11]. However, the development of sensitive diagnostic tools for the accurate measurement of EV miRNAs is challenging because of the low abundance of miRNAs and high sequence similarity between miRNA family groups. Recently, microfluidics coupled with SERS has been developed for multiplex detection of exosomal miRNAs with high sensitivity and specificity (Fig. 9.9a) [40]. This SERS sensor contains a substrate with highly uniform plasmonic gold nanopillars. The nanopillars are self-closed in a solvent driven (with a gap of less than 10 nm) by capillary force to create multiple hotspots for enhanced coupling of local surface plasmons [44]. For detection of the target miRNA, a sandwich hybridization method using two short locked nucleic acid (LNA) probes was proposed. One LNA probe was modified on the SERS substrate for capture of the target miRNA, and the other was conjugated with a Cy3 fluorophore for signal readout. The sandwiched hybridization was performed at

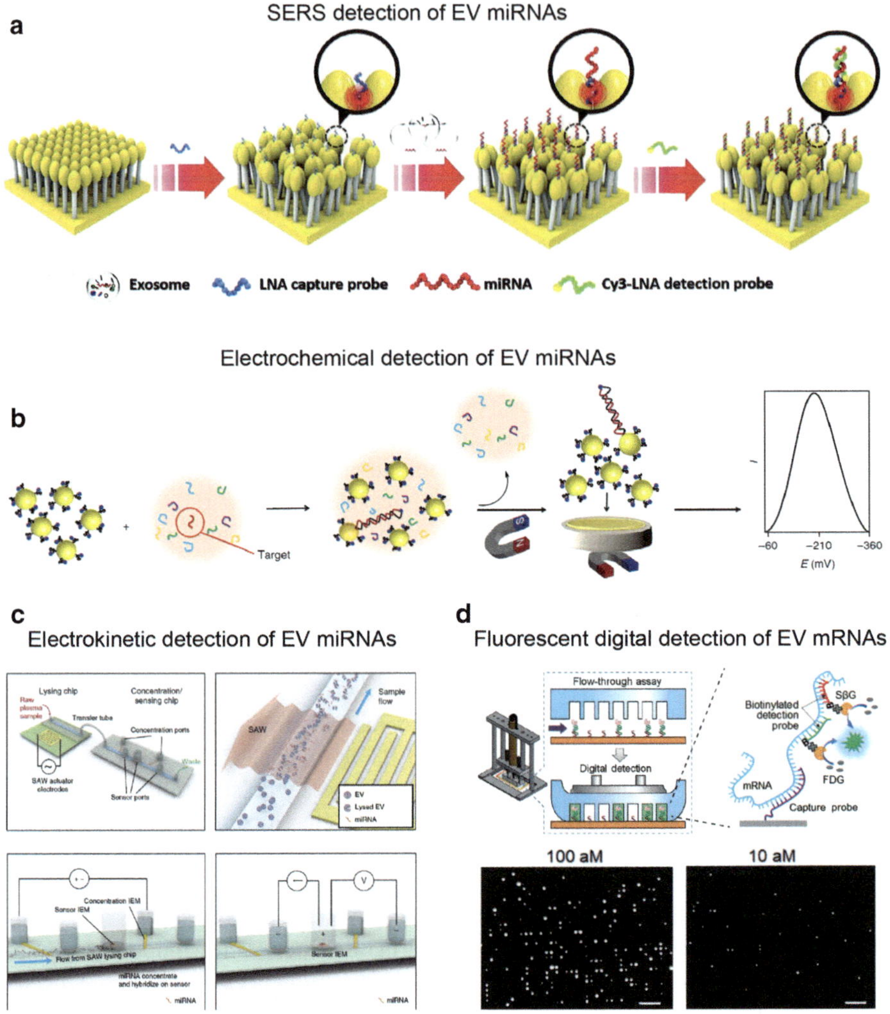


**Fig. 9.9** Ex situ detection of EV RNAs using microfluidic platforms. (**a**) Microfluidics coupled with surface-enhanced Raman scattering (SERS) for detecting exosomal miRNAs. Reprinted with permission [40]. *(Copyright 2019, Wiley-VCH)*. (**b**) Electrochemical detection of exosomal miRNAs based on an electrically reconfigurable network of gold-coated magnetic nanoparticles. Reprinted with permission [41]. *(Copyright 2018, Nature Publishing Group)*. (**c**) EV miRNA detection using a microfluidic platform consisting of a surface acoustic wave (SAW) EV lysing chip and an electrokinetic membrane sensing chip. Reprinted with permission [42]. *(Copyright 2019, Nature Publishing Group)*. (**d**) Microfluidic digital chip for absolute quantification of mutant mRNAs in tumor-derived EVs. [Reprinted with permission [43]. *(Copyright 2018, Royal Society of Chemistry)*]

64 °C to increase target selectivity, enabling precise discrimination among single-base mismatch-containing miRNAs. The SERS sensor showed a limit of detection (LoD) of 1 aM with a wide dynamic range from 1 aM to 100 nM. Because of the

unique fingerprinting of Raman spectra of individual molecules [11], simultaneous detection of multiple miRNA analytes was achieved. The expression levels of exosomal *miR-21*, *miR-222*, and *miR-200c* detected by SERS were used to discriminate different molecular subtypes of breast cancer cell lines.

#### 9.3.1.2 SPR Sensing

SPR imaging (SPRi) is a sensitive, high-throughput biosensing technique that can detect various analytes of clinical interest in real time [11]. Recently, an SPRi sensor was developed for simultaneous detection of exosomal miRNAs derived from clinical samples [45]. The substrate of the SPRi sensor was functionalized with a DNA tetrahedral framework (DTF) that contained a single-stranded DNA (ssDNA) capture arm showing partial complementarity to the target miRNA. Notably, the DTF-functionalized substrate exhibited good capacity for resisting nonspecific absorption of complex physiological matrices compared with its ssDNA-functionalized counterpart [46]. After the capture of target miRNA on the sensor substrate, silver nanocubes (AgNCs) functionalized with ssDNA were used to hybridize with the other part of the miRNA, followed by assembly of ssDNA-coated Au nanoparticles (AuNPs) on the surface of AgNCs to form Au-on-Ag heterostructures for SPR signal amplification. With the aid of the Au-on-Ag heterostructure and DTF, this SPRi sensor achieved a wide linear range from 2 fM to 20 nM with a low LoD of 1.68 fM. The use of different DTFs on multiple detection sites of one gold substrate allows for multiplex detection of exosomal miRNAs by a single sensor. Applying this technique to exosomes purified from the plasma of patients with NSCLC and healthy donors, four miRNAs (*miR-21*, *miR-378*, *miR-200*, and *miR-139*) were found to be overexpressed in patients with NSCLC, suggesting that these miRNAs could be used for accurate discrimination of these patients.

#### 9.3.1.3 Electrochemical Sensing

Electrochemical assays have emerged as a versatile platform for direct analysis of RNAs with low detection limits. For example, an electrically reconfigurable DNA-gold-coated MNP (DNA-Au@MNP) network has been developed for detecting exosomal miRNAs with high sensitivity (Fig. 9.9b) [41]. DNA-Au@MNPs are functionalized with DNA probes that are complementary to the target miRNA and labeled with methylene blue as the electrochemical transduction element. The DNA-Au@MNPs then act as dispersible electrodes to capture target miRNAs in a homogeneous solution of the exosome RNA extract, allowing for more efficient target recognition and mass transfer compared with conventional assays, which rely on analyte diffusion to the sensor interface. Using an external magnetic field, DNA-Au@MNPs with a captured miRNA target can be separated from the complex sample matrix and collected onto a gold electrode for electrical reconfiguration through square-wave voltammetry. The formation of a double helix on the target-captured DNA-Au@MNP surface can increase the interparticle distance, resulting in an increased electron transfer length during the redox reaction to impede electron tunneling through the formed Au@MNP network [47]. Therefore, ultrasensitive quantification of miRNAs can be achieved by measuring changes in the current.

This sensor shows a wide linear range across eight orders of concentrations (10 aM to 1 nM). Moreover, the sensor was able to detect changes in the concentrations of A549 exosomal *miR-21* upon transfection of a *miR-21* single-strand oligonucleotide inhibitor into parental A549 cells.

#### 9.3.1.4 An Integrated Platform with Acoustic EV Lysis and Electrokinetic Sensing

For one-step quantification of EV miRNAs, an integrated microfluidic platform consisting of a surface acoustic wave (SAW) lysing chip and an miRNA concentration/sensing chip has been developed (Fig. 9.9c) [42]. An acoustofluidic chip was used for reagent-free EV lysis upon mechanical EV membrane rupture [48]. This chip was constructed by building a polycarbonate microchannel on the surface of a lithium niobate ($LiNbO_3$) piezoelectric substrate that was patterned with interdigitated electrodes. The generated SAWs refracted into the microchannel, leading to EV lysis by inducing acoustic pressure [11]. The lysed samples were subsequently transferred to the second chip containing two sets of ion-exchange membranes (IEMs). By applying an electric field along the microchannel, the first set of IEMs (concentration IEMs) with negative surface charge could continuously intercept negatively charged nucleic acids from lysed samples, resulting in an increase in the nucleic acid concentration by up to three orders of magnitude [49]. The second set of IEMs (sensing IEMs) functionalized with an oligonucleotide antisense probe was arranged right above the concentration position. Hybridization of the target miRNA with the probe reduced the ion depletion action and resulted in a voltage shift in the current response curve, enabling the quantification of target miRNA. This integrated platform achieved an LoD of 1 pM from a small sample volume (20 μL). The platform identified the overexpression of *miR-21* in untreated human plasma samples from patients with liver cancer ($n = 3$), revealing a nearly 13-fold increase in *miR-21* expression compared with that in healthy donors ($n = 3$).

#### 9.3.1.5 Microfluidic Digital Detection

Digital PCR (dPCR) is a powerful technology for sensitive, absolute quantification of nucleic acids by partitioning a sample into many individual, parallel PCR assays. For high-throughput, on-chip analysis of EV lncRNAs, a multicolor fluorescence dPCR-based EV lncRNA (miDER) assay has been developed [50]. Multitarget detection was accomplished in a single amplification, and different target cDNAs produced varied fluorescence signals during PCR amplification. The miDER assay displayed a wide linear range from 1 to $2.9 \times 10^4$ copies/μL, with an LoD of 10 copies/μL of the target gene from peripheral blood. The quantification results by miDER were consistent with those obtained by qRT-PCR. Using this multiplex assay, the expression levels of two lung cancer-related genes (*SLC9A3-AS1* and *PCAT6*) were detected in patients with lung cancer ($n = 32$) and in healthy controls ($n = 30$), and a significant difference was observed between the two groups. The combination of two lncRNAs in the miDER assay yielded a high area under the curve (AUC) value of 0.811 (95% confidence interval, 0.705–0.918).

The absolute quantification of mutant mRNAs in tumor-derived EVs has also been explored using microfluidic digital chips in a PCR-free manner (Fig. 9.9d) [43]. The EV lysate was injected into the microfluidic chip, in which the target mRNA was captured by functionalization of the substrate with complementary DNA probes. Reporter probes conjugated with β-galactosidase were used to detect the target mRNA by catalyzing the conversion of the nonfluorescent FDG substrate into fluorescein. For digital detection, the top slide of the microfluidic chip was tightly sealed with the substrate, creating multiple isolated microreactors (with a volume in the fL range). This integrated microfluidic platform was able to quantify as little as 6.5 and 0.3 copies of *EWS-FLI1* fusion transcripts per $10^5$ EVs derived from two Ewing sarcoma cell lines (CHLA-9 and CHLA-258 cells).

## 9.3.2 In Situ Fluorescence Detection of EV Nucleic Acids

In addition to ex situ analysis of RNA from EV lysates, microfluidic platforms complemented with sophisticated DNA probes and advanced signal amplification/detection strategies have been applied for in situ detection of EV-associated RNAs without the need for EV lysis or elimination of interference from nonvesicular RNAs.

### 9.3.2.1 Detection of EV RNAs Using Virus-Mimicking Fusogenic Vesicles

To enable the in situ detection of target EV RNAs, fusogenic vesicles have been used for efficient delivery of fluorescence probes into EVs [11]. Recent work has shown that a biochip combined with cationic lipid-polymer hybrid nanoparticles (LPHNs) can be used for in situ detection of EV mRNAs in a single step (Fig. 9.10a) [51]. A gold substrate immobilized with LPHNs could capture negatively charged EVs via electrostatic interactions to form LPHN-EV complexes. Moreover, a catalyzed hairpin DNA circuit (CHDC) preloaded into the LPHNs could be transferred into the EV lumen for mRNA detection. The CHDC consisted of two hairpin DNAs (H1 and H2) for target mRNA recognition and catalytic hybridization and a reporter with self-quenched fluorescence for signal readout. One target mRNA can catalyze multiple cycles of H1–H2 hybridization through toehold-mediated strand displacement reactions and further destabilize multiple reporters to generate an amplified fluorescence signal readout. This LPHN-CHDC biochip achieved a low LoD of 6.60 pg (298 amol) for the detection of the *GPC-1* ssDNA oligo. This assay was used to detect EV *GPC-1* mRNA levels directly in serum samples from patients with pancreatic ductal adenocarcinoma (PDAC) at stages I–II ($n$ = 111) and III–IV ($n$ = 55), benign pancreatic disease (BPD, $n$ = 23; patients with pancreatitis), and healthy donors ($n$ = 75). Analysis of *GPC-1* mRNA levels enabled good discrimination of patients with PDAC (stages I–IV) from healthy donors and patients with BPD with 100% sensitivity and specificity and an AUC of 1.0.

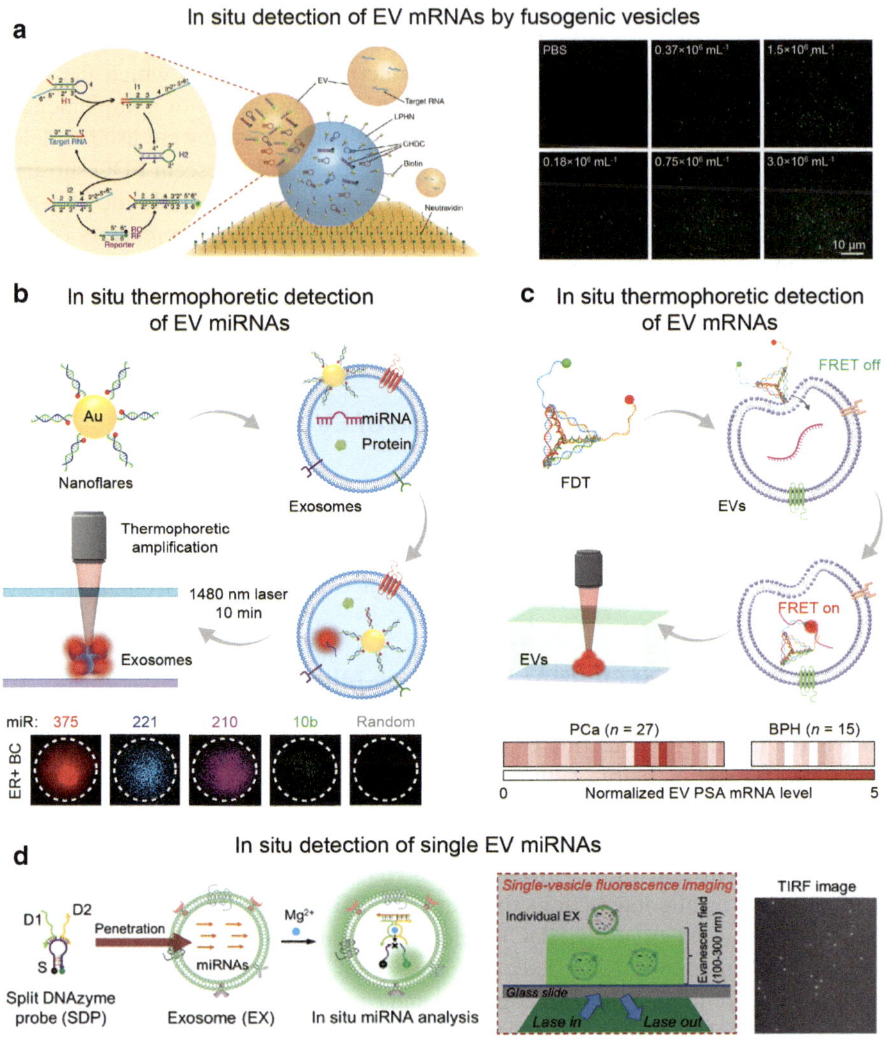


**Fig. 9.10** In situ detection of EV RNAs using microfluidic platforms. (**a**) In situ detection of EV mRNAs using microfluidic biochips combined with cationic lipid-polymer hybrid nanoparticles (LPHNs) and a catalyzed hairpin DNA circuit (CHDC). Reproduced with permission [51]. *(Copyright 2017, Nature Publishing Group)*. (**b**) Thermophoretic sensor combined with nanoflares (TSN) for in situ detection of exosomal miRNAs. Reproduced with permission [52]. *(Copyright 2020, American Chemical Society)*. (**c**) Thermophoretic sensor combined with a FRET-based DNA tetrahedron (FDT) for in situ detection of EV mRNAs. Reproduced with permission [53]. *(Copyright 2021, Elsevier)*. (**d**) In situ analysis of exosomal miRNAs using a split DNAzyme probe (SDP) and TIRF imaging. [Reproduced with permission [54]. *(Copyright 2019, Ivyspring International Publisher)*]

#### 9.3.2.2 Thermophoretic Detection of EV RNAs

Thermophoresis refers to directed migration of particles in a temperature gradient, and thermophoretic detection has emerged as an efficient strategy for manipulation and identification of EVs and viruses [11]. Sun et al. developed a thermophoretic sensor combined with nanoflares (TSN) for in situ detection of exosomal miRNAs with high sensitivity and specificity (Fig. 9.10b) [52]. Nanoflares were constructed by functionalizing a 13-nm spherical AuNP with antisense ssDNA complementary to the target miRNA. The ssDNA was prehybridized to a Cy5-modified reporter flare whose fluorescence was quenched due to the proximity to AuNPs [55]. Nanoflares can be passively transported into exosomes by incubation, generating a fluorescence signal upon displacement of the reporter flare by the target miRNA. Under concomitant thermophoresis and thermal convection, exosomes encapsulated with nanoflares can be significantly enriched toward the laser spot center at the bottom of a microchamber, resulting in an amplified fluorescence signal. Notably, thermophoretic accumulation of EVs is size-dependent, whereas circulating small miRNAs cannot be enriched to generate a fluorescence signal. Therefore, the fluorescence intensity is exclusively related to the expression level of the target miRNA in exosomes. The TSN showed a good linear response to exosomal *miR-375* in the range between 0.17 and 170 fM, with an LoD of 0.36 fM, comparable to that of qRT-PCR. Moreover, exosomal *miR-375* achieved an accuracy of 85% for detecting patients with estrogen receptor-positive breast cancer at early stages (stages I, II).

For in situ measurement of EV mRNAs, an ultrasensitive DNA tetrahedron-based thermophoretic assay has been developed (Fig. 9.10c) [53]. To increase the EV uptake efficiency of nanoprobes, a fluorescence resonance energy transfer (FRET)-based DNA tetrahedron (FDT) with high structural rigidity was constructed, showing a high internalization rate of approximately 50%. After EV uptake of FDTs, the two Cy3- or Cy5-modified recognition sequences of the FDT could bind to target mRNA, leading to the generation of a FRET signal owing to the close proximity of the two fluorophores. The FRET signal in EVs was significantly amplified by size-selective thermophoretic accumulation, yielding an LoD of 14 aM for detecting *PSA* mRNA in serum EVs. The level of EV *PSA* mRNA was used to discriminate patients with prostate cancer ($n = 27$) from those with benign prostatic hyperplasia ($n = 15$) with an AUC of 0.93. This was superior to serum PSA protein (AUC = 0.74), which is the current gold standard in prostate cancer screening.

#### 9.3.2.3 Detection of Single EV miRNAs

A total internal reflection fluorescence (TIRF)-based single-vesicle imaging assay was developed for direct visualization and quantification of exosomal miRNAs at the single-vesicle level in serum samples (Fig. 9.10d) [54]. Exosomes were first treated with streptolysin O for membrane permeabilization, followed by delivery of a split DNAzyme probe (SDP, consisting of two divided DNAzyme fragments and a ribonucleobase-containing substrate) into the exosomes. The hybridization of target miRNA with SDP activated the DNAzyme for catalytic cleavage of the substrate with the aid of $Mg^{2+}$, producing a fluorescence signal readout that could be monitored under a TIRF imaging system. Because the evanescent field generated by TIRF showed a penetration depth of approximately 100–300 nm, only the exosomes entering the confined evanescent field layer could be excited, thus eliminating the background signal from bulk solution. Each individual bright spot in TIRF images represented a single fluorescent exosome, allowing for in situ quantitative and stoichiometric analysis of exosomal miRNAs by simply counting the number and fluorescence intensity of spots in the sample region. This TIRF imaging assay exhibited a linear detection range from $3 \times 10^4$ to $10^7$ copies/μL and an LoD of 378 copies/μL of exosomal *miR-21*. This method was then used to detect exosomal *miR-21* directly from serum samples, and the expression of this miRNA was able to distinguish patients with cancer from healthy donors. Additionally, the exosomal *miR-21* level in serum was also informative for monitoring tumor progression and responses to treatment in patients with different cancer types (e.g., melanoma, breast cancer, and cervical cancer).

Quantitative analysis of EV-associated RNAs may be a promising approach for noninvasive, precise, and sensitive cancer diagnostics. Microfluidic platforms combined with state-of-the-art signal transduction and amplification strategies provide a versatile toolbox for EV RNA profiling with high sensitivity and specificity. Moreover, microfluidics can be used to analyze single EVs, revealing their heterogeneity and facilitating the development of new strategies for cancer diagnostics. Before being translated into real clinical practice, additional studies are needed to validate these microfluidic sensing platforms using large clinical cohorts.

## References

1. Chen C, Skog J, Hsu C-H, Lessard RT, Balaj L, Wurdinger T, et al. Microfluidic isolation and transcriptome analysis of serum microvesicles. Lab Chip. 2010;10(4):505–11.
2. Kanwar SS, Dunlay CJ, Simeone DM, Nagrath S. Microfluidic device (ExoChip) for on-chip isolation, quantification and characterization of circulating exosomes. Lab Chip. 2014;14(11):1891–900.
3. Lo T-W, Zhu Z, Purcell E, Watza D, Wang J, Kang Y-T, et al. Microfluidic device for high-throughput affinity-based isolation of extracellular vesicles. Lab Chip. 2020;20(10):1762–70.
4. Zhang P, He M, Zeng Y. Ultrasensitive microfluidic analysis of circulating exosomes using a nanostructured graphene oxide/polydopamine coating. Lab Chip. 2016;16(16):3033–42.
5. Zhao Z, Yang Y, Zeng Y, He M. A microfluidic ExoSearch chip for multiplexed exosome detection towards blood-based ovarian cancer diagnosis. Lab Chip. 2016;16(3):489–96.
6. Tayebi M, Zhou Y, Tripathi P, Chandramohanadas R, Ai Y. Exosome purification and analysis using a facile microfluidic hydrodynamic trapping device. Anal Chem. 2020;92(15):10733–42.
7. Woo H-K, Sunkara V, Park J, Kim T-H, Han J-R, Kim C-J, et al. Exodisc for rapid, size-selective, and efficient isolation and analysis of nanoscale extracellular vesicles from biological samples. ACS Nano. 2017;11(2):1360–70.
8. Kim D, Woo H-K, Lee C, Min Y, Kumar S, Sunkara V, et al. EV-Ident: identifying tumor-specific extracellular vesicles by size fractionation and single-vesicle analysis. Anal Chem. 2020;92(8):6010–8.
9. Yeh Y-T, Zhou Y, Zou D, Liu H, Yu H, Lu H, et al. Rapid size-based isolation of extracellular vesicles by three-dimensional carbon nanotube arrays. ACS Appl Mater Interfaces. 2020;12(11):13134–9.

10. Wang Z, Wu H-J, Fine D, Schmulen J, Hu Y, Godin B, et al. Ciliated micropillars for the microfluidic-based isolation of nanoscale lipid vesicles. Lab Chip. 2013;13(15):2879–82.
11. Wu M, Ouyang Y, Wang Z, Zhang R, Huang P-H, Chen C, et al. Isolation of exosomes from whole blood by integrating acoustics and microfluidics. Proc Natl Acad Sci U S A. 2017;114(40):10584–9.
12. Lee K, Shao H, Weissleder R, Lee H. Acoustic purification of extracellular microvesicles. ACS Nano. 2015;9(3):2321–7.
13. Ibsen SD, Wright J, Lewis JM, Kim S, Ko S-Y, Ong J, et al. Rapid isolation and detection of exosomes and associated biomarkers from plasma. ACS Nano. 2017;11(7):6641–51.
14. Lewis JM, Vyas AD, Qiu Y, Messer KS, White R, Heller MJ. Integrated analysis of exosomal protein biomarkers on alternating current electrokinetic chips enables rapid detection of pancreatic cancer in patient blood. ACS Nano. 2018;12(4):3311–20.
15. Zhang H, Lyden D. Asymmetric-flow field-flow fractionation technology for exomere and small extracellular vesicle separation and characterization. Nat Protoc. 2019;14(4):1027–53.
16. Wunsch BH, Smith JT, Gifford SM, Wang C, Brink M, Bruce RL, et al. Nanoscale lateral displacement arrays for the separation of exosomes and colloids down to 20 nm. Nat Nanotechnol. 2016;11(11):936–40.
17. Liu C, Guo J, Tian F, Yang N, Yan F, Ding Y, et al. Field-free isolation of exosomes from extracellular vesicles by microfluidic viscoelastic flows. ACS Nano. 2017;11(7):6968–76.
18. Huang LR, Cox EC, Austin RH, Sturm JC. Continuous particle separation through deterministic lateral displacement. Science. 2004;304(5673):987–90.
19. Liu C, Zhao J, Tian F, Chang J, Zhang W, Sun J. λ-DNA- and aptamer-mediated sorting and analysis of extracellular vesicles. J Am Chem Soc. 2019;141(9):3817–21.
20. Vaidyanathan R, Naghibosadat M, Rauf S, Korbie D, Carrascosa LG, Shiddiky MJA, et al. Detecting exosomes specifically: a multiplexed device based on alternating current electrohydrodynamic induced nanoshearing. Anal Chem. 2014;86(22):11125–32.
21. Cho S, Jo W, Heo Y, Kang JY, Kwak R, Park J. Isolation of extracellular vesicle from blood plasma using electrophoretic migration through porous membrane. Sens Actuators B Chem. 2016;233:289–97.
22. Mark D, Haeberle S, Roth G, von Stetten F, Zengerle R. Microfluidic lab-on-a-chip platforms: requirements, characteristics and applications. Chem Soc Rev. 2010;39(3):1153–82.
23. Im H, Shao H, Park YI, Peterson VM, Castro CM, Weissleder R, et al. Label-free detection and molecular profiling of exosomes with a nano-plasmonic sensor. Nat Biotechnol. 2014;32(5):490–5.
24. Kang Y-T, Purcell E, Hadlock T, Lo T-W, Mutukuri A, Jolly S, et al. Multiplex isolation and profiling of extracellular vesicles using a microfluidic DICE device. Analyst. 2019;144(19):5785–93.
25. Shao H, Chung J, Balaj L, Charest A, Bigner DD, Carter BS, et al. Protein typing of circulating microvesicles allows real-time monitoring of glioblastoma therapy. Nat Med. 2012;18(12):1835–40.
26. Kabe Y, Suematsu M, Sakamoto S, Hirai M, Koike I, Hishiki T, et al. Development of a highly sensitive device for counting the number of disease-specific exosomes in human sera. Clin Chem. 2018;64(10):1463–73.
27. Sakamoto S, Kabe Y, Hatakeyama M, Yamaguchi Y, Handa H. Development and application of high-performance affinity beads: toward chemical biology and drug discovery. Chem Rec. 2009;9(1):66–85.
28. Daaboul GG, Gagni P, Benussi L, Bettotti P, Ciani M, Cretich M, et al. Digital detection of exosomes by interferometric imaging. Sci Rep. 2016;6(1):37246.
29. Friedrich R, Block S, Alizadehheidari M, Heider S, Fritzsche J, Esbjörner EK, et al. A nano flow cytometer for single lipid vesicle analysis. Lab Chip. 2017;17(5):830–41.
30. Lee K, Fraser K, Ghaddar B, Yang K, Kim E, Balaj L, et al. Multiplexed profiling of single extracellular vesicles. ACS Nano. 2018;12(1):494–503.

31. Beekman P, Enciso-Martinez A, Rho HS, Pujari SP, Lenferink A, Zuilhof H, et al. Immunocapture of extracellular vesicles for individual multi-modal characterization using AFM, SEM and Raman spectroscopy. Lab Chip. 2019;19(15):2526–36.
32. Raghu D, Christodoulides JA, Christophersen M, Liu JL, Anderson GP, Robitaille M, et al. Nanoplasmonic pillars engineered for single exosome detection. PLoS One. 2018;13(8):e0202773.
33. Liu C, Xu X, Li B, Situ B, Pan W, Hu Y, et al. Single-exosome-counting immunoassays for cancer diagnostics. Nano Lett. 2018;18(7):4226–32.
34. Tian Q, He C, Liu G, Zhao Y, Hui L, Mu Y, et al. Nanoparticle counting by microscopic digital detection: selective quantitative analysis of exosomes via surface-anchored nucleic acid amplification. Anal Chem. 2018;90(11):6556–62.
35. Wu D, Yan J, Shen X, Sun Y, Thulin M, Cai Y, et al. Profiling surface proteins on individual exosomes using a proximity barcoding assay. Nat Commun. 2019;10(1):3854.
36. Mazutis L, Gilbert J, Ung WL, Weitz DA, Griffiths AD, Heyman JA. Single-cell analysis and sorting using droplet-based microfluidics. Nat Protoc. 2013;8(5):870–91.
37. Ko J, Wang Y, Carlson JCT, Marquard A, Gungabeesoon J, Charest A, et al. Single extracellular vesicle protein analysis using immuno-droplet digital polymerase chain reaction amplification. Adv Biosyst. 2020;4(12):1900307.
38. Stoorvogel W. Functional transfer of microRNA by exosomes. Blood. 2012;119(3):646–8.
39. Möller A, Lobb RJ. The evolving translational potential of small extracellular vesicles in cancer. Nat Rev Cancer. 2020;20(12):697–709.
40. Lee JU, Kim WH, Lee HS, Park KH, Sim SJ. Quantitative and specific detection of exosomal miRNAs for accurate diagnosis of breast cancer using a surface-enhanced Raman scattering sensor based on plasmonic head-flocked gold nanopillars. Small. 2019;15(17):1804968.
41. Tavallaie R, McCarroll J, Le Grand M, Ariotti N, Schuhmann W, Bakker E, et al. Nucleic acid hybridization on an electrically reconfigurable network of gold-coated magnetic nanoparticles enables microRNA detection in blood. Nat Nanotechnol. 2018;13(11):1066–71.
42. Ramshani Z, Zhang C, Richards K, Chen L, Xu G, Stiles BL, et al. Extracellular vesicle microRNA quantification from plasma using an integrated microfluidic device. Commun Biol. 2019;2(1):189.
43. Zhang P, Crow J, Lella D, Zhou X, Samuel G, Godwin AK, et al. Ultrasensitive quantification of tumor mRNAs in extracellular vesicles with an integrated microfluidic digital analysis chip. Lab Chip. 2018;18(24):3790–801.
44. Ding S-Y, Yi J, Li J-F, Ren B, Wu D-Y, Panneerselvam R, et al. Nanostructure-based plasmon-enhanced Raman spectroscopy for surface analysis of materials. Nat Rev Mater. 2016;1(6):16021.
45. Wu W, Yu X, Wu J, Wu T, Fan Y, Chen W, et al. Surface plasmon resonance imaging-based biosensor for multiplex and ultrasensitive detection of NSCLC-associated exosomal miRNAs using DNA programmed heterostructure of Au-on-Ag. Biosens Bioelectron. 2021;175:112835.
46. Nie W, Wang Q, Zou L, Zheng Y, Liu X, Yang X, et al. Low-fouling surface plasmon resonance sensor for highly sensitive detection of microRNA in a complex matrix based on the DNA tetrahedron. Anal Chem. 2018;90(21):12584–91.
47. LMH L, Goon IY, Chuah K, Lim M, Braet F, Amal R, et al. The biochemiresistor: an ultrasensitive biosensor for small organic molecules. Angew Chem Int Ed. 2012;51(26):6456–9.
48. Liu C, Zhang W, Li Y, Chang J, Tian F, Zhao F, et al. Microfluidic sonication to assemble exosome membrane-coated nanoparticles for immune evasion-mediated targeting. Nano Lett. 2019;19(11):7836–44.
49. Slouka Z, Senapati S, Chang H-C. Microfluidic systems with ion-selective membranes. Annu Rev Anal Chem. 2014;7(1):317–35.
50. Bai Y, Qu Y, Wu Z, Ren Y, Cheng Z, Lu Y, et al. Absolute quantification and analysis of extracellular vesicle lncRNAs from the peripheral blood of patients with lung cancer based on multi-colour fluorescence chip-based digital PCR. Biosens Bioelectron. 2019;142:111523.
51. Hu J, Sheng Y, Kwak KJ, Shi J, Yu B, Lee LJ. A signal-amplifiable biochip quantifies extracellular vesicle-associated RNAs for early cancer detection. Nat Commun. 2017;8(1):1683.

52. Zhao J, Liu C, Li Y, Ma Y, Deng J, Li L, et al. Thermophoretic detection of exosomal microRNAs by nanoflares. J Am Chem Soc. 2020;142(11):4996–5001.
53. Han Z, Wan F, Deng J, Zhao J, Li Y, Yang Y, et al. Ultrasensitive detection of mRNA in extracellular vesicles using DNA tetrahedron-based thermophoretic assay. Nano Today. 2021;38:101203.
54. He D, Wang H, Ho S-L, Chan H-N, Hai L, He X, et al. Total internal reflection-based single-vesicle in situ quantitative and stoichiometric analysis of tumor-derived exosomal microRNAs for diagnosis and treatment monitoring. Theranostics. 2019;9(15):4494–507.
55. Zhai L-Y, Li M-X, Pan W-L, Chen Y, Li M-M, Pang J-X, et al. In situ detection of plasma exosomal microRNA-1246 for breast cancer diagnostics by a Au nanoflare probe. ACS Appl Mater Interfaces. 2018;10(46):39478–86.

# Biosensing Technologies for Detection of Extracellular Vesicles

# 10

Bo Li, Rongrong Huang, Xinyu Li, Ye Zhang, Boyue Fan, Kaili Di, Zhiyang Li, Wei Cheng, and Shijia Ding

## 10.1 Introduction

With advancements in biomedical technologies, the requirements for high-sensitivity and high-specificity detection of tumor-derived exosomes are becoming increasingly rigorous owing to the essential roles of these approaches in the early diagnosis of cancer. Some traditional detection methods described in previous chapters, such as western blotting, nanoparticle tracking analysis (NTA), enzyme-linked immunosorbent assay (ELISA), and flow cytometry [1], have several disadvantages. Thus, a specific, sensitive, and rapid detection platform is essential for clinical point-of-care diagnosis [2]. In recent years, researchers have developed novel technologies for the detection of exosomes. Among these, biosensor technology is

---

Bo Li, Rongrong Huang and Xinyu Li contributed equally to this work.

B. Li · Y. Zhang
Department of Laboratory Medicine, Nanfang Hospital, Southern Medical University, Guangzhou, China

R. Huang · B. Fan · K. Di · Z. Li (✉)
Department of Clinical Laboratory, The Affiliated Drum Tower Hospital of Nanjing University Medical School, Nanjing, China
e-mail: lizhiyang@nju.edu.cn

X. Li · S. Ding
Key Laboratory of Clinical Laboratory Diagnostics (Ministry of Education), College of Laboratory Medicine, Chongqing Medical University, Chongqing, China

W. Cheng (✉)
The Center for Clinical Molecular Medical detection, The First Affiliated Hospital of Chongqing Medical University, Chongqing, China
e-mail: chengwei@cqmu.edu.cn



Q. Wang, L. Zheng (eds.), *Extracellular Vesicles*,
https://doi.org/10.1007/978-981-99-8365-0_10

favored by researchers because of its unique advantages. In this chapter, we introduce biosensor technology and its applications in exosome detection.

Biosensor technology is an analytical tool or system that combines biologically active materials as recognition elements, with physical transducers and signal amplifiers. The bioreceptor specifically recognizes the target analyte, and the transducer converts this response into different types of energy that are amplified, processed, and converted into the desired signal format [3]. A major challenge in biosensor applications is achieving effective signal capture of biometric transduction events. The recognition element is critical for the validity of the capture. Typically, recognition elements include enzymes, proteins, DNA, antibodies, antigens, aptamers, and biofilms [4]. Aptamers are short single-stranded oligonucleotides with unique tertiary structures that bind to the target molecule with high affinity and specificity. Compared with antibodies, aptamers are superior in terms of easy synthesis, simple modification, low cost, and high stability in body fluids. Moreover, aptamers have unique properties, such as small size (2–3 nm), favorable affinity, and good specificity, making them ideal recognition ligands for exosomes [5, 6]. Biosensor methods have been favored by researchers because they offer an approach for simple, rapid, and low-cost diagnostic testing in environmental monitoring, food, medicine, and military medicine. In particular, the application of biosensors in clinical cancer diagnosis has several considerable advantages, including higher assay speed and flexibility, multitarget analysis capability, and low cost [7]. Biosensors have been developed for several clinical applications owing to their potential to promote point-of-care testing (POCT) [8].

In recent years, many approaches based on electrochemistry, chemiluminescence (CL), fluorescence, colorimetry, surface plasmon resonance (SPR), surface-enhanced Raman spectroscopy (SERS), and other biosensors have been used to sense exosomes and exosomal biomarkers [9]. Electrochemistry biosensors are promising tools that use electrodes (solid electrodes, ion-selective electrodes, gas-sensitive electrodes, etc.) as conversion elements for the detection of molecular reactions by measuring solely the current and/or voltage. Electrochemical methods have shown advantages such as simple operation, cost-effectiveness, rapid analysis, and easy miniaturization for exosome detection [10–12]. CL is a trace analysis method that identifies a substance based on light radiation produced by the redox reaction. CL has many advantages in the detection of exosomes; for example, the approach is light source-free and highly sensitive and can be performed using common instruments [13]. Another highly sensitive readout approach for biosensors is fluorescence [14]. Fluorescent nanomaterials, such as metal nanoparticles (NPs), quantum dots (QDs), upconversion NPs, and other small fluorescent molecules, have been used in biosensors based on fluorescence resonance energy transfer and CL resonance energy transfer [15, 16]. Because of its advantages of high sensitivity and ease of operation, fluorescence-based biosensors have been extensively applied in the field of exosomal detection and imaging [9]. The results obtained from colorimetric biosensors can be instantly observed with the naked eye without additional analytical equipment [17, 18], a promising approach in POCT. Therefore, researchers have developed many biosensors based on colorimetry for detecting exosomes

[19]. SPR is a biosensor based on the principle that there is a direct correlation between the change in the interface refractive index and the attenuation degree of reflected light formed by incident light of a certain wavelength during biomolecule recognition [20, 21]. SPR shows high sensitivity and can be used in real-time without the need for labeling or the use of excessive reagents; therefore, this approach is thought to be suitable for exosome analysis [22]. SERS is another popular POCT method. The enhancement of Raman intensity relies on localized SPR, which can generate a strong electromagnetic field among metallic NPs. In SERS-based strategies, Raman spectroscopy is used to analyze the chemical composition of exosomes because of its remarkable potential to enhance signals from negligible or low amounts of biomarkers on the exosomal surface [23].

Although biosensors have been used for several clinical applications, few have been developed for exosome-related clinical detection. Harnessing the potential of biosensors is challenging because of the complexity and diversity of tumor-derived exosomes. Successful development of biosensor-based exosome detection will require continued advances in recognition ligands, including non-antibody recognition ligands, advances in exosome isolation and enrichment, and integration of microfluidics or other advanced manufacturing techniques, as well as the development of new, more sensitive, less costly multichannel transducers.

## 10.2 Electrochemical Biosensors

A typical electrochemical biosensor consists of a biological recognition element and electrochemical transducer that converts biological interactions into electrical signals. Owing to their portability, self-containment, and low cost, electrochemical biosensors have been widely applied in the quantitative analysis of EVs in recent years (Table 10.1) [5, 14, 42, 43]. In this section, we discuss research in the fields of proteins, nucleic acids, and other biomarker-based biosensors. In addition, perspectives related to clinical applications are highlighted to inspire readers with broader interests in diverse areas.

### 10.2.1 Protein Biomarker-Based Biosensors

EV membrane surface proteins are common targets for EV electrochemical sensors. Zhou et al. constructed a label-free electrochemical detection system for EVs based on a CD63 aptamer-immobilized gold electrode [24]. Redox moiety-labeled beacons were first hybridized with the aptamer to be anchored on the electrode surface, which was then released in the presence of EVs and simultaneously caused a decrease in the electrochemical signal. Yin et al. also used the CD63 aptamer to capture EVs, and by combining an aptamer recognition-triggered amplification strategy, the detection limit of their sensor was found to be considerably lower than that of the abovementioned assay ($1.2 \times 10^4$/mL compared with $1.2 \times 10^6$/mL) [25]. They designed an aptamer probe (P1), a trigger DNA probe (P2), and a hairpin

**Table 10.1** Comparison of different EV electrochemical biosensors

| Methods | Technique | Maker | Detection range | Detection time | Detection limit | References |
|---|---|---|---|---|---|---|
| Release of redox moieties labeled beacons in the presence of EVs and caused a decrease of electrochemical signal | Electrochemical biosensors | CD63 | $10^6$–$10^9$ particles/mL | ~15 min | $1.2 \times 10^6$ particles/mL | [24] |
| Aptamer recognition-trigged enzyme digestion leading to the release of redox-active indicator | Electrochemical biosensors | CD63 | $3.4 \times 10^1$–$3.4 \times 10^5$ particles/mL | 2 h | $1.2 \times 10^4$ particles/mL | [25] |
| CD9 antibody-EVs-HRP conjugated second antibody sandwich structure | Electrochemical biosensors | CD9 | $1 \times 10^2$–$1 \times 10^6$ particles/μL | 1 h | $2 \times 10^2$ particles/μL | [26] |
| CTSDR triggered by CD63 aptamer binding with EVs and served as catalytic molecule machine | Electrochemical biosensors | CD63 | $1 \times 10^5$–$5 \times 10^7$ particles/mL | 2 h | $1.72 \times 10^4$ particles/mL | [1] |
| Two-stage microfluidic | | CD63 | m $7.61 \times 10^4$ particles/mL–$7.61 \times 10^8$ particles/mL | 1 h | $4.39 \times 10^3$ particles/mL | [27] |
| platform integrated on-chip isolation and in situ electrochemical analysis of EVs from serum | Electrochemical biosensors | Programmed death ligand-1 (PD L1) | $1 \times 10^3$–$1 \times 10^{10}$ particles/mL | >4 h | $3.34 \times 10^2$ particles/mL | [28] |
| Catalytic hairpin assembly-triggered DNA walker sensitized with Ag@C core-shell nanocomposites | Electrochemical biosensors | CD63, EpCAM | 100–75,000 particles/μL | >4 h | 40 particles/μL | [29] |
| Metal-organic frameworks encapsulated with methylene blue absorbed on exosome surface | Electrochemical biosensors | Epidermal growth factor receptor (EGFR) | $9.5 \times 10^3 - 1.9 \times 10^7$ particles/μL | 3 h | $7.38 \times 10^3$ particles/μL | [30] |

**Table 10.1** (continued)

| Methods | Technique | Maker | Detection range | Detection time | Detection limit | References |
|---|---|---|---|---|---|---|
| MB-EVs-SiO2 NPs probe separated by magnetic beads followed by the release of ferrocene carboxamide, causing oxidation current signals | Electrochemical biosensors | MUC1, HER2, EpCAM, and CEA | $1.2 \times 10^3$ to $1.2 \times 10^7$ particles/μL | >6 h | – | [31] |
| Gold-loaded nanoporous ferric oxide nanozymes were used to accomplish ELISA-based sensing | Electrochemical biosensors | CD63, placenta alkaline phosphatase | $10^3$–$10^7$ particles/mL | 1–2 h | $10^2$ particles/mL | [32] |
| Electrical current generates when TMB encountered with HRP-labeled EVs | Electrochemical biosensors | CD63, EpCAM, CD24, CA125, HER2, MUC18, EGFR | $10^4$–$10^8$ particles/mL | 1 h | $3 \times 10^4$ EVs | [33] |
| QD nanoparticles in combination with SWASV readout | Electrochemical biosensors | CD63, HER2, and FAM134B | – | 1–2 h | $10^2$ EVs/μL | [34] |
| Detachable microfluidic device implemented with electrochemical apt sensor | Electrochemical biosensors | EpCAM | $1 \times 10^2$–$1 \times 10^9$ EVs/μL | – | 17 EVs/μL | [35] |
| Magnetic EVs-enrichment combined HRP antibody labeling | Electrochemical biosensors | CD9, CD63, CD81, CD24, CD44, CD54, CD326 and CD340 | – | >3 h | $10^5$/μL | [36] |
| EVs captured by aptamer modified circular gold electrodes, and aptamer conjugated AgNPs and CuNPs were applied for electrochemical readout | Electrochemical biosensors | EpCAM, PSMA | – | 2 h | – | [37] |

**Table 10.1** (continued)

| Methods | Technique | Maker | Detection range | Detection time | Detection limit | References |
|---|---|---|---|---|---|---|
| Aptamer recognition-induced multi-DNA release and cyclic enzymatic amplification | Electrochemical biosensors | PSMA | $10^3$–$10^5$ EVs/μL | 6 h | $7 \times 10^2$ EVs/μL | [10] |
| Incorporating an expanded nucleotide-containing aptamer into a DNA NTH structure | electrochemical Biosensors | Hepatocarcinoma cell specific protein | $1 \times 10^4$–$10^{13}$ EVs/mL | 30 min | $2.09 \times 10^4$ EVs/mL | [38] |
| Hemin/G-quadruplex-assisted signal amplification | Electrochemical biosensors | MUC1 | $4.8 \times 10^3$–$4.8 \times 10^6$ EVs/mL | 4 h | $9.54 \times 10^2$ EVs/mL | [11] |
| Concentration-normalized dual-marker analysis | Electrochemical biosensors | CD81, Syntenin-1 protein | – | – | $1.9 \times 10^5$ EVs/mL | [39] |
| Magnetic EVs-enrichment platform combined Ru(bpy)32+-polymer amplification strategy | Electrochemical biosensors | miRNA 21 | $10^3$–$10^6$ EVs/μL | 3–5 h | $10^2$ EVs/μL | [40] |
| Binding of hpDNA and miR-122 triggered HCR and led to the capture of large amounts of electroactive [Ru(NH3)6]3+ | Electrochemical biosensors | miRNA 122 | $100^{-1} \times 10^{11}$ aM | >6 h | 53 aM | [41] |

probe (HP). In the absence of EVs, the electrochemical redox-active indicator anthracycline doxorubicin (DOX) was intercalated into P1–P2 and HP. P2 was released after P1 bound to EVs, yielding the P2–HP duplex, which could be digested by Exo III. The cycle of the digestion procedure caused a large number of free DOX molecules in the solution to produce a strong current signal.

Antibodies have also been used to immobilize EVs. Doldán reported ananti-CD9 antibody EV horseradish peroxidase (HRP)-conjugated secondary antibody sandwich structure for the detection of EVs [26]. The signal amplification of the sensor was achieved by binding multiple detector antibodies to each EV. The sensor exhibited a dynamic range spanning almost four orders of magnitude, with a limit of detection of $2 \times 10^2$/mL.

Detection of specific transmembrane and cytosolic protein biomarkers can be used to identify disease-related EVs and assist with efficient treatment monitoring [30–32]. Boriachek and coworkers reported a nano-immunosensor based on square-wave anodic stripping voltammetry for detection of cancer EVs using QDs as a signal amplifier. EVs were captured using anti-CD63 antibody-modified magnetic beads [34]. Biotinylated monoclonal antibodies targeting family with sequence similarity 134 member B (a colon cancer-specific protein) and human epidermal growth factor receptor-2 (HER2; a breast cancer-specific protein) were functionalized with streptavidin-coated CdSe QDs. The presented method showed a detection limit of 100 particles/μL with a relative standard deviation of less than 5.5% in cancer cell lines.

Wang reported a sensor incorporating an expanded nucleotide-containing aptamer into a DNA nanotetrahedron (NTH) structure for enhanced electrochemical detection of cancer-related EVs [38]. The aptamer was generated via modified in vitro selection of liver cancer cells. The results showed that the NTH-assisted sensor could detect EVs with 100-fold higher sensitivity ($2.09 \times 10^4$/mL) than the single-stranded aptamer-functionalized sensor. In another cancer EV-specific detection method, Huang first demonstrated the overexpression of Mucin1 in gastric cancer cell-derived EVs and developed a hemin/G-quadruplex-assisted signal amplification strategy for the detection of target EVs [11]. Jiang used DNA NTH to profile cancerous exosomal proteins [44]. In their experiment, aptamer-modified DNA NTHs were used as the recognition and capture units, and Au nanoparticle (NP)-DNA conjugates coupled with HRP were used to realize signal amplification. The sensor could profile four types of proteins using the corresponding aptamer.

In addition, microfluidic chips that integrate electrochemical methods with a platform for sample manipulation devices can be advantageous for EV detection [27, 45]. Zhou and coworkers first demonstrated the overexpression of epithelial cell adhesion molecule (EpCAM) and prostate-specific membrane antigen (PSMA) in prostate cancer serum-derived EVs over healthy controls. They then fabricated a sensor chip containing 11 EpCAM aptamer-modified circular gold electrodes capable of specifically capturing cancer EVs [37]. EpCAM- and PSMA-aptamer-conjugated AgNPs and CuNPs were used for the electrochemical readout. The sensor may be able to provide multiplex profiling of surface markers associated with prostate cancer in a minimally invasive manner.

In addition to the external markers (e.g., tetraspanin and EV membrane proteins) described above, internal EV-specific biomarker detection can also be applied in clinical analysis and associated care. However, robust evaluation of EV internal biomarkers has been hindered by issues with sample preparation and assay sensitivity. To solve this problem, Li performed CD81 electroanalysis prior to lysis to normalize exosomal syntenin quantification [39]. Exosomal internal proteins are released prior to interfacial immunocapture, which may expose the sensory interface to growth factors and other proteins. The sensor exhibited a detection limit of $1.9 \times 10^5$ particles/mL for intact exosomes and 3–5 pM for internal exosomal syntenin levels with almost a linear dynamic range of almost five orders.

To be integrated into clinical practice, a method must (1) contain both EV isolation and detection steps [35]; (2) be compatible with multiplexed screening; (3) be fast, cost-effective, and simple; and (4) show good reproducibility and stability. For example, Cao and coworkers enriched EVs on anti-CD63 antibody-functionalized immunobeads and then detected the beads using a DNA chain containing the CD63 aptamer region, which subsequently initiated a catalytic molecule machine that relied on a cascade toehold-mediated strand displacement reaction [1]. Owing to the high efficiency of the amplification strategy, the method showed a detection limit of $1.72 \times 10^4$ particles/mL and high specificity in serum samples.

Jeong and coworkers developed an eight-channel device to profile multiple EV protein markers [33]. The sensor contained two orthogonal modalities: an isolation element using immunomagnetic beads for the capture and HRP enzyme labeling of EVs and an electrochemical detection element to provide information for high-throughput analysis. They applied this device to screen plasma EVs from patients with ovarian cancer. The expression levels of CD63, EpCAM, CD24, and CA125 were measured. The results demonstrated the potential of the sensor to accelerate effective cancer management and provide insights into tumor biology.

Similarly, Moura and coworkers reported an electrochemical sensor for detecting epithelial cell-derived EVs in serum [36]. In their experiment, EVs were preconcentrated using magnetic particles modified with antibodies against general tetraspanin proteins as well as specific receptors of cancer (CD24, CD44, CD54, CD326, and CD340), followed by anti-HRP antibody labeling. The electrochemical sensor showed reliable results for the differentiation of healthy donors from individuals with breast cancer based on specific epithelial biomarkers.

Dong and coworkers reported an original sensor using aptamer-magnetic beads to capture tumor EVs and convert EV detection to nucleic acid detection [10]. Unlike other EV detection methods that also utilize aptamers, in their experiment, three types of messenger DNAs (mDNAs) were hybridized with PSMA aptamers modified on magnetic beads. In the presence of target EVs, the aptamer bound to the EV membrane, leading to the release of mDNAs. Furthermore, they combined an exonuclease III-assisted target recycling strategy with mDNAs to amplify the signal.

### 10.2.2 Nucleic Acid Biomarker-Based Biosensors

EV-containing nucleic acid biomarkers, such as microRNAs (miRNAs) [46–48], circulating tumor DNA, and circular RNA, can shed light on the molecular diagnosis and monitoring of diseases [49, 50]. miRNAs are small, endogenous, noncoding, single-stranded RNA molecules containing approximately 20–24 nucleotides that regulate gene expression post-transcriptionally by binding with protein-coding mRNAs [51]. Evidence has shown that regulation of miRNA expression plays important roles in various diseases, including cancer [52]. To analyze EVs containing *miR-21*, Luo and coworkers designed a locked nucleic acid (LNA)-modified Y shape-like structure. In the presence of miRNA, an LNA-assisted strand displacement reaction occurred, and the "Y" shape-like structure transformed into a hairpin structure, which turned the methylene blue (MB) signal on and the ferrocene signal off. Notably, this biosensor exhibited high accuracy and sensitivity with a limit of detection as low as 2.3 fM. Using another *miR-21*-based sensor, Ghazizadeh and coworkers covalently connected p19 RNA-binding protein with EVs to switch off electrochemical activity. An*miR-21*/probe duplex was added to shift the sensor, providing rapid, label-free detection of EV miRNAs.

Additionally, Guo and coworkers developed a sensitive assay for evaluating *miR-122* concentrations in different cancer-derived exosomes [41]. First, hairpin DNA (hpDNA) probes are immobilized on the electrode surface. Then, binding of hpDNA and *miR-122* triggers a hybridization chain reaction (HCR). Using electroactive $[Ru(NH_3)^6]^{3+}$, which showed a strong binding affinity to DNA duplexes, as a signal reporter, a significant increase in the electrochemical signal was achieved. Exonuclease I was used to cleave the curly single-stranded hpDNA prior to detection. The results showed that the addition of exonuclease I greatly improved the hybridization efficiency, and the detection limit was estimated to be 500 exosomes when using MCF-7 human breast cancer cells.

An EV isolation step can also be included in nucleic acid-based sensors. Fan and coworkers integrated a rapid magnetic EV-enrichment platform and $Ru(bpy)_3^{2+}$-polymer amplification strategy [40]. First, cancer EVs were captured using a biological-affinity identification platform with anti-EpCAM antibodies. After combining the recognition domain of the polymer-amplified probe with *miR-21*, an electrochemiluminescence (ECL) signal was generated by the linear $Ru(bpy)_3^{2+}$-polymer domain of the probe. This strategy may provide a new direction for the nondestructive detection of liquid biopsy for tumors.

### 10.2.3 Other Biomarker-Based Biosensors

Size distribution measurement can be used not only to assess the quality of EV isolation techniques but also to compare samples of different cellular origins. The size of EVs may be reflective of their cellular origin [53–55]. For example, studies have shown that salivary exosomes from patients with oral cancer are larger and more prone to aggregation than those from healthy controls [56]. Such features have been used in several EV isolation [57] and quantification methods [58], and in electrochemical sensors, EV size has also been considered. Mathew and coworkers reported an ultrasensitive tumor-derived EV detection assay at clinically relevant concentrations using a double amplification mechanism combining redox cycling and an enzymatic reaction, along with a sandwich immunoassay ensuring two-level selectivity [59]. Relatively small EVs with an estimated diameter of 30–150 nm were found on anti-EpCAM antibody-functionalized electrodes. These EVs were the least susceptible to shear forces when flushing the microchannel to remove unbound species because the Stokes drag force linearly scales with object size.

## 10.3 Fluorescent Biosensors

Fluorescent biosensors have been widely used to analyze biomarkers owing to their simple operation and flexibility. Moreover, fluorescent biosensing technology has developed rapidly owing to the use of nanomaterials, which endow fluorescent methods with excellent specificity and sensitivity. Therefore, fluorescent biosensors are regarded as promising candidates for EV detection. According to their reaction phases and characteristics, biosensors can be divided into three categories (Table 10.2): paper-based sensors [60], solution-based fluorescent sensors [61–67, 72, 73], and microfluidic-based sensors [74–77].

### 10.3.1 Paper-Based Fluorescent Biosensors

Paper-based fluorescent biosensors, such as paper-based analytical devices (PADs), are simple, fast, and low-cost analytical platforms that rapidly detect clinical samples. Chen et al. reported fluorescent PADs using upconversion nanoparticles (UCNPs) and gold nanorods (AuNRs) as a fluorescence-quenching pair to detect exosomes based on upconversion luminescence resonance energy transfer (LRET) between UCNP and AuNRs [60]. First, the CD63 protein DNA aptamer was divided into two different fragments (CP and DP), and the UCNP at the CP end was then used to modify the filter paper, whereas the DP end modified the AuNRs. When mixed with exosomes and dropped onto filter paper, the CD63 protein on the surface of exosomes can bind CP to DP, thereby reducing the distance between AuNRs and paper UCNP and inducing LRET. The fluorescence quenching rate can be used to accurately calculate the concentration of exosomes, achieving high-sensitivity detection of exosomes derived from hepatocellular carcinoma cells.

### 10.3.2 Solution-Based Fluorescent Biosensors

Solution-based fluorescent sensors have been widely applied in exosome detection owing to their homogeneous phase reaction properties, which endow the sensors with a rapid response and high signal output. Zhang's group reported a simple exosome detection platform exploiting the advantages of two-dimensional (2D) nanomaterial $Ti_3C_2$MXenes and fluorescent aptamers [61]. When the target exosomes were introduced into the system, the fluorescent aptamer showed a higher preference for binding to exosomes than $Ti_3C_2$MXenes. Thus, the fluorescent aptamers detached from the $Ti_3C_2$MXene surface, followed by the appearance of a "turn-on" fluorescent signal. This competitive strategy is so universal that many nanomaterials, particularly graphene oxide (GO), have been applied to detect target exosomes [72]. For example, Kong et al. designed exosome detection nanosensors using GO and aptamers [62, 63]. Zheng et al. also employed a new signal report or, aggregation-induced emission luminogen (AIEgen), a chemical molecule that binds to and aggregates aptamers and exhibits high-intensity fluorescence, to detect tumor-derived exosomes [64]. By adopting AIEgen, the established aptasensor could show

**Table 10.2** Fluorescent biosensors for EVs

| Method | Substrate | Target | Detection range (particles/μL) | Reaction time (min) | LOD (particles/μL) | References |
|---|---|---|---|---|---|---|
| UCNPs@Split-Apt-Au NRs | Paper-based biosensor | CD63 | $1.0 \times 10^4$–$10^8$ | ~30 | $1.1 \times 10^3$ | [60] |
| Cy3-CD63 aptamer/$Ti_3C_2$ MXenes nanocomplex | Solution-based biosensor | CD63 | $10^1$–$10^6$ | 60 | 1.4 | [61] |
| ExoAPP | Solution-based biosensor | CD63, EpCAM, CEA, PTK-7, AFP, PSMA and PDGF | $1.6 \times 10^2$–$10^5$ | 40 | $1.6 \times 10^2$ | [62] |
| Aptamer/GO/DNase I detection system | Solution-based biosensor | CD63, EpCAM | $10^3$–$10^6$ | ~30 | $2.1 \times 10^4$ | [63] |
| TPE-TAs/aptamer/GO aptasensor | Solution-based biosensor | PSMA, HER2, EGFR, EpCAM | $4.07 \times 10^5$–$1.83 \times 10^7$ | 25 | $3.43 \times 10^5$ | [64] |
| MB-exosomes-$Zr^{4+}$-liposomes sandwich complex | Solution-based biosensor | CD63 | $1.68 \times 10^4$–$4.2 \times 10^7$ | ~140 | $7.6 \times 10^3$ | [65] |
| hMFEX | Solution-based biosensor | GPC-1 | $7.8 \times 10^4$–$3.9 \times 10^9$ | ~60 | $6.56 \times 10^4$ | [66] |
| ABDNs-based TIRF assay | Solution-based biosensor | PTK7 | $10^3$–$10^7$ | 120 | $10^3$ | [67] |
| Cy3-CD63 aptamer/$Ti_3C_2$ MXenes nanocomplex | Solution-based biosensor | CD63 | $10^1$–$10^6$ | 60 | 1.4 | [68] |
| IC on GO/PDA nano-interface | Microfluidic-based biosensor | CD9, CD63, CD81, and EpCAM | $10^3$–$10^6$ | – | ~50 (80 aM) | [69] |
| Microchip digital detection @RIDA | Microchip-based biosensor | GPC-1 and membrane | – | 90 | Single-particle level | [70] |
| Droplet digital ExoELISA | Microfluidic droplet-based biosensor | CD63, GPC-1 | $10^1$–$10^5$ | 30 | 10 | [10, 70, 71] |

reduced background noise, resolve the limitations of enzymes, and thus improve detection performance.

Because of the relatively low concentrations of specific subpopulations of exosomes and complicated environments within biological fluids, integrated biosensing platforms have been developed for the isolation and detection of exosome subpopulations. Li et al. demonstrated that immunomagnetic beads could capture specific exosomes; thus, the beads were labeled with fluorescent liposomes to detect exosomes [65]. Furthermore, Zheng et al. reported a homogenous magneto-fluorescent nanosensor for tumor-derived exosome isolation and analysis. The analysis was conducted via the transformation of exosome signals into DNA nanostructures for amplification [66]. In addition, Li's group applied a glass-based platform to capture CD63-positive exosomes, introducing an aptamer-induced specific HCR for signal amplification [67]. Lee et al. also reported a solution-based signal-amplifiable biochip using HCR for exosomal biomarker detection with anionic liposomes as a carrier and delivered the detection materials into exosomes to realize the in situ detection of exosomal nucleic acids [73]. With numerous choices of nanomaterials and flexible reaction strategies, solution-based fluorescent sensors have shown great promise for exosome detection in biofluids.

### 10.3.3 Microplatform-Based Fluorescent Biosensors

Microplatform-based fluorescent biosensors have the characteristics of automation, integration, and miniaturization; these properties can significantly improve the detection sensitivity of EVs [68]. Zhang et al. developed a microfluidic exosome analysis platform based on a GO/polydopamine (PDA) nano-interface [69]. GO-induced three-dimensional (3D) nanoporous PDA layer-by-layer microfluidic deposition could greatly improve the efficiency of exosome capture while effectively inhibiting the adsorption of nonspecific exosomes. By modifying different exosomal membrane protein antibodies to capture exosomes, only 2 μL of plasma was needed to quantify exosomes for diagnosis in patients with ovarian cancer.

In recent years, microfluidic technology has gradually been applied to high-sensitivity EV detection. Tian et al. introduced a quantitative exosome analysis method based on a rapid isothermal nucleic acid detection assay on a digital microchip [70]. DNA oligonucleotides were self-assembled in the lipid bilayer membrane on the surface of exosomes to represent total exosome analysis. Specific antibody-DNA conjugates are used to label specific exosome proteins. DNA-anchored exosomes are then allocated to microchip chambers, and according to the Poisson distribution, each chamber has one or fewer exosomes. The chamber containing exosomes is recorded as "1" following detection of a fluorescence signal, whereas the chamber without exosomes is recorded as "0." This method provides a universal platform for the quantitative detection of various NPs. Liu et al. developed a digital detection methodology for breast cancer exosomes based on droplet microfluidic technology [71]. This method integrates the advantages of traditional ExoELISA for the specific detection of exosomes with the advantages of ultrasensitivity and integration of droplet microfluidic technology. In this droplet microfluidic platform, the detection limit for exosomes is as low as 10/μL, which allows digital absolute quantitative detection of breast cancer serum exosomes and effectively distinguishes tumor samples from normal samples.

In summary, fluorescent biosensors can recognize EV molecules and convert the response into a fluorescence signal that is proportional to the number of EVs in the system, enabling flexible and sensitive analysis of EVs. However, despite the advantages described above, it is necessary to address the challenges related to application of fluorescent biosensors in clinical laboratories. For example, the recognition element may nonspecifically adhere or form aggregates to produce an interference signal, thereby disrupting quantitative analysis. Moreover, fluorescent materials with different spectra may cause mutual interference, limiting their application in the detection of multiple markers for EVs. Therefore, efforts should be made to develop novel triggering strategies with high specificity and fluorescent dyes with high fluorescence quantum yield and narrow emission spectra to promote the clinical application of EV fluorescent biosensors.

## 10.4 CL-Based Biosensors

CL-based biosensors play important roles in biosensing and have been widely applied for biomedical detection [78]. CL includes ECL and direct CL (DC). ECL involves both electrochemical and CL reactions. Recently, a growing number of ECL biosensors have been proposed for the detection of nucleic acids, proteins, and cells. Many technologies have been proposed to integrate ECL [79]. As an immunoassay, DC combines labeling substrate and luminescence techniques, which are simpler and more sensitive than other immunoassay technologies [80]. As a simple and versatile analytical platform, CL has emerged as a satisfactory tool for the detection of EV-associated biomarkers (Table 10.3). Studies have demonstrated that CL can be adapted to many technologies, including nanomaterial, enzyme, and DNA nanotechnologies [82].

### 10.4.1 Nanomaterial-Assisted CL Assays

Nanomaterials are a class of materials with nanoscale sizes, e.g., metals, microgels, and inorganic NPs [89]. These materials could have important applications in CL assays owing to their specific surface area, good conductivity, and stability [9]. Furthermore, the biological properties of nanomaterials, including specific recognition, enzyme-mimic activity, and biocompatibility, greatly increase their value in CL assays [90]. Owing to their superior properties, nanomaterials are often adopted

**Table 10.3** CL-based biosensors for EVs

| Method | Technique | Target | Detection range (particles/μL) | Reaction time (min) | LOD (particles/μL) | References |
|---|---|---|---|---|---|---|
| Anti-GPC1-g-$C_3N_4$@Galinstan-PDA | ECL | GPC1, AFP, CEA, CD9 | $50–10^5$ | 120 | 31 | [81] |
| AuNPs-MXenes-Apt/exosomes/Apt/SAPAM/GCE | ECL | CD63 | $10^2–10^5$ | 120 | 30 | [82] |
| MXenes-BPQDs@$Ru(dcbpy)_3^{2+}$-PEI-$Ab_{CD63}$ | ECL | CD63 | $1.1 \times 10^2–10^7$ | 130 | 37 | [83] |
| CuS-MG | CL | CD63, HER2 | $10–10^5$ | – | 10 | [84] |
| MPA-CdS:Eu NCs | ECL | CD63 | $3.4 \times 10^2–1.7 \times 10^5$ | 120 | 74.1 | [85] |
| CdS QDs@$CaCO_3$ | ECL | EpCAM | $5 \times 10–1 \times 10^5$ | 120 | 21 | [86] |
| DNA-GOD | ECL | CD63 | $2 \times 10^2–7.5 \times 10^4$ | – | 60 | [87] |
| $Ru(bpy)_3^{2+}$-polymer | ECL | EpCAM, miR-21 | – | 180–300 | 100 | [88] |

as stowage, catalytic interfaces, and highly electrocatalytic materials for EV detection [91].

AuNPs, which are frequently used for CL assays, have good performance for loading nucleic acids [92]. Liu et al. constructed an ECL biosensor to detect exosomes and their surface proteins by combining electrode interface-modified multivalent poly(amidoamine) dendrimers AuNPs and g$C_3N_4$-coated Galinstan liquid metal nanohybrids nanoprobes [81]. The PAMAM-AuNPs were first connected with a specific antibody and then dropped onto the electrode interface, which provided a large area for exosome capture. At the same time, the Galinstan NPs were adopted as the signal probe. The g-$C_3N_4$@Galinstan-PDA-modified antibody could be effectively combined with exosomes to yield powerful signals. This is because the Galinstan NPs can facilitate electron transfer and suppress g-$C_3N_4$ passivation. The proposed platform can be used to identify multiple exosome surface biomarkers (e.g., carcinoembryonic antigen [CEA], alpha-fetoprotein, glypican 1 [GPC1], and CD9) derived from HeLa, OVCAR-3, and BT474 cells with a limit of detection of 31 particles/μL.

Two-dimensional nanomaterials have been widely used in CL biosensors because of their intrinsic features, including good conductivity, hydrophilicity, and easy film formation [93]. Dai's group developed an ECL and photothermal dual-mode biosensor for exosome detection using MXenes and black phosphorus QDs (BPQDs) [83]. BPQDs act as coreactants to catalyze the oxidization of $Ru(dcbpy)_3^{2+}$; the fabricated $Ru(dcbpy)_3^{2+}$@BPQDs can reduce energy loss and shorten the electron transfer distance, which generates an enhanced ECL signal. Because of their large surface area and good conductivity, MXenes have been adopted as carriers to load large amounts of $Ru(dcbpy)_3^{2+}$ and BPQDs. The proposed ECL platform exhibits high sensitivity for exosome detection, with a limit of detection of 37.0 particles/μL.

In addition, diverse nanomaterials have emerged as alternatives for exosome detection [8]. Zhong et al. reported a CuS-enclosed microgel-based CL assay for rapid quantification of EVs [84]. The proposed CuS-enclosed microgels could enrich specific EVs from complex bimatrices and produce strong CL signals. These antibody-fabricated microgels can recognize specific target exosomes. The limit of detection of this platform is $10^4$ EV particles/mL. Moreover, the proposed CL assay can directly detect specific EVs in human serum and cell culture medium without tedious sample preparation.

### 10.4.2 Enzyme-Assisted CL Assays

Enzymes are important biocatalysts because of their high efficiency and specificity. Diverse enzymes that can be triggered at a constant temperature have been adopted for ECL assays [94]. Shen et al. proposed mercaptopropionic acid-modified $Eu^{3+}$-doped CdS nanocrystals (MPA-CdS:Eu NCs) as ECL emitters [85]. First, the MPA-CdS:Eu NCs were dropped onto the GCE surface. The CD63 aptamer modified on the CdS:Eu NCs could recognize and capture exosomes to form a DNAzyme structure. In addition to hemin, the DNAzyme/hemin complex could efficiently catalyze the decomposition of $H_2O_2$, leading to a decreased signal. This ECL biosensor can detect exosomes in the range from $3.4 \times 10^5$ to $1.7 \times 10^8$ particles/mL, with a limit of detection of $7.41 \times 10^4$ particles/mL. To improve the sensitivity of exosome detection, Feng et al. proposed an enzymatic target-recycling amplification method based on a homogeneous ECL sensing system. Stimulus-responsive DNA microcapsule-loaded CdSQDs were adopted as ECL emitters [86]. Upon the addition of exosomes, DNA-modified exosomes were formed and hybridized with the DNA shells. Then, the DNA microcapsules were unfolded to release the encapsulated CdS QDs, which generated an ECL signal. Using the Nt.BbvCl, the hybridized DNA shells could be digested, and modified DNA could be released to hybridize with other DNA shells. The proposed ECL platform exhibits a wide detection range, high sensitivity, and good selectivity using enzymatic target recycling amplification and CdSQD-loaded DNA microcapsules.

### 10.4.3 DNA Nanotechnology-Assisted CL Assays

DNA nanotechnology is an attractive tool in biosensing and involves the use of various nanodevices, including walkers, tetrahedrons, and intelligent robots, which facilitate the intrinsic addressable, programmable, and precise Watson–Crick base pairing of the approach [87]. Xu et al. developed an ECL aptasensor for exosome quantitation using a DNA walker [88]. The DNA walker consisted of a RuSiNP-modified electrode. Two types of DNA (anchor DNA and double-stranded DNA) were modified on the surface of RuSiNPs. Anchor DNA was used as the template. Double-stranded DNA was formed by hybridization with the arm and aptamer. In the absence of the target exosomes, the walking strand was effectively blocked by CD63. In the presence of target exosomes, the aptamer combined specifically with the exosome surface. The walking strand then hybridized with the anchor DNA, generating the cleavage site. Anchor DNA was released in the presence of Nb.BbvCI. Simultaneously, the walking strand is hybridized with another anchor DNA for a new round. Finally, the glucose oxidase-modified DNA probe was captured by the abundant DNA anchor DNA, generating an amplified signal for exosome detection.

In general, the high heterogeneity and relatively small size of EVs pose technical challenges that preclude seamless investigation of their molecular features and interactions. Emerging platforms have been proposed and optimized for convenient and sensitive detection of EVs. CL-based biosensors show great potential for EV-associated biomarker detection. Many studies have been conducted to improve the sensitivity and specificity of CL. CL-based biosensors with inherent properties have been used for exosome-associated biomarker detection, with promising performance and convenience. Despite obvious progress, further improvements are still needed to transition CL-based biosensors to the clinical setting, as follows. First, the results obtained using CL biosensors are greatly affected by sample pretreatment. Suitable pretreatment is required to obtain reliable, repeatable results from CL biosensors for clinical applications. Second, it is essential to set baselines for disease status because physiological factors can affect the amounts and contents of EVs. Third, analysis of individual EVs is critical for identifying specific EVs, and this should accelerate the clinical applications of CL-based biosensors.

## 10.5 SPR-Based Biosensors

### 10.5.1 Fundamentals of SPR Biosensors

SPR is a plasmonic biosensor method that has emerged in recent years as an efficient and reliable platform for detecting the interactions of biological molecules [95]. This label-free technology does not require tedious sample handling steps [14]. SPR sensors are thin-film refractometers that measure changes in the refractive index occurring at the surface of a metal film [96]. The SPR effect occurs at the metal-dielectric interface when the energy of the incident light is coupled into surface plasmons, which causes a weakening of the total internal reflection of the light beam, thereby leading to changes in the refractive index. Quantification of the analytes can be completed by measuring these changes. The total internal reflection effect occurs at a specific incident angle and wavelength [97].

### 10.5.2 Patterns and Characteristics of SPR Signal Amplification

SPR alone is not very sensitive for the detection of analyte molecules, particularly small molecules ($\leq$800 kDa); thus, the scope of SPR for biological applications is limited. Owing to the extremely low content of exosomes in plasma, amplification of the analytical signal is required. In the past few decades, different methodologies have been used to increase the sensitivity of SPR, and signal amplification carried out on SPR platforms can be divided into two categories.

The first is induction of SPR signal amplification by the introduction of metallic NPs [34, 98]. The unique features of metallic NPs, such as controlled shape and size, high surface area to volume ratio, adjustable absorption and emission attributes, and biocompatibility, make them ideal for designing efficient biosensors. AuNPs are among the most commonly used metallic NPs in the field of biosensors [99]. The main advantages of AuNPs are their excellent catalytic and electrical properties, potent absorption in the near-infrared and visible wavelength regions, enhancement of the SPR electromagnetic field, and improvement of signal amplification for analytes, particularly those present at low concentrations [100, 101]. Secondary labeling of captured exosomes with metallic NPs results in larger shifts in the SPR angle or changes in the reflective intensity [102].

Another method involves stacking 2D substances on the surface of the chip. Typically, this is used in conjunction with the first method and involves substances such as GO and $MoS_2$. These 2D substances are unique owing to their strong molecule adsorption capacity, optical signal amplification, high carrier mobility, electronic bridging, and simple fabrication; therefore, these molecules may facilitate enhancement of the sensitivity of substrates in SPR. Graphene/$MoS_2$-based nanocomposites provide sensitivity enhancement by inducing a large field enhancement at the substrate interface [103]. Hence, the signal is amplified by the plasmon material, and the sensitivity of molecular detection increased up to the femto to atto molar level.

### 10.5.3 Application of SPR Biosensors in the Detection of EVs

SPR biosensing is one of the most effective methods for monitoring the affinity of biomolecules and preliminary screening of drug molecules [104]. SPR technology is increasingly being used to study various biological molecules, including DNA, RNA, proteins, and EVs, in biomedical fields [104]. Among these molecules, EVs (having particle sizes of 50–200 nm) have emerged as tumor detection markers in recent years, and their particle size aligns perfectly with the sensitive detection domain of SPR (approximately 200 nm) [104]. Therefore, SPR biosensors are suitable for detecting EVs.

#### 10.5.3.1 Detection of Exosomes Using Traditional SPR Sensors

The detection mechanism of traditional SPR biosensing is mainly based on antigen–antibody binding, ligand–receptor kinetics, and enzyme substrate reactions [104] to achieve relatively simple and rapid detection of analytes. The most common detection mechanism is based on the specific recognition of antigens and antibodies. The primary antibody is first modified on the gold chip through gold-amino bonds. After the target substance is added, the primary antibody specifically recognizes and captures the target to increase the upload molecular weight on the chip. Moreover, the SPR refractive index changes, and CCD detector captures the change in real time to reflect the existence of the target. The change in the magnitude of the refractive index is then compared to achieve qualitative and quantitative detection of targets [104]. As shown in Fig. 10.1, exosomes are detected by antigen-antibody binding. The anti-CD63 antibody specifically captures the CD63 protein on human mast cell-derived exosomes. The detection efficiency of this method is comparable to that of ELISA [104].

#### 10.5.3.2 Detection and Classification of Exosomes Using Signal-Enhanced SPR Sensors

The main drawback of traditional SPR biosensing is that it is difficult to achieve sensitive detection of trace substances; therefore, it is necessary to further amplify the response signal of SPR [104]. As a result, methods for the secondary amplification or indirect detection of bound antibodies have emerged. Many polymer nanomaterials, such as NP plasma nanostructures [104] with high refraction sensitivity and nanostructures [104] with adjustable plasma domains, can improve the detection sensitivity of SPR and realize detection signal amplification. Secondary amplification detection (NPs, polymer materials, etc.) is the most common signal amplification method used in clinical analyses. The greatest advantage of this

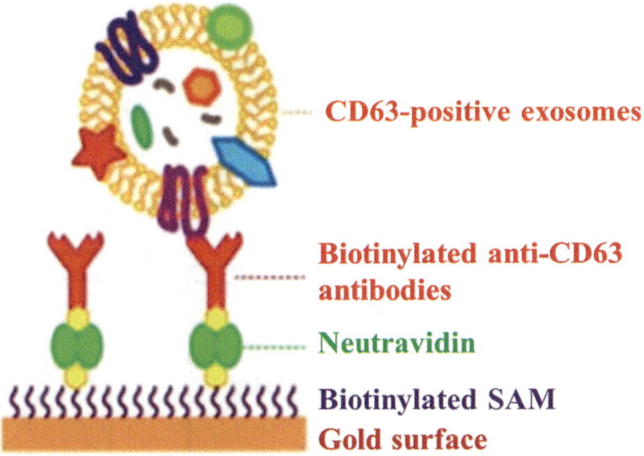


**Fig. 10.1** Schematic diagram of an antigen–antibody binding-based SPR biosensor for the detection of exosomes. [Reproduced with permission [105]. (*Copyright 2014, American Chemical Society*)]

method is that the analyte in the biological fluid is captured by the secondary antibody or aptamer strand to reduce nonspecific background and improve the signal-to-noise ratio [104]. Through proper elution and addition of antifouling materials, nonspecific adsorption can be minimized. Finally, quantitative detection of trace analytes in concentrated biological fluids was achieved.

**NP Enhancement**

In recent years, metal NPs, particularly AuNPs [104], have been widely used in the signal amplification of SPR-based sensors (Fig. 10.2). These metal NPs serve as signal amplification beacons and can be used to enhance the sensing signal. The principle of NP signal amplification is based on the fact that the introduction of functionalized AuNPs can increase the mass of the analyte on the chip, thereby significantly changing the refractive index of SPR. For example, Lyon et al. used various types of colloidal AuNPs (30–59 nm in diameter) to cover a 47-nm-thick gold film to reduce the detection limit [104]. They also labeled AuNPs with anti-human IgG and detected h-IgG using the triple-hybrid immunological method. As shown in Fig. 10.2, after adding colloidal AuNPs functionalized with antibody, the SPR signal was increased by 25-fold, and the minimum detection limit was reduced to 6.7 pM. In addition to AuNPs, silver NPs and silica NPs can also be used to enhance the SPR detection signal.

As shown in Fig. 10.3, AuNPs can be used to enhance the SPR signal to achieve sensitive detection of exosomes from multiple tumor cells [104]. The SPR signal was successfully amplified by adding AuNPs modified by the secondary antibody, and sensitive detection of non-small cell lung cancer-derived exosomes was realized. The limit of detection was as low as $10^4$ particles/mL. Moreover, by comparing signal differences, the method could also be used for the detection and classification of multicomponent exosomes.

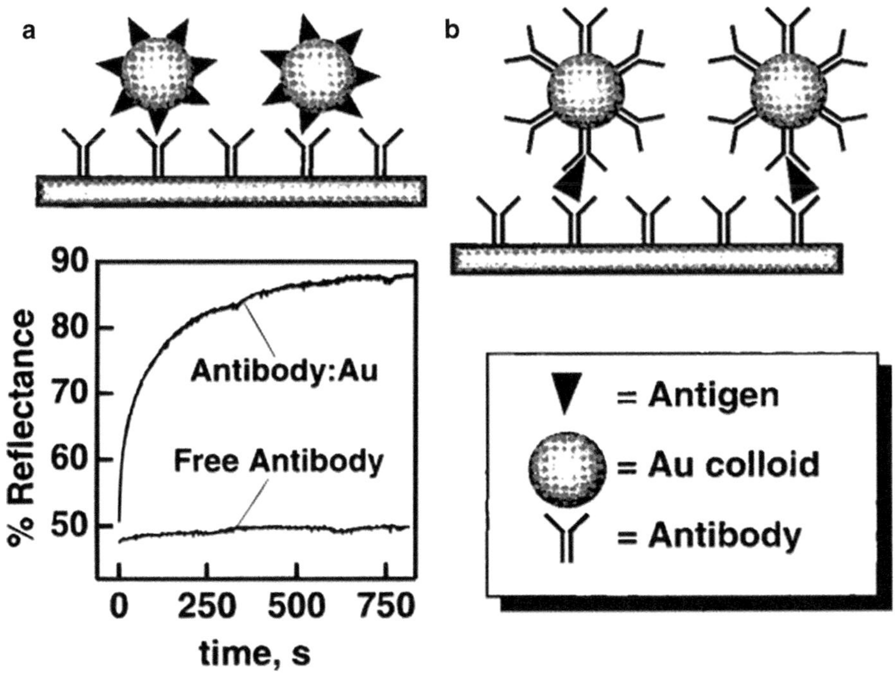


**Fig. 10.2** Schematic diagram of an AuNP-enhanced SPR biosensor. [Reproduced with permission [106]. *(Copyright 1998, American Chemical Society)*]

### Polymer Material Enhancement

The emergence of polymer nanomaterials provides more possibilities for SPR signal enhancement [104]. In recent years, an increasing number of researchers have chosen polymer nanomaterials, such as DNA hydrogels, polydopamine hydrogels, and DNA branches [104], to further enhance the SPR detection signal. First, compared with other inorganic nanomaterials (metal particles, semiconductor materials, etc.), the DNA nanostructure design is more flexible, with a strictly controllable self-assembly process; these nanostructures are also easy to synthesize and biocompatible. These advantages can compensate for the deficiencies of inorganic materials. Among these materials, DNA hydrogels are widely used, as shown in Fig. 10.4. In this method, AuNPs are embedded in loose and porous DNA hydrogels, and this alteration greatly enhances the SPR signal and enables the detection of prostate cancer cell-derived exosomes with a limit of detection of $1.00 \times 10^5$ particles/mL.

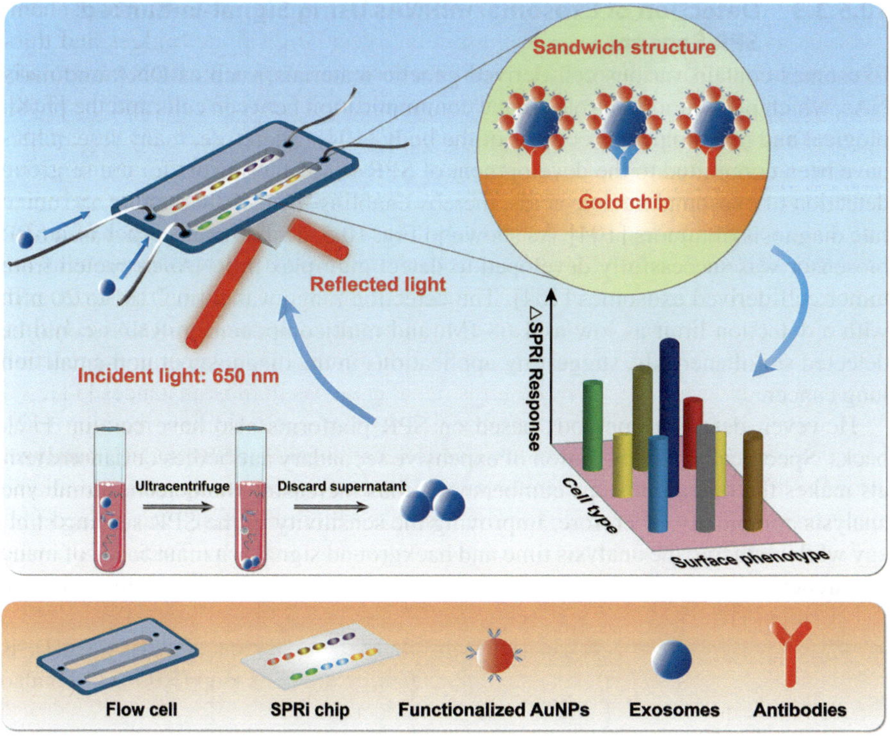


**Fig. 10.3** AuNP-based SPR biosensor for the detection of tumor cell-derived exosomes

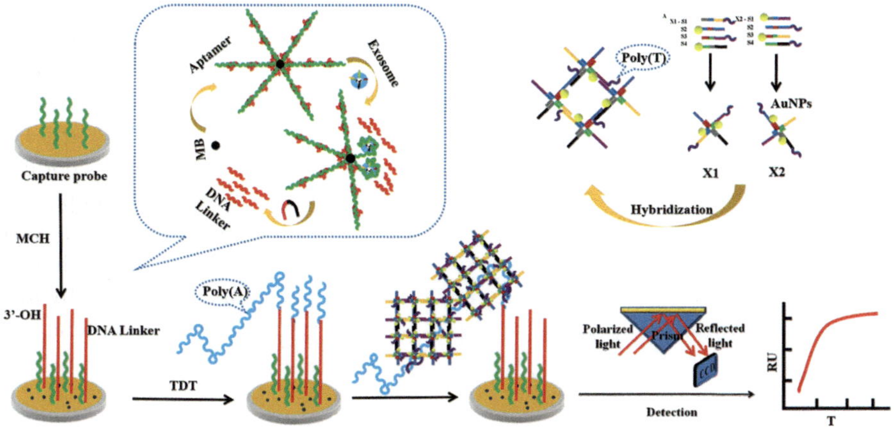


**Fig. 10.4** Hydrogel-AuNP-based SPR biosensor for the detection of prostate cancer cell-derived exosomes

### 10.5.3.3 Detection of Exosomal miRNAs Using Signal-Enhanced SPR Sensors

Exosomes contain various cell-derived genetic materials, such as DNA and miRNAs, which play important roles in the communication between cells and the physiological and pathological processes of the body [104]. Therefore, many researchers have been committed to the development of SPR-based biosensors for the sensitive detection of exosomal nucleic acids, thereby enabling the early screening and accurate diagnosis of tumors [104]. As shown in Fig. 10.5, a silver nanocage-based SPR biosensor was successfully developed to detect multiplex miRNAs extracted from tumor cell-derived exosomes [104]. The detection range was from 2 fM to 20 nM, with a detection limit as low as 1.68 fM, and multicomponent miRNAs could be detected simultaneously, suggesting applications in the diagnosis of non-small cell lung cancer.

However, detection methods based on SPR platforms also have certain drawbacks. Specifically, the utilization of expensive secondary antibodies or nanomaterials makes the operation steps cumbersome, thus increasing the detection time and analysis complexity. Therefore, improving the sensitivity of the SPR sensing strategy while reducing the analysis time and background signal is a main focus of many researchers.

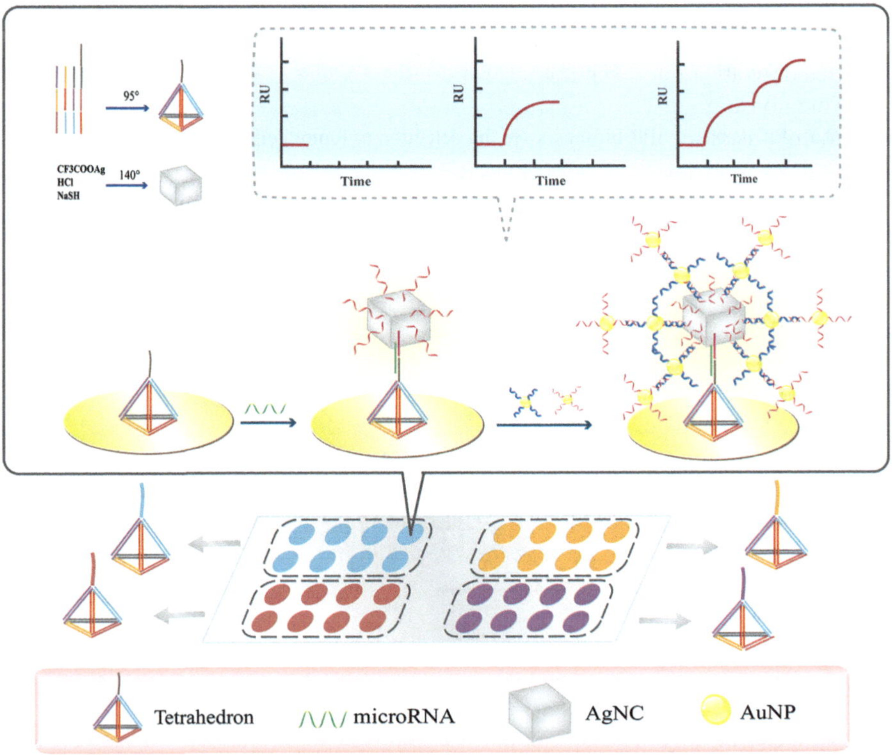


**Fig. 10.5** A silver nanocube-based SPR biosensor for the detection of multiplex exosomal miRNAs

### 10.5.4 Outlook

Major advancements in SPR biosensors have been made in the past few decades, and these biosensors were initially limited to salt solutions and simple bioassays for the detection of nucleic acids, proteins, and other biological macromolecules. Current research focuses on the development of portable SPR biosensors that can be used for analysis of clinical samples, such as tumor cell-derived exosomes. Further studies are expected to reveal many promising applications in clinical tumor diagnosis and treatment.

## 10.6 Raman-Based Biosensors

Raman spectroscopy is a molecular spectroscopy technology based on inelastic scattering that can identify proteins, nucleic acids, lipids, and other biochemical substances. In 1974, Fleischmann et al. [107] found that the Raman signal was greatly enhanced when molecules were adsorbed on a rough metal surface. By 1977, this surface-enhancement effect associated with rough surfaces was further confirmed and was designated the SERS effect [108]. SERS overcomes the weakness of the poor signals inherent in traditional Raman spectra and can increase the Raman signal intensity by several orders of magnitude.

Existing methods for the characterization of exosomes, such as western blotting, NTA, electron microscopy, and flow cytometry, are cumbersome or can only provide limited information. In recent years, ultra-high-sensitivity SERS technology has been used to detect exosomes and facilitate the development of novel, efficient, nondestructive liquid biopsy technologies. Exosome detection strategies based on SERS technology can be divided into two types. The first is a strategy to directly obtain the Raman signals of the exosomes themselves using markless SERS technology. The advantages of this method are that it is simple and fast and can more directly analyze the components of exosomes; however, various biomolecular signals may overlap with each other. The second is SERS labeling using Raman probe molecules. The presence and concentration of exosomes can be determined indirectly by detecting the Raman signals of the labeled molecules. This method often has a higher detection sensitivity and specificity [85].

### 10.6.1 Protein Biomarker-Based SERS Biosensors

The detection of exosome protein biomarkers relies mainly on SERS tags. SERS tags are NPs with a strong Raman signal. By binding specific molecules, such as antibodies or aptamers, SERS tags can combine with target exosomes to form a complex and can be used to indirectly characterize the distributions and concentrations of target molecules by detecting the Raman strength of the Raman tag molecule of this complex. The common complex structure is referred to as Raman tag-exosome-magnetic bead/chip.

Zong et al. [109] developed a method to identify and capture target exosomes using magnetic nanobeads and core-shell structure nanoprobes and subsequently used these nanomaterials for qualitative and quantitative detection. The center of the nanoprobe contained gold core-silver shell NRs(Au@Ag NRs) as the SERS active core, which was then modified with a dye as a Raman reporter molecule, and the outer layer was covered with silica as the protective layer. Finally, an anti-HER2 antibody was attached to the outermost layer. Magnetic NPs were prepared by coating $Fe_3O_4$ NPs with silicon dioxide and attaching an anti-CD63 antibody to the silicon shell. In the presence of target exosomes, magnetic nanobeads and SERS nanoprobes can capture exosomes by forming sandwich-type immune complexes. Magnets can be used to precipitate immune complexes; therefore, SERS signals can be detected in the sediments. Exosomes from SKBR3 human breast cancer cells and MRC5 human fetal lung fibroblasts were directly detected using this method in cell culture medium.

The same research group proposed an improved method that could simultaneously detect multiple exosomes [110]. Magnetic beads modified with CD63 aptamers were used to capture exosomes. SERS probes corresponding to three different tumor protein biomarkers (HER2, CEA, and PSMA) were designed. When the target exosomes were added, they are specifically combined with the corresponding magnetic beads and probes to form complexes, which aggregated to the bottom of the tube under the action of magnets. At this time, the corresponding SERS probe signals in the supernatant decreased, whereas the signals of other unrelated SERS probes remained unchanged. Using this method, a minimum of 32, 73, and 203 exosomes/μL secreted from SKBR3, TB4, and LNCaP cells, respectively, were detected. Finally, blood samples from patients with breast, colorectal, and prostate cancer were tested to verify the usefulness of the method.

Li et al. [111] developed an ultrasensitive SERS detection method using chips that could detect single exosomes in 2 μL cell culture medium. Moreover, by analyzing 2-μL serum samples, researchers found that this method could distinguish patients with pancreatic cancer from healthy individuals, identify whether the tumor was metastatic, and determine the tumor stage with a sensitivity of 95.7%. In this method, PDA is self-assembled on glass slides and coated with a layer of specific antibodies (anti-macrophage inhibitory factor, anti-GPC1, anti-CD63, or anti-epidermal growth factor receptor [EGFR]) to capture and concentrate exosomes. The SERS probes used to detect signals have a multilayer structure, that is, Au, reporter Ag, and PDA-encapsulated antibodies, from inside to outside.

Also using the chip, Kwizera et al. [112] applied 3D printing technology to fix an antibody array on the surface of gold-plated glass slides. The Raman probe used gold NRs coated with the QSY21 Raman reporter gene to quantitatively detect the expression of several surface protein markers on exosomes derived from breast cancer cell culture medium and samples from patients with HER2-positive breast cancer. The detection limit was $2 \times 10^6$ exosomes/mL, and the expression profiles of eight proteins (EpCAM, CD44, HER2, EGFR, insulin-like growth factor receptor 1, CD81, CD63, and CD9) in more than 80 samples were analyzed within 2 h. The results showed that exosomes derived from the breast cancer cell lines MDA-MB-231,

MDA-MB-468, and SKBR3 had different protein profiles than normal breast cancer cells (MCF12A cells) and that plasma exosomes from patients with HER2-positive breast cancer had higher levels of HER2 and EpCAM than those of healthy donors.

Zhang et al. [113] developed a novel Raman probe with intense electromagnetic hot spots that could provide dramatically enhanced Raman scattering. This novel probe structure consisted of AuNPs bound to triangular pyramid DNA (TP-AuNPs). The authors used this probe to detect EpCAM on exosomes and effectively distinguish exosomes extracted from the plasma of healthy individuals and patients with breast cancer.

Immunological oncology therapy has become an essential treatment modality in patients with advanced malignant tumors, and anti-programmed cell death 1 (PD-1)/ programmed cell death ligand 1 (PD-L1) antibody drugs are effective tumor immunotherapy agents. Blelloch et al. [114] reported that the PD-L1 levels of circulating exosomes in patients are associated with drug treatment. Therefore, the level of exosomal PD-L1 can be used as a predictor of the clinical outcome of anti-PD-1/ PD-L1 therapy and as a biomarker of tumor burden. In 2020, Pang et al. [115] proposed a method that could capture and analyze PD-L1 exosomes directly from the serum. The magnetic beads they used contained $Fe_3O_4$ on the inside and $TiO_2$ on the outside; this structure was expected to facilitate combination with the phospholipid layer of exosomes and then efficiently and rapidly enrich exosomes. Subsequently, the Au@Ag@MBA SERS probe modified with an anti-PD-L1 antibody was added to exosomal PD-L1 for quantification. As few as one exosome/μL can be detected in 40 min, and the sample size required for analysis in the clinical setting is only 4 μL.

### 10.6.2 Nucleic Acid Biomarker-Based SERS Biosensors

In recent years, SERS methods have attracted much interest for DNA and RNA detection [116, 117]. SERS has many unique advantages, including high sensitivity, small sample volume, rapidity, stability, and availability of specific signals. The detection of nucleic acids by SERS can be carried out using two strategies. First, label-free detection and Raman reporter molecules can be used. Label-free SERS detection facilitates the detection of the intrinsic Raman spectra of an analyte without the use of Raman labels. The main advantage of label-free detection is the simplicity of the experiments [118]. However, for complex samples, the signal may be indistinguishable, and the specificity is insufficient. The use of Raman reporter molecules allows for indirect detection of DNA. The advantage of this method is that, compared with DNA, it has a higher Raman scattering cross-section for the Raman reporter molecule [119].

SERS is a promising ultrasensitive method that can be used to sensitively and accurately identify specific DNA fragments. Many different methods can be used to detect specific DNA using SERS spectroscopy. Raman spectroscopy provides fingerprint vibrational spectra of molecules and allows the identification of individual components in the mixture. However, Raman signals are usually very weak because the incidence of inelastic scattering events is low. Therefore, researchers have begun to use rough metals to enhance Raman signals. For label-free analysis, many studies have reported increased sensitivity through the continuous improvement of materials. A recent study introduced dichloromethane as the interface reagent. The internal standard renders SERS a universal method for label-free detection of single-stranded DNA with different chain lengths and high-sensitivity A/G quantification [120]. The low signal intensity and heterogeneity of EVs hinder their detection by unlabeled SERS. To overcome these problems, researchers have exploited various chemical and physical properties to fix EVs on the surface of plasma NPs or nanostructures and have then performed SERS detection. Jiang et al. established a platform for directly extracting exosomes from serum samples and used SERS to detect miRNAs in situ. They synthesized LNA-modified Au@DTNB as a SERS tag to enter exosomes and assemble with target miRNAs, resulting in the induction of hot-spot SERS signals; they also added $Fe_3O_4$@$TiO_2$NPs to enrich exosomes and further detect signals from SERS [121].

Compared with label-free analysis, Raman reporter molecules have been more frequently employed for ultrasensitive and multiplexed detection of circulating nucleic acids. The following three methods have been used: formation of a sandwich structure, turning a signal on or off, and HCR. Raman reporter molecules approach plasmonic nanostructures through hybridization or the formation or decomposition of hairpin strands. When the Raman reporter is close to the plasma nanostructure, the signal is very strong; by contrast, when the Raman reporter is close to the plasma nanostructure, the signal is very weak [122]. The sandwich structure usually consists of a detection probe, target molecule, and capture unit. Compared with traditional polymerase chain reaction-based methods, SERS-based methods provide rapid, selective, sensitive, and simple detection of miRNAs [123]. A recent work described an SERS-active AgNR array with highly sensitive and specific SERS sensing of DNAs. This repeatable, reliable method uses a double amplification strategy and exhibits a linear detection range for human serum samples, with a limit of detection of up to 40.4 aM [124]. Another SERS-based detection method utilizes dual-SERS enhancement of $Fe_3O_4$@Ag-SERS tag assemblies and DSN-assisted recycling signal amplification for one-step detection of *miR-10b* in exosomes of blood samples to diagnose pancreatic cancer. A detection limit of 1 aM was achieved for single-base mismatch recognition in the diagnosis of patients with pancreatic ductal adenocarcinoma [125].

This SERS technique has been used to detect target nucleic acids and may have promising applications as a label-free homogeneous biosensor for cancer diagnosis without the need for target amplification.

### 10.6.3 Other Biomarker-Based SERS Biosensors

In addition to proteins and nucleic acids, Raman signals can also provide structural information on lipid molecules in membranes [126]. However, it is challenging to study the basic properties of lipid membranes based on lipid information. Raman probes are usually used to localize the probe in the membrane and emit a strong signal when excited by a laser, enabling researchers to ensure insertion into the lipid membrane. Carbonate-terminated AuNPs have been used as artificial ion transporters across biological membranes. Furthermore, synthetic NPs can interact with lipid membranes. In a previous study, AuNPs were used to determine the surface charge of the lipid film, and Raman spectroscopy was used to continuously monitor the oxidation of phosphatidylcholine [127]. Additionally, Czamara et al. analyzed the Raman spectra of 35 lipids, including saturated/unsaturated fatty acids, triacylglycerols, cholesterol, cholesterol esters, and phospholipids, to screen for key lipids involved in the development of cardiovascular diseases [126]. Although few studies have reported using SERS to detect exosomal lipids, the membrane structure of exosomes is similar to that of cells; therefore, the technologies used for cell lipid detection may also have applications in exosomal detection.

The SERS signal depends on the interactions between the sample, the metal substrate, and the laser. Even for exosomes from the same cell line, the intensity and shape of the SERS spectrum may change when different parts of exosomes are located in the plasmon-active region. Principle component analysis of exosomes is usually used to classify the source of exosomes; alternatively, by training with the Raman spectra of pure samples, models can be built and used to distinguish different types of vesicles [85]. Overall, although SERS may have applications in the detection of exosomes, there are still many problems to overcome for the widespread application of this technique.

## 10.7 Other Biosensors

### 10.7.1 Piezoelectric Biosensors

Piezoelectric sensors, which are based on the principle of oscillation change due to a mass bound on the piezoelectric crystal surface, have attracted much interest as biosensors for label-free detection, easy operation, and real-time measurement [128, 129]. In recent years, piezoelectric biosensors have been used in the field of exosome detection. Jugal and coworkers designed a label-free immunosensing strategy using quartz crystal microbalance and dissipation monitoring to detect the surface protein profile of exosomes [130]. Similarly, Wang et al. reported an AuNP-amplified surface acoustic wave sensor for exosome detection based on the mass increasing effect of AuNPs [131] (Fig. 10.6a). In their design, AuNPs were connected to the trapped exosomes via a secondary antibody, which increased the mass of the exosomes to realize surface acoustic wave signal amplification. Accordingly, the sensitivity was two orders higher than that when signal amplification was not used (i.e., as low as $1.1 \times 10^3$ particles/mL). Furthermore, the deformation caused by the mass was more directly reflected by the degree of deformation. Etayash and coworkers fabricated cantilever arrays for the simultaneous detection of multiple exosomal surface antigens [134]. In their experiment, each cantilever was functionalized with antibodies from different tumor-derived exosome biomarkers to capture the corresponding targets. Subsequently, antibody-coated AuNPs were

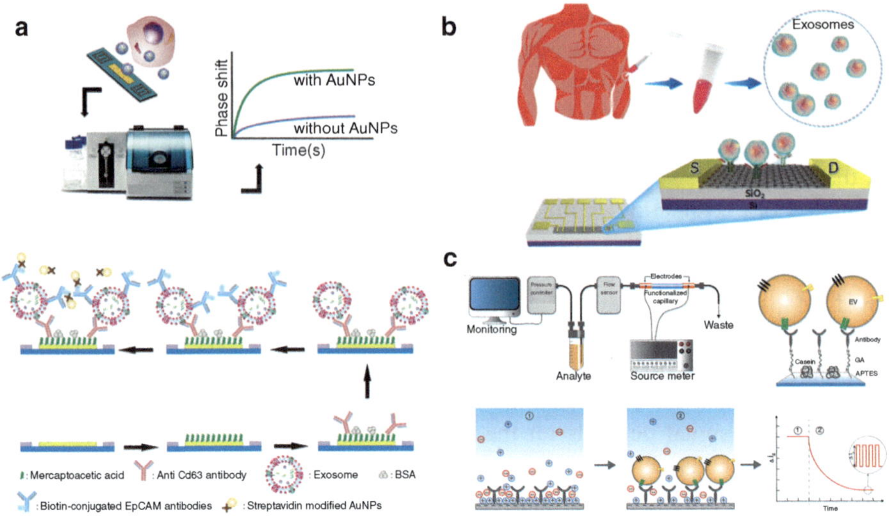


**Fig. 10.6** Other representative types of biosensors. (**a**) Quantitative detection of exosomes based on AuNP-amplified surface acoustic wave. Reproduced with permission [131]. *(Copyright 2020, American Chemical Society)*. (**b**) Quantification of exosomes based on the reduced graphene oxide field effect transistor. Reproduced with permission [132]. *(Copyright 2019, American Chemical Society)*. (**c**) Surface protein profiling of exosomes based on the electrokinetic principle. [Reproduced with permission [133]. *(Copyright 2019, American Chemical Society)*]

combined with the captured exosomes to increase their mass. Different cantilever bending signals were observed after injecting exosomes with bending amplitudes scaled according to the targeted antigens. The simple technique was used to detect exosomes with ultra-low concentrations (approximately 200 exosomes/mL, 0.1 pg/mL), compare the expression levels of exosomal-surface antigens, and distinguish between tumorigenic and nontumorigenic exosomes in real time.

### 10.7.2 Reduced Graphine Oxide (RGO) Field Effect Transistor Biosensors

The field-effect transistor biosensor based on RGO is a widely studied biosensor based on 2D nanomaterials. Through the field-effect transistor, the micro-electric signal generated by the interaction of biomolecules on the sensing interface is converted into a readable electric signal and amplified, resulting in high sensitivity and specificity [135]. Additionally, Yu et al. developed an RGO field-effect transistor biosensor [132] (Fig. 10.6b). An anti-CD63 antibody was covalently immobilized on the field-effect transistor surface. When the exosomes were captured by the specific antibody, the net carrier density on the chip surface changed owing to the contribution of the negative charges of the exosomes, resulting in a left shift of the Dirac point. This method achieved a low detection limit of 33 particles/μL and was employed to detect exosomes in clinical serum samples.

### 10.7.3 Electrokinetic Biosensors

Electrokinetic measurement is an accurate and effective method for measuring the deposition of particles and proteins. The most common method is to change the streaming potential caused by the deposition of particles at the solid/liquid interface [136]. Cavallaro and coworkers proposed a simple and inexpensive electronic method for detecting and analyzing the size range of exosomes [133] (Fig. 10.6c). When surface markers of small EVs interact with the affinity reagent fixed on the inner surface of the silica gel microcolumn, the change in flow current can be monitored, and this approach has been shown to have a sensitivity of $1.75 \times 10^5$small EVs/mL. Furthermore, this electrokinetic method may have applications in the label-free detection and profiling of small EVs in the size range of exosomes for the first time.

## 10.8 Perspectives

Individualized diagnosis and treatment are considered the hallmarks of precision medicine [137]. EVs have been extensively accepted by the scientific community in terms of their function in transferring biological molecules between cells as well as their potential to become biomarkers for various diseases. Because analysis of EVs in liquid biopsy samples may have applications in the early diagnosis of various diseases, the development of EV detection technologies is essential for future clinical purposes. EVs not only contain proteins and nucleic acids but also carry information on multilevel targets, such as liposomes, protein phosphorylation, and nucleic acid methylation. Research focused on the development of methods that can simultaneously detect EV multilevel targets is necessary for accurate diagnoses and successful treatments [138]. Moreover, artificial intelligence (AI) provides major opportunities for improvement of the detection and statistical analysis of multilevel targets [139]. AI is expected to undoubtedly be an important component in the biomedical field.

Although EVs may have promising applications in the precise screening of diseases, the progress of EV research depends on the development of separation and detection methods; however, these methods still require further improvement and optimization. For EVs to become routine testing items in clinical practice, EV detection methods must have the following characteristics: high-throughput, rapid assay time, operability, and high sensitivity, specificity, and stability [140]. Furthermore, the discovery of novel EV biomarkers is urgently needed. Most existing protein biomarkers are limited to the membrane surface of EVs, whereas inner protein markers remain unexplored.

In the future, data on the combination of multilevel targets and AI for the detection and analysis of EVs will greatly improve efficiency. To accomplish this goal, joint endeavors by multidisciplinary researchers are required [138].

## References


1. Cao Y, Li L, Han B, et al. A catalytic molecule machine-driven biosensing method for amplified electrochemical detection of exosomes. Biosens Bioelectron. 2019;141:111397.
2. An Y, Jin T, Zhu Y, et al. An ultrasensitive electrochemical aptasensor for the determination of tumor exosomes based on click chemistry. Biosens Bioelectron. 2019;142:111503.
3. Qureshi A, Gurbuz Y, Niazi JH. Biosensors for cardiac biomarkers detection: a review. Sens Actuators B Chem. 2012;171:62–76.
4. Daniels JS, Pourmand N. Label-free impedance biosensors: opportunities and challenges. Electroanalysis. 2007;19(12):1239–57.
5. Bagheri Hashkavayi A, Cha BS, Lee ES, et al. Advances in exosome analysis methods with an emphasis on electrochemistry. Anal Chem. 2020;92(19):12733–40.
6. Wu M, Chen Z, Xie Q, et al. One-step quantification of salivary exosomes based on combined aptamer recognition and quantum dot signal amplification. Biosens Bioelectron. 2021;171:112733.
7. Rasooly A, Jacobson J. Development of biosensors for cancer clinical testing. Biosens Bioelectron. 2006;21(10):1851–8.
8. Syedmoradi L, Norton ML, Omidfar K. Point-of-care cancer diagnostic devices: from academic research to clinical translation. Talanta. 2021;225:122002.
9. Shao B, Xiao Z. Recent achievements in exosomal biomarkers detection by nanomaterials-based optical biosensors—a review. Anal Chim Acta. 2020;1114:74–84.
10. Dong H, Chen H, Jiang J, et al. Highly sensitive electrochemical detection of tumor exosomes based on aptamer recognition-induced multi-DNA release and cyclic enzymatic amplification. Anal Chem. 2018;90(7):4507–13.
11. Huang R, He L, Xia Y, et al. A sensitive Aptasensor based on a hemin/G-Quadruplex-assisted signal amplification strategy for electrochemical detection of gastric cancer exosomes. Small. 2019;15(19):e1900735.
12. Zhou YG, Mohamadi RM, Poudineh M, et al. Interrogating circulating microsomes and exosomes using metal nanoparticles. Small. 2016;12(6):727–32.
13. Cai S, Ye J, Al-Maskri AAA, et al. A conformational switch-based aptasensor for the chemiluminescence detection of microRNA. Luminescence. 2019;34(8):823–9.
14. Xu L, Shoaie N, Jahanpeyma F, et al. Optical, electrochemical and electrical (nano)biosensors for detection of exosomes: a comprehensive overview. Biosens Bioelectron. 2020;161:112222.
15. Freeman R, Liu X, Willner I. Chemiluminescent and chemiluminescence resonance energy transfer (CRET) detection of DNA, metal ions, and aptamer-substrate complexes using hemin/G-quadruplexes and CdSe/ZnS quantum dots. J Am Chem Soc. 2011;133(30):11597–604.
16. Okumoto S, Jones A, Frommer WB. Quantitative imaging with fluorescent biosensors. Annu Rev Plant Biol. 2012;63:663–706.
17. Xia Y, Liu M, Wang L, et al. A visible and colorimetric aptasensor based on DNA-capped single-walled carbon nanotubes for detection of exosomes. Biosens Bioelectron. 2017;92:8–15.
18. Zhang Y, Wang D, Yue S, et al. Sensitive multicolor visual detection of exosomes via dual signal amplification strategy of enzyme-catalyzed metallization of Au nanorods and hybridization chain reaction. ACS Sens. 2019;4(12):3210–8.
19. Jiang Y, Shi M, Liu Y, et al. Aptamer/AuNP biosensor for colorimetric profiling of exosomal proteins. Angew Chem Int Ed Engl. 2017;56(39):11916–20.
20. Brolo AG. Plasmonics for future biosensors. Nat Photonics. 2012;6(11):709–13.
21. Rojalin T, Phong B, Koster HJ, et al. Nanoplasmonic approaches for sensitive detection and molecular characterization of extracellular vesicles. Front Chem. 2019;7:279.

22. Chin LK, Son T, Hong JS, et al. Plasmonic sensors for extracellular vesicle analysis: from scientific development to translational research. ACS Nano. 2020;14(11):14528–48.
23. Hao N, Pei Z, Liu P, et al. Acoustofluidics-assisted fluorescence-SERS bimodal biosensors. Small. 2020;16(48):e2005179.
24. Zhou Q, Rahimian A, Son K, et al. Development of an aptasensor for electrochemical detection of exosomes. Methods. 2016;97:88–93.
25. Yin X, Hou T, Huang B, et al. Aptamer recognition-trigged label-free homogeneous electrochemical strategy for an ultrasensitive cancer-derived exosome assay. Chem Commun. 2019;55(91):13705–8.
26. Doldán X, Fagúndez P, Cayota A, et al. Electrochemical sandwich immunosensor for determination of exosomes based on surface marker-mediated signal amplification. Anal Chem. 2016;88(21):10466–73.
27. Xu H, Liao C, Zuo P, et al. Magnetic-based microfluidic device for on-chip isolation and detection of tumor-derived exosomes. Anal Chem. 2018;90(22):13451–8.
28. Cao Y, Wang Y, Yu X, et al. Identification of programmed death ligand-1 positive exosomes in breast cancer based on DNA amplification-responsive metal-organic frameworks. Biosens Bioelectron. 2020;166:112452.
29. Guo Y, Cao Q, Feng Q. Catalytic hairpin assembly-triggered DNA walker for electrochemical sensing of tumor exosomes sensitized with Ag@C core-shell nanocomposites. Anal Chim Acta. 2020;1135:55–63.
30. Sun Z, Wang L, Wu S, et al. An electrochemical biosensor designed by using Zr-based metal-organic frameworks for the detection of glioblastoma-derived exosomes with practical application. Anal Chem. 2020;92(5):3819–26.
31. An Y, Li R, Zhang F, et al. Magneto-mediated electrochemical sensor for simultaneous analysis of breast cancer exosomal proteins. Anal Chem. 2020;92(7):5404–10.
32. Boriachek K, Masud MK, Palma C, et al. Avoiding pre-isolation step in exosome analysis: direct isolation and sensitive detection of exosomes using gold-loaded nanoporous ferric oxide nanozymes. Anal Chem. 2019;91(6):3827–34.
33. Jeong S, Park J, Pathania D, et al. Integrated magneto-electrochemical sensor for exosome analysis. ACS Nano. 2016;10(2):1802–9.
34. Boriachek K, Islam MN, Gopalan V, et al. Quantum dot-based sensitive detection of disease specific exosome in serum. Analyst. 2017;142(12):2211–9.
35. Kashefi-Kheyrabadi L, Kim J, Chakravarty S, et al. Detachable microfluidic device implemented with electrochemical aptasensor (DeMEA) for sequential analysis of cancerous exosomes. Biosens Bioelectron. 2020;169:112622.
36. Moura SL, Mart NCG, Mart M, et al. Electrochemical immunosensing of nanovesicles as biomarkers for breast cancer. Biosens Bioelectron. 2020;150:111882.
37. Zhou YG, Mohamadi RM, Poudineh M, et al. Interrogating circulating microsomes and exosomes using metal nanoparticles. Small (Weinheim an der Bergstrasse, Germany). 2016;12(6):727–32.
38. Wang S, Zhang L, Wan S, et al. Aptasensor with expanded nucleotide using DNA nanotetrahedra for electrochemical detection of cancerous exosomes. ACS Nano. 2017;11(4):3943–9.
39. Li Q, Tofaris GK, Davis JJ. Concentration-normalized electroanalytical assaying of exosomal markers. Anal Chem. 2017;89(5):3184–90.
40. Fan Z, Yu J, Lin J, et al. Exosome-specific tumor diagnosis via biomedical analysis of exosome-containing microRNA biomarkers. Analyst. 2019;144(19):5856–65.
41. Guo Q, Yu Y, Zhang H, et al. Electrochemical sensing of exosomal microRNA based on hybridization chain reaction signal amplification with reduced false-positive signals. Anal Chem. 2020;92(7):5302–10.
42. Soda N, Rehm BH, Sonar P, et al. Advanced liquid biopsy technologies for circulating biomarker detection. J Mater Chem B. 2019;7(43):6670–704.
43. Wang W, Luo J, Wang S. Recent progress in isolation and detection of extracellular vesicles for cancer diagnostics. Adv Healthc Mater. 2018;7(20):1800484.

44. Jiang J, Yu Y, Zhang H, et al. Electrochemical aptasensor for exosomal proteins profiling based on DNA nanotetrahedron coupled with enzymatic signal amplification. Anal Chim Acta. 2020;1130:1–9.
45. Ortega FG, Piguillem SV, Messina GA, et al. EGFR detection in extracellular vesicles of breast cancer patients through immunosensor based on silica-chitosan nanoplatform. Talanta. 2019;194:243–52.
46. Boriachek K, Umer M, Islam MN, et al. An amplification-free electrochemical detection of exosomal miRNA-21 in serum samples. Analyst. 2018;143(7):1662–9.
47. Liu L, Lu H, Shi R, et al. Synergy of peptide-nucleic acid and spherical nucleic acid enabled quantitative and specific detection of tumor exosomal microRNA. Anal Chem. 2019;91(20):13198–205.
48. Li X, Li X, Li D, et al. Electrochemical biosensor for ultrasensitive exosomal miRNA analysis by cascade primer exchange reaction and MOF@Pt@MOF nanozyme. Biosens Bioelectron. 2020;168:112554.
49. Cheng W, Ma J, Cao P, et al. Enzyme-free electrochemical biosensor based on double signal amplification strategy for the ultra-sensitive detection of exosomal microRNAs in biological samples. Talanta. 2020;219:121242.
50. Wang LL, Chen WQ, Wang YR, et al. Numerous long single-stranded DNAs produced by dual amplification reactions for electrochemical detection of exosomal microRNAs. Biosens Bioelectron. 2020;169:112555.
51. Zhang J, Wang LL, Hou MF, et al. A ratiometric electrochemical biosensor for the exosomal microRNAs detection based on bipedal DNA walkers propelled by locked nucleic acid modified toehold mediate strand displacement reaction. Biosens Bioelectron. 2018;102:33–40.
52. Miao P, Tang Y. Dumbbell hybridization chain reaction based electrochemical biosensor for ultrasensitive detection of exosomal miRNA. Anal Chem. 2020;92(17):12026–32.
53. Colombo M, Moita C, Van Niel G, et al. Analysis of ESCRT functions in exosome biogenesis, composition and secretion highlights the heterogeneity of extracellular vesicles. J Cell Sci. 2013;126(Pt 24):5553–65.
54. Yáñez-Mó M, Siljander PR-M, Andreu Z, et al. Biological properties of extracellular vesicles and their physiological functions. J Extracellular Vesicles. 2015;4(1):27066.
55. Caponnetto F, Manini I, Skrap M, et al. Size-dependent cellular uptake of exosomes. Nanomed Nanotechnol Biol Med. 2017;13(3):1011–20.
56. Sharma S, Gillespie BM, Palanisamy V, et al. Quantitative nanostructural and single-molecule force spectroscopy biomolecular analysis of human-saliva-derived exosomes. Langmuir ACS J Surf Colloids. 2011;27(23):14394–400.
57. Liu F, Vermesh O, Mani V, et al. The exosome total isolation chip. ACS Nano. 2017;11(11):10712–23.
58. van der Pol E, Coumans F, Varga Z, et al. Innovation in detection of microparticles and exosomes. J Thromb Haemostasis JTH. 2013;11(Suppl 1):36–45.
59. Mathew DG, Beekman P, Lemay SG, et al. Electrochemical detection of tumor-derived extracellular vesicles on nanointerdigitated electrodes. Nano Lett. 2020;20(2):820–8.
60. Chen X, Lan J, Liu Y, et al. A paper-supported aptasensor based on upconversion luminescence resonance energy transfer for the accessible determination of exosomes. Biosens Bioelectron. 2018;102:582–8.
61. Zhang Q, Wang F, Zhang H, et al. Universal $Ti_3C_2$ MXenes based self-standard ratiometric fluorescence resonance energy transfer platform for highly sensitive detection of exosomes. Anal Chem. 2018;90(21):12737–44.
62. Jin D, Yang F, Zhang Y, et al. ExoAPP: exosome-oriented, aptamer nanoprobe-enabled surface proteins profiling and detection. Anal Chem. 2018;90(24):14402–11.
63. Wang H, Chen H, Huang Z, et al. DNase I enzyme-aided fluorescence signal amplification based on graphene oxide-DNA aptamer interactions for colorectal cancer exosome detection. Talanta. 2018;184:219–26.

64. Li B, Liu C, Pan W, et al. Facile fluorescent aptasensor using aggregation-induced emission luminogens for exosomal proteins profiling towards liquid biopsy. Biosens Bioelectron. 2020;168:112520.
65. Wang L, Yang Y, Liu Y, et al. Bridging exosome and liposome through zirconium–phosphate coordination chemistry: a new method for exosome detection. Chem Commun (Camb). 2019;55(18):2708–11.
66. Li B, Pan W, Liu C, et al. Homogenous magneto-fluorescent nanosensor for tumor-derived exosome isolation and analysis. ACS Sens. 2020;5(7):2052–60.
67. He D, Ho SL, Chan HN, et al. Molecular-recognition-based DNA nanodevices for enhancing the direct visualization and quantification of single vesicles of tumor exosomes in plasma microsamples. Anal Chem. 2019;91(4):2768–75.
68. Liu C, Feng Q, Sun J. Lipid nanovesicles by microfluidics: manipulation, synthesis, and drug delivery. Adv Mater. 2019;31(45):e1804788.
69. Zhang P, He M, Zeng Y. Ultrasensitive microfluidic analysis of circulating exosomes using a nanostructured graphene oxide/polydopamine coating. Lab Chip. 2016;16(16):3033–42.
70. Tian Q, He C, Liu G, et al. Nanoparticle counting by microscopic digital detection: selective quantitative analysis of exosomes via surface-anchored nucleic acid amplification. Anal Chem. 2018;90(11):6556–62.
71. Liu C, Xu X, Li B, et al. Single-exosome-counting immunoassays for cancer diagnostics. Nano Lett. 2018;18(7):4226–32.
72. Li F, Liu X, Zhao B, et al. Graphene nanoprobes for real-time monitoring of isothermal nucleic acid amplification. ACS Appl Mater Interfaces. 2017;9(18):15245–53.
73. Hu J, Sheng Y, Kwak KJ, et al. A signal-amplifiable biochip quantifies extracellular vesicle-associated RNAs for early cancer detection. Nat Commun. 2017;8(1):1683.
74. Lewis JM, Vyas AD, Qiu Y, et al. Integrated analysis of exosomal protein biomarkers on alternating current electrokinetic chips enables rapid detection of pancreatic cancer in patient blood. ACS Nano. 2018;12(4):3311–20.
75. Zhang P, Zhou X, Zeng Y. Multiplexed immunophenotyping of circulating exosomes on nano-engineered ExoProfile chip towards early diagnosis of cancer. Chem Sci. 2019;10(21):5495–504.
76. Re Tegui E, van der Vos KE, Lai CP, et al. Engineered nanointerfaces for microfluidic isolation and molecular profiling of tumor-specific extracellular vesicles. Nat Commun. 2018;9(1):175.
77. Liu C, Zhao J, Tian F, et al. Low-cost thermophoretic profiling of extracellular-vesicle surface proteins for the early detection and classification of cancers. Nat Biomed Eng. 2019;3(3):183–93.
78. Roda A, Mirasoli M, Michelini E, et al. Progress in chemical luminescence-based biosensors: a critical review. Biosens Bioelectron. 2016;76:164–79.
79. Zhang J, Zou H, Lei J, et al. Multifunctional Au(I)-based AIEgens: manipulating molecular structures and boosting specific cancer cell imaging and theranostics. Angew Chem Int Ed Engl. 2020;59(18):7097–105.
80. Yan Y, Shi P, Song W, et al. Chemiluminescence and bioluminescence imaging for biosensing and therapy: in vitro and in vivo perspectives. Theranostics. 2019;9(14):4047–65.
81. Zhang Y, Wang F, Zhang H, et al. Multivalency Interface and g-$C_3N_4$ coated liquid metal nanoprobe signal amplification for sensitive electrogenerated chemiluminescence detection of exosomes and their surface proteins. Anal Chem. 2019;91(18):12100–7.
82. Zhang H, Wang Z, Wang F, et al. In situ formation of gold nanoparticles decorated $Ti_3C_2$ MXenes nanoprobe for highly sensitive electrogenerated chemiluminescence detection of exosomes and their surface proteins. Anal Chem. 2020;92(7):5546–53.
83. Fang D, Zhao D, Zhang S, et al. Black phosphorus quantum dots functionalized MXenes as the enhanced dual-mode probe for exosomes sensing. Sens Actuators B Chem. 2020;305:127544.
84. Jiang Q, Liu Y, Wang L, et al. Rapid enrichment and detection of extracellular vesicles enabled by CuS-enclosed microgels. Anal Chem. 2019;91(24):15951–8.

85. Qiao B, Guo Q, Jiang J, et al. An electrochemiluminescent aptasensor for amplified detection of exosomes from breast tumor cells (MCF-7 cells) based on G-quadruplex/hemin DNAzymes. Analyst. 2019;144(11):3668–75.
86. Guo Y, Cao Q, Zhao C, et al. Stimuli-responsive DNA microcapsules for homogeneous electrochemiluminescence sensing of tumor exosomes. Sens Actuators B Chem. 2021;329:129136.
87. Tiwari A, Dhoble SJ. Recent advances and developments on integrating nanotechnology with chemiluminescence assays. Talanta. 2018;180:1–11.
88. Feng QM, Ma P, Cao QH, et al. An aptamer-binding DNA walking machine for sensitive electrochemiluminescence detection of tumor exosomes. Chem Commun (Camb). 2019;56(2):269–72.
89. Pirzada M, Altintas Z. Nanomaterials for healthcare biosensing applications. Sensors (Basel). 2019;19(23):5311.
90. Zhong Y, Li J, Lambert A, et al. Expanding the scope of chemiluminescence in bioanalysis with functional nanomaterials. J Mater Chem B. 2019;7(46):7257–66.
91. Zhang L, Gu C, Wen J, et al. Recent advances in nanomaterial-based biosensors for the detection of exosomes. Anal Bioanal Chem. 2021;413(1):83–102.
92. Liu X, Wang Q, Chen J, et al. Ultrasensitive electrochemiluminescence biosensor for the detection of tumor exosomes based on peptide recognition and luminol-AuNPs@g-$C_3N_4$ nanoprobe signal amplification. Talanta. 2021;221:121379.
93. Li R, An Y, Jin T, et al. Detection of MUC1 protein on tumor cells and their derived exosomes for breast cancer surveillance with an electrochemiluminescence aptasensor. J Electroanal Chem. 2021;882:115011.
94. Hvastkovs EG, Schenkman JB, Rusling JF. Metabolic toxicity screening using electrochemiluminescence arrays coupled with enzyme-DNA biocolloid reactors and liquid chromatography-mass spectrometry. Annu Rev Anal Chem (Palo Alto, Calif). 2012;5(1):79–105.
95. Fathi F, Sharifi M, Jafari A, et al. Kinetic and thermodynamic insights into interaction of albumin with piperacillin: spectroscopic and molecular modeling approaches. J Mol Liq. 2019;296:111770.
96. Homola J. Surface plasmon resonance sensors for detection of chemical and biological species. Chem Rev. 2008;108(2):462–93.
97. Falkowski P, Lukaszewski Z, Gorodkiewicz E. Potential of surface plasmon resonance biosensors in cancer detection. J Pharm Biomed Anal. 2021;194:113802.
98. Mitchell JS, Wu Y, Cook CJ, et al. Sensitivity enhancement of surface plasmon resonance biosensing of small molecules. Anal Biochem. 2005;343(1):125–35.
99. Hasanzadeh M, Tagi S, Solhi E, et al. An innovative immunosensor for ultrasensitive detection of breast cancer specific carbohydrate (CA 15-3) in unprocessed human plasma and MCF-7 breast cancer cell lysates using gold nanospear electrochemically assembled onto thiolated graphene quantum dots. Int J Biol Macromol. 2018;114:1008–17.
100. Saha K, Agasti SS, Kim C, et al. Gold nanoparticles in chemical and biological sensing. Chem Rev. 2012;112(5):2739–79.
101. Su H, Li S, Jin Y, et al. Nanomaterial-based biosensors for biological detections. Adv Health Care Technol. 2017;3:19–29.
102. He L, Musick MD, Nicewarner SR, Salinas FG, Benkovic SJ, Natan MJ, Keating CD. Colloidal Au-enhanced surface plasmon resonance for ultrasensitive detection of DNA hybridization. J Am Chem Soc. 2000;122:9071–7.
103. Cittadini M, Bersani M, Perrozzi F, et al. Graphene oxide coupled with gold nanoparticles for localized surface plasmon resonance based gas sensor. Carbon. 2014;69:452–9.
104. Patil PO, Pandey GR, Patil AG, et al. Graphene-based nanocomposites for sensitivity enhancement of surface plasmon resonance sensor for biological and chemical sensing: a review. Biosens Bioelectron. 2019;139:111324.
105. Rupert DL, Lässer C, Eldh M, et al. Determination of exosome concentration in solution using surface plasmon resonance spectroscopy. Anal Chem. 2014;86(12):5929–36.
106. Lyon LA, Musick MD, Natan MJ. Colloidal Au-enhanced surface plasmon resonance immunosensing. Anal Chem. 1998;70(24):5177–83.

107. Fleischmann MP, Hendra PJ, McQuillan AJ. Raman spectra of pyridine adsorbed at a silver electrode. Chem Phys Lett. 1974;26(2):163–6.
108. Jeanmaire DL, Van Duyne RP. Surface Raman spectroelectrochemistry: Part I. Heterocyclic, aromatic, and aliphatic amines adsorbed on the anodized silver electrode. J Electroanal Chem Interfacial Electrochem. 1977;84(1):1–20.
109. Zong SF, Wang L, Chen C, et al. Facile detection of tumor-derived exosomes using magnetic nanobeads and SERS nanoprobes. Anal Methods. 2016;8(25):5001–8.
110. Wang Z, Zong S, Wang Y, et al. Screening and multiple detection of cancer exosomes using an SERS-based method. Nanoscale. 2018;10(19):9053–62.
111. Li TD, Zhang R, Chen H, et al. An ultrasensitive polydopamine bi-functionalized SERS immunoassay for exosome-based diagnosis and classification of pancreatic cancer. Chem Sci. 2018;9(24):5372–82.
112. Kwizera EA, O'Connor R, Vinduska V, et al. Molecular detection and analysis of exosomes using surface-enhanced Raman scattering gold nanorods and a miniaturized device. Theranostics. 2018;8(10):2722–38.
113. Zhang XR, Liu C, Pei YJ, et al. Preparation of a novel Raman probe and its application in the detection of circulating tumor cells and exosomes. ACS Appl Mater Interfaces. 2019;11(32):28671–80.
114. Poggio M, Hu T, Pai CC, et al. Suppression of exosomal PD-L1 induces systemic anti-tumor immunity and memory. Cell. 2019;177(2):414–27.e13.
115. Pang YF, Shi JM, Yang XS, et al. Personalized detection of circling exosomal PD-L1 based on $Fe_3O_4$@$TiO_2$ isolation and SERS immunoassay. Biosens Bioelectron. 2020;148:9.
116. Driskell JD, Seto AG, Jones LP, et al. Rapid microRNA (miRNA) detection and classification via surface-enhanced Raman spectroscopy (SERS). Biosens Bioelectron. 2008;24(4):917–22.
117. Crawford BM, Wang HN, Stolarchuk C, et al. Plasmonic nanobiosensors for detection of microRNA cancer biomarkers in clinical samples. Analyst. 2020;145(13):4587–94.
118. Morla-Folch J, Gisbert-Quilis P, Masetti M, et al. Conformational SERS classification of K-Ras point mutations for cancer diagnostics. Angew Chem Int Ed Engl. 2017;56(9):2381–5.
119. Pyrak E, Krajczewski J, Kowalik A, et al. Surface enhanced Raman spectroscopy for DNA biosensors-how far are we? Molecules. 2019;24(24):4423.
120. Li Y, Gao T, Xu G, et al. Direct approach toward label-free DNA detection by surface-enhanced Raman spectroscopy: discrimination of a single-base mutation in 50 base-paired double helixes. Anal Chem. 2019;91(13):7980–4.
121. Jiang S, Li Q, Wang C, et al. In situ exosomal microRNA determination by target-triggered SERS and $Fe_3O_4$@$TiO_2$-based exosome accumulation. ACS Sens. 2021;6(3):852–62.
122. Ni J, Lipert RJ, Dawson GB, et al. Immunoassay readout method using extrinsic Raman labels adsorbed on immunogold colloids. Anal Chem. 1999;71(21):4903–8.
123. Guven B, Dudak FC, Boyaci IH, et al. SERS-based direct and sandwich assay methods for mir-21 detection. Analyst. 2014;139(5):1141–7.
124. Zhang J, Yang Y, Jiang X, et al. Ultrasensitive SERS detection of nucleic acids via simultaneous amplification of target-triggered enzyme-free recycling and multiple-reporter. Biosens Bioelectron. 2019;141:111402.
125. Pang Y, Wang C, Lu L, et al. Dual-SERS biosensor for one-step detection of microRNAs in exosome and residual plasma of blood samples for diagnosing pancreatic cancer. Biosens Bioelectron. 2019;130:204–13.
126. Czamara K, Majzner K, Pacia MZ, et al. Raman spectroscopy of lipids: a review. J Raman Spectrosc. 2015;46(1):4–20.
127. Xiang S, Xu Y, Liao X, et al. Dynamic monitoring of the oxidation process of phosphatidylcholine using SERS analysis. Anal Chem. 2018;90(22):13751–8.
128. Pohanka M. Overview of piezoelectric biosensors, immunosensors and DNA sensors and their applications. Materials (Basel, Switzerland). 2018;11(3):448.
129. Románszki L, Varga Z, Mihály J, et al. Electromagnetic piezoelectric acoustic sensor detection of extracellular vesicles through interaction with detached vesicle proteins. Biosensors. 2020;10(11):173.

130. Suthar J, Parsons ES, Hoogenboom BW, et al. Acoustic immunosensing of exosomes using a quartz crystal microbalance with dissipation monitoring. Anal Chem. 2020;92(5):4082–93.
131. Wang C, Wang C, Jin D, et al. AuNP-amplified surface acoustic wave sensor for the quantification of exosomes. ACS Sens. 2020;5(2):362–9.
132. Yu Y, Li YT, Jin D, et al. Electrical and label-free quantification of exosomes with a reduced graphene oxide field effect transistor biosensor. Anal Chem. 2019;91(16):10679–86.
133. Cavallaro S, Horak J, Hååg P, Gupta D, et al. Label-free surface protein profiling of extracellular vesicles by an electrokinetic sensor. ACS Sens. 2019;4(5):1399–408.
134. Etayash H, McGee AR, Kaur K, et al. Nanomechanical sandwich assay for multiple cancer biomarkers in breast cancer cell-derived exosomes. Nanoscale. 2016;8(33):15137–41.
135. Mao S, Chang J, Pu H, et al. Two-dimensional nanomaterial-based field-effect transistors for chemical and biological sensing. Chem Soc Rev. 2017;46(22):6872–904.
136. Adamczyk Z, Sadlej K, Wajnryb E, et al. Streaming potential studies of colloid, polyelectrolyte and protein deposition. Adv Colloid Interface Sci. 2010;153(1–2):1–29.
137. König IR, Fuchs O, Hansen G, et al. What is precision medicine? Eur Respir J. 2017;50(4):1700391.
138. Wang S, Khan A, Huang R, et al. Recent advances in single extracellular vesicle detection methods. Biosens Bioelectron. 2020;154:112056.
139. Chen C, Zong S, Liu Y, et al. Profiling of exosomal biomarkers for accurate cancer identification: combining DNA-PAINT with machine-learning-based classification. Small (Weinheim an der Bergstrasse, Germany). 2019;15(43):e1901014.
140. Shen M, Di K, He H, et al. Progress in exosome associated tumor markers and their detection methods. Mol Biomed. 2020;1:1.

# Technologies for the Surface Functionalization of Extracellular Vesicles

11

Gang Han, Mengyuan Geng, and HaiFang Yin

## 11.1 Introduction

Extracellular vesicles (EVs) are highly heterogeneous bilayer lipid vesicles with an average diameter of 30–1000 nm, primarily including exosomes and microvesicles. Exosomes are formed via fusion of endosome-derived multivesicular bodies with the plasma membrane and microvesicles that directly bud from the plasma membrane [1, 2]. EVs are exquisite natural nanovesicles characterized by a sophisticated membrane structure and distinct nucleic acid, protein, and lipid constituents [3]; thus, synthetic or artificial membranous nanoparticles cannot recapitulate EVs. Notably, the compositions of EVs originating from different sources vary greatly, and therefore EVs can either function as a "friend" or "foe" depending on their cellular origin. Mesenchymal stem cell-derived EVs have been tested in clinical trials for diabetes, chronic kidney diseases, and macular degeneration, with promising results [4], representing an effective example of well-tolerated EVs. Heat shock protein 70-enriched EVs have also been shown to mediate protection of auditory hair cells from ototoxic insults [5]. By contrast, EVs have been shown to promote tumor metastasis or dissemination of prion proteins in tumors or neurodegenerative disorders [6, 7]. Thus, the origins of EVs should be carefully considered prior to application in the clinical setting. As natural drug carriers, EVs exhibit favorable biocompatibility, low immunogenicity, and well-established safety profiles in different species [8]. The presence of CD47, a do-not-eat signal [9], on the surface of EVs was shown to enable EVs to escape from the surveillance of macrophages and thus extend the circulation time [10]. These favorable attributes have attracted much interest in the possibility of using EVs for diagnostic and therapeutic purposes; thus,

G. Han · M. Geng · H. Yin (✉)
School of Medical Technology & School of Medical Sciences, Tianjin Medical University, Tianjin, China
e-mail: haifangyin@tmu.edu.cn



Q. Wang, L. Zheng (eds.), *Extracellular Vesicles*,
https://doi.org/10.1007/978-981-99-8365-0_11

EVs are being actively pursued by biotech companies and clinicians. Despite these intrinsic advantages, most EVs, irrespective of origin and administration route, preferentially accumulate in the liver and spleen [11]. Accordingly, EVs should be modified with tissue- or cell-targeting moieties. Notably, the ability to load exogenous cargo is a prerequisite for the therapeutic application of EVs. Moreover, the development of EV surface bioengineering technologies has improved the applicability and effectiveness of EVs in the biomedical field. In this chapter, we will summarize recent developments on EV surface functionalization technologies and carefully weigh the advantages and limitations of each approach. We also add our perspectives on the clinical feasibility of these approaches and other future research directions.

## 11.2 Surface Modification of EVs Via Genetic Approaches

Cell- or tissue-specific delivery and external drug loading are essential for the therapeutic applications of EVs. To confer the specificity of EVs to target cells or tissues and equip EVs with external functional moieties, genetic approaches have been extensively employed by expressing tissue-targeting peptides and therapeutic peptides/proteins fused to the EV transmembrane- or membrane-associated proteins. Below, we summarize different EV surface functionalization approaches via genetic engineering based on the type of proteins used (Fig. 11.1).

### 11.2.1 Tetraspanin

EVs are highly enriched in tetraspanin, a superfamily of proteins with four transmembrane domains involved in cell signaling, signal transduction, and cell adhesion [12], particularly CD63, CD81, and CD9 [13, 14]. Knockout of CD63 expression impairs the secretion of small EVs, suggesting that CD63 may have critical roles in EV biogenesis [15]. These tetraspanin proteins have been defined as EV biomarkers based on Minimal Information for Studies of Extracellular Vesicles 2018 (MISEV2018) [16] and are therefore the preferred molecular scaffolds for customizing the EV surface given the relative abundance per EV. Corso et al. fused green fluorescence protein (GFP) tags to 12 EV-associated proteins, including transmembrane proteins (CD63, CD9, CD81, syndecan, lysosome-associated membrane protein 2b [Lamp2b]), membrane-associated proteins (flotillin-2, C1C2 domain of lactadherin, and myristoylation tag), and soluble proteins (syntenin, ALG-2-interacting protein X, small integral membrane protein of the lysosome/late endosome, and ubiquitin C), and compared the relative loading efficiency and specificity of these proteins for HEK293T EVs. The results showed that the transmembrane proteins were more efficient than cytosolic proteins for vesicular loading, particularly the tetraspanin proteins CD63, CD9, and CD81. Ligation of GFP to CD63 is robust and results in minimal vesicle perturbation, irrespective of the integration site [17]. This study was also corroborated with N-terminal fusion of tetraspanins and

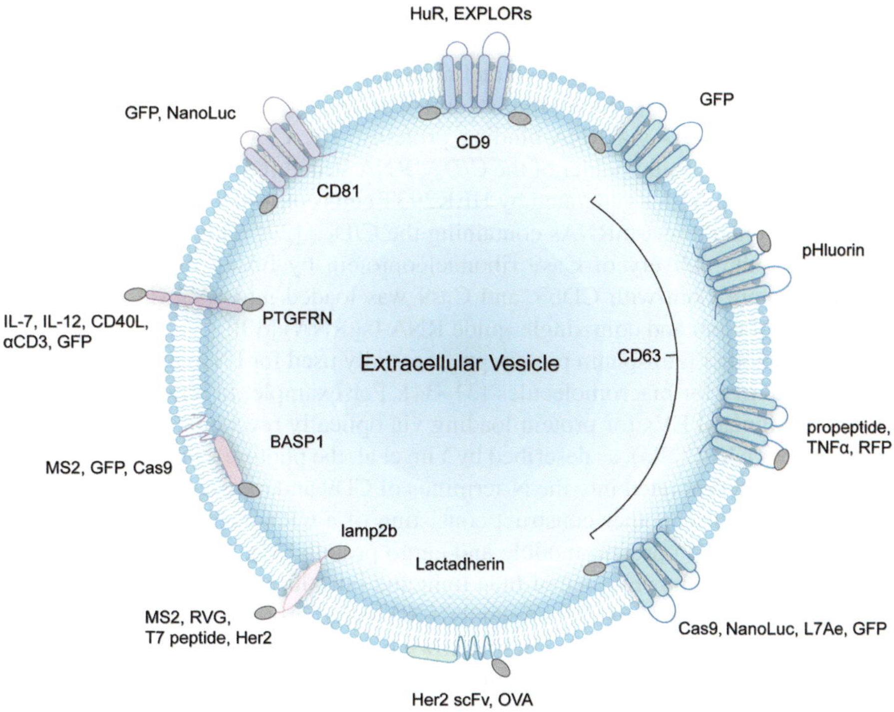


**Fig. 11.1** Schematic for surface functionalization of EVs via genetic approaches

luciferase, as described by Gupta et al. [18]. Thus, CD63 has been widely implemented for surface engineering of EVs in vitro and in vivo, such as for tracking EVs with fluorescent proteins [19, 20] or the nano-luciferase reporter gene. In particular, labeled CD63 controlled by tissue-specific promoters can be used to track cell-type specific EVs in vivo [21]. To improve our ability to observe fusion events of multivesicular bodies with the plasma membrane and EV secretion, Sung et al. adopted a pH-sensitive GFP derivative (pHluorin) and displayed it on the first extracellular loop of CD63 [22, 23]. This pH-responsive reporter system enabled real-time intracellular visualization of EV maturation. Through sequential truncation of fluorescent protein-tagged CD63, Curley et al. discovered that transmembrane helix 3 is critical for surface engineering of EVs [24], and this may facilitate the design of more compact fusion proteins for EV functionalization.

CD63 has also been used for cargo loading and decoration of the EV surface with targeting ligands or functional moieties other than visualization and imaging. Duong et al. designed decoy EVs by fusing the tumor necrosis factor alpha (TNFα) binding domain of human TNF receptor-1 (hTNFR1) to the second extracellular loop of CD63 and stably transfecting the construct into HEK293 cells, resulting in the generation of EVs with hTNFR1 projected on the external surface. These hTNFR1-functionalized EVs inhibited TNFα activation in vitro [25]. Similarly, Ran et al. expressed myostatin propeptide, a natural inhibitor of myostatin [26], on the outer surface of EVs via the fusion of propeptide with the second extracellular loop of CD63. Propeptide-enriched EVs bound to mature myostatin in circulation and contributed to increased muscle growth and mass and phenotypic rescue of dystrophic mice [27]. In addition to these proteins and peptides, CD63 has also been used for the loading of RNAs via RNA-binding proteins. The CD63-fused RNA-binding domain L7Ae, a specific binder of the $C/D_{box}$ RNA structure [28], effectively loaded target mRNAs into EVs generated by HEK293T cells cotransfected with the CD63-L7Ae fusion and target mRNAs containing the $C/D_{box}$ [29]. This strategy was also adopted for the delivery of Cas9 ribonucleoprotein by fusing the aptamer com-binding protein Com with CD63, and Cas9 was loaded into EVs via the specific interaction of Com and com-single-guide RNA (sgRNA) [30].

CD9 is another tetraspanin protein preferentially used for EV internal drug loading, particularly for macromolecules [31–34]. For example, in an optogenetic EV system designated EVs for protein loading via optically reversible protein–protein interactions (EXPLORs), as described by Yim et al. the photoreceptor cryptochrome 2 (CRY2) is incorporated into the N-terminus of CD9 and transfected into HEK293 cells together with another construct consisting of a CRY-interacting basic-helix–loop–helix 1 (CIB1) protein module and cargo protein [31]. CIB1-CD9 interacted with CRY only in the presence of blue light and introduced CRY-cargo proteins to EVs during EV biogenesis. Subsequently, CRY-cargo proteins detach from the EV membrane in the absence of blue light, allowing cytosolic delivery of cargo proteins [31]. This system was employed to load super-repressor IκBα protein into EVs, and IκBα-loaded EVs were found to attenuate mortality and systemic inflammation in septic mouse models [32]. To facilitate RNA drug loading, RNA-binding proteins, such as human antigen R (HuR), were incorporated into the C-terminus of CD9. Cotransfection of target microRNAs (e.g., *miR-155*) or RNAs may permit the binding of HuR to *miR-155* or sgRNAs prior to sorting to EVs inside cells. Subsequently, HuR-associated *miR-155* was delivered to recipient cells via EVs in vitro and in vivo [33].

In summary, tetraspanin proteins have been extensively applied as molecular scaffolds for surface functionalization of EVs via genetic engineering and show promise in EV imaging, visualization, and drug loading. The development of innovative tools, such as EXPLOR, supports the use of genetically modified tetraspanins for EV engineering, although the sorting efficiency is highly variable depending on the origin of the transfected cells; this may limit the application of this approach.

### 11.2.2 Lamp2b

In addition to tetraspanin family proteins, Lamp2b is another transmembrane protein enriched on EVs, particularly in dendritic cell (DC)-derived EVs [35]. The function of Lamp2b as a scaffold was demonstrated in our seminal study in 2011. To target EVs to neurons, the rabies virus-derived acetylcholine receptor-specific RVG peptide [36] was fused to the N-terminus of Lamp2b and expressed in immature DCs. EVs were shown to cross the blood–brain barrier and mediate neuron-specific delivery of small interfering RNAs (siRNAs) in Alzheimer's disease mouse models [36]. Subsequently, the same strategy was adopted for neuron-specific delivery of α-synuclein siRNA [37], *miR-124* [38], nerve growth factor protein and mRNA [39], and *miR-193b-3p* [40] to mouse models of Parkinson's disease, cerebral ischemia, and subarachnoid hemorrhage. However, Huang et al. reported that small RNAs actively loaded into EVs via the fusion of RNA-binding MS2 loops and Lamp2b were subjected to rapid degradation in recipient cells [41]; this result was not entirely unsurprising because Lamp2b is known to be associated with lysosomes in cells. In addition to RVG, the transferrin receptor-binding T7 peptide [42] has been used to decorate EVs and target the delivery of anti-*miR-21* oligonucleotides to glioblastoma in a rat model [43]. Similarly, Lamp2b fusions with cardiac-specific [44], ischemic myocardium-targeting [45], and chondrocyte-binding peptides [46] have also been implemented for specific delivery of EVs to the heart, ischemic myocardium, or deep cartilage regions in mouse and rat models. In addition, peptides such as α-integrin-specific RGD [47], tLyp-1 peptide [48], and human epidermal growth factor receptor 2 (HER2) [49] were used for delivery of EVs to tumors. However, the target moieties enriched on EVs via Lamp2b were degraded rapidly. Although glycosylation signals at various positions in the fusion protein can improve stability and production yield [50], the need for dramatic optimization of stability makes Lamp2b less attractive than tetraspanins.

### 11.2.3 Lactadherin

Lactadherin is a secreted, membrane-associated protein that consists of two functional domains, i.e., $NH_2$-terminal epidermal growth factor (EGF)-like and COOH-terminal (i.e., C1C2) domains [51]. C1C2 shows strong binding affinity to phosphatidylserine-enriched lipid membranes [52] and thus is able to facilitate the sorting or enrichment of functional moieties on membranous vesicles, such as EVs. Notably, as a molecular scaffold for the surface functionalization of EVs, C1C2 has been primarily used for tumor antigen display and tumor immunotherapy [53–56]. To validate whether EV can be a source of tumor antigens, Zeelenberg et al. targeted chicken egg ovalbumin (OVA) to EVs via fusion with the C1C2 domain of lactadherin and demonstrated that EV-associated OVA induced more potent antitumor immune responses than soluble OVA in vitro and in vivo [53]. Similarly, C1C2 was adopted for the display of prostate cancer antigens on EVs for prostate cancer immunotherapy [54]. C1C2 can also anchor tumor-targeting moieties, such as HER2 single-chain variable fragment (scFv), on the surface of EVs [57, 58]. HER2 scFv-decorated EVs direct tumor-specific delivery of the prodrug/enzyme regimen to HER2-positive human breast cancer mouse models [58]. However, lactadherin is a membrane-associated protein rather than a transmembrane protein; thus, tethering of C1C2 with functional moieties may result in less stable interactions with the exosomal membrane [59].

### 11.2.4 Other Moieties

Glycosylphosphatidylinositol (GPI)-anchored proteins show strong targeting to the lipid bilayer membrane [60], and GPI-anchored protein decay-accelerating factor (DAF) has been shown to selectively sort to EVs during reticulocyte maturation [61]. Based on these observations, Kooijmans et al. anchored EGF receptor (EGFR) nanobodies on the surfaces of EVs by fusing an EGFR nanobody to the C-terminus of the DAF signal peptide, thereby enabling localization of the nanobody to the GPI-rich lipid rafts of EV membranes.

EGFR nanobodies displayed on EVs showed enhanced binding to EGFR-overexpressing cancer cells, although no increase in cellular uptake was detected [62]. In addition to GPI-anchored proteins, Dooley et al. used proteomic analyses to identify two scaffold membrane proteins, prostaglandin F2 receptor negative regulator (PTGFRN) and brain acid soluble protein 1 (BASP1), from EVs derived from HEK293 cells [63]. In these cells, PTGFRN is a single transmembrane protein [64], and BASP1 is a membrane-bound protein [65]. The results showed that these two proteins are preferentially sorted into EVs and enable surface display and luminal loading of cytokines, RNA-binding proteins, antibody fragments, antigens, or Cas9 proteins by fusing either to the N- or C-terminus of PTGFRN and the C-terminus of BASP1 [63]. Other than utilizing endogenous membrane or membrane-associated proteins, Ohno et al. adopted a commercial pDisplay vector, which enables the display of target moieties on the plasma membrane via fusion with the transmembrane domain of platelet-derived growth factor receptor (PDGFR), for the surface display of the EGFR-binding peptide GE11. GE11 peptides were presented on the surfaces of EVs derived from transfected HEK293 cells and GE11-modified EV-directed specific delivery of *let-7a* miRNA to EGFR-expressing xenograft breast cancer tissues in immunodeficient mice [66]. The strategy was also utilized to anchor anti-CD3 and anti-EGFR monoclonal antibodies on the surface of EVs, as described by Cheng et al. for targeted immunotherapy in EGFR-expressing breast cancer models in vitro and in vivo [67]. Interestingly, it is not clear why PDGFR is preferentially localized to the EV surface and whether this mechanism is conserved in all cell types.

Overall, cellular engineering has been exploited to achieve surface functionalization of EVs. However, this strategy faces several challenges, as follows: (1) only genetically encodable moieties without potential hazards, such as intracellular triggers of proliferation, can be used; (2) cell transfection efficiency varies dramatically and depends on the cells used and cargo to be expressed and therefore may not present a manufacturable option; (3) sorting efficiency is generally low with the possibility of loss when the cells are at a high passage number; (4) the method is time-consuming and labor-intensive because the target cell line needs to be established whenever a new target is identified. Therefore, genetically encoded moieties will likely be a niche approach to clinical deployment, and direct, simpler, and scalable EV surface functionalization will be the more popular approach.

## 11.3 Direct Surface Functionalization of EVs

A parallel approach, direct surface functionalization, may have applications in the preparation of smart nanodelivery systems based on ease, high performance, enhanced efficacy, safety, and cost. We describe these approaches, which have been extensively tested in vitro and in vivo (Fig. 11.2), in the sections below.

### 11.3.1 Covalent Conjugation

Chemical methods, particularly copper-catalyzed or copper-free azide-alkyne cycloaddition (click chemistry), have been applied to directly attach imaging or targeting molecules to the surface of EVs via covalent bonds for visualization and targeted delivery of EVs in vitro and in vivo [68–73]. To achieve simultaneous imaging and treatment of glioma, Jia et al. loaded superparamagnetic iron oxide nanoparticles (SPIONs) and curcumin into EVs, followed by conjugation of neuropilin-1-specific peptide (RGE) to EV membranes with click chemistry. RGE-modified EVs could cross the blood–brain barrier and elicit synergistic effects for SPION-mediated imaging and chemotherapy of glioma in mice bearing orthotopic tumors [74]. To endow EVs with the ability to release cargo in response to pH changes, Nie et al. designed responsive EV nanobioconjugates by modifying M1

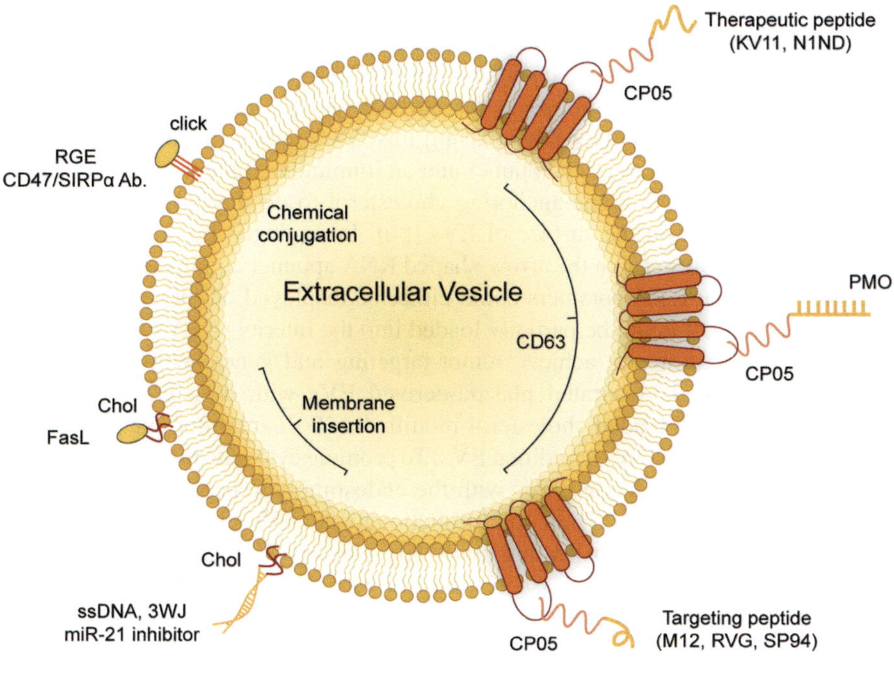


**Fig. 11.2** Illustration of the direct surface functionalization of EVs

macrophage membranes with azide groups, thereby generating azide-functionalized EVs. These EVs were simultaneously conjugated with dibenzocyclooctyne-modified anti-CD47 (aCD47) and anti-signal regulatory protein alpha (SIRPα) antibodies (aSIRPα) linked with pH-sensitive benzoic-imine bonds. EVs equipped with aCD47 and aSIRPα showed specific accumulation in CD47-overexpressing tumor tissues. Selective cleavage of the benzoic-imine bond in response to the acidic tumor microenvironment led to the release of aCD47 and aSIRPα, which consequently abolished the "don't eat me" signal mediated by CD47 and SIRPα, improved phagocytosis by macrophages, and enhanced the antitumor effect in vivo [75]. This strategy is generally applicable to other therapeutic or targeting moieties as long as the moiety can be modified to react with azide-functionalized EVs.

Although nonspecific covalent modification of EVs is a straightforward method for surface modification, such bioconjugation strategies could impair EV functions by altering or obscuring functional sites of surface proteins. Additionally, although chemical modification enables the surface display of different moieties via conjugation reactions, the exquisite surface structure of EVs may limit the conjugation owing to a lack of site specificity.

### 11.3.2 Membrane Insertion

Cholesterol is capable of spontaneous insertion into the bilayer lipid membrane via hydrophobic interactions, which result in stable binding to the membrane [76]. Thus, cholesterol has been used as an exosomal anchor to load different moieties, primarily nucleic acids, on the surface or the interior of EVs [77–81]. Yerneni et al. designed cholesterol-conjugated single-stranded DNA (ssDNA) for EV display and attached other functional groups to the surface of EVs via interaction with the ssDNA or its complementary DNA. Using this DNA tether, the authors equipped EVs with a nucleolin-targeting aptamer and an immunomodulatory protein [79]. Pi et al. further used membrane-anchoring cholesterol to display three-way junction RNA nanoparticles on the surface of EVs [80]. Intriguingly, by manipulating the orientation of cholesterol in the arrow-shaped RNA aptamer conjugates, the authors demonstrated that RNA aptamers could either be displayed on the surface of EVs for cell targeting or could be partially loaded into the interior of EVs for drug loading [80]. Furthermore, to achieve tumor-targeting and gene/chemo combination therapy, Zhan et al. decorated plasma-derived EVs with transferrin-conjugated magnetic beads and loaded cholesterol-modified *miR-21* inhibitor and unmodified doxorubicin into transferrin-modified EVs. To promote cytosolic release, the authors also modified plasma-derived EVs with the endosomolytic peptide L17E [82] via electrostatic interactions. These multifunctionalized EVs elicited combinatorial therapeutic effects in tumor-bearing mice [83]. Notably, although cholesterol is an effective linker for membrane association or hydrophobic insertion, one practical challenge to conjugating cholesterol to any functional moiety is the need to identify an appropriate solvent to dissolve amphiphilic molecules because cholesterol-containing compounds have a strong tendency to self-assemble [84]. Moreover, external introduction of cholesterol may result in side effects under certain pathophysiological conditions [85].

### 11.3.3 Noncovalent Modifications

Noncovalent binding via specific interactions is another strategy for EV surface functionalization. A striking example is the EV-anchoring peptide CP05, which was identified via phage screening; this peptide shows strong binding capacity to CD63 on EVs. As a proof-of-concept study, Gao et al. painted EVs from different origins with targeting moieties via CP05 without compromising the integrity and function of the EVs. Importantly, CP05 enabled simultaneous decoration of EVs with muscle-targeting peptides and phosphorodiamidate morpholino oligomers (PMOs) on single EVs, which resulted in targeted delivery of PMOs to muscle and restoration of functional dystrophin protein in body-wide muscles of dystrophic mice [86]. In addition to tissue-targeting peptides and nucleic acid analogs, immunomodulatory, antiangiogenic, and neuroprotective peptides were also painted on the surfaces of EVs via CP05 [87–89]. Furthermore, Zha et al. used CP05 as a flexible linker to decorate polycaprolactone three-dimensional-printed porous bone scaffolds with EVs encapsulating vascular endothelial growth factor for an acellular sustainable and stable treatment approach in rats with large segmental bone defects [90]. In addition, CP05 can be used for capturing circulatory EVs and thus serves as a handy tool for efficient tissue-targeting, drug loading, and capture of EVs without compromising the intrinsic functionality of EVs [86]. Thus, noncovalent direct surface functionalization of EVs represents a versatile platform for engineering EVs with defined therapeutic properties, although more comprehensive studies are warranted to investigate the loading efficiencies of these proteins.

## 11.4 Future Perspectives

In summary, EVs show strong potential for both diagnostic and therapeutic applications. The development of isolation and functionalization technologies has created many paths for the clinical utility of EVs. However, there are still barriers to the widespread clinical application of EVs. Surface functionalization technologies endow EVs with tissue/cell selectivity and the ability to add functional moieties to improve therapeutic efficacy, although such targeting technologies have intrinsic physical limitations, such as the inability to cross the blood–brain and heart barriers. There is also room for improving the loading efficiency, particularly for large molecules, including proteins such as Cas9 and nucleic acids, and this should not be overlooked when designing novel gene therapies. Because surface functionalization exposes the attached moieties to the environment, not all molecules are suitable for display. Those suitable include targeting molecules, immunomodulatory moieties (cytokines, antibodies, and small compounds), inhibitors, and antisense oligonucleotides with relatively stable backbones. Beyond therapeutic delivery, surface functionalization of EVs can also capture and remove hazardous biomolecules, thereby depleting tumor-promoting EVs or controlling the release of captured EVs containing clinically beneficial moieties. Thus, we are enthusiastic for the future development and clinical prospects of EV surface functionalization.

## References

1. van Niel G, D'Angelo G, Raposo G. Shedding light on the cell biology of extracellular vesicles. Nat Rev Mol Cell Biol. 2018;19(4):213–28.
2. Biressi S, Filareto A, Rando TA. Stem cell therapy for muscular dystrophies. J Clin Invest. 2020;130(11):5652–64.
3. Kalluri R, LeBleu VS. The biology, function, and biomedical applications of exosomes. Science. 2020;367(6478):eaau6977.
4. Mendt M, Rezvani K, Shpall E. Mesenchymal stem cell-derived exosomes for clinical use. Bone Marrow Transplant. 2019;54(Suppl 2):789–92.
5. Breglio AM, May LA, Barzik M, et al. Exosomes mediate sensory hair cell protection in the inner ear. J Clin Invest. 2020;130(5):2657–72.
6. Azmi AS, Bao B, Sarkar FH. Exosomes in cancer development, metastasis, and drug resistance: a comprehensive review. Cancer Metastasis Rev. 2013;32(3–4):623–42.
7. Hartmann A, Muth C, Dabrowski O, et al. Exosomes and the prion protein: more than one truth. Front Neurosci. 2017;11:194.
8. Mendt M, Kamerkar S, Sugimoto H, et al. Generation and testing of clinical-grade exosomes for pancreatic cancer. JCI Insight. 2018;3(8):e99263.
9. Brightwell RM, Grzankowski KS, Lele S, et al. The CD47 "don't eat me signal" is highly expressed in human ovarian cancer. Gynecol Oncol. 2016;143(2):393–7.
10. Kamerkar S, LeBleu VS, Sugimoto H, et al. Exosomes facilitate therapeutic targeting of oncogenic KRAS in pancreatic cancer. Nature. 2017;546(7659):498–503.
11. Wiklander OP, Nordin JZ, O'Loughlin A, et al. Extracellular vesicle in vivo biodistribution is determined by cell source, route of administration and targeting. J Extracell Vesicles. 2015;4:26316.
12. Charrin S, Jouannet S, Boucheix C, et al. Tetraspanins at a glance. J Cell Sci. 2014;127(Pt 17):3641–8.
13. Andreu Z, Yanez-Mo M. Tetraspanins in extracellular vesicle formation and function. Front Immunol. 2014;5:442.
14. Escola JM, Kleijmeer MJ, Stoorvogel W, et al. Selective enrichment of tetraspan proteins on the internal vesicles of multivesicular endosomes and on exosomes secreted by human B-lymphocytes. J Biol Chem. 1998;273(32):20121–7.
15. Hurwitz SN, Conlon MM, Rider MA, et al. Nanoparticle analysis sheds budding insights into genetic drivers of extracellular vesicle biogenesis. J Extracell Vesicles. 2016;5:31295.
16. Thery C, Witwer KW, Aikawa E, et al. Minimal information for studies of extracellular vesicles 2018 (MISEV2018): a position statement of the International Society for Extracellular Vesicles and update of the MISEV2014 guidelines. J Extracell Vesicles. 2018;7(1):1535750.
17. Corso G, Heusermann W, Trojer D, et al. Systematic characterization of extracellular vesicle sorting domains and quantification at the single molecule—single vesicle level by fluorescence correlation spectroscopy and single particle imaging. J Extracell Vesicles. 2019;8(1):1663043.
18. Gupta D, Liang X, Pavlova S, et al. Quantification of extracellular vesicles in vitro and in vivo using sensitive bioluminescence imaging. J Extracell Vesicles. 2020;9(1):1800222.
19. Stickney Z, Losacco J, McDevitt S, et al. Development of exosome surface display technology in living human cells. Biochem Biophys Res Commun. 2016;472(1):53–9.
20. Joshi BS, de Beer MA, Giepmans BNG, et al. Endocytosis of extracellular vesicles and release of their cargo from endosomes. ACS Nano. 2020;14(4):4444–55.

21. Luo W, Dai Y, Chen Z, et al. Spatial and temporal tracking of cardiac exosomes in mouse using a nano-luciferase-CD63 fusion protein. Commun Biol. 2020;3(1):114.
22. Miesenbock G, De Angelis DA, Rothman JE. Visualizing secretion and synaptic transmission with pH-sensitive green fluorescent proteins. Nature. 1998;394(6689):192–5.
23. Sung BH, von Lersner A, Guerrero J, et al. A live cell reporter of exosome secretion and uptake reveals pathfinding behavior of migrating cells. Nat Commun. 2020;11(1):2092.
24. Curley N, Levy D, Do MA, et al. Sequential deletion of CD63 identifies topologically distinct scaffolds for surface engineering of exosomes in living human cells. Nanoscale. 2020;12(22):12014–26.
25. Duong N, Curley K, Brown A, et al. Decoy exosomes as a novel biologic reagent to antagonize inflammation. Int J Nanomed. 2019;14:3413–25.
26. Collins-Hooper H, Sartori R, Macharia R, et al. Propeptide-mediated inhibition of myostatin increases muscle mass through inhibiting proteolytic pathways in aged mice. J Gerontol A Biol Sci Med Sci. 2014;69(9):1049–59.
27. Ran N, Gao X, Dong X, et al. Effects of exosome-mediated delivery of myostatin propeptide on functional recovery of mdx mice. Biomaterials. 2020;236:119826.
28. Saito H, Kobayashi T, Hara T, et al. Synthetic translational regulation by an L7Ae-kink-turn RNP switch. Nat Chem Biol. 2010;6(1):71–8.
29. Kojima R, Bojar D, Rizzi G, et al. Designer exosomes produced by implanted cells intracerebrally deliver therapeutic cargo for Parkinson's disease treatment. Nat Commun. 2018;9(1):1305.
30. Yao X, Lyu P, Yoo K, et al. Engineered extracellular vesicles as versatile ribonucleoprotein delivery vehicles for efficient and safe CRISPR genome editing. J Extracell Vesicles. 2021;10(5):e12076.
31. Yim N, Ryu SW, Choi K, et al. Exosome engineering for efficient intracellular delivery of soluble proteins using optically reversible protein–protein interaction module. Nat Commun. 2016;7:12277.
32. Choi H, Kim Y, Mirzaaghasi A, et al. Exosome-based delivery of super-repressor IkappaBalpha relieves sepsis-associated organ damage and mortality. Sci Adv. 2020;6(15):eaaz6980.
33. Li Z, Zhou X, Wei M, et al. In vitro and in vivo RNA inhibition by CD9-HuR functionalized exosomes encapsulated with miRNA or CRISPR/dCas9. Nano Lett. 2019;19(1):19–28.
34. Li Z, Zhou X, Gao X, et al. Fusion protein engineered exosomes for targeted degradation of specific RNAs in lysosomes: a proof-of-concept study. J Extracell Vesicles. 2020;9(1):1816710.
35. Simhadri VR, Reiners KS, Hansen HP, et al. Dendritic cells release HLA-B-associated transcript-3 positive exosomes to regulate natural killer function. PLoS One. 2008;3(10):e3377.
36. Alvarez-Erviti L, Seow Y, Yin H, et al. Delivery of siRNA to the mouse brain by systemic injection of targeted exosomes. Nat Biotechnol. 2011;29(4):341–5.
37. Cooper JM, Wiklander PB, Nordin JZ, et al. Systemic exosomal siRNA delivery reduced alpha-synuclein aggregates in brains of transgenic mice. Mov Disord. 2014;29(12):1476–85.
38. Yang J, Zhang X, Chen X, et al. Exosome mediated delivery of miR-124 promotes neurogenesis after ischemia. Mol Ther Nucleic Acids. 2017;7:278–87.
39. Yang J, Wu S, Hou L, et al. Therapeutic effects of simultaneous delivery of nerve growth factor mRNA and protein via exosomes on cerebral ischemia. Mol Ther Nucleic Acids. 2020;21:512–22.
40. Lai N, Wu D, Liang T, et al. Systemic exosomal miR-193b-3p delivery attenuates neuroinflammation in early brain injury after subarachnoid hemorrhage in mice. J Neuroinflamm. 2020;17(1):74.
41. Hung ME, Leonard JN. A platform for actively loading cargo RNA to elucidate limiting steps in EV-mediated delivery. J Extracell Vesicles. 2016;5:31027.
42. Han L, Huang R, Liu S, et al. Peptide-conjugated PAMAM for targeted doxorubicin delivery to transferrin receptor overexpressed tumors. Mol Pharm. 2010;7(6):2156–65.
43. Kim G, Kim M, Lee Y, et al. Systemic delivery of microRNA-21 antisense oligonucleotides to the brain using T7-peptide decorated exosomes. J Control Release. 2020;317:273–81.

44. Kim H, Yun N, Mun D, et al. Cardiac-specific delivery by cardiac tissue-targeting peptide-expressing exosomes. Biochem Biophys Res Commun. 2018;499(4):803–8.
45. Wang X, Chen Y, Zhao Z, et al. Engineered exosomes with ischemic myocardium-targeting peptide for targeted therapy in myocardial infarction. J Am Heart Assoc. 2018;7(15):e008737.
46. Liang Y, Xu X, Li X, et al. Chondrocyte-targeted microRNA delivery by engineered exosomes toward a cell-free osteoarthritis therapy. ACS Appl Mater Interfaces. 2020;12(33):36938–47.
47. Tian Y, Li S, Song J, et al. A doxorubicin delivery platform using engineered natural membrane vesicle exosomes for targeted tumor therapy. Biomaterials. 2014;35(7):2383–90.
48. Bai J, Duan J, Liu R, et al. Engineered targeting tLyp-1 exosomes as gene therapy vectors for efficient delivery of siRNA into lung cancer cells. Asian J Pharm Sci. 2020;15(4):461–71.
49. Liang G, Zhu Y, Ali DJ, et al. Engineered exosomes for targeted co-delivery of miR-21 inhibitor and chemotherapeutics to reverse drug resistance in colon cancer. J Nanobiotechnol. 2020;18(1):10.
50. Hung ME, Leonard JN. Stabilization of exosome-targeting peptides via engineered glycosylation. J Biol Chem. 2015;290(13):8166–72.
51. Stubbs JD, Lekutis C, Singer KL, et al. cDNA cloning of a mouse mammary epithelial cell surface protein reveals the existence of epidermal growth factor-like domains linked to factor VIII-like sequences. Proc Natl Acad Sci U S A. 1990;87(21):8417–21.
52. Andersen MH, Graversen H, Fedosov SN, et al. Functional analyses of two cellular binding domains of bovine lactadherin. Biochemistry. 2000;39(20):6200–6.
53. Zeelenberg IS, Ostrowski M, Krumeich S, et al. Targeting tumor antigens to secreted membrane vesicles in vivo induces efficient antitumor immune responses. Cancer Res. 2008;68(4):1228–35.
54. Rountree RB, Mandl SJ, Nachtwey JM, et al. Exosome targeting of tumor antigens expressed by cancer vaccines can improve antigen immunogenicity and therapeutic efficacy. Cancer Res. 2011;71(15):5235–44.
55. Morishita M, Takahashi Y, Matsumoto A, et al. Exosome-based tumor antigens-adjuvant co-delivery utilizing genetically engineered tumor cell-derived exosomes with immunostimulatory CpG DNA. Biomaterials. 2016;111:55–65.
56. Hartman ZC, Wei J, Glass OK, et al. Increasing vaccine potency through exosome antigen targeting. Vaccine. 2011;29(50):9361–7.
57. Longatti A, Schindler C, Collinson A, et al. High affinity single-chain variable fragments are specific and versatile targeting motifs for extracellular vesicles. Nanoscale. 2018;10(29):14230–44.
58. Wang JH, Forterre AV, Zhao J, et al. Anti-HER2 scFv-directed extracellular vesicle-mediated mRNA-based gene delivery inhibits growth of HER2-positive human breast tumor xenografts by prodrug activation. Mol Cancer Ther. 2018;17(5):1133–42.
59. Yang Y, Hong Y, Cho E, et al. Extracellular vesicles as a platform for membrane-associated therapeutic protein delivery. J Extracell Vesicles. 2018;7(1):1440131.
60. Heider S, Dangerfield JA, Metzner C. Biomedical applications of glycosylphosphatidylinositol-anchored proteins. J Lipid Res. 2016;57(10):1778–88.
61. Rabesandratana H, Toutant JP, Reggio H, et al. Decay-accelerating factor (CD55) and membrane inhibitor of reactive lysis (CD59) are released within exosomes during in vitro maturation of reticulocytes. Blood. 1998;91(7):2573–80.
62. Kooijmans SA, Aleza CG, Roffler SR, et al. Display of GPI-anchored anti-EGFR nanobodies on extracellular vesicles promotes tumour cell targeting. J Extracell Vesicles. 2016;5:31053.
63. Dooley K, McConnell RE, Xu K, et al. A versatile platform for generating engineered extracellular vesicles with defined therapeutic properties. Mol Ther. 2021;29(5):1729–43.
64. Aguila B, Morris AB, Spina R, et al. The Ig superfamily protein PTGFRN coordinates survival signaling in glioblastoma multiforme. Cancer Lett. 2019;462:33–42.
65. Hartl M, Nist A, Khan MI, et al. Inhibition of Myc-induced cell transformation by brain acid-soluble protein 1 (BASP1). Proc Natl Acad Sci U S A. 2009;106(14):5604–9.
66. Ohno S, Takanashi M, Sudo K, et al. Systemically injected exosomes targeted to EGFR deliver antitumor microRNA to breast cancer cells. Mol Ther. 2013;21(1):185–91.

67. Cheng Q, Shi X, Han M, et al. Reprogramming exosomes as nanoscale controllers of cellular immunity. J Am Chem Soc. 2018;140(48):16413–7.
68. Smyth T, Petrova K, Payton NM, et al. Surface functionalization of exosomes using click chemistry. Bioconjug Chem. 2014;25(10):1777–84.
69. Wang M, Altinoglu S, Takeda YS, et al. Integrating protein engineering and bioorthogonal click conjugation for extracellular vesicle modulation and intracellular delivery. PLoS One. 2015;10(11):e0141860.
70. Xu L, Faruqu FN, Liam-Or R, et al. Design of experiment (DoE)-driven in vitro and in vivo uptake studies of exosomes for pancreatic cancer delivery enabled by copper-free click chemistry-based labelling. J Extracell Vesicles. 2020;9(1):1779458.
71. Song S, Shim MK, Lim S, et al. In situ one-step fluorescence labeling strategy of exosomes via bioorthogonal click chemistry for real-time exosome tracking in vitro and in vivo. Bioconjug Chem. 2020;31(5):1562–74.
72. Zhang H, Wu J, Wu J, et al. Exosome-mediated targeted delivery of miR-210 for angiogenic therapy after cerebral ischemia in mice. J Nanobiotechnol. 2019;17(1):29.
73. Tian T, Zhang HX, He CP, et al. Surface functionalized exosomes as targeted drug delivery vehicles for cerebral ischemia therapy. Biomaterials. 2018;150:137–49.
74. Jia G, Han Y, An Y, et al. NRP-1 targeted and cargo-loaded exosomes facilitate simultaneous imaging and therapy of glioma in vitro and in vivo. Biomaterials. 2018;178:302–16.
75. Nie W, Wu G, Zhang J, et al. Responsive exosome nano-bioconjugates for synergistic cancer therapy. Angew Chem Int Ed Engl. 2020;59(5):2018–22.
76. Bunge A, Loew M, Pescador P, et al. Lipid membranes carrying lipophilic cholesterol-based oligonucleotides—characterization and application on layer-by-layer coated particles. J Phys Chem B. 2009;113(51):16425–34.
77. O'Loughlin AJ, Mager I, de Jong OG, et al. Functional delivery of lipid-conjugated siRNA by extracellular vesicles. Mol Ther. 2017;25(7):1580–7.
78. Salmaninejad A, Valilou SF, Bayat H, et al. Duchenne muscular dystrophy: an updated review of common available therapies. Int J Neurosci. 2018;128(9):854–64.
79. Yerneni SS, Lathwal S, Shrestha P, et al. Rapid on-demand extracellular vesicle augmentation with versatile oligonucleotide tethers. ACS Nano. 2019;13(9):10555–65.
80. Pi F, Binzel DW, Lee TJ, et al. Nanoparticle orientation to control RNA loading and ligand display on extracellular vesicles for cancer regression. Nat Nanotechnol. 2018;13(1):82–9.
81. Didiot MC, Hall LM, Coles AH, et al. Exosome-mediated delivery of hydrophobically modified siRNA for huntingtin mRNA silencing. Mol Ther. 2016;24(10):1836–47.
82. Akishiba M, Takeuchi T, Kawaguchi Y, et al. Cytosolic antibody delivery by lipid-sensitive endosomolytic peptide. Nat Chem. 2017;9(8):751–61.
83. Zhan Q, Yi K, Qi H, et al. Engineering blood exosomes for tumor-targeting efficient gene/chemo combination therapy. Theranostics. 2020;10(17):7889–905.
84. Hosta-Rigau L, Zhang Y, Teo BM, et al. Cholesterol—a biological compound as a building block in bionanotechnology. Nanoscale. 2013;5(1):89–109.
85. Tabas I. Consequences of cellular cholesterol accumulation: basic concepts and physiological implications. J Clin Invest. 2002;110(7):905–11.
86. Gao X, Ran N, Dong X, et al. Anchor peptide captures, targets, and loads exosomes of diverse origins for diagnostics and therapy. Sci Transl Med. 2018;10(444):eaat0195.
87. Dong X, Lei Y, Yu Z, et al. Exosome-mediated delivery of an anti-angiogenic peptide inhibits pathological retinal angiogenesis. Theranostics. 2021;11(11):5107–26.
88. Wang T, Li Y, Guo M, et al. Exosome-mediated delivery of the neuroprotective peptide PACAP38 promotes retinal ganglion cell survival and axon regeneration in rats with traumatic optic neuropathy. Front Cell Dev Biol. 2021;9:659783.
89. Zuo B, Qi H, Lu Z, et al. Alarmin-painted exosomes elicit persistent antitumor immunity in large established tumors in mice. Nat Commun. 2020;11(1):1790.
90. Zha Y, Li Y, Lin T, et al. Progenitor cell-derived exosomes endowed with VEGF plasmids enhance osteogenic induction and vascular remodeling in large segmental bone defects. Theranostics. 2021;11(1):397–409.

# Technologies for Visualization and Tracking of Extracellular Vesicles

# 12

Zi-Li Yu, He-Jing Zhang, Xu Duan, Bo-Lin Xiao, Hou-Fu Xia, Zhuo-Kun Chen, Qi-Hui Xie, Jue Huang, Yin-Xue Chen, Gao-Hong Chen, and Gang Chen

## 12.1 Overview of Technologies for the Visualization and Tracking of Extracellular Vesicles (EVs)

The visualization and tracking of EVs involves labeling of EVs with specific substances and subsequently observing or measuring them through optical imaging, magnetic resonance imaging (MRI), radionuclide imaging, and other technologies. Owing to the ultrasmall size of EVs, appropriate methods for the visualization and

Zi-Li Yu, He-Jing Zhang and Xu Duan contributed equally with all other contributors.

Z.-L. Yu · H.-F. Xia
State Key Laboratory of Oral & Maxillofacial Reconstruction and Regeneration, Key Laboratory of Oral Biomedicine Ministry of Education, Hubei Key Laboratory of Stomatology, School & Hospital of Stomatology, Wuhan University, Wuhan, China

Department of Oral and Maxillofacial Surgery, School and Hospital of Stomatology, Wuhan University, Wuhan, China

H.-J. Zhang · X. Duan · B.-L. Xiao · Z.-K. Chen · Q.-H. Xie · J. Huang · Y.-X. Chen · G.-H. Chen
State Key Laboratory of Oral & Maxillofacial Reconstruction and Regeneration, Key Laboratory of Oral Biomedicine Ministry of Education, Hubei Key Laboratory of Stomatology, School & Hospital of Stomatology, Wuhan University, Wuhan, China

G. Chen (✉)
State Key Laboratory of Oral & Maxillofacial Reconstruction and Regeneration, Key Laboratory of Oral Biomedicine Ministry of Education, Hubei Key Laboratory of Stomatology, School & Hospital of Stomatology, Wuhan University, Wuhan, China

Department of Oral and Maxillofacial Surgery, School and Hospital of Stomatology, Wuhan University, Wuhan, China

TaiKang Center for Life and Medical Sciences, Wuhan University, Wuhan, China

Frontier Science Center for Immunology and Metabolism, Wuhan University, Wuhan, China
e-mail: geraldchan@whu.edu.cn


Q. Wang, L. Zheng (eds.), *Extracellular Vesicles*,
https://doi.org/10.1007/978-981-99-8365-0_12

tracking of these vesicles are essential for studying the biological behaviors or functions of EVs at the subcellular, cellular, and whole-body levels.

EVs, which are important carriers of biological information, participate in the regulation of pathophysiological processes by mediating intercellular communication. However, the limitations of EV research, particularly EV visualization and tracking, have hampered in-depth studies of the biological behaviors and functions of EVs. Therefore, there is an urgent need to establish stable and reliable technologies for visualizing and tracking EVs. In addition, owing to their biosafety and natural transport properties, EVs have recently been used to construct drug delivery vehicles. However, further research is required to study the in vivo distribution and kinetic characteristics of EVs as drug carriers.

There are still many challenges associated with studying the unique structures and physicochemical properties of EVs. First, EVs are submicron or nanoscale double-layer membrane structures. The ultrasmall size of EVs necessitates extremely high spatial resolution during imaging and tracing. Moreover, EVs transmit biological information and interact with recipient cells in a dynamic process; this highlights the requirements for high stability of the tracer signals and time resolution during imaging. Tissue penetration of tracer signals is also crucial for in vivo tracing of EVs. However, given that EVs are biologically active, ideal labeling materials and strategies should have no major impact on the natural physicochemical properties and biological functions of EVs.

The main steps in the visualization and tracking of EVs include labeling, visualization, tracking, and data analysis. The requirements for the observation level and target indicators are specific. It is essential to select suitable labeling schemes and construct corresponding imaging platforms to obtain reliable data. Previous studies have attempted to trace EVs both in vitro and in vivo. Organic fluorescent dyes, fluorescent proteins, luciferase, inorganic fluorescent nanomaterials such as quantum dots (QDs), and other optical imaging-based markers to label EVs have been used for tracking and visualizing EVs in vitro. Markers based on optical imaging, MRI, and radionuclide imaging are often used to track EVs in vivo. However, optical, magnetic resonance, and radionuclide imaging platforms, which are presently widely used, are all designed for cell imaging or medical imaging, and there is still a lack of super-resolution imaging platforms dedicated to the dynamic tracking of EVs.

Organic fluorescent dyes, including PKH26 and PKH67, were the earliest fluorescent probes used to label EVs. Although these compounds have played important roles in exploring the biological functions of EVs, with the deepening and refinement of recent researches, defects such as nonspecific staining, low fluorescence intensity, easy light bleaching, and long half-life have made these dyes insufficient for some experimental requirements. Therefore, inorganic fluorescent nanomaterials with high fluorescence intensity, excellent photostability, and satisfactory tissue penetration ability are gradually replacing traditional organic fluorescent dyes. A typical example is the use of QDs. Presently, most labeling and tracing imaging of EVs is aimed at a single EV component. However, many present studies have focused on the information-delivery function of EVs, and it is important to realize multicomponent labeling and tracing of EVs as well as analysis of their key information molecules. Simultaneously, continuous dynamic tracking based on time series analysis is gradually replacing traditional static displays and is expected to become a powerful tool for elucidation of the biological behaviors of EVs. Therefore, multicomponent labeling and dynamic tracking will be promising approaches in the field of EV visualization and tracking.

## 12.2 EV Labeling Strategies

Commonly used EV labeling strategies for EVs can be classified as indirect or direct labeling according to the labeling mechanisms used in their preparation (Fig. 12.1). In indirect labeling strategies, EV donor cells are first labeled with appropriate labeling agents, and EVs secreted from these cells then inherit the

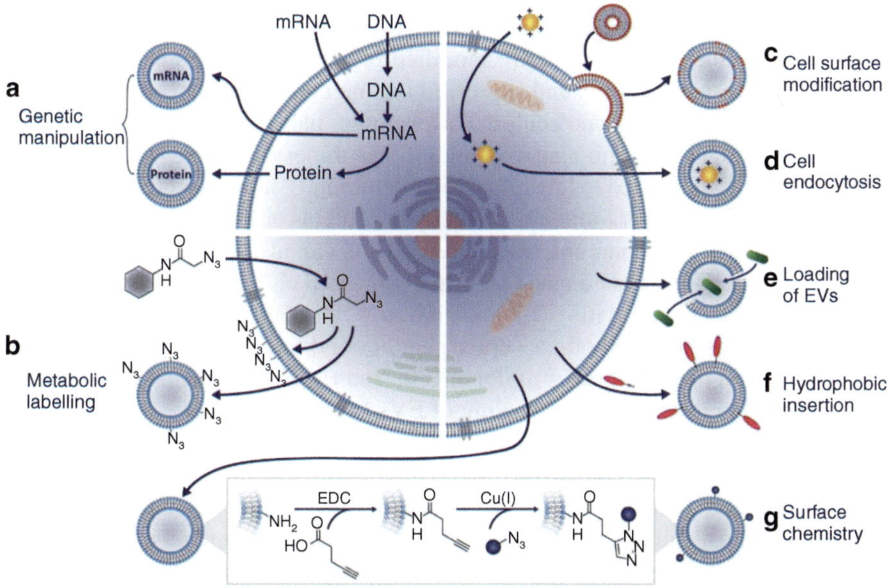


**Fig. 12.1** EV labeling strategies. (**a**) Genetic manipulation: mRNAs and proteins, whose expression is controlled using recombinant plasmids, are secreted into EVs. (**b**) Metabolic labeling: metabolite analogs are introduced into the cell's biosynthesis mechanism to label the EV donor cells, and EVs secreted from the cells then inherit the labeling agents from the donor cells. (**c**) Cell modification: exogenous labeling materials, such as liposomes or natural phospholipids, are fused into the membranes of the EV donor cells. (**d**) Cell endocytosis: exogenous materials are endocytosed by donor cells and then inherited by the secreted EVs. (**e**) Loading of EVs: membrane-permeable labeling materials directly enter EVs to achieve labeling. (**f**) Hydrophobic insertion: lipophilic dye molecules form hydrophobic bonds with EV membranes to achieve labeling. (**g**) Surface chemistry: EVs are labeled with chemical substances via a chemical reaction occurring at the EV surface [1]. [*Figure reprinted with permission (Re-engineering extracellular vesicles as smart nanoscale therapeutics. American Chemical Society, 2017)*]

labeling agents from donor cells. In direct labeling strategies, EVs are labeled using fluorescent dyes or nanomaterials directly. In this section, we review presently available strategies for EV labeling according to their mechanisms and efficiency.

## 12.2.1 Indirect EV Labeling Strategies Via Modification of Parent Cells

### 12.2.1.1 Genetic Manipulation

Genetic manipulation recombines target genes encoding label proteins, such as fluorescent proteins or luciferases, to plasmids (such as lentiviral vectors), which are then transfected into the parent cells. Consequently, EVs carrying fluorescent proteins emit fluorescent light of a specific wavelength under excitation, and EVs carrying luciferase emit visible light by reacting with substrates.

Presently, the most commonly used recombinant plasmids are those expressing fusion proteins, which combine label proteins and EV surface markers (e.g., CD63 and lectin). These fusion proteins can be used to label EV subgroups [2, 3]. Additionally, label proteins that are modified to carry specific molecular binding sites by genetic manipulation and expressed in donor cells can then be encapsulated into EVs to label specific molecules. For example, palmitoylated fluorescent proteins expressed by recombinant plasmids form thioester bonds with sulfhydryl groups, which can be used to label the inner membranes of donor cells and EVs [4].

Label proteins are not limited to fluorescent proteins or luciferase, which can be directly detected, and some intermediate products can also be used. For example, the fusion proteins streptavidin and lactectin expressed on the surface of EVs can be conjugated and labeled with $^{125}$I-bound biotin ($^{125}$I-IBB) [5]. After biotinylation, label proteins fused with biotin acceptor peptides and transmembrane domains can be expressed on the membranes of donor cells and EVs. Thus, the biotin–streptavidin system can realize multiple labeling of EVs [6].

### 12.2.1.2 Metabolic Labeling

By exploiting the metabolic processes of donor cells, metabolic materials, such as nonnatural lipids, amino acids, and saccharides, can be incorporated into the metabolome of donor cells and then packed into EVs. For example, through lipid metabolism in donor cells, fluorescent fatty acid analogs with BODIPY fluorophores can participate in intracellular lipid metabolism and the formation of an EV lipid bilayer without disrupting cellular homeostasis [7]. L-Azidoalanine, a methionine analog containing azide, can be incorporated into amino acid metabolism and substituted for methionine in cell membranes [8]. Additionally, $Ac_4ManNAz$, a monosaccharide containing azido groups, can be incorporated into the polysaccharide synthesis mechanism to modify the surface of EVs. Alkynyl-containing fluorescent dyes, such as ADIBO fluorescent dyes, can be used to generate click chemical reactions with the azido groups on EV membranes for fast, mild, and efficient labeling of EVs [9].

Furthermore, given the continuous natural exchange of phospholipids between cell membranes and the extracellular environment, our group established an EV labeling method based on biotinylation of the parent cell membrane [10]. First, biotinylated phosphatidyl ethanolamine (DSPE-PEG-biotin) is added to the culture medium at a final concentration of 10 μg/mL. The cells are then cultured in conventional medium for 3–5 days and starved for 24 h in serum-free medium. EVs in cell supernatants are harvested by ultracentrifugation and incubated with QDs and 2 nM streptavidin at 37 °C for 30 min. Owing to the high affinity of streptavidin for biotin, biotinylated EVs are specifically labeled with streptavidin-containing QDs. Flow cytometry analyses demonstrated that the labeling efficiency of this approach is greater than 80%.

#### 12.2.1.3 Endocytosis

Exogenous labeling materials can be endocytosed by donor cells and secreted into EVs. For example, after incubation with ultrasmall superparamagnetic iron oxide nanoparticles (USPIOs, 4–6 nm, 200 μg Fe/mL) for 72 h, all adipogenic stem cells were labeled by USPIOs, and cell activity was not affected. Transmission electron microscopy analysis revealed that secreted exosomes were loaded with USPIOs, and spectrophotometry analysis further revealed that 0.643 μg Fe was present in 100 μg exosomes [11].

Given that macrophages can take up large amounts of exogenous materials through phagocytosis, the concentration and incubation time of labeling nanoparticles can be reduced to achieve optimal labeling efficiency when macrophages are employed as donor cells. Therefore, some researchers have co-incubated THP-1 human mononuclear cells with 5 mM iron oxide nanoparticles and a photosensitizer at 37 °C for 2 h to harvest EVs with magnetic targeting and optical imaging properties [12].

The indirect EV labeling strategies described above have good specificity under mild conditions (Table 12.1). However, genetic manipulation is a laborious and time-consuming process, and the synthesis of nonnatural metabolic materials requires days of cell culture. Moreover, because the modification efficiency of donor cells and EVs may be disproportional, the number of successfully modified EVs may not be as high as expected.

**Table 12.1** Indirect EV labeling strategies

| Labeling methods | Labeling materials | Characteristics | |
|---|---|---|---|
| | | Advantages | Disadvantages |
| Genetic manipulation | Fluorescent proteins or luciferases | Stable expression; sustainability | Step-cumbersome and time-consuming |
| Metabolic labeling | Azido groups or biotin | Fast, mild and efficient | Time-consuming and unsustainability |
| Endocytosis | USPIOs | Few steps; easy; multiple additional ability | Low conversion efficiency and unsustainability |

### 12.2.2 Direct EV Labeling Strategies

EV membranes are composed of phospholipid bilayers and proteins, such as tetraspanins (CD9, CD63, CD81, and CD82) and lectin. Proteins on the EV membrane can be direct labeling sites for dyes or probes. Alternatively, labeling materials can be packed into EVs in an active or passive manner. Compared with indirect labeling strategies, direct EV labeling strategies are faster, cheaper, and more efficient (Table 12.2). Moreover, direct labeling strategies can also be applied to label EVs that have already been released into the extracellular space, such as EVs from human bodily fluids (e.g., blood, saliva, etc.).

#### 12.2.2.1 Direct Labeling Strategies Targeting EV Membranes

The phospholipid bilayer of EVs can be labeled with lipophilic dyes, such as PKH dyes (e.g., PKH26, PKH67, etc.), Di series dyes (e.g., DiI, DiD, etc.), and R18. The hydrophobic structures of lipophilic dyes can stably form nonspecific bonds with the phospholipid bilayer when co-incubated with EVs. Owing to the lateral diffusion of dye molecules, the outer membrane of the EVs can be evenly colored. However, some lipophilic dye molecules tend to aggregate into clusters. Clusters of dye molecules are difficult to distinguish from the labeled EVs. Therefore, it is necessary to use sucrose density gradient centrifugation to remove excess dye from EVs after incubation [13].

In addition to phospholipid bilayers, proteins on EV membranes are commonly used as labeling sites. Direct labeling targeting the membrane proteins of EVs usually does not have specificity. For example, maleimide with different fluorescent groups can react with membrane proteins containing sulfhydryl groups (such as a tetraspan network) to form stable thioether bonds [14]. Such a reaction can realize simple and rapid labeling. The tricarbonyl technetium complex ($^{99}$mTc-tricarbonyl complex) can also be used to label EV membrane proteins via chelation [15]. Moreover, the labeling of EV membranes with high specificity was realized by receptor–ligand interactions. For example, both superparamagnetic nanoparticles with transferrin-binding ability and antibodies modified with gold-carbon QDs targeting EV surface markers can effectively bind to the surfaces of EVs isolated from blood [16, 17]. Notably, antibodies modified by QDs can also function as fluorescent nanoprobes [16].

#### 12.2.2.2 Direct Labeling Strategies Targeting EV Cargo

Exogenous markers with membrane permeability can be passively loaded into EVs. For example, succinimide ester (CFSE) can penetrate the membrane and be hydrolyzed by nonspecific esterases to emit fluorescence. Hydrolyzed CFSE can then covalently bind with amine groups and remain in the EVs. After chelation with radioactive tracers, such as $^{99}$mTc, the lipophilic chelating agent HMPAO can be loaded into EVs via permeation [18]. Furthermore, with the help of transporters on the membrane, EVs can actively take up modified labeling materials. For example, EVs derived from mesenchymal stem cells can actively take up glucose-coated gold nanoparticles via glucose transporter-1 [19].

**Table 12.2** Direct EV labeling strategies

| Labeling sites | | Labeling materials | Labeling principles | Characteristics | |
|---|---|---|---|---|---|
| | | | | Advantages | Disadvantages |
| Membrane | Phospholipid bilayer | Lipophilic dyes | Hydrophobic binding | Variety for different requirements; easily availability; few steps | Easy to aggregate and escape to interfere distinguishment |
| | Proteins | Maleimide-based fluorescent dyes | Thioether linkage | Stability; few steps; fast and easy | |
| | | Tricarbonyl technetium complex | Chelation effect | Tracking with radioactivity | |
| | | Surface-modified nanoparticles | Receptor–ligand interactions | High specificity; high cost | |
| Intracellular molecules | Amino group | CFSE | Covalent binding | Widely applied; easy to use; stability | |
| | Nucleic acids | Nucleic acid dyes | Embedding charge interaction | Specific for nucleic acids | |
| | None | Inorganic nanoparticles | Electroporation | Various for different imaging platforms | Strict loading conditions; effect on the integrity of membrane |

Electroporation can instantaneously increase the permeability of the membrane. Electroporation creates small holes (10 nm in diameter); consequently, labeling materials in the surrounding medium, such as QDs (1–6 nm in diameter) and USPIOs (5 nm in diameter), can be loaded into EVs. In our previous research, we loaded $Ag_2Se$@Mn QDs (approximately 1.8 nm in diameter) into EVs. First, 50 μL EVs and 50 μL QDs (1 mg/mL) were added to a 0.4 cm electroporation cuvette containing 150 μL electroporation buffer (Bio-Rad Laboratories, Hercules, CA, USA) and mixed well. Second, the cuvette was placed in a Gene Pulser Xcell electroporation system (Bio-Rad Laboratories), and 250 V/350 μF voltage/capacitance was applied for a single pulse with a duration of less than 1 s. The conversion was completed after the electric shock. Excessive QDs were removed using sucrose density gradient (15–60%) ultracentrifugation for 1 h. Through flow analysis, this one-step method can quickly introduce ultrasmall-diameter QDs with efficiencies as high as 90% [20].

## 12.3 Technologies for Tracking of EVs

Imaging methods for EVs, such as fluorescence imaging, bioluminescence imaging, MRI, and radionuclide imaging, are summarized in Table 12.3.

### 12.3.1 Optical Imaging of EVs

Optical imaging is a technique for imaging cells, tissues, organs, or living bodies using an optical detection system. This approach can realize the real-time observation of labeled objects without producing any ionizing radiation. Presently, optical imaging of EVs mainly includes fluorescence imaging and bioluminescent imaging. Fluorescence imaging is an imaging technique that detects light emitted from fluorescent substances excited by a specific wavelength of light. Fluorescent proteins, fluorescent dyes, and inorganic fluorescent nanomaterials, such as fluorescent QDs, can all be used in fluorescence imaging.

Bioluminescence imaging mainly detects light signals generated by a specific enzymatic reaction between luciferase and luciferin. Commonly used luciferase enzymes include firefly luciferase, Renilla luciferase, Gaussia luciferase, and NanoLuc derived from deep-sea shrimp. Compared with fluorescence imaging, bioluminescence imaging can effectively avoid the interference of biological spontaneous illumination, and quantitative analysis can be completed during imaging. However, bioluminescence has strict requirements with regard to imaging instruments. Additionally, the reaction requires the addition of luciferin as a reaction substrate, which is affected by the biological activities of cells. Therefore, bioluminescence is often used for imaging in vivo.

Compared with bioluminescence imaging, fluorescence imaging is more commonly used for the dynamic tracking of EVs. Presently, fluorescence imaging of EVs mainly focuses on cellular-level analyses using fluorescence microscopy and

**Table 12.3** EV imaging methods

| Imaging methods | Labeling materials | Imaging planes | Research purposes | Detection systems |
|---|---|---|---|---|
| Fluorescence imaging | Dye, fluorescent group, fluorescent protein, quantum dot | Cell | Release, transfer, uptake, and other physiological processes of EVs | Confocal fluorescence microscope |
| | Di series dyes, near-infrared quantum dots, BODPIY fluorophore | Cells, isolated organs, living bodies | The physiological process of EVs and its biological distribution in the body | Confocal fluorescence microscope, live small animal optical imager |
| Bioluminescence imaging | Luciferase | Living bodies | Biological distribution of EVs | Live small animal optical imager |
| Magnetic resonance imaging | Superparamagnetic iron oxide, superparamagnetic iron oxide nanoparticles | Isolated organs, living bodies | Biological distribution of EVs | Magnetic resonance imager |
| Radionuclide imaging | Indium-111 oxine complexes, biotin combined with $^{125}I$ | Isolated organs, living bodies | The biological distribution or tumor targeting of EVs was quantitatively studied | γ-Counter |
| | Technetium-tricarbonyl complex | Living bodies | The biological distribution of EVs was quantitatively studied | SPECT/CT |

whole animal-level analyses using in vivo imaging systems. By analyzing the location of EVs in parent cells or receptor cells, we can elucidate the biological processes of EVs, including their production, secretion, and uptake. For example, if we directly labeled EVs with QDs or organic fluorescent dyes (such as CFSE and PKH67), confocal fluorescence microscopy of receptor cells can enable analysis of the uptake of EVs according to the location of EVs in cells at different time points. In addition, EVs can be indirectly labeled via parent cells to analyze their production and release.

When EVs derived from different cell sources are labeled simultaneously, we can study the transfer of EVs among cells [4]. Researchers have used C5-maleimide-Alexa488 to directly label EVs. The nucleus and intracellular lysosomes of recipient cells were also labeled with fluorescence simultaneously. Confocal fluorescence microscopy imaging revealed that the endocytosed EVs gradually aggregated around the nucleus after uptake, and their colocalization with lysosomes gradually increased [2]. In another study, exchange of EVs was observed between different cells by constructing cells expressing green fluorescent protein or tdTomato to identify EVs derived from different cells. Researchers also simultaneously labeled mRNAs to observe their transportation via EVs [4].

Near-infrared (NIR) fluorescence imaging, which has high sensitivity and spatial resolution, is another popular approach for tracking EVs in vivo. Indeed, NIR imaging shows lower absorption, scattering, and emission of NIR light when applied to biological components, and the small amount of spontaneous NIR fluorescence inside biological tissues can cause lower background noise. The most commonly used NIR probes include Di dyes and QDs. EVs labeled with NIR probes can be injected into animals, and images of EVs in organs or living animals can be captured using an in vivo imaging system (IVIS). The metabolism or distribution of EVs in vivo can be analyzed according to the fluorescence intensity of different organs at different times. Some researchers have labeled EVs derived from different cells with DIR and injected the EVs into mice using various approaches. Images of living mice and organs obtained using IVIS at different times after injection have shown that the cell source and injection mode of EVs can affect the distribution of EVs in vivo [3]. Our group used $Ag_2Se$@Mn QDs to label EVs derived from tumor cells and observed their biological distributions in nude mice using an NIR imager (Fig. 12.2).

Fluorescence imaging is used in a wide range of applications. Different labeling probes and methods can be used to obtain images of cells, tissues, and living bodies. Probes with different emitted light can be used in combination to achieve multicolor imaging. Fluorescence imaging has advantages, such as simple operation, relatively low equipment requirements, high sensitivity, and good time resolution. However, the absorption and scattering of light in the tissues and the interference of spontaneous light increase the background noise of fluorescence imaging and the requirement for a good signal-to-noise ratio. This leads to a low spatial resolution and limited depth of in vivo imaging.

### 12.3.2 MRI for EVs

Under an external magnetic field, the spin of the atomic nucleus changes from a disordered to an ordered pattern. Simultaneously, a certain frequency of radiofrequency pulses excites the nucleus, which causes resonance. When the radiofrequency pulse stops, the nucleus returns to the equilibrium state from the excited state with the released energy and generates a radio signal. MRI detects these radio signals and analyzes them using mathematical algorithms, thereby obtaining images of the moving nucleus. The process through which the nucleus returns from the excited state to the equilibrium state is called the relaxation process. The required time, which is called the relaxation time, is divided into $T_1$ (spin–lattice or longitudinal relaxation time) and $T_2$ (spin–spin or transverse relaxation time).

Contrast agents are usually required for MRI. Commonly used MRI contrast agents include paramagnetic gadolinium contrast agents, superparamagnetic iron oxide (SPIOs), and USPIOs. Paramagnetic gadolinium comparators enhance the signals and yield brighter images mainly by shortening the $T_1$ of the nucleus,

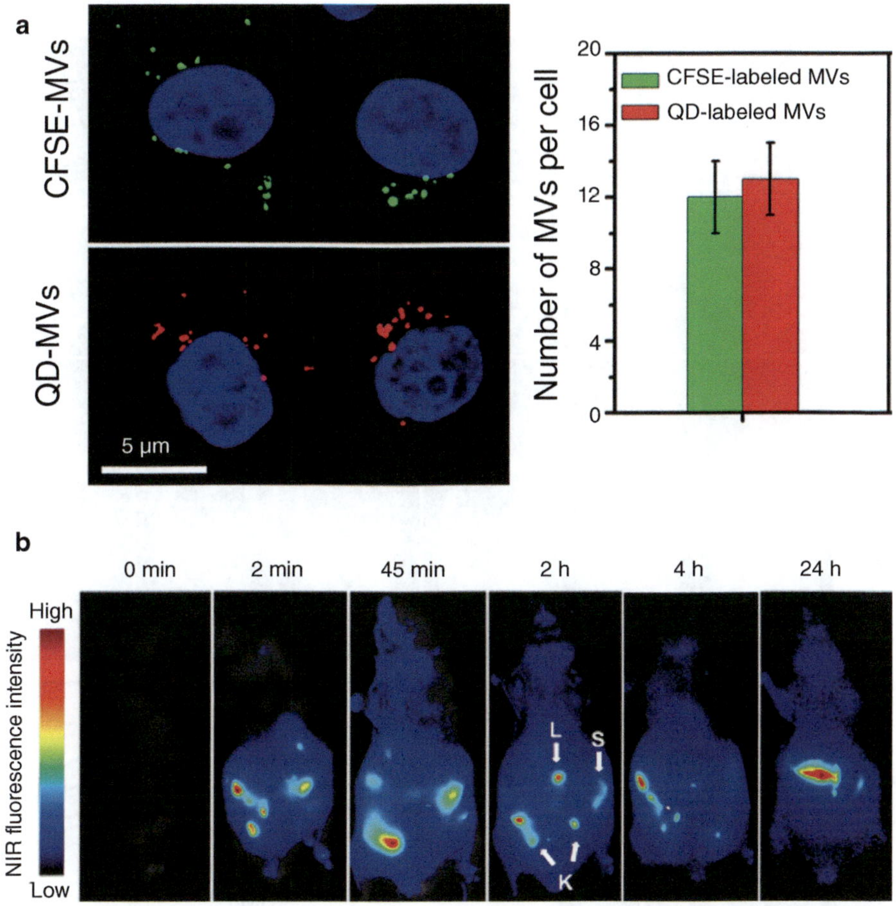


**Fig. 12.2** Fluorescence labeling and imaging of EVs. (**a**) Tumor cells ingest EVs labeled with CFSE or $Ag_2Se$ QDs. (**b**) Near-infrared imaging of EVs labeled with $Ag_2Se$ QDs in living mice [21]. [*Figure reprinted with permission (Ultrasmall magnetically engineered* $Ag_2Se$ *quantum dots for instant efficient labeling and whole-body high-resolution multimodal real-time tracking of cell-derived microvesicles. American Chemical Society, 2016)*]

whereas SPIOs and USPIOs mainly rely on shortening the $T_2$ and weakening the signals to obtain darker images. Although gadolinium contrast agents are widely used clinically, SPIOs and USPIOs are the main contrast agents used by researchers for MRI of EVs. USPIOs can be loaded into EVs by electroporation, and $T_2$-weighted images (often used in vitro) and $T_1$-weighted images (often used in vivo) of labeled EVs can then be obtained.

USPIOs can also be loaded into EVs through endocytosis and secretion by the parental cells. $T_2$-weighted images of USPIOs loaded in EVs can be obtained in vitro or in vivo [11, 22]. Researchers have used USPIOs to label EVs derived from B16F10 cells and have detected their distribution in lymph nodes. The findings

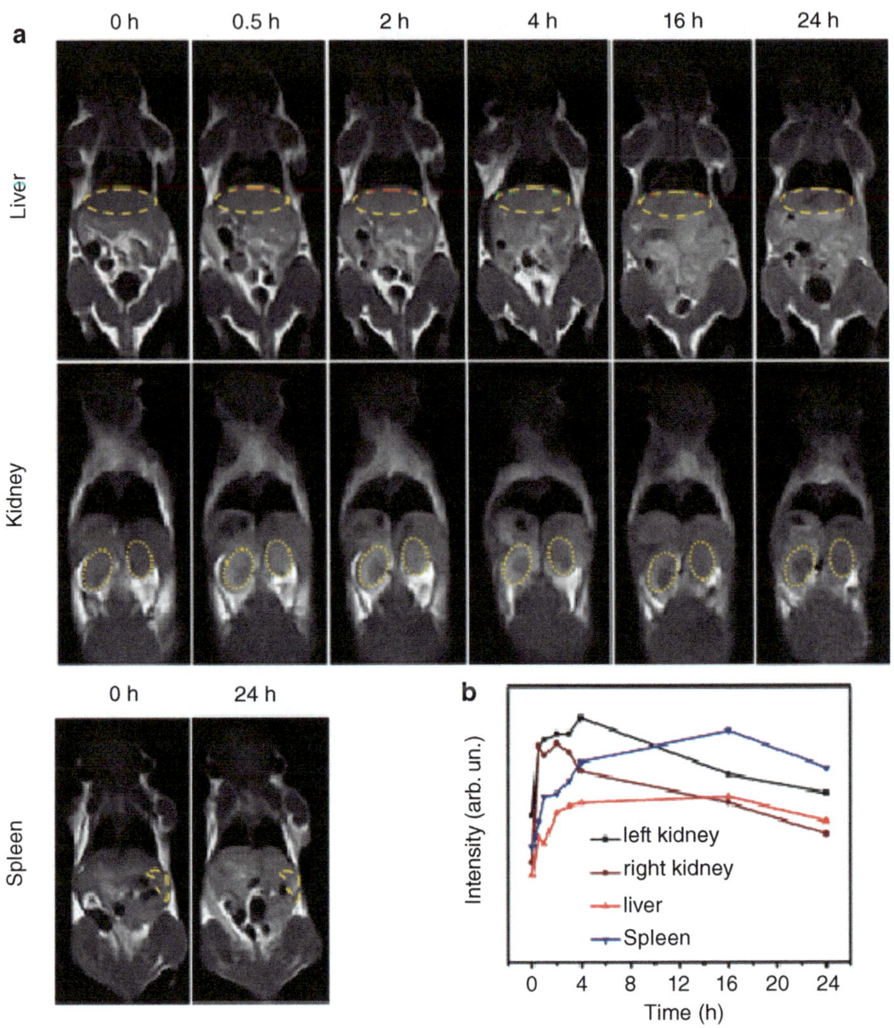


**Fig. 12.3** Magnetic resonance imaging of EVs labeled with $Ag_2Se$@Mn-QDs. (**a**) $T_1$-weighted images of the mouse liver, kidney, and spleen at 0–24 h after injection of $Ag_2Se$@Mn-QD-labeled EVs. (**b**) Quantitative analysis of the distribution of EVs labeled with $Ag_2Se$@Mn-QDs in the mouse liver, kidney, and spleen by MRI [21]. [*Figure reprinted with permission (Ultrasmall magnetically engineered Ag2Se quantum dots for instant efficient labeling and whole-body high-resolution multimodal real-time tracking of cell-derived microvesicles. American Chemical Society, 2016)*]

showed that EVs derived from B16F10 cells can gather in a specific region in the popliteal lymph nodes of mice [22]. Additionally, other studies have shown that the relaxation time of SPIO-labeled EVs derived from human bone marrow mesenchymal stem cells is significantly shorter than that of unlabeled EVs [23]. In a previous study, MRI was used to observe $Ag_2Se$@Mn-QD-labeled EVs in nude mice (Fig. 12.3).

MRI has obvious advantages over optical imaging techniques in vivo. The limited tissue penetration of fluorescence signals makes optical imaging suitable for the imaging of superficial tissues, whereas MRI can be used to obtain signals from deeper tissues and shows higher spatial resolution than optical imaging. Thus, MRI is more effective for detecting the biological distribution of EVs in tissues and organs. However, MRI is relatively expensive and requires more contrast agents owing to its low sensitivity.

### 12.3.3 Radionuclide Imaging of EVs

Radionuclide imaging is an imaging technology in which radionuclides or their markers are introduced into the body to participate in tissue metabolism; metabolic processes are then identified and quantitatively assessed by measurement of emitted gamma photons that can penetrate the tissue. A radioactive detector is used for body surface detection. The primary radionuclide imaging technologies include gamma cameras, single-photon emission computed tomography (SPECT), and positron emission computed tomography (PET). SPECT and PET are also known as emission computed tomography. Commonly used radionuclides include $^{99m}$Tc, iodine ($^{123}$I, $^{124}$I, $^{125}$I, $^{131}$I), and $^{111}$In. $^{99}$mTc, $^{123}$I, $^{125}$I, $^{131}$I, and $^{111}$In can be detected using γ-counters or SPECT/CT imaging systems, and $^{124}$I is recommended for PET. Presently, γ-counter and SPECT/CT imaging systems are commonly used in EV tracking imaging studies, and $^{99m}$Tc, $^{125}$I, and $^{111}$In are commonly used marker elements.

By directly labeling the outer membrane of EVs with $^{111}$In oxine complexes or biotinylated membranes with $^{125}$I-IBB, the biodistribution of EVs in vivo could be quantitatively analyzed using γ cameras [5, 24]. The membrane proteins of EVs can also be labeled with $^{99m}$Tc, and the biological distribution of EVs in vivo can be quantitatively analyzed using SPECT/CT [15]. In some studies, $^{99m}$Tc-HMPAO was used to label macrophage-derived EVs ($^{99m}$Tc-HMPAO-EVs), after which a SPECT/CT imaging system was used to observe the distribution of EVs in mouse organs. The results showed that macrophage-derived EVs were mainly distributed in the liver, with lower accumulation in the spleen and kidney [18] (Fig. 12.4).

The γ photons emitted by radionuclides are empowered by strong penetration; therefore, radionuclide imaging is not limited by tissue depth and has a higher resolution than optical imaging. These advantages render it suitable for imaging in vivo. In addition, radionuclide imaging can quantitatively analyze the biological distributions of markers in real time, which is difficult to achieve using MRI. At the same time, the sensitivity of radionuclide imaging is higher than that of MRI, enabling researchers to avoid disruption of the structures and biological behaviors of labeled EVs by the addition of an excessive amount of labeled materials. However, the cost of radionuclide imaging is higher than that of optical imaging and MRI. Radionuclide labeling also carries the risk of ionizing radiation. The use of radioactive substances requires experimenters to be well trained and may be affected by various regulatory policies.

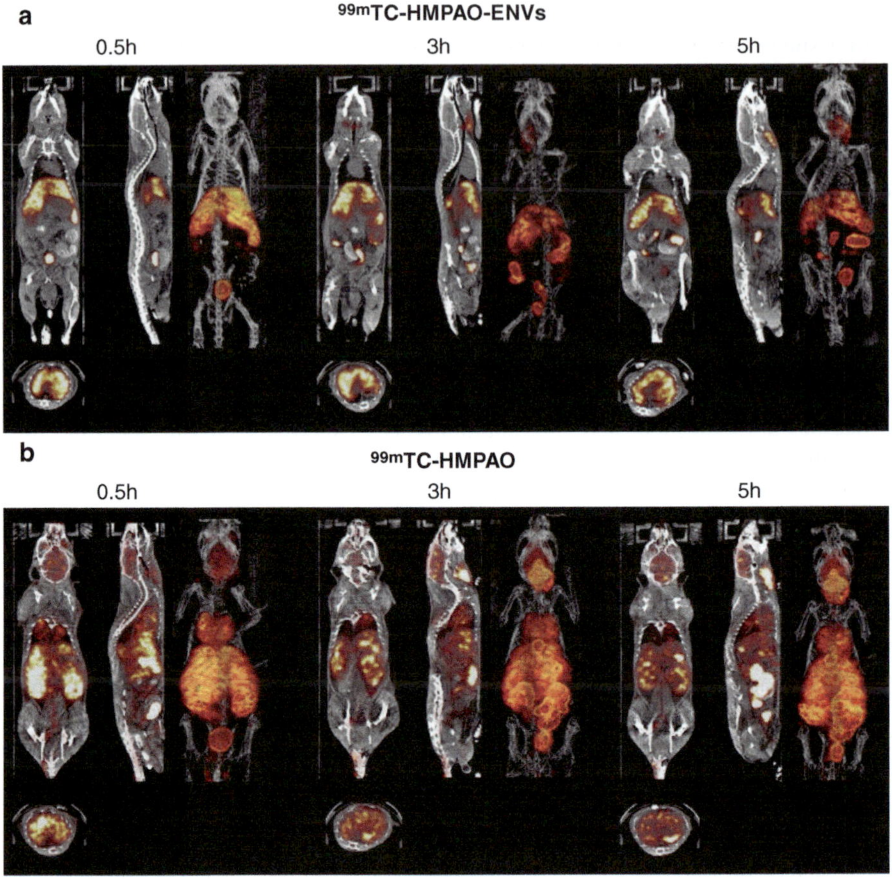


**Fig. 12.4** Analysis of the biodistribution of $^{99}$mTc-hmpao-EVs in mice using SPECT/PET imaging. (**a**) SPECT/CT imaging showed that $^{99}$mTc-HMPAO-EVs were mainly distributed in the liver, salivary gland, and intestinal tract at 5 h after administration. (**b**) Compared with $^{99}$mTc-HMPAO-EVs, $^{99}$mTc-HMPAO was mainly distributed in the brain tissue [18]. [*Figure reprinted with permission (Noninvasive imaging of radiolabeled exosome-mimetic nanovesicle using (99m) Tc-HMPAO. Springer Nature, 2015)*]

## 12.4 Future Perspectives for the Tracking and Imaging of EVs

Present strategies for studying the biological behaviors and functions of EVs fall into two categories. The first is "indirect inference," in which the possible biological functions of EVs are inferred based on indirect results. For example, polymerase chain reaction and Western blotting are widely used to study EVs carrying nucleic acids and proteins, respectively. Coculture experiments with EVs and recipient cells have been performed to study the related functions of EVs. The second strategy is "static display," which enables the visualization of EVs using static imaging platforms. Static displays include observing the morphological characteristics of EVs using transmission electron microscopy, studying the particle size of EVs using dynamic light scattering, and verifying the uptake of EVs by recipient cells using fluorescence microscopy. However, both indirect inference and static display ignore the dynamic nature of EVs. Therefore, realizing sensitive, nondestructive, in situ, real-time, and dynamic visualization of the pathophysiological processes regulated by EVs has become a hotspot in recent research efforts. With the development of new research strategies, dynamic tracking has attracted worldwide attention. Dynamic tracking combines the advantages of two existing strategies, which are capable of presenting EVs from the macro perspective of heterogeneous populations and the micro perspective of specific cargo. Moreover, continuous dynamic imaging based on time series analysis has been developed to track the complex movements of EVs. However, dynamic tracking research on EVs is still in its early stages, and further breakthroughs are necessary to realize multicomponent labeling, multimodal imaging, and single-particle tracking of EVs. In the near future, dynamic tracking is expected to provide important experimental evidence for EV research.

### 12.4.1 Multicomponent Labeling and Imaging of EVs

In addition to being natural carriers of biological information in vivo, EVs play a vital role in drug delivery. There are two main applications of dynamic EV tracking: (1) elucidation of the dynamic process of EV cargo delivery in the field of basic biological research and (2) evaluating the effectiveness and safety of EVs as drug carriers in vivo. In the former case, present strategies for single-label tracking of EVs are insufficient to study key biological molecules in EVs. Similarly, in the latter case, presently available strategies for single-component tracking of EVs fail to enable visualization of the active components of EVs. Therefore, multicomponent dynamic tracking is essential for developing an appropriate technique for EV tracking and imaging. The purpose of dynamic tracking is to analyze the nature and function of EVs on a multidimensional space-time scale from a time series. The multicomponent dynamic tracking strategy provides a powerful means for in-depth studies of EVs. However, the main goal of this approach is the effective, nondestructive visualization of EV cargo with high spatial and temporal resolution. First, multicomponent dynamic tracking must be biologically friendly and must allow elucidation of the roles of EV subcomponents in a high-fidelity manner. The impact of the tracking and imaging processes on biological activities should be minimized. Second, optimization of spatial resolution should be performed based on signal strength, the signal-to-noise ratio, distinction of adjacent EVs (e.g., the Abbe diffraction limit), the spatial localization of a single component, and other factors. Optimizing the spatial resolution is also important to ensure tracking quality. To date, our team has made some progress in the study of multicomponent dynamic tracking of EVs. To explore the roles of microRNAs (miRNAs) in tumor angiogenesis, we developed a multicolor and multicomponent visualization strategy for tumor-derived EVs. First, we labeled the outer membrane of EVs by biotinylating the cell membrane with fluorescent QDs. Then, we electroporated EVs to enable a variety of fluorescently labeled miRNAs to enter EVs. Thus, we dynamically monitored the real-time uptake of EVs into vascular endothelial cells using confocal microscopy based on a coculture platform in living cells.

### 12.4.2 Multimodal Tracking and Imaging of EVs

Recent studies have shown that the clinical application of EVs mainly depends on dynamic tracking and imaging of EVs in vivo. Presently, few imaging platforms have been developed for the in vivo imaging of EVs; existing optical, magnetic resonance, and radionuclide imaging platforms have been designed for cell imaging or even medical imaging, and the parameters of these platforms can be adjusted to be compatible with EVs. The experimental designs are limited by the imaging threshold. With the horizontal and vertical expansion of research on EVs, it is necessary to establish a multimodal and dynamic strategy for the imaging and tracking of EVs. Considering the above requirements of multicomponent dynamic tracking, multimodal tracking platforms with the capacity for multiple imaging modes should also be developed for mainstream use in the future. Indeed, single-mode imaging strategies may not meet the requirements for EV imaging and tracking. The imaging channel overlap or mismatch, response delay, continuous tracking time, and other issues should also be considered. Thus, it is important to develop a multimodal dynamic tracking technology for EVs for the following reasons: (1) expanding the combination of labeling, imaging, and tracking methods may be more conducive to realizing multicomponent simultaneous in situ dynamic tracking; (2) one tracking mode can provide information for another tracking mode to improve image accuracy and reconstruction quality; and (3) various modes can complement each other to expand the applications of the dynamic tracking system. To realize multimodal and dynamic tracking of EVs, we can combine different labeling strategies or develop new labeling materials with different imaging modes simultaneously. In our previous work on multimodal visualization strategies, we synthesized ultrasmall QDs and demonstrated the capacity for dual-mode imaging in both magnetic resonance and near-infrared fluorescence; the imaging results of bimodal tracking were consistent with each other. Thus, we achieved dual-mode visualization of EVs in vivo, thereby enabling us to monitor the dynamic distribution of EVs in various tissues.

### 12.4.3 Dynamic Tracking of Single Particles

As a representative method for studying the properties or movement characteristics of individual particles in a heterogeneous medium, dynamic tracking of single particles differs from traditional methods for uniformly analyzing a large number of particles. Dynamic tracking of a single particle does not conceal the unique features of single particles and can reveal the essential characteristics of heterogeneous particles on a multidimensional scale of time and space. This is realized by analysis of the highly complex and dynamic activities of EVs. Hence, dynamic tracking of single particles is expected to lead to major advances in the field of EV tracking and imaging.

Visualization strategies, imaging platforms, and data analyses are the core techniques that must be developed for single-particle tracking. For visualization strategies, the rapid response and signal-to-noise ratio of the imaging signal are important for providing continuous and dynamic monitoring and high spatial resolution. Present visualization strategies are based on accumulated experience with optical imaging platforms and are used for single-particle tracking in the fields of virology, subcellular structure analysis, and molecular biology. Based on an extensive comparison of the labeling strategies of EVs, our team clarified the advantages of QDs in single-particle tracking. Owing to the small particle size, high fluorescence intensity, satisfactory optical stability, wide range of excitation light, and narrow range of emission light, QDs are suitable for the development of multicolor and multicomponent strategies for single-particle tracking of EVs. However, interference associated with imaging of the particles must be minimized when a tracking signal with sufficient strength is detected. High-sensitivity imaging platforms with low detection delays and good biocompatibility are required. Several imaging platforms meet the above requirements, including optical microscopy and PET-CT, the former of which can be divided into confocal microscopy and total internal reflection microscopy. Given the inherent limitations of these platforms and imaging strategies, researchers have mainly used optical microscopy to conduct single-particle tracking of EVs in vitro. However, PET-CT is more commonly used in single-particle tracking of EVs in vivo owing to its good spatial positioning characteristics. Regardless of which visualization strategy and imaging platform are applied, the key to single-particle tracking lies in data analysis, which can be divided into three steps. First, the coordinates of the target particles in each frame of the images, namely, particle positioning, are determined based on the detected signals. Second, the positions of the same particles in each frame are connected according to time series analyses (i.e., trajectory reconstruction). The third step is analysis of the obtained features. Thus, accurate trajectory reconstruction depends on the coordination of multiple factors, including signal strength, signal-to-noise ratio, imaging time, and trajectory algorithm.

## References

1. Armstrong JPK, Holme MN, Stevens MM. Re-engineering extracellular vesicles as smart nanoscale therapeutics. ACS Nano. 2017;11(1):69–83.
2. Suetsugu A, Honma K, Saji S, Moriwaki H, Ochiya T, Hoffman RM. Imaging exosome transfer from breast cancer cells to stroma at metastatic sites in orthotopic nude-mouse models. Adv Drug Deliv Rev. 2013;65(3):383–90.
3. Takahashi Y, Nishikawa M, Shinotsuka H, Matsui Y, Ohara S, Imai T, et al. Visualization and in vivo tracking of the exosomes of murine melanoma B16-BL6 cells in mice after intravenous injection. J Biotechnol. 2013;165(2):77–84.
4. Lai CP, Kim EY, Badr CE, Weissleder R, Mempel TR, Tannous BA, et al. Visualization and tracking of tumour extracellular vesicle delivery and RNA translation using multiplexed reporters. Nat Commun. 2015;6(1):7029.

5. Morishita M, Takahashi Y, Nishikawa M, Sano K, Kato K, Yamashita T, et al. Quantitative analysis of tissue distribution of the B16BL6-derived exosomes using a streptavidin-lactadherin fusion protein and iodine-125-labeled biotin derivative after intravenous injection in mice. J Pharm Sci. 2015;104(2):705–13.
6. Lai CP, Mardini O, Ericsson M, Prabhakar S, Maguire C, Chen JW, et al. Dynamic biodistribution of extracellular vesicles in vivo using a multimodal imaging reporter. ACS Nano. 2014;8(1):483–94.
7. Coscia C, Parolini I, Sanchez M, Biffoni M, Boussadia Z, Zanetti C, et al. Generation, quantification, and tracing of metabolically labeled fluorescent exosomes. Methods Mol Biol (Clifton, NJ). 2016;1448:217–35.
8. Wang M, Altinoglu S, Takeda YS, Xu Q. Integrating protein engineering and bioorthogonal click conjugation for extracellular vesicle modulation and intracellular delivery. PLoS One. 2015;10(11):e0141860.
9. Lee TS, Kim Y, Zhang W, Song IH, Tung CH. Facile metabolic glycan labeling strategy for exosome tracking. Biochim Biophys Acta. 2018;1862(5):1091–100.
10. Chen G, Zhu J-Y, Zhang Z-L, Zhang W, Ren J-G, Wu M, et al. Transformation of cell-derived microparticles into quantum-dot-labeled nanovectors for antitumor siRNA delivery. Angew Chem Int Ed. 2015;54(3):1036–40.
11. Busato A, Bonafede R, Bontempi P, Scambi I, Schiaffino L, Benati D, et al. Magnetic resonance imaging of ultrasmall superparamagnetic iron oxide-labeled exosomes from stem cells: a new method to obtain labeled exosomes. Int J Nanomed. 2016;11:2481–90.
12. Silva AKA, Luciani N, Gazeau F, Aubertin K, Bonneau S, Chauvierre C, et al. Combining magnetic nanoparticles with cell derived microvesicles for drug loading and targeting. Nanomedicine. 2015;11(3):645–55.
13. Puzar Dominkus P, Stenovec M, Sitar S, Lasic E, Zorec R, Plemenitas A, et al. PKH26 labeling of extracellular vesicles: characterization and cellular internalization of contaminating PKH26 nanoparticles. Biochim Biophys Acta. 2018;1860(6):1350–61.
14. Roberts-Dalton HD, Cocks A, Falcon-Perez JM, Sayers EJ, Webber JP, Watson P, et al. Fluorescence labelling of extracellular vesicles using a novel thiol-based strategy for quantitative analysis of cellular delivery and intracellular traffic. Nanoscale. 2017;9(36):13693–706.
15. Varga Z, Gyurkó I, Pálóczi K, Buzás EI, Horváth I, Hegedűs N, et al. Radiolabeling of extracellular vesicles with 99mTc for quantitative in vivo imaging studies. Cancer Biother Radiopharm. 2016;31(5):168–73.
16. Jiang X, Zong S, Chen C, Zhang Y, Wang Z, Cui Y. Gold-carbon dots for the intracellular imaging of cancer-derived exosomes. Nanotechnology. 2018;29(17):175701.
17. Qi H, Liu C, Long L, Ren Y, Zhang S, Chang X, et al. Blood exosomes endowed with magnetic and targeting properties for cancer therapy. ACS Nano. 2016;10(3):3323–33.
18. Hwang DW, Choi H, Jang SC, Yoo MY, Park JY, Choi NE, et al. Noninvasive imaging of radiolabeled exosome-mimetic nanovesicle using (99m)Tc-HMPAO. Sci Rep. 2015;5:15636.
19. Betzer O, Perets N, Angel A, Motiei M, Sadan T, Yadid G, et al. In vivo neuroimaging of exosomes using gold nanoparticles. ACS Nano. 2017;11(11):10883–93.
20. Yu Z-L, Zhang W, Zhao J-Y, Zhong W-Q, Ren J-G, Wu M, et al. Development of a dual-modally traceable nanoplatform for cancer theranostics using natural circulating cell-derived microparticles in oral cancer patients. Adv Funct Mater. 2017;27(40):1703482.
21. Zhao JY, Chen G, Gu YP, Cui R, Zhang ZL, Yu ZL, et al. Ultrasmall magnetically engineered $Ag_2Se$ quantum dots for instant efficient labeling and whole-body high-resolution multimodal real-time tracking of cell-derived microvesicles. J Am Chem Soc. 2016;138(6):1893–903.
22. Hu L, Wickline SA, Hood JL. Magnetic resonance imaging of melanoma exosomes in lymph nodes. Magn Reson Med. 2015;74(1):266–71.
23. Dabrowska S, Del Fattore A, Karnas E, Frontczak-Baniewicz M, Kozlowska H, Muraca M, et al. Imaging of extracellular vesicles derived from human bone marrow mesenchymal stem cells using fluorescent and magnetic labels. Int J Nanomed. 2018;2018(13):1653–64.
24. Smyth T, Kullberg M, Malik N, Smith-Jones P, Graner MW, Anchordoquy TJ. Biodistribution and delivery efficiency of unmodified tumor-derived exosomes. J Control Release. 2015;199:145–55.

# Methods for Analysis of Extracellular Vesicle Functions

# 13

Shijing Yue, Haobin Wang, Kun Zhao, Wei Mu, and Zhe Wang

## 13.1 Introduction

In recent years, research on extracellular vesicles (EVs) has progressed rapidly, with many studies providing evidence on the mechanisms, functions, and applications of these particles [1]. In this chapter, we summarize and discuss the main methods developed for studies of EV function. EVs may have applications as a potential drug delivery system to target specific organs or recipient cells. To alter the phenotypes of recipient cells, EV cargo must be modified in such a way as to regulate gene expression in the target cells. Nucleic acid modification allows EVs to target specific organs or recipient cells. Moreover, EVs can also be loaded with specific regulatory nucleic acid molecules to modulate gene expression in recipient cells. Modifying EVs with proteins can also alter the surface molecules and enable the targeting of specific organs or recipient cells. EV cargo-specific proteins may

Shijing Yue, Haobin Wang, Kun Zhao, Wei Mu, and Zhe Wang contributed equally to this work.

S. Yue (✉)
School of Medicine, Nankai University, Tianjin, China
e-mail: shijingyue@nankai.edu.cn

H. Wang
The Third People's Hospital of Chengdu, Chengdu, Sichuan, China

K. Zhao
The First Affiliated Hospital of Zhengzhou University, Zhengzhou, Henan, China

W. Mu
School of Public Health, Shanghai JiaoTong University School of Medicine, Shanghai, China

Z. Wang
The First Affiliated Hospital of Guangdong Pharmaceutical University, Guangzhou, Guangdong, China



Q. Wang, L. Zheng (eds.), *Extracellular Vesicles*,
https://doi.org/10.1007/978-981-99-8365-0_13

regulate the activation of distinct signaling pathways in recipient cells. The double-layer phospholipid molecular membrane allows EVs to be easily modified by lipids, thereby changing their properties and characteristics. Lipid components may provide adaptors to facilitate EV labeling or detection. Additionally, coculture with recipient cells has been widely applied for functional research on EVs in vitro. The uptake and internalization of EVs by recipient cells are affected by various membrane proteins. In vivo functional investigations of EVs have been performed using multiple routes of administration in small animal models. The effects of EV uptake in vivo are related to the injection method and specific EV modifications. Owing to the capillary system present in tissues, the liver and spleen are common sites showing the highest retention of EVs following systemic injection. Recent results from in vitro and in vivo studies have enabled us to build a systematic view of EV modifications according to particle size and other parameters.

## 13.2 EV Modifications

Cell-derived EVs can be used as therapeutic drugs for precise treatment and as molecular imaging agents for diagnosis. Because of their unique membrane structure, consisting of a water core and highly enriched lipid rafts, EVs are ideal nanoparticles for loading hydrophilic and lipophilic reagents [2]. However, present knowledge on EVs has revealed the major challenges that must be overcome in order to facilitate the application of EVs in the clinical setting including the lack of a unified characterization method for EVs from different sources, the low yield and encapsulation efficiency of EVs, and the lack of high-efficiency advanced purification technologies. Based on these major challenges, EVs engineered with multiple modifications have been proposed. EVs can be modified with nucleic acids, proteins, and lipids using various methods. To further improve the applicability of EVs in precision therapy, novel strategies have been developed to modify EVs and engineered EVs as drug delivery platforms with high loading efficiencies, targeting abilities, and controllable drug release capabilities (Table 13.1).

### 13.2.1 EV Modification with Nucleic Acids

EVs have been developed as a new-generation targeted delivery system that can be modified using nucleic acids. Nucleic acid aptamer modification facilitates the propagation of EVs through the microvasculature and interstitial tissue to bind to specific tissues. As an aptamer-nanoparticle drug delivery system, nonspecific uptake by phagocytes in the reticuloendothelial system must be resolved. Engineered EVs guided by aptamers can provide higher specificity and efficacy in targeted therapies. In addition, studies have demonstrated that cell-derived EVs can evade phagocytosis, exhibiting a longer circulating half-life and lower immunogenicity. The superior performance of nucleic acid aptamers and EVs has resulted in the development of targeted drug delivery systems [17].

**Table 13.1** Strategies for EV modification

| Modification types | Molecular targets | Names of modification | Mechanisms | Disease or application | References |
|---|---|---|---|---|---|
| DNA aptamer | Nucleolin | AS1411 | Binding to cancer hallmark nucleolin for targeting | Cancer | [3] |
| | Mucin 1 | AptA, AptB | Binding to cancer hallmark mucin 1 for targeting | Cancer | [4] |
| | EpCAM | SYL3C, Ep1 | Binding to cancer hallmark EpCAM for targeting | Cancer | [4] |
| | HER2 | Herceptamers | Binding to cancer hallmark HER2 for targeting | Breast cancer | [4] |
| mRNA | PTEN | PTEN mRNA | Restore PTEN expression | Cancer | [5] |
| siRNA | Survivin | Survivin siRNA | Inhibit survivin gene expression | Cancer | [6] |
| | Kras | $Kras^{G12D}$ siRNA or shRNA | Suppress $Kras^{G12D}$ expression and inactivate Kras signaling | Cancer | [7] |
| miRNA | PTEN/PI3K/AKT signaling | miR-126 | Interrupt the PTEN/PI3K/AKT signaling pathway | Cancer | [8] |
| | miR-155 | Anti-miR-155 | Inhibit miR-155 expression | Cancer | [9] |
| Proteins | HER2 | DARPin | Target recognition of HER2-positive cancer cells | Breast cancer | [10] |
| | CD95/death receptor | TRAIL | Induce apoptosis and necrosis in the TRAIL-sensitive cells | Cancer | [11] |
| | TNFR | CTNF-α | Enhance cancer cell growth inhibition and apoptosis | Cancer | [12] |
| Lipids | EV membrane | DSPE-PEG | EV labelling or targeting | Diagnosis | [13, 14] |
| | EV membrane | Cholesterol | Drug delivery or EV detection | Diagnosis and therapy | [15, 16] |

#### 13.2.1.1 EV Surface Modification with Nucleic Acid Aptamers

Aptamers, also known as chemical antibodies, are single-stranded DNA or RNA molecules that fold into a three-dimensional (3D) structure and bind to targets with high affinity and specificity. Aptamers are produced by polymerase chain reaction-based in vitro selection using systematic evolution of ligands by exponential enrichment, a tool for deciphering protein DNA-binding sequence specificity [18]. Single-stranded nucleic acids adaptively fold via the pairing of some complementary bases in the chain, electrostatic interactions, and hydrogen bonding to form stable 3D structures, such as hairpins, pseudo knots, stem loops, and G-quartets. The stable 3D structure of nucleic acids can bind to a target to form a stable complex. Nucleic acid aptamers have low immunogenicity, high specificity and sensitivity, and strong tumor penetration depth. EVs combined with aptamers for targeting effects are expected to become intelligent engineered nanovesicles for the next generation of precision medicine [7].

#### RNA Aptamers

Aptamers have been developed based on the discovery that RNA molecules fold into "pockets" for specific binding with small molecules to form a stable surface. Specific RNA molecules have been developed into scaffolds that can recognize and bind to specific molecules. For example, an RNA aptamer was screened to target vascular endothelial growth factor (VEGF), which is commonly used as a therapeutic agent in the clinical setting. The RNA aptamer used as a VEGF antagonist, called pegaptanib (trade name Macugen), was approved by the US Food and Drug Administration in 2004 for the treatment of angiogenesis and age-related macular degeneration. In addition, nuclease-resistant RNA aptamers (A9 and A10) against prostate-specific membrane antigen (PSMA) have been extensively studied in preclinical models [19].

#### DNA Aptamers

Compared with RNA aptamers, DNA aptamers are more stable and have been widely applied. A DNA aptamer was discovered accidentally based on its ability to be absorbed by cancer cells, and this aptamer has since been evaluated in clinical trials for cancer treatment. The molecule is a G-rich DNA nucleolar aptamer (AS1411) composed of 26 nucleotides and has direct antiproliferative effects in vitro. The target of AS1411 is nucleolin, which is expressed in the nuclei of all cells. Nucleolin is overexpressed in the cytoplasm and plasma membranes of cancer cells [3]. Moreover, DNA aptamers targeting hallmark cancer molecules, such as human epidermal growth factor receptor-2 (HER-2), epithelial cell adhesion molecule (EpCAM), and mucin-1, have been screened and are expected to have applications in clinical treatment [4].

#### Methods for Modification by Nucleic Acid Aptamers

EVs can be modified with nucleic acid aptamers using various methods. Nucleic acid aptamers can be fixed onto the EV membrane via click chemistry. Furthermore, EVs can be fused to nucleic acid-modified liposomes. However, these two methods involve multiple steps that may be difficult to complete accurately at the nanometer level. Based on the characteristics of distearoylphosphatidylethanolamine (DSPE), C18, cholesterol, and other lipids that can be directly inserted into the plasma membrane, nucleic acid aptamers can be modified on EV surfaces using specific lipids. Owing to its amphiphilic properties, effective labeling, and excellent retention ability at low temperatures, cholesterol has been widely applied to modified EVs with nucleic acid aptamers [20].

#### 13.2.1.2 Nucleic Acids as EV Cargo

Small RNAs, including mRNAs, microRNAs (miRNAs), small interfering RNAs (siRNAs), and PIWI-interacting RNAs (piRNAs), are involved in the regulation of cell proliferation, apoptosis, migration, and many other cellular events via the silencing of gene expression. A study indicated that EVs containing phosphatase and tensin homolog (*PTEN*) mRNA can effectively restore the functions of tumor suppressors to enhance the inhibition of tumor growth and increase survival in a PTEN-deficient glioma mouse model [5]. siRNAs containing 20–25 nucleotides (nt) prevent target gene expression by interfering with mRNA. The mechanism involves the degradation of mRNA after the transcription of specific target genes by binding complementary nucleotide sequences [21]. Additionally, miRNAs are non-coding single-stranded small RNA molecules (19–23 nt) encoded by endogenous genes. miRNAs can regulate the stability of mRNAs in various types of cells to regulate gene expression levels and transcription efficiency. piRNAs are also non-coding RNAs (24–32 nt) that exist primarily in the reproductive system. piRNAs serve as PIWI guides to regulate target gene expression, transcription, and posttranscriptional modifications [22].

Owing to their instability, small RNAs are extremely difficult to translate into the clinical setting. Therefore, it is important to identify a suitable RNA transport carrier that can effectively deliver the RNAs to the target tissues and cells. Traditional small RNA delivery systems include viral and nonviral vectors. Viral vectors have higher transfection efficiency and can cause an immune response. Nonviral vectors include liposomes, cationic polymers, and inorganic nanomaterials. Although nonviral vectors have several advantages, including a wide range of sources, controllable structure, easy preparation, and low toxicity, they can be easily removed using antibodies. Furthermore, liposomes are prone to electrostatic binding of negatively charged substances in the plasma causing aggregation. EVs are endogenous biologically active microvesicle structures released by cells; they can function by transporting small RNA molecules between cells and can therefore be used as potential transport carriers for small RNAs [23].

EVs have several advantages as potential small RNA transport carriers. For example, EVs are produced by endogenous cells and show high biocompatibility and low biological immune clearance. Furthermore, modified EVs can deliver siRNAs across the blood-brain barrier. EVs can be produced by all cell types, have a wide range of sources, act as convenient storage molecules, and can be transported throughout the body. EVs can also be modified by surface particles to produce active targeting substances, thereby improving encapsulation efficiency; this feature may confer EVs with important roles in tumor treatment. siRNAs are promising gene therapy drugs that can show good stability when packaged as EV cargo. Studies have demonstrated that EV siRNAs can effectively inhibit the occurrence and development of pancreatic cancer in mice. Indeed, the delivery of siRNA targeting transforming growth factor β1 as EV cargo successfully inhibits tumor growth. The siRNA can also integrate with the backbone of highly enriched pre-miRNAs for endogenous expression, and packaging into EVs increases the therapeutic effects and reduces the side effects of the siRNA [24]. EVs can also be used as carriers to deliver siRNAs for treatment using electroporation to load the siRNAs. Faruqu et al. used electroporation to encapsulate siRNAs into HEK-293 cell-derived exosomes for transport to PANC-1 human pancreatic cancer cells in vitro [25]. EVs transfer miRNAs to recipient cells, inducing the activation of signaling pathways by regulating specific genes to alter their phenotypes. For example, EVs loaded with *miR-126* can strongly inhibit the proliferation and migration of A549 lung cancer cells by interrupting the PTEN/phosphatidylinositol 3-kinase/AKT signaling pathway. In a lung metastasis model in nude mice, systemic administration of EVs loaded with *miR-126* showed anticancer effects [8]. Additionally, miRNA inhibitors delivered by EVs can neutralize distant metastases caused by cancer. EVs loaded with *miR-155* inhibitors act on macrophages to reduce lipopolysaccharide-induced tumor necrosis factor (TNF)-α production, with a reduced inflammatory response in the body [9]. Moreover, fibroblast-like mesenchymal cell (MSC)-derived EVs carry siRNAs or short hairpin RNAs specific to oncogenic $Kras^{G12}$ to target oncogenic KRAS with enhanced efficacy [7]. EVs containing survivin siRNA with PSMA aptamer modification effectively inhibit prostate cancer xenograft formation [6].

### 13.2.2 EV Modification with Proteins

EVs contain many types of proteins, including biogenesis-related and cell type-specific proteins. The production of EVs varies in different cell types and under certain conditions. Cell-derived and cell-engineered EVs are potential drug delivery systems, particularly for cancer therapy. Cancer-derived EVs play important roles in tumor escape from immune surveillance [7]. Many studies have demonstrated that EVs are involved in gastrointestinal cancer metastasis. Cancer metastasis is regulated by EVs via the epithelial-mesenchymal transition, angiogenesis, and the tumor microenvironment [26].

#### 13.2.2.1 Protein Modifications for EV Targeting

EVs have been used as drug delivery systems. The specific targeting of cell-derived or cell-engineered EVs is a key point that may facilitate the application of EVs in the clinical setting. Protein modification for EV targeting has been a focus in the field.

##### Genetic Engineering

A designed ankyrin repeat protein (DARPin) was expressed on the surface of EVs using genetic engineering. DARPin recognizes and binds specifically to HER-2. EVs modified with DARPin were used to treat and target HER2-positive breast cancer cells [10].

#### O-Glycosylation for Modification of Glycosylphosphatidylinositol (GPI)-Anchored Proteins

GPI-anchored proteins are macromolecular complexes of proteins, lipids, and sugar chains that interact via covalent bonds. These proteins are involved in various physiological processes, including immune recognition, cell communication, and signal transduction. GPI-anchored proteins on the surface regulate the functions of EVs, including immune responses and tumor growth. Cancer cells alter the expression of short O-glycans by interfering with the O-glycosylation pathway. Glycosylation is one of the most abundant posttranslational modifications in cells. However, malignant tumors can interfere with the O-glycosylation pathway and alter the expression of short O-glycans, a common and important feature. Exosomal glypican 1 has been reported as a biomarker for the early diagnosis of pancreatic cancer.

#### Peptide Modification for Targeting

The short peptide Arg-Gly-Asp (RGD) is an important ligand that can be recognized by integrins. Matrix proteins, including fibronectin, fibrinogen, and collagen, containing the RGD tripeptide structural unit (which is responsible for integrin binding), can affect cell adhesion, proliferation, differentiation, and migration. Specific integrins are expressed at high levels on the surfaces of tumor cells, and EVs may be modified with RGD peptides using genetic engineering methods [27]. EVs have also been modified with RGD peptides that competitively bind to integrin receptors to interfere with the proliferation and metastasis of cancer cells.

#### Folic Acid Modification for Targeting

The folic acid receptor (FR) is anchored to the cell membrane surface via GPI. It has high affinity for folic acid binding and transport. Owing to metabolic reprogramming and the sustained proliferation of cancer cells, the demand for folic acid is relatively higher than that of normal cells. FRs are expressed at high levels in cancer cells and can modify EVs with folic acid to target cancer cells for efficient drug delivery [28].

### 13.2.2.2 EV Cargo Proteins for Signaling Activation

#### EV Delivery of Functional Molecules Using Biotin

Biotin and streptavidin, which have high affinity for each other, are widely used in the field of EV modification. Indeed, the biotin-streptavidin system can be used for EV modification with cancer immunotherapy molecules or biomarker molecules for quantification. In addition, chemical editing approaches can facilitate the preparation of EVs with specific targeting abilities, i.e., EVs chemically labeled with dual ligands (biotin and avidin) and packed drugs into the cytosol via a 3D nanostructured microfluidic chip [29].

#### EV Modification with Chimeric Proteins of Receptors and Ligands

EVs enriched with chimeric programmed death ligand-1 (PD-L1) can mediate programmed death-1 (PD1) cross-linking and immunosuppression. Furthermore, EVs can also enrich chimeric MHC molecules that are essential for T-cell signal transduction. EVs derived from EL4 leukemia/lymphoma cells and containing TNF-related apoptosis-inducing ligand (TRAIL) as cargo can induce apoptosis and necrosis in TRAIL-sensitive cells [11]. Furthermore, MSCs expressing TRAIL/green fluorescent protein (GFP)-containing EVs can induce apoptosis in lymphoma and INT12 melanoma. EVs modified with a targeting peptide (CPP) and TNF-α fusion protein (CTNF-α) combined with superparamagnetic iron oxide nanoparticles (SPIONs) were found to significantly enhance cancer cell growth inhibition and the TNF receptor I-mediated apoptosis pathway [12]. EVs modified with the neuropilin-1 targeting peptide RGERPPR by click chemistry for delivery of SPIONs and curcumin have been applied in the imaging and treatment of glioma. The fusion of target proteins with EV membrane proteins improves their specificity for loading into EVs. For example, anti-epidermal growth factor receptor ligands on EV membranes act through a GPI anchor for cancer cell targeting. Moreover, CD63 and lysosome-associated membrane protein-2b (Lamp2b), as the fusion product of a targeting peptide with EV marker proteins, have been commonly used to acquire targeting EVs [30].

#### EVs as Carriers for Protein Delivery

To avoid protein degradation and poor cellular uptake during the delivery of therapeutic proteins through systemic administration, EVs have been suggested as ideal delivery vectors for functional proteins [31]. For example, melanoma cells were transfected with a dominant-negative mutant (survivin-T34A) to produce survivin-T34A-containing EVs. Another study reported the modification of EVs by anchoring targeting ligands on their membranes; the findings showed that EVs engineered to express targeting ligands (iRGD) showed strong and efficient targeting to αV integrin-positive breast cancer cells.

### 13.2.2.3 Enrichment of Functional Molecules in EVs

#### Enrichment by Genetic Engineering

Proteins fused with the lactadherin (LA) C1C2 domain can be enriched on EVs via lipid binding. The biogenesis of EVs can be enhanced by selective enrichment of transmembrane proteins, including Lamp2b, tetraspanin membrane protein superfamily (CD63, CD9, CD81), and platelet growth factor receptor. Presently, the most widely used method for enrichment of molecular EVs is to generate fusion proteins with specific proteins and selectively enriched transmembrane proteins through genetic engineering technology [32].

#### Enrichment by Ubiquitination

Proteins can be enriched in EVs via the ubiquitination modification pathway. E3 ubiquitin ligases can be divided into three families: the homologous to the E6-AP carboxyl terminus (HECT)-family protein E6-associated protein E3 ligase, the ring E3 ligase, and the ring-in-between-ring (RBR) E3 ligase. The enzymatic activities of HECT and RBR E3 ligase directly catalyze the attachment of ubiquitin to the substrate, whereas ring E3 ligase transfers ubiquitin to the target protein. Ubiquitin fusion proteins can be developed to enrich the target proteins in EVs for treatment. The antigen protein Ag85B generates a fusion protein with ubiquitin-bound enhanced GFP via transient transfection in HEK293 cells to produce EVs for immune therapy [33].

**Modification with Surface Proteins for EV Tracking**

Modification with surface proteins can be used to track EV localization. CD63-fused light-activated fluorescent protein (PAmcherry) has been used in cancer imaging research. PAmcherry was shuttled into the 3′ end of the *CD63* gene fragment using genetic engineering technology to express the CD63-fused fluorescent protein, enabling the tracking of cancer cells. In addition, EVs are usually tracked using fluorescently labeled tetraspanin proteins, which are associated with membranes, even after EV cargo delivery. Assessment of EV cargo release and transfer should involve the use of cytosolic EV markers [34].

### 13.2.3 EV Modification with Lipids

EV membranes contain phosphatidylcholine (PS), phosphatidylethanolamine (PE), phosphatidylinositol, phosphatidic acid (PA), cholesterol, ceramide, sphingomyelin, and glycosphingomyelin. The proportions and concentrations of lipids in purified EVs differ from those in plasma and other cell membranes. These differences may provide clues regarding the mechanisms of biogenesis. Studies have shown that ceramides can stimulate EV production. EVs contain high levels of cone and reverse cone lipids, including PE, PS, lysophospholipids, PA, and ceramide, which cause the strong membrane curvature observed in EVs [35]. Lipids play important structural roles in the membranes of EVs and are essential for EV biogenesis and release into the extracellular environment. Through recent technological developments and investigations, elucidation of the importance of lipids in the biology of EVs will be possible [36].

#### 13.2.3.1 DSPE Modification

**EV Protein Analysis**

EVs are promising biomarkers for liquid biopsy in cancer. The levels of specific EV proteins have been shown to correlate with the disease state and can therefore be used as indicators in cancer diagnosis. However, because of the small sizes and complex chemical compositions of EVs, EV protein analysis is challenging. Importantly, DSPE-polyethylene glycol (PEG)-SH can be used to engineer EVs through hydrophobic interactions, and assembly with gold nanoparticles (2 nm) can be used to prepare Exo@Au nanozymes. The proteins on Exo@Au nanozymes can be selectively captured using specific antibodies inoculated into 96-well plates. The immobilized Exo@Au nanozymes can then react with 3,3′,5,5′-tetramethylbenzidine and $H_2O_2$, exhibiting peroxidase-like activity, and can be colorimetrically assessed to determine EV protein levels using a microplate reader [13].

**Targeting Immune Cells**
Targeted delivery of antigens and drugs to lymph nodes is essential for successful immunotherapy. Specific antigens and drugs targeting lymph nodes should be taken up by dendritic cells (DCs) and macrophages to effectively activate T-cells. Moreover, mannose ligands can be selectively taken up by DCs through the mannose receptor CD206, which is a transmembrane protein mainly expressed in antigen-presenting cells (APCs). Mannose-conjugated PEG-lipid (DSPE-PEG-man) was synthesized using isothiocyanate-amine coupling chemistry and incorporated into EVs to effectively target DCs. PEG has been used to inhibit nonspecific adsorption, prolong drug circulation time, and reduce the immunogenicity of PEG-conjugated biomaterials [14].

#### 13.2.3.2 Cholesterol Modification

Cholesterol is a lipid sterol that plays key roles in normal cellular functions. As the main component of the eukaryotic cell membrane, cholesterol regulates the fluidity and permeability of the cell membrane. Owing to its high affinity for the double-layer membrane structure, cholesterol can be efficiently inserted into the double-layer membranes of EVs without destroying the original membrane structure. Therefore, cholesterol has unique advantages for targeted drug delivery systems and tumor detection [20].

**Targeted Drug Delivery**
Glucocorticoids (GCs) have been shown to exert therapeutic effects in rheumatoid arthritis. However, owing to the nonspecific distribution of GCs following systemic administration and the severe adverse reactions associated with long-term administration, the clinical application of GCs is limited. Studies have established biomimetic EVs through encapsulation of dexamethasone sodium phosphate and surface modification with folic acid-PEG-Chol for targeted drug delivery. Multidrug resistance is a major challenge in clinical chemotherapy, and in one study, the nucleolin-targeting aptamer AS1411 was connected to EVs through a cholesterol linker, cargo siRNA, and the chemotherapeutic drug doxorubicin to target multidrug-resistant tumor cells [15].

**EV Detection**
EVs are potential biomarkers for many diseases. However, reliable and convenient quantitative methods remain technically challenging. Immunomagnetic beads that capture EVs by spontaneously inserting a DNA anchor labeled with high-affinity bivalent cholesterol (B-Chol) can be directly and effectively separated and quantified. This method can sensitively detect $2.2 \times 10^3$ EVs/μL with a relative standard deviation of less than 5.6% and a sensitivity 100 times that of enzyme-linked immunosorbent assay.

In addition, a convenient and highly sensitive surface-enhanced Raman scattering (SERS)-based method can be used for specific separation and accurate quantification of EVs. SERS nanoprobes are immobilized on the surface of EVs through hydrophobic interactions between cholesterol and lipid membranes to form sandwich-type immune complexes. Immune complexes can then be magnetically captured to generate enhanced SERS signals. The degree of immune complex assembly and the corresponding SERS signal are positively correlated with EV concentration in a wide linear range from 40 to $4 \times 10^7$ particles/μL, with a detection limit of 27 particles/μL [16].

#### 13.2.3.3 Hybrid Nanovesicles of EVs and Liposomes

Liposomal formulations have been extensively explored owing to standardized manufacturing processes, simple surface modification, and high drug-loading efficiency. Owing to similarities in the lipid bilayers of cell-derived EVs and liposomes, many recent studies have reported the generation of hybrid nanovesicles by fusing liposomes with cell-derived EVs using different methods. These hybrid nanovesicles exhibit the advantages of both liposomes and cell-derived EVs. More importantly, hybrid nanovesicles that fuse immunogenic EVs with liposomes to encapsulate chemotherapeutics have many applications in immunochemotherapy. EVs are endogenous communication vesicles that represent an attractive strategy for advanced drug delivery. However, efficient and reproducible loading of EVs with therapeutic or imaging agents remains a bottleneck for their use as a drug delivery system. Many technologies have been developed to modify EVs through fusion with liposomes containing both membranes and soluble cargo. The fusion of EVs with functionalized liposomes can be triggered using PEG to create a smart biosynthetic hybrid carrier. This versatile method has been shown to efficiently enrich EVs with exogenous lipophilic or hydrophilic compounds while preserving their intrinsic content and biological properties. Hybrid EVs have been shown to improve the cellular delivery efficiency of chemotherapeutic compounds compared with the free drug or drug-loaded liposome precursor. This method allows the biocamouflage of liposomes by enriching their lipid bilayer and inner compartment with biogenic molecules. Thus, the proposed fusion strategy enables efficient EV loading and may facilitate the pharmaceutical development of EVs with adaptable activities and drug delivery capacity.

The composition of hybrid nanovesicles can be designed by varying the liposome-to-EV ratio and the composition of the precursor liposomes. This method was demonstrated by the fusion of MSC-derived EVs with PEGylated liposomes, enabling lower internalization by macrophages in situ. In addition, MSC-derived EVs generated upon fusion with commercial therapeutic liposomes display a three- to fourfold increased ability to deliver antitumor drugs to tumor cells compared with precursor liposomes. Interestingly, MSC-derived EVs retain their proliferative effects, even after fusion, indicating that their intrinsic properties are maintained after the fusion process. Therefore, the PEG-mediated fusion enables the engineering of hybrid nanovesicles with tunable properties while preserving their inner cargo and biological activity. This method can be applied to load EVs of any cellular origin with potential compounds associated with synthetic liposomes, from hydrophilic compounds encapsulated in their inner core (siRNAs, miRNAs, chemotherapeutic agents, proteins, nanoparticles, and imaging contrast agents) to lipophilic compounds residing at the bilayer surface (membrane proteins, targeting agents, PEGylated lipids). For example, we anticipate that PEGylation of EVs will increase their circulation time, resulting in improved targeting and enhanced permeability and retention effects, and that their intrinsic biocompatibility will be advantageous for avoiding vector-mediated toxicity. This fusion method also enables the addition of biogenic functionalities, such as enhanced cellular uptake, to liposomes. This is a promising approach for enhancing the properties of several liposomal formulations, particularly those that have already received market authorization.

### 13.2.4 Summary

Owing to the broad sources of EVs, functional studies have shown that EVs lack specific targeting capacity and defined cargo. To address these challenges, cell-derived EVs can be modified using biological molecules. Nucleic acids, proteins, and lipids can be engineered to modify EVs using multiple methods. EV surface modification with specific molecules mediates their association with recipient cell surface proteins for targeting. Modification of EV cargo with specific molecules can regulate gene expression and signaling pathways in recipient cells. Nucleic acid modification of EVs effectively facilitates targeting and regulation of recipient cells. Engineered EVs with nucleic acid aptamers clearly increase the specificity and efficacy of targeting. EVs modified with specific nucleic acids may effectively regulate the gene expression patterns of recipient cells. Protein modifications have been widely applied for EV targeting and EV-dependent regulation of gene expression in recipient cells. Lipid modification of EVs can improve detection sensitivity and can facilitate drug loading to achieve targeted drug delivery. Lipid modification methods are constantly being improved; for example, methods using cholesterol, which can be randomly inserted into the membrane, have been shown to enhance drug targeting.

## 13.3 Coculture of EVs with Target Cells

Donor cell-derived EVs with genetic engineering modifications can be used as drug carriers. Signaling molecules can be taken up by target cells and play functional roles in regulating intercellular communication. The uptake of EVs by target cells is a complex process mediated by EV membrane proteins, lipid molecules, and other structures. structures. Studies have shown that various types of endocytosis, such as macropinocytosis, phagocytosis, clathrin-mediated endocytosis, vesicle-dependent endocytosis, lipid raft-dependent endocytosis (the main clathrin-independent mechanism), and clathrin/vesicle protein-independent endocytosis." to "Studies have shown that various types of endocytosis have been identified as the intercellular transport mechanism of EV contents, such as macropinocytosis, phagocytosis, clathrin-mediated, vesicle-dependent, lipid raft dependent endocytosis (the main clathrin independent mechanism) and clathrin/vesicle protein independent endocytosis.

### 13.3.1 Interaction of EV Membrane Proteins and Recipient Cells

EVs selectively enrich specific proteins, including tetraspanin family proteins, cytoskeletal structure proteins, membrane transport proteins, cell adhesion molecules, lectins, and heparin sulfate proteoglycans.

#### 13.3.1.1 Tetraspanin Family Proteins and EV Uptake

The tetraspanin family, which includes CD9, CD63, CD82, CD81, CD151, Tspan9, and Tspan8, plays key roles in various physiological processes, including cell adhesion and viral infection. Proteins in the tetraspanin family contain four transmembrane domains and two extracellular loops. The large extracellular loop of tetraspanins combines with a wide range of functional proteins that regulate cellular function. Tetraspanins can form tetraspanin-enriched microdomains (TEMs) through a combination of tetraspanin proteins and/or cell surface molecules, such as integrins, immunoglobulin superfamily proteins, proteases, growth factor receptors, and intracellular signaling molecules. TEMs are involved in many physiological processes, including cell signal transduction, endocytosis, cell adhesion, migration, proliferation, and differentiation. Notably, knockdown of CD29 or CD81 significantly inhibits EV uptake [37]. Tetraspanins may be used for visualizing EV endocytosis. Stickney et al. developed a new EV surface display technology for mammalian cell-derived EVs using tetraspanin molecules. The fluorescent reporter genes were inserted into the two selected sites of tetraspanin to be expressed on the internal and external surfaces of EVs; this approach was used to successfully track EVs after gene transfection. The EV surface protein lymphocyte function-associated antigen-1 (LFA-1) may interact with integrin $\alpha4\beta7$ in recipient cells to increase EV uptake. Macrophage-derived EVs can be effectively taken up via the interaction between LFA-1 and intercellular adhesion molecule-1 (ICAM-1) expressed in human brain endothelial cells. Inflammation in the brain can increase the uptake of macrophage-derived EVs by 3.1-fold via induction of ICAM-1 expression in brain endothelial cells [38].

#### 13.3.1.2 Endocytosis by Membrane Proteins and Membrane Fusion

The particle size, surface membrane charge, and surrounding ionic strength of EVs play important roles in communication with target cells [39]. EVs can establish contact with target cells in vivo through various mechanisms, including membrane proteins, clathrin-mediated structures, and lipid membrane structures. Notably, the macrophage receptor (MARCO) promotes EV uptake via kinetic-dependent endocytosis and macrocytosis. MARCO-mediated actin reorganization and dynein activity are involved in EV internalization. Endocytosis may occur through clathrin-dependent and/or clathrin-independent mechanisms. In a previous study, clathrin-mediated endocytosis was induced using a cell permeability inhibitor of dynein; the treatment inhibited hypoxia-damaged cardiomyocytes by EVs in a concentration-dependent manner. Adaptor protein-2 and dynamin are also involved in the regulation of clathrin-dependent pathways [40]. Membrane fusion is a common method mediating the interactions of EVs with target cells. Quinn et al.

demonstrated that human epidermal growth factor receptor-2 (HER-2)-overexpressing BT-474 cell-derived EVs could successfully introduce HER-2 onto the surface of MDA-MB-231 triple-negative breast cancer cells via membrane fusion [41].

#### 13.3.1.3 Lectin and Heparin Sulfate Proteoglycan (HSPG) Contribute EV Uptake

Lectin is a soluble membrane receptor that recognizes and binds to sugar groups. The lectin protein family is involved in various cell-to-cell communication events, including adhesion and intracellular transport. Most lectins can be classified as transmembrane C-type lectins and selectins, transmembrane agglutinins (combined with sialic acid), or cytoplasmic galectins (combined with galactosidase). Studies have indicated that EV cargo galectin interacts with T-cell immunoglobulin mucin 1 (TIM1). TIM1 is a membrane receptor in phagocytes that interacts with PS and plays key roles in the uptake of EVs. HSPG is a ubiquitous glycoprotein that is involved in many cellular functions. These proteoglycans are hybrid receptors that bind various ligands through heparin sulfate chains. HSPG is involved in the interaction between EVs and cells and promotes extracellular interactions, including EV attachment and movement. Moreover, HSPG also regulates the binding and cargo delivery of EVs into hepatic satellite cells [42].

#### 13.3.1.4 Integrin-Dependent Regulation of the Interactions Between EVs and Target Cells

Integrins determine the tissue distribution of EVs to facilitate cancer metastatic niche formation. Integrins support EV internalization, membrane protein expression, and the activation of related kinase-mediated signaling pathways in target cells. EV cargo integrins bind to specific ligands on target cells and promote EV internalization. The interaction of integrin $\alpha4\beta7$ with the ligand mucosal addressin cell adhesion molecule 1 contributes to EV uptake by endothelial cells [43]. Subramanian et al. found that neutrophil-derived EVs adhere to endothelial cells in an integrin $\beta2$-dependent manner. The interaction between EVs and endothelial cells locally releases leukotriene B4, which plays pivotal roles in neutrophil recruitment and the inflammatory signal response. Furthermore, the Tspan8-$\alpha4\beta1$ complex is necessary for the interaction and internalization of EVs with endothelial cells to enhance tumor angiogenesis. In addition, integrin adhesion receptors are enriched in EVs that promote cancer cell migration and mediate organ-specific metastasis [44].

Integrin binding to Tspan8 promotes the uptake of EVs into target cells. Tspan8 is involved in the integrin-linked kinase signaling pathway and downstream molecules that regulate the invasive ability of melanoma cells. Additionally, Tspan8 promotes tumorigenesis and development by binding to and activating integrins. Tspan8 can also bind to a variety of integrins, including a3$\beta$1, a6$\beta$1, a6$\beta$4, av$\beta$3, and av$\beta$6, which activate downstream extracellular signal-regulated kinase, AKT, and c-Jun N-terminal kinase signaling; inhibit tumor cell apoptosis; and promote the

EMT in tumor cells. The mechanisms through which Tspan8 promotes tumor cell metastasis include activation of the VEGF signaling pathway to induce tumor angiogenesis, promotion of vascular endothelial barrier penetration by tumor cells, induction of tumor cell spreading, and reduction of cell adhesion to laminin [45].

### 13.3.2 EV Cargo Regulates Gene Expression Patterns in Recipient Cells

EVs play key roles in various physiological and pathological processes, including the induction of stimulatory and inhibitory immune responses, establishment of tumor metastatic niches, promotion of tumorigenesis, enhancement of viral spread, and progression of neurodegenerative diseases. EV cargo proteins, mRNAs, miRNAs, and lipids may reprogram the gene expression patterns of target cells, thereby regulating the phenotypes and physiological activities of target cells. EV cargo nucleic acids (mRNAs and noncoding regulatory RNAs) contribute to the reprogramming of target cells.

#### 13.3.2.1 Reprogramming of Target Cells by EV Cargo Proteins

EV cargo proteins are involved in carcinogenesis and cancer development. Metastatic cancer cell-derived EV cargo proteins promote cancer cell proliferation, migration, invasion, and angiogenesis. Metastatic melanoma secretes EVs expressing PD-L1, which can interact with the PD-1 receptor on T-cells to initiate an immune checkpoint response and is an indicator of the adaptive response of tumor cells to T-cell resurrection [46].

EV cargo proteins can regulate the functions and physiological states of target cells through receptor-ligand interactions. Chandra et al. demonstrated that human immunodeficiency virus (HIV)-infected cell-derived EVs induce cell/mitochondrial superoxide production and reduce mitochondrial membrane potential to cause HIV-related endothelial dysfunction. EVs contain the HIV-Tat protein, which reduces the expression of phosphorylated endothelial nitric oxide synthase and induces cells and mitochondria to produce superoxide to accelerate brain endothelial dysfunction. Furthermore, lipopolysaccharide binding protein and soluble E-cadherin expression levels in EVs have been used as biomarkers for metastatic non-small cell lung cancer and ovarian cancer cells [47].

#### 13.3.2.2 Reprogramming of Target Cells by EV Cargo miRNAs

MSC-derived EVs can transfer miRNAs, such as *miR-125a* and *miR-30b*, to promote tube formation in human umbilical vein endothelial cells by blocking the angiogenesis inhibitor delta-L4. Interestingly, MSC-derived EVs contain antiangiogenic miRNAs, such as *miR-16* and *miR-100*. Alteration of the target cell phenotype can be affected by the efficiency of miRNA interference with target gene expression in recipient cells. EV cargo miRNAs targeting VEGF in the tumor microenvironment can inhibit angiogenesis in breast cancer cells [48].

#### 13.3.2.3 Reprogramming of Target Cells by EV Cargo Circular RNAs (circRNAs)

CircRNAs have a closed continuous loop structure without a 5′-3′ polarity and a poly (tail). CircRNAs are desorbed and combine with miRNAs to release target genes. CircRNAs are abundant and stable in EVs and can sustain EV uptake by target cells. Moreover, circRNAs are more abundant in EVs than in cells, and the levels of circRNAs vary with the mutation status of KRAS (proto-oncogene).

#### 13.3.2.4 Reprogramming of Target Cells by EV Cargo Long Noncoding RNAs (lncRNAs)

EV cargo lncRNAs have recently been shown to regulate tumor cell apoptosis, proliferation, and migration and to induce angiogenesis. EV-enriched lncRNAs have potential applications in cancer diagnosis and treatment [49]. Additionally, EV cargo lncRNAs can affect the tumor microenvironment and regulate tumor development. Hypoxic bladder cancer cell-derived EVs enriched with carcinogenic *lncRNA-UCA1* were found to reshape the tumor microenvironment and promote tumor growth and development. The levels of human serum-derived EV cargo *lncRNA-UCA1* can be measured as a bladder cancer diagnostic biomarker [50]. *LncARSR* is highly expressed in sunitinib-tolerant renal cell carcinoma cells and can be transported by EVs to recipient cells for survival. In addition, EV cargo lncRNAs, including *lincUCA1* and *lincROR*, play key roles in cancer metastasis [51].

### 13.3.3 Summary

EV functions are primarily regulated by interactions of EV surface molecules with target cells and regulation of target cell gene expression by EV cargo. Proteins are major EV surface molecules that regulate the interactions between EVs and recipient cells. EVs contain proteins, mRNAs, miRNAs, and lipids that regulate gene expression and signal transduction. However, the major EV surface and cargo molecules in recipient cells are heterogeneous. The various sources of EVs determine their effects on the regulation of target cells. A major challenge in studies of EV function in vitro is the identification of key molecules that regulate phenotypic changes in recipient cells. Multiple -omics analyses of EVs and target cells could provide insights into these mechanisms and could facilitate identification of EV cargo-specific molecules involved in the observed phenotypic changes in recipient cells.

## 13.4 EV Treatment in Small Animal Models

Systematic methods for exploring the unknown components of EVs and their potential biological functions in vitro have been described. Animal experimentation with EV treatment is a promising approach for observing biological behaviors in the natural environment. Studying the biological processes of EVs, such as EV distribution, target recognition, and functional activity, in small animal models in vivo could help to elucidate the physiopathological changes that occur during the disease process and could provide new therapies to cure disease. Thus, it is necessary to investigate the potential clinical applications of EVs.

### 13.4.1 Research Objectives and Selection of Animal Models

EVs facilitate tissue regeneration, tumor diagnosis, and therapy. Benefiting from animal models, researchers have identified many unexpected functions of EVs in the occurrence, progression, and treatment of specific diseases using optical in vivo imaging to study circulation times, tissue distributions, target recognition, and EV depletion in different disease models. Although mice (e.g., nude, C57BL/6, and BALB/c mice) are the most commonly used animal models in studies of EV treatment, EVs may be applied to all animal models. Normally, a specific animal model is chosen based on research objectives; therefore, the delivery method, EV dose, and treatment frequency need to be adjusted when using animal models other than mice.

### 13.4.2 Administration Methods

Different types of administration methods are used to deliver EVs based on specific research objectives and animal models. The injection type and administration route affect absorption efficiency. The most common route is intravenous injection because this approach is similar to that used in the treatment of human patients. The caudal vein is typically chosen for intravenous injection because it is convenient to find and puncture without necessitating surgical exposure. EVs enter the body from the tail vein, drain into the inferior vena cava, similar to the human vascular system, and finally distribute throughout the whole organ system. This route of administration represents the fastest, most accurate, and most controlled method for tracking the movement of EVs in real time. However, intravenous injection is a double-edged sword; it has the advantage that the EVs can enter the circulation system directly, thereby bypassing absorption barriers and the first-pass effect, but also has the disadvantage that the EVs are relatively rapidly cleared via the blood circulation. According to the guidelines for the injection of substances by most Institutional Animal Care and Use Committees, the maximum intravenous injection bolus amount for mice is 5 mL/kg, and the recommended location of intravenous administration is the lateral tail.

Intraperitoneal injection is another common means of administering EVs and is performed by injection into the peritoneum in the lower right quadrant of the animal's abdomen. EVs are absorbed by the veins of the mesentery and then gathered into the portal vein of the liver. Although the bioavailability of EVs administered via intraperitoneal injection is generally low in laboratory animals, this administration route is effective for absorption because the peritoneum is full of blood vessels and lymphatic vessels. The maximum administration volume for intraperitoneal injection in mice is 10 mL/kg. Instead of accumulating in the liver, spleen, lungs, and gastrointestinal tract, as is observed after intravenous injection, EV accumulation after intraperitoneal injection has been reported in the liver, pancreas, and gastrointestinal tract [52].

In solid tumor models, intratumoral injection, which is quite an invasive approach, delivers the "living drug" directly to the tumor microenvironment, yielding higher concentrations of EVs to facilitate appropriate interactions. Intramyocardial injection, another direct injection method, has been used in models of myocardial infarction. Oral administration is another noninvasive method; however, it is difficult to control the dose of drugs administered via this route. EVs may elicit functional responses in the digestive system. Intraventricular injection and intranasal administration have also been used to treat brain disease. Intramuscular, intradermal, and subcutaneous injections are additional approaches that have been used to explore the potential applications of EVs as vaccines. Subcutaneous injection of EVs has also been used to study second-degree skin burns.

### 13.4.3 Injection Dose and Frequency

For any clinical applications, determination of the appropriate dose and administration frequency is necessary, and evaluation of the side effects of EV injection based on its biodistribution is also required. Based on the recommendations of the Minimal Information for Studies of Extracellular Vesicles 2018 document, protein amount, particle number, and lipid amount are used to measure the EV amounts in many laboratories. Total protein amount is the most commonly used parameter for EV dose determination, followed by particle number [53]. Given the heterogeneity of EV particle size, there is no clear linear relationship between the quantity of EVs and their total protein mass. As described above, volume should be considered for every injection, and injections should be based on the number of EVs per unit body weight (ranging from $0.25 \times 10^{10}$ particles/g body weight [p/g] to $1.5 \times 10^{10}$ p/g) or the total protein amount per unit body weight (2–10 mg/kg) [54]. Because EV secretion is affected by different health conditions (e.g., the concentration of EVs in tumor-bearing mice is much higher than that in peripheral blood form healthy mice) [55], the original level of EVs secreted by the experimental subjects should be taken into account. Accordingly, in studies of the biological functions of EVs, if the quantity of EVs injected into small animals exceeds that naturally secreted in the body, the results may be unreliable. Additionally, when the injection dose is higher than that able to be handled by the mononuclear phagocytic system, the EVs that escape may distribute to other organs, resulting in relatively lower amounts of EVs in the liver. Given that EVs are rapidly distributed throughout the body following intravenous and intraperitoneal injection, the local concentrations of EVs and the delivery method should be taken into consideration when the injection dose is calculated.

The injection frequency is also related to the research objective. A single shot is used for real-time tracking from the injection site until the signal is completely quenched for detection of the circulation time, tissue distribution, and clearance mechanism, thereby avoiding unexpected interference from multiple injections. Multiple injections are required to investigate disease progression. Considering the injection dose, circulation time, and other influencing factors, many studies have used regimens in which injections are performed three to four times/week for 2–3 weeks.

### 13.4.4 Observation Index

To date, studies of EV treatment in small animals have mainly focused on biosafety assessments, biological activities, and the functions of EVs in various pathophysiological processes. Body weight analyses, complete blood counts, liver function tests, kidney function tests, image examinations, and histological assessments of the main organs as well as analyses of oncogenicity, tumor-promoting ability, cytotoxicity, and cell viability have been performed in safety analyses. Using the different injection methods described above, various bioluminescence and fluorescence signals in EVs derived from different host cells have been tracked to measure the circulation time, tissue distribution, and clearance rules via fluorescent imaging. Body weights, tumor sizes, cell categories and activities, protein and gene expression patterns in tumor tissues, body immune status, and microenvironment have been evaluated during analyses of the functional mechanisms of EVs in tumor progression. The specific observation indices are related to the disease characteristics, which are in turn affected by the animal model used to build the disease model. These are all crucial factors for determining the functions of EVs in disease progression.

### 13.4.5 Other Factors That Influence EV Treatment with Small Animals

EVs released by different donor cells have different half-lives and tissue distributions. Because the concentration of EVs is decreased to half the initial concentration in only 2 min, EVs derived from B cells are cleared from the blood 5 min after circulation, whereas EVs secreted by human renal epithelial cells require 6 h to be cleared after injection into mice. EVs carry different markers that can target many recipient cells. EVs derived from DCs and endothelial cells are more likely to accumulate in the spleen.

The luminescence properties of a material are critical for imaging. Bioluminescence, fluorescence, and MRI have been used for optical in vivo imaging. In general, bioluminescence imaging using luciferase has the highest sensitivity, whereas nuclear imaging has the highest penetration but requires substrates for imaging. Fluorescence imaging using fluorescent proteins has the highest spatial resolution (nm), which is limited by the lack of a noninvasive in vivo imaging platform. MRI has the highest penetration (m) with the highest temporal resolution (min–h); however, MRI shows low sensitivity [53]. Moreover, suitable labeling methods for EVs, including genetic modification, covalent binding, membrane integration, encapsulation (or internalization), and metabolic labeling, are important for imaging. Genetic modification can be used to avoid surface modification; however, genetic changes in cells may also lead to changes in the properties of cells and EVs and uneven loading into EVs. The advantage of covalent binding is the tight binding of probes to proteins; however, this approach cannot distinguish between EV and non-EV proteins [56]. With regard to membrane integration, which is easy to implement, lipophilic dyes have a longer half-life than EVs and can exhibit nonspecific binding [57]. Thus, dye binding to nonspecific targets may be analyzed rather than dye bound to real EVs during the late tracking period. Encapsulation is achievable and may be used to avoid surface modification; however, this method may cause aggregation or fusion of EVs or background signals [57]. Metabolic labeling is also simple and easy to perform during the cell culture process, although the addition of extra substances during metabolic labeling may alter the characteristics of the cells or EVs [58].

Overall, the biological behaviors and functions of EVs (e.g., circulation time, tissue distribution, target recognition, depletion, etc.) in physiological and pathological processes can be affected by the EV source, delivery method, and injection dose as well as the labeling materials and methods used and the state of the EV receptor. More studies are needed to provide high-quality guidance for designing EV treatment strategies in patients with various diseases. These studies will also facilitate the selection of an appropriate animal model and the comprehensive analysis of other factors that may influence the results.

### 13.4.6 Summary

Small animals are promising models for investigation of the functions of EVs in vivo. However, because of the characteristics of EVs, the selection of the appropriate animal model should be based on the study's specific research objectives. Methods for EV administration in animal models have been reported, and intravenous injection remains the most common method for direct distribution through the circulatory system. The major challenge of intravenous injection is the relatively rapid clearance by blood circulation and the retention of EVs in the liver and spleen. The detailed mechanisms of EV retention in the liver and spleen should be investigated in greater detail in further studies. Targeting tumor tissue should be the focus of clinical applications of EVs in the treatment of tumors. The dose and frequency of EV administration are important for achieving accurate results. Presently, intravenous injection performed once every 3 days is a common dosing regimen. However, because not all of the characteristics of EVs have been elucidated, the effects of EV treatment are not unique to specific cargo. Common labeling methods have been applied for the monitoring of EV treatment and to obtain data for various observation indices. Furthermore, several other factors affect EV treatment in small animals. One major challenge limiting the application of EVs in vivo is the heterogeneity of the cargo, which should be addressed in future studies.

## References


1. Dinh NTH, Lee J, Lee J, Kim SS, Go G, Bae S, et al. Indoor dust extracellular vesicles promote cancer lung metastasis by inducing tumour necrosis factor-alpha. J Extracell Vesicles. 2020;9(1):1766821.
2. Yong T, Zhang X, Bie N, Zhang H, Zhang X, Li F, et al. Tumor exosome-based nanoparticles are efficient drug carriers for chemotherapy. Nat Commun. 2019;10(1):3838.
3. Nimjee SM, White RR, Becker RC, Sullenger BA. Aptamers as therapeutics. Annu Rev Pharmacol. 2017;57:61–79.
4. Alves LN, Missailidis S, Lage CAS, CEB DEA. Anti-MUC1 aptamer as carrier tool of the potential radiosensitizer 1,10 phenanthroline in MCF-7 breast cancer cells. Anticancer Res. 2019;39(4):1859–67.
5. Yang Z, Shi J, Xie J, Wang Y, Sun J, Liu T, et al. Large-scale generation of functional mRNA-encapsulating exosomes via cellular nanoporation. Nat Biomed Eng. 2020;4(1):69–83.
6. Pi F, Binzel DW, Lee TJ, Li Z, Sun M, Rychahou P, et al. Nanoparticle orientation to control RNA loading and ligand display on extracellular vesicles for cancer regression. Nat Nanotechnol. 2018;13(1):82–9.
7. Kamerkar S, LeBleu VS, Sugimoto H, Yang S, Ruivo CF, Melo SA, et al. Exosomes facilitate therapeutic targeting of oncogenic KRAS in pancreatic cancer. Nature. 2017;546(7659):498–503.
8. Nie HF, Xie XD, Zhang DD, Zhou Y, Li BF, Li FQ, et al. Use of lung-specific exosomes for miRNA-126 delivery in non-small cell lung cancer. Nanoscale. 2020;12(2):877–87.
9. Tian YH, Li SP, Song J, Ji TJ, Zhu MT, Anderson GJ, et al. A doxorubicin delivery platform using engineered natural membrane vesicle exosomes for targeted tumor therapy. Biomaterials. 2014;35(7):2383–90.
10. Limoni SK, Moghadam MF, Moazzeni SM, Gomari H, Salimi F. Engineered exosomes for targeted transfer of siRNA to HER2 positive breast cancer cells. Appl Biochem Biotechnol. 2019;187(1):352–64.
11. Tavakkoli S, Sotoodehnejadnematalahi F, Fathollahi A, Bandehpour M, Haji Molla Hoseini M, Yeganeh F. EL4-derived exosomes carry functional TNF-related apoptosis-inducing ligand that are able to induce apoptosis and necrosis in the target cells. Int J Mol Cell Med. 2020;9(3):207–15.
12. Zhuang M, Chen X, Du D, Shi J, Deng M, Long Q, et al. SPION decorated exosome delivery of TNF-alpha to cancer cell membranes through magnetism. Nanoscale. 2020;12(1):173–88.
13. Wan Y, Cheng G, Liu X, Hao SJ, Nisic M, Zhu CD, et al. Rapid magnetic isolation of extracellular vesicles via lipid-based nanoprobes. Nat Biomed Eng. 2017;1:0058.
14. Duan HX, Liu YH, Gao ZG, Huang W. Recent advances in drug delivery systems for targeting cancer stem cells. Acta Pharmacol Sin B. 2021;11(1):55–70.
15. Wang TF, Luo Y, Lv HY, Wang J, Zhang Y, Pei RJ. Aptamer-based erythrocyte-derived mimic vesicles loaded with siRNA and doxorubicin for the targeted treatment of multidrug-resistant tumors. Acs Appl Mater Inter. 2019;11(49):45455–66.
16. Zhang Y, Wang D, Yue S, Lu Y, Yang C, Fang J, et al. Sensitive multicolor visual detection of exosomes via dual signal amplification strategy of enzyme-catalyzed metallization of au Nanorods and hybridization chain reaction. ACS Sens. 2019;4(12):3210–8.
17. Wang T, Luo Y, Lv H, Wang J, Zhang Y, Pei R. Aptamer-based erythrocyte-derived mimic vesicles loaded with siRNA and doxorubicin for the targeted treatment of multidrug-resistant tumors. ACS Appl Mater Interfaces. 2019;11(49):45455–66.
18. Tran PH, Xiang D, Nguyen TN, Tran TT, Chen Q, Yin W, et al. Aptamer-guided extracellular vesicle theranostics in oncology. Theranostics. 2020;10(9):3849–66.
19. Zhu GZ, Chen XY. Aptamer-based targeted therapy. Adv Drug Deliver Rev. 2018;134:65–78.
20. Wan Y, Wang LX, Zhu CD, Zheng Q, Wang GX, Tong JL, et al. Aptamer-conjugated extracellular nanovesicles for targeted drug delivery. Cancer Res. 2018;78(3):798–808.
21. Nikam RR, Gore KR. Journey of siRNA: clinical developments and targeted delivery. Nucleic Acid Ther. 2018;28(4):209–24.

22. Huang XW, Toth KF, Aravin AA. piRNA biogenesis in drosophila melanogaster. Trends Genet. 2017;33(11):882–94.
23. Wu QB, Liu WR, Wang J, Zhu L, Wang ZQ, Peng Y. Exosomal noncoding RNAs in colorectal cancer. Cancer Lett. 2020;493:228–35.
24. Reshke R, Taylor JA, Savard A, Guo HS, Rhym LH, Kowalski PS, et al. Reduction of the therapeutic dose of silencing RNA by packaging it in extracellular vesicles via a pre-microRNA backbone. Nat Biomed Eng. 2020;4(1):52–68.
25. Faruqu FN, Xu LZ, Al-Jamal KT. Preparation of exosomes for siRNA delivery to cancer cells. Jove-J Vis Exp. 2018:(142).
26. Kahroba H, Hejazi MS, Samadi N. Exosomes: from carcinogenesis and metastasis to diagnosis and treatment of gastric cancer. Cell Mol Life Sci. 2019;76(9):1747–58.
27. Xin L, Yuan YW, Liu C, Zhou LQ, Liu L, Zhou Q, et al. Preparation of internalizing RGD-modified recombinant methioninase exosome active targeting vector and antitumor effect evaluation. Dig Dis Sci. 2021;66(4):1045–53.
28. Ojha T, Pathak V, Shi Y, Hennink WE, Moonen CTW, Storm G, et al. Pharmacological and physical vessel modulation strategies to improve EPR-mediated drug targeting to tumors. Adv Drug Deliver Rev. 2017;119:44–60.
29. Wang J, Li W, Zhang L, Ban L, Chen P, Du W, et al. Chemically edited exosomes with dual ligand purified by microfluidic device for active targeted drug delivery to tumor cells. ACS Appl Mater Interfaces. 2017;9(33):27441–52.
30. Gao X, Ran N, Dong X, Zuo B, Yang R, Zhou Q, et al. Anchor peptide captures, targets, and loads exosomes of diverse origins for diagnostics and therapy. Sci Transl Med. 2018;10(444).
31. Zhang X, Zhang H, Gu J, Zhang J, Shi H, Qian H, et al. Engineered extracellular vesicles for cancer therapy. Adv Mater. 2021;33:e2005709.
32. Armstrong JP, Holme MN, Stevens MM. Re-engineering extracellular vesicles as smart nanoscale therapeutics. ACS Nano. 2017;11(1):69–83.
33. Cheng Y, Schorey JS. Targeting soluble proteins to exosomes using a ubiquitin tag. Biotechnol Bioeng. 2016;113(6):1315–24.
34. Murphy DE, de Jong OG, Brouwer M, Wood MJ, Lavieu G, Schiffelers RM, et al. Extracellular vesicle-based therapeutics: natural versus engineered targeting and trafficking. Exp Mol Med. 2019;51(3):1–12.
35. Pegtel DM, Gould SJ. Exosomes. Annu Rev Biochem. 2019;88:487–514.
36. Guo Q, Jiang C. Delivery strategies for macromolecular drugs in cancer therapy. Acta Pharm Sin B. 2020;10(6):979–86.
37. Hu Q, Su H, Li J, Lyon C, Tang W, Wan M, et al. Clinical applications of exosome membrane proteins. Precis Clin Med. 2020;3(1):54–66.
38. Soe ZY, Prajuabjinda O, Myint PK, Gaowa A, Kawamoto E, Park EJ, et al. Talin-2 regulates integrin functions in exosomes. Biochem Bioph Res Co. 2019;512(3):429–34.
39. Tabak S, Schreiber-Avissar S, Beit-Yannai E. Influence of anti-glaucoma drugs on uptake of extracellular vesicles by trabecular meshwork cells. Int J Nanomedicine. 2021;16:1067–81.
40. Mettlen M, Chen PH, Srinivasan S, Danuser G, Schmid SL. Regulation of clathrin-mediated endocytosis. Annu Rev Biochem. 2018;87:871–96.
41. Quinn Z, Mao W, Xia Y, John R, Wan Y. Conferring receptors on recipient cells with extracellular vesicles for targeted drug delivery. Bioact Mater. 2021;6(3):749–56.
42. Chen L, Brigstock DR. Integrins and heparan sulfate proteoglycans on hepatic stellate cells (HSC) are novel receptors for HSC-derived exosomes. FEBS Lett. 2016;590(23):4263–74.
43. Park EJ, Prajuabjinda O, Soe ZY, Darkwah S, Appiah MG, Kawamoto E, et al. Exosomal regulation of lymphocyte homing to the gut. Blood Adv. 2019;3(1):1–11.
44. Altei WF, Pachane BC, dos Santos PK, Ribeiro LNM, Sung BH, Weaver AM, et al. Inhibition of alpha v beta 3 integrin impairs adhesion and uptake of tumor-derived small extracellular vesicles. Cell Commun Signal. 2020;18(1):158.
45. Wang Z, Zoller M. Exosomes, metastases, and the miracle of cancer stem cell markers. Cancer Metastasis Rev. 2019;38(1–2):259–95.

46. Chen G, Huang AC, Zhang W, Zhang G, Wu M, Xu W, et al. Exosomal PD-L1 contributes to immunosuppression and is associated with anti-PD-1 response. Nature. 2018;560(7718):382.
47. Chandra PK, Rutkai I, Kim H, Braun SE, Abdel-Mageed AB, Mondal D, et al. Latent HIV-exosomes induce mitochondrial hyperfusion due to loss of phosphorylated dynamin-related protein 1 in brain endothelium. Mol Neurobiol. 2021;58:2974.
48. Tomasetti M, Lee W, Santarelli L, Neuzil J. Exosome-derived microRNAs in cancer metabolism: possible implications in cancer diagnostics and therapy. Exp Mol Med. 2017;49(1):e285.
49. Zhao R, Zhang YL, Zhang X, Yang YM, Zheng X, Li XH, et al. Exosomal long noncoding RNA HOTTIP as potential novel diagnostic and prognostic biomarker test for gastric cancer. Mol Cancer. 2018;17:68.
50. Xue M, Chen W, Xiang A, Wang RQ, Chen H, Pan JJ, et al. Hypoxic exosomes facilitate bladder tumor growth and development through transferring long non-coding RNA-UCA1. Mol Cancer. 2017;16:143.
51. Zhang P, Zhou HX, Lu KF, Lu YN, Wang Y, Feng TB. Exosome-mediated delivery of MALAT1 induces cell proliferation in breast cancer. Oncotargets Ther. 2018;11:291–9.
52. Mentkowski KI, Snitzer JD, Rusnak S, Lang JK. Therapeutic potential of engineered extracellular vesicles. AAPS J. 2018;20(3):50.
53. Yi YW, Lee JH, Kim S-Y, Pack C-G, Ha DH, Park SR, et al. Advances in analysis of biodistribution of exosomes by molecular imaging. Int J Mol Sci. 2020;21(2):665.
54. Zhao J-Y, Chen G, Gu Y-P, Cui R, Zhang Z-L, Yu Z-L, et al. Ultrasmall magnetically engineered Ag2Se quantum dots for instant efficient labeling and whole-body high-resolution multimodal real-time tracking of cell-derived microvesicles. J Am Chem Soc. 2016;138(6):1893–903.
55. Erb U, Zhao K, Wang Z, Xiao L, Zöller M. Murine and human pancreatic tumor exosome recovery in mouse serum: diagnostic and prognostic potential and target cell delivery. Cancer Lett. 2017;403:1–12.
56. Royo F, Cossío U, de Angulo AR, Llop J, Falcon-Perez JM. Modification of the glycosylation of extracellular vesicles alters their biodistribution in mice. Nanoscale. 2019;11(4):1531–7.
57. Gangadaran P, Hong CM, Oh JM, Rajendran RL, Kalimuthu S, Son SH, et al. In vivo non-invasive imaging of radio-labeled exosome-mimetics derived from red blood cells in mice. Front Pharmacol. 2018;9:817.
58. Lee H, Park H, Noh GJ, Lee ES. pH-responsive hyaluronate-anchored extracellular vesicles to promote tumor-targeted drug delivery. Carbohydr Polym. 2018;202:323–33.

# Chemical Biology of Extracellular Vesicles

# 14

Gaoge Sun and Hang Yin

## 14.1 Bioorthogonal Click Chemistry and Introduction of Unnatural Amino Acids in Studies of EVs

Conventional strategies of using biomolecules, dyes, and encoded reporters (e.g., fluorescent proteins or fusion protein tags) to investigate proteins have developed rapidly and maturely. Green fluorescent protein (GFP), which is composed of 238 amino acid residues, was originally derived from the jellyfish *Aequorea victoria* in 1978 [1, 2]. The discovery and reconstruction of GFP by three scientists led them to receive the Nobel Prize in Chemistry in 2008. Aided by cellular RNA polymerase and tRNAs, fusion proteins are synthesized in vivo. However, other macromolecules, including nucleic acids, lipids, and glycans, cannot genetically encode tracers or tags, such as proteins. Moreover, using conventional genetic biology approaches to explore posttranslational modifications (PTMs) in living systems is not mature [3]. Furthermore, identification of biofunctional components that can be used in conjunction with chemical methods has become increasingly important in chemical biology. The reaction should be easy to perform, provide a high yield without side

G. Sun
School of Pharmaceutical Sciences, Key Laboratory of Bioorganic Phosphorous Chemistry and Chemical Biology (Ministry of Education), Department of Chemistry, Tsinghua University, Beijing, China

H. Yin (✉)
School of Pharmaceutical Sciences, Key Laboratory of Bioorganic Phosphorous Chemistry and Chemical Biology (Ministry of Education), Department of Chemistry, Tsinghua University, Beijing, China

Beijing Advanced Innovation Center for Structural Biology, Tsinghua University, Beijing, China

Tsinghua-Peking Joint Center for Life Sciences, Tsinghua University, Beijing, China
e-mail: yin_hang@mail.tsinghua.edu.cn



Q. Wang, L. Zheng (eds.), *Extracellular Vesicles*,
https://doi.org/10.1007/978-981-99-8365-0_14

products under physiological conditions, and have little effect on natural biological processes. Under these circumstances, chemical bonding is the optimal scheme. The term "bioorthogonal" was proposed by Bertozzi in 2003 [4]. The discovery of bioorthogonal reactions has prompted biologists to probe biomolecules in living systems and to establish connections between two proteins, reporters, or functional elements using selective, fast, and stable methods in aqueous living systems. Studies of EVs via bioorthogonal reactions targeting membrane components, such as phospholipids or glycans, have also become popular [5, 6].

Protein modification methods have become an important research topic in chemical biology. Cysteine residues forming intramolecular disulfide bridges and thioether alkylation in thiol groups can be utilized for selective modification, particularly single-site modification, owing to the low proportion of cysteine in proteins [7, 8]. By contrast, lysine residues, which show a high frequency in nature, are more popular than cysteine in multiple modification methods because of the amine group ($-NH_2$) [3]. The end products could be amide, urea, thiourea, sulfonamide, or amine following reductive amination from a Schiff base [9]. The other 18 types of amino acids can be modified by classic conjugation methods [10], in which proteins are bound to fluorophore and biotin small-molecule probes or fixed on the surface. In addition to these classic methods, novel modification schemes for cysteine, lysine, tyrosine, and tryptophan have emerged in recent years [11]. In general, the reaction occurs at the N-terminal, which requires metal catalysis in most instances [11].

In 1963, Huisgen et al. reported a 1,3-dipolar cycloaddition reaction [12]. Sharpless and Meldal found that this reaction was accelerated by a Cu(I) catalyzer through Cu-acetylide intermediate formation, named copper-catalyzed azide-alkyne cycloaddition [13–15]. Notably, before the term "bioorthogonal" was coined [11, 16], the concept of biorthogonal reactions was introduced by Tsien et al., who observed the reaction of fluorophores with tetra-cysteine inside live cells with high selectivity [17].

Azide ($R\text{-}N_3$), one of the substrates in click chemistry catalyzed by Cu(I), is small and stable in biosystems. The other substrate alkyne group ($R\text{-}C \equiv C\text{-}R$) also satisfies the requirements of a bioorthogonal reaction.

In one example of a bioorthogonal reaction mediated by EVs, Santamaría et al. integrated the targetability of EVs and the preferential tropism of EVs for their parental cells as therapeutic drug carriers and demonstrated that the Pd-catalyzed reaction could transform prodrugs into toxic species in situ [18]. The bioorthogonal reaction was facilitated by the EVs and Pd nanoparticles. The EV-catalyst hybrids retained both targetability and reactivity, and reduction of $Pd^{2+}$ cations with CO treatment occurred at 30 °C and 6 bar for 40 min, followed by the assembly of nanosheets into EVs (Pd-exosomes). Pd-exosomes have a strong catalytic activity that catalyzes the conversion of nonfluorescent molecules into fluorescent reporters in vitro. Similarly, prodrug probes can be transformed into drugs and fluorophores in cells. Similarly, another chemical biology method attempts to deliver drugs to tumors via EVs using Pd catalyst bioorthogonal chemistry [19]. Like a Trojan horse, a prodrug and Pd nanosheets can trigger dealkylation reactions inside cells by targeting potential EVs to create opportunities for novel targeted anticancer therapy.

To monitor EV events and positions in vitro and in vivo, chemical tracing modification of EV membranes shows more selectivity than direct observation by microscopy or fluorescent staining using lipophilic dyes, including PKH67, PKH26, DiI, and DiD [20–27]. Moreover, compared with bioluminescent imaging using luciferase, EV labeling via the bioorthogonal reaction can be regulated without a substrate [28–30]. Bioorthogonal conjugation is an effective method for promoting the development of EV tracking strategies. Moreover, combining a label with EVs using bioorthogonal conjugation with specific groups may help to distinguish different subtypes of EVs that do not bind with nonspecific dyes. Aided by phospholipid optical probe labeling, we visualized the dynamic distribution and uptake of EVs in vivo [31]. Cancer cells are fed fatty acids containing azide groups ($-N_3$), resulting in the production of azide groups on EV membranes via metabolic and vesicle secretion pathways. Fluorescent dyes are linked to azide-containing EVs via copper-free click chemistry. This provides a new strategy for monitoring biological events related to EVs. Another similar example using carbohydrate metabolism has also been reported for the tracking of EVs in vitro and in vivo [32].

Bioorthogonal click chemistry of EVs shows great potential for applications in drug delivery and labeling. Future investigations of bioorthogonal chemistry in EVs will focus on tracing EVs in vivo, -omics or PTMs in EVs, and in situ drug delivery to address biological and clinical problems [33]. Given the rapid advances in the compatibility and diversity of bioorthogonal approaches, chemical biology remains an exciting approach to studies of EVs.

## 14.2 Small-Molecule Probes for EVs

EVs are desirable natural drug delivery carriers; therefore, the development of regulatory tools for EVs is valuable. However, studies on small-molecule EV probes are still in their infancy. The development of specific small-molecule probes for EVs is expected to be difficult. EVs have been shown to be associated with various diseases and to function as activators or inhibitors of disease development and progression. Identification of compounds with therapeutic potential related to EVs from numerous molecules according to the Lipinski rule of five [34] (molecular mass < 500 Da, $\leq$5 hydrogen bond donors, $\leq$10 hydrogen bond acceptors, partition coefficient $\log P < 5$) may limit the applications of candidate compounds. Furthermore, to find lead compounds in chemical biology, chemists have attempted high-throughput chemical genetic screening in vitro and in vivo using natural products or drug-like libraries [35, 36]. Small-molecule probes are rationally designed to show better affinity and selectivity based on the structure of compounds [37, 38]. To find a key to "lock" or "unlock" EVs directly, the small-molecule probe has to interact with the core binding domains of phospholipids, which have a negative charge, and the high curvature or high membrane protein content on EVs, which is difficult to realize because EVs are much larger and more flexible than proteins. Additionally, EVs may be regulated indirectly. The binding sites and mechanisms of action of small-molecule probes need to be elucidated. In addition to the pharmaceutical potential

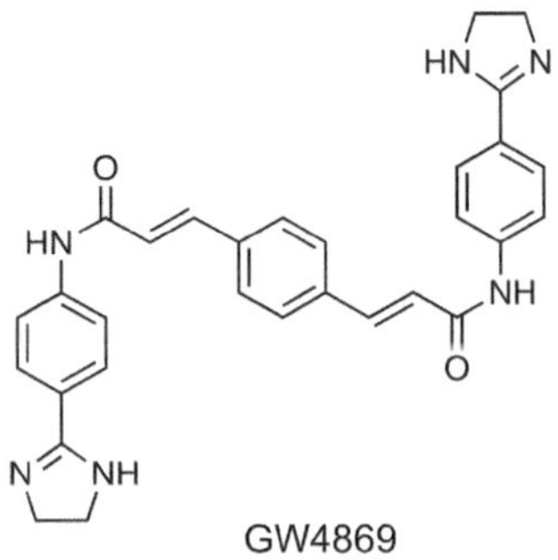


**Fig. 14.1** The molecular structure of GW4869 [40]

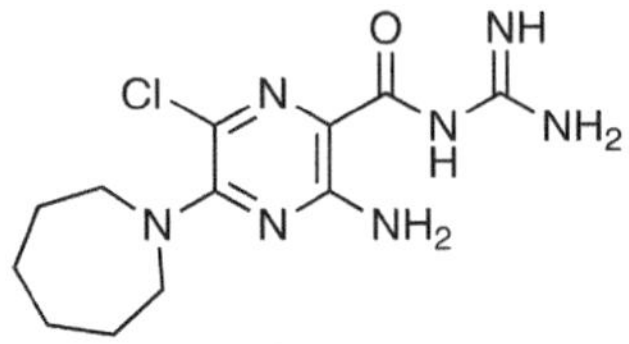


**Fig. 14.2** The molecular structure of DMA [43]

of candidate compounds related to EVs, it is important to identify small-molecule probes to target or regulate EVs using chemical biology methods.

One commonly recognized EV-related small molecule is GW4869 (Fig. 14.1). In 2008, Simons et al. reported that GW4869 is a neutral sphingomyelinase (nSMase) inhibitor that decreases EV secretion by preventing ceramide formation [39]. In 2002, a high-throughput screening assay showed that GW4869 is a neutral, magnesium-dependent sphingomyelinase inhibitor [40–42]. The half-maximal inhibitory concentration of the nSMase noncompetitive inhibitor with an asymmetric structure was 1 μM in vitro. This study showed that GW4869 inhibited the production of ceramide, which participates in the endosomal sorting complexes required for transport (ESCRT)-independent secretory machinery [39]. Another EV secretion inhibitor, dimethyl amiloride (DMA), blocks the calcium-dependent protein transferrin (Fig. 14.2) [43–45].

Notably, no specific small-molecule probes have been shown to act directly on the surfaces of EVs. The abovementioned compounds DMA and GW4869, as well as dozens of species related to the inhibition or stimulation of EV secretion and biosynthesis, act upstream of the signaling pathway. Several of these molecules are involved in autophagy, apoptosis, and phospholipase activity. EVs show complex mechanisms of intracellular biogenesis, exhibit different trafficking patterns, and have complex secretory pathways [46]. Upstream regulatory manipulations may affect various outcomes, making it difficult to develop small-molecule probes to regulate EVs [46–48]. Nevertheless, the size and heterogeneity of EVs makes it difficult for them to interact directly with small-molecule probes. With the rapid development of technologies, particularly artificial intelligence, computer-assisted drug discovery may facilitate the development of small-molecule probes for EVs in the near future [49–51]. In addition, phenotypic readout reporters for probe screening are crucial [52]. For example, identification of EV inhibitors using high-throughput screening from compound libraries necessitates a rational and sensitive readout. Based on the screening results, identification of targets of the small-molecule probes will be necessary. Understanding the binding mechanism may also help to enhance the effects of probes owing to structural modification. Thus, further studies are needed to identify small-molecule regulators of EVs. We are looking forward to developing specific probes that can promote or inhibit EV secretion or uptake. These major demands for EV regulation using probes have attracted the interest of chemical biologists in the field.

## 14.3 Peptide Probes for EVs

### 14.3.1 Peptide Synthesis and Integration of EVs

In 1994, native chemical ligation (NCL) was a promising technique devised for the synthesis of two unprotected peptides into native backbone proteins [53]. NCL has promoted the generation of protein products to a greater extent than solid-phase peptide synthesis and isotopic labeling [54, 55]. Because solid-phase peptide synthesis is inexpensive and controllable, this approach can be combined with the bioorthogonal click chemistry reaction onto functional chips or small-molecule probes. EVs conjugate peptides by affinity binding, coculture fusion, or chemical bonding.

The sizes of EVs are similar to those of oligopeptides and polypeptides. Therefore, dimensional matching of peptide probes and EVs is theoretically feasible. In addition, the amphiphilic membranes of EVs are suitable for binding with the effector domains of peptides located on the membrane or transmembrane. Consequently, further studies of EVs may necessitate the development of sensing peptide probes. Owing to the diagnostic and therapeutic potential of EVs, it may be beneficial to identify EVs modified with targeting peptides or peptide-drug conjugates. In the following two subsections, we describe the field of EV peptide chemistry.

### 14.3.2 EV-Binding Peptides

Peptide probes for EVs can be established via the application of novel techniques, including machine-dependent peptide synthesis, computational stimulation, and observation of the interfacial binding mechanisms between peptide probes and EVs. Probes are designed to sense and bind to phospholipid bilayer membranes via functional peptide domains.

In the double-layer phospholipid membranes of EVs, the formation of a lipid bilayer determines the shape and properties of the membrane structure. Compared with other membranous organelles, nanosized EVs show prominent features in curvature and phospholipid components. Curvature plays an important role in cellular processes, e.g., the biogenesis and vesicular trafficking of EVs by ESCRT proteins from endosomal membranes to produce the high curvature preferable for plasma membrane fusion and secretion to the extracellular space [46]. Phospholipids constitute a major component of membranes. The properties and functions of lipids are critical in studies of EVs owing to the large surface area of EVs. EVs undergo processing changes from multivesicular bodies to the extracellular environment. Phospholipid processing requires the activity of the lipid processing protein ARF6, leading to asymmetric distribution and turnover of phosphatidylserine (PS) [56].

A highly curved membrane exposes more package defects and gaps in the outer leaflet [57–59]. The outer membrane is rich in negative phospholipids, such as PS and ceramide. Ideally, the EV-binding probes generate an interaction force between EVs and probes or sense defects on the EV membrane, as shown in Fig. 14.3. Phospholipid-recognizing probes may also have important applications.

Small molecules usually bind to the pockets of specific proteins or biomacromolecules but not to a larger structure. Under these circumstances, peptide probes are more suitable for direct binding to EVs. There are several domains of proteins that detect curvature and reflection in physiological processes. Based on the natural curvature-sensing protein function, we can use chemical and biological approaches to develop EV-binding peptide probes in vivo. Peptide enantiomers consist of artificial D-form amino acids that exhibit stability in vivo and ex vivo. The mirror image of D-peptides shows a strong resistance to proteases. Unlike covalent attachment, peptide-lipid interactions are too weak to affect the original structure of the membrane.

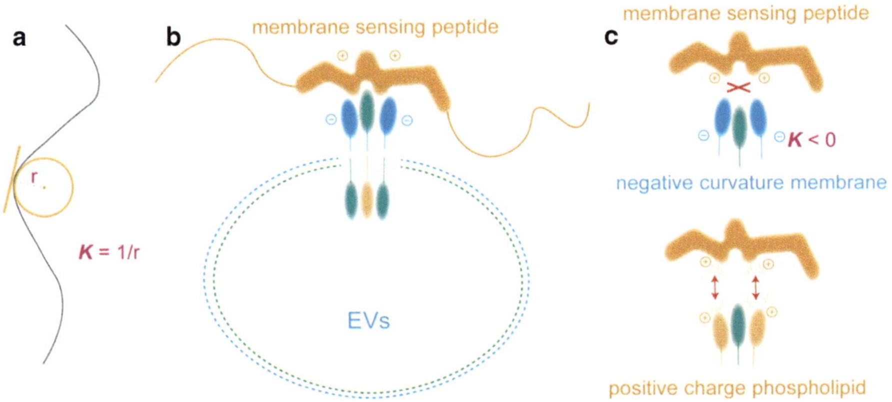


**Fig. 14.3** (**a**) The membrane curvature reflects the degree of curving of the membrane, which is a three-dimensional definition. The best approximating circle is the osculating (kissing) circle. The curvature of the curve is the reciprocal of the radius ($K = 1/r$). The curvature of the surface is measured by the planes at a small distance; this plane cuts off the surface as well as a circumscribing cylinder. The curvature can be considered the reciprocal of the radius of the cylinder. (**b**) The membrane-sensing peptide binding mode. Three factors promote the binding from peptides and curved EV membranes, as follows: suitable curvature; a negative charge (usually PS with a negative charge) in the outer leaflet of the EV membrane and a positive charge from the peptide (lysine, histidine, and arginine); suitable defects in the EV membrane can be optimized as insertion sites for peptide probes. (**c**) The membrane-sensing peptide has a low interaction force with the membrane, resulting in mismatches in the curvature or defects in assembly. Importantly, the same electrical charges repel each other

Transmembranes are easy for small molecules; however, peptide probes exist in the extracellular environment, providing ideal conditions for interactions with EVs. In summary, peptide recombination as a sensor in protein engineering is a promising method for the capture and manipulation of EVs.

In many physiological processes, sensing the curvature of the membrane is used for the dynamic regulation of membrane proteins [60] and vice versa [61]. Protein-lipid interactions result in curvature transformation. Annexin A5 (annexin V) is a classic PS tracer [62] that can be used to determine whether the inner plasma membrane leaflet has undergone an inside-out flip during cellular apoptosis [63]. Alternatively, PS is abundant on the EV surface, indicating that lipid-sensing proteins offer one route to target EV membranes. Natural peptide sequences, such as the ArfGAP1 lipid packing sensor motif or Bin/amphiphysin/Rvs domain, were found to sense curved membranes as well [64, 65].

Since 2002, researchers have studied the effector domains of myristoylated alanine-rich C-kinase substrate (MARCKS) [66]. MARCKS is a substrate of protein kinase C, which dissociates from $PI(4,5)P_2$. The effector domain (MARCKS-ED) contains a lysine-rich motif that interacts with PS, which is a negative phospholipid. The package defects form an adaptation pocket that is suitable for interactions with serine and phenylalanine residues in MARCKS-ED. Basic-aromatic clusters assembled by lysine and phenylalanine residues are normally present in lipid-interacting membrane proteins [67].

MARCKS-ED also shows better affinity for apoptotic cells than annexin A5 in *Caenorhabditis elegans* [68]. The binding modes for different membranous targets are related to reverse curvature. PS in the outer leaflet of EVs has a positive curvature, whereas PS is in the introversion membrane of apoptotic cells does not. Liposome models identified two binding modes for MARCKS-ED in vitro [69]. PS and lysine residues generate electrostatic interactions; phenylalanine residues, which have hydrophobic aromatic groups, are inserted into packing defects on high-curvature membranes. Subsequently, both L-MARCKS-ED and D-MARCKS-ED had the same binding mode and affinity for EVs [70].

Combining chemical and biological methods, the peptide probe MARCKS-ED was shown to link with the photoactive ligand photodoxaz; these results may have practical applications [71]. As shown in Fig. 14.4, MARCKS-ED-photodoxaz releases the cytotoxic agent doxaz under ultraviolet light. MARCKS-ED binds EVs that selectively target cancerous cells in situ. EV-sensing peptides integrated with prodrugs can enhance and expand upon the applications of EV-mediated drug delivery.

Other EV-binding peptides, such as bradykinin-derived trimer peptides or C2BL3L, have also been used for EV profiling [72, 73]. Bradykinin (BK, RPPGFSPFR) is a cationic peptide that interacts with a bilayer lipid membrane [74, 75]. As shown in Fig. 14.5, using click chemistry, an 11-residue precursor of BK can yield trimerized ε-azidolysine (Lys $\varepsilon$-$N_3$) and tripropargylamine. Glycine residues are spacers for the import of Lys($\varepsilon$-$N_3$) and the reporter molecule "(N-(7-nitrobenz-2-oxa-1,3-diazol-4-yl) (NBD) aminohexanoic acid (Ahx) in fluorescence assays.

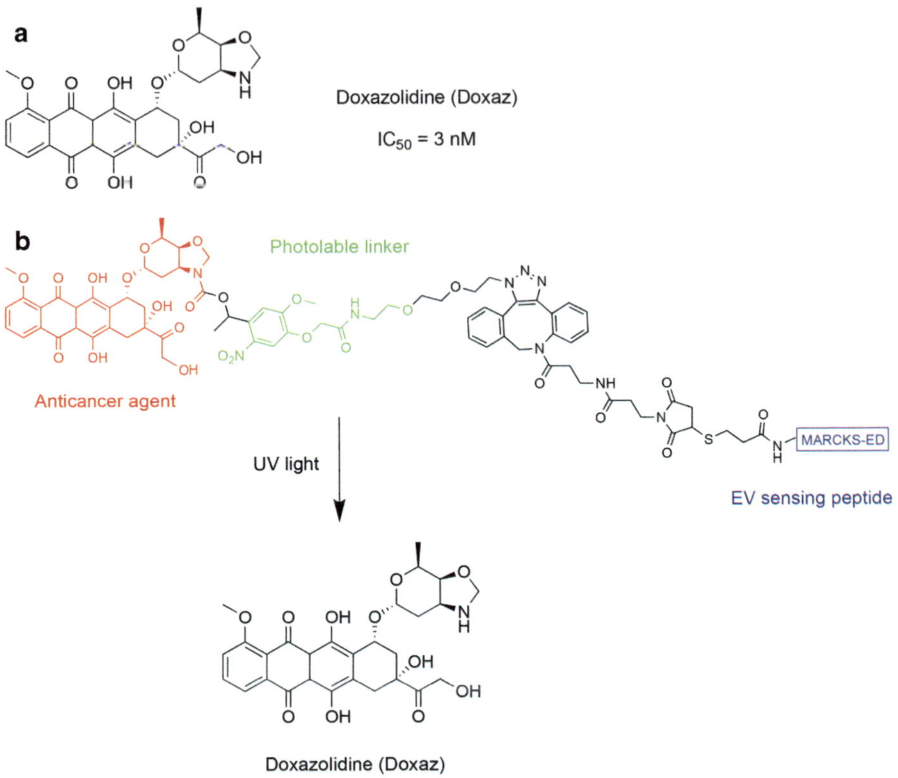


**Fig. 14.4** Application of MARCKS-ED with anticancer drugs. (**a**) Doxazolidine (doxaz). Half-maximal inhibitor concentrations were obtained using a DOX-resistant NCI-RES/Adr cell line. (**b**) MARCKS-ED-photodoxazolidine consists of three parts, i.e., an anticancer agent (doxaz), a photolabile linker, and the MARCKS-ED peptide, and the mechanism involves release of doxaz with near UV light. After sensing and entering tumor cells via EVs, UV light exposure results in doxaz release. Furthermore, the compound can combine with UV light to inhibit tumor growth in situ

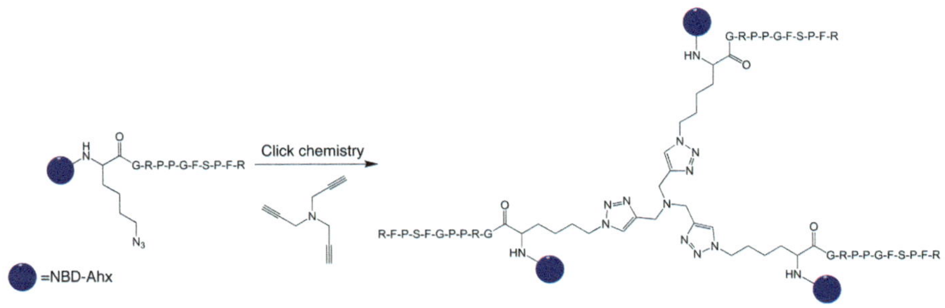


**Fig. 14.5** Bradykinin is trimerized by click chemistry. The ε-amino group of the lysine residue is converted to an azide by solid-phase Cu(II)-catalyzed amine-to-azide conversion to produce the Lys(ε-$N_3$) residue using standard Fmoc chemistry and microwave-assisted solid-phase peptide synthesis. The bioorthogonal strategy for copper-catalyzed azide-alkyne 1,3-dipolar cycloaddition is a crucial step for trimeric peptide formation [76]

C2BL3L is a functional cyclic peptide mimic derived from synaptotagmin-1 (Syt1). As shown in Fig. 14.6, a novel artificial cyclic peptide was obtained from Syt1 membrane penetration at C2B loop 3 (DYDKIGKNDA) [77], which was formed by click chemistry to achieve 10-residue cyclic orthogonal ligation via the alkyne-functionalized L-propargylglycine (Pra) group and azide-functionalized ε-azido norleucine (Nle) residue cyclization [78]. An additional -GG- dipeptide linker at the N-terminus was used as a functional handle. For example, engineered peptides can be labeled as C2BL3C-Trp and C2BL3C-NBD following the introduction of Trp or NBD (Table 14.1).

EV sensing and separating platforms, such as microarray chips or microfluid chips, are promising and novel analytical toolboxes. Recently, researchers have reported the first use of a microarray instrument for EV profiling using membrane-sensing peptides [79]. Peptide-based profiling chips are more sensitive and operational for functional modification than antibody-based chips. Overall, these reports

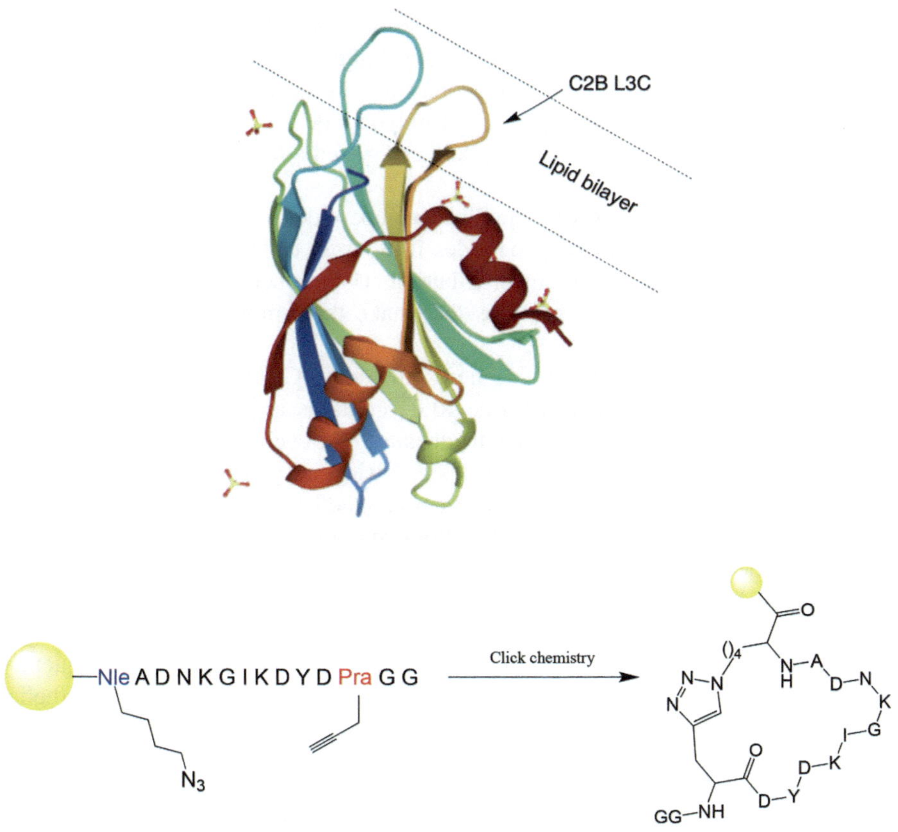


**Fig. 14.6** Crystal structure of the Syt1 C2B domain (PDB: 6 U41). C2B looped three inserts in the lipid bilayer by binding to membranes and sensing curvature. The precursor is GGPraDYDKIGKNDANle (ε-N3). Design and synthesis of C2BL3C as a structural and functional mimic of Syt1 C2B loop 3. Solid-phase synthesis of designed cyclic C2BL3C peptides

**Table 14.1** Typical EV-sensing peptides

| Name | Main binding mechanism | References |
|---|---|---|
| MARCKS-ED | Negatively charged PS attraction and hydrophobic Phe interactions | [68] |
| C2BL3C | Highly curved membrane sensing | [72] |
| Bradykinin | Negatively charged PS attraction | [73] |
| BAR dimers | Electrostatic interaction on concave | [64] |

have greatly increased our confidence in the development and application of specific peptide probes for EVs and related derivatives.

### 14.3.3 EV Targeting and Conjugation of Therapeutic Peptides with EVs

In addition to EV-specific peptides, multifunctional peptides have a broad developmental history and many fields of application. Integration of functional peptides in EVs has been thoroughly explored, and many studies have demonstrated that conjugation of EVs with one type of peptide or polypeptide drug may have applications in the development of improved drug delivery systems for disease-specific targeting of EVs.

The first step involves packing the functional peptides into EVs. From this study, coupling of EVs with delivering peptides into target cells or tissues can prevent protease degradation and random distribution. Using chemical biology methods, scientists can further modify peptides that make them more suitable for loading into EVs.

Common functional peptide fragments include RGD and cyclo(RGD-f/y-K) peptides. The integrin-binding peptide RGD contains arginine-glycine-aspartate, which can target tumors and trigger cell adhesion. The cyclic peptides can be further modified to azide groups (Lys-$N_3$) on lysine residues and can participate in click chemistry. Curcumin-loaded, c(RGDyK)-conjugated EVs have been shown to block the inflammatory response and cellular apoptosis in an ischemic brain model [80]. Similarly, the designed RGD-EVs produced by arginyl-glycyl-aspartic acid-functionalized 1,2-dioleoyl-glycero-3-phosphoethanolamine-poly(ethylene glycol), combined with gold nanoshells, gold nanorods, and gold nanocages, can target tumors and then absorb light near the 700–900 nm wavelength range, subsequently transferring the heat from the light to cancer cells [81]. A c-mesenchymal-epithelial transition factor-binding peptide engineered in macrophage-derived EVs showed targeting preference for breast cancer, which is promising for the delivery of drug in breast cancer treatment [82]. Engineered EVs harboring functional or tumor-targeting peptides could be used for further studies of the chemical biology of EVs.

## 14.4 Chemical Biology Approaches for Investigation of Posttranslational and Epigenetic Modifications of EVs

Since the Human Genome Project was completed in 2003, we have identified over 20,000 genes encoded in the three billion base pairs in human DNA. However, these thousands of human genes cannot explain the functional diversity of human physiological processes. Many functional proteins are produced by RNA splicing and PTMs. During protein processing, various protein assemblies and modifications function, and these processes are of great interest in cell biology and genetics. Protein PTMs occur throughout the entire life cycle of the protein and can alter the structure and chemical activity of proteins; this further affects enzyme activity, structural stability, and subcellular localization. There are many types of PTMs (Fig. 14.7), ranging from simple chemical groups (methylation, acetylation, phosphorylation, etc.) to complex polypeptides (ubiquitination, SUMOylation, etc.). PTMs regulate protein functions and are related to certain diseases. Some studies of EV PTMs are promising but challenging, e.g., clarifying the machinery required for dynamic PTMs to regulate EV biogenesis, cargo loading, secretion, and uptake and distinguishing the differences in PTMs between EVs to recognize and sort specific EV subtypes, which necessitates the development of precise EV and PTM detection methods. Notably, lipids, carbohydrates, and amino acids, which are crucial in many metabolic processes, can also modify proteins. PTM of EV proteins may

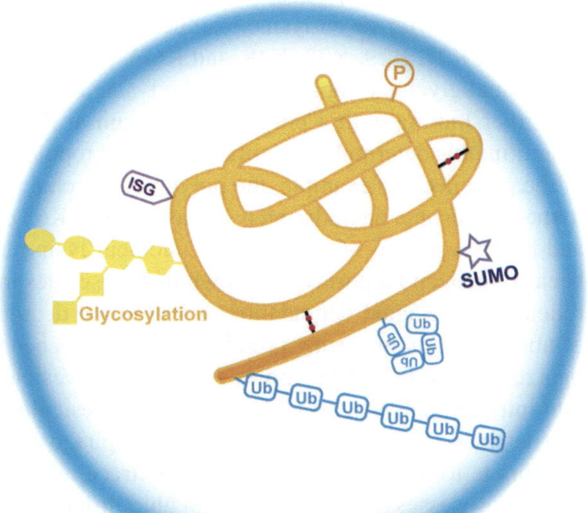


**Fig. 14.7** Various PTMs are present in EVs. Phosphorylation occurs on serine, threonine, and tyrosine residues. The glycosylation of proteins (usually in transmembrane proteins) affects the specific distribution of the protein in EVs. Ubiquitination of proteins can affect the fate of the protein, and EVs have been shown to be rich in ubiquitinated proteins. Furthermore, the other PTMs have been shown to occur in EV proteins

reflect the extracellular or parent cell internal environment, providing insights into the use of liquid biopsy for disease diagnosis and prognosis. Analysis of EV protein PTMs may be helpful for analysis of disease status. Chemical biology approaches are also useful tools for studying PTMs in EVs. For example, mass spectroscopy can be used to determine the specific molecular weight of a molecule, thereby enabling the identification of various types of PTM. In addition, small-molecule probes can not only regulate PTM but also act as prodrugs in PTM-related diseases.

Before discussing PTMs of EV proteins, it is necessary to understand the basic concepts of chemical biology. The writer-eraser-reader model regulates the epigenetic modification system. Briefly, writer and eraser proteins add or remove PTMs on amino acid residues. Reader proteins perform this function by recognizing and binding to PTMs. Classic biochemical approaches can be used for the detection and identification of PTMs; for example, phosphorylated antibodies and acetylated antibodies can be applied in Western blot analyses. However, this approach is limited by the specificity and sensitivity of these specific antibodies, and further development of mass spectrometry technology has enabled the identification of novel PTM modes, substrates, and sites through chemical biology investigations. In this section, some PTMs related to EVs are introduced, and the technical principles and methods are briefly discussed.

Proteins in EVs may have multiple PTMs and therefore display different functions and sorting preferences in EVs [83]. Some PTMs are related to vesicular sorting mechanisms, such as ubiquitination and SUMOylation. EVs contain ubiquitinated proteins [84]. Ubiquitin is a conserved protein consisting of 76 amino acid residues with seven lysine residues mediating the formation of the polychain (K6, K11, K27, K29, K33, K48, and K63). Mono-ubiquitination via ligation of the C-terminal glycine of ubiquitin and the lysine residue of the target protein can lead to further linking with a polyubiquitin chain at the seven different lysine sites in ubiquitin; this determines the different spatial topologies and the complex functions of the proteins. For example, the polyubiquitin chain at Ub-K48 targets the protein to the proteasome degradation pathway; the Ub-K63 chain is related to endocytosis and autophagy [85, 86]. Other noncanonical lysine ubiquitination sites have also been studied, and reports have shown that these sites may be related to endoplasmic reticulum-associated degradation or lysosomal degradation [86–90]. The function of ubiquitination in sorting proteins into EVs has been studied. Ubiquitination recognized by the ESCRT complex is crucial for vesicle budding and protein sorting into intraluminal vesicles (ILVs) [91]. Multivesicular bodies containing ILVs have two destinies: fusion with lysosomes or autophagosomes and fusion with the plasma membrane to release vesicles into the extracellular environment. These two competitive processes are also involved in ubiquitination. Similarly, some ubiquitin-like modifications, such as SUMOylation by a small ubiquitin-like modifier (SUMO), and some other modifiers, such as neural precursor cell expressed developmentally downregulated 8 or ISG15, have also been identified. The sorting of miRNAs and proteins into EVs has been reported to be controlled by SUMOylated proteins [92, 93]. Furthermore, ISGylation, which is similar to SUMOylation, inhibits EV secretion by inducing TSG101 protein aggregation and degradation [94]. EV proteins can also undergo ubiquitination in an independent manner [95].

PTMs in EVs can also be detected by mass spectrometry or by using specific antibodies. Aberrant phosphorylation of Thr181 in tau causes the protein to aggregate in EVs secreted by M1C cells and causes elevated cerebrospinal fluid tau levels in early Alzheimer's disease [96]. Glycosylation affects EV cargo sorting by galectins and galectin-associated proteins. Moreover, mannose, α-2,6-sialic acid, polylactosamine, sialoglycoproteins, and N-glycans are abundant and have potential functions in EVs [97, 98]. PTMs are crucial for regulating EV secretion and function. Indeed, many recent reports have illustrated the importance of EVs carrying modified proteins and delivering substances as information to their recipient cells. These EVs can mediate the transfer of lactylated and acetylated high-mobility group box 1 protein from macrophages, thereby causing polymicrobial sepsis [99].

Because of the existence of thousands of proteins, which can be modified by PTMs in many different ways, proteins can have diverse cellular functions. It is important to identify influential PTMs in studies of EVs. Research topics should be related to EV biogenesis and cargo. With the development of -omics studies and big data analysis techniques, EV protein PTMs can be studied in sufficient detail. In this section, we will discuss how to identify PTM types and interaction sites using chemical biology techniques. After protein isolation and accumulation, the proteins in a solution or gel can be digested into polypeptide fragments using a protease (typically trypsin). The modified peptides can then undergo an enrichment process before separation by high-performance liquid chromatography. The types of PTMs in EVs can be recognized based on variations in mass determined by tandem mass spectrometry analysis. By using these chemical biology methods, we can determine the functions of PTMs in EVs or the classifications for EV clusters. Some add-on PTM markers of EVs can also provide guidance for the diagnosis and prognosis [100] of related diseases (Table 14.2).

Chemical biologists use chemical tools to study and regulate protein activity. Moreover, combined with the bioorthogonal reaction and small-molecule rational design, as described above, EV protein PTMs can be studied in detail. For example, a small molecule can be designed as a PTM mimic or inhibitor. Bioorthogonal reactions can capture and enrich EVs with specific PTMs. Chemical biology tools are expected to be important for EV research.

**Table 14.2** PTMs in EVs

| PTM type | Function | References |
|---|---|---|
| Phosphorylation | Protein aggregation | [96] |
| Glycosylation | Cargo sorting | [97, 98] |
| Ubiquitinoylation | Cargo loading, discharge | [91] |
| SUMOylation | Control miRNAs and protein loading | [92, 93] |
| ISGylation | Inhibit EV secretion | [94] |
| Lactylation/acetylation | Increase endothelium permeability | [99] |

## References


1. Prendergast, FG, Mann KG. Chemical and physical properties of aequorin and the green fluorescent protein isolated from Aequorea forskålea. Biochemistry. 1978;17:3448–53.
2. Shimomura O, Johnson FH, Saiga Y. Extraction, purification and properties of aequorin, a bioluminescent protein from the luminous hydromedusan, Aequorea. J Cell Comp Physiol. 1962;59:223–39.
3. Ramil CP, Lin Q. Bioorthogonal chemistry: strategies and recent developments. Chem Commun (Camb). 2013;49(94):11007–22.
4. Prescher JA, Bertozzi CR. Chemistry in living systems. Nat Chem Biol. 2005;1(1):13–21.
5. Lee TS, Kim Y, Zhang W, Song IH, Tung CH. Facile metabolic glycan labeling strategy for exosome tracking. Biochim Biophys Acta Gen Subj. 2018;1862(5):1091–100.
6. Lai CP, Kim EY, Badr CE, Weissleder R, Mempel TR, Tannous BA, et al. Visualization and tracking of tumour extracellular vesicle delivery and RNA translation using multiplexed reporters. Nat Commun. 2015;6(1):7029.
7. Hermanson GT. Chapter 1—Introduction to bioconjugation. In: Hermanson GT, editor. Bioconjugate techniques. 3rd ed. Boston: Academic Press; 2013. p. 1–125.
8. Glazer AN. Specific chemical modification of proteins. Annu Rev Biochem. 1970;39:101–30.
9. Francis SDTNSJMB. New methods for protein bioconjugation. Chem Biol. 2007:593–634.
10. Francis MB. Proteins: chemistry and chemical reactivity. In: Wiley encyclopedia of chemical biology; 2008. p. 1–16.
11. Sletten EM, Bertozzi CR. Bioorthogonal chemistry: fishing for selectivity in a sea of functionality. Angew Chem Int Ed. 2009;48(38):6974–98.
12. Huisgen R. 1,3-dipolar cycloadditions. Past and future. Angew Chem Int Ed Engl. 1963;2(10):565–98.
13. Rostovtsev VV, Green LG, Fokin VV, Sharpless KB. A stepwise huisgen cycloaddition process: copper(I)-catalyzed regioselective "ligation" of azides and terminal alkynes. Angew Chem Int Ed Engl. 2002;41(14):2596–9.
14. Tornøe CW, Christensen C, Meldal M. Peptidotriazoles on solid phase: [1,2,3]-triazoles by regiospecific copper(i)-catalyzed 1,3-dipolar cycloadditions of terminal alkynes to azides. J Org Chem. 2002;67(9):3057–64.
15. Himo F, Lovell T, Hilgraf R, Rostovtsev VV, Noodleman L, Sharpless KB, et al. Copper(I)-catalyzed synthesis of azoles. DFT study predicts unprecedented reactivity and intermediates. J Am Chem Soc. 2005;127(1):210–6.
16. Bertozzi CR. A decade of bioorthogonal chemistry. Acc Chem Res. 2011;44(9):651–3.
17. Griffin BA, Adams SR, Tsien RY. Specific covalent labeling of recombinant protein molecules inside live cells. Science. 1998;281(5374):269–72.
18. Sebastian V, Sancho-Albero M, Arruebo M, Pérez-López AM, Rubio-Ruiz B, Martin-Duque P, et al. Nondestructive production of exosomes loaded with ultrathin palladium nanosheets for targeted bio-orthogonal catalysis. Nat Protoc. 2020;16:131.
19. Sancho-Albero M, Rubio-Ruiz B, Pérez-López AM, Sebastián V, Martín-Duque P, Arruebo M, et al. Cancer-derived exosomes loaded with ultrathin palladium nanosheets for targeted bioorthogonal catalysis. Nat Catal. 2019;2(10):864–72.
20. Koga K, Matsumoto K, Akiyoshi T, Kubo M, Yamanaka N, Tasaki A, et al. Purification, characterization and biological significance of tumor-derived exosomes. Anticancer Res. 2005;25(6a):3703–7.
21. Miyauchi K, Kim Y, Latinovic O, Morozov V, Melikyan GB. HIV enters cells via endocytosis and dynamin-dependent fusion with endosomes. Cell. 2009;137(3):433–44.
22. Tian T, Wang Y, Wang H, Zhu Z, Xiao Z. Visualizing of the cellular uptake and intracellular trafficking of exosomes by live-cell microscopy. J Cell Biochem. 2010;111(2):488–96.
23. Morelli AE, Larregina AT, Shufesky WJ, Sullivan ML, Stolz DB, Papworth GD, et al. Endocytosis, intracellular sorting, and processing of exosomes by dendritic cells. Blood. 2004;104(10):3257–66.

24. Näslund TI, Paquin-Proulx D, Paredes PT, Vallhov H, Sandberg JK, Gabrielsson S. Exosomes from breast milk inhibit HIV-1 infection of dendritic cells and subsequent viral transfer to CD4+ T cells. AIDS. 2014;28(2):171–80.
25. Franzen CA, Simms PE, Van Huis AF, Foreman KE, Kuo PC, Gupta GN. Characterization of uptake and internalization of exosomes by bladder cancer cells. Biomed Res Int. 2014;2014:619829.
26. Pužar Dominkuš P, Stenovec M, Sitar S, Lasič E, Zorec R, Plemenitaš A, et al. PKH26 labeling of extracellular vesicles: characterization and cellular internalization of contaminating PKH26 nanoparticles. Biochim Biophys Acta Biomembr. 2018;1860(6):1350–61.
27. Takahashi Y, Nishikawa M, Shinotsuka H, Matsui Y, Ohara S, Imai T, et al. Visualization and in vivo tracking of the exosomes of murine melanoma B16-BL6 cells in mice after intravenous injection. J Biotechnol. 2013;165(2):77–84.
28. Roberts-Dalton HD, Cocks A, Falcon-Perez JM, Sayers EJ, Webber JP, Watson P, et al. Fluorescence labelling of extracellular vesicles using a novel thiol-based strategy for quantitative analysis of cellular delivery and intracellular traffic. Nanoscale. 2017;9(36):13693–706.
29. Coscia C, Parolini I, Sanchez M, Biffoni M, Boussadia Z, Zanetti C, et al. Generation, quantification, and tracing of metabolically labeled fluorescent exosomes. Methods Mol Biol. 2016;1448:217–35.
30. Suetsugu A, Honma K, Saji S, Moriwaki H, Ochiya T, Hoffman RM. Imaging exosome transfer from breast cancer cells to stroma at metastatic sites in orthotopic nude-mouse models. Adv Drug Deliv Rev. 2013;65(3):383–90.
31. Zhang P, Dong B, Zeng E, Wang F, Jiang Y, Li D, et al. In vivo tracking of multiple tumor exosomes labeled by phospholipid-based bioorthogonal conjugation. Anal Chem. 2018;90(19):11273–9.
32. Song S, Shim MK, Lim S, Moon Y, Yang S, Kim J, et al. In situ one-step fluorescence labeling strategy of exosomes via bioorthogonal click chemistry for real-time exosome tracking in vitro and in vivo. Bioconjug Chem. 2020;31(5):1562–74.
33. Devaraj NK. The future of bioorthogonal chemistry. ACS Cent Sci. 2018;4(8):952–9.
34. Lipinski CA, Lombardo F, Dominy BW, Feeney PJ. Experimental and computational approaches to estimate solubility and permeability in drug discovery and development settings1PII of original article: S0169-409X(96)00423-1. The article was originally published in advanced drug delivery reviews 23 (1997) 3–25.1. Adv Drug Deliv Rev. 2001;46(1):3–26.
35. Dallakyan S, Olson AJ. Small-molecule library screening by docking with PyRx. Methods Mol Biol. 2015;1263:243–50.
36. Wenderski TA, Stratton CF, Bauer RA, Kopp F, Tan DS. Principal component analysis as a tool for library design: a case study investigating natural products, brand-name drugs, natural product-like libraries, and drug-like libraries. In: Hempel JE, Williams CH, Hong CC, editors. Chemical biology: methods and protocols. New York, NY: Springer New York; 2015. p. 225–42.
37. Hempel JE, Hong CC. Practical strategies for small-molecule probe development in chemical biology. In: Hempel JE, Williams CH, Hong CC, editors. Chemical biology: methods and protocols. New York, NY: Springer New York; 2015. p. 209–23.
38. Vendrell-Navarro G, Brockmeyer A, Waldmann H, Janning P, Ziegler S. Identification of the targets of biologically active small molecules using quantitative proteomics. In: Hempel JE, Williams CH, Hong CC, editors. Chemical biology: methods and protocols. New York, NY: Springer New York; 2015. p. 263–86.
39. Trajkovic K, Hsu C, Chiantia S, Rajendran L, Wenzel D, Wieland F, et al. Ceramide triggers budding of exosome vesicles into multivesicular endosomes. Science. 2008;319(5867):1244–7.
40. Luberto C, Hassler DF, Signorelli P, Okamoto Y, Sawai H, Boros E, et al. Inhibition of tumor necrosis factor-induced cell death in MCF7 by a novel inhibitor of neutral sphingomyelinase. J Biol Chem. 2002;277(43):41128–39.
41. Marchesini N, Luberto C, Hannun YA. Biochemical properties of mammalian neutral sphingomyelinase 2 and its role in sphingolipid metabolism. J Biol Chem. 2003;278(16):13775–83.

42. Kolmakova A, Kwiterovich P, Virgil D, Alaupovic P, Knight-Gibson C, Martin SF, et al. Apolipoprotein C-I induces apoptosis in human aortic smooth muscle cells via recruiting neutral sphingomyelinase. Arterioscler Thromb Vasc Biol. 2004;24(2):264–9.
43. Savina A, Furlán M, Vidal M, Colombo MI. Exosome release is regulated by a calcium-dependent mechanism in K562 cells. J Biol Chem. 2003;278(22):20083–90.
44. Chalmin F, Ladoire S, Mignot G, Vincent J, Bruchard M, Remy-Martin JP, et al. Membrane-associated Hsp72 from tumor-derived exosomes mediates STAT3-dependent immunosuppressive function of mouse and human myeloid-derived suppressor cells. J Clin Invest. 2010;120(2):457–71.
45. Lyu L, Wang H, Li B, Qin Q, Qi L, Nagarkatti M, et al. A critical role of cardiac fibroblast-derived exosomes in activating renin angiotensin system in cardiomyocytes. J Mol Cell Cardiol. 2015;89(Pt B):268–79.
46. Colombo M, Raposo G, Théry C. Biogenesis, secretion, and intercellular interactions of exosomes and other extracellular vesicles. Annu Rev Cell Dev Biol. 2014;30:255–89.
47. Pegtel DM, Gould SJ. Exosomes. Annu Rev Biochem. 2019;88:487–514.
48. Kalluri R, LeBleu VS. The biology, function, and biomedical applications of exosomes. Science. 2020;367(6478).
49. Yang X, Wang Y, Byrne R, Schneider G, Yang S. Concepts of artificial intelligence for computer-assisted drug discovery. Chem Rev. 2019;119(18):10520–94.
50. Schneider P, Walters WP, Plowright AT, Sieroka N, Listgarten J, Goodnow RA Jr, et al. Rethinking drug design in the artificial intelligence era. Nat Rev Drug Discov. 2020;19(5):353–64.
51. Jing Y, Bian Y, Hu Z, Wang L, Xie XQ. Deep learning for drug design: an artificial intelligence paradigm for drug discovery in the big data era. AAPS J. 2018;20(3):58.
52. Kulak O, Yamaguchi K, Lum L. Identification of therapeutic small-molecule leads in cultured cells using multiplexed pathway reporter readouts. In: Hempel JE, Williams CH, Hong CC, editors. Chemical biology: methods and protocols. New York, NY: Springer New York; 2015. p. 3–14.
53. Dawson PE, Muir TW, Clark-Lewis I, Kent SB. Synthesis of proteins by native chemical ligation. Science. 1994;266(5186):776–9.
54. Dawson PE, Kent SB. Synthesis of native proteins by chemical ligation. Annu Rev Biochem. 2000;69:923–60.
55. Camarero JA, Shekhtman A, Campbell EA, Chlenov M, Gruber TM, Bryant DA, et al. Autoregulation of a bacterial sigma factor explored by using segmental isotopic labeling and NMR. Proc Natl Acad Sci U S A. 2002;99(13):8536–41.
56. Muralidharan-Chari V, Clancy J, Plou C, Romao M, Chavrier P, Raposo G, et al. ARF6-regulated shedding of tumor cell-derived plasma membrane microvesicles. Curr Biol. 2009;19(22):1875–85.
57. Bigay J, Casella JF, Drin G, Mesmin B, Antonny B. ArfGAP1 responds to membrane curvature through the folding of a lipid packing sensor motif. EMBO J. 2005;24(13):2244–53.
58. Antonny B. Mechanisms of membrane curvature sensing. Annu Rev Biochem. 2011;80:101–23.
59. Kastelowitz N, Yin H. Exosomes and microvesicles: identification and targeting by particle size and lipid chemical probes. Chembiochem. 2014;15(7):923–8.
60. McMahon HT, Gallop JL. Membrane curvature and mechanisms of dynamic cell membrane remodelling. Nature. 2005;438(7068):590–6.
61. Tonnesen A, Christensen SM, Tkach V, Stamou D. Geometrical membrane curvature as an allosteric regulator of membrane protein structure and function. Biophys J. 2014;106(1):201–9.
62. Tait JF, Gibson D. Phospholipid binding of annexin V: effects of calcium and membrane phosphatidylserine content. Arch Biochem Biophys. 1992;298(1):187–91.
63. Andree HA, Stuart MC, Hermens WT, Reutelingsperger CP, Hemker HC, Frederik PM, et al. Clustering of lipid-bound annexin V may explain its anticoagulant effect. J Biol Chem. 1992;267(25):17907–12.

64. Bhatia VK, Madsen KL, Bolinger PY, Kunding A, Hedegård P, Gether U, et al. Amphipathic motifs in BAR domains are essential for membrane curvature sensing. EMBO J. 2009;28(21):3303–14.
65. Mesmin B, Drin G, Levi S, Rawet M, Cassel D, Bigay J, et al. Two lipid-packing sensor motifs contribute to the sensitivity of ArfGAP1 to membrane curvature. Biochemistry. 2007;46(7):1779–90.
66. Rauch ME, Ferguson CG, Prestwich GD, Cafiso DS. Myristoylated alanine-rich C kinase substrate (MARCKS) sequesters spin-labeled phosphatidylinositol 4,5-bisphosphate in lipid bilayers. J Biol Chem. 2002;277(16):14068–76.
67. Zhang W, Crocker E, McLaughlin S, Smith SO. Binding of peptides with basic and aromatic residues to bilayer membranes: phenylalanine in the myristoylated alanine-rich C kinase substrate effector domain penetrates into the hydrophobic core of the bilayer. J Biol Chem. 2003;278(24):21459–66.
68. Morton LA, Yang H, Saludes JP, Fiorini Z, Beninson L, Chapman ER, et al. MARCKS-ED peptide as a curvature and lipid sensor. ACS Chem Biol. 2013;8(1):218–25.
69. Morton LA, Saludes JP, Yin H. Constant pressure-controlled extrusion method for the preparation of Nano-sized lipid vesicles. J Vis Exp. 2012:(64).
70. Yan L, de Jesus AJ, Tamura R, Li V, Cheng K, Yin H. Curvature sensing MARCKS-ED peptides bind to membranes in a stereo-independent manner. J Pept Sci. 2015;21(7):577–85.
71. Tamura R, Balabanova A, Frakes SA, Bargmann A, Grimm J, Koch TH, et al. Photoactivatable prodrug of Doxazolidine targeting exosomes. J Med Chem. 2019;62(4):1959–70.
72. Saludes JP, Morton LA, Ghosh N, Beninson LA, Chapman ER, Fleshner M, et al. Detection of highly curved membrane surfaces using a cyclic peptide derived from synaptotagmin-I. ACS Chem Biol. 2012;7(10):1629–35.
73. Saludes JP, Morton LA, Coulup SK, Fiorini Z, Cook BM, Beninson L, et al. Multivalency amplifies the selection and affinity of bradykinin-derived peptides for lipid nanovesicles. Mol BioSyst. 2013;9(8):2005–9.
74. Lopez JJ, Shukla AK, Reinhart C, Schwalbe H, Michel H, Glaubitz C. The structure of the neuropeptide bradykinin bound to the human G-protein coupled receptor bradykinin B2 as determined by solid-state NMR spectroscopy. Angew Chem Int Ed Engl. 2008;47(9):1668–71.
75. Bonechi C, Ristori S, Martini G, Martini S, Rossi C. Study of bradykinin conformation in the presence of model membrane by nuclear magnetic resonance and molecular modelling. Biochim Biophys Acta. 2009;1788(3):708–16.
76. Alleti R, Rao V, Xu L, Gillies RJ, Mash EA. A solanesol-derived scaffold for multimerization of bioactive peptides. J Org Chem. 2010;75(17):5895–903.
77. Cheng Y, Sequeira SM, Malinina L, Tereshko V, Söllner TH, Patel DJ. Crystallographic identification of Ca2+ and Sr2+ coordination sites in synaptotagmin I C2B domain. Protein Sci. 2004;13(10):2665–72.
78. Punna S, Kuzelka J, Wang Q, Finn MG. Head-to-tail peptide cyclodimerization by copper-catalyzed azide-alkyne cycloaddition. Angew Chem Int Ed Engl. 2005;44(15):2215–20.
79. Gori A, Romanato A, Bergamaschi G, Strada A, Gagni P, Frigerio R, et al. Membrane-binding peptides for extracellular vesicles on-chip analysis. J Extracell Vesicles. 2020;9(1):1751428.
80. Tian T, Zhang HX, He CP, Fan S, Zhu YL, Qi C, et al. Surface functionalized exosomes as targeted drug delivery vehicles for cerebral ischemia therapy. Biomaterials. 2018;150:137–49.
81. Wang J, Dong Y, Li Y, Li W, Cheng K, Qian Y, et al. Designer exosomes for active targeted chemo-Photothermal synergistic tumor therapy. Adv Funct Mater. 2018;28(18):1707360.
82. Li S, Wu Y, Ding F, Yang J, Li J, Gao X, et al. Engineering macrophage-derived exosomes for targeted chemotherapy of triple-negative breast cancer. Nanoscale. 2020;12(19):10854–62.
83. Moreno-Gonzalo O, Villarroya-Beltri C, Sánchez-Madrid F. Post-translational modifications of exosomal proteins. Front Immunol. 2014;5:383.
84. Buschow SI, Liefhebber JM, Wubbolts R, Stoorvogel W. Exosomes contain ubiquitinated proteins. Blood Cells Mol Dis. 2005;35(3):398–403.

85. Komander D, Rape M. The ubiquitin code. Annu Rev Biochem. 2012;81(1):203–29.
86. Swatek KN, Komander D. Ubiquitin modifications. Cell Res. 2016;26(4):399–422.
87. Chastagner P, Israël A, Brou C. Itch/AIP4 mediates Deltex degradation through the formation of K29-linked polyubiquitin chains. EMBO Rep. 2006;7(11):1147–53.
88. Matsumoto ML, Wickliffe KE, Dong KC, Yu C, Bosanac I, Bustos D, et al. K11-linked polyubiquitination in cell cycle control revealed by a K11 linkage-specific antibody. Mol Cell. 2010;39(3):477–84.
89. Huang H, Jeon M-s, Liao L, Yang C, Elly C, Yates JR, et al. K33-linked polyubiquitination of T cell receptor-ζ regulates proteolysis-independent T cell signaling. Immunity. 2010;33(1):60–70.
90. Peng DJ, Zeng M, Muromoto R, Matsuda T, Shimoda K, Subramaniam M, et al. Noncanonical K27-linked polyubiquitination of TIEG1 regulates Foxp3 expression and tumor growth. J Immunol. 2011;186(10):5638–47.
91. Raiborg C, Stenmark H. The ESCRT machinery in endosomal sorting of ubiquitylated membrane proteins. Nature. 2009;458(7237):445–52.
92. Villarroya-Beltri C, Gutiérrez-Vázquez C, Sánchez-Cabo F, Pérez-Hernández D, Vázquez J, Martin-Cofreces N, et al. Sumoylated hnRNPA2B1 controls the sorting of miRNAs into exosomes through binding to specific motifs. Nat Commun. 2013;4(1):2980.
93. Kunadt M, Eckermann K, Stuendl A, Gong J, Russo B, Strauss K, et al. Extracellular vesicle sorting of α-synuclein is regulated by sumoylation. Acta Neuropathol. 2015;129(5):695–713.
94. Villarroya-Beltri C, Baixauli F, Mittelbrunn M, Fernández-Delgado I, Torralba D, Moreno-Gonzalo O, et al. ISGylation controls exosome secretion by promoting lysosomal degradation of MVB proteins. Nat Commun. 2016;7(1):13588.
95. Villarroya-Beltri C, Baixauli F, Gutiérrez-Vázquez C, Sánchez-Madrid F, Mittelbrunn M. Sorting it out: regulation of exosome loading. Semin Cancer Biol. 2014;28:3–13.
96. Saman S, Kim W, Raya M, Visnick Y, Miro S, Saman S, et al. Exosome-associated tau is secreted in tauopathy models and is selectively phosphorylated in cerebrospinal fluid in early Alzheimer disease*. J Biol Chem. 2012;287(6):3842–9.
97. Batista BS, Eng WS, Pilobello KT, Hendricks-Muñoz KD, Mahal LK. Identification of a conserved glycan signature for microvesicles. J Proteome Res. 2011;10(10):4624–33.
98. Escrevente C, Grammel N, Kandzia S, Zeiser J, Tranfield EM, Conradt HS, et al. Sialoglycoproteins and N-glycans from secreted exosomes of ovarian carcinoma cells. PLoS One. 2013;8(10):e78631.
99. Yang K, Fan M, Wang X, Xu J, Wang Y, Tu F, et al. Lactate promotes macrophage HMGB1 lactylation, acetylation, and exosomal release in polymicrobial sepsis. Cell Death Differ. 2021;29:133.
100. Moreno-Gonzalo O, Fernandez-Delgado I, Sanchez-Madrid F. Post-translational add-ons mark the path in exosomal protein sorting. Cell Mol Life Sci. 2018;75(1):1–19.

# Part III

# EVs Biomarker in Diseases

# Screening and Clinical Verification of Extracellular Vesicle Biomarkers

15

Bo Li, Yiyao Huang, Bodeng Wu, Huixian Lin, Qi Xiu, Qian Wang, and Lei Zheng

## 15.1 Introduction

Biomarkers, including gene expression products, metabolites, polysaccharides, and proteins [1], refer to a class of indicators found in the blood, body fluids, and tissues that can be used in clinical diagnosis and treatment because they reflect patient physiology, disease pathology, or drug treatment effects [2]. As shown in Table 15.1, there are six types of clinical biomarkers, categorized based on their applications.

Based on our increased understanding of many diseases, we expect that early diagnosis will lead to more rapid cures for various diseases [3]. However, the complexities of body fluid samples have impeded biomarker analysis because potential disease biomarkers are often present in low abundance and are readily masked by high-abundance proteins in body fluids. Indeed, low-abundance proteins can be difficult to detect because 99% of plasma protein mass is made up of 22 abundant proteins [4]. Therefore, alternative solutions are needed to facilitate biomarker discovery and application.

---

Bo Li, Yiyao Huang, and Bodeng Wu contributed equally to this work.

B. Li · Y. Huang · B. Wu · H. Lin · Q. Xiu · L. Zheng (✉)
Department of Laboratory Medicine, Nanfang Hospital, Southern Medical University, Guangzhou, China

Guangdong Engineering and Technology Research Center for Rapid Diagnostic Biosensors, Nanfang Hospital, Southern Medical University, Guangzhou, China
e-mail: nfyyzhenglei@smu.edu.cn

Q. Wang (✉)
Laboratory Medicine Center, Zhujiang Hospital, Southern Medical University, Guangzhou, Guangdong, China
e-mail: wangqian@smu.edu.cn



Q. Wang, L. Zheng (eds.), *Extracellular Vesicles*,
https://doi.org/10.1007/978-981-99-8365-0_15

**Table 15.1** Types of clinical biomarkers

| Types | Applications |
|---|---|
| Screening | Early screening of diseases |
| Diagnostic | Definitely diagnosis of diseases |
| Prognostic | Predict disease outcomes of treatment |
| Predictive | Predict the effect of treatment |
| Monitoring | Evaluate the effect of treatment and predict the recurrence and progression |
| Toxicity | Reflect on the side effects of drugs |

Extracellular vesicles (EVs) are secreted by a wide range of cells, including immune, neuronal, and cancer cells, and can be found circulating in almost all biological fluids [4–7]. EVs are filled with a cytoplasmic fluid that contains many bioactive substances, including lipids, proteins, and nucleic acids [8]; thus, these vesicles play important roles in intercellular communication and have enormous diagnostic potential. Moreover, because EVs circulating in the blood or other body fluids are likely to have been derived from various tissues, the isolation of cell-specific EVs can provide a wealth of information on specific pathological conditions. Therefore, EVs may represent the global landscape of tumor heterogeneity, which cannot be fully appreciated using traditional methods of mutation analyses [9]. In 2020, nearly 35,000 EV-related studies had been published in PubMed-indexed journals. For example, the levels of phosphorylated EV proteins in the blood are well correlated with tumor progression [10]. Furthermore, glypican 1 [11], integrin [12], and programmed death-1 [13] expression in EVs can be used in the early diagnosis of tumors, detection of tumor metastasis, and treatment decisions in patients with cancer. Furthermore, using deep sequencing of exosomal RNA species, It was concluded that microRNAs (miRNAs), abundant functional RNA species in exosomes, could be used for cancer diagnosis; in particular, *miR-373*, *miR-200a*, *miR-200b*, and *miR-200c* were shown to have potential applications as diagnostic and prognostic biomarkers in ovarian cancer, and *miR-372* was shown to be a biomarker of colorectal cancer [14, 15].

Of the 100,000 publications describing candidate biomarkers, only 1% of the literature represents clinical studies [16]. Indeed, not only should biomarkers be correlated with outcomes but changes in biomarker expression must also explain the clinical outcomes [17]. Several approaches may improve the diagnostic accuracy and clinical translation of EV-based biomarkers. First, high-throughput technology can be used to perform a preliminary screening of numerous target molecules in parallel. Second, the reproducibility of the results must also be improved. Finally, it is necessary to establish a laboratory-based quality-control system for EV biomarkers. EVs containing various biomolecules play crucial roles in the transmission of information between different types of cells. Thus, the stability, specific cargo capacity, and noninvasive sampling procedures of EVs make them promising clinical biomarkers [4]. This chapter will focus on the above topics.

## 15.2 Experimental Design

Approaches for the screening and verification of potential biomarkers mainly include five steps, i.e., study design, screening, verification, validation, and clinical application (Fig. 15.1). Potential biomarkers can only be used in clinical diagnosis after rigorous clinical verification to demonstrate their high sensitivity, specificity, predictive ability, risk/benefit ratio, and clinical feasibility.

### 15.2.1 Pre-analysis Factors

Owing to the extensive sources and complex heterogeneity of EVs [19], it is necessary for researchers to fully understand disease diagnosis and treatment strategies as well as clinical mechanisms to ensure that the selected EV biomarkers can improve

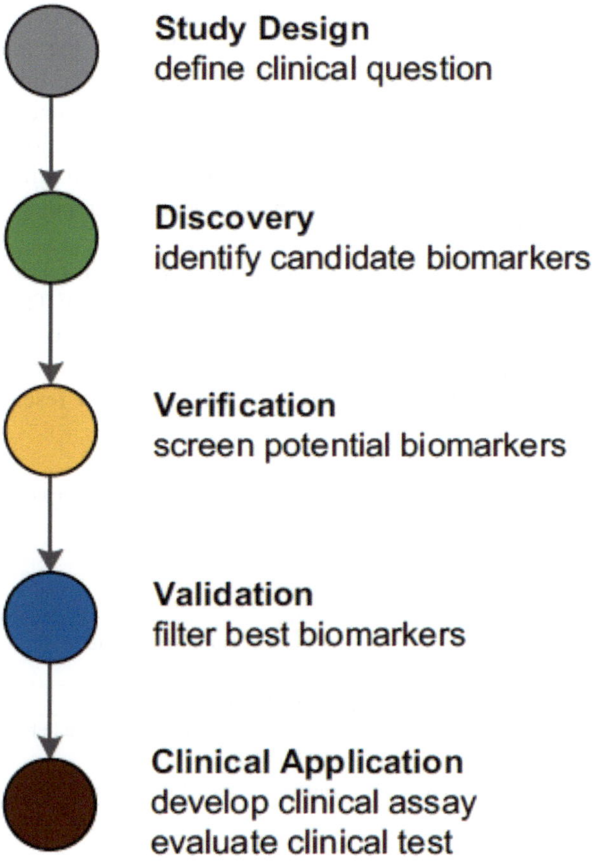


**Fig. 15.1** Workflow overview for biomarker development (*Comprehensive biomarker discovery and validation for clinical application.* Reprinted with permission from *Royal Society of Chemistry*) [18]

available methods for diagnosis and treatment to enhance clinical outcomes. The experimental design should include study objectives, biological sample database establishment methods, a research program, and feasibility analyses. The research program should include at least the following information: (1) sample size, (2) sample inclusion/exclusion criteria, (3) sample collection and handling requirements, (4) limitations of the analytical methods, (5) statistical methods, (6) independent database and evaluation, and (7) complete and transparent experimental results.

Therefore, it is necessary to standardize specimen collection and preparation methods for EV biomarker research. Pre-analysis factors are a major cause of unrepeatable and biased results. The International Society for Extracellular Vesicles (ISEV) issued guidelines for the collection and storage of EV samples in 2013 [20] and 2015 [21] and launched the Minimum Requirements for Extracellular Vesicle Research in 2018 (MISEV2018) [22] to standardize and improve EV research repeatability. In addition, MISEV2018 recommends that at least one protein in categories one to three must be evaluated in any EV preparation: (1) transmembrane or glycosylphosphatidylinositol-anchored proteins localized at the external membrane of prokaryotic cells, and plasma membrane and/or endosomes of eukaryotic cells represent hallmarks of any type of EV; (2) the presence of cytosolic proteins (eukaryotic cells and gram-positive bacteria) or periplasmic proteins (gram-negative bacteria) to support the structure of lipid bilayers enclosing intracellular material, as expected for any EV; and (3) some proteins are major constituents of non-EV structures often co-isolated with EVs. In addition, two different methods, e.g., transmission electron microscopy and nanoparticle tracking analysis, should be used to characterize EVs [22].

Table 15.2 lists the pre-analytical specimen processing factors that may affect plasma EVs. Moreover, ISEV established the EV-TRACK platform (www.evtrack.org) and recommends that EV researchers upload experimental details to provide a reference for subsequent researchers, improve the transparency of EV research, and gradually establish standardized and mature EV research methods.

**Table 15.2** Pre-analysis factors influencing biomarker detection

| Physiological factors | Specimen collection factors | Other interference factors |
|---|---|---|
| Age<br>Gender<br>Weight | Anticoagulant<br>Stabilizer<br>Tourniquet compression time | Circulating antibodies<br>Drug<br>Endogenous complement |
| Height | Needle model | Hemolysis |
| Lifestyle | Specimen temperature and time before centrifugation | Lipemia |
| Circadian rhythm changes | Centrifugal force, time, temperature | Jaundice |
| Blood pressure | Storage time, temperature | Chronic underlying disease |
| Pregnancy/ menstruation | Freeze-thaw times | Surgery or infection |

### 15.2.2 Strategies for Biomarker Selection

According to the ISEV 2018 recommendations, to ensure the consistency of EV purity and amount, EVs should be quantified in terms of total protein, total lipids, and the number of particles. Presently, for EV nucleic acids, there are two technologies commonly used for detection: next-generation sequencing and gene chip technology. By contrast, EV proteins and lipids are mostly monitored using various mass spectrometry methods. After obtaining high-throughput data, researchers should first check the quality of sequencing or mass spectrometry data. For EV RNA markers, researchers can perform quality control and quantitative polymerase chain reaction (qPCR) verification to ensure that the RNA sequencing results are credible according to the guidelines proposed by the ISEV in 2013 [23]. The quality of the sequencing results can also be checked individually based on the "Experts Consensus of Second-Generation Sequencing in Clinical Molecular Pathology Laboratory" and "Consensus on the Application of NGS Technology in Clinical Tumor Precision Diagnosis" documents. However, for EV protein spectrum results, there are still no relevant authoritative guidelines for reference. Researchers can preliminarily judge data quality based on the false discovery rate, segment matching error distribution, peptide segment quantity or length distribution, and protein molecule mass distribution [24].

When the quality of data is confirmed, researchers can analyze the data and screen for potential biomarkers. Initially, potential biomarkers are characterized statistically based on a cutoff expression level, significant *P* value, and large distribution difference between groups. Based on these findings, researchers can utilize bioinformatics methods, such as Gene Ontology, Kyoto Encyclopedia of Genes and Genomes, and protein interaction networks to analyze protein expression, function, localization, and other information from a systems biology perspective, enabling the identification of related key molecules. For EVs, there are also some specific databases, including exoRBase, ExoCarta, and Vesiclepedia, that can be used. If there is little information about the corresponding disease in the above three databases, large-scale platforms, such as The Cancer Genome Atlas, Gene Expression Omnibus, and Proteome Xchange, can also be used.

## 15.3 Verification of Biomarkers

In biomarker screening research, validation experiments are an indispensable step to determine whether the target biomarkers meet clinical needs [25]. In general, biomarker validation can be roughly divided into preliminary verification and clinical verification. Specifically, three steps are required for validation experiments. First, researchers can use the specimens during the screening stage for verification, eliminating the interference factors associated with the specimen source. Then, a suitable verification method is established for the initial screen of potential biomarkers. Sequentially, the established verification method can be utilized to detect new specimens and verify the repeatability and diagnostic efficacy of the identified markers.

However, biomarker validation often fails for various reasons, including variations among samples and detection methods. Furthermore, small sample volumes and insufficient statistical power can also cause verification failure [26]. Only biomarkers with well-validated clinical benefits can be used in the clinical setting to improve diagnostic and therapeutic capabilities, thereby enhancing patient health management and medical decision-making.

### 15.3.1 Preliminary Verification

In preliminary verification, a small sample volume is needed to establish appropriate detection methods (including fitting a combined diagnostic model) for further screening of multiple potential markers and preliminary verification. During this process, detection variations, poor-quality samples, sample variance, and other factors can affect the screening and verification of potential biomarkers. Therefore, researchers should fully understand the advantages and disadvantages of their chosen methods before screening and should select sensitive, specific techniques. Additionally, clinical samples that can represent the target population while controlling for age, sex, and other factors (common biases, Table 15.3) should also be considered to reduce interference, improve comparisons, and obtain answers to the clinical questions being asked. In addition to healthy controls, the control group should also include patients with other diseases who have similar biomarker expression profiles and/or clinical symptoms in order to improve the specificity of disease diagnosis.

**Table 15.3** Common biases in biomarker research

| Bias | Sample collection | | Examples |
|---|---|---|---|
| | Before | After | |
| Samples (ages, gender, drug) | √ | | The experimental group is all men, and the control group is all women<br>Bias: Gender differences |
| Sample collection process | √ | √ | The experimental group and control group come from different clinical research centers<br>Bias: Differences in diagnosis and sample processing |
| Sample storage and pretreatment | | √ | The experimental group collects samples for 10 years, and the control group costs 1 year because of different incidence rates<br>Bias: Differences in storage time and repeated freeze-thaw |
| Sample analysis | | √ | The experimental group and control group processed samples on different dates<br>Bias: Differences in analytical time and processing |

### 15.3.2 Clinical Verification

After preliminary verification, the potential biomarkers need to be clinically confirmed in a larger population to evaluate the sensitivity, specificity, detection limit, and other indicators. Through these indicators, the overall efficacy of potential biomarkers can be evaluated, and their potential applications in clinical diagnosis can be verified.

However, most biomarker studies have not included clinical specimen verification, and only a few researchers have reported large-sample verification [27] and multicenter independent verification [28]. The clinical reproducibility of potential biomarkers is the biggest bottleneck for their development. Among the various factors that interfere with the diagnostic efficacy of biomarkers, the most influential factors are pre-analysis factors, such as physiological status, specimen collection method, and disease-related factors (Table 15.3). Moreover, the absence of standard operating procedure documents to strictly control pre-analysis factors can cause specimen heterogeneity in various experimental stages and between research teams.

Accordingly, many researchers are calling for the following protocols to be in place: (1) strict management of specimen banks for biomarker research; (2) in biomarker studies, specimen collection and experimental methods should be clearly explained; and (3) third-party laboratory verification should be provided to ensure the repeatability of the results [29]. Furthermore, researchers can improve the accuracy of confirmation trials by setting appropriate inclusion criteria, observing a large number of samples, conducting multistage and multicenter verification, and evaluating retrospective and prospective results.

### 15.3.3 EV Detection Technology in Biomarker Validation

Suitable methods are essential for the verification of screened biomarkers. Differences in the methods used for the screening and validation of biomarkers may result in considerable system errors, which could affect the validation. EV biomarkers mainly include nucleic acids and proteins that are present in EVs.

In the detection of EV nucleic acids, qPCR or digital PCR is commonly used. The ISEV suggested using multiple RNAs together as references for both internal and external references [22]. During amplification, the external reference, e.g., *cel-miR-39*, *ath-miR-159a*, and *cel-line-4*, or an internal reference is recommended for calibration [30]. However, there is no recognized internal reference. Researchers can choose a suitable internal reference according to their experiment design after verifying its stability in each group [31]. Furthermore, in recent years, some new nucleic acid detection technologies, such as real-time fluorescent nucleic acid constant temperature amplification detection technology, local surface plasmon resonance-based nucleic acid analysis, and fluorescent nano-biosensing technology, have emerged and have resulted in improved EV nucleic acid detection. Zhang et al. proposed a robust electrochemical biosensor using multifunctional DNA tetrahedron-assisted catalytic hairpin assembly, which can used for quantification of exosome miRNAs down to 7.2 aM [32].

For detection of EV proteins, labeled/unlabeled protein mass spectrometry and protein chip technology are commonly applied to screen for differentially expressed EV proteins, followed by Western blotting, high-sensitivity flow technology, enzyme linked immunosorbent assay (ELISA), and parallel reaction monitoring. Although high-sensitivity flow cytometry technology has many advantages in the clinical detection of EV biomarkers, such as its high rapidity, sensitivity, and stability, this approach only targets EV membrane proteins, resulting in limited applications. Thus, Western blotting and ELISA are the most commonly used methods for verification of EV protein markers.

Microfluidic technology has recently been applied in various fields. Because of its high sensitivity, short analysis time, high throughput, strong controllability, and small sample volume [33], microfluidic technology has become popular in the field of laboratory diagnosis. Professor Zheng Lei from Southern Medical University and Professor Yao Shuhuai from Hong Kong University of Science and Technology reported for the first time that microdroplet technology can be used to achieve quantitative detection of breast cancer EVs harboring specific markers [34]. This method not only significantly improves the detection limit of EV biomarkers but also provides insights into single EV detection.

Although these new technologies can improve the sensitivity and specificity of EV biomarker detection, there are still many problems to overcome in order to apply these new technologies to EV biomarker validation. First, the clinical requirements and potential applications of the test should be clear. Second, selection of appropriate biomarkers and detection methods, comparison with classical methods (i.e., the gold standard method), and large-sample multicenter clinical trials can be used to certify the diagnostic efficacy of the method. Finally, a clinical diagnosis quality management system (e.g., ISO 15189) should be gradually established to further improve the clinical applications of the approach [35].

## 15.4 Conclusion and Perspectives

As an important feature of liquid biopsy analysis, EVs have potential applications in the diagnosis and treatment of diseases. EVs act as carriers of various molecules, thereby playing important physiological roles in cell information exchange and signal transduction. Their stability, high cargo content, and noninvasiveness sampling make EVs promising new circulating biomarkers for disease diagnosis. With improvements in EV detection technology, a multitude of disease-related EV biomarkers have been discovered; however, relatively few have been translated into the clinical setting. Additionally, several issues still need to be overcome in order to facilitate the clinical application of EV biomarkers. Specifically, no rapid, stable, high-yield, and clinically applicable EV purification methods have been developed for use with clinical specimens. This has greatly restricted the clinical application of EV biomarkers. Moreover, additional work is required for standardization of specimen storage, purification, and subsequent EV identification methods. Furthermore, because almost all cells in the body can secrete EVs, screening for tissue-specific EV biomarkers from complex body fluids or tissues is a major challenge. Finally, further studies are needed to elucidate the intracellular sorting mechanisms of EV molecules. Despite these limitations, we believe that additional developments in basic research on EVs and improvements in EV detection technology will allow EV detection to be applied in the clinical diagnosis and treatment of diseases.

## References


1. Nallagangula KS, Shashidhar KN, Lakshmaiah V, Muninarayana. Evolution of proteomic biomarker for chronic liver disease: promise into reality. Circ Biomark. 2018;7:1849454418777186.
2. Hoshino A, Kim HS, Bojmar L, et al. Extracellular vesicle and particle biomarkers define multiple human cancers. Cell. 2020;182(4):1044–61.
3. Gardner CC Jr. Pheochromocytoma: early diagnosis usually leads to cure. Am Pract Dig Treat. 1961;12:337–44.
4. Boukouris S, Mathivanan S. Exosomes in bodily fluids are a highly stable resource of disease biomarkers. Proteomics Clin Appl. 2015;9(3–4):358–67.
5. Wolfers J, Lozier A, Raposo G, et al. Tumor-derived exosomes are a source of shared tumor rejection antigens for CTL cross-priming. Nat Med. 2001;7(3):297–303.
6. Fevrier B, Vilette D, Archer F, et al. Cells release prions in association with exosomes. Proc Natl Acad Sci U S A. 2004;101(26):9683–8.
7. Escola JM, Kleijmeer MJ, Stoorvogel W, et al. Selective enrichment of tetraspan proteins on the internal vesicles of multivesicular endosomes and on exosomes secreted by human B-lymphocytes. J Biol Chem. 1998;273(32):20121–7.
8. Shao H, Im H, Castro CM, et al. New technologies for analysis of extracellular vesicles. Chem Rev. 2018;118(4):1917–50.
9. Sumrin A, Moazzam S, Khan AA, et al. Exosomes as biomarker of cancer. BRAZILIAN ARCHIVES OF BIOLOGY AND TECHNOLOGY; 2018.
10. Chen IH, Xue L, Hsu CC, et al. Phosphoproteins in extracellular vesicles as candidate makers for breast cancer. Proc Natl Acad Sci U S A. 2017;114(12):3175–80.
11. Melo SA, Luecke LB, Kahlert C, et al. Glypican-1 identifies cancer exosomes and detects early pancreatic cancer. Nature. 2015;523(7559):177–82.
12. Hoshino A, Costa-silv AB, Shen TL, et al. Tumour exosome integrins determine organotropic metastasis. Nature. 2015;527(7578):329–35.
13. Chen G, Huang AC, Zhang W, et al. Exosomal PD-L1 contributes to immunosuppression and is associated with anti-PD-1 response. Nature. 2018;560(7718):382–6.
14. Meng X, Müller V, Milde-Langosch K, et al. Diagnostic and prognostic relevance of circulating exosomal miR-373, miR-200a, miR-200b and miR-200c in patients with epithelial ovarian cancer. Oncotarget. 2016;7(13):16923–35.
15. Yu J, Jin L, Jiang L, Gao L, et al. Serum miR-372 is a diagnostic and prognostic biomarker in patients with early colorectal cancer. Anti Cancer Agents Med Chem. 2016;16(4):424–31.
16. Barker AD, Compton CC, Poste G. The National Biomarker Development Alliance accelerating the translation of biomarkers to the clinic. Biomark Med. 2014;8(6):873–6.
17. Califf RM. Biomarker definitions and their applications. Exp Biol Med (Maywood). 2018;243(3):213.
18. Horvatovich P, Bischoff R. Comprehensive biomarker discovery and validation for clinical application. Royal Society of Chemistry; 2013. ISBN: 978-1-84973-422-6.
19. Pegtel DM, Gould SJ. Exosomes. Annu Rev Biochem. 2019;88:487–514.

20. Witwer KW, Buzas EI, Bemis LT, et al. Standardization of sample collection, isolation and analysis methods in extracellular vesicle research. J Extracell Vesicles. 2013;2(1):20360.
21. Yuana Y, Boding AN, Grootemaat AE, et al. Handling and storage of human body fluids for analysis of extracellular vesicles. J Extracell Vesicles. 2015;4:29260.
22. Thery C, Witwer KW, Aikawa E, et al. Minimal information for studies of extracellular vesicles 2018 (MISEV2018):a position statement of the International Society for Extracellular Vesicles and update of the MISEV2014 guidelines. J Extracell Vesicles. 2018;7(1):1535750.
23. Hill AF, Pegtel DM, Lambertz U, et al. ISEV position paper: extracellular vesicle RNA analysis and bioinformatics. J Extracell Vesicles. 2013;2(1):22859.
24. Geyer PE, Holdt LM, Teupser D, et al. Revisiting biomarker discovery by plasma proteomics. Mol Syst Biol. 2017;13(9):942.
25. Rayyan M, Zheutlin A, Byrd JB. Clinical research using extracellular vesicles: insights from the International Society for Extracellular Vesicles 2018 annual meeting. J Extracell Vesicles. 2018;7(1):1535744.
26. Ransohoff DF. Lessons from controversy: ovarian cancer screening and serum proteomics. J Natl Cancer Inst. 2005;97(4):315–9.
27. Mensa E, Guescini M, Giuliani A, et al. Small extracellular vesicles deliver miR-21 and miR-217 as pro-senescence effectors to endothelial cells. J Extracell Vesicles. 2020;9(1):1725285.
28. Woo J, Santasusagna S, Banks J, et al. Urine extracellular vesicle GATA2 mRNA discriminates biopsy result in men with suspicion of prostate cancer. J Urol. 2020;204(4):691–700.
29. Barker AD, Compton CC, Poste G. The national biomarker development alliance accelerating the translation of biomarkers to the clinic. Biomark Med. 2014;8(6):873–6.
30. Bustin SA, Benes V, Garson JA, et al. The MIQE guidelines: minimum information for publication of quantitative real-time PCR experiments. Clin Chem. 2009;55(4):611–22.
31. Shimomura A, Shino S, Kawauchi J, et al. Novel combination of serum microRNA for detecting breast cancer in the early stage. Cancer Sci. 2016;107(3):326–34.
32. Zhang Y, Zhang X, Situ B, et al. Rapid electrochemical biosensor for sensitive profiling of exosomal microRNA based on multifunctional DNA tetrahedron assisted catalytic hairpin assembly. Biosens Bioelectron. 2021;183:113205.
33. Iliescu FS, Vrtacnik D, Neuzil P, et al. Microfluidic technology for clinical applications of exosomes. Micromachines (Basel). 2019;10(6):392.
34. Liu C, Xu X, Li B, et al. Single-exosome-counting immunoassays for cancer diagnostics. Nano Lett. 2018;18(7):4226–32.
35. Ayers L, Pink R, Carter DRF, et al. Clinical requirements for extracellular vesicle assays. J Extracell Vesicles. 2019;8(1):1593755.

# Extracellular Vesicles as Biomarkers for Cancer

# 16

Shuhong Wang, Yusheng Lin, Jia Hu, Weixiao Fan,
Lijuan Yu, Xiaoke Hao, Shenglin Huang, and Hao Zhang

## 16.1 Extracellular Vesicle (EV) Biomarkers for Cancers of the Digestive Tract

Owing to the minimally invasive nature of biofluid collection, liquid biopsy has promising applications in the diagnosis, management, surveillance, and prognosis of malignancies. EVs derived from biofluids have emerged as potential noninvasive

S. Wang · H. Zhang (✉)
Institute of Precision Cancer Medicine and Pathology, Jinan University Medical College, Guangzhou, Guangdong, People's Republic of China

Y. Lin
Institute of Precision Cancer Medicine and Pathology, Jinan University Medical College, Guangzhou, Guangdong, People's Republic of China

Department of Hematology, University Medical Center Groningen, University of Groningen, Groningen, the Netherlands

J. Hu · S. Huang (✉)
Fudan University Shanghai Cancer Center and Institutes of Biomedical Sciences, Fudan University, Shanghai, People's Republic of China
e-mail: slhuang@fudan.edu.cn

W. Fan
Department of Clinical Microbiology, Greater Xi'an Area Medical Diagnostic Lab Center, Xi'an, Shaanxi, People's Republic of China

L. Yu
Department of Clinical Laboratory Medicine, Xijing Hospital, Fourth Military Medical University (Air Force Medical University), Xi'an, Shaanxi, People's Republic of China

X. Hao (✉)
Division of Life Sciences and Medicine, Northwest University, Xi'an, Shaanxi, People's Republic of China

Greater Xi'an Area Medical Diagnostic Lab Center, Xi'an, Shaanxi, People's Republic of China



Q. Wang, L. Zheng (eds.), *Extracellular Vesicles*,
https://doi.org/10.1007/978-981-99-8365-0_16

cancer biomarkers that can reflect their cellular origins. Notably, the presence of a membrane ensures that the EV molecular cargo is highly stable, thereby avoiding variations caused by exogenous factors [1]. EVs contain different forms of RNA, including mRNAs, chimeric RNAs, rRNA, and long and small noncoding RNA (ncRNAs), such as microRNAs (miRNAs), tRNA fragments, piwi-interacting RNAs, vault RNAs, and Y RNAs [2, 3]. Additionally, EVs are also enriched in tumor-derived DNA and proteins [3]. Many molecules in EVs differ significantly in biofluids from patients with cancer compared with those from healthy individuals, allowing for biomarker identification in many cancer types [4–10].

Commonly used biomarkers for cancers of the digestive tract, such as carbohydrate antigen 125 (CA125), CA 19-9, squamous cell carcinoma antigen, alpha fetoprotein (AFP), tissue inhibitor of metalloproteinases 1, cytokeratin 19 fragments, and carcinoembryonic antigen, have unsatisfactory specificity and sensitivity and thus interfere with diagnostic and prognostic accuracy. Present studies support that EV-based approaches show increased detection sensitivity and specificity compared with traditional tests. In this chapter, we describe and discuss EV biomarkers for digestive tract cancers.

### 16.1.1 Esophageal Cancer (EC)

EC is one of the most common malignancies of the digestive tract [11]. EC can be classified as esophageal squamous cell carcinoma (ESCC) or esophageal adenocarcinoma (EA) according to the pathological characteristics of cancer. In 2018, more than 572,000 people were newly diagnosed with EC worldwide, and ESCC still accounts for the vast majority of all EC cases [11]. Because outcomes are better when EC is treated early, much effort has been devoted to early detection and surveillance. However, EC diagnosis and prognostic prediction are faced with several challenges, including late appearance of symptoms, costly and invasive endoscopy procedures, radiation risk, and relatively insensitive and inaccurate biomarkers [8, 12]. Therefore, it is important to identify new, sensitive, less invasive, and cost-effective biomarkers for EC diagnosis and prognosis. Increasing evidences suggest that EV RNAs and proteins can be applied clinically as noninvasive biomarkers for prognosis prediction, early diagnosis, and treatment response monitoring in patients with EC (Table 16.1), as discussed below.

#### 16.1.1.1 EV RNAs in EC

Tanaka et al. provided the first evidence of *miR-21* enrichment in EVs in the serum of patients with ESCC. Their study demonstrated that EV *miR-21* may participate in ESCC progression, revealing an association between circulatory exosomal *miR-21* and ESCC diagnosis and progression [13]. *MiR-1246*, one of the most widely studied oncogenic miRNAs, was estimated in the peripheral blood of patients with ESCC, showing good performance in diagnostic and prognostic evaluations [14]. Furthermore, *miR-93-5p* is another cancer-associated miRNA that is dysregulated in many malignant tumors and can be transferred between EC9706 cells by

**Table 16.1** Potential EV biomarkers in EC

| Cancer type | Cargoes | EV source | Patient cohorts | Detection method | Application | Tendency | Year | References |
|---|---|---|---|---|---|---|---|---|
| EC | RNA | | | | | | | |
| | miR-21 | Serum | 51 EC patients and 41 patients with benign diseases | qRT-PCR | Diagnostic and prognostic biomarker | Up | 2013 | [13] |
| | MiR-1246 | Serum | 46 healthy controls and 101 ESCC patients | miRNA microarray and qRT-PCR | Diagnostic and prognostic biomarker | Up | 2013 | [14] |
| | MiR-93-5p | Serum | 84 healthy controls and 84 ESCC patients | qRT-PCR | Diagnostic and prognostic biomarker | Up | 2018 | [15] |
| | LncRNA PCAT1 | Serum | 39 paired ESCC samples and adjacent normal tissues of patients with ESCC | RT-qPCR and semiquantitative PCR | Diagnostic biomarker | Up | 2019 | [16] |
| | chr8-23234-3p,chr1-17695-5p, chr 8-2743-5p, and miR-432-5p | Serum | Discovery phase ($N = 20$); training phase ($N = 178$); validation phase ($n = 188$) | RNA sequencing and qRT-PCR | LN metastasis prediction biomarker | Up | 2020 | [17] |
| | PROTEIN stathmin-1 | Serum | 364 ESCC; 53 HCC; 59 GC; 57 CRC; 48 HNSCC; 53 LAC; 52LSCC patients | Western blotting and AlphaLISA | Diagnostic and lymph node metastases | Up | 2018 | [18] |

EVs, thereby promoting the proliferation of recipient EC9706 cells [15]. The expression of *miR-93-5p* was significantly higher in patients with EC than in healthy participants [15]. Upregulation of plasma *miR-93-5p* in patients with ESCC increases the risk of EC [15]. Therefore, *miR-93-5p* could be identified as a plasma EV biomarker for the diagnosis and prognosis of EC. Serum EV *miR-182* and *miR-766-3p* are two potential biomarkers for predicting the unfavorable prognosis of ESCC [19, 20]. The long ncRNA (lncRNA) *PCAT1*, present in ESCC-derived EVs, showed higher expression in the serum of patients with ESCC and may serve as a noninvasive biomarker for ESCC [16]. According to a recent study, a four-lncRNA panel (*UCA1*, *POU3F3*, *ESCCAL-1*, and *PEG10*) in EVs has considerable value for ESCC diagnosis, and *POU3F3* may serve as a novel and independent prognostic predictor in clinical applications [21]. A four-stage study, including screening, training, testing, and validation, identified that plasma EV lncRNAs NR_039819, NR_036133, NR_003353, ENST00000 442416.1, and ENST00000416100.1 may serve as useful tumor biomarkers for ESCC diagnosis [22]. Instead of blood tests, salivary EV chimeric RNA was first proposed as a disease biomarker by Lin et al. [8] *GOLM1-NAA35* expression in salivary EVs can serve as a noninvasive and convenient marker for ESCC detection, even during the early stages of the disease. To identify patients with ESCC with lymph node (LN) metastasis, four differentially expressed miRNAs (*chr 8-23234-3p*, *chr 1-17695-5p*, *chr 8-2743-5p*, and *miR-432-5p*) were tested in serum EVs [17]. Subsequently, an optimized four-EV miRNA model was constructed, which could effectively identify patients with ESCC with LN metastasis and was significantly superior to the preoperative computed tomography report [17]. Plasma EV levels of *hsa_circ_00021946* and *hsa_circ_0043603* were used as diagnostic biomarkers for ESCC, revealing the roles of EV circular RNAs (circRNAs) in ESCC [12].

EVs are also important biomarkers for the diagnosis and progression of EA. An EV multi-biomarker panel consisting of five specific miRNA ratios (*RNU6-1/miR-16-5p*, *miR-25-3p/miR-320a*, *let-7e-5p/miR-15b-5p*, *miR-30a-5p/miR-324-5p*, and *miR-17-5p/miR-194-5p*) was proposed by Chiam et al. to distinguish EA from controls and Barrett's esophagus [23]. The potential advantage of combining multiple biomarkers with a single biomarker was demonstrated in this study. Warnecke-Eberz et al. reported that overexpression of serum EV *miR-223-5p* and *miR-483-5p* in patients with EA and *miR-223-5p* expression was associated with tumor development stage [24]. In 2017, Chen et al. isolated EVs from serum samples of patients with EA and normal controls and reported that *miR-21-5p*, *miR-16-5p*, *miR-25-3p*, and *miR-155-5p* were significantly upregulated, whereas *miR-192-5p* was significantly downregulated in patients with EA, suggesting the potential applications of these miRNAs as biomarkers [25]. Furthermore, there is still debate on whether plasma or serum is a better source of EVs for clinical applications. Chiam et al. reported that EVs isolated from serum are less contaminated with blood cells and protein-associated miRNAs, indicating that serum is preferable for studies of cancer-associated EV biomarkers [26]. They also compared the diagnostic performance of the previously identified multi-miRNA biomarker panel for EA and produced a higher cross-validated accuracy for discriminating patients with EA from healthy controls using serum EV preparations compared with plasma EV preparations [23, 26].

#### 16.1.1.2 EV Proteins in EC

Using nearly 1000 clinical samples, Yan et al. showed that serum stathmin-1 levels were dramatically elevated in patients with squamous cell carcinoma (SCC), particularly ESCC, with a sensitivity and specificity of 81% and 94%, respectively [18]. Serum stathmin-1 achieved an area under the receiver operating characteristic (ROC) curve (AUC) of 0.88 for early-stage ESCC, and higher stathmin-1 expression was shown to be related to LN metastasis. Furthermore, stathmin-1 is abundant in EVs and can be released into circulation via EVs [18]. Serum stathmin-1 levels were found to exhibit excellent diagnostic capability for ESCC and may serve as a risk assessment marker for LN metastasis [18]. In addition to the EV protein content, EV-specific proteins can be used to evaluate the concentration of EVs. Indeed, Zhao et al. evaluated CD9 expression in EVs isolated from serum samples of patients with ESCC and revealed that the serum concentration of circulating EVs was significantly higher in patients with ESCC than in healthy controls and could distinguish patients with ESCC with a sensitivity of 75% and a specificity of 85%; moreover, increased levels of circulating exosomes were also correlated with poor overall survival and progression-free survival [27].

Large-scale validation studies of EV biomarkers for EC have not yet been conducted. Prospective studies on EV biomarkers are needed to substantiate the diagnostic and prognostic value of these biomarkers in the clinical setting.

### 16.1.2 Laryngeal SCC (LSCC)

LSCC is one of the most common forms of head and neck cancers [28]. Present therapeutic strategies, including surgery, chemoradiation, gene therapy, and immunotherapy, have a moderate effect on early-stage cases; however, the therapeutic outcomes are still unsatisfactory, and the overall 5-year survival rate remains poor for patients with advanced LSCC. Early detection of LSCC is challenging because of the anatomical location of the larynx, providing the opportunity for a tumor to be more hidden. Furthermore, the traditional biopsy procedure for LSCC is invasive, and novel, minimally invasive, and highly sensitive markers are attractive alternatives. Here, we present an assessment of EV-based biomarkers for LSCC, with a focus on techniques available in the clinical setting (Table 16.2).

#### 16.1.2.1 EV RNA in LSCC

Wang et al. demonstrated that the expression of EV *miR-21* and *HOTAIR* was markedly higher in patients with LSCC than in those with benign laryngeal disease and was significantly associated with the clinical parameters of LSCC [29]. They also showed that the combined evaluation of *miR-21* and *HOTAIR* from serum EVs was superior to that of EV *miR-21* or EV *HOTAIR* alone in screening for LSCC, suggesting a promising tool for predicting prognosis in patients with LSCC [29]. In

**Table 16.2** Potential EV biomarkers in LSCC

| Cancer type | Cargoes | EV source | Patient cohorts | Detection method | Application | Tendency | Year | References |
|---|---|---|---|---|---|---|---|---|
| **LSCC** | **RNA** | | | | | | | |
| | miR-21 and HOTAIR | Serum | 52 LSCC patients and 49 patients with polyps | qRT-PCR | Diagnostic and prognostic biomarker | Up | 2014 | [29] |
| | miR-941 | Plasma | 59 LSCC patients and 31 HC subjects | RNA-seq analysis and qRT-PCR | Diagnostic biomarker and therapeutic target | Up | 2020 | [30] |

addition to *miR-21*, EV *miR-941* was also shown to be highly expressed in the plasma of patients with LSCC compared with that in controls, and ROC curve analysis showed that *miR-941* has an AUC of 0.797 for distinguishing patients with LSCC from controls [30]. They further demonstrated the functional role of *miR-941* in promoting cell proliferation and invasion in LSCC [30]. Intriguingly, a recent study demonstrated that EV *circRASSF2* promoted LSCC progression, and its expression in serum was significantly higher in patients with LSCC than that in control groups, indicating the potential role of EV *circRASSF2* as a novel clinical molecular marker for the diagnosis of LSCC [28].

### 16.1.3 Tongue SCC (TSCC)

TSCC, the most common malignancy of the oral cavity, is known for its high mortality rate, and many patients develop recurrence and metastasis [31]. There are typically no alarming symptoms during TSCC progression, resulting in delays in diagnosis and thereby leading to a poor prognosis. Thus, there is a need for robust diagnostic and prognostic biomarkers that can accurately identify patients with early TSCC and predict its behavior.

In this section, we describe the characteristics and potential clinical implications of each EV biomarker in TSCC (Table 16.3).

#### 16.1.3.1 EV RNA in TSCC

In 2017, Rabinowits et al. presented the first study comparing the expression patterns of miRNAs in TSCC; they matched benign tissue and plasma samples (both free and within EVs) from the same patients to identify a signature pattern of miRNAs that could serve as a "liquid biopsy" for TSCC [33]. They identified an aberrantly expressed pattern of miRNAs in both the tumor and plasma of patients with TSCC relative to that in matched benign tissue. In addition, compared with circulating free miRNA, the EV miRNA expression pattern was much more consistent with the tumor miRNA expression pattern. Although the applicability of this EV miRNA expression pattern as a promising biomarker for TSCC remains unclear, the findings support the potential utility of EV miRNAs in evaluation of the diagnosis and treatment response in patients with TSCC [33]. A recent study focusing on chemoresistance in TSCC showed that EV *miR-200c* effectively enhances the chemosensitivity of TSCC cells to docetaxel, thus providing a potential diagnostic marker for TSCC [32].

#### 16.1.3.2 EV Proteins in TSCC

In a study by Sanada et al., lysyl oxidase like 2 (LOXL2) protein levels in serum EVs from patients with TSCC and low-grade TSCC were elevated compared with those in healthy controls; however, the value of serum EV LOXL2 as a promising diagnostic marker still needs to be explored in a large cohort study [34].

**Table 16.3** Potential EV biomarkers in TSCC

| Cancer type | Cargoes | EV source | Patient cohorts | Detection method | Application | Tendency | Year | Reference |
|---|---|---|---|---|---|---|---|---|
| **TSCC** | **RNA** | | | | | | | |
| | miR-200c | Serum | | RNA-Seq and qRT-PCR | Regulator of chemoresistance | Up | 2020 | [32] |

### 16.1.4 Gingival Cancer and Buccal Carcinoma

Gingival cancers and buccal carcinomas are rare diseases worldwide. They most commonly affect elderly populations. Biomarker studies on these cancers have not yielded conclusive information.

### 16.1.5 Colorectal Cancer (CRC)

CRC is one of the leading causes of cancer-related deaths worldwide. Recent studies have indicated that exosomes may be important sources of CRC biomarkers. Moreover, some miRNAs, lncRNAs, and proteins in EVs isolated from the blood are significantly different between patients with CRC and normal controls. These exosomal molecules may serve as diagnostic indicators for CRC.

Exploring the correlations between exosomes and CRC may be beneficial for developing novel biomarkers for CRC diagnosis (Table 16.4).

#### 16.1.5.1 EV Proteins in CRC

Many studies have demonstrated differential expression of proteins in the EVs of patients with CRC. In one study, 36 proteins were found to be upregulated, whereas 22 were downregulated in the serum exosomes of patients with CRC [39]. Campanella et al. reported that heat shock protein 60 (HSP60) in plasma exosomes of patients with CRC is a promising candidate for CRC diagnosis [35]. Another study showed that the percentage of glypican 1 $(GPC1)^+$ exosomes and the level of GPC1 protein expression in exosomes from tumor tissues and plasma of patients with CRC before surgical treatment were significantly higher than those in peritumoral tissues and the plasma of healthy controls. Increased plasma $GPC1^+$ exosomes are specific markers for the diagnosis of CRC and may be therapeutic targets in patients with CRC [36]. A similar study showed that the levels of transmembrane protein CD147-positive exosomes were significantly elevated in the plasma of patients with CRC, and ROC curve analysis illustrated that the concentration of CD147-positive exosomes in the plasma showed excellent diagnostic ability, with an AUC of 0.932 [40]. Thus, these blood exosomal proteins may be potential diagnostic candidates in CRC.

#### 16.1.5.2 EV RNAs in CRC

ncRNAs play vital roles in the regulation of CRC development and drug resistance and could serve as biomarkers for CRC diagnosis and prognostic prediction.

Serum exosomal levels of seven miRNAs (*let-7a*, *miR-1229*, *miR-1246*, *miR-150*, *miR-21*, *miR-223*, and *miR-23a*) are significantly higher in patients with primary CRC, even in the early stages of the disease, compared with that in healthy controls and are significantly downregulated after surgical resection of tumors. These miRNAs are also secreted at significantly higher levels in CRC cell lines than in normal colon cancer cell lines. The high sensitivities of the seven selected exosomal miRNAs were confirmed using ROC analysis [41]. These results indicate that several

**Table 16.4** Potential EV biomarkers in CRC

| Cancer type | Cargoes | EV source | Patient cohorts | Detection method | Application | Tendency | Year | References |
|---|---|---|---|---|---|---|---|---|
| **CRC** | **PROTEIN** | | | | | | | |
| | HSP60 | Plasma | 57 CRC patients 40 matched subjects | Western blot and IHC | Diagnosis biomarker and therapeutic target | Up | 2019 | [35] |
| | GPC1 | Plasma | 102 tumor tissues and 89 normal colon tissues | Western blot | Diagnosis biomarker and therapeutic target | Up | 2017 | [36] |
| | miR-19a-3p, miR-21–5p, and miR-425–5p | Serum | Training (30 CRC vs. 30 NCs) and testing stages (136 CRC vs. 90 NCs) | qRT-PCR | Diagnostic biomarker | Up | 2020 | [37] |
| | lncRNA LINC02418 | Serum or plasma | 155 CRC patients and from 145 healthy volunteers | qRT-PCR | Diagnostic biomarker | Up | 2019 | [38] |

miRNAs are promising biomarkers for CRC and can be evaluated using a noninvasive diagnostic approach.

Zhu et al. performed miRNA microarray analyses to profile exosomal RNAs from the serum and showed that the serum levels of three miRNAs (*miR-19a-3p*, *miR-21-5p*, and *miR-425-5p*) were significantly higher in patients with CRC than in healthy controls. The ROCs of the three-miRNA panel were 0.86, 0.74, and 0.87 for the training, testing, and external validation stages, respectively, confirming the applicability of this serum three-miRNA panel for the diagnosis of CRC [37].

A study of six lncRNAs (*LNCV6_116109*, *LNCV6_98390*, *LNCV6_38772*, *LNCV_108266*, *LNCV6_84003*, and *LNCV6_98602*) indicated that the expression of these lncRNAs was significantly upregulated in patients with CRC compared with that in healthy individuals. Furthermore, the AUCs of these lncRNAs were 0.770, 0.7500, 0.6500, 0.6900, 0.7500, and 0.7200, respectively, suggesting that they may serve as potential noninvasive biomarkers for early diagnosis of CRC [42]. In another study, Zhao et al. reported that *LINC02418* was highly expressed in CRC tissues and cell lines. More importantly, the AUC of exosomal *LINC02418* (0.897) was higher than that of cell-free *LINC02418* (0.678), suggesting that exosomal ncRNAs may be more suitable as diagnostic biomarkers in patients with CRC [38].

In addition to miRNAs and lncRNAs, exosomal circRNAs in peripheral blood are considered emerging diagnostic biomarkers of cancers. Pan et al. reported that exosomal *hsa_circ_0004771* could distinguish patients with CRC at stage I/II from healthy individuals (AUC, 0.86; sensitivity, 81.43%; and specificity, 80%), supporting potential applications as an early detection biomarker for CRC [43].

Although exosomes have many advantages as novel tumor markers, including their sensitivity and specificity, abundance in various body fluids, and noninvasive sampling procedures, it is necessary to develop more efficient and uniform separation methods to facilitate the translation of exosome detection to clinical applications. More effort is needed to resolve all possible sources of variation when attempting to utilize exosomes as biomarkers.

### 16.1.6 Gastric Cancer (GC)

GC is one of the leading causes of cancer-related deaths in both men and women owing to delayed diagnosis and high metastatic frequency. EVs deliver biomolecules, such as proteins, nucleic acids, metabolites, and pharmacological compounds, that can participate in the physiological and pathological processes of GC, including tumor growth, metastasis, immune response and evasion, chemoresistance, and treatment response [43]. Furthermore, EVs derived from patients could serve as prognostic and diagnostic biomarkers for GC (Table 16.5).

#### 16.1.6.1 EV Proteins in GC

Fu et al. found that the expression levels of tripartite motif-containing 3 (TRIM3) mRNA and protein were decreased in GC tissues compared with those in matched

**Table 16.5** Potential EV biomarkers in GC

| Cancer type | Cargoes | EV source | Patient cohorts | Detection method | Application | Tendency | Year | References |
|---|---|---|---|---|---|---|---|---|
| **GC** | **Protein** | | | | | | | |
| | TRIM3 | Serum | 80 gastric cancer patients and 80 healthy volunteers | ELISA and Western blot | Diagnostic biomarker and therapy target | | 2018 | [44] |
| | **RNA** | | | | | | | |
| | circSHKBP1 | Serum | Blood samples of 20 GC patients and 20 health controls; tumors and paired normal tissues of 72 cases were frozen tissues | qRT-PCR | Diagnosis and progression biomarker and therapeutic target | | 2020 | [45] |
| | miR-374a-5p | Serum | 59 GC patients and 17 gastritis patients, and 17 healthy donors | qRT-PCR | Diagnosis and drug resistance therapy target | | 2015 | [46] |
| | circKIAA1244 | Plasma | 10 GC patients and 5 healthy controls | RNA-FISH and RIP | Diagnosis biomarker | | 2018 | [47] |
| | lncHOTTIP | Serum | 126 GC patients and 120 healthy people | qRT-PCR | Diagnosis and prognosis biomarker | | 2018 | [48] |

control tissues. In addition, TRIM3 protein levels in the serum exosomes of patients with GC were lower than those in healthy controls, suggesting that exosomal TRIM3 may serve as a biomarker for GC diagnosis [44].

Yoon et al. reported that gastrokine 1 (GKN1) was secreted and internalized in the gastric epithelium by exosome-driven transfer, which inhibited gastric tumorigenesis, supporting the clinical application of GKN1 protein in GC diagnosis and treatment [49].

#### 16.1.6.2 EV RNAs in GC

Exosomal miRNAs are considered promising biomarkers for cancer diagnosis. Wei et al. found that *exo-miR-15b-3p* promotes GC development and GES-1 cell malignant transformation. Furthermore, serum *exo-miR-15b-3p* may be a potential biomarker for GC diagnosis and prognosis and can be used in precise targeted GC therapy [46]. *MiR-374a-5p* promotes GC chemoresistance and could therefore be a novel biomarker for GC diagnosis and prognosis [46].

CircRNAs play vital roles in human cancers by serving as miRNA sponges and modulating the expression of protein-encoding genes that regulate RNA-binding proteins. A study of *circSHKBP1* found that exosomal *circSHKBP1* regulates the *miR-582-3p*/Hu antigen R/vascular endothelial growth factor pathway, suppresses HSP90 degradation, and promotes GC progression, suggesting that *circSHKBP1* is a promising circulating biomarker for GC diagnosis and prognosis and an exceptional candidate for further therapeutic exploration [45]. Tang et al. demonstrated that decreased expression of *circ-KIAA1244* in the plasma was negatively correlated with TNM stage, lymphatic metastasis, and shorter overall survival in patients with GC. Moreover, *circ-KIAA1244* could be detected in GC plasma exosomes and showed no significant difference compared with the expression levels observed in corresponding control plasma samples. Thus, GC-tissue-derived *circ-KIAA1244* could serve as a novel circulating biomarker for detection of GC [47].

LncRNAs, such as miRNAs, function as important regulators of gene regulatory networks and play crucial roles in cancer progression [50]. Guo et al. revealed that circulating exosomal GC-associated lncRNA1 is a biomarker for early detection by determining its diagnostic efficiency based on the AUC, specificity, and sensitivity [51]. Additionally, Lin et al. evaluated the diagnostic accuracy of exosomal *lncUEGC1* in discriminating patients with early GC (EGC) from healthy individuals and those with premalignant chronic atrophic gastritis. They found that exosomal *lncUEGC1* may have applications as a highly sensitive, stable, and noninvasive biomarker for EGC diagnosis [52]. Zhao et al. found a correlation between increased exosomal *HOTTIP* levels and poor overall survival. Further univariate and multivariate Cox analyses revealed that overexpression of the exosomal lncRNA *HOTTIP* is an independent prognostic factor in patients with GC. These findings demonstrated that exosomal *HOTTIP* may be a potential biomarker for GC diagnosis and prognosis [48].

#### 16.1.6.3 EV DNA in GC

A study on BarH-like 2 homeobox protein (BARHL2) showed that BARHL2 methylation generated an AUC of 0.923, with 90% sensitivity and 100% specificity for distinguishing patients with GC from healthy controls when analyzing gastric juice-derived exosome DNA samples, indicating that methylation analysis of BARHL2 from gastric juice-derived exoDNA could be useful for early detection of GC in the clinical setting [53].

Despite increased focus on EVs used as diagnostic biomarkers in GC, there are still several challenging problems to be addressed, including the establishment of standardized methods for the isolation, quantification, and analysis of exosomes in the body fluids of patients with GC; the detection of GC-specific exosomal DNA, RNA species, or proteins in the circulation; and the identification of exosomes from the optimal donor cells for drug loading and large-scale preparation, storage, and formulation. Therefore, more effort is required to address this issue.

### 16.1.7 Pancreatic Cancer

Pancreatic cancer arises when cells in the pancreas, a glandular organ behind the stomach, begin to multiply out of control and form a mass. Pancreatic cancer is aggressive, and few symptoms are detectable until the cancer is already at an advanced stage; thus, the survival rate in patients with pancreatic cancer is poor, and novel molecularly targeted therapies and companion diagnostics are required. EVs, including exosomes and microvesicles, play crucial roles in pancreatic cancer progression and chemoresistance (Table 16.6). Furthermore, these EVs could be promising biomarkers, therapeutic targets, and tools for treating pancreatic cancer.

#### 16.1.7.1 EV Proteins in Pancreatic Cancer

Pancreatic ductal adenocarcinoma (PDAC) is a lethal malignancy with a poor prognosis owing to late diagnosis and a lack of reliable prognostic biomarkers. Kimura et al. found that cytoskeleton-associated protein 4 (CKAP4) is secreted by small EVs (sEVs) from PDAC cells, and these sEVs exhibit the characteristics of exosomes. Furthermore, serum CKAP4 levels are higher in patients with PDAC than in healthy controls. CKAP4 is highly expressed in serum samples from mice bearing pancreatic tumor xenografts and from patients with PDAC, whereas CKAP4 is barely detectable in serum samples from normal mice and from patients who have undergone surgery. Accordingly, CKAP4 secreted in exosomes may represent a biomarker for PDAC, and anti-CKAP4 antibodies could contribute to the development of novel diagnostic methods and therapeutics [54].

#### 16.1.7.2 EV RNAs in Pancreatic Cancer

Xu et al. found that *miR-196a* and *miR-1246* are highly enriched in pancreatic cancer exosomes. Consistent with these findings, plasma exosome *miR-196a* and *miR-1246* levels are significantly elevated in patients with pancreatic cancer

**Table 16.6** Potential EV biomarkers in pancreatic cancer

| Cancer type | Cargoes | EV source | Patient cohorts | Detection method | Application | Tendency | Year | References |
|---|---|---|---|---|---|---|---|---|
| **PC** | **Protein** | | | | | | | |
| | CKAP4 | Serum | | ELISA | Diagnosis biomarker and therapy target | Up | 2019 | [54] |
| | **RNA** | | | | | | | |
| | miR-196a and miR-1246 | Plasma | 15 cancer patients and 15 healthy subjects | RNA sequencing and qRT-PCR | Diagnosis biomarker | Up | 2017 | [55] |
| | miR-155, 181a,181b, 196a | Plasma | 140 PCa patients and 111 chronic pancreatitis patients | qRT-PCR | Diagnostic biomarker | Up | 2012 | [56] |
| | miR-1246 and miR-4644 | Salivary | 12 pancreatobiliary tract cancer patients and 13 healthy control participants | qRT-PCR | Diagnostic biomarker | Up | 2016 | [57] |

compared with those in healthy individuals, suggesting that these miRNAs may have the potential to be utilized in the diagnosis of pancreatic cancer [55].

In another study, Goto et al. found that the expression of exosomal *miR-191*, *miR-21*, and *miR-451a* was significantly higher in patients with pancreatic cancer and intraductal papillary mucinous neoplasm (IPMN) compared with that in healthy donors. Furthermore, ROC curve analysis showed that the AUC and diagnostic accuracy of ExmiRs were 5–20% superior to those of three serum bulky circulating miRNAs (e.g., *ExmiR-21*, AUC: 0.826, accuracy: 80.8%; circulating *miR-21*, AUC: 0.653, accuracy: 62.3%). These results indicated that the three ExmiRs could serve as markers for early diagnosis and progression in patients with pancreatic cancer and IPMN and could be more useful markers than circulating miRNAs (limited to these three miRNAs) [58].

Liu et al. extracted plasma RNAs from 140 patients with pancreatic cancer, 111 patients with chronic pancreatitis (CP), and 68 healthy controls; the relative abundances of seven miRNAs (*miR-16*, *miR-21*, *miR-155*, *miR-181a*, *miR-181b*, *miR-196a*, and *miR-210*) were measured. All seven miRNAs were significantly upregulated in patients with pancreatic cancer compared with those in the CP and normal groups, among which only four miRNAs (*miR-155*, *miR-181a*, *miR-181b*, and *miR-196a*) were significantly different. Furthermore, only *miR-16* and *miR-196a* played independent roles in discriminating patients with pancreatic cancer from normal controls and patients with CP. The combination of *miR-16*, *miR-196a*, and CA19-9 was more effective at discriminating pancreatic cancer. Moreover, the combination was effective in the identification of stage 1 tumors, indicating that the combination of *miR-16*, *miR-196a*, and CA19-9 may be effective for pancreatic cancer diagnosis, particularly in early tumor screening [56].

In addition to exosomes extracted from serum, unstimulated whole saliva samples collected from 12 patients with pancreatobiliary tract cancer and 13 healthy control participants were analyzed as cancer and control groups, respectively. The relative expression ratios of *miR-1246* and *miR-4644* in salivary exosomes were significantly higher in the cancer group than in the control group. For *miR-1246*, the results yielded an AUC of 0.814 ($P = 0.008$). For *miR-4644*, the results yielded an AUC of 0.763 ($P = 0.026$). The combination of *miR-1246* and *miR-4644* yielded an increased AUC of 0.833 ($P = 0.005$). This pilot study suggested that *miR-1246* and *miR-4644* in salivary exosomes could be candidate biomarkers for pancreatobiliary tract cancer [57].

Kawamura et al. focused on exosomal miRNAs in the portal vein blood (PVB) and peripheral blood (PB) as potential biomarkers to identify patients at high risk for recurrence and poor postoperative outcomes. They found that exosomal *miR-4525*, *miR-451a*, and *miR-21* levels were higher in the PVB than in the PB. High expression of *miR-4525*, *miR-451a*, and *miR-21* in the PVB was associated with recurrence with higher sensitivity, specificity, and accuracy than that in the PB. Furthermore, *miR-4525*, *miR-451a*, and *miR-21* levels in the PVB were independent prognostic factors for overall survival and disease-free survival, indicating that *miR-451a* and *miR-21* in the PVB are potential biomarkers for identifying patients at high risk for recurrence and poor survival following PDAC resection [59]. Another study on exosomal *miR-550* extracted from PANC1 cells revealed the diagnostic potential of miRNAs in PDAC [60].

## 16.2 EV Biomarkers for Cancers of the Mammary Reproductive System

### 16.2.1 Breast Cancer (BC)

BC is one of the most common malignant tumors that endanger women's health and is reported to be a leading cause of cancer-related deaths globally, with a high incidence in women ages 20–59 years. Although progress has been made in the treatment of BC, the diagnosis and prognosis of BC remain challenging.

#### 16.2.1.1 EV RNAs in BC

EV miRNAs are potentially effective candidates for the diagnosis and monitoring of BC (Table 16.7). Eichelser et al. found that exosomal *miR-373* was obviously upregulated in patients with triple-negative BC [61]. Hannafon et al. found that plasma exosomal *miR-21* and *miR-1246* levels were significantly elevated in patients with BC [62], suggesting that these miRNAs may be potential novel biomarkers for the diagnosis of BC; the combination of the two miRNAs was also proven to be more effective than analysis of each miRNA alone. Zhou et al. reported that serum exosomal *miR-105* levels were higher in animals in the precancerous and metastatic stages than in tumor-free controls [63], suggesting that *miR-105* may be a diagnostic candidate for metastatic progression in BC. Di Modica et al. [64] observed higher expression levels of exosomal *miR-939* in basal-like tumor subtypes and proposed that this upregulation was associated with worse prognosis in triple-negative BC.

#### 16.2.1.2 EV Proteins in BC

Fang et al. quantified the expression levels of circulating human epidermal growth factor receptor 2 (HER2)-positive exosomes in 19 patients with BC and revealed that exosomal HER2 protein could be indicative of BC diagnosis. Kibria et al. observed the differential expression of circulating exosomal CD47 in serum derived from patients with BC when compared with that in healthy individuals and further demonstrated that exosomal CD47 may serve as a potential biomarker. Furthermore, based on the findings of Chae et al., the levels of exosomal Del-1 were significantly upregulated in patients with BC at all stages compared with that in other groups [65], indicating that exosomal Del-1 could also be a novel diagnostic biomarker for patients with BC.

**Table 16.7** Potential EV biomarkers in BC

| Cancer type | Cargoes | EV source | Patient cohorts | Detection method | Application | Tendency | Year | References |
|---|---|---|---|---|---|---|---|---|
| **BC** | **RNA** | | | | | | | |
| | miR-373 | Serum | 168 BC patients, 19 patients with benign breast diseases, and 28 healthy controls | Quantitative TaqMan MicroRNA assays | Diagnostic biomarker | Up | 2014 | [61] |
| | miR-21 and miR-1246 | Plasma | 16 patients with BC and 16 healthy controls | Small RNA sequencing and qRT-PCR | Diagnostic biomarkers | Up | 2016 | [62] |
| | miR-105 | Serum | The MDA-MB-231 metastatic BC line and the MCF-10A noncancerous mammary epithelial line | RT-PCR | Predictive or early diagnostic biomarker and therapeutic target | Up | 2014 | [63] |
| | miR-939 | N.D | 63 TNBC patients | qRT-PCR | Prognostic biomarker | Up | 2017 | [64] |
| | **Proteins** | | | | | | | |
| | Del-1 | Plasma | 22 patients with ductal carcinomas or ductal carcinomas in situ and 111 female BC patients | Enzyme-linked immunosorbent assay | Diagnostic biomarker | Up | 2017 | [65] |

*N.D* not determined

### 16.2.2 Cervical Cancer (CC)

CC is the fourth most common malignant tumor in women, with approximately 313,365 deaths and 569,000 newly diagnosed cases annually worldwide; in particular, the death rate in developing or undeveloped countries is as high as 85%. Accurate, effective screening methods for detection and prevention of early cervical cancer are needed to improve survival rates in patients. The most commonly used screening methods in clinical practice are human papilloma virus detection and Bai's new liquid-based cytology test; however, these methods are technically challenging and can be easily affected by factors related to the technician's skill. Therefore, there is an urgent demand for effective new treatment methods and diagnostic biomarkers for rapid screening and detection of CC.

#### 16.2.2.1 EV RNAs in CC

Zheng et al. found that plasma exosomal *let-7d-3p* and *miR-30d-5p* were differentially expressed between cervical tumors and adjacent normal tissues and were further validated as valuable diagnostic biomarkers for the noninvasive screening of CC and its precursors [66]. In addition, exosomal *miR-125a-5p* has been identified as a potential biomarker for CC diagnosis. Lv et al. assessed the expression levels of exosomal *miR-125a-5p* in 22 healthy individuals and 38 patients with CC and found that exosomal *miR-125a-5p* was significantly downregulated in patients with CC compared with that in healthy controls, confirming that exosomal *miR-125a-5p* could be used as a potential biomarker for CC diagnosis [67]. In addition, Wu et al. reported that exosomal *miR-221-3p* could act as a potentially novel diagnostic biomarker and therapeutic target for cervical SCC [68].

Zhang et al. found that the exosomal lncRNAs *HOTAIR*, *MALAT1*, and *MEG3* were significantly differentially expressed in patients with CC compared with those in normal controls and further proposed that exosomal lncRNAs may serve as potential biomarkers for the early detection and diagnosis of CC [69]. Furthermore, Guo et al. identified exosomal *lncRNA-EXOC7* as an important biomarker for the diagnosis and detection of recurrent CC [70]. These potential biomarkers are summarized in Table 16.8.

### 16.2.3 Ovarian Cancer (OC)

OC is the most lethal malignant tumor of the mammary reproductive system, causing serious harm to women's health. Epithelial OC (EOC) accounts for approximately 90% of all OC cases. Owing to a lack of typical symptoms and effective diagnostic methods, approximately 70% of OC cases are not diagnosed until an advanced stage, resulting in poor prognoses for patients.

#### 16.2.3.1 EV RNAs in OC

The roles of EV miRNAs in OC have attracted much attention (Table 16.9). The first study on circulating exosomal miRNAs in OC was performed in 2008 [71]. The authors found that compared with patients with benign disease, patients with OC

**Table 16.8** Potential EV biomarkers in CC

| Cancer type | Cargoes | EV source | Patient cohorts | Detection method | Application | Tendency | Year | References |
|---|---|---|---|---|---|---|---|---|
| **CC** | **RNA** | | | | | | | |
| | let-7d-3p and miR-30d-5p | Plasma | 121 samples from healthy volunteers, CC patients, and cervical intraepithelial neoplasia patients | miRNA sequencing and qRT-PCR | Diagnostic biomarkers | Up | 2019 | [66] |
| | miR-125a-5p | Plasma | 44 patients with CC and 28 healthy controls | RNA sequencing and qRT-PCR | Diagnostic biomarker | Down | 2021 | [67] |
| | miR-221-3p | Serum | Human CSCC cell lines SiHa and C33a, human umbilical vein endothelial cells, and human embryonic kidney 293T | qRT-PCR | Diagnostic biomarker | Up | 2019 | [68] |
| | lncRNAs HOTAIR, MALAT1, and MEG3 | Cervicovaginal lavage samples (ultracentrifugation) | 30 CC patients and 60 healthy controls | qRT-PCR | Diagnostic biomarkers | Up | 2016 | [69] |

**Table 16.9** Potential EV biomarkers in OC

| Cancer type | Cargoes | EV source | Patient cohorts | Detection method | Application | Tendency | Year | References |
|---|---|---|---|---|---|---|---|---|
| **OC** | **RNA** | | | | | | | |
| | miR-21, miR-141, miR-200a, miR-200b, miR-200c, miR-203, miR-205, and miR-214 | Serum | 50 OC patients, 10 patients with benign ovarian adenoma, and 10 healthy controls | N.D | Diagnostic biomarkers | Up | 2008 | [71] |
| | miR-373 and miR-200a | Serum | 163 EOC patients, 20 patients with benign ovarian diseases, and 32 healthy controls | qRT-PCR | Diagnostic biomarkers | Up | 2016 | [72] |
| | miR-200b and miR-200c | Serum | 163 EOC patients, 20 patients with benign ovarian diseases, and 32 healthy controls | qRT-PCR | Diagnostic and prognostic biomarkers | Up | 2016 | [72] |
| | miR-222-3p | Serum | 6 EOC patients and 6 healthy controls | qRT-PCR | Diagnostic biomarker | Up | 2016 | [73] |
| | miR-101 | Serum | 36 OC patients and 20 healthy controls | qRT-PCR | Diagnostic biomarker | Down | 2017 | [74] |
| | miRNA-484 | Serum | 113 OC patients and 60 healthy controls | qRT-PCR | Prognostic biomarker | Down | 2020 | [75] |
| | miR-34a | Serum | 58 EOC patients | qRT-PCR | Diagnostic and prognostic biomarker | Up | 2020 | [76] |
| | **Proteins** | | | | | | | |
| | CA125 | Serum | 55 OC patients, 33 patients with malignant tumors (non-OC), and 55 healthy controls | N.D | Diagnostic biomarker | Up | 2020 | [77] |
| | TGF-β1 and MAGE3/6 | Plasma | 22 OC patients, 10 patients with benign tumors, and 10 healthy controls | Western blot assays | Prognostic biomarkers | Up | 2013 | [78] |

*N.D* not determined

showed significant differential expression of eight miRNAs, i.e., *miR-21*, *miR-141*, *miR-200a*, *miR-200b*, *miR-200c*, *miR-203*, *miR-205*, and *miR-214*; these miRNAs were not detected in healthy individuals [71]. Accordingly, exosomal miRNAs may serve as promising biomarkers for the early diagnosis of OC. In another study, four miRNAs (*miR-373*, *miR-200a*, *miR-200b*, and *miR-200c*) were significantly overexpressed in patients with OC compared with those in healthy controls, and three of these (*miR-200a*, *miR-200b*, and *miR-200c*) were significantly upregulated in patients with OC compared with those in patients with benign disease [72], supporting the diagnostic value of the above circulating exosomal miRNAs in OC. Furthermore, *miR-200b* and *miR-200c* were found to be significantly associated with worse overall survival [72], revealing their potential prognostic value. Ying et al. also reported significant upregulation of serum-derived circulating exosomal *miR-222-3p* in patients with OC compared with that in healthy controls and subsequently suggested a potential role of exosomal *miR-222-3p* in OC [73]. Based on a study by Xu et al., the expression level of serum circulating exosomal *miR-101* was markedly lower in patients with OC than in healthy controls [74], revealing the potential diagnostic value of this miRNA. Moreover, Zhang et al. found that decreased exosomal *miR-484* expression in the serum of patients with OC was correlated with worse clinical outcomes, shorter overall survival, and reduced progression-free survival, indicating that exosomal *miR-484* could be used as a potential biomarker for the diagnosis of OC [75]. Maeda et al. [76] demonstrated that the expression levels of serum-derived exosomal *miR-34a* were significantly increased in patients with early-stage OC compared with that in patients with advanced-stage disease. Additionally, exosomal *miR-34a* levels were lower in the recurrence group than in the recurrence-free group. These findings provided evidence for the potential diagnostic and prognostic value of EV RNAs in EOC.

#### 16.2.3.2 EV Proteins in OC

According to a study published by Chen et al., the expression level of CA125 is markedly higher in exosomes than in serum, suggesting that exosomal CA125 could improve the diagnostic sensitivity of OC. Furthermore, a combination of serum $HE4^+$ and exosomal CA125 significantly improved the diagnosis of OC [77]. Compared with benign ovarian lesion-derived exosomes, higher expression levels of transforming growth factor β1, melanoma-associated antigen (MAGE) 3, and MAGE6 were clearly observed in exosomes from patients with OC, suggesting their potential for distinguishing malignancy from benign disease [78].

## 16.3 EV Biomarkers for Hematologic Tumors

### 16.3.1 Multiple Myeloma (MM)

MM is the second most common hematological cancer, and some clinical characteristics are used in the diagnosis of MM, including the accumulation and clonal growth of malignant plasma cells (PCs) within the bone marrow (BM), as well as monoclonal immunoglobulin presence in the urine and serum [79]. PCs, primarily produced by MM cells, are thought to enable the induction of a series of alterations in the BM microenvironment (BMM) because they not only stimulate the production of inflammatory cytokines and activate the expression of osteoclasts (OCs) but also impair the balance between OCs and osteoblasts; this has serious implications in bone tissues [80]. However, with the transfer of clonal PCs from the BM to another site through the peripheral blood, genetic and epigenetic differences can be clearly observed in the development of MM. Multiple clinical studies have shown that in patients with MM, the numbers of circulating PCs and monoclonal free light chains in the sera are increased in cases exhibiting some clinical symptoms, such as anemia, dehydration, renal failure, proteinuria, and osteolytic bone lesions in their bone scans [81–83]. To date, the main treatments for MM include chemotherapy, immunotherapy, and autologous stem cell transplantation; however, the involvement of BMM easily leads to complications and MM therapy failure. Hence, it is important to determine the mechanisms of interaction between MM cells and BMM and to explore novel biomarkers for MM diagnosis and prognosis (Table 16.10).

#### 16.3.1.1 EV RNAs in MM

Recently, circulating exosomal miRNAs have also been shown to have great prognostic value for patients with MM. In one study, the authors found that *let-7* family members, i.e., *let-7c*, *miR-20a*, *miR-103a*, *miR-140*, and *miR-185a*, were markedly downregulated in MM, whereas *miR-4505* and *miR-4741* showed obviously higher expression, suggesting that the above serum exosomal miRNAs may have great potential as novel biomarkers for disease progression [87]. Additionally, Manier et al. analyzed 22 biologically related miRNAs from 156 patients with MM using a quantitative reverse transcription polymerase chain reaction (qRT-PCR) array and found that two of these miRNAs, i.e., *let-7b* and *miR-18a*, were significantly associated with both progression-free survival and overall survival; these miRNAs remained statistically significant, even after adjustment for the International Staging System for MM scoring and adverse cytogenetics in multivariate analysis [84]. In another study, in comparison with patients without osteoporosis, serum-derived

**Table 16.10** Potential EV biomarkers in MM

| Cancer type | Cargoes | EV source | Patient cohorts | Detection method | Application | Tendency | Year | References |
|---|---|---|---|---|---|---|---|---|
| **MM** | **RNA** | | | | | | | |
| | let-7b and miR-18a | Serum | 156 MM patients and 5 healthy controls | Small RNA sequencing and qRT-PCR | Prognostic biomarkers | Down | 2017 | [84] |
| | miR-214 | Serum | 42 osteoporotic and 23 non-osteoporotic patients | qRT-PCR | Diagnostic biomarkers | Up | 2016 | [85] |
| | miR-16-5p, miR-15a-5p, miR-20a-5p, and miR-17-5p | Serum | 204 MM patients with Bz, thalidomide, and lenalidomide-based regimens | RT-PCR | Drug resistance biomarkers | Down | 2016 | [81] |
| | **Proteins** | | | | | | | |
| | DIS3 | Serum | | N.D | Diagnostic biomarkers | Up | 2018 | [86] |

circulating *miR-214* levels were found to be significantly upregulated in patients with osteoporosis, thus suggesting its potential value as a biomarker for MM-related bone disease [85].

Furthermore, exosomal miRNAs can be used as biomarkers for drug resistance. For example, Zhang et al. found that four exosomal miRNAs (i.e., *miR-16-5p*, *miR-15a-5p*, *miR-20a-5p*, and *miR-17-5p*) were significantly downregulated in bortezomib-resistant patients when compared with bortezomib-sensitive patients [81].

#### 16.3.1.2 EV Proteins in MM

DIS3, an exosomal protein complex with two isoforms, has been shown to exist as two isoforms and can be used as a diagnostic candidate for MM and other hematological cancers [86].

### 16.3.2 Lymphoma

Lymphomas are a group of hematopoietic-derived cancers characterized by the infiltration of lymphoid tissues [88]. According to the latest World Health Organization lymphoma classification, the majority of lymphomas can be categorized as B-cell lymphomas and are further classified into two main subtypes: Hodgkin lymphoma (HL, approximately 10% of cases) and non-Hodgkin lymphoma (NHL, approximately 90% of cases) [89]. HL is a rare hematopoietic malignancy and can be further divided into nodular lymphocyte-predominant HL (5%) and classical HL (95%), whereas NHL mainly includes acute lymphoblastic lymphoma (ALL), Burkitt lymphoma, follicular lymphoma, diffuse large B-cell lymphoma (DLBCL), and anaplastic large cell lymphoma [90]. In recent years, great advances have been made in NHL treatment. For example, the combination of rituximab- and anthracycline-containing chemotherapy can significantly improve therapeutic effects [91]. However, the subsequent prognosis was so poor that it led to refractory disease or recurrence in NHL-postoperative patients owing to toxicity and chemotherapy resistance. Therefore, it is imperative to explore effective therapeutic methods and novel biomarkers for the clinical management of lymphoma.

#### 16.3.2.1 EVs RNA in Lymphoma

Exosomal miRNAs can act as potential monitoring, prognostic, and diagnostic biomarkers in lymphoma (Table 16.11). Van et al. found that five EV miRNAs (i.e., *miR24-3p*, *miR127-3p*, *miR21-5p*, *miR155-5p*, and *let7a-5p*) had markedly higher expression levels in patients with untreated classical HL (cHL) than in patients with cHL after treatment and in healthy controls. In particular, EV miRNA levels increased in patients who experienced relapse, suggesting that these miRNAs may serve as monitoring biomarkers [92]. In another study, the authors identified 37 significantly upregulated and 17 downregulated exosomal miRNAs derived from serum and further found that *miR-99a-5p* and *miR-125b-5p* could be used as potential prognostic predictors of DLBCL [93]. Inada et al. reported that serum-derived exosomal *miR-15a-3p*, *miR-21-5p*, and *miR-181a-5p* were significantly

**Table 16.11** Potential EV biomarkers in lymphoma

| Cancer type | Cargoes | EV source | Patient cohorts | Detection method | Application | Tendency | Year | References |
|---|---|---|---|---|---|---|---|---|
| **Lymphomas** | **RNA** | | | | | | | |
| | miR24-3p, miR127-3p, miR21-5p, miR155-5p, and let7a-5p | Plasma | 20 cHL patients, 4 patients with systemic lupus erythematosus, and 9 healthy controls | TaqMan qRT-PCR | Monitoring biomarkers | Up | 2016 | [92] |
| | miR-99a-5p and miR-125b-5p | Serum | 83 DLBCL patients with chemosensitive and 33 DLBCL patients with chemoresistant | qRT-PCR | Prognostic biomarkers | Up | 2019 | [93] |
| | miR-15a-3p, miR-21-5p, and miR-181a-5 | Serum | N.D | | Diagnostic biomarkers | | 2014 | [94] |

differentially expressed between patients with DLBCL and healthy individuals, and ROC curve analysis further revealed that these three miRNAs were valuable biomarkers for DLBCL diagnosis [94].

### 16.3.3 Leukemia

Leukemia is a hematological malignancy of the blood and BM caused by the unusual proliferation of abnormal white blood cells [95]. Leukemias can be broadly divided into several subtypes, i.e., acute myeloid leukemia (AML), acute lymphoblastic leukemia (ALL), chronic myeloid leukemia, and chronic lymphocytic leukemia. ALL is the most common type of blood cancer worldwide [96]. To date, the biological mechanisms and etiologies of leukemia remain unclear; however, some common factors increase disease risk, including chemicals, smoking, ionizing radiation, and family history. In clinical practice, traditional treatment approaches include chemotherapy, radiotherapy, BM transplantation, palliative care, and combination therapy [97]. Nonetheless, whether the therapy regimens succeed depends not only on the patient age and leukemia type but also on communication between leukemia cells and the surrounding BMM, which is a major challenge in leukemia treatment [98]. Therefore, understanding the interactions between leukemia cells and the BMM will be very useful for elucidating therapeutic effects and drug resistance.

#### 16.3.3.1 EV RNAs in Leukemia

Some studies have shown that exosomal miRNAs may serve as biomarkers for clinical diagnosis (Table 16.12). Parikh et al. found that the expression levels of *miR-155* were significantly higher in patients with CLL, AML, and Waldenström's macroglobulinemia than those in controls, suggesting that these miRNAs could be effective candidates as prognostic biomarkers [103]. Moreover, in a study by Manier et al., the expression levels of EV *miR-155* were assessed in patients with AML, and the results showed that *miR-155* expression was related to white blood cell count; from these findings, the authors concluded that EV *miR-155* may be a novel noninvasive diagnostic biomarker for leukemia [104]. In addition, Huan et al. indicated that AML-derived exosomes express abundant amounts of nucleophosmin, fms-like tyrosine kinase 3, matrix metalloproteinase (MMP) 9, insulin-like growth factor-1R, and C-X-C motif chemokine receptor 4, which are involved in AML pathogenesis and can act as potential prognostic biomarkers [100]. Additionally, using a mouse model, Hornick et al. reported that serum exosomal *miR-150*, *miR-155*, and *miR-1246* could be applied as potential biomarkers for the early detection of AML recurrence [101]. In another study, Zhao et al. indicated that exosomal *miR-4532* derived from AML could prevent normal hematopoiesis, suggesting its potential prognostic value for AML [102].

**Table 16.12** Potential EV biomarkers in leukemia

| Cancer type | Cargoes | EV source | Patient cohorts | Detection method | Application | Tendency | Year | References |
|---|---|---|---|---|---|---|---|---|
| **Leukemia** | **RNA** | | | | | | | |
| | miR-155 | Plasma | Group A (63 MBL patients, 70 CLL patients, and 8 healthy controls), group B (48 MBL patients, 23 CLL patients, and 12 healthy controls) | TaqMan miRNA qRT-PCR | Prognostic biomarkers | Up | 2013 | [99] |
| | NPM 1, FLT 3, MMP 9, IGF-1R, and CXCR 4 | N.D | N.D | qRT-PCR | Prognostic biomarkers | Up | 2013 | [100] |
| | miR-150, miR-155, and miR-1246 | Serum | | qRT-PCR | Diagnostic biomarkers | | 2015 | [101] |
| | miR-4532 | N.D | | RT-qPCR | Prognostic biomarker | Up | 2019 | [102] |

*N.D* not determined

## 16.4 EV Biomarkers for Cancers of the Lungs and Liver

### 16.4.1 Lung Cancer (LC)

LC is the leading cause of fatal malignant tumors worldwide, accounting for 13.2% and 25.9% of all new cancer cases and all cancer-related deaths, respectively [105]. LC can be divided pathologically into two subsets: small cell LC (SCLC, approximately 15% of cases) and non-small cell LC (NSCLC, approximately 85% of cases); the former is thought to be very difficult to treat and intermittently metastatic, whereas the latter is less destructive but more widespread [106]. NSCLC comprises three subgroups: SCC (25%), large cell carcinoma (10%), and adenocarcinoma (40%) [107]. To date, surgical excision therapy is effective for nonmetastatic lung tumors in the early stages, whereas chemotherapy is urgently needed for patients with advanced or metastatic LC. Previous studies have revealed that biomarkers and targeted therapies for NSCLC involve many signal transduction mechanisms, including epidermal growth factor receptor (EGFR), KRAS, anaplastic lymphoma kinase, HER2, BRAF, phosphatidylinositol-4,5-bisphosphate 3-kinase catalytic subunit alpha, AKT1, mitogen-activated protein kinase kinase 1, and MET [108, 109]. However, the mean 5-year survival rate for patients still remains low (approximately 18.1%), mainly because of poor early diagnosis [105]. Therefore, additional studies are needed to elucidate the related molecular mechanisms and explore novel biomarkers and effective treatment methods.

#### 16.4.1.1 EV RNAs in LC

Circulating exosomal miRNAs play important roles in the diagnosis and prognosis of LC (Table 16.13). For example, Rabinowits et al. investigated 27 patients with lung adenocarcinoma and nine healthy controls and confirmed that, compared with healthy controls, the total expression levels of 12 specific miRNAs, i.e., *hsa-miR-17-3p*, *hsa-miR-21*, *hsa-miR-106a*, *hsa-miR-146*, *hsa-miR-155*, *hsa-miR-191*, *hsa-miR-192*, *hsa-miR-203*, *hsa-miR-205*, *hsa-miR-210*, *hsa-miR-212*, and *hsa-miR-214*, were notably elevated in patients with NSCLC, suggesting that these miRNAs could be potential diagnostic candidates for lung adenocarcinoma [110]. The seven exosomal miRNAs derived from the serum of patients with LC (i.e., *miR-451a*, *miR-486-5p*, *miR-363-3p*, *miR-660-5p*, *miR-15b-5p*, *miR-25-3p*, and *miR-16-2-3p*) were specifically distinguished with a mean specificity and sensitivity of 87.55% and 83.33%, respectively [111], thus indicating that these miRNAs may serve as candidates for the diagnosis of LC. Furthermore, the authors identified three specific miRNAs (i.e., *hsa-miR-221-3p*, *hsa-miR-224-5p*, and *hsa-miR-125b-5p*) as potential factors for monitoring treatment responses in patients with SCLC [111]. Jin et al. reported that the expression levels of exosomal miRNAs, i.e., *let-7b-5p*, *miR-24-3p*, and *miR-486-5p*, which were identified in LC serum, were higher than those in the healthy group; they further identified 43 cases of LC from 60 undiagnosed patients based on the differential expression of these exosomal miRNAs [112].

**Table 16.13** Potential EV biomarkers in LC

| Cancer type | Cargoes | EV source | Patient cohorts | Detection method | Application | Tendency | Year | References |
|---|---|---|---|---|---|---|---|---|
| **LC** | **RNA** | | | | | | | |
| | miR-17-3p, miR-21, miR-106a, miR-146, miR-155, miR-191, miR-192, miR-203, miR-205, miR-210, miR-212, and miR-214 | Plasma | 27 patients with LC and 9 healthy controls | | Diagnostic biomarkers | Up | 2009 | [110] |
| | miR-451a, miR-486-5p, miR-363-3p, miR-660-5p, miR-15b-5p, miR-25-3p, and miR-16-2-3p | Serum | 9 SCLC patients, 11 NSCLC patients, 10 and healthy controls | miRNA sequencing and qPCR | Diagnostic biomarkers and monitoring biomarkers | Up | 2018 | [111] |
| | miR-16-5p, miR-15a-5p, miR-20a-5p, and miR-17-5p | Plasma | 46 NSCLC patients, 60 individuals with the possibility of suffering from NSCLC, and 42 healthy controls | miRNA sequencing and qPCR | Diagnostic biomarkers | Up | 2017 | [112] |
| | **Proteins** | | | | | | | |
| | NY-ESO-1 | Plasma | 276 NSCLC patients (198 adenocarcinoma, 69 SCC, and 9 other NSCLC) | N.D | Prognostic biomarker | Up | 2016 | [113] |
| | SRGN, TPM3, THBS1, and HUWE1 | Plasma | | Mass-spectrometry analysis | | | 2017 | [114] |

*N.D* not determined

#### 16.4.1.2 EV Proteins in LC

A prospective study of 276 patients with LC showed overexpression of exosomal NY-ESO-1, suggesting that exosomal NY-ESO-1 could serve as a sensitive biomarker for the prognosis of NSCLC [113]. Vykoukal et al. found that the exosomal proteins serglycin, tropomyosin 3, thrombospondin 1, and HUWE domain containing E3 ubiquitin protein ligase 1, identified by mass spectrometry, were specifically expressed in patients with adenocarcinoma [114].

### 16.4.2 Liver Cancer

Hepatocellular carcinoma (HCC), also called primary liver cancer, is one of the five most common malignant tumors worldwide, with 840,000 new cases and at least 780,000 deaths each year [115]. The genesis and development of HCC are associated with hepatitis B virus (HBV) infection in China [116]. The main treatment approaches for patients with HCC are surgical resection, liver transplantation, local administration of radiation or chemical drugs, and combined therapy [117]. However, commonly used radiation therapies, such as radiofrequency ablation and transcatheter arterial chemoembolization, can generally kill part of tumor cells, and chemotherapy treatments can easily cause drug resistance over time and as the dose increases [118]. Other commonly used molecular markers for the diagnosis of HCC, such as AFP [119], desaturation-γ-lock-up-thrombin [120], and phosphatidylinositol proteoglycan-3 [121], are often unable to diagnose HCC accurately at an early stage owing to disease heterogeneity; this results in low survival rates (<20%) in patients with advanced HCC. Therefore, there is an urgent need to identify promising biomarkers for early diagnosis and develop novel effective treatment methods for clinical practice.

#### 16.4.2.1 EV RNAs in Liver Cancer

Several studies have recommended exosome-derived RNAs as screening and prognostic biomarkers for HCC (Table 16.14). According to a study by Won Sohn et al., the expression of HCC-derived exosomal miRNAs (i.e., *miR-18a*, *miR-221*, *miR-222*, and *miR-224*) was much lower than that of HBV-derived miRNAs, supporting the use of these miRNAs as potential serum markers for diagnosis [122]. A study using patient-derived cells as a model reported that three exosomal miRNAs (i.e., *miR-30d*, *miR-140*, and *miR-29b*) have significant effects on survival in patients with HCC, indicating that exosomes may be prognostic biomarkers [123]. In addition, Xu et al. indicated that two lncRNAs (*ENSG00000258332.1* and *LINC00635*) showed notably higher expression levels in patients with HCC than in the other groups (all $P < 0.05$) and that the combination of the two lncRNAs and serum AFP had an AUC of 0.894. Taken together, these findings suggested that the combination of these two exosomal lncRNAs and AFP could be used as potential biomarkers for the diagnosis and prognosis of HCC [124]. Another study suggested that circulating exosomal ncRNAs, such as *miR-21* and *lncRNA-ATB*, may serve as novel prognostic biomarkers and therapeutic targets for HCC [125]. Moreover, based on their

**Table 16.14** Potential EV biomarkers in HCC

| Cancer type | Cargoes | EV source | Patient cohorts | Detection method | Application | Tendency | Year | References |
|---|---|---|---|---|---|---|---|---|
| **HCC** | **RNA** | | | | | | | |
| | miR-18a, miR-221, miR-222, and miR-224 | Serum | 20 CHB patients with chronic hepatitis B, 20 patients with liver cirrhosis, and 20 HCC patients | qRT-PCR | Serological biomarker | Up | 2015 | [122] |
| | miR-140-3p, miR-30d-5p, miR-29b-3p, miR-130b-3p, and miR-330-5p | N.D | 36 HCC patients | microRNA-sequencing | Prognostic biomarkers | Down | 2019 | [123] |
| | miR-296-3p | N.D | 36 HCC patients | microRNA-sequencing | Prognostic biomarkers | Up | 2019 | [123] |
| | ENSG00000258332.1 and LINC00635 | Serum | 115 HCC patients, 85 patients with liver cirrhosis, 156 patients with chronic hepatitis B, and 120 healthy controls | TaqMan real-time PCR | Diagnostic and prognostic biomarkers | Up | 2018 | [124] |
| | miRNA-21 and lncRNA-ATB | Serum | 79 HCC patients | TaqMan miRNA assays | Prognostic biomarkers and therapeutic targets | Down | 2019 | [125] |
| | miR-718 | Serum | 39 HCC patients with hepatitis C infection, 12 HCC patients with hepatitis B infection, and 8 HCC patients with other causes | qRT-PCR | Prognostic biomarkers | Down | 2015 | [126] |

*N.D* not determined

findings, Sugimachi et al. proposed that exosome-derived circulating *miR-718* may be a novel biomarker for prognosis after treatment in patients with HCC [126].

#### 16.4.2.2 EV Proteins in Liver Cancer

In a previous study, exosomal *circUHRF1* was shown to exhibit higher expression in HCC tissues than in matched adjacent nontumor tissues, indicating its prognostic value in HCC [127].

## 16.5 Bladder Cancer

### 16.5.1 EV Proteins in Bladder Cancer

Because the bladder is anatomically located in the lower urinary tract, studies on bladder cancer biomarkers have mostly focused on EVs derived from the urine, which can be easily and noninvasively collected. Many EVs are present in the urine, and the detection of proteins and RNAs in urine EVs is critical for the early diagnosis and prognosis of bladder cancer (Table 16.15).

Beckham et al. found that the levels of epidermal growth factor-like repeats and discoidin 1-like domains-containing protein 3 (EDIL3) were much higher in urine EVs from patients with bladder cancer than in those from healthy controls. EV EDIL3 could promote the angiogenesis and migration of bladder cancer cells and endothelial cells by activating EGFR signaling [128]. Moreover, Lin et al. used matrix-assisted laser desorption ionization time-of-flight mass spectrometry to analyze the proteomics of urine EVs from 129 patients with bladder cancer and 62 healthy individuals; they found that α-1-antitrypsin and histone H2B1K could be used for the diagnosis of bladder cancer, and this result was verified by immunohistochemistry in tissue samples [129].

### 16.5.2 EV RNAs in Bladder Cancer

In contrast to the large number of studies on EV miRNAs in pancreatic cancer, more studies have evaluated EV lncRNAs as diagnostic markers in bladder cancer. Through comparative analysis of EV lncRNAs in the urine of patients with bladder cancer and healthy controls, Zhan et al. found that the combined detection of the urine EV lncRNAs *MALAT1*, *PCAT-1*, and *SPRY4-IT1* had high diagnostic value, suggesting applications in the noninvasive diagnosis of bladder cancer [130]. *HOTAIR* and some tumor-related lncRNAs are enriched in the urine EVs of patients with bladder transitional cell carcinoma (UBC), which is a highly muscle invasive disease. Downregulation of *HOTAIR* in UBC cell lines reduces in vitro migration and invasion. Importantly, lack of *HOTAIR* expression in UBC cell lines also alters the expression of epithelial-mesenchymal transition-related genes, including *SNAI1*, *TWIST1*, *ZEB1*, *ZO1*, *MMP1*, *LAMB3*, and *LAMC2* [131]. Chen et al. found that *LNMAT2* was overexpressed in urine EVs and serum EVs from patients with

**Table 16.15** Potential EV biomarkers in bladder cancer

| Cancer type | Cargoes | EV source | Patient cohorts | Detection method | Application | Tendency | Year | References |
|---|---|---|---|---|---|---|---|---|
| Bladder cancer | Protein EDIL3 | Urine | 12 patients with BC and 12 healthy controls | Ultracentrifugation | Diagnosis | Up | 2014 | [128] |
| Bladder cancer | Protein Alpha 1-antitrypsin H2B1K | Urine | 129 UC patients and 62 controls | Ultracentrifugation | Diagnosis and prognosis | Up | 2016 | [129] |
| | RNAs | | | | | | | |
| Bladder cancer | lncRNA MALAT1, PCAT-1, and SPRY4-IT1 | Urine | 184 BC patients and 184 healthy controls | Urine exosome RNA isolation kit (NORGEN) | Diagnosis | Up | 2018 | [130] |
| Bladder cancer | lnc RNA HOTAIR | Urine | 8 UBC patients and 5 healthy controls | Ultracentrifugation | Diagnosis | Up | 2016 | [131] |
| Bladder cancer | lncRNA-LNMAT2 | Urine | 154 bladder cancer patients without LN metastasis and 52 patients with LN metastasis | Ultracentrifugation | Diagnosis and prognosis | Up | 2020 | [132] |
| Bladder cancer | lncRNA PTENP1 | Plasma | 50 bladder cancer patients and 60 healthy controls | ExoQuick exosome precipitation solution (system biosciences) | Diagnosis | Down | 2018 | [133] |
| | lncRNA-UCA1 | Plasma | 30 bladder cancer patients and 30 healthy controls | ExoQuick solution (system biosciences) | Diagnosis | Up | 2017 | [134] |

bladder cancer and that both urine and serum EV *LNMAT2* were positively correlated with LN metastasis in patients with bladder cancer; these findings indicated that analysis of EV *LNMAT2* could be used to detect early LN metastasis. Their results also demonstrated that EV *LNMAT2* promotes angiogenesis and LN metastasis in bladder cancer, suggesting that *LNMAT2* may represent a potential molecular target for clinical intervention in patients with bladder cancer with LN metastasis [133].

*PTENP1* levels were also shown to be downregulated in bladder cancer tissues and plasma EVs from patients with bladder cancer. EV *PTENP1* derived from normal cells suppresses tumor growth in vitro and in vivo and enhances the expression of the tumor-suppressor gene *PTEN*. Therefore, EV *PTENP1* is a promising new biomarker that can be used for the clinical detection of bladder cancer [134].

In addition, the expression level of *lncRNA-UCA1* is higher in human serum EVs from patients with bladder cancer than in those from healthy controls, suggesting potential applications as a diagnostic biomarker for bladder cancer [135].

A study on miRNA expression profiles found that there are correlations among bladder cancer tumor tissues, white blood cells, and urine EVs and that some miRNAs are upregulated in bladder cancer tumor tissues, urine EVs, and white blood cells. Furthermore, miRNAs in urine EVs may be potential biomarkers for bladder cancer and other malignant tumors, and noninvasive detection methods may facilitate the diagnosis of bladder cancer [136].

Bladder cancer is the most common malignant tumor of the urinary system in China and is second only to prostate cancer in the Western world. However, existing tumor markers have limited diagnostic efficacy. Because of the unique anatomical location of bladder cancer, urine EV markers may be potential biomarkers for bladder cancer; however, there is still much work to be done to translate the results from bench to bedside.

## 16.6 Thyroid Cancer

Specific tumor markers are critical for the early diagnosis and identification of malignant thyroid nodules and for monitoring recurrence and assessing prognosis. However, the pathogenesis of thyroid cancer has not yet been fully elucidated, and analysis of a single tumor marker alone has limited applications. EVs reflect the overall genetic characteristics of the source cell, and detection of EV tumor markers combined with advances in monitoring methods may facilitate future research efforts. Moreover, additional studies are needed to assess the clinical applications of the identified tumor markers.

The miRNA profiles of EVs derived from the plasma of patients with thyroid cancer change after surgery, indicating that EV miRNA profiles could reflect different states of the disease. Through further cohort studies, plasma EV *miR-31* levels were shown to distinguish papillary thyroid cancer from benign thyroid nodules, whereas *miR-21* levels were shown to distinguish follicular thyroid cancer from benign thyroid nodules [137].

For patients undergoing total thyroidectomy and/or ablation and radioactive iodine therapy, elevated urine EV thyroglobulin has been shown to perform better than serum thyroglobulin in predicting the recurrence of thyroid cancer. Indeed, detection of EV thyroglobulin in the urine is expected to replace the detection of serum thyroglobulin and can be used for rapid detection of recurrence [138].

Plasma EV *miR-485-3p* and *miR-4433a-5p* may be used as biomarkers for diagnosing papillary thyroid carcinoma (PTC). Plasma EV *miR-485-3p* can distinguish between high- and low-risk PTC [139].

LN metastasis is associated with an increased risk of recurrence and a poor prognosis in patients with PTC. By analyzing plasma-derived EV miRNAs in 49 patients with thyroid cancer with LN metastasis and 15 patients with thyroid cancer without LN metastasis, researchers found that *miR-146b-5p* and *miR-222-3p* were both significantly upregulated in patients with LN metastasis and that EV-derived *miR-146b-5p* and *miR-222-3p* could serve as potential biomarkers for LN metastasis in patients with PTC [140].

Researchers from Shandong University have demonstrated that *circFNDC3B* is upregulated in serum EVs from patients with PTC, suggesting that *circFNDC3B* may regulate PTC progression through the *miR-1178*/Toll-like receptor 4 pathway. Accordingly, EV-derived *circFNDC3B* could be a promising target for the treatment of patients with PTC [141].

## 16.7 Pituitary Tumors

The EV lncRNA *H19* can inhibit the growth of remote tumor cells and can be transported across the cell membrane to exert its inhibitory effects. Cabergoline treatment increases the expression of *H19* and exerts synergistic therapeutic effects with EV *H19*. Additionally, EV *H19* inhibits the phosphorylation of the mammalian target of rapamycin complex 1 substrate 4E-BP1. Among patients with all subtypes of pituitary tumors, the expression level of EV *H19* is significantly lower than that in healthy individuals. Changes in plasma EV *H19* levels may be related to patient prognosis or drug response [142].

Cyclin-dependent kinase (*CDK*) 6 and *RhoU* mRNA in serum EVs can be used as markers to predict the aggressiveness of nonfunctioning pituitary adenoma (NF-PA). The combination of these two genes performs better for distinguishing between invasive NF-PA and noninvasive NF-PA. Thus, *CDK6* and *RhoU* play important roles in the invasion of NF-PA, and established diagnostic methods are important for guiding clinical screening and postoperative treatment [143].

Pituitary tumors are mostly benign, and the main applications of EV markers in pituitary tumors would be related to assessment of tumor malignancy (Table 16.16).

**Table 16.16** Potential EV biomarkers in pituitary cancer

| Cancer type | Cargoes | EV source | Patient cohorts | Detection method | Application | Tendency | Year | References |
|---|---|---|---|---|---|---|---|---|
| Thyroid cancer | MiRNA MiRNA-21 | Plasma | 42 thyroid cancer patients and 8 benign thyroid adenomas | Ultracentrifugation | Diagnosis | Up | 2016 | [136] |
| | MiRNA-31 | Plasma | Follicular (n.8) vs. papillary (n.34) thyroid cancer | Ultracentrifugation | Diagnosis | Up | 2016 | [136] |
| | Protein Thyroglobulin | Urine | 5 thyroid cancer patients (T1 and T2) and 9 (T3 or LN) | ExoQuick-TC (System Biosciences, Palo Alto, CA, USA) | Diagnosis and prognosis | Up | 2020 | [137] |
| | miR-485-3p and miR-4433a-5p | Plasma | 136 patients with PTC and 82 patients with benign thyroid nodules | Ultracentrifugation | Diagnosis | Up | 2020 | [138] |
| | miR-146b-5p and miR-222-3p | Plasma | 49 thyroid cancer patients with LNM and 15 patients without LNM | exoRNeasy Serum/Plasma Maxi Kit (QIAGEN, Germany) | Diagnosis | Up | 2020 | [139] |
| Pituitary tumor | circFNDC3B | Plasma | 42 PTC patients and 40 healthy controls | Ultracentrifugation | Diagnosis | Up | 2020 | [140] |
| | lncRNA H19 | Serum | 200 healthy subjects and 206 patients with pituitary tumors | Serial ultracentrifugation | Diagnosis | Down | 2019 | [141] |
| | CDK6 and RHOU mRNA | Serum | 32 NNF-PAs and 14 INF-PAs | Total Exosome Isolation Reagent (Invitrogen) | Diagnosis | Up | 2019 | [142] |

## 16.8 Prostate Cancer

### 16.8.1 EV Proteins in Prostate Cancer

Proteomic studies in patients with prostate cancer have recently suggested that EVs derived from prostate tissue, prostate cancer cells, blood (serum and plasma), urine, and seminal plasma may serve as biomarkers for predicting the occurrence and development of prostate cancer (Table 16.17).

By comparing EV proteomics for PC3 and LNCaP cells, researchers showed that ITGA3 and ITGB1 were associated with prostate cancer metastasis and that the expression levels of these two proteins were significantly higher in urine EVs from patients with metastasis than that those from patients without metastasis and benign prostatic hyperplasia (BPH), highlighting the potential applications of EV proteins in the diagnosis of prostate cancer [144]. Park reported that the concentration of prostate-specific membrane antigen (PSMA)-positive EVs could distinguish patients with prostate cancer from those with BPH. Patients with lower plasma PSMA-positive EV concentrations also had lower Gleason scores and lower probability of biochemical recurrence during the 18-month follow-up period [145].

However, despite other reports on EV protein biomarkers, unstandardized isolation methods, different sample sources, limited case numbers, tumor heterogeneity, and the low reproducibility of protein identification/quantification techniques have limited the development of this field. Further studies are needed to discover clinically valuable EV protein biomarkers.

### 16.8.2 EV RNAs in Prostate Cancer

EVs contain many types of RNAs, including miRNAs, lncRNAs, and mRNAs; these molecules may be derived from different prostate cancer specimens and could be used as biomarker candidates for prostate cancer diagnosis. The relationships between miRNAs and tumorigenesis have been extensively studied, and therefore miRNAs are the most extensively studied type of EV RNA biomarker.

An early study compared the expression of 742 miRNAs in plasma-derived EVs from 78 patients with prostate cancer and 28 healthy controls; through this analysis, researchers identified 11 miRNAs with significantly elevated expression; *miR-574-3p* and *miR-107* were identified by qRT-PCR in the validation cohort. By comparing plasma EV miRNAs from patients with and without metastasis, they found that the expression levels of *miR-141* and *miR-375* were significantly upregulated in patients with metastasis compared with those in patients without metastasis. The increased expression in EVs, combined with the known roles of these two miRNAs in the metastasis of prostate cancer, led to the conclusion that detection of EV *miR-141* and/or *miR-375* from blood samples may have applications in the identification of metastatic prostate cancer [146]. Furthermore, Huang et al. evaluated the

**Table 16.17** Potential EV biomarkers in prostate cancer

| Cancer type | Cargoes | EV source | Patient cohorts | Detection method | Application | Tendency | Year | References |
|---|---|---|---|---|---|---|---|---|
| Prostate cancer | Proteins<br>ITGA3 and ITGB1 | Urine | 5 BPH patients, 5 PCa patients, and 3 metastatic PCa patients | Ultracentrifugation | Diagnosis | Up | 2013 | [143] |
| | PSMA | Plasma | 28 BPH patients and 82 PCa patients | Polyethylene glycol (PEG)/dextran (DEX) aqueous two-phase system | Diagnosis | Up | 2016 | [144] |
| | RNAs<br>MiRNA-141 and miRNA-375 | Serum | 47 patients with prostate cancer metastases and 72 patients with nonrecurrent prostate cancer | Filtration | Diagnosis | Up | 2012 | [145] |
| | miR-1290 and miR-375 | Plasma | 23 CRPC patients in screening cohort and 100 CRPC patients in follow-up cohort | ExoQuick Exosome precipitation solution (System Biosciences) | Prognosis | Up | 2015 | [146] |
| | miR-141 | Serum | 51 PCa patients and 40 healthy controls | ExoQuick Exosome Precipitation Solution (System Biosciences) | Diagnosis | Up | 2015 | [153] |
| | MiR-1246 | Serum | 44 PCa patients, 8 race-matched normal, and 4 BPH patients | Total exosome isolation reagent (Life Technologies) | Diagnosis and prognosis | Up | 2018 | [147] |
| | EPI (ERG, PCA3, and SPDEF) | Urine | 229 patients: a prior negative biopsy and "gray zone" serum PSA levels (2–10 ng/mL) | Ultrafiltration centrifugation | Diagnosis and prognosis | Up | 2016<br>2018<br>2020 | [149, 150, 154] |

**Table 16.17** (continued)

| Cancer type | Cargoes | EV source | Patient cohorts | Detection method | Application | Tendency | Year | References |
|---|---|---|---|---|---|---|---|---|
| | ERG, PCA3, and TMPRSS2:ERG genes | Urine | 163 men who were stratified into 4 groups: biopsy negative (Bx Neg, $n = 39$), biopsy positive (Bx Pos, $n = 47$), post-radical prostatectomy (RP, $n = 37$), and controls (males <35 years, $n = 40$) | Filtration | Diagnosis | Up | 2012 | [152] |
| | Lipids phosphatidylserine (PS)18:1/18:1 and lactosylceramide (d18:1/16:0) | Urine | 15 prostate cancer patients and 13 healthy controls | Ultracentrifugation | Diagnosis | Up | 2017 | [155] |
| | Glucuronate, D-ribose5-phosphate and isobutyryl-L-carnitine | Urine | 3 healthy controls, 3 pre-prostatectomy patients, and 3 post-prostatectomy patients | Ultracentrifugation | Diagnosis | Down | 2017 | [156] |

clinical use of plasma EV miRNAs using RNA sequencing and qRT-PCR in patients with castration-resistant prostate cancer. Cox regression analysis revealed that *miR-1290*, *miR-1246*, and *miR-375* were correlated with overall survival. Among these miRNAs, higher levels of *miR-1290* and *miR-375* were significantly associated with a poorer overall survival rate in the follow-up cohort. Additionally, in patients with castration-resistant prostate cancer, incorporating *miR-1290*/*miR-375* into a hypothetical model based on clinical prognostic factors significantly improved prediction performance [147]. Li et al. reported that serum-derived EV *miR-141* could be used as a diagnostic marker for prostate cancer to distinguish prostate cancer from BPH and healthy controls [154]. Another study showed that EV *miR-1246* is a potential EV biomarker with diagnostic and predictive/prognostic potential in patients with prostate cancer. Importantly, high levels of EV *miR-1246* were found to indicate a higher pathological grade, high probability of LN metastasis and distant metastasis, and poor prognosis [148]. Yu et al. also found that osteogenic tumor-derived exosomes promote osteoclast differentiation and inhibit osteoblastogenesis by transferring miRNAs, thus directly degrading the bone extracellular matrix. Prostate cancer exosomes and the transferred miRNA *miR-92a-1-5p* are critical mediators that disrupt bone homeostasis, degrade the bone matrix, induce pathological bone remodeling, and ultimately create an osteoclastic premetastatic niche for tumor growth. The findings of their study not only provided insights into tumor bone metastasis, in which exosomes derived from osteoblastic tumors induce osteoclast differentiation, but also suggested potential therapeutic targets for prostate cancer bone metastasis [149].

In addition to miRNAs, EV mRNA markers have also been widely used as diagnostic and prognostic markers. In particular, Exosome Diagnostics launched a diagnostic kit (ExoDx Prostate IntelliScore [EPI]) based on previous research and has progressed into clinical applications. The EPI is a completely noninvasive urine EV detection method launched by Exosome Diagnostics. This test examines the gene expression levels of three RNA markers associated with high-grade prostate cancer (*ERG*, *PCA3*, and *SPDEF*) and aims to help doctors identify whether a male patient with a prostate-specific antigen (PSA) level of 2–10 ng/mL needs a prostate biopsy. Through two independently published clinical validation studies, the researchers proved that the EPI helps avoid unnecessary initial prostate biopsies by up to 30%. In May 2019, the National Comprehensive Cancer Network included the EPI in the "Prostate Cancer Screening Guidelines," and in June 2019, the EPI was granted a breakthrough medical device designation by the US Food and Drug Administration. This is the first liquid biopsy product of EVs to be certified as a breakthrough medical device [150, 151, 155].

In addition, Nilsson et al. analyzed the relationships of urine EV *PCA3* and *TMPRSS2*:*ERG* expression with different subgroups of prostate cancer. The results also confirmed the potential of EV mRNA derived from urine as a diagnostic marker for prostate cancer and suggested that EV mRNAs may be used to predict the occurrence of metastasis and the effects of androgen deprivation therapy [152].

Scholars from Colombia have examined *PSA*, *PCA3*, *AR*, survivin, *NCOA2*, *RAD21*, *TMPRSS2*, *ERG*, and *TMPRSS2:ERG* in urine EVs from patients with negative and positive biopsies, patients who have undergone radical prostatectomy, and healthy controls. Compared with the negative biopsy group, the expression levels of EV *ERG*, *PCA3*, *TMPRSS2:ERG*, *TMPRSS2*, and survivin were significantly higher in the positive biopsy group. Moreover, the incidence rates of the *TMPRSS2:ERG* fusion gene were 68% and 44%, respectively, in the positive and negative biopsy groups and only 5% in the control group. *TMPRSS2:ERG* was not detected in the radical surgery group. The loss of *TMPRSS2:ERG* expression in four patients with positive biopsy results after prostatectomy further supported this finding, indicating the specificity of the fusion event for prostate cancer diagnosis. Overall, the findings of this study confirmed that mRNAs are stably expressed in EVs and may be exploited for simple, accurate liquid biopsy for prostate cancer diagnosis [153].

Compared with protein and lipid molecules, RNAs are more abundant in EVs and easier to detect, and research on EV RNA markers is extensive. Additionally, many studies on the mechanisms of EV miRNAs in castration-resistant prostate cancer and prostate cancer with distant metastasis have led to the discovery of prostate cancer-specific EV miRNAs and EV miRNA panels. With advancements in detection technologies and the implementation of large-scale clinical research, RNAs derived from EVs may play important roles in the early diagnosis, prognosis, and monitoring of the minimal change residual of prostate cancer.

### 16.8.3 EV Lipids in Prostate Cancer

Lipid molecules in urine EVs have potential applications as diagnostic markers for prostate cancer. Skotland et al. conducted a quantitative analysis of 36 lipid types in urine EVs from patients with prostate cancer and healthy controls. They found that nine lipid molecules showed differential levels in EVs derived from the urine of patients with prostate cancer and healthy controls; among these lipids, phosphatidylserine (PS) 18:1/18:1 and lactosylceramide (d18:1/16:0) showed the most significant differences. After combining these two molecules and PS 18:0–18:2 to distinguish patients with prostate cancer from healthy controls, the sensitivity and specificity reached 93% and 100%, respectively [156].

Researchers from the University of Helsinki in Finland explored the metabolomics of EVs derived from the urine as well as the metabolomics of urine samples and platelets from patients with prostate cancer before and after prostatectomy. The results showed that glucuronate, D-ribose 5-phosphate, and isobutyryl-L-carnitine levels were 2- to 26-fold lower in patients before prostatectomy than those in patients after prostatectomy and in healthy controls; thus, EV metabolomics could be used to probe novel biomarkers [157].

In addition to participating in the biosynthesis and uptake of EVs, lipids also participate in various biological processes, such as immune surveillance, tumor microenvironment modification, and inflammation regulation. Therefore, the lipid molecules carried by EVs not only reflect the biological characteristics of the cells from which they originate but also have the potential to change the microenvironment of the tumor metastasis site and promote the formation of metastatic lesions. With the development of lipidomics and metabolomics, EV lipid molecules and small molecule metabolites can be identified as novel tumor markers.

## 16.9 Osteosarcoma

Most studies on EV biomarkers have focused on adult cancers, and progress in childhood and adolescent cancers is very limited. However, despite the low incidence rates of tumors that tend to occur in adolescents, such as osteosarcoma and Ewing's sarcoma, the fatality rates of these cancers are very high, and treatment options are often limited. Therefore, studies of EV markers have clinical value in the early diagnosis of such tumors (Table 16.18).

Brady et al. analyzed the protein contents of EVs in the serum of patients with different stages of osteosarcoma and in healthy volunteers; they found that the accuracy of more than ten proteins for distinguishing patients with osteosarcoma from healthy volunteers was 85%. Further research showed that serine protease inhibitory protein D1 and major histocompatibility complex type III complement component C6 have high identification accuracy rate and may be used to assess the stage of osteosarcoma [158].

Serum EV programmed death ligand-1 (PD-L1) and N-cadherin levels are significantly higher in patients with osteosarcoma than in healthy controls. Additionally, serum EV PD-L1 and N-cadherin levels were also high in patients with osteosarcoma with lung metastasis than in those without metastasis; EV PD-L1 expression was decreased significantly after patients underwent treatment [159].

Another study revealed that many miRNAs and mRNAs were present in blood EVs from patients with osteosarcoma. These RNAs are differentially expressed in the EVs of patients with osteosarcoma; moreover, serum EV miRNAs may be reliable diagnostic biomarkers for differential responses to chemotherapy in these patients. Overall, these results provide evidence of the potential value of these biomarkers for the diagnosis of osteosarcoma and for monitoring treatment outcomes following chemotherapy. The findings also provide insights into the mechanisms of abnormal regulation of EV RNAs in patients with osteosarcoma in whom the tumor is poorly responsive to chemotherapy [160].

**Table 16.18** Potential EV biomarkers in osteosarcoma and Ewing's sarcoma

| Cancer type | Cargoes | EV source | Patient cohorts | Detection method | Application | Tendency | Year | References |
|---|---|---|---|---|---|---|---|---|
| Osteosarcoma | Proteins<br>Serine protease inhibitory protein D1 and major histocompatibility complex type III complement component C6 | Serum | Five osteosarcoma patients at three time points over their disease course | Total Exosome Isolation (Invitrogen) | Diagnosis | Up | 2018 | [158] |
| | PD-L1 and N-CADHERIN | Serum | 70 osteosarcoma patients and 9 benign tumor patients and 22 healthy donors | Ultracentrifugation | Diagnosis and prognosis | Up | 2020 | [159] |
| | A profile of RNAs | Serum | 48 OS patients with poor chemotherapeutic response, 45 OS patients with good chemotherapeutic response, and 51 healthy controls | Ultracentrifugation | Diagnosis and prognosis | Up and down | 2017 | [160] |
| Ewing's sarcoma | EWS-FLI1 fusion gene | Plasma | ES cell line-derived exosomes were serially diluted in 10 ml of plasma of a healthy donor | Ultracentrifugation | Diagnosis | Up | 2013 | [161] |
| | EWS/FLI-1 fusion gene | Plasma | Plasma from ES cell/ xenografted mice | Centrifugation and filtration | Diagnosis | Up | 2013 | [162] |

## 16.10 Ewing's Sarcoma

The pathogenesis of Ewing's sarcoma is closely related to the occurrence of chromosomal translocations and the formation of fusion genes. Therefore, most studies have focused on determining the expression levels of characteristic fusion genes in EVs and exploring the roles of EV fusion genes in tumorigenesis.

Miller et al. demonstrated the presence of EVs in an Ewing's sarcoma cell line for the first time and detected the expression of the specific fusion gene *EWS-FLI1* in tumor-derived EVs. Further analyses revealed that EV mRNAs mainly participate in G-protein-coupled signaling pathways, neurotransmitter signaling, and cell stemness-related signaling pathways. These findings provided a basis for exploring how Ewing's sarcoma cells affect the tumor microenvironment and established an approach for marker screening [161]. Another study also detected the *EWS/FLI-1* fusion gene in EVs derived from an Ewing's sarcoma cell line carrying t(11; 22)(q24; 12). *EWS-FLI1* mRNA has also been detected in EVs derived from mice inoculated with an Ewing's sarcoma cell line [162]. Zeng showed that EVs derived from Ewing's sarcoma cells could be stably enriched by combining a microfluidic chip with immunocapture, making it possible to routinely detect the *EWS-ELI1* fusion gene in EVs from clinical samples [163].

The unique genetic characteristics of Ewing's sarcoma provide a direction for the screening of EV biomarkers. Additionally, the detection of pathogenic genes in EVs may have clinical applications in early diagnosis.

## References

1. Russell AE, Sneider A, Witwer KW, Bergese P, Bhattacharyya SN, Cocks A, et al. Biological membranes in EV biogenesis, stability, uptake, and cargo transfer: an ISEV position paper arising from the ISEV membranes and EVs workshop. J Extracell Vesicles. 2019;8(1):1684862.
2. Chu YL, Li H, Ng PLA, Kong ST, Zhang H, Lin Y, et al. The potential of circulating exosomal RNA biomarkers in cancer. Expert Rev Mol Diagn. 2020;20(7):665–78.
3. O'Brien K, Breyne K, Ughetto S, Laurent LC, Breakefield XO. RNA delivery by extracellular vesicles in mammalian cells and its applications. Nat Rev Mol Cell Biol. 2020;21(10):585–606.
4. Pang B, Zhu Y, Ni J, Thompson J, Malouf D, Bucci J, et al. Extracellular vesicles: the next generation of biomarkers for liquid biopsy-based prostate cancer diagnosis. Theranostics. 2020;10(5):2309–26.
5. Best MG, Wesseling P, Wurdinger T. Tumor-educated platelets as a noninvasive biomarker source for cancer detection and progression monitoring. Cancer Res. 2018;78(13):3407–12.
6. Melo SA, Luecke LB, Kahlert C, Fernandez AF, Gammon ST, Kaye J, et al. Glypican-1 identifies cancer exosomes and detects early pancreatic cancer. Nature. 2015;523(7559):177–82.
7. San Lucas FA, Allenson K, Bernard V, Castillo J, Kim DU, Ellis K, et al. Minimally invasive genomic and transcriptomic profiling of visceral cancers by next-generation sequencing of circulating exosomes. Ann Oncol. 2016;27(4):635–41.
8. Lin Y, Dong H, Deng W, Lin W, Li K, Xiong X, et al. Evaluation of salivary exosomal chimeric GOLM1-NAA35 RNA as a potential biomarker in esophageal carcinoma. Clin Cancer Res. 2019;25(10):3035–45.

9. Trino S, Lamorte D, Caivano A, De Luca L, Sgambato A, Laurenzana I. Clinical relevance of extracellular vesicles in hematological neoplasms: from liquid biopsy to cell biopsy. Leukemia. 2021;35(3):661–78.
10. Xu R, Rai A, Chen M, Suwakulsiri W, Greening DW, Simpson RJ. Extracellular vesicles in cancer—implications for future improvements in cancer care. Nat Rev Clin Oncol. 2018;15(10):617–38.
11. Thrift AP. Global burden and epidemiology of Barrett oesophagus and oesophageal cancer. Nat Rev Gastroenterol Hepatol. 2021;18:432.
12. Fan L, Cao Q, Liu J, Zhang J, Li B. Circular RNA profiling and its potential for esophageal squamous cell cancer diagnosis and prognosis. Mol Cancer. 2019;18(1):16.
13. Tanaka Y, Kamohara H, Kinoshita K, Kurashige J, Ishimoto T, Iwatsuki M, et al. Clinical impact of serum exosomal microRNA-21 as a clinical biomarker in human esophageal squamous cell carcinoma. Cancer. 2013;119(6):1159–67.
14. Takeshita N, Hoshino I, Mori M, Akutsu Y, Hanari N, Yoneyama Y, et al. Serum microRNA expression profile: miR-1246 as a novel diagnostic and prognostic biomarker for oesophageal squamous cell carcinoma. Br J Cancer. 2013;108(3):644–52.
15. Liu MX, Liao J, Xie M, Gao ZK, Wang XH, Zhang Y, et al. miR-93-5p transferred by exosomes promotes the proliferation of esophageal cancer cells via intercellular communication by targeting PTEN. Biomed Environ Sci. 2018;31(3):171–85.
16. Huang L, Wang Y, Chen J, Wang Y, Zhao Y, Wang Y, et al. Long noncoding RNA PCAT1, a novel serum-based biomarker, enhances cell growth by sponging miR-326 in oesophageal squamous cell carcinoma. Cell Death Dis. 2019;10(7):513.
17. Liu T, Du LT, Wang YS, Gao SY, Li J, Li PL, et al. Development of a novel serum exosomal microRNA nomogram for the preoperative prediction of lymph node metastasis in esophageal squamous cell carcinoma. Front Oncol. 2020;10:573501.
18. Yan L, Dong X, Gao J, Liu F, Zhou L, Sun Y, et al. A novel rapid quantitative method reveals stathmin-1 as a promising marker for esophageal squamous cell carcinoma. Cancer Med. 2018;7(5):1802–13.
19. Qiu ML, Li X, Lin JB, Luo RG, Liu B, Feng Z. Serum exosomal miR-182 upregulation predicts unfavorable prognosis of esophageal squamous cell carcinoma. Eur Rev Med Pharmacol Sci. 2020;24(10):5412–8.
20. Liu S, Lin Z, Zheng Z, Rao W, Lin Y, Chen H, et al. Serum exosomal microRNA-766-3p expression is associated with poor prognosis of esophageal squamous cell carcinoma. Cancer Sci. 2020;111(10):3881–92.
21. Yan S, Du L, Jiang X, Duan W, Li J, Xie Y, et al. Evaluation of serum exosomal lncRNAs as diagnostic and prognostic biomarkers for esophageal squamous cell carcinoma. Cancer Manag Res. 2020;12:9753–63.
22. Jiao Z, Yu A, Rong W, He X, Zen K, Shi M, et al. Five-lncRNA signature in plasma exosomes serves as diagnostic biomarker for esophageal squamous cell carcinoma. Aging (Albany NY). 2020;12(14):15002–10.
23. Chiam K, Wang T, Watson DI, Mayne GC, Irvine TS, Bright T, et al. Circulating serum exosomal miRNAs as potential biomarkers for esophageal adenocarcinoma. J Gastrointest Surg. 2015;19(7):1208–15.
24. Warnecke-Eberz U, Chon SH, Holscher AH, Drebber U, Bollschweiler E. Exosomal onco-miRs from serum of patients with adenocarcinoma of the esophagus: comparison of miRNA profiles of exosomes and matching tumor. Tumour Biol. 2015;36(6):4643–53.
25. Chen H, Zhou X. Serum exosomal miRNAs expression as novel biomarkers for detection of esophageal adenocarcinoma. United Eur Gastroenterol. 2017;5:A340.
26. Chiam K, Mayne GC, Wang T, Watson DI, Irvine TS, Bright T, et al. Serum outperforms plasma in small extracellular vesicle microRNA biomarker studies of adenocarcinoma of the esophagus. World J Gastroenterol. 2020;26(20):2570–83.
27. Zhao A, Guo L, Xu J, Zheng L, Guo Z, Ling Z, et al. Identification and validation of circulating exosomes-based liquid biopsy for esophageal cancer. Cancer Med. 2019;8(7):3566–74.

28. Tian L, Cao J, Jiao H, Zhang J, Ren X, Liu X, et al. CircRASSF2 promotes laryngeal squamous cell carcinoma progression by regulating the miR-302b-3p/IGF-1R axis. Clin Sci (Lond). 2019;133(9):1053–66.
29. Wang J, Zhou Y, Lu J, Sun Y, Xiao H, Liu M, et al. Combined detection of serum exosomal miR-21 and HOTAIR as diagnostic and prognostic biomarkers for laryngeal squamous cell carcinoma. Med Oncol. 2014;31(9):148.
30. Zhao Q, Zheng X, Guo H, Xue X, Zhang Y, Niu M, et al. Serum exosomal miR-941 as a promising oncogenic biomarker for laryngeal squamous cell carcinoma. J Cancer. 2020;11(18):5329–44.
31. Almangush A, Heikkinen I, Makitie AA, Coletta RD, Laara E, Leivo I, et al. Prognostic biomarkers for oral tongue squamous cell carcinoma: a systematic review and meta-analysis. Br J Cancer. 2017;117(6):856–66.
32. Cui J, Wang H, Zhang X, Sun X, Zhang J, Ma J. Exosomal miR-200c suppresses chemoresistance of docetaxel in tongue squamous cell carcinoma by suppressing TUBB3 and PPP2R1B. Aging (Albany NY). 2020;12(8):6756–73.
33. Rabinowits G, Bowden M, Flores LM, Verselis S, Vergara V, Jo VY, et al. Comparative analysis of microRNA expression among benign and malignant tongue tissue and plasma of patients with tongue cancer. Front Oncol. 2017;7:191.
34. Sanada T, Islam A, Kaminota T, Kirino Y, Tanimoto R, Yoshimitsu H, et al. Elevated exosomal lysyl oxidase like 2 is a potential biomarker for head and neck squamous cell carcinoma. Laryngoscope. 2020;130(5):E327–34.
35. Campanella C, Rappa F, Sciumè C, Marino Gammazza A, Barone R, Bucchieri F, et al. Heat shock protein 60 levels in tissue and circulating exosomes in human large bowel cancer before and after ablative surgery. Cancer. 2015;121(18):3230–9.
36. Li J, Chen Y, Guo X, Zhou L, Jia Z, Peng Z, et al. GPC1 exosome and its regulatory miRNAs are specific markers for the detection and target therapy of colorectal cancer. J Cell Mol Med. 2017;21(5):838–47.
37. Zhu M, Huang Z, Zhu D, Zhou X, Shan X, Qi LW, et al. A panel of microRNA signature in serum for colorectal cancer diagnosis. Oncotarget. 2017;8(10):17081–91.
38. Zhao Y, Du T, Du L, Li P, Li J, Duan W, et al. Long noncoding RNA LINC02418 regulates MELK expression by acting as a ceRNA and may serve as a diagnostic marker for colorectal cancer. Cell Death Dis. 2019;10(8):568.
39. Chen Y, Xie Y, Xu L, Zhan S, Xiao Y, Gao Y, et al. Protein content and functional characteristics of serum-purified exosomes from patients with colorectal cancer revealed by quantitative proteomics. Int J Cancer. 2017;140(4):900–13.
40. Tian Y, Ma L, Gong M, Su G, Zhu S, Zhang W, et al. Protein profiling and sizing of extracellular vesicles from colorectal cancer patients via flow cytometry. ACS Nano. 2018;12(1):671–80.
41. Ogata-Kawata H, Izumiya M, Kurioka D, Honma Y, Yamada Y, Furuta K, et al. Circulating exosomal microRNAs as biomarkers of colon cancer. PLoS One. 2014;9(4):e92921.
42. Hu D, Zhan Y, Zhu K, Bai M, Han J, Si Y, et al. Plasma exosomal long non-coding RNAs serve as biomarkers for early detection of colorectal cancer. Cell Physiol Biochem. 2018;51(6):2704–15.
43. Pan B, Qin J, Liu X, He B, Wang X, Pan Y, et al. Identification of serum exosomal hsa-circ-0004771 as a novel diagnostic biomarker of colorectal cancer. Front Genet. 2019;10:1096.
44. Fu H, Yang H, Zhang X, Wang B, Mao J, Li X, et al. Exosomal TRIM3 is a novel marker and therapy target for gastric cancer. J Exp Clin Cancer Res. 2018;37(1):162.
45. Xie M, Yu T, Jing X, Ma L, Fan Y, Yang F, et al. Exosomal circSHKBP1 promotes gastric cancer progression via regulating the miR-582-3p/HUR/VEGF axis and suppressing HSP90 degradation. Mol Cancer. 2020;19(1):112.
46. Ji R, Zhang X, Gu H, Ma J, Wen X, Zhou J, et al. miR-374a-5p: a new target for diagnosis and drug resistance therapy in gastric cancer. Mol Ther Nucleic Acids. 2019;18:320–31.
47. Tang W, Fu K, Sun H, Rong D, Wang H, Cao H. CircRNA microarray profiling identifies a novel circulating biomarker for detection of gastric cancer. Mol Cancer. 2018;17(1):137.

48. Zhao R, Zhang Y, Zhang X, Yang Y, Zheng X, Li X, et al. Exosomal long noncoding RNA HOTTIP as potential novel diagnostic and prognostic biomarker test for gastric cancer. Mol Cancer. 2018;17(1):68.
49. Yoon JH, Ham IH, Kim O, Ashktorab H, Smoot DT, Nam SW, et al. Gastrokine 1 protein is a potential theragnostic target for gastric cancer. Gastric Cancer. 2018;21(6):956–67.
50. Zhang H, Chen Z, Wang X, Huang Z, He Z, Chen Y. Long non-coding RNA: a new player in cancer. J Hematol Oncol. 2013;6:37.
51. Guo X, Lv X, Ru Y, Zhou F, Wang N, Xi H, et al. Circulating exosomal gastric cancer-associated long noncoding RNA1 as a biomarker for early detection and monitoring progression of gastric cancer: a multiphase study. JAMA Surg. 2020;155(7):572–9.
52. Lin LY, Yang L, Zeng Q, Wang L, Chen ML, Zhao ZH, et al. Tumor-originated exosomal lncUEGC1 as a circulating biomarker for early-stage gastric cancer. Mol Cancer. 2018;17(1):84.
53. Yamamoto H, Watanabe Y, Oikawa R, Morita R, Yoshida Y, Maehata T, et al. BARHL2 methylation using gastric wash DNA or gastric juice exosomal DNA is a useful marker for early detection of gastric cancer in an H. pylori-independent manner. Clin Transl Gastroenterol. 2016;7(7):e184.
54. Kimura H, Yamamoto H, Harada T, Fumoto K, Osugi Y, Sada R, et al. CKAP4, a DKK1 receptor, is a biomarker in exosomes derived from pancreatic cancer and a molecular target for therapy. Clin Cancer Res. 2019;25(6):1936–47.
55. Xu YF, Hannafon BN, Zhao YD, Postier RG, Ding WQ. Plasma exosome miR-196a and miR-1246 are potential indicators of localized pancreatic cancer. Oncotarget. 2017;8(44):77028–40.
56. Liu J, Gao J, Du Y, Li Z, Ren Y, Gu J, et al. Combination of plasma microRNAs with serum CA19-9 for early detection of pancreatic cancer. Int J Cancer. 2012;131(3):683–91.
57. Machida T, Tomofuji T, Maruyama T, Yoneda T, Ekuni D, Azuma T, et al. miR-1246 and miR-4644 in salivary exosome as potential biomarkers for pancreatobiliary tract cancer. Oncol Rep. 2016;36(4):2375–81.
58. Goto T, Fujiya M, Konishi H, Sasajima J, Fujibayashi S, Hayashi A, et al. An elevated expression of serum exosomal microRNA-191, -21, -451a of pancreatic neoplasm is considered to be efficient diagnostic marker. BMC Cancer. 2018;18(1):116.
59. Kawamura S, Iinuma H, Wada K, Takahashi K, Minezaki S, Kainuma M, et al. Exosome-encapsulated microRNA-4525, microRNA-451a and microRNA-21 in portal vein blood is a high-sensitive liquid biomarker for the selection of high-risk pancreatic ductal adenocarcinoma patients. J Hepatobiliary Pancreat Sci. 2019;26(2):63–72.
60. Taller D, Richards K, Slouka Z, Senapati S, Hill R, Go DB, et al. On-chip surface acoustic wave lysis and ion-exchange nanomembrane detection of exosomal RNA for pancreatic cancer study and diagnosis. Lab Chip. 2015;15(7):1656–66.
61. Eichelser C, Stückrath I, Müller V, Milde-Langosch K, Wikman H, Pantel K, et al. Increased serum levels of circulating exosomal microRNA-373 in receptor-negative breast cancer patients. Oncotarget. 2014;5(20):9650–63.
62. Hannafon BN, Trigoso YD, Calloway CL, Zhao YD, Lum DH, Welm AL, et al. Plasma exosome microRNAs are indicative of breast cancer. Breast Cancer Res. 2016;18(1):90.
63. Zhou W, Fong MY, Min Y, Somlo G, Liu L, Palomares MR, et al. Cancer-secreted miR-105 destroys vascular endothelial barriers to promote metastasis. Cancer Cell. 2014;25(4):501–15.
64. Di Modica M, Regondi V, Sandri M, Iorio MV, Zanetti A, Tagliabue E, et al. Breast cancer-secreted miR-939 downregulates VE-cadherin and destroys the barrier function of endothelial monolayers. Cancer Lett. 2017;384:94–100.
65. Lee SJ, Lee J, Jung JH, Park HY, Lee C-H, Moon P-G, et al. Exosomal Del-1 as a potent diagnostic marker for breast cancer: a prospective cohort study. J Clin Oncol. 2017;35(15_suppl):11548.
66. Zheng M, Hou L, Ma Y, Zhou L, Wang F, Cheng B, et al. Exosomal let-7d-3p and miR-30d-5p as diagnostic biomarkers for non-invasive screening of cervical cancer and its precursors. Mol Cancer. 2019;18(1):76.

67. Lv A, Tu Z, Huang Y, Lu W, Xie B. Circulating exosomal miR-125a-5p as a novel biomarker for cervical cancer. Oncol Lett. 2021;21(1):54.
68. Wu XG, Zhou CF, Zhang YM, Yan RM, Wei WF, Chen XJ, et al. Cancer-derived exosomal miR-221-3p promotes angiogenesis by targeting THBS2 in cervical squamous cell carcinoma. Angiogenesis. 2019;22(3):397–410.
69. Zhang J, Liu SC, Luo XH, Tao GX, Guan M, Yuan H, et al. Exosomal long noncoding RNAs are differentially expressed in the cervicovaginal lavage samples of cervical cancer patients. J Clin Lab Anal. 2016;30(6):1116–21.
70. Guo Y, Wang X, Wang K, He Y. Appraising the value of serum and serum-derived exosomal LncRNA-EXOC7 as a promising biomarker in cervical cancer. Clin Lab. 2020;66(7).
71. Taylor DD, Gercel-Taylor C. MicroRNA signatures of tumor-derived exosomes as diagnostic biomarkers of ovarian cancer. Gynecol Oncol. 2008;110(1):13–21.
72. Meng X, Müller V, Milde-Langosch K, Trillsch F, Pantel K, Schwarzenbach H. Diagnostic and prognostic relevance of circulating exosomal miR-373, miR-200a, miR-200b and miR-200c in patients with epithelial ovarian cancer. Oncotarget. 2016;7(13):16923–35.
73. Ying X, Wu Q, Wu X, Zhu Q, Wang X, Jiang L, et al. Epithelial ovarian cancer-secreted exosomal miR-222-3p induces polarization of tumor-associated macrophages. Oncotarget. 2016;7(28):43076–87.
74. Xu Y, Xu L, Zheng J, Geng L, Zhao S. MiR-101 inhibits ovarian carcinogenesis by repressing the expression of brain-derived neurotrophic factor. FEBS Open Bio. 2017;7(9):1258–66.
75. Zhang W, Su X, Li S, Liu Z, Wang Q, Zeng H. Low serum exosomal miR-484 expression predicts unfavorable prognosis in ovarian cancer. Cancer Biomark. 2020;27(4):485–91.
76. Maeda K, Sasaki H, Ueda S, Miyamoto S, Terada S, Konishi H, et al. Serum exosomal microRNA-34a as a potential biomarker in epithelial ovarian cancer. J Ovarian Res. 2020;13(1):47.
77. Chen Z, Liang Q, Zeng H, Zhao Q, Guo Z, Zhong R, et al. Exosomal CA125 as a promising biomarker for ovarian cancer diagnosis. J Cancer. 2020;11(21):6445–53.
78. Szajnik M, Derbis M, Lach M, Patalas P, Michalak M, Drzewiecka H, et al. Exosomes in plasma of patients with ovarian carcinoma: potential biomarkers of tumor progression and response to therapy. Gynecol Obstet (Sunnyvale). 2013;Suppl 4:3.
79. Pawlyn C, Davies FE. Toward personalized treatment in multiple myeloma based on molecular characteristics. Blood. 2019;133(7):660–75.
80. Yaccoby S. Advances in the understanding of myeloma bone disease and tumour growth. Br J Haematol. 2010;149(3):311–21.
81. Zhang L, Pan L, Xiang B, Zhu H, Wu Y, Chen M, et al. Potential role of exosome-associated microRNA panels and in vivo environment to predict drug resistance for patients with multiple myeloma. Oncotarget. 2016;7(21):30876–91.
82. Raimondi L, De Luca A, Amodio N, Manno M, Raccosta S, Taverna S, et al. Involvement of multiple myeloma cell-derived exosomes in osteoclast differentiation. Oncotarget. 2015;6(15):13772–89.
83. Wang J, Hendrix A, Hernot S, Lemaire M, De Bruyne E, Van Valckenborgh E, et al. Bone marrow stromal cell-derived exosomes as communicators in drug resistance in multiple myeloma cells. Blood. 2014;124(4):555–66.
84. Manier S, Liu CJ, Avet-Loiseau H, Park J, Shi J, Campigotto F, et al. Prognostic role of circulating exosomal miRNAs in multiple myeloma. Blood. 2017;129(17):2429–36.
85. Sun W, Zhao C, Li Y, Wang L, Nie G, Peng J, et al. Osteoclast-derived microRNA-containing exosomes selectively inhibit osteoblast activity. Cell Discov. 2016;2:16015.
86. Robinson SR, Viegas SC, Matos RG, Domingues S, Bedir M, Stewart HJS, et al. DIS3 isoforms vary in their endoribonuclease activity and are differentially expressed within haematological cancers. Biochem J. 2018;475(12):2091–105.
87. Zhang ZY, Li YC, Geng CY, Zhou HX, Gao W, Chen WM. Serum exosomal microRNAs as novel biomarkers for multiple myeloma. Hematol Oncol. 2019;37(4):409–17.
88. Fernandes M, Teixeira AL, Medeiros R. The opportunistic effect of exosomes on non-Hodgkin lymphoma microenvironment modulation. Crit Rev Oncol Hematol. 2019;144:102825.

89. Elenitoba-Johnson KSJ, Lim MS. New insights into lymphoma pathogenesis. Annu Rev Pathol. 2018;13:193–217.
90. Galardy PJ, Bedekovics T, Hermiston ML. Targeting childhood, adolescent and young adult non-Hodgkin lymphoma: therapeutic horizons. Br J Haematol. 2016;173(4):625–36.
91. Armitage JO, Gascoyne RD, Lunning MA, Cavalli F. Non-Hodgkin lymphoma. Lancet. 2017;390(10091):298–310.
92. van Eijndhoven MA, Zijlstra JM, Groenewegen NJ, Drees EE, van Niele S, Baglio SR, et al. Plasma vesicle miRNAs for therapy response monitoring in Hodgkin lymphoma patients. JCI Insight. 2016;1(19):e89631.
93. Feng Y, Zhong M, Zeng S, Wang L, Liu P, Xiao X, et al. Exosome-derived miRNAs as predictive biomarkers for diffuse large B-cell lymphoma chemotherapy resistance. Epigenomics. 2019;11(1):35–51.
94. Inada K, Okoshi Y, Cho Y, Saito H, Iijima T, Hori M, et al. Availability of circulating microRNAs as a biomarker for the early diagnosis of diffuse large B-cell lymphoma. Blood. 2014;124(21):2988.
95. Stansly PG. Progress in virus research. Oncogenic viruses causing neoplasms of reticular tissue in the mouse: a comparative study. Prog Exp Tumor Res. 1963;3:216–53.
96. Global, regional, and national incidence, prevalence, and years lived with disability for 310 diseases and injuries, 1990-2015: a systematic analysis for the global burden of disease study 2015. Lancet. 2016;388(10053):1545–1602.
97. Dombret H, Gardin C. An update of current treatments for adult acute myeloid leukemia. Blood. 2016;127(1):53–61.
98. Behrmann L, Wellbrock J, Fiedler W. Acute myeloid leukemia and the bone marrow niche-take a closer look. Front Oncol. 2018;8:444.
99. Ferrajoli A, Shanafelt TD, Ivan C, Shimizu M, Rabe KG, Nouraee N, et al. Prognostic value of miR-155 in individuals with monoclonal B-cell lymphocytosis and patients with B chronic lymphocytic leukemia. Blood. 2013;122(11):1891–9.
100. Huan J, Hornick NI, Shurtleff MJ, Skinner AM, Goloviznina NA, Roberts CT Jr, et al. RNA trafficking by acute myelogenous leukemia exosomes. Cancer Res. 2013;73(2):918–29.
101. Hornick NI, Huan J, Doron B, Goloviznina NA, Lapidus J, Chang BH, et al. Serum exosome MicroRNA as a minimally-invasive early biomarker of AML. Sci Rep. 2015;5:11295.
102. Zhao C, Du F, Zhao Y, Wang S, Qi L. Acute myeloid leukemia cells secrete microRNA-4532-containing exosomes to mediate normal hematopoiesis in hematopoietic stem cells by activating the LDOC1-dependent STAT3 signaling pathway. Stem Cell Res Ther. 2019;10(1):384.
103. Parikh SA, Shanafelt TD. Prognostic factors and risk stratification in chronic lymphocytic leukemia. Semin Oncol. 2016;43(2):233–40.
104. Mineo M, Garfield SH, Taverna S, Flugy A, De Leo G, Alessandro R, et al. Exosomes released by K562 chronic myeloid leukemia cells promote angiogenesis in a Src-dependent fashion. Angiogenesis. 2012;15(1):33–45.
105. Siegel RL, Miller KD, Jemal A. Cancer statistics, 2017. CA Cancer J Clin. 2017;67(1):7–30.
106. Iqbal MA, Arora S, Prakasam G, Calin GA, Syed MA. MicroRNA in lung cancer: role, mechanisms, pathways and therapeutic relevance. Mol Aspects Med. 2019;70:3–20.
107. Herbst RS, Heymach JV, Lippman SM. Lung cancer. N Engl J Med. 2008;359(13):1367–80.
108. Park K, Tan EH, O'Byrne K, Zhang L, Boyer M, Mok T, et al. Afatinib versus gefitinib as first-line treatment of patients with EGFR mutation-positive non-small-cell lung cancer (LUX-Lung 7): a phase 2B, open-label, randomised controlled trial. Lancet Oncol. 2016;17(5):577–89.
109. Kwak EL, Bang YJ, Camidge DR, Shaw AT, Solomon B, Maki RG, et al. Anaplastic lymphoma kinase inhibition in non-small-cell lung cancer. N Engl J Med. 2010;363(18):1693–703.
110. Rabinowits G, Gerçel-Taylor C, Day JM, Taylor DD, Kloecker GH. Exosomal microRNA: a diagnostic marker for lung cancer. Clin Lung Cancer. 2009;10(1):42–6.
111. Poroyko V, Mirzapoiazova T, Nam A, Mambetsariev I, Mambetsariev B, Wu X, et al. Exosomal miRNAs species in the blood of small cell and non-small cell lung cancer patients. Oncotarget. 2018;9(28):19793–806.

112. Jin X, Chen Y, Chen H, Fei S, Chen D, Cai X, et al. Evaluation of tumor-derived exosomal miRNA as potential diagnostic biomarkers for early-stage non-small cell lung cancer using next-generation sequencing. Clin Cancer Res. 2017;23(17):5311–9.
113. Sandfeld-Paulsen B, Aggerholm-Pedersen N, Bæk R, Jakobsen KR, Meldgaard P, Folkersen BH, et al. Exosomal proteins as prognostic biomarkers in non-small cell lung cancer. Mol Oncol. 2016;10(10):1595–602.
114. Vykoukal J, Sun N, Aguilar-Bonavides C, Katayama H, Tanaka I, Fahrmann JF, et al. Plasma-derived extracellular vesicle proteins as a source of biomarkers for lung adenocarcinoma. Oncotarget. 2017;8(56):95466–80.
115. Chen R, Xu X, Tao Y, Qian Z, Yu Y. Exosomes in hepatocellular carcinoma: a new horizon. Cell Commun Signal. 2019;17(1):1.
116. Zheng C, Zheng L, Yoo JK, Guo H, Zhang Y, Guo X, et al. Landscape of infiltrating T cells in liver cancer revealed by single-cell sequencing. Cell. 2017;169(7):1342–1356.e16.
117. Forner A, Reig M, Bruix J. Hepatocellular carcinoma. Lancet. 2018;391(10127):1301–14.
118. Kuczynski EA, Lee CR, Man S, Chen E, Kerbel RS. Effects of Sorafenib dose on acquired reversible resistance and toxicity in hepatocellular carcinoma. Cancer Res. 2015;75(12):2510–9.
119. Marrero JA, Feng Z, Wang Y, Nguyen MH, Befeler AS, Roberts LR, et al. Alpha-fetoprotein, des-gamma carboxyprothrombin, and lectin-bound alpha-fetoprotein in early hepatocellular carcinoma. Gastroenterology. 2009;137(1):110–8.
120. Mitchell PS, Parkin RK, Kroh EM, Fritz BR, Wyman SK, Pogosova-Agadjanyan EL, et al. Circulating microRNAs as stable blood-based markers for cancer detection. Proc Natl Acad Sci U S A. 2008;105(30):10513–8.
121. Li J, Wang T, Jin B, Li W, Wang Z, Zhang H, et al. Diagnosis accuracy of serum glypican-3 level in patients with hepatocellular carcinoma: a systematic review with meta-analysis. Int J Biol Markers. 2018;33(4):353–63.
122. Sohn W, Kim J, Kang SH, Yang SR, Cho JY, Cho HC, et al. Serum exosomal microRNAs as novel biomarkers for hepatocellular carcinoma. Exp Mol Med. 2015;47(9):e184.
123. Yu LX, Zhang BL, Yang Y, Wang MC, Lei GL, Gao Y, et al. Exosomal microRNAs as potential biomarkers for cancer cell migration and prognosis in hepatocellular carcinoma patient-derived cell models. Oncol Rep. 2019;41(1):257–69.
124. Xu H, Chen Y, Dong X, Wang X. Serum exosomal long noncoding RNAs ENSG00000258332.1 and LINC00635 for the diagnosis and prognosis of hepatocellular carcinoma. Cancer Epidemiol Biomark Prev. 2018;27(6):710–6.
125. Lee YR, Kim G, Tak WY, Jang SY, Kweon YO, Park JG, et al. Circulating exosomal non-coding RNAs as prognostic biomarkers in human hepatocellular carcinoma. Int J Cancer. 2019;144(6):1444–52.
126. Sugimachi K, Matsumura T, Hirata H, Uchi R, Ueda M, Ueo H, et al. Identification of a bona fide microRNA biomarker in serum exosomes that predicts hepatocellular carcinoma recurrence after liver transplantation. Br J Cancer. 2015;112(3):532–8.
127. Zhang PF, Gao C, Huang XY, Lu JC, Guo XJ, Shi GM, et al. Cancer cell-derived exosomal circUHRF1 induces natural killer cell exhaustion and may cause resistance to anti-PD1 therapy in hepatocellular carcinoma. Mol Cancer. 2020;19(1):110.
128. Beckham CJ, Olsen J, Yin PN, Wu CH, Ting HJ, Hagen FK, et al. Bladder cancer exosomes contain EDIL-3/Del1 and facilitate cancer progression. J Urol. 2014;192(2):583–92.
129. Lin SY, Chang CH, Wu HC, Lin CC, Chang KP, Yang CR, et al. Proteome profiling of urinary exosomes identifies alpha 1-antitrypsin and H2B1K as diagnostic and prognostic biomarkers for urothelial carcinoma. Sci Rep. 2016;6:34446.
130. Zhan Y, Du L, Wang L, Jiang X, Zhang S, Li J, et al. Expression signatures of exosomal long non-coding RNAs in urine serve as novel non-invasive biomarkers for diagnosis and recurrence prediction of bladder cancer. Mol Cancer. 2018;17(1):142.
131. Berrondo C, Flax J, Kucherov V, Siebert A, Osinski T, Rosenberg A, et al. Expression of the long non-coding RNA HOTAIR correlates with disease progression in bladder cancer and is contained in bladder cancer patient urinary exosomes. PLoS One. 2016;11(1):e0147236.

132. Chen C, Luo Y, He W, et al. Exosomal long noncoding RNA LNMAT2 promotes lymphatic metastasis in bladder cancer. J Clin Invest. 2020;130(1):404–21. https://doi.org/10.1172/JCI130892.
133. Sun X, Li CW, Wang WJ, Chen MK, Li H, Lai YJ, et al. Inhibition of c-MET upregulates PD-L1 expression in lung adenocarcinoma. Am J Cancer Res. 2020;10(2):564–71.
134. Zheng R, Du M, Wang X, Xu W, Liang J, Wang W, et al. Exosome-transmitted long non-coding RNA PTENP1 suppresses bladder cancer progression. Mol Cancer. 2018;17(1):143.
135. Xue M, Chen W, Xiang A, Wang R, Chen H, Pan J, et al. Hypoxic exosomes facilitate bladder tumor growth and development through transferring long non-coding RNA-UCA1. Mol Cancer. 2017;16(1):143.
136. Armstrong DA, Green BB, Seigne JD, Schned AR, Marsit CJ. MicroRNA molecular profiling from matched tumor and bio-fluids in bladder cancer. Mol Cancer. 2015;14:194.
137. Samsonov R, Burdakov V, Shtam T, Radzhabova Z, Vasilyev D, Tsyrlina E, et al. Plasma exosomal miR-21 and miR-181a differentiates follicular from papillary thyroid cancer. Tumour Biol. 2016;37(9):12011–21.
138. Huang T-Y, Wang C-Y, Chen K-Y, Huang L-T. Urinary exosomal thyroglobulin in thyroid cancer patients with post-ablative therapy: a new biomarker in thyroid cancer. Front Endocrinol (Lausanne). 2020;11:382.
139. Dai D, Tan Y, Guo L, Tang A, Zhao Y. Identification of exosomal miRNA biomarkers for diagnosis of papillary thyroid cancer by small RNA sequencing. Eur J Endocrinol. 2020;182(1):111–21.
140. Jiang K, Li G, Chen W, Song L, Wei T, Li Z, et al. Plasma exosomal miR-146b-5p and miR-222-3p are potential biomarkers for lymph node metastasis in papillary thyroid carcinomas. Onco Targets Ther. 2020;13:1311–9.
141. Wu G, Zhou W, Pan X, Sun Z, Sun Y, Xu H, et al. Circular RNA profiling reveals exosomal circ_0006156 as a novel biomarker in papillary thyroid cancer. Mol Ther Nucl Acids. 2020;19:1134–44.
142. Zhang Y, Liu YT, Tang H, Xie WQ, Yao H, Gu WT, et al. Exosome-transmitted lncRNA H19 inhibits the growth of pituitary adenoma. J Clin Endocrinol Metab. 2019;104(12):6345–56.
143. Yu S, Wang XS, Cao KC, Bao XJ, Yu J. Identification of CDK6 and RHOU in serum exosome as biomarkers for the invasiveness of non-functioning pituitary adenoma. Chin Med Sci J. 2019;34(3):168–76.
144. Bijnsdorp IV, Geldof AA, Lavaei M, Piersma SR, van Moorselaar RJA, Jimenez CR. Exosomal ITGA3 interferes with non-cancerous prostate cell functions and is increased in urine exosomes of metastatic prostate cancer patients. J Extracell Vesicles. 2013;2(1):22097.
145. Park YH, Shin HW, Jung AR, Kwon OS, Choi YJ, Park J, et al. Prostate-specific extracellular vesicles as a novel biomarker in human prostate cancer. Sci Rep. 2016;6:30386.
146. Bryant RJ, Pawlowski T, Catto JWF, Marsden G, Vessella RL, Rhees B, et al. Changes in circulating microRNA levels associated with prostate cancer. Br J Cancer. 2012;106(4):768–74.
147. Huang X, Yuan T, Liang M, Du M, Xia S, Dittmar R, et al. Exosomal miR-1290 and miR-375 as prognostic markers in castration-resistant prostate cancer. Eur Urol. 2015;67(1):33–41.
148. Bhagirath D, Yang TL, Bucay N, Sekhon K, Majid S, Shahryari V, et al. microRNA-1246 is an exosomal biomarker for aggressive prostate cancer. Cancer Res. 2018;78(7):1833–44.
149. Yu L, Sui B, Fan W, Lei L, Zhou L, Yang L, et al. Exosomes derived from osteogenic tumor activate osteoclast differentiation and concurrently inhibit osteogenesis by transferring COL1A1-targeting miRNA-92a-1-5p. J Extracell Vesicles. 2021;10(3):e12056.
150. McKiernan J, Donovan MJ, O'Neill V, Bentink S, Noerholm M, Belzer S, et al. A novel urine exosome gene expression assay to predict high-grade prostate cancer at initial biopsy. JAMA Oncol. 2016;2(7):882–9.
151. McKiernan J, Noerholm M, Tadigotla V, Kumar S, Torkler P, Sant G, et al. A urine-based exosomal gene expression test stratifies risk of high-grade prostate cancer in men with prior negative prostate biopsy undergoing repeat biopsy. BMC Urol. 2020;20(1):138.

152. Nilsson J, Skog J, Nordstrand A, Baranov V, Mincheva-Nilsson L, Breakefield XO, et al. Prostate cancer-derived urine exosomes: a novel approach to biomarkers for prostate cancer. Br J Cancer. 2009;100(10):1603–7.
153. Russo LM, Bate K, Motamedinia P, Salazar G, Scott A, Lipsky M, et al. Urinary exosomes as a stable source of mRNA for prostate cancer analysis. J Clin Oncol. 2012;30(5_suppl):174.
154. Li Z, Ma YY, Wang J, Zeng XF, Li R, Kang W, et al. Exosomal microRNA-141 is upregulated in the serum of prostate cancer patients. Onco Targets Ther. 2016;9:139–48.
155. McKiernan J, Donovan MJ, Margolis E, Partin A, Carter B, Brown G, et al. A prospective adaptive utility trial to validate performance of a novel urine exosome gene expression assay to predict high-grade prostate cancer in patients with prostate-specific antigen 2-10ng/ml at initial biopsy. Eur Urol. 2018;74(6):731–8.
156. Skotland T, Ekroos K, Kauhanen D, Simolin H, Seierstad T, Berge V, et al. Molecular lipid species in urinary exosomes as potential prostate cancer biomarkers. Eur J Cancer. 2017;70:122–32.
157. Puhka M, Takatalo M, Nordberg M-E, Valkonen S, Nandania J, Aatonen M, et al. Metabolomic profiling of extracellular vesicles and alternative normalization methods reveal enriched metabolites and strategies to study prostate cancer-related changes. Theranostics. 2017;7(16):3824–41.
158. Brady JV, Troyer RM, Ramsey SA, Leeper H, Yang L, Maier CS, et al. A preliminary proteomic investigation of circulating exosomes and discovery of biomarkers associated with the progression of osteosarcoma in a clinical model of spontaneous disease. Transl Oncol. 2018;11(5):1137–46.
159. Wang J, Zhang H, Sun X, Wang X, Ren T, Huang Y, et al. Exosomal PD-L1 and N-cadherin predict pulmonary metastasis progression for osteosarcoma patients. J Nanobiotechnol. 2020;18(1):151.
160. Xu JF, Wang YP, Zhang SJ, Chen Y, Gu HF, Dou XF, et al. Exosomes containing differential expression of microRNA and mRNA in osteosarcoma that can predict response to chemotherapy. Oncotarget. 2017;8(44):75968–78.
161. Miller IV, Raposo G, Welsch U, Prazeres da Costa O, Thiel U, Lebar M, et al. First identification of Ewing's sarcoma-derived extracellular vesicles and exploration of their biological and potential diagnostic implications. Biol Cell. 2013;105(7):289–303.
162. Tsugita M, Yamada N, Noguchi S, Yamada K, Moritake H, Shimizu K, et al. Ewing sarcoma cells secrete EWS/Fli-1 fusion mRNA via microvesicles. PLoS One. 2013;8(10):e77416.
163. Zhang P, Samuel G, Crow J, Godwin AK, Zeng Y. Molecular assessment of circulating exosomes toward liquid biopsy diagnosis of Ewing sarcoma family of tumors. Transl Res. 2018;201:136–53.

# Extracellular Vesicle Biomarkers for Diagnosis of Cardiovascular Diseases

# 17

Lu Peng and Huina Zhang

## 17.1 Introduction

The cardiovascular system comprises the heart, arteries, veins, and capillaries, among which the heart is of paramount significance as a powerful pump [1]. Circulating blood transfers oxygen and nutrients to different organs and removes metabolic waste derived from cells to maintain the stability of whole body function. Obesity, cigarette smoking, high blood glucose levels, and elevated cholesterol levels are common risk factors for the development of cardiovascular diseases (CVDs) [2]. With improvements in living standards, the prevalence of and mortality rates associated with CVDs are increasing annually, thereby affecting the health and financial stability of patients suffering from cardiovascular events [3]. As key mediators of intercellular communication, extracellular vesicles (EVs) play active roles in the initiation and progression of cardiovascular lesions and have promising value in the diagnosis of CVDs owing to their specific compositions [4].

## 17.2 Current Status of CVD Diagnosis

CVD, a life-threatening illness, is the leading cause of mortality worldwide [5]. Traditional diagnostic methods, such as cardiovascular imaging and biochemical detection, are usually used for the diagnosis of patients with impaired cardiovascular structure and function [6]. However, these methods are not applicable in patients with acute and dangerous diseases, such as acute coronary syndrome and aortic aneurysm. Consequently, there is an urgent need to identify novel biomarkers of CVDs, which will help identify cardiovascular events in the early stages prior to the

L. Peng · H. Zhang (✉)
Beijing Institute of Heart, Lung and Blood Vessel Disease, Beijing Anzhen Hospital, Capital Medical University, Beijing, China



Q. Wang, L. Zheng (eds.), *Extracellular Vesicles*,
https://doi.org/10.1007/978-981-99-8365-0_17

occurrence of obvious clinical symptoms. In recent years, the number of emerging cardiovascular biomarkers has continued to increase with the development of -omics technologies. Several protein biomarkers, including myeloperoxidase and leptin, identified by proteomic platforms, have been reported to predict cardiovascular outcomes and all-cause mortality [7]. Circulating microRNAs (miRNAs) have also been shown to be associated with CVDs [8]. Future efforts should be made to explore more accurate and convenient biomarkers in CVD.

## 17.3 Diagnostic Value of EVs in CVDs

EVs carry specific parental cell-derived cargo that reflects the cell's physiological and pathological features and may therefore have applications as diagnostic biomarkers [9]. An increasing number of researchers are focusing on the potential clinical applications of EVs. The results of experimental and clinical research on EVs have provided clues for the translation of EV biomarkers from bench to bedside. EV-based analyses have conformed with the "less is more" principle, characterized by a smaller but more defined sample highly enriched with specific biomarkers [10]. Unlike biomacromolecules circulating in biofluids, proteins and nucleic acids encapsulated in EVs can stably exist in the extracellular microenvironment for a longer time because of the protection provided by the lipid bilayer [11]. In addition, the number of EVs is associated with the presence and severity of CVDs [12]. These characteristics make EVs ideal biomarker carriers. Early warning information from EVs will assist clinicians in the prevention and treatment of CVDs.

## 17.4 Associations of EV Biomarkers with CVDs

### 17.4.1 EVs as Potential Biomarkers in Vascular Diseases

#### 17.4.1.1 EVs and Atherosclerosis

Atherosclerosis is a chronic, multifocal, immunoinflammatory disease of the large and moderate arteries [13]. The main features of atherosclerosis are stimulated endothelium, lipid disposition, leukocyte recruitment, and the fibroproliferation response mediated by smooth muscle cells [14]. Acute myocardial infarction attributed to acute thrombotic occlusion of an artery is one of the main causes of death in patients with atherosclerosis [15]. EVs derived from various cell types can have both beneficial and detrimental effects on the development of atherosclerosis by regulating certain physiological and pathological processes [16]. A previous study found that exosomal *miR-155* derived from smooth muscle cells promotes atherosclerosis formation by destroying the integrity of endothelial barriers [17]. Thrombin-stimulated platelets secrete exosomes enriched with *miR-223*, *miR-339*, and *miR-21*, which inhibit the proliferation of vascular smooth muscle cells induced by platelet-derived growth factor [18]. In patients with atherosclerosis, exosomal *miR-92a* and *miR-30e* levels are significantly elevated [19]. Annexin V+ circulating microvesicles released by platelets, neutrophils, and endothelial cells can discriminate coronary atherosclerotic plaques and calcifications in asymptomatic patients with familial hypercholesterolemia [20]. These findings indicate that EV-encapsulated cargo and EV amount can provide messages related to atherosclerosis and may be useful biomarkers for detection of early pathological changes associated with atherosclerosis.

#### 17.4.1.2 EVs and Hypertension

Hypertension, characterized by endothelial dysfunction, is a common disease that can seriously damage the heart, brain, and kidneys and is associated with metabolic changes in the body [21]. Because the molecular pathogenesis of hypertension is sophisticated and has not yet been fully elucidated, it is necessary to gain insights into the pathogenic mechanisms of hypertension and to identify easy-to-measure markers for the optimization of clinical treatment. Exosomal *miR-17* levels are elevated in serum from hypertensive rats and activate pro-inflammatory signaling pathways in endothelial cells [22]. The concentration of circulating endothelial microparticles, defined by membrane-specific antigens, is significantly higher in hypertensive patients than in control subjects [23]. Moreover, *miR-27a* and *miR-181a* expression levels are increased in plasma EVs of patients with hypertension, and monocyte-derived *miR-27a* in EVs participates in impaired vasodilation [24]. In patients diagnosed with obstructive sleep apnea (OSA), the biological functions and miRNA components of plasma exosomes differ in hypertensive individuals with normal dipping blood pressure and reverse dipping blood pressure, which explains the unnatural circadian patterns of blood pressure and provides evidence for individual treatment of hypertension in patients with OSA [25]. These findings suggest that circulating EVs and their specific cargo are likely novel biomarkers that can aid in the diagnosis and management of different types of hypertension [26].

#### 17.4.1.3 EVs and Venous Thromboembolism

Venous thromboembolism, including deep venous thrombosis and pulmonary embolism, is associated with slow blood flow, injured venous endothelium, and hypercoagulation [27]. Venous thromboembolism is related to the prothrombotic effects of lifestyle, medical interventions, and cancer, which can explain the higher prevalence of venous thromboembolism in the growing population [28]. Tissue factor-exposed microparticles are involved in the pathogenic mechanism of venous thromboembolism and are increased in patients with unprovoked venous thromboembolism, although the results remain conflicting; these features provide important clues for the diagnosis of venous thromboembolism [29, 30]. Additionally, the procoagulant activity of plasma EVs was found to be higher in patients with acute leukemia who experienced thrombotic events, supporting the potential applications of EVs as biomarkers to distinguish patients with or without thrombotic risk [31]. In the context of cancer, EV-exposing tissue factor was found to be positively associated with the risk of thrombosis and could be used as a biomarker for evaluating cancer-related venous thromboembolism, as demonstrated in a cohort study [32]. In addition, tissue factor-positive EVs may act as biomarkers for predicting survival in patients with several types of cancer, including pancreatic cancer [33]. In a small case-control study conducted by Jamaly et al., patients with unprovoked venous thromboembolism had higher levels of plasma P-selectin glycoprotein ligand-1-bearing microvesicles than healthy subjects [34]. Moreover, circulating levels of microparticles shedding from platelets were found to increase significantly in patients with pulmonary thromboembolism and to exhibit high accuracy, similar to that of D-dimer, in the diagnosis of pulmonary thromboembolism [35]. Most studies evaluating the relationships between EVs and venous thrombosis have focused on tissue factor-positive EVs and cancer-related venous thrombosis. Larger population-based clinical studies and exploration of venous thromboembolism risk in other diseases are required to uncover the latent value of EVs in the diagnosis of venous thromboembolism.

#### 17.4.1.4 EVs and Aortic Aneurysm

Aortic aneurysm is a life-threatening CVD characterized by focal pathological enlargement and structural abnormalities of the arterial wall [36]. Ruptured aortic aneurysms are extremely dangerous events that may cause sudden death and are a challenge to manage [37]. Accordingly, a comprehensive understanding of the mechanisms underlying aortic aneurysm is essential to identify and confirm effective biomarkers that can indicate the risk of aneurysm rupture and to distinguish patients with aortic aneurysm from healthy subjects. A previous differential proteomic analysis of plasma-derived microvesicles detected the levels of proteins related to the pathological mechanisms of abdominal aortic aneurysm and observed upregulation of ferritin, small integral membrane protein 1, platelet factor 4, and mitochondrial heat shock protein 60 in samples from patients with abdominal aortic aneurysm, providing informative data for exploring protein biomarkers of abdominal aortic aneurysm [38]. Another recent study found that *miR-574-5p* levels were upregulated in serum samples from patients with thoracic aortic aneurysm as well as in plasma EVs from Marfan $Fbn1^{C1041G/+}$ mice, supporting the diagnostic potential of *miR-574-5p* in circulating EVs for thoracic aortic aneurysm [39]. In patients with ascending aortic aneurysms, circulating exosomal expression of *miR-133a*, *miR-320a*, and *miR-34a* differs depending on the aortic valve morphotype [40]. Furthermore, elevated wall tension associated with aneurysm disease increases the expression of *miR-133a* in exosomes derived from thoracic aortic fibroblasts, suggesting that exosomal *miR-133a* may be a sensitive biomarker for aortic aneurysm [41]. This evidence indicates that EV-containing proteins and miRNAs may become important biomarkers in the initiation and progression of aortic aneurysms.

### 17.4.2 EVs as Potential Biomarkers in Heart Diseases

#### 17.4.2.1 EVs and Coronary Artery Disease (CAD)

CAD, an atherosclerotic disease, is the leading cause of mortality globally and manifests as stable angina, unstable angina, myocardial infarction, or sudden cardiac death [42]. Coronary angiography, intravascular ultrasound, optical coherence tomography, and functional assessment are common invasive diagnostic methods that can offer invaluable information to aid the diagnosis and management of CAD [43]. EVs exert crucial effects on the development and progression of CAD as mediators of intercellular communication [44]. Moreover, EVs in numerous liquid biopsies contain abundant biological information and have the potential to serve as diagnostic biomarkers [45]. A previous study demonstrated that the expression of serum exosomal *miR-208a* was increased in patients with acute coronary syndrome (ACS), suggesting that exosomal *miRNA-208a* may be critical for the timely recognition of ACS [46]. Additionally, the total number of plasma EVs is increased in patients with ACS, and this increase is mainly due to the increased number of EVs expressing surface CD41a antigen derived from platelets, indicating that platelet-derived EVs may act as early indicators of ACS [47, 48]. Compared with patients with stable angina, those with ACS have higher levels of leukocyte-derived microparticles [49]. Interestingly, endothelial microvesicle numbers are increased in patients with CAD compared with controls, patients with ACS have higher numbers of endothelial microvesicles than patients with stable CAD, and elevated numbers of endothelial microvesicles are positively and strongly associated with central systolic blood pressure [50]. Circulating microvesicle-incorporated *miR-92a-3p*, predominantly derived from endothelial cells, is increased in patients with CAD compared with that in patients without CAD; therefore, further exploration of the potential diagnostic roles of endothelial EVs in CAD is necessary [51]. The presence of ubiquitinated adenosine $A_{2A}$ receptors in plasma EVs has been observed in patients with CAD and is related to hyperhomocysteinemia [52]. Overall, analysis of EVs as potential circulating biomarkers in patients with CAD may have important clinical applications and should be evaluated in additional large-scale clinical studies.

#### 17.4.2.2 EVs and Myocardial Infarction

Myocardial infarction is a devastating syndrome characterized by cardiac cell death resulting from a sharp reduction in or disruption of the coronary blood supply [53]. Although therapeutic strategies that can reduce ischemic injury of the myocardium have been identified, accessible biomarkers beneficial for rapid and correct diagnosis of acute myocardial infarction are still lacking [54]. Circulating cardiac-specific miRNAs, such as *miR-499*, produced by cardiomyocytes are dramatically increased in patients with acute myocardial infarction, suggesting the diagnostic potential of these miRNAs as attractive myocardial infarction biomarkers [55]. Exosomes enriched with cardiac-specific miRNAs can reflect cardiomyocyte damage in response to ischemia and may serve as valuable reservoirs of available biomarkers for diagnostic purposes [56]. A proteomic analysis compared the protein cargo of plasma EVs between patients with myocardial infarction and control patients with stable angina and obtained a panel of six novel EV markers of myocardial injury associated with complement activation, lipid metabolism, and platelet activation. The results validated by the antibody-based assay also demonstrated that effective EV-based diagnostic strategies can be used to inform clinical decisions [57]. The quantity of circulating plasma EVs mainly originating from endothelial cells was reported to increase in patients diagnosed with ST-segment elevation acute myocardial infarction and was positively correlated with the severity of myocardial injury [58]. Another study revealed that the number of $CD31^{+}/CD42b^{-}$ microparticles secreted by endothelial cells was higher in patients with acute myocardial infarction than that in patients with stable coronary heart disease and healthy individuals [59]. Although plasma and EVs isolated from plasma shared common differentially expressed proteins in patients with myocardial infarction and controls, chemokine ligand 17, chymotrypsin C, and proto-oncogene tyrosine-protein kinase SRC were dysregulated only in the EV fraction, indicating that analyzing the proteomic profiles of circulating EVs can provide additional diagnostic information [60]. Taken together, these findings suggest that the number of EVs and their molecular cargo can change significantly in conditions of myocardial infarction and may be attractive tools for the diagnosis of myocardial infarction [61].

#### 17.4.2.3 EVs and Heart Failure

Heart failure is a debilitating condition in which the heart is too weak to pump enough blood to satisfy the body's metabolic needs and is the final outcome of various CVDs [62]. Heart failure is a prevalent disease in modern society, affecting approximately 40 million people worldwide [63]. To assist in the diagnosis of heart failure, multiple studies have focused on finding reliable biomarkers primarily from serum [64]; however, novel biomarkers in EVs are also expected to be beneficial. A dyspnea cohort study observed that the levels of SerpinG1, cystatin C, and CD14 packaged in total plasma EVs were significantly higher in patients with heart failure than in controls [65]. A study initiated by Beg et al. reported that the circulating exosomal *miR-146a/miR-16* ratio, which is closely related to inflammation, was higher in patients with heart failure than in controls, whereas the plasma *miR-146a/miR-16* ratio did not differ significantly between the two groups [66]. In children with univentricular heart disease, the concentrations of *miR129-5p* in plasma microvesicles are negatively associated with the degree of clinical heart failure, demonstrating that *miR129-5p* contained in microvesicles may act as a specific biomarker for the presence and stratification of heart failure in pediatric univentricular patients [67]. The levels of plasma exosomes and plasma-derived exosomal mitochondrial DNA (mtDNA) copy numbers were reported to increase distinctly in patients with chronic heart failure, revealing a new mechanism of exosome-mediated inflammatory process and establishing an intriguing research direction in exosomal mtDNA-related fields [68]. Reductions in plasma exosomal *miR-425* and *miR-744* have been identified in patients with heart failure and have been shown to be linked to fibrogenesis of cardiac fibroblasts [69]. Moreover, a recent study utilizing RNA sequencing analysis showed that the expression levels of serum exosomal *hsa-piR-020009* and *hsa-piR-006426* were dramatically downregulated in patients with heart failure compared with those in healthy volunteers, providing insights into available biomarker selection for heart failure [70]. Furthermore, serum *exo-miR-92b-5p* levels were confirmed to be elevated in patients with heart failure with reduced ejection fraction and in patients with acute heart failure caused by dilated cardiomyopathy, indicating a promising diagnostic role of *exo-miR-92b-5p* [71, 72]. Additionally, a recent study found that subpopulations of apoptotic endothelial cell-derived microvesicles were associated with different phenotypes of heart failure and may improve the predictive capacity of established chronic heart failure biomarkers [73]. Overall, circulating EVs have been extensively studied in heart failure and may have applications as diagnostic biomarkers for heart failure and progression.

#### 17.4.2.4 EVs and Cardiomyopathy

Cardiomyopathy, a pathological condition characterized by impaired structure and function of the myocardium, can be categorized as primary or secondary cardiomyopathy [74]. The process of primary cardiomyopathy, including hypertrophic cardiomyopathy, dilated cardiomyopathy (DCM), restrictive cardiomyopathy, arrhythmogenic right ventricular cardiomyopathy, and left ventricular noncompaction, is predominantly confined to the heart, whereas secondary cardiomyopathy usually occurs subsequent to infection, metabolic disease, or other systemic conditions [75, 76]. The underlying pathogenesis of cardiomyopathy is not well defined, whereas genetic testing, which can decipher the molecular etiology of the disease, provides guidance on the identification of patients with cardiomyopathy and the recognition of individuals vulnerable to cardiomyopathy [77]. Due to the complex heterogeneity cardiomyopathy pathogenesis, more informative detection methods are required. A previous study used proteome analysis to compare the proteomic signature of plasma-derived EVs from patients with DCM and healthy controls and found that 51 proteins mainly linked to stress and protein activation were unique to the DCM group, revealing that differentially expressed proteins in plasma EVs may mirror the morbid state of DCM and have the potential to be diagnostic biomarkers of DCM [78]. However, another study observed that circulating levels of low-density lipoprotein receptor-related protein 1 were increased in the plasma rather than in plasma-derived EVs from patients with idiopathic DCM compared with healthy controls [79]. Therefore, EVs may contain partial disease-specific information about DCM, and further investigations are needed. A recent study found that exosomal *miR-29a* participates in obesity-related cardiomyopathy by dampening mitochondrial function [80]. In obese individuals, the level of plasma exosomal *miR-122* is significantly increased and positively correlated with cardiac dysfunction. These findings suggest that exosomal miRNAs may hold promise as diagnostic markers of metabolic cardiomyopathy [81]. In the field of cardiomyopathy, ongoing biomarker research is still in its infancy, and the applicability of EVs as diagnostic biomarkers needs to be verified through more in-depth studies.

#### 17.4.2.5 EVs and Myocarditis

Myocarditis is clinically and pathologically viewed as an inflammatory disease of the myocardium, mainly attributed to viral infections, particularly the Coxsackie B virus [82]. Presently, the most common methods applied for the diagnosis of myocarditis are echocardiography, cardiac magnetic resonance imaging, and endomyocardial biopsy [83]. However, studies on EV biomarkers in myocarditis have attracted much attention. Coxsackie B virus infection can boost the release of *miR-590-5p* from EVs, illustrating a potential mechanism through which Coxsackie B virus facilitates viral infection and organ damage [84]. Notably, $CD144^{+}$ microparticles secreted by endothelial cells are markedly increased in patients diagnosed with myocarditis induced by parvovirus B19 compared with that in patients with myocarditis who are negative for parvovirus B19 and that in healthy controls [85]. In a study of patients with acute myocarditis, five miRNAs encapsulated in serum exosomes, i.e., *miR-30a*, *miR-181d*, *miR-125a*, *miR-155*, and *miR-21*, were found to be significantly altered, revealing the roles of these miRNAs as potential candidates for viral myocarditis diagnosis [86]. A recent study found that *miR-142* expression in circulating exosomes was higher in mice with experimental autoimmune myocarditis than in control groups and was involved in the development of myocarditis by inducing immune-metabolic reprogramming [87]. Thus, EVs can not only mediate mechanisms related to pathological damage but also act as possible biomarkers for the diagnosis and classification of myocarditis.

#### 17.4.2.6 EVs and Arrhythmia

Arrhythmia is generally considered a frequency and/or rhythm abnormality of the cardiac beat ascribed to an aberrant origin and/or cardiac conduction disorders [88]. Several protein factors reflecting inflammation, fibrosis, and autoimmunity have been reported as available biomarkers for the early recognition of cardiac arrhythmia, particularly atrial fibrillation (AF), the most common type of arrhythmia [89]. A clinical study conducted by Pourtau et al. found that the levels of tissue factor-dependent procoagulant activity of plasma microparticles were strikingly decreased in patients with paroxysmal or persistent AF compared with that in controls, supporting the local action of procoagulant microparticles [90]. Another study observed higher levels of plasma EVs exposing tissue factor in nonvalvular AF patients than in patients without AF [91]. In addition, the plasma levels of platelet microparticle CD62P were significantly higher in patients with AF in a hypercoagulable state than in patients with non-hypercoagulable AF and healthy volunteers [92]. In patients with nonvalvular AF, total circulating phosphatidylserine (PS)-bearing microparticles mostly derived from platelets were increased and played a pivotal role in the hypercoagulable state [93]. By contrast, Siwaponanan et al. found that serum PS-expressing microparticle levels were similar between patients with AF and healthy controls [94]. There were no significant differences in mtDNA copy numbers encapsulated in plasma exosomes between patients with AF and healthy controls, according to a recent study [95]. Additionally, microarray analysis of miRNAs determined the changes in miRNA expression in serum exosomes from patients with AF and showed that five miRNAs, i.e., *miR-107*, *miR-320d*, *miR-103a*, *miR-486*, and *let-7b*, had a more than 4.5-fold increase in expression in patients with persistent AF compared with controls with supraventricular tachycardia, suggesting the potential role of serum exosomal miRNAs as novel biomarkers of AF identification and progression [96]. Presently, EV biomarker studies in arrhythmia mainly focus on AF and have provided insights into novel diagnostic strategies for AF; hence, there is an urgent need for further exploration of other types of arrhythmia.

#### 17.4.2.7 EVs and Other Relatively Rare Heart Diseases

Plasma endothelial microparticles reflecting endothelial dysfunction are significantly increased in patients with mitral valve disease and are associated with impairment of mitral valve endothelial function by reducing nitric oxide production and increasing $O_2^-$ generation [97]. In children with congenital heart disease (CHD), platelet microparticles are significantly elevated compared with those in healthy children controls, and patients with cyanotic CHD show more obvious elevation of platelet microparticle levels than patients with acyanotic CHD [98]. The levels of plasma endothelial microparticles are higher in adults with atrial septal defects or ventricular septal defects than in healthy subjects, and these microparticles may play pivotal roles in endothelial dysfunction via p38 mitogen-activated protein kinase-dependent pathways [99]. A recent clinical study revealed that the numbers of plasma microvesicles released from various blood cells were significantly higher in patients suffering from infective endocarditis than in patients with other bacterial infections, demonstrating that plasma EVs may have potential applications as useful diagnostic biomarkers for patients with suspected infective endocarditis [100]. Overall, these seminal studies have established a basis for the future application of EVs as diagnostic biomarkers in CVDs.

Based on present experimental and clinical studies, the roles of EVs in detection of the initiation and development of CVDs are unclear. Furthermore, bioactive molecules packaged in EVs have promising potential as diagnostic biomarkers for CVDs and also have diagnostic value for discriminating patients from healthy subjects. Improving our understanding of the molecular architecture of EVs in patients with CVDs and in healthy individuals could lead to clinical advances in the diagnosis and treatment of CVDs. We look forward to an era when EVs can be utilized as effective and accurate tools in the diagnosis of CVDs to achieve early prevention (Tables 17.1 and 17.2).

**Table 17.1** Summary of EV contents as potential biomarkers in CVDs

| Disease | Sample size ($n$) | Source | Method of isolation | Cargoes | Quantification method | Trend | Potential application | References |
|---|---|---|---|---|---|---|---|---|
| Coronary atherosclerosis | 42 | Plasma | Precipitation | miR-92a, miR-30e | qRT-PCR | Increase | Diagnosis | [19] |
| Hypertension | 6 | Plasma | Ultracentrifugation | miR-27a, miR-181a | qRT-PCR | Increase | Diagnosis | [24] |
| Abdominal aortic aneurysms | 10 | Plasma | Ultracentrifugation | Ferritin, small integral membrane protein 1, platelet factor 4, mitochondrial HSP60 | Proteomic analysis | Increase | Diagnosis | [38] |
| Acute coronary syndrome | 500 | Serum | Precipitation | miR-208a | qRT-PCR | Increase | Diagnosis | [46] |
| Coronary artery disease | 139 | Plasma | Ultracentrifugation | miR-92a-3p | qRT-PCR | Increase | Diagnosis | [51] |
| | 8 | Plasma | Precipitation | $A_{2A}$ receptor | Western blotting | Increase | Diagnosis | [52] |
| Myocardial infarction | 17 | Plasma | Precipitation | 6 proteins | Luminex assay and ELISA | Increase | Diagnosis, therapeutic targets | [57] |
| ST-segment elevation myocardial infarction | 60 | Plasma | Acoustic seed trapping | Chemokine ligand 17, chymotrypsin C, and proto-oncogene tyrosine-protein kinase SRC | Proteomic analysis | Decrease | Diagnosis | [60] |
| Acute decompensated heart failure | 141 | Plasma | Precipitation | SerpinG1, Cystatin C, CD14 | Luminex | Increase | Diagnosis | [65] |
| Heart failure with reduced ejection fraction | 40 | Plasma | Precipitation | miR-146a | qRT-PCR | Increase | Diagnosis | [66] |

| | | | | | | | | |
|---|---|---|---|---|---|---|---|---|
| Heart failure in univentricular heart disease | 71 | Plasma | Precipitation | miR129-5p | qRT-PCR | Decrease | Diagnosis | [67] |
| Chronic heart failure | 20 | Plasma | Precipitation | mtDNA | Droplet digital PCR | Increase | Diagnosis | [68] |
| Acute heart failure | 31 | Serum | Precipitation | miR-425, miR-744 | qRT-PCR | Decrease | Prediction | [69] |
| Heart failure | 3 | Serum | Ultracentrifugation | piR-020009, piR-006426 | RNA sequencing | Decrease | Diagnosis | [70] |
| Acute heart failure with reduced ejection fraction | 28 | Serum | Precipitation | miR-92b-5p | qRT-PCR | Increase | Diagnosis | [71] |
| Acute heart failure caused by dilated cardiomyopathy | 43 | Serum | Precipitation | miR-92b-5p | qRT-PCR | Increase | Diagnosis | [72] |
| Dilated cardiomyopathy | 20 | Plasma | Size-exclusion chromatography | 51 proteins | Proteomic analysis | – | Diagnosis | [78] |
| Acute myocarditis | 23 | Serum | Ultracentrifugation | miR-30a, miR-181d, miR-125a, increase; miR-155, miR-21, decrease | qRT-PCR | – | Diagnosis | [86] |
| Persistent atrial fibrillation | 40 | Serum | Precipitation | miR-107, miR-320d, miR-103a, miR-486, let-7b | qRT-PCR | Increase | Progression | [96] |

**Table 17.2** Summary of EVs with surface-specific markers as potential biomarkers in CVDs

| Disease | Sample size ($n$) | Source | Parental cell | Method of isolation | Surface-specific marker | Quantification method | Trend | Potential application | References |
|---|---|---|---|---|---|---|---|---|---|
| Atherosclerosis with familial hypercholesterolemia | 48 | Plasma | Platelet, granulocyte, and neutrophil | Centrifugation | Annexin $V^+$ | Flow cytometry | Increase | Diagnosis | [20] |
| Hypertension | 21 | Plasma | Endothelial cell | – | $CD31^+/41^-$, $CD62e^+$, $CD144^+$ | Flow cytometry | Increase | Diagnosis | [23] |
| Venous thromboembolism | 30 | Plasma | Endothelial cell, platelet | – | Tissue factor, CD146, CD61 | Flow cytometry | Increase | Diagnosis | [30] |
| | 20 | Plasma | Monocyte, endothelial cell | Centrifugation | P-selectin glycoprotein ligand-$1^+$ | Flow cytometry | Increase | Diagnosis | [34] |
| Pulmonary thromboembolism | 102 | Plasma | Platelet | – | CD41a, Annexin V | Flow cytometry | Increase | Diagnosis | [35] |
| Acute coronary syndrome | 13 | Plasma | Platelet | Magnetic separation | CD41a | Flow cytometry | Increase | Diagnosis | [47] |
| | 12 | Plasma | Platelet | – | CD41a | Flow cytometry | Increase | Diagnosis | [48] |
| | 37 | Plasma | Neutrophil, monocyte | Centrifugation | $AnV^+CD42a^-CD66b^+$, $AnV^+CD14^+TF^+$ | Flow cytometry | Increase | Diagnosis | [49] |
| | 38 | Plasma | Endothelial cell | – | CD105, CD144 | Flow cytometry | Increase | Diagnosis | [50] |
| ST-segment elevation acute myocardial infarction | 22 | Plasma | Endothelial cell | Centrifugation | P selectin, VCAM-1 | Nanoparticle tracking analysis | Increase | Diagnosis | [58] |

| | | | | | | | | | |
|---|---|---|---|---|---|---|---|---|---|
| Acute myocardial infarction | 37 | Plasma | Endothelial cell | – | $CD31^+/CD42b^-$ | Flow cytometry | Increase | Diagnosis | [59] |
| Heart failure with reduced ejection fraction | 85 | Plasma | Apoptotic endothelial cell | Centrifugation | $CD144^+$/Annexin $V^+$ | Flow cytometry | Increase | Prediction | [73] |
| Parvovirus B19+ myocarditis | 23 | Plasma | Apoptotic endothelial cell | – | $CD31^+AV^+$ | Flow cytometry | Increase | Diagnosis | [85] |
| Nonvalvular atrial fibrillation | 13 | Plasma | Monocyte-macrophage lineage | – | Tissue factor | Protein microarray | Increase | Diagnosis | [91] |
| Atrial fibrillation | 60 | Plasma | Platelet | – | CD62P | Protein microarray | Increase | Diagnosis | [92] |
| Nonvalvular atrial fibrillation | 60 | Plasma | Platelet, leukocyte, erythrocyte, endothelial cell | Centrifugation | Phosphatidylserine | Flow cytometry | Increase | Diagnosis, therapeutic targets | [93] |
| Mitral valve disease | 81 | Plasma | Endothelial cell | – | $CD31^+/CD42b^-$ | Flow cytometry | Increase | Therapeutic target | [97] |
| Cyanotic congenital heart disease | 23 | Plasma | Platelet | – | $CD41b^+$ | Flow cytometry | Increase | Diagnosis | [98] |
| Atrial septal defect; ventricular septal defect | 20; 23 | Plasma | Endothelial cell | – | $CD31^+/CD42b^-$ | Flow cytometry | Increase | Therapeutic target | [99] |
| Infective endocarditis | 57 | Plasma | Platelet; leukocyte; neutrophil; monocytes; lymphocyte | Centrifugation | CD41a; CD45; CD66b; CD14; CD3 | Flow cytometry | Increase | Diagnosis | [100] |

## 17.5 Associations of EV Biomarkers with Cardiovascular-Related Diseases

Many other metabolic diseases or abnormalities, such as diabetes, obesity, and OSA, are independent risk factors for CVDs. Owing to the characteristics of metabolic diseases, EV biomarkers related to metabolic disease-associated CVDs also have unique properties. A previous study found that plasma exosomal *miR-630* expression was reduced in obese children and children with OSA with endothelial dysfunction compared with their counterparts without endothelial dysfunction. Whole-genome transcriptomic analyses revealed that NRF2-mediated oxidative stress responses and AMP kinase and tight junction signaling pathways in endothelial cells are the major canonical pathways regulated by *miR-630*, suggesting the potential role of exosomal *miR-630* as an indicator of CVD risk [101]. Healthy men exposed to intermittent hypoxia, a major factor contributing to the pathogenesis of OSA-related cardiovascular comorbidities, have altered exosomal miRNAs that can disrupt the endothelial cell monolayer barrier and reduce the expression of endothelial nitric oxide synthase [102]. Moreover, four differentially expressed exosomal miRNAs, i.e., *miR-16-5p*, *miR-4459*, *miR-451a*, and *miR-6510-5p*, were identified in pretreated and post-positive airway pressure subjects diagnosed with obesity hypoventilation syndrome, confirming the role of exosomal cargo in assessing response to therapy [103]. In diabetic patients, the concentration of serum exosomes and the content of arginase 1 delivered by serum exosomes were significantly higher than those in nondiabetic controls, thereby contributing to impaired endothelial function; these findings suggested the potential applications of serum exosomal arginase 1 in diabetes-associated vasculopathy [104]. In addition, Sonic Hedgehog-positive exosomes were reported to be increased in patients with type 2 diabetes [105]. Compared with that in patients with obesity, circulating exosomal *miR-23a-5p* and *miR-6087* expression levels were more than twofold higher in patients with obesity combined with diabetes mellitus, whereas circulating exosomal *miR-6751-3p* levels showed the opposite tendency, providing candidates as biomarkers for obesity-related diabetes mellitus and CVD [106]. Additionally, urinary uromodulin (*UMOD*) mRNA levels are significantly elevated in urinary EVs from patients with diabetic kidney disease (DKD), and the change is correlated with the disease severity; thus, urinary EV *UMOD* mRNA may be a novel biomarker for DKD detection and stratification [107].

In general, EVs provide seminal opportunities for identifying clinical diagnostic biomarkers. Numerous efforts have been made to identify EV biomarkers in CVDs. The analysis of dysregulated EV cargo may lead to the development of new diagnostic tools. There is no doubt that these studies have paved the way for further clinical applications. However, large-scale clinical studies are urgently required to evaluate whether EV analysis could be of sufficient clinical significance and thus dictate a more accurate treatment strategy in patients with altered EV levels or cargo. We expect that EVs will soon have applications as diagnostic biomarkers in the field of CVDs.

## References


1. Granger A, Emambokus N. Cell metabolism cardiovascular biology. Cell Metab. 2015;21(2):151.
2. Chandra P, Bhave R. Assessing the risk factors associated with cardiovascular disease. Eur J Prev Cardiol. 2018;25(9):932–3.
3. Shen C, Ge J. Epidemic of cardiovascular disease in China. Circulation. 2018;138(4):342–4.
4. Jansen F, Nickenig G, Werner N. Extracellular vesicles in cardiovascular disease. Circ Res. 2017;120(10):1649–57.
5. Townsend N, Wilson L, Bhatnagar P, et al. Cardiovascular disease in Europe: epidemiological update 2016. Eur Heart J. 2016;37(42):3232–45.
6. Kramer CK, Shaw LJ, Chandrashekhar Y. Progress in cardiovascular imaging. JACC Cardiovasc Imaging. 2018;11(12):1883–914.
7. Ho JE, Lyass A, Courchesne P, et al. Protein biomarkers of cardiovascular disease and mortality in the community. J Am Heart Assoc. 2018;7(14):e008108.
8. Barwari T, Joshi A, Mayr M. MicroRNAs in cardiovascular disease. J Am Coll Cardiol. 2016;68(23):2577–84.
9. Doyle L, Wang M. Overview of extracellular vesicles, their origin, composition, purpose, and methods for exosome isolation and analysis. Cells. 2019;8(7):727.
10. Momen-Heravi F, Getting SJ, Moschos SA. Extracellular vesicles and their nucleic acids for biomarker discovery. Pharmacol Ther. 2018;192:170–87.
11. Chung IM, Rajakumar G, Venkidasamy B, et al. Exosomes: current use and future applications. Clin Chim Acta. 2020;500:226–32.
12. Sluijter JPG, Davidson SM, Boulanger CM, et al. Extracellular vesicles in diagnostics and therapy of the ischaemic heart: position paper from the working group on cellular biology of the heart of the European Society of Cardiology. Cardiovasc Res. 2018;114(1):19–34.
13. Falk E. Pathogenesis of atherosclerosis. J Am Coll Cardiol. 2006;47(8):C7–C12.
14. Zhu Y, Xian X, Wang Z, et al. Research Progress on the relationship between atherosclerosis and inflammation. Biomolecules. 2018;8(3):80.
15. Muscogiuri G, Annweiler C, Duval G, et al. Vitamin D and cardiovascular disease: from atherosclerosis to myocardial infarction and stroke. Int J Cardiol. 2017;230:577–84.
16. Deng W, Tang T, Hou Y, et al. Extracellular vesicles in atherosclerosis. Clin Chim Acta. 2019;495:109–17.
17. Zheng B, Yin W-N, Suzuki T, et al. Exosome-mediated miR-155 transfer from smooth muscle cells to endothelial cells induces endothelial injury and promotes atherosclerosis. Mol Ther. 2017;25(6):1279–94.
18. Tan M, Yan H-B, Li J-N, et al. Thrombin stimulated platelet-derived exosomes inhibit platelet-derived growth factor receptor-Beta expression in vascular smooth muscle cells. Cell Physiol Biochem. 2016;38(6):2348–65.
19. Wang Z, Zhang J, Zhang S, et al. MiR-30e and miR-92a are related to atherosclerosis by targeting ABCA1. Mol Med Rep. 2019;19(4):3298–304.
20. Chiva-Blanch G, Padró T, Alonso R, et al. Liquid biopsy of extracellular microvesicles maps coronary calcification and atherosclerotic plaque in asymptomatic patients with familial hypercholesterolemia. Arterioscler Thromb Vasc Biol. 2019;39(5):945–55.
21. Khaddaj Mallat R, Mathew John C, Kendrick DJ, et al. The vascular endothelium: a regulator of arterial tone and interface for the immune system. Crit Rev Clin Lab Sci. 2017;54(7–8):458–70.
22. Osada-Oka M, Shiota M, Izumi Y, et al. Macrophage-derived exosomes induce inflammatory factors in endothelial cells under hypertensive conditions. Hypertens Res. 2017;40(4):353–60.
23. Sansone R, Baaken M, Horn P, et al. Endothelial microparticles and vascular parameters in subjects with and without arterial hypertension and coronary artery disease. Data Brief. 2018;19:495–500.

24. Zou X, Wang J, Chen C, et al. Secreted monocyte miR-27a, via mesenteric arterial mas receptor-eNOS pathway, causes hypertension. Am J Hypertens. 2020;33(1):31–42.
25. Khalyfa A, Gozal D, Chan WC, et al. Circulating plasma exosomes in obstructive sleep apnoea and reverse dipping blood pressure. Eur Respir J. 2020;55(1):1901072.
26. La Salvia S, Gunasekaran PM, Byrd JB, et al. Extracellular vesicles in essential hypertension: hidden messengers. Curr Hypertens Rep. 2020;22(10):76.
27. Battinelli EM, Murphy DL, Connors JM. Venous thromboembolism overview. Hematol Oncol Clin North Am. 2012;26(2):345–67.
28. Schulman S, Ageno W, Konstantinides SV. Venous thromboembolism: past, present and future. Thromb Haemost. 2017;117(7):1219–29.
29. Nieuwland R, Sturk A, Bleker S, et al. Clinical significance of tissue factor–exposing microparticles in arterial and venous thrombosis. Semin Thromb Hemost. 2015;41(7):718–27.
30. Campello E, Spiezia L, Radu CM, et al. Endothelial, platelet, and tissue factor-bearing microparticles in cancer patients with and without venous thromboembolism. Thromb Res. 2011;127(5):473–7.
31. Gheldof D, Haguet H, Dogné J-M, et al. Procoagulant activity of extracellular vesicles as a potential biomarker for risk of thrombosis and DIC in patients with acute leukaemia. J Thromb Thrombolysis. 2017;43(2):224–32.
32. Van Es N, Hisada Y, Di Nisio M, et al. Extracellular vesicles exposing tissue factor for the prediction of venous thromboembolism in patients with cancer: a prospective cohort study. Thromb Res. 2018;166:54–9.
33. Hisada Y, Mackman N. Cancer cell-derived tissue factor-positive extracellular vesicles. Curr Opin Hematol. 2019;26(5):349–56.
34. Jamaly S, Basavaraj MG, Starikova I, et al. Elevated plasma levels of P-selectin glycoprotein ligand-1-positive microvesicles in patients with unprovoked venous thromboembolism. J Thromb Haemost. 2018;16(8):1546–54.
35. Wang M, Fu Y, Xu L, et al. Diagnostic value of platelet-derived microparticles in pulmonary thromboembolism: a population-based study. Exp Ther Med. 2018;16(4):3099–106.
36. Quintana RA, Taylor WR. Introduction to the compendium on aortic aneurysms. Circ Res. 2019;124(4):470–1.
37. Fraga-Silva RA, Trachet B. Editorial: novel insights on aortic aneurysm. Curr Pharm Des. 2015;21(28):3993–5.
38. Martinez-Pinna R, De Peredo AG, Monsarrat B, et al. Label-free quantitative proteomic analysis of human plasma-derived microvesicles to find protein signatures of abdominal aortic aneurysms. Proteomics Clin Appl. 2014;8(7–8):620–5.
39. Boileau A, Lino Cardenas CL, Courtois A, et al. MiR-574-5p: a circulating marker of thoracic aortic aneurysm. Int J Mol Sci. 2019;20(16):3924.
40. Pasta S, Agnese V, Gallo A, et al. Shear stress and aortic strain associations with biomarkers of ascending thoracic aortic aneurysm. Ann Thorac Surg. 2020;110(5):1595–604.
41. Akerman AW, Blanding WM, Stroud RE, et al. Elevated wall tension leads to reduced miR-133a in the thoracic aorta by exosome release. J Am Heart Assoc. 2019;8(1):e010332.
42. Malakar AK, Choudhury D, Halder B, et al. A review on coronary artery disease, its risk factors, and therapeutics. J Cell Physiol. 2019;234(10):16812–23.
43. Pyxaras SA, Wijns W, Reiber JHC, et al. Invasive assessment of coronary artery disease. J Nucl Cardiol. 2017;25(3):860–71.
44. Gao X-F, Wang Z-M, Wang F, et al. Exosomes in coronary artery disease. Int J Biol Sci. 2019;15(11):2461–70.
45. Boulanger CM, Loyer X, Rautou P-E, et al. Extracellular vesicles in coronary artery disease. Nat Rev Cardiol. 2017;14(5):259–72.
46. Bi S, Wang C, Jin Y, et al. Correlation between serum exosome derived miR-208a and acute coronary syndrome. Int J Clin Exp Med. 2015;8(3):4275–80.
47. Vagida MS, Arakelyan A, Lebedeva AM, et al. Analysis of extracellular vesicles using magnetic nanoparticles in blood of patients with acute coronary syndrome. Biochemistry (Mosc). 2016;81(4):382–91.

48. Vagida M, Arakelyan A, Lebedeva A, et al. Flow analysis of individual blood extracellular vesicles in acute coronary syndrome. Platelets. 2017;28(2):165–73.
49. Mavroudis CA, Eleftheriou D, Hong Y, et al. Microparticles in acute coronary syndrome. Thromb Res. 2017;156:109–16.
50. Gkaliagkousi E, Gavriilaki E, Vasileiadis I, et al. Endothelial microvesicles circulating in peripheral and coronary circulation are associated with central blood pressure in coronary artery disease. Am J Hypertens. 2019;32(12):1199–205.
51. Liu Y, Li Q, Hosen MR, et al. Atherosclerotic conditions promote the packaging of functional MicroRNA-92a-3p into endothelial microvesicles. Circ Res. 2019;124(4):575–87.
52. Ruf J, Vairo D, Paganelli F, et al. Extracellular vesicles with ubiquitinated adenosine A receptor in plasma of patients with coronary artery disease. J Cell Mol Med. 2019;23(10):6805–11.
53. Thygesen K, Alpert JS, Jaffe AS, et al. Fourth universal definition of myocardial infarction (2018). J Am Coll Cardiol. 2018;72(18):2231–64.
54. Sayed ASM, Xia K, Yang T-L, et al. Circulating microRNAs: a potential role in diagnosis and prognosis of acute myocardial infarction. Dis Markers. 2013;35:561–6.
55. Chen X, Zhang L, Su T, et al. Kinetics of plasma microRNA-499 expression in acute myocardial infarction. J Thorac Dis. 2015;7(5):890–6.
56. Chistiakov D, Orekhov A, Bobryshev Y. Cardiac extracellular vesicles in normal and infarcted heart. Int J Mol Sci. 2016;17(1):63.
57. Cheow ESH, Cheng WC, Lee CN, et al. Plasma-derived extracellular vesicles contain predictive biomarkers and potential therapeutic targets for myocardial ischemic (MI) injury. Mol Cell Proteomics. 2016;15(8):2628–40.
58. Akbar N, Digby JE, Cahill TJ, et al. Endothelium-derived extracellular vesicles promote splenic monocyte mobilization in myocardial infarction. JCI Insight. 2017;2(17):e93344.
59. Zhang Y, Cheng J, Chen F, et al. Circulating endothelial microparticles and miR-92a in acute myocardial infarction. Biosci Rep. 2017;37(2):BSR20170047.
60. Gidlöf O, Evander M, Rezeli M, et al. Proteomic profiling of extracellular vesicles reveals additional diagnostic biomarkers for myocardial infarction compared to plasma alone. Sci Rep. 2019;9(1):8991.
61. Li N, Rochette L, Wu Y, et al. New insights into the role of exosomes in the heart after myocardial infarction. J Cardiovasc Transl Res. 2018;12(1):18–27.
62. Tanai E, Frantz S. Pathophysiology of heart failure. Compr Physiol. 2015;6:187–214.
63. Baman JR, Ahmad FS. Heart failure. JAMA. 2020;324(10).
64. Pourafkari L, Tajlil A, Nader ND. Biomarkers in diagnosing and treatment of acute heart failure. Biomark Med. 2019;13(14):1235–49.
65. Lionetti V, Zhang Y-N, Vernooij F, et al. Extracellular vesicle proteins associated with systemic vascular events correlate with heart failure: an observational study in a Dyspnoea cohort. PLoS One. 2016;11(1):e0148073.
66. Beg F, Wang R, Saeed Z, et al. Inflammation-associated microRNA changes in circulating exosomes of heart failure patients. BMC Res Notes. 2017;10(1):751.
67. Fan G-C, Ramachandran S, Lowenthal A, et al. Plasma microvesicle analysis identifies microRNA 129-5p as a biomarker of heart failure in univentricular heart disease. PLoS One. 2017;12(8):e0183624.
68. Ye W, Tang X, Yang Z, et al. Plasma-derived exosomes contribute to inflammation via the TLR9-NF-κB pathway in chronic heart failure patients. Mol Immunol. 2017;87:114–21.
69. Wang L, Liu J, Xu B, et al. Reduced exosome miR-425 and miR-744 in the plasma represents the progression of fibrosis and heart failure. Kaohsiung J Med Sci. 2018;34(11):626–33.
70. Yang J, Xue FT, Li YY, et al. Exosomal piRNA sequencing reveals differences between heart failure and healthy patients. Eur Rev Med Pharmacol Sci. 2018;22(22):7952–61.
71. Wu T, Chen Y, Du Y, et al. Circulating exosomal miR-92b-5p is a promising diagnostic biomarker of heart failure with reduced ejection fraction patients hospitalized for acute heart failure. J Thorac Dis. 2018;10(11):6211–20.
72. Wu T, Chen Y, Du Y, et al. Serum exosomal MiR-92b-5p as a potential biomarker for acute heart failure caused by dilated cardiomyopathy. Cell Physiol Biochem. 2018;46(5):1939–50.

73. Berezin AE, Kremzer AA, Samura TA, et al. Altered signature of apoptotic endothelial cell-derived microvesicles predicts chronic heart failure phenotypes. Biomark Med. 2019;13(9):737–50.
74. Mckenna WJ, Maron BJ, Thiene G. Classification, epidemiology, and global burden of cardiomyopathies. Circ Res. 2017;121(7):722–30.
75. Brieler J, Breeden MA, Tucker J. Cardiomyopathy: an overview. Am Fam Physician. 2017;96(10):640–6.
76. Dadson K, Hauck L, Billia F. Molecular mechanisms in cardiomyopathy. Clin Sci (Lond). 2017;131(13):1375–92.
77. Burke MA, Cook SA, Seidman JG, et al. Clinical and mechanistic insights into the genetics of cardiomyopathy. J Am Coll Cardiol. 2016;68(25):2871–86.
78. Roura S, Gámez-Valero A, Lupón J, et al. Proteomic signature of circulating extracellular vesicles in dilated cardiomyopathy. Lab Invest. 2018;98(10):1291–9.
79. Roura S, Gálvez-Montón C, De Gonzalo-Calvo D, et al. Extracellular vesicles do not contribute to higher circulating levels of soluble LRP1 in idiopathic dilated cardiomyopathy. J Cell Mol Med. 2017;21(11):3000–9.
80. Li F, Zhang K, Xu T, et al. Exosomal microRNA-29a mediates cardiac dysfunction and mitochondrial inactivity in obesity-related cardiomyopathy. Endocrine. 2019;63(3):480–8.
81. Wang Y, Jin P, Liu J, et al. Exosomal microRNA-122 mediates obesity-related cardiomyopathy through suppressing mitochondrial ADP-ribosylation factor-like 2. Clin Sci. 2019;133(17):1871–81.
82. Leone O, Pieroni M, Rapezzi C, et al. The spectrum of myocarditis: from pathology to the clinics. Virchows Arch. 2019;475(3):279–301.
83. Tschöpe C, Cooper LT, Torre-Amione G, et al. Management of myocarditis-related cardiomyopathy in adults. Circ Res. 2019;124(11):1568–83.
84. Germano JF, Sawaged S, Saadaeijahromi H, et al. Coxsackievirus B infection induces the extracellular release of miR-590-5p, a proviral microRNA. Virology. 2019;529:169–76.
85. Bachelier K, Biehl S, Schwarz V, et al. Parvovirus B19-induced vascular damage in the heart is associated with elevated circulating endothelial microparticles. PLoS One. 2017;12(5):e0176311.
86. Fan KL, Li MF, Cui F, et al. Altered exosomal miR-181d and miR-30a related to the pathogenesis of CVB3 induced myocarditis by targeting SOCS3. Eur Rev Med Pharmacol Sci. 2019;23(5):2208–15.
87. Sun P, Wang N, Zhao P, et al. Circulating exosomes control CD4+ T cell immunometabolic functions via the transfer of miR-142 as a novel mediator in myocarditis. Mol Ther. 2020;28:2605.
88. Fu D-G. Cardiac arrhythmias: diagnosis, symptoms, and treatments. Cell Biochem Biophys. 2015;73(2):291–6.
89. Hammerer-Lercher A, Namdar M, Vuilleumier N. Emerging biomarkers for cardiac arrhythmias. Clin Biochem. 2020;75:1–6.
90. Pourtau L, Sellal JM, Lacroix R, et al. Platelet function and microparticle levels in atrial fibrillation: changes during the acute episode. Int J Cardiol. 2017;243:216–22.
91. Mørk M, Andreasen JJ, Rasmussen LH, et al. Elevated blood plasma levels of tissue factor-bearing extracellular vesicles in patients with atrial fibrillation. Thromb Res. 2019;173:141–50.
92. Zhang XZ, Liu AG, Guo ML, et al. Relationship of platelet microparticle CD62P and activated GP IIb/IIIa with hypercoagulable state after atrial fibrillation radiofrequency catheter ablation. Eur Rev Med Pharmacol Sci. 2018;22(12):3919–24.
93. Wang L, Bi Y, Yu M, et al. Phosphatidylserine-exposing blood cells and microparticles induce procoagulant activity in non-valvular atrial fibrillation. Int J Cardiol. 2018;258:138–43.
94. Siwaponanan P, Keawvichit R, Udompunturak S, et al. Altered profile of circulating microparticles in nonvalvular atrial fibrillation. Clin Cardiol. 2019;42(4):425–31.

95. Soltész B, Urbancsek R, Pös O, et al. Quantification of peripheral whole blood, cell-free plasma and exosome encapsulated mitochondrial DNA copy numbers in patients with atrial fibrillation. J Biotechnol. 2019;299:66–71.
96. Mun D, Kim H, Kang J-Y, et al. Expression of miRNAs in circulating exosomes derived from patients with persistent atrial fibrillation. FASEB J. 2019;33(5):5979–89.
97. Ci H-B, Ou Z-J, Chang F-J, et al. Endothelial microparticles increase in mitral valve disease and impair mitral valve endothelial function. Am J Physiol Endocrinol Metab. 2013;304(7):E695–702.
98. Ismail EAR, Youssef OI. Platelet-derived microparticles and platelet function profile in children with congenital heart disease. Clin Appl Thromb Hemost. 2012;19(4):424–32.
99. Lin Z-B, Ci H-B, Li Y, et al. Endothelial microparticles are increased in congenital heart diseases and contribute to endothelial dysfunction. J Transl Med. 2017;15(1):4.
100. Guimarães Júnior MH, Ferrari TCA, Teixeira-Carvalho A, et al. Cell-derived microvesicles in infective endocarditis: role in diagnosis and potential for risk stratification at hospital admission. J Infect. 2019;79(2):101–7.
101. Khalyfa A, Kheirandish-Gozal L, Khalyfa AA, et al. Circulating plasma extracellular microvesicle microRNA cargo and endothelial dysfunction in children with obstructive sleep apnea. Am J Respir Crit Care Med. 2016;194(9):1116–26.
102. Khalyfa A, Zhang C, Khalyfa AA, et al. Effect on intermittent hypoxia on plasma exosomal micro RNA signature and endothelial function in healthy adults. Sleep. 2016;39(12):2077–90.
103. Bhattacharjee R, Khalyfa A, Khalyfa AA, et al. Exosomal cargo properties, endothelial function and treatment of obesity hypoventilation syndrome: a proof of concept study. J Clin Sleep Med. 2018;14(5):797–807.
104. Zhang H, Liu J, Qu D, et al. Serum exosomes mediate delivery of arginase 1 as a novel mechanism for endothelial dysfunction in diabetes. Proc Natl Acad Sci U S A. 2018;115(29):E6927–36.
105. Song M, Han L, Chen F-F, et al. Adipocyte-derived exosomes carrying Sonic Hedgehog mediate M1 macrophage polarization-induced insulin resistance via Ptch and PI3K pathways. Cell Physiol Biochem. 2018;48(4):1416–32.
106. Kim H, Bae Y-U, Lee H, et al. Effect of diabetes on exosomal miRNA profile in patients with obesity. BMJ Open Diabetes Res Care. 2020;8(1):e001403.
107. Yamamoto CM, Murakami T, Oakes ML, et al. Uromodulin mRNA from urinary extracellular vesicles correlate to kidney function decline in type 2 diabetes mellitus. Am J Nephrol. 2018;47(5):283–91.

# Extracellular Vesicle Biomarkers for Immune Disorders

# 18

Yiqin Ge, Yuji Huang, Yanning Li, Huanjin Liao, Xuchu Wang, Zhihua Tao, and Li Li

## 18.1 Extracellular Vesicle (EV) Biomarkers for the Diagnosis of Immune Disorders

The immune system is an interactive network of lymphoid organs, cells, humoral factors, and cytokines that contributes to immune defense, immune surveillance, and homeostasis. The essential function of the immune system is best illustrated in the context of disease; that is, underactivity results in severe infections, tumor development, immunodeficiency-related diseases, such as complement deficiency, T-cell deficiency, B-cell deficiency, and severe combined immunodeficiency, whereas overactivity results in allergic and autoimmune diseases and unnecessary immunological rejection related to organ and hematopoietic stem cell transplantation. Extracellular vesicles (EVs) are a heterogeneous group of cell-derived membranous structures consisting of exosomes and microvesicles (MVs) containing complex proteins, lipids, and various enclosed RNAs that originate from the endosomal system or are shed from the plasma membrane. EVs are considered a mechanism for intercellular communication, allowing cells to exchange proteins, lipids, and genetic materials implicated in immune regulation and killing [1]. EVs are also involved in several physiological processes that maintain homeostasis and immune balance,

Zhihua Tao and Li Li contributed equally with all other contributors.

Y. Ge · Y. Huang · Y. Li · H. Liao
Department of Laboratory Medicine, Shanghai General Hospital, Shanghai Jiao Tong University School, Shanghai, People's Republic of China

X. Wang · Z. Tao (✉) · L. Li (✉)
Department of Laboratory Medicine, The Second Affiliated Hospital of Zhejiang University School of Medicine, Hangzhou, Zhejiang, People's Republic of China
e-mail: zrtzhi@zju.edu.cn



Q. Wang, L. Zheng (eds.), *Extracellular Vesicles*,
https://doi.org/10.1007/978-981-99-8365-0_18

such as blood coagulation, stem cell expansion, and neuronal communication [2, 3], as well as pathological states that promote disease development, such as inflammation and tumorigenesis [4]. In recent years, EVs have been used as biomarkers to predict the progression of various pathological conditions and to assess the risk of disease, suggesting important clinical applications. This chapter focuses on EV biomarkers and their implications in infections, autoimmune diseases, allergies, organ transplantation, and other immune-related diseases.

## 18.2 EV Biomarkers for the Diagnosis of Autoimmune Diseases

Autoimmune diseases result from an abnormal immune response that leads to the self-damage of tissues and organs. Typically, various tolerance mechanisms play protective roles in preventing the autoreactive lymphocyte response, whereas in autoimmune diseases, such mechanisms become problematic, resulting in overreactivity of immune cells. Large quantity of inflammatory cytokines and chemokines contribute to the imbalance between the stimulatory and inhibitory effects of immune cells as well as the clearance and presentation of autoantigens. Immune tolerance, abnormal immune regulation, changes in the self-components of tissues, and foreign antigens mimicking the structure of self-tissues cause the system to mistake its members for foreign antigens and continue to attack its own tissues through humoral or cellular immunity, thereby causing autoimmune disease [5].

EVs have been identified in biological fluids, including plasma, urine, and blood. They have been shown to contain proteins, mRNAs, microRNAs (miRNAs), and other substances, highlighting the systemic spread of these nanocarriers in vivo. Growing evidence indicates that circulating EVs contribute to pathological processes [6–8]. In autoimmune diseases, EVs play essential roles in crosstalk pathways between multiple immune cells. For example, the balance between Th17 cells and regulator T cells (Tregs) is critical for maintaining immunological tolerance and monitoring unwanted or exaggerated immune responses. Although immunomodulatory cytokines (interleukin [IL]-10, transforming growth factor [TGF]-β, IL-35) released by Tregs and the expression of some inhibitory proteins on Tregs are engaged in preventing T cells from activation, miRNAs (e.g., *let-7d*) in Treg cell-derived EVs are also essential mediators that facilitate the suppression of T cells [9]. Moreover, the expression of ecto-5′-nucleotidase (CD73) on Treg cell-derived EVs has been shown to contribute to their suppressive function by converting extracellular adenosine-5-monophosphate to adenosine, which leads to immune modulation via the interaction with adenosine receptors expressed on target cells [10]. EVs also play functional roles in communication between antigen-presenting cells (APCs) and T cells. Multiple groups have reported that EVs from APCs can present peptide/major histocompatibility complex (MHC) complexes to T cells [11–13].

Specifically, EVs can directly bind to the surface of APCs and interact with T cells [14]. Under pathological conditions, endogenous EVs are the primary source of autoantigens and can activate autoreactive T cells involved in forming immune complexes, thereby leading to autoimmune diseases [7, 15].

During the progression of autoimmune diseases, the content of EVs reflect the physiological and pathological state of the source cells. Thus, EVs are potential biomarkers for noninvasive disease diagnosis. In this chapter, we review recent advances of EVs in autoimmune diseases and highlight state-of-the-art studies on the functions of EVs in autoimmune disease development.

### 18.2.1 Current Status of Autoimmune Disease Diagnosis

Autoimmune factors are involved in the occurrence and development of many diseases, which leads to complex clinical characteristics and causes tremendous difficulties in disease diagnosis.

Autoantibody detection is essential for the diagnosis and treatment of autoimmune diseases. With advanced in early diagnosis and standardized treatment, autoantibody detection has attracted much attention. The normal immune response depends on the cooperation and mutual restriction of various immune cells, particularly T-cell subgroups, which mediates cellular immunity and participate in humoral immunity. Simultaneously, autoimmune disorders result from abnormal changes in immune cells. Hence, we usually detect immune cell subsets to monitor the development and progression of autoimmune disease. In addition, autoimmune disease is often accompanied by abnormal expression of cytokines and antigen-antibody complexes. Antigen-antibody complexes formed by autoantibodies activate the complement system. Therefore, immunoglobulins, complements, cytokines, and chemokines are commonly used as serum immunological indicators for autoimmune disease.

EVs play essential roles in maintaining and regulating autoimmune homeostasis. Recently, EVs have aroused attention as new circulating biomarkers and have gradually become a research hotspot. Compared with traditional circulating markers, such as cytokines and hormones, EVs can stably exist in body fluids and have a long half-life. Furthermore, owing to the protective effects of the lipid bilayer, they can transport biologically active substances, such as RNA and proteins, to target organs. Compared with fine-needle aspiration and histopathological examination, liquid biopsy is easier to perform and much less invasive, making specimens obtained by this method more acceptable for the diagnosis of patients who favor real-time, dynamic, and individualized clinical testing. Alterations in the number, membrane molecules, and contents of EVs may also be used as biomarkers in patients. Unfortunately, current studies are insufficient because of small sample sizes and a lack of mature EV diagnostic markers. The following analysis summarizes research progress on EVs in autoimmune disease, infections, and other inflammation-related conditions.

### 18.2.2 Functions of EVs in Autoimmunity

EVs play various roles in cell-to-cell communication, regulation of cell signals, and initiation of enzymatic reactions [6] based on their origins and microenvironments, as follows. First, the self-tissue components carried by EVs can be used as self-antigen proteins and peptides to form complexes with MHC-I/II. As sources of autoantigens, EVs and MHC-I/II are recognized by dendritic cells or autoreactive T cells, inducing Th0 polarization toward Th1 or Th2 cells. These cells then activate B cells to produce autoantibodies against self-tissue components. For example, salivary gland exosomes from patients with systemic lupus erythematosus (SLE) carry autoantigens, such as SS-A, SS-B, and Sm. Second, EVs may be involved in the formation of immune complexes that exacerbate hypersensitivity reactions. The synovial fluid of patients with rheumatoid arthritis (RA) contains immune complexes consisting of platelet-derived MVs and anti-citrullinated peptide autoantibodies. Third, EVs promote the presentation of autoantigens to T cells. APCs secrete EVs that express MHC-I, MHC-II, and costimulatory molecules, forming the material basis for antigen presentation and activation of autoreactive T cells. Fourth, EVs can be used as adjuvants in APCs to trigger and enhance autoimmune responses. They also carry pathogen-associated molecular pattern and damage-associated molecular pattern molecules. After being recognized by the pattern recognition receptors carried on the surface of APCs, These molecules can promote the maturation of APCs, the activation of lymphocytes, and the release of pro-inflammatory factors. Fifth, biologically active molecules carried and secreted by EVs are involved in the maintenance of inflammation and immune responses. In RA, platelet-derived EVs release large amounts of IL-1β, which participates in the inflammatory reaction of synovial fibroblasts. Synovial fibroblasts release IL-6 and IL-8, which exacerbate joint inflammation. Finally, EVs can inhibit and regulate the immune response. This effect is mainly reflected in the pathological and physiological processes of tumors, allergies, and pregnancies. For example, in the early stages of inflammation, neutrophil-derived EVs can inhibit the release of the pro-inflammatory factors IL-8, IL-10, and tumor necrosis factor (TNF)-α from macrophages and trigger the release of the anti-inflammatory cytokine TGF-β. EVs can also enhance the functions of Tregs, inhibit the cytotoxic effects of natural killer and $CD8^+$ cells, and suppress the differentiation of monocytes and the maturation of dendritic cells.

## 18.2.3 EVs and Autoimmune Diseases

### 18.2.3.1 EVs and RA

RA is an autoimmune disease with a high incidence, characterized by the infiltration of lymphocytes and macrophages in the synovial tissue and the proliferation of synovial cells. These features lead to immune reactions in the joint, joint inflammation, and erosive destruction. The proteome and transcriptome of EVs (mRNAs, miRNAs, and long noncoding RNAs [lncRNAs]) in the body fluids of patients with autoimmune diseases are different from those of healthy individuals. The differences are the basis for exploring EVs as diagnostic disease markers. Yoo et al. [16] used mass spectrometry and enzyme-linked immunosorbent assay (ELISA) to compare differences in the proteomes of serum EVs from patients with RA with or without clinical remission. The results showed that serum amyloid A and hyaluron from lymphatic endothelial cells expressing lymphatic endothelial hyaluronan receptor-1 are associated with RA disease grade. EVs can also improve the inflammatory environment and increase the survival rate of autoreactive T cells by secreting or expressing pro-inflammatory factors. Zhang et al. also found that the TNF-α level in supernatants from synovial fibroblast cultures and serum exosomes from patients with RA was significantly higher than that in normal joint synovium and serum exosomes. These exosomes have the capacity to inhibit apoptosis in activated T cells and may be related to the pathogenesis of T cell-mediated RA [17].

LncRNAs are also involved in the occurrence and development of many diseases via regulation of pluripotency, differentiation, and embryo development. HOX transcript antisense RNA (*HOTAIR*) was the first lncRNA shown to have trans-transcriptional regulatory effects. Song et al. concluded that the *HOTAIR* levels of exosomes derived from peripheral blood mononuclear cells and serum exosomes in patients with RA were significantly higher than those in healthy individuals [18]. Moreover, Zhang et al. identified lncRNAs that were differentially expressed in the EVs of patients with RA and normal human synovial cells. They found that *ENST00000483588*, *ENST00000438399*, *uc004afb.1*, and *ENST00000452247* may be involved in the pathological process of RA and that these lncRNAs may have potential applications in the diagnosis and evaluation of active RA [19]. Yuan et al. [20] analyzed lncRNAs in exosomes derived from peripheral blood mononuclear cells from patients with RA and healthy individuals using RNA-seq and bioinformatics and identified four differentially expressed lncRNAs using quantitative polymerase chain reaction for verification. The expression of *ENST00000456270* was comparable to that in patients with RA, and serum IL-6 and TNF-α levels were shown to be closely related to disease activity.

Multiple studies have shown that the numbers of platelet-derived MVs in the blood circulation of patients with RA are increased and are related to disease progression. The concentrations of platelet MVs in the synovial fluid of patients are significantly higher than those of circulating MVs, indicating that platelet MVs are produced locally and may exhibit priority positioning.

#### 18.2.3.2 EVs and SLE

SLE is a typical systemic autoimmune disease that is characterized by the production of a series of pathogenic autoantibodies. Lupus nephritis (LN) is the leading cause of death and complications in patients with SLE. Lee et al. found that the number of serum exosomes in patients with SLE was significantly higher than that in healthy controls and was positively correlated SLE [21]. EVs from SLE serum can induce peripheral blood mononuclear cells (PBMCs) to secrete high levels of interferon (IFN)-α, TNF-α, IL-1, and IL-6β. Additionally, Toll-like receptor (TLR) 4 antagonists can significantly inhibit EV-mediated TNF-α, IL-1β, and IL-6 production, whereas TLR1/2, TLR7, and TLR9 antagonists can significantly inhibit IFN-α in EVs.

Kidney biopsy is the "gold standard" for diagnosing clinical kidney disease; however, this approach is invasive. Accordingly, analysis of body fluids, such as urine, may be more appropriate for analysis of biomarkers owing to the noninvasiveness of sample collection and the abundance of samples. Research on the correlation between urine EVs and diseases have also provided a direction for clinical medicine and basic medical research in recent years. Urine EV *miR-29c* in patients with SLE is related to the chronic progression of LN but is unrelated to renal function [22]. Perez-Hernandez found that urinary exosome *miR-146a* levels in patients experiencing active periods of LN were significantly higher than those in healthy controls [23]. Platelets, white blood cells, and endothelial-derived EVs in the serum of patients with SLE were also significantly higher than those in healthy controls. Some researchers have examined the potential relationships between EV concentrations, autoantibody profiles, and disease progression. They concluded that platelet EVs and anti-double-stranded DNA antibody levels were negatively correlated. In 2012, Nielsen showed that the numbers of plasma platelets, white blood cells, and endothelial-derived EVs decreased in patients with SLE, presumably because of increased clearance rates of changes in EV formation in these patients.

#### 18.2.3.3 EVs and Sjogren's Syndrome (SS)

SS is a chronic autoimmune disease that mainly involves the lacrimal, salivary, and other exocrine glands. EVs derived from salivary gland epithelial cells contain the autoantigens Ro/SSA, La/SSB, Sm/RNP, and epithelial cell-specific cytokeratin. Michael et al. compared saliva-derived exosomes from patients with SS with those from healthy controls and identified many differentially expressed miRNAs [24, 25].

Alevizos et al. found that the overexpression of *miR-181a* and *miR-16* in salivary gland tissues of patients is associated with salivary gland function attenuation [26]. Gallo et al. found that the specific miRNA *EBV-miR-BART13-3p* produced by Epstein-Barr virus-infected B cells is transferred from B cells to salivary gland cells through EVs. This miRNA targets the calcium ion sensing pathway through 1 stromal interaction molecule 1, resulting in loss of salivary gland function, which may be related to the pathogenesis of SS caused by EB infection [27]. The study of miRNAs in readily accessible and noninvasive salivary EVs is an integral part of exploring disease pathogenesis, prediction, and diagnostic markers. Research on EVs and SS is still in its infancy, and currently available reports are still insufficient. The contents of EVs derived from platelets, endothelial cells, monocytes, and T lymphocytes in patients with SS and changes in the proteome and transcriptome content should be evaluated. Additionally, research on diseases and exploration of diagnostic markers are essential. However, unification and standardization of technical methods and differences in patient individualization are the first issues that need to be resolved.

#### 18.2.3.4 EVs and Multiple Sclerosis (MS)

MS is an autoimmune disease characterized by inflammatory demyelination of the white matter in the central nervous system. During MS progression, the contents of vesicles derived from skin cells, white blood cells, and platelets increase significantly. Accordingly, the activity of matrix metalloproteinases in vesicles also increases. These vesicles can cause blood-brain barrier damage in MS, promote the survival of immune cells, and penetrate the central nervous system. Astrocytes are the most abundant glial cells in the central nervous system, and glial fibrillary acidic protein (GFAP) is a marker of astrocyte activation. Willis et al. [28] found that the content of GFAP was significantly increased in peripheral blood EVs of patients with MS, suggesting that EVs are involved in the central nervous system inflammatory response of MS. Myelin oligodendrocyte glycoprotein was found at a substantially higher volume in T-cell EVs isolated from the serum, cerebrospinal fluid, and PBMCs of patients with relapsing-remitting MS or secondary progressive MS than that in healthy subjects. Nine miRNAs were expressed differently in patients with MS than in healthy individuals by sequencing serum-derived EV miRNAs. EVs may be used as a basis for MS classification and have potential applications as MS diagnostic markers. Sáenz-Cuesta evaluated EVs derived from platelets, leukocytes, and monocytes in patients with MS; explored the effects of treatment on the numbers of these EVs; and found that the numbers of platelet-derived EVs in treated patients with MS were significantly lower than those in untreated patients.

#### 18.2.3.5 EVs and Other Autoimmune Diseases

Inflammatory bowel diseases (IBDs), including ulcerative colitis and Crohn's disease, are intestinal immune function disorders that are increasing in prevalence in Asia. In vitro experiments have shown that EVs derived from patients with IBDs can promote the inflammatory response of intestinal epithelial cells and macrophages. The mRNA and protein profiles of EVs may serve as potential markers for the clinical diagnosis of IBDs. Although IBDs are chronic gastrointestinal diseases, they can also cause oral lesions. Zheng et al. [29] used liquid chromatography-mass spectrometry to compare salivary exosome contents and proteins in patients with IBDs with those of healthy controls and found that the expression of proteasome subunit alpha 7 (PSMA7) was significantly higher in patients with IBDs than in healthy individuals. Gene Ontology analysis showed that this exosomal protein was related to proteasome activity and inflammation. Animal experiments confirmed that PSMA7 is highly expressed in the colon and oral tissues of IBD model mice compared with that in healthy controls. They proposed that PSMA7 carried by salivary exosomes could replace colonoscopy examination, reduce patient suffering, and serve as a new diagnostic marker for IBDs.

The communication function of EVs between cells enables them to exert antiviral and immunomodulatory effects during viral infections. Viral myocarditis (VMC) is a type of myocardial inflammation caused by various viral infections. Clinical diagnosis of this disease is difficult. To date, endocardial biopsy remains the gold standard for diagnosing VMC. Owing to invasiveness and complexity of myocardial biopsy, this approach is not preferred in the clinical setting. Zhao et al. [30] used two-dimensional electrophoresis analysis to compare plasma exosomal proteins between the VMC and control groups. After identification by mass spectrometry, the upregulated protein retinol binding protein 4 (RBP4) was selected and verified by ELISA. The results showed that the expression of RBP4 protein was increased in patients with VMC, which may be related to the antiviral response in the body. Thus, RBP4 may be a candidate for the early diagnosis of VMC as a plasma exosomal biomarker; however, multicenter and large-sample studies are still required to confirm these results.

Most liver diseases have no significant symptoms in the early stages and are often detected only in the late stages. Routine imaging and laboratory tests have limited value for the early diagnosis of the disease, and liver biopsy is still the gold standard. Therefore, the development of reliable and noninvasive diagnostic methods is urgently needed. Recent studies have shown that EVs may participate in the pathogenesis of viral hepatitis by facilitating the transfer of viruses between liver stem cells through genetic materials containing viruses or by inhibiting the killing functions of immune cells. Dreux et al. [31] found that EVs containing hepatitis C virus RNA can induce plasmacytoid dendritic cells to secrete IFN-α, thereby inhibiting the viral life cycle and blocking viral spread. Alcohol and its metabolites have been shown to cause liver damage; however, we lack clinical markers for monitoring the severity of alcoholic liver damage. Momen-Heravi et al. analyzed the serum EVs of alcohol-fed mice and the plasma of patients with alcoholic hepatitis. The results showed that the content of EVs in mouse models and patient samples was significantly increased and that *miR-192* and *miR-30a* were both highly expressed. Receiver operating characteristic curve analysis showed that *miR-192* could be used for the diagnosis of alcoholic hepatitis. The area under the curve (AUC) was 0.95 ($P < 0.001$), whereas the AUC of *miR-30a* was only 0.58 ($P = 0.56$). The Sukriti study found that the levels of EVs derived from T cells, macrophages, hematopoietic stem cells, and hepatocytes were higher in patients with alcoholic hepatitis who did not respond to corticosteroids than in those who did respond to corticosteroids. The degree of exosome elevation was shown to be related to the amount and duration of alcohol intake and the severity of hepatitis, supporting the value of EVs for guiding the treatment of patients with alcoholic hepatitis using cortisol.

Several studies have shown that EV contents in body fluids are closely related to disease, thereby supporting the importance of EVs in the autoimmune process and the pathogenesis of autoimmune disease. However, research on EVs remains insufficient. The structures, functions, and roles of these vesicles in diseases and their application as biomarkers and therapeutic agents require further discussion. EV separation, detection technology, and quality control standards are key focuses of EV research. A comprehensive understanding of EV contents, functions, and activities, along with further clarification of their exact mechanisms in the pathogenesis of autoimmune disease, is essential for improving diagnostic tools for autoimmune disease and other diseases.

## 18.3 EV Biomarkers for the Diagnosis of Allergic Diseases

EVs derived from multiple cell types play crucial roles in the persistence and development of allergic diseases. Researchers have found many proteins and RNAs in EVs from clinical specimens of patients with asthma, such as bronchoalveolar lavage fluid, body fluid, and sputum, supporting the establishment of new strategies for clarifying the mechanism of disease. As messengers, EVs contain intracellular proteins, inflammatory mediators, and nucleic acids that have emerged as key players in regulating immune responses. This section briefly describes the roles of EVs secreted from different immune cells in modulating allergic diseases and emphasizes the diagnostic value and therapeutic prospects of EV biomarkers.

### 18.3.1 Functions of EVs in Allergic Diseases

EVs participate in cell-cell communication and immune regulation by transferring biological contents in a paracrine manner in allergic diseases.

#### 18.3.1.1 EVs from Dendritic Cells/Monocyte-Derived Macrophages (MDMs)

Dendritic cell-derived EVs expressing MHC/peptide complexes take up inhaled allergens and then activate T cells. Théry et al. [32] found that EVs containing allergens or peptides can induce an allergen-specific Th2 cell response in vivo; however, they failed in vitro. Cytokines such as TGF-β, IL-1β, and IFN-γ from airway smooth muscle affect the rate of EV generation and delivery from peripheral blood monocytes and alveolar macrophages [33].

#### 18.3.1.2 EVs from Eosinophils

Eosinophils are essential adaptive immune cells in asthma and spirochete infection that release cytokines, such as IL-2, IL-4, IL-5, IL-10, IL-12, IL-13, IL-16, IL-18, TGF-α/β, lipid mediators, platelet-activating factor, eosinophil peroxidase, and regulated on activation, normal T-cell expressed and secreted.

Eosinophil-derived EVs contain enzymes, including eosinophil peroxidase and major basic protein. They also induce the production of reactive oxygen species and nitric oxide and augment the chemotaxis of eosinophils and adhesion. Furthermore, these EVs are involved in protein degradation and tissue damage, which exacerbate asthma.

#### 18.3.1.3 EVs from Lymphocytes

Hough et al. [34] demonstrated that EVs from dendritic cells and B cells induce T-cell proliferation, activation, and Th2 immune responses via various mechanisms. $CD8^+$ T cell-derived EVs exhibit cytotoxic functions. B cell-derived EVs contain the costimulatory factors CD40, CD80, and CD86 as well as integrins (α1 and α2), which play roles in T-cell activation. Admyre et al. found that B cells secrete EVs containing MHC, bind Betv1-derived peptides, and subsequently induce dose-dependent T-cell proliferation, Th2 cytokine release, and IFN-γ and TNF-α inhibition [35].

#### 18.3.1.4 EVs from Mast Cells and Basophils

Mast cells in humans and mice release EVs with various functions, including the transfer of proteins and nucleic acids to interact with different types of immune cells and induce Th2 inflammation; modulation of cell differentiation, maturation, and activity through membrane surface molecules expressed on EVs, such as CD40 (costimulatory molecule), FcεRI (IgE high-affinity receptor), KIT, four transmembrane proteins, lymphocyte function-associated antigen-1, and intercellular adhesion molecule-1; modulation of immune reactions by activating T and B cells and IgE production; and encapsulation of a variety of inflammatory mediators, cytokines, and chemokines to induce allergies.

Mayoral et al. [36] proposed a model in which *miR-221* overexpression performs dual roles in mast cells, i.e., contributing to cell cycle regulation and cytoskeletal function in the resting state. However, in response to stimulation through IgE-antigen complexes, the effects of *miR-221* are mast cell specific and activation dependent, contributing to the regulation of degranulation, cytokine production, and cell adherence.

Basophils are rarely found in the blood circulation, accounting for 0–1% of all blood-derived cells. They migrate from the bloodstream to sites of inflammation, thereby affecting the body's response to infection. Dvorak et al. reported that EVs derived from basophils were similar to granules in basophils, which may be associated with mediator exocytosis. To date, we have few reports on EVs from basophils [37].

#### 18.3.1.5 Novel Functions of EVs

In addition to transferring proteins and RNAs, EVs play roles in mitochondrial transfer. Bronchoalveolar lavage fluid from both healthy volunteers and asthmatics contained EVs with encapsulated mitochondria. Moreover, the levels of mitochondrial DNA within the EVs were significantly higher in the asthmatic group. Hough et al. [38] reported that myeloid-derived regulatory cells transfer EVs containing mitochondria to peripheral T cells and regulate T-cell responses in asthma.

Mesenchymal stromal cells (MSCs) promote an anti-inflammatory and highly phagocytic macrophage phenotype (M2 type) through EV-mediated mitochondrial transfer in acute respiratory distress syndrome [39].

In general, EVs play important roles in novel cell-to-cell communication involving exosomal transfer of mitochondria and bioenergetic regulation of target cells.

### 18.3.2 EVs as Potential Biomarkers in Allergic Diseases

EVs are considered biomarkers for clinical diagnosis because of their intracellular proteins and nucleic acids, which are closely related to physiological and pathological conditions. They can be isolated from the blood, urine, saliva, breast milk, ascitic fluid, and bronchoalveolar lavage fluid (BALF), which can be obtained noninvasively. EVs have been extensively studied and are associated with various diseases, including cancer. Thus, EVs have attracted the attention of researchers in the fields of allergic and respiratory diseases.

MiRNAs are essential regulators of gene expression in the immune system (Table 18.1). EV-packaging miRNAs play critical roles in the modulation of post-transcriptional allergic diseases. Indeed, EVs in the BALF of experimental mice with bronchial asthma (BA) suppress airway inflammation. Administration of BALF-derived EVs from experimental mice with BA inhibits the IgE response,

**Table 18.1** EV as biomarkers in allergic diseases

| Biomarker | Function | Disease |
|---|---|---|
| miRNA-34/449 | Regulate ciliated cell differentiation | Asthma [42] |
| miRNA-19 | (1) Promote epithelial cell proliferation; (2) Regulate type 2 inflammation | Asthma [42] |
| miRNA-21 | (1) Suppress IL12p35 gene expression and decrease the IL-12 level in serum (2) Positive relation with severity of asthma; (3) Modulate airway remodeling | Asthma [43] |
| miRNA-455 | Upregulation in the lungs of allergen-exposed mice | Asthma [44] |
| Let7-miRNA | (1) Improve AHR and mucus secretion; (2) Repress IL-13 expression | Asthma [45] |
| miRNA-126 | Upregulate PU.1 and Myb, inhibit IL-1β and TNF-α expression | Asthma [46] |
| miRNA-24/27 | Inhibit IL-4 production in T cells | Asthma [47] |
| miRNA-21 | ActivateIL-12/IFN-gamma pathway | Delayed-type hypersensitivity [48] |
| miRNA-150 | Inhibit T cell-mediated contact dermatitis | Atopic dermatitis & allergic rhinitis [49] |

blocks Th2 cytokine production, and suppresses airway inflammation. Levänen et al. [40] first confirmed the existence of miRNAs in EVs from the BALF. They found that EV miRNA profiles in asthma were different from those in healthy controls; for example, the *let-7* family and *miR-200* family were downregulated, and this effect was shown to be significantly associated with airway inflammation and remodeling. Activation by dust mites also leads to unique miRNA subset expression in patients with allergic diseases. Gon et al. [41] analyzed miRNAs and mRNAs derived from airway-secreted EVs in house dust mite (HDM)-exposed mice. They found that HDM exposure induced significant changes in the expression of 139 miRNAs in EVs and upregulated 31 genes, including *IL-13* and *IL-5Ra*, which are putative targets of *miR-346* and *miR-574-5p*. Administration of the sphingomyelinase inhibitor GW4869 also reduces EV secretion and alleviates BA in mice. Currently, two safety studies of MSCs and their trophic factors in patients with BA are in progress (NCT02192736 and NCT03137199).

#### 18.3.2.1 Roles of MiRNAs in Asthma

Several miRNAs, including *miR-34/449*, *miR-19*, *miR-21*, and *miR-455*, have been shown to have roles in asthma [42–49]. The IL-13-induced reduction in *miR-34/449* expression in the bronchial epithelium contributes to alterations in epithelial differentiation in asthma. Moreover, *miR-21* plays roles in Th2 cell polarization through negative regulation of IL-12p35 in models of allergic airway inflammation [43]. When exposed to TGF-β, human lung fibroblasts show increased expression of *miR-21* and *miR-455* [44]. Additionally, Garbacki et al. showed that *miR-455* was upregulated in the lungs of allergen-exposed mice.

*MiR-126* was shown to link innate and adaptive immune responses in an HDM mouse model of allergic airway inflammation. Antagonism of *miR-126* significantly suppresses the effector functions of Th2 cells via the negative regulation of GATA3 expression, resulting in diminished Th2 cell-mediated inflammation [46]. Pua et al. [47] found that *miR-24* and *miR-27* both inhibit IL-4 production in T cells in vitro and that mice lacking the expression of these miRNAs in T cells show enhanced allergic airway hypersensitivity inflammatory responses.

#### 18.3.2.2 Roles of MiRNAs in Allergic Rhinitis

*MiR-155* targets c-Maf and promotes the secretion of the Th2 cytokines IL-4, IL-5, and IL-10, which play roles in cell infiltration and airway remodeling. *MiR-155*-deficient mice show diminished eosinophilic inflammation, mucus hypersecretion, and Th2 cytokine levels in the lungs after allergen exposure [50]. Recent studies have demonstrated that *miR-155* is overexpressed in the nasal mucosa of patients with allergic rhinitis and the skin of patients with atopic dermatitis, suggesting important roles in the pathogenesis of allergic diseases [51].

#### 18.3.2.3 Roles of MiRNAs in Delayed Hypersensitivity

Lu et al. [48] showed that *miR-21* plays central roles in the Th1/Th2 balance. In ovalbumin-challenge models, *miR-21*-deficient mice showed reduced eosinophilic inflammation and IL-4 expression in the lungs and BALF; however, the levels of the Th1 cytokine IFN-γ increased. In addition, loss of *miR-21* significantly enhances Th1-associated delayed-type hypersensitivity cutaneous responses.

Cell-mediated allergic reactions to haptens can be inhibited by EVs carrying *miR-150*. Bryniarski et al. [49] demonstrated a mechanism of T-cell tolerance in mice mediated by exosome-like nanovesicles carrying *miR-150* produced by $CD8^+$ T cells. Nanovesicles carrying *miR-150* inhibit contact sensitivity in sensitized mice at the peak response.

Hence, miRNAs in EVs have been extensively evaluated as potential biomarkers of allergic diseases. Most of these studies were performed in vitro and in animal models; therefore, more robust studies are needed to provide better insights into the roles of miRNAs in human cells and tissues.

### 18.3.3 EVs as Potential Therapeutic Targets for Allergic Diseases

Many studies have explored the application of EVs in the diagnosis, prevention, and treatment of allergic diseases. EVs not only promote immune responses but also induce allergen tolerance. Intranasal administration of EVs in the BALF collected from experimental mice with pollinosis induces immune tolerance and suppresses allergic reactions. Additionally, EVs are valuable tools for drug delivery. Mast cell- and dendritic cell-derived EVs display a specific lipid composition that is stable in circulation and is protected from complement-mediated lysis via the expression of CD55 and CD59. Moreover, EVs can deliver various therapeutic biomolecules, ranging from nucleic acids to small molecules.

FcεRI expressed on EVs derived from mast cells can bind to free IgE in serum and reduce IgE-FcεRI conjunction on mast cells, thereby alleviating allergic reactions [52]. Moreover, researchers have found that loading low-dose allergens on EVs reduces airway inflammation and induces Th1 responses in ovalbumin challenge models.

Given their multiple functions in antigen presentation, immune modulation, and allergen tolerance, EVs may be potential tools for vaccine preparation to achieve desensitization dependent on multiple low-dose treatments.

## 18.4 EV Biomarkers for the Diagnosis of Organ Transplant Rejection

Clinical organ transplantation is one of the most important medical advances of the century, providing a means of giving the gift of life to patients with terminal failure of vital organs via donation of organs from deceased or living individuals. The risk of allograft rejection is a significant hurdle in transplant medicine. The allogeneic response is mediated by both cells and antibodies that infiltrate the graft and cause allograft injury and loss of function [53]. This is mainly driven by the recognition of MHC proteins called human leukocyte antigens (HLAs), which are expressed on graft tissue cells. After organ transplantation, circulating HLA antibodies bind to HLA molecules expressed in the endothelium of the graft and induce type II allergic reactions [54]. Immune complexes are deposited in blood vessels and glomerular basement membranes, causing vasculitis, thrombosis, necrosis of transplanted tissue, and organ failure, which ultimately leads to transplantation failure [55].

The identification of allograft injury, including acute clinical and subclinical injury, is vital for increasing the longevity of the transplanted organ. Newly identified relationships between allograft injury and the levels of chemokines, cytokines, HLA class II proteins, CD4/CD8 cell ratio, complement and adhesion molecule testing, and nHLA-related proteins in various biofluids offer noninvasive methods of immune analysis. These approaches have facilitated the monitoring of organ transplant rejection but cannot meet the requirements for prediction and prevention. Therefore, identifying early biomarkers that can be used for organ transplant rejection is crucial.

EVs play critical roles as carriers of molecular information during the signal transmission process. As biological information molecules that are widely distributed in human body fluids, EVs are biologically stable and can be collected in a noninvasive manner, making EVs promising circulating biomarkers with broad application prospects. Studies have evaluated the presence of EVs in tissues after pancreatic islet transplantation and have shown that these EVs may serve as biomarkers for noninvasive monitoring of immune reperfusion. Compared with traditional circulating markers, such as cytokines and hormones, EVs can exist stably in body fluids and have a long half-life.

This section briefly summarizes the roles of EVs in organ transplantation and their possible applications as new molecular markers in organ transplant rejection.

### 18.4.1 EVs and Lung Transplant Rejection

Human lung transplantation (LTx) is a viable treatment option for patients with end-stage lung diseases, such as idiopathic pulmonary fibrosis, chronic obstructive pulmonary disease, and cystic fibrosis [56, 57]. Short-term survival rates have increased; however, acute rejection (AR) and chronic rejection (CR), in the form of chronic lung allograft dysfunction, result in morbidity and mortality. Both AR and bronchiolitis obliterans (BOS) are primarily caused by immunological insults to transplanted lungs. Some studies have reported a significant correlation between BOS and the de novo development of antibodies to mismatched donor HLA (donor-specific antibodies [DSAs]) and antibodies to lung-associated self-antigens (SAgs), K-alpha 1 tubulin (Ka1T), and collagen V (Col-V) after LTx [58, 59]. Recent studies have demonstrated that exosomes isolated from the sera and BALF of LTx recipients diagnosed with AR or CR contain DSAs and the lung SAgs Col-V and Ka1T. These results suggest that EVs released from the transplanted lungs during rejection can induce immune responses, thereby increasing the risk of BOS development. Therefore, detecting EVs carrying autoantigens from the graft can be used as biomarkers of noninvasive rejection.

EVs, which are membrane vesicles released from cells within the lung allograft, contain a diverse array of biomolecules that closely reflect the biological state of the cells and tissues from which they were released. EV transcriptomes may provide a better understanding of the rejection process. Furthermore, biomarkers originating from this transcriptome could provide timely and sensitive detection of acute cellular rejection, reduce the incidence of severe AR and chronic lung allograft dysfunction, and improve patient outcomes. AR is an extremely inflammatory environment involved in innate and adaptive immune responses, and identification of novel upstream regulators may facilitate the establishment of potential therapeutic targets. Many olfactory receptor genes, including *OR2T3*, *OR8B3*, *OR2A4*, *OR1N1*, *OR8U1*, *OR2W1*, and *OR51A7*, are elevated in the lung allograft transcriptome during AR [60]. Novel upstream pathways associated with rejection could provide an early warning mechanism to prevent rejection and identify targets for pharmacological intervention in acute lung allograft rejection.

### 18.4.2 EVs and Heart Transplant Rejection

Biomarkers are critically needed to monitor cardiac allografts for injury and immunologic rejection after heart transplantation. EVs are tissue-specific nanovesicles released into the circulation by many cell types. Their profiles are dynamic and reflect the conditional changes imposed on their tissue counterparts. Transplanted hearts may release donor-specific EVs into the recipient's circulation, and these EVs may be conditionally altered during immunological rejection.

A recent study by Dewi et al. found a significant increase in exosomal *miR-142-3p* content in the total serum exosome pool of heart transplant recipients under conditions of acute cellular rejection versus no rejection. Further analysis suggested that *miR-142-3p*, a hematopoietic tissue-specific miRNA, is released from T cells upon stimulation, utilizing exosomal transport, and can modify the expression of its target mRNA, *RAB11FIP2*, in endothelial cells [61]. Furthermore, *miR-142-3p* expression was significantly increased in the exosomal fraction but not in the supernatants of the T-cell culture. This T cell/endothelial cell axis via an exosomal shuttle may contribute to T cell-mediated graft rejection. Transplant heart exosome profiling enables noninvasive monitoring of early acute rejection with high accuracy.

### 18.4.3 EVs and Kidney Transplant Rejection

Kidney allograft biopsy is currently considered the gold standard for diagnosing kidney rejection. However, renal biopsy is invasive and can be inconclusive. The main concerns of clinicians during the early posttransplantation period are delayed graft function (DGF) and AR. Early detection of DGF and AR could have both short- and long-term prognostic implications for transplant survival.

EVs have also been investigated as potential biomarkers for kidney transplantation. Plasma and urinary EVs have been studied in various transplant settings.

In a study of 231 kidney transplant patients, circulating endothelial microparticle levels in plasma were analyzed before and periodically after kidney transplantation (up to 2 months). Plasma levels were found to increase during antibody-mediated rejection episodes and decrease after therapy, with a slower decline in patients with peritubular capillary C4d staining [62]. In another study, quantification of plasma $C4d^{+}CD144^{+}$ EVs released from endothelial cells associated with antibody-mediated rejection demonstrated an 11-fold increase in concentration compared with that in patients without rejection and decreased more than 70% after successful antirejection therapy [63].

In one study, increased expression of 17 urinary EV proteins was found in patients with T cell-mediated rejection, and two proteins, i.e., tetraspanin-1 and hemopexin, were proposed as biomarkers [64].

In addition to participating in transplant rejection of the lungs, heart, and kidneys, EVs also play essential roles in the rejection of other organ transplants. CD4, CD8, and CD31 are highly expressed in the peripheral blood microparticles of patients with AR after liver transplantation, although the numbers of total microparticles and the expression levels of CD25 and MHC-I were not different. The levels of EVs obtained from the serum of patients undergoing gland transplantation are closely associated with rejection.

When rejection occurs, the EV level can reach the peak value; therefore, the detection of EV level has predictive value for the prognosis of transplant rejection.

Detecting EV levels after organ transplantation and specific markers carried by EVs may be the best approach for exploring the mechanisms of transplant rejection. These EVs may be the best source of predictive and diagnostic markers for organ transplant rejection. Further in-depth research on EVs may provide insights into the mechanisms of organ transplant rejection and thereby facilitate early prediction and diagnosis.

## References


1. van Niel G, D'Angelo G, Raposo G. Shedding light on the cell biology of extracellular vesicles. Nat Rev Mol Cell Biol. 2018;19(4):213–28.
2. Schorey JS, Cheng Y, Singh PP, Smith VL. Exosomes and other extracellular vesicles in host-pathogen interactions. EMBO Rep. 2015;16(1):24–43.
3. Deatherage BL, Cookson BT. Membrane vesicle release in bacteria, eukaryotes, and archaea: a conserved yet underappreciated aspect of microbial life. Infect Immun. 2012;80(6):1948–57.
4. Lo Cicero A, Stahl PD, Raposo G. Extracellular vesicles shuffling intercellular messages: for good or for bad. Curr Opin Cell Biol. 2015;35:69–77.
5. Qi S, Chen Q, Xu D, Xie N, Dai Y. Clinical application of protein biomarkers in lupus erythematosus and lupus nephritis. Lupus. 2018;27(10):1582–90.
6. Turpin D, Truchetet ME, Faustin B, et al. Role of extracellular vesicles in autoimmune diseases. Autoimmun Rev. 2016;15(2):174–83.
7. Buzas EI, Gyorgy B, Nagy G, Falus A, Gay S. Emerging role of extracellular vesicles in inflammatory diseases. Nat Rev Rheumatol. 2014;10(6):356–64.
8. Piszczatowska K, Czerwaty K, Cyran AM, et al. The emerging role of small extracellular vesicles in inflammatory airway diseases. Diagnostics (Basel). 2021;11(2):222.
9. Okoye IS, Coomes SM, Pelly VS, et al. MicroRNA-containing T-regulatory-cell-derived exosomes suppress pathogenic T helper 1 cells. Immunity. 2014;41(3):503.
10. Smyth LA, Ratnasothy K, Tsang JY, et al. CD73 expression on extracellular vesicles derived from CD4+ CD25+ Foxp3+ T cells contributes to their regulatory function. Eur J Immunol. 2013;43(9):2430–40.
11. Montecalvo A, Shufesky WJ, Stolz DB, et al. Exosomes as a short-range mechanism to spread alloantigen between dendritic cells during T cell allorecognition. J Immunol. 2008;180(5):3081–90.
12. Segura E, Nicco C, Lombard B, et al. ICAM-1 on exosomes from mature dendritic cells is critical for efficient naive T-cell priming. Blood. 2005;106(1):216–23.
13. Utsugi-Kobukai S, Fujimaki H, Hotta C, Nakazawa M, Minami M. MHC class I-mediated exogenous antigen presentation by exosomes secreted from immature and mature bone marrow derived dendritic cells. Immunol Lett. 2003;89(2–3):125–31.
14. Wakim LM, Bevan MJ. Cross-dressed dendritic cells drive memory CD8+ T-cell activation after viral infection. Nature. 2011;471(7340):629–32.
15. Robbins PD, Dorronsoro A, Booker CN. Regulation of chronic inflammatory and immune processes by extracellular vesicles. J Clin Invest. 2016;126(4):1173–80.
16. Yoo J, Lee SK, Lim M, Sheen D, Choi EH, Kim SA. Exosomal amyloid A and lymphatic vessel endothelial hyaluronic acid receptor-1 proteins are associated with disease activity in rheumatoid arthritis. Arthritis Res Ther. 2017;19(1):119.
17. Zhang HG, Liu C, Su K, et al. A membrane form of TNF-alpha presented by exosomes delays T cell activation-induced cell death. J Immunol. 2006;176(12):7385–93.
18. Song J, Kim D, Han J, Kim Y, Lee M, Jin EJ. PBMC and exosome-derived Hotair is a critical regulator and potent marker for rheumatoid arthritis. Clin Exp Med. 2015;15(1):121–6.
19. Zhang Y, Xu YZ, Sun N, et al. Long noncoding RNA expression profile in fibroblast-like synoviocytes from patients with rheumatoid arthritis. Arthritis Res Ther. 2016;18(1):227.

20. Yuan M, Wang S, Yu L, et al. Long noncoding RNA profiling revealed differentially expressed lncRNAs associated with disease activity in PBMCs from patients with rheumatoid arthritis. PLoS One. 2017;12(11):e0186795.
21. Lee JY, Park JK, Lee EY, Lee EB, Song YW. Circulating exosomes from patients with systemic lupus erythematosus induce an proinflammatory immune response. Arthritis Res Ther. 2016;18(1):264.
22. Sole C, Cortes-Hernandez J, Felip ML, Vidal M, Ordi-Ros J. miR-29c in urinary exosomes as predictor of early renal fibrosis in lupus nephritis. Nephrol Dial Transplant. 2015;30(9):1488–96.
23. Perez-Hernandez J, Forner MJ, Pinto C, Chaves FJ, Cortes R, Redon J. Increased urinary exosomal microRNAs in patients with systemic lupus erythematosus. PLoS One. 2015;10(9):e0138618.
24. Michael A, Bajracharya SD, Yuen PS, et al. Exosomes from human saliva as a source of microRNA biomarkers. Oral Dis. 2010;16(1):34–8.
25. Carr PR, Jansen L, Walter V, et al. Associations of red and processed meat with survival after colorectal cancer and differences according to timing of dietary assessment. Am J Clin Nutr. 2016;103(1):192–200.
26. Alevizos I, Alexander S, Turner RJ, Illei GG. MicroRNA expression profiles as biomarkers of minor salivary gland inflammation and dysfunction in Sjogren's syndrome. Arthritis Rheum. 2011;63(2):535–44.
27. Gallo A, Jang SI, Ong HL, et al. Targeting the Ca(2+) sensor STIM1 by exosomal transfer of Ebv-miR-BART13-3p is associated with Sjogren's syndrome. EBioMedicine. 2016;10:216–26.
28. Saenz-Cuesta M, Osorio-Querejeta I, Otaegui D. Extracellular vesicles in multiple sclerosis: what are they telling us? Front Cell Neurosci. 2014;8:100.
29. Zheng X, Chen F, Zhang Q, et al. Salivary exosomal PSMA7: a promising biomarker of inflammatory bowel disease. Protein Cell. 2017;8(9):686–95.
30. 赵宗磊, 杜松, 沈淑馨, et al. 病毒性心肌炎患者血浆外泌体生物标志物蛋白质组学筛选.
31. Dreux M, Garaigorta U, Boyd B, et al. Short-range exosomal transfer of viral RNA from infected cells to plasmacytoid dendritic cells triggers innate immunity. Cell Host Microbe. 2012;12(4):558–70.
32. Vallhov H, Gutzeit C, Hultenby K, Valenta R, Gronlund H, Scheynius A. Dendritic cell-derived exosomes carry the major cat allergen Fel d 1 and induce an allergic immune response. Allergy. 2015;70(12):1651–5.
33. Kulshreshtha A, Ahmad T, Agrawal A, Ghosh B. Proinflammatory role of epithelial cell-derived exosomes in allergic airway inflammation. J Allergy Clin Immunol. 2013;131(4):1194–203, 1203.e1191–e1114.
34. Hough KP, Chanda D, Duncan SR, Thannickal VJ, Deshane JS. Exosomes in immunoregulation of chronic lung diseases. Allergy. 2017;72(4):534–44.
35. Admyre C, Bohle B, Johansson SM, et al. B cell-derived exosomes can present allergen peptides and activate allergen-specific T cells to proliferate and produce TH2-like cytokines. J Allergy Clin Immunol. 2007;120(6):1418–24.
36. Xia YC, Harris T, Stewart AG, Mackay GA. Secreted factors from human mast cells trigger inflammatory cytokine production by human airway smooth muscle cells. Int Arch Allergy Immunol. 2013;160(1):75–85.
37. Dvorak AM. Degranulation and recovery from degranulation of basophils and mast cells. Chem Immunol Allergy. 2005;85:205–51.
38. Hough KP, Trevor JL, Strenkowski JG, et al. Exosomal transfer of mitochondria from airway myeloid-derived regulatory cells to T cells. Redox Biol. 2018;18:54–64.
39. Morrison TJ, Jackson MV, Cunningham EK, et al. Mesenchymal stromal cells modulate macrophages in clinically relevant lung injury models by extracellular vesicle mitochondrial transfer. Am J Respir Crit Care Med. 2017;196(10):1275–86.
40. Levanen B, Bhakta NR, Torregrosa Paredes P, et al. Altered microRNA profiles in bronchoalveolar lavage fluid exosomes in asthmatic patients. J Allergy Clin Immunol. 2013;131(3):894–903.

41. Gon Y, Maruoka S, Inoue T, et al. Selective release of miRNAs via extracellular vesicles is associated with house-dust mite allergen-induced airway inflammation. Clin Exp Allergy. 2017;47(12):1586–98.
42. van den Berge M, Tasena H. Role of microRNAs and exosomes in asthma. Curr Opin Pulm Med. 2019;25(1):87–93.
43. 王湘云, 陈乾, 孙亚红, 白冲, 第二军医大学学报 商J. 不同严重程度哮喘患者血清外泌体中微RNA-21的表达水平及其诊断价值. 2018;39(347):50–54.
44. Ong J, Timens W, Rajendran V, et al. Identification of transforming growth factor-beta-regulated microRNAs and the microRNA-targetomes in primary lung fibroblasts. PLoS One. 2017;12(9):e0183815.
45. Kumar M, Ahmad T, Sharma A, et al. Let-7 microRNA-mediated regulation of IL-13 and allergic airway inflammation. J Allergy Clin Immunol. 2011;128(5):1077–85. e1071–e1010.
46. Mattes J, Collison A, Plank M, Phipps S, Foster PS. Antagonism of microRNA-126 suppresses the effector function of TH2 cells and the development of allergic airways disease. Proc Natl Acad Sci U S A. 2009;106(44):18704–9.
47. Pua HH, Steiner DF, Patel S, et al. MicroRNAs 24 and 27 suppress allergic inflammation and target a network of regulators of T helper 2 cell-associated cytokine production. Immunity. 2016;44(4):821–32.
48. Lu TX, Hartner J, Lim EJ, et al. MicroRNA-21 limits in vivo immune response-mediated activation of the IL-12/IFN-gamma pathway, Th1 polarization, and the severity of delayed-type hypersensitivity. J Immunol. 2011;187(6):3362–73.
49. Bryniarski K, Ptak W, Jayakumar A, et al. Antigen-specific, antibody-coated, exosome-like nanovesicles deliver suppressor T-cell microRNA-150 to effector T cells to inhibit contact sensitivity. J Allergy Clin Immunol. 2013;132(1):170–81.
50. Malmhäll C, Alawieh S, Lu Y, et al. MicroRNA-155 is essential for TH2-mediated allergen-induced eosinophilic inflammation in the lung. J Allergy Clin Immunol. 2014;133(5):1429–38, 1438.e1–7.
51. Suojalehto H, Toskala E, Kilpelinen M, Majuri ML, Alenius H. MicroRNA profiles in nasal mucosa of patients with allergic and nonallergic rhinitis and asthma. Paper presented at: John Wiley & Sons, Ltd.2013, 3, 612.
52. Xie G, Yang H, Peng X, et al. Mast cell exosomes can suppress allergic reactions by binding to IgE. J Allergy Clin Immunol. 2018;141(2):788–91.
53. Nankivell BJ, Alexander SI. Rejection of the kidney allograft. N Engl J Med. 2010;363:1451–62.
54. Nasr M, Sigdel T, Sarwal M. Advances in diagnostics for transplant rejection. Expert Rev Mol Diagn. 2016;16(10):1121–32.
55. Alegre ML, Lakkis FG, Morelli AE. Antigen presentation in transplantation. Trends Immunol. 2016;37(12):831–43.
56. Afonso Junior JE, Werebe Ede C, Carraro RM, et al. Lung transplantation. Einstein (Sao Paulo). 2015;13(2):297–304.
57. Lund LH, Edwards LB, Kucheryavaya AY, et al. The registry of the International Society for Heart and Lung Transplantation: thirty-second official adult heart transplantation report—2015; focus theme: early graft failure. J Heart Lung Transplant. 2015;34(10):1244–54.
58. Jaramillo A, Smith MA, Phelan D, et al. Development of ELISA-detected anti-HLA antibodies precedes the development of bronchiolitis obliterans syndrome and correlates with progressive decline in pulmonary function after lung transplantation. Transplantation. 1999;67(8):1155–61.
59. Hachem RR, Yusen RD, Meyers BF, et al. Anti-human leukocyte antigen antibodies and preemptive antibody-directed therapy after lung transplantation. J Heart Lung Transplant. 2010;29(9):973–80.
60. Gregson AL, Hoji A, Injean P, et al. Altered exosomal RNA profiles in bronchoalveolar lavage from lung transplants with acute rejection. Am J Respir Crit Care Med. 2015;192(12):1490–503.
61. Dewi IS, et al. Exosomal miR-142-3p is increased during cardiac allograft rejection and augments vascular permeability through down-regulation of endothelial RAB11FIP2 expression. Cardiovasc Res. 2017;113(5):440–52.

62. Qamri Z, Pelletier R, Foster J, et al. Early posttransplant changes in circulating endothelial microparticles in patients with kidney transplantation. Transpl Immunol. 2014;31(2):60–4.
63. Tower CM, Reyes M, Nelson K, et al. Plasma C4d+ endothelial microvesicles increase in acute antibody-mediated rejection. Transplantation. 2017;101(9):2235–43.
64. Lim JH, Lee CH, Kim KY, et al. Novel urinary exosomal biomarkers of acute T cell-mediated rejection in kidney transplant recipients: a cross-sectional study. PLoS One. 2018;13(9):e0204204.

# Extracellular Vesicle Biomarkers for Neurological Diseases

# 19

Yanli Tan, Lei Xin, Jianglong Xu, Chuan Fang, and Chunsheng Kang

## 19.1 Introduction

Extracellular vesicles (EVs) are membrane-bound nanoparticles that can be classified as small extracellular vesicles (sEVs, popularly known as "exosomes") and large extracellular vesicles (lEVs, commonly known as "microvesicles") based on their size, biogenesis, and biophysical characteristics [1–3]. sEVs are approximately 30–200 nm in size and are released when multivesicular bodies (MVBs) fuse with the plasma membrane [4]. MVBs are formed by inward budding of the cell membrane. lEVs, which have diameters of approximately 200–1000 nm, are small vesicles that detach from cell membranes after activation, injury, or apoptosis [5]. Different central nervous system (CNS) cells, such as neurons, astrocytes, microglia, and oligodendrocytes, have been shown to produce and release EVs [6, 7]. These EVs carry important substances, such as DNA, mRNAs, microRNAs (miRNAs), other noncoding RNAs, proteins, and lipids from parent cells; these cargoes

Yanli Tan and Lei Xin contributed equally with all other contributors.

Y. Tan · L. Xin
Hebei University, Baoding, China

Hebei Key Laboratory of Precise Diagnosis and Treatment of Glioma, Baoding, China

J. Xu · C. Fang (✉)
Hebei Key Laboratory of Precise Diagnosis and Treatment of Glioma, Baoding, China

Department of Neurosurgery, Affiliated Hospital of Hebei University, Baoding, China
e-mail: chuanfang@hub.edu.cn

C. Kang (✉)
Laboratory of Neuro-oncology, Tianjin Neurological Institute, Tianjin Medical University General Hospital, Tianjin, China
e-mail: kang97061@tmu.edu.cn



Q. Wang, L. Zheng (eds.), *Extracellular Vesicles*,
https://doi.org/10.1007/978-981-99-8365-0_19

enter recipient cells through endocytosis or receptor-dependent cell membrane activity. Thus, EVs participate in cell-to-cell communication and play important roles in the physiological and pathological conditions of the CNS. For example, sEVs from oligodendrocytes nourish and support healthy and diseased tissues when absorbed by neurons [8].

EVs can cross the blood-brain barrier and enter circulation. EVs have been successfully isolated from human serum, plasma, cerebrospinal fluid (CSF), saliva, and urine. sEVs can be efficiently labeled using surface proteins, including transmembrane 4 superfamily members (e.g., CD9, CD63, CD81, and CD82), ALG-2 interacting protein X, tumor susceptibility gene 101 protein, 70-kDa heat shock protein, guanosine triphosphatase, and major histocompatibility complex molecules [9, 10]. Adenosine diphosphate-ribosylation factor 6 and vesicle-associated membrane protein 3 have also been identified as potential lEV markers [11, 12]. However, these proteins are also present in other types of vesicles and in cell membranes. EVs secreted by neurons have been found to carry glycosylphosphatidylinositol-anchored proteins, L1 cell adhesion molecule, and glutamate receptors 2 and 3. Thus, brain-derived EVs can be isolated based on these characteristics [13, 14].

These EVs can protect carried substances from being hydrolyzed due to their structures, and compared with traditional biomarkers, EV biomarkers are more stable, abundant, and noninvasive, with great potential for the diagnosis and treatment of CNS diseases. In this chapter, we summarize research progress on EV biomarkers in CNS diseases and review existing problems and potential prospects.

## 19.2 Neoplastic Diseases

CNS tumors include primary and metastatic tumors. Prior studies have shown that tumor cells secrete more EVs than normal cells and that EV contents may affect the immune response to the host, alter chemical drug resistance, enhance tumor proliferation, and promote tumor invasion [15]. Therefore, EVs are a potential biomarker source that can provide information about tumor malignancy and prognosis as well as insights into diagnosis and individualized treatment for patients with CNS tumors.

In 2016, molecular biomarkers, such as isocitrate dehydrogenase (*IDH*) gene mutation status [16], were included in the World Health Organization classification of CNS tumors for the first time. *IDH* mutations have been detected in EVs from the CSF and plasma of patients with glioma. Furthermore, EV-*IDH* mutations play important roles in the diagnosis of glioma [17]. In addition, epidermal growth factor receptor (EGFR)-positive EVs are effective diagnostic and prognostic markers of glioma, and EGFR levels in EVs are positively correlated with Ki-67 levels in tumor tissues [18]. The EGFR levels in serum EVs can be used to identify patients with different grades of glioma. In addition, serum $CD9^+$/glial fibrillary acidic protein $(GFAP)^+$/$survivin^+$ sEVs have been shown to be of great value in early glioma diagnosis [19]. Polymerase I and transcriptional releasing factor (PTRF) can be detected in the blood sEVs of patients with glioma and can be used as biomarkers for glioma. Indeed, PTRF-containing sEVs are also associated with poor prognosis and recurrence. PTRF enhances intercellular communication with sEVs and affects the malignant phenotype of glioma cells. Higher PTRF/CD63 ratios in serum sEVs from patients with glioma after surgery are associated with poor prognosis, and this factor can also be used to evaluate the postoperative efficacy of glioma [20]. This evidence suggests that components in circulating or local EVs can be used as biomarkers to evaluate patient response to treatment and prognosis. In addition, *miR-21* in EVs derived from the CSF has predictive value for glioma recurrence [21]. Moreover, *miR-10b* is also expressed in nervous system tumor cells but not in normal brain tissue [22]. Circulating sEV *miR-10b* is positively correlated with tumor malignancy, indicating that *miR-10b* can be used as an important tumor marker to evaluate prognosis [23].

Glioblastoma (GBM) is a highly heterogeneous and malignant type of glioma, and EGFR variant III (EGFRvIII), the most common mutant form of EGFR, is closely associated with GBM prognosis. EGFRvIII has been detected in CSF-derived EVs from patients with EGFRvIII-positive or EGFRvIII-negative GBM. The sensitivity and specificity of EGFRvIII in CSF-derived EVs for diagnosing EGFRvIII-positive GBM were 61% and 98%, respectively. EGFRvIII mutation-guided targeted therapy for GBM also seems feasible [24]. One small RNA (*RNU6-1*) and two miRNAs (*miR-320* and *miR-574-3p*) were found in sEVs isolated from the plasma of patients with GBM. Additionally, *RNU6-1* or *miR-320/miR-574-3p/RNU6-1* binding can identify both healthy controls and patients with GBM [25]. As a biomarker for the diagnosis of GBM, *RNU6-1* expression in sEVs can be used to differentiate GBM from nonneoplastic lesions (including stroke, hemorrhage, and demyelinating lesions) and primary CNS lymphoma (PCNSL) [26].

Due to the rapid development of gene chips and high-throughput sequencing technology, EV miRNAs from CSF can be examined (e.g., using the TaqMan OpenArray Real-Time PCR System). EV *miR-21* levels are significantly higher in cancer patients than in controls [27]. Additionally, proteomic analysis of intraoperative suction-isolated EVs showed that all eight subunit EV levels of the key molecular chaperone T-complex protein 1 (TCP1) ring complex, including CCT2, CCT3, CCT5, CCT6A, CCT7, and TCP1, were higher in patients with GBM than in patients with glioma grades II and III. Immunohistochemical staining confirmed that CCT6A levels may be associated with EGFR expression [28]. Another study examined serum-derived EVs using liquid-chromatography mass spectrometry in four groups of patients with GBM, including patients with non-small cell lung cancer with brain metastasis, meningioma, and lumbar disc hernia (as a control). In total, 10 whole serum proteins (i.e., matrix metalloproteinase [MMP] 9, heat-shock protein B1 [HSPB1], caspase 14 [CASP14], hemoglobin subunit gamma 1 [HBG1], immunoglobulin heavy constant gamma 4, defensin alpha 1, von Willebrand factor, heterogeneous nuclear ribonucleoprotein A1, S100A8, and talin 1) and 17 EV proteins (MMP9, HSPB1, CASP14, HBG1, fibrinogen B, gamma-glutamylcyclotransferase, platelet factor 4, S100A7, fibronectin 1, alanyl aminopeptidase, filaggrin 2, HSPA8, immunoglobulin lambda-like polypeptide 1, multimerin 1, S100A14, suprabasin, small proline-rich protein 2E) showed significant differences, and principal component analysis suggested that EVs have great potential for monitoring tumors of the CNS [29].

Temozolomide (TMZ) is the first-line chemotherapeutic agent for glioma treatment. O6-methylguanine-DNA methyltransferase (MGMT) is a key enzyme with DNA repair activity, and the levels of these enzymes are used as predictors of response to TMZ therapy. The level of MGMT in sEVs correlates with the number of tumor cells. Furthermore, miRNAs associated with clinical responses to chemotherapy, including *miR-21* and *miR-9*, have been detected in CSF-isolated EVs from patients with GBM [30, 31]. In addition, sEVs from GBM cells carry *miR-221*, which targets Dynamin3, leading to increased TMZ resistance [32]. The expression level of *miR-1238* in circulating sEVs has been shown to be higher in patients with GBM than in healthy controls. Through sEVs, *miR-1238* can induce the upregulation of *miR-1238* in TMZ-sensitive cells, activate the phosphatidylinositol 3-kinase (PI3K)/Akt/mammalian target of rapamycin (mTOR) pathway, significantly reduce the expression levels of Caveolin-1 in receptor cells, and eventually lead to the development of drug resistance in TMZ-sensitive cells [33]. Similar to radiotherapy, GBM-derived sEV *miR-301a* can specifically inhibit the expression of transcription elongation factor A-like 7 (TCEAL7) in recipient cells in the hypoxic tumor microenvironment. During this process, TCEAL7 also prevents the transport of β-catenin from the cytoplasm to the nucleus, which negatively regulates the Wnt/β-catenin signaling pathway and promotes radiotherapy resistance in GBM cells [34].

## 19.3 Neurodegeneration

Neurodegenerative diseases are characterized by the accumulation of intracellular and extracellular insoluble proteins with similar clinical manifestations. Thus, neurodegenerative diseases are also referred to as brain protein-related diseases. As potential vectors of intercellular interactions, EVs may play roles in the pathogenesis of these diseases. Moreover, because EVs are released into body fluids, they may be promising noninvasive peripheral biomarkers for the prediction and diagnosis of these diseases.

### 19.3.1 Alzheimer's Disease (AD)

AD, a progressive neurodegenerative neuropathy and the most common form of dementia, is characterized by the aggregation of beta-amyloid and the microtubule-associated protein tau [35]. The proteins in blood sEVs can serve as biomarkers of AD. Neurogenic sEVs isolated from the plasma of patients with AD showed higher expression levels of Aβ1–42, total tau, phospho-T181 tau, and phospho-S396 tau, thus providing higher disease predictability during the preclinical phase [36]. The levels of phosphorylated insulin receptor substrate 1 in neurogenic sEVs can accurately predict the development of AD in patients with diabetes [37, 38]. In another study, researchers isolated serum neuron-derived sEVs from patients with AD and from healthy controls and showed that synaptosomal-associated protein 25 kDa (SNAP-25) levels could also distinguish the two groups; indeed, SNAP-25 levels in serum neuron-derived sEVs have been shown to correlate with Mini-Mental State Examination scores and cognitive status [39]. In addition, the expression levels of C-type lectin domain family 1 member B and C-C motif chemokine ligand 11 (CCL11) proteins in plasma EVs are associated with nerve and inflammatory processes and have diagnostic power [40].

In addition to proteins, miRNAs are also effective biomarkers. Analysis of blood-derived EVs from patients with AD with different levels of dementia severity (mild, moderate, and severe) revealed that myelin oligodendrocyte glycoprotein (MOG), *let-7G-5p*, *miR-126-3p*, *miR-142-3p*, *miR-146a-5p*, and *miR-223-3p* could be used as biomarkers of dementia severity [41]. In addition, *miR-135a* and *miR-384* were upregulated in serum sEVs from patients with AD, whereas *miR-193b* was downregulated. Combining all three miRNAs allows for early screening in patients with mild cognitive impairment (MCI) [42]. Moreover, blood sEV *miR-193b* levels were lower in patients with AD than in those with MCI [43]. Therefore, sEVs could serve as biomarkers for the early diagnosis of AD and provide insights into the recognition and prevention of AD.

In another study, researchers used next-generation deep sequencing to analyze blood sEV miRNAs in patients with AD and in healthy, age- and sex-matched controls. Screenings showed significant differences in 24 miRNAs, among which three miRNAs (*miR-30b-5p*, *miR-22-3p*, and *miR-378a-3p*) could better predict AD status (area under the curve of 0.880) [44]. In addition, integration of miRNA expression data with other data could provide useful and reliable biomarkers for AD [45].

### 19.3.2 Parkinson's Disease (PD)

PD is a common neurodegenerative disease characterized by the accumulation of α-synuclein (α-syn), a terminal protein of neurons. Serum sEV α-synuclein can be used as a biomarker for the identification of PD [46]. In addition, plasma sEV α-synuclein levels have been shown to be significantly increased in patients with PD [47]; the ratio of α-syn to total α-syn is also significantly increased and is negatively correlated with disease severity [48]. sEVs can transfer α-syn between cells and induce apoptosis; this process is correlated with the severity of cognitive impairment and promotes the progression of PD [35]. In addition, α-syn levels in salivary EVs can be used to distinguish patients with PD from healthy controls [49].

Another study analyzed the protein profiles of plasma-derived sEVs from patients with stage II and III PD using Hoehn classification-based proteomic analysis. The results indicated that plasma-derived sEV proteins, including clusterin, the complement C1R subcomponent, apolipoprotein A1, and fibrinogen γ chain, are potential biomarker candidates for PD diagnosis [50]. In addition, SNAP23 and calbindin levels were found to be increased in patients with PD, and their success rates for disease diagnosis and prediction within the Discovery Replication cohort were 86% and 76%, respectively [51]. sEVs also contain abundant miRNAs, including *miR-29c*, *miR-30c*, *miR-146a-5p*, *miR-185*, *miR-19b*, *miR-214*, and *miR-222*. *miR-19b* and *miR-29c* may be correlated with disease severity, thus providing a basis for early clinical diagnosis of PD [52].

### 19.3.3 Amyotrophic Lateral Sclerosis (ALS)

ALS is a neurodegenerative disease characterized by chronic degeneration of upper and lower motor neurons and the accumulation of phosphorylated TDP43 and superoxide dismutase 1. Thus, CSF sEV TDP43 may be a potential diagnostic biomarker in patients with ALS [53]. A comparison of the mRNA expression profiles of CSF sEVs in four patients with sporadic ALS and four normal healthy donors using high-sensitivity next-generation sequencing revealed significant changes in 543 genes, including *CUEDC2* [54]. *miR-124-3p* is also highly abundant in CSF sEVs, and there is a significant correlation between *miR-124-3p* levels in CSF sEVs and disease stage (as defined by ALSFRS-R scores) in male patients with ALS [55]. Therefore, miRNA biomarkers may be helpful for the early diagnosis of ALS.

### 19.3.4 Huntington's Disease (HD)

HD is caused by the accumulation of mutated Huntington proteins. A previous study found that 13 miRNAs in plasma-derived sEVs from patients with HD, including *miR-877-5p*, *miR-223-3p*, *miR-30D-5p*, *miR-128*, *miR-22-5p*, *miR-223-5p*, *miR-338-3p*, *miR-130b-3p*, *miR-628-3p*, *miR-361-5p*, *miR-425-5p*, and *miR-942*, were significantly upregulated [56]. Thus, these disease-associated amplified RNAs in the body fluids of patients with HD may be good candidate biomarkers for monitoring disease progression and treatment.

### 19.3.5 Multiple Sclerosis (MS)

MS is characterized by T-cell targeting of myelin antigens as well as destruction of the protective myelin sheath around neurons. sEVs mainly express three types of myelin proteins: myelin basic protein, lipoprotein, and MOG. MOG levels are increased in patients with MS and are closely associated with disease activity [57]. Notably, *let-7i* is significantly increased in patients with MS. Induction of circulating sEVs in human T cells is inhibited in patients with MS, and sEV *let-7i* regulates MS progression by blocking the IGF1R/transforming growth factor (TGF) β receptor 1 pathway [58]. The miRNA profiles of sEVs in patients with MS are different from those in healthy controls. Additionally, *miR-122-3p* and *miR-200a-5p* have been shown to be enriched in serum sEVs of patients with relapse-remission MS (RRMS) and neuromyelitis optic spectrum disease. These miRNAs are also positively correlated with disease severity [59]. In another study, researchers found that *miR-326* expression levels were significantly increased in T-cell-derived sEVs from patients with RRMS [60]. Thus, all of these components may be biomarkers for MS.

Experimental allergic encephalomyelitis (EAE) is an ideal animal model for MS. The miRNAs in urinary sEVs, plasma, and spinal cords were systematically analyzed during the pre-onset, onset, and peak stages of an EAE model established in chronic B6 mice. Urine sEVs can serve as a specific and sensitive source of miRNA biomarkers for all three stages of EAE. For example, *miR-155-5p* expression has been shown to be increased in urinary sEVs, plasma, and spinal cords 6 days before the onset of EAE, suggesting its early involvement in the progression of EAE [61]. In addition, sEV circRNA expression can be used to diagnose immune-mediated demyelinating diseases, such as MS; *hsa_circ_0087862* and *hsa_circ_0012077* have been validated as biomarkers in independent subject cohorts. Receiver operating characteristic curves showed that both *hsa_circ_0087862* and *hsa_circ_0012077* have 100% diagnostic accuracy in identifying immune-mediated demyelinating diseases [62]. With the development of various omics tools, more potential markers for EVs are likely to be identified. However, further validation will also be needed.

## 19.4 Infectious Diseases

### 19.4.1 Prion Disease

Prion diseases, also known as transmissible spongiform encephalopathies, are characterized by spongy vacuolation and the progressive loss of neuronal structure and function. Pathological features also include accumulation of $PrP^{Sc}$, which results from the misfolding of normal viral proteins ($PrP^{C}$). $PrP^{C}$ is highly expressed in sEVs, and disease-associated forms of $PrP^{Sc}$ have also been found in sEVs, which are taken up by immature cells [63]. In addition, as EV cargo, miRNAs can be identified as valuable biomarkers using high-throughput approaches, although they may also be altered in other neurodegenerative diseases.

### 19.4.2 Other Infectious Diseases

EVs can also serve as potential biomarkers for diseases outside the CNS. For example, macrophages and dendritic cells release more EVs after the onset of human immunodeficiency virus (HIV) infection [64, 65]. The HIV critical protein Nef has been detected in plasma EVs of patients [66]. Examination of miRNA expression in serum sEVs revealed that four candidate miRNAs (*miR-671-5p*, *miR-16-5p*, *miR-150-3p*, and *miR-4281*) showed differential expression levels across three groups: patients with mild hand, foot, and mouth disease, patients with severe hand, foot, and mouth disease, and healthy controls [67].

## 19.5 Cerebrovascular Diseases

Stroke is a complex and life-threatening cerebrovascular disease. There are two subtypes of stroke, i.e., ischemic and hemorrhagic. After stroke, sEVs are synthesized and released from the brain cells. They can pass through the blood-brain barrier and play roles in intercellular communication. sEVs can also be detected in the peripheral blood or CSF and are involved in the occurrence and development of stroke. In 2014, Lausted et al. identified biomarkers related to stroke treatment in bodily fluids; however, these biomarkers had the disadvantages of low abundance and poor heterogeneity [68].

### 19.5.1 Cerebral Ischemic Stroke

After ischemia, astrocytes release more sEVs, indicating that the viral protein (PrP) in sEVs increases. In addition, caspase-1 protein levels in the serum EVs of patients with ischemic stroke can be used as potential biomarkers [69]. miRNAs have also become reliable biomarkers for the diagnosis, treatment, and prognosis of ischemic stroke. One study showed that *miR-223* was significantly upregulated in blood samples from patients with acute ischemic stroke within 72 h, as demonstrated using reverse transcription polymerase chain reaction, and that sEV *miR-223* levels were positively correlated with NIH Stroke Scale/Score and poor prognosis [70]. The combined application of *miR-30a-5p* and *miR-21-5p* in plasma-derived sEVs can be used to differentiate ischemic stroke from the hyperacute phase (HIS) and subacute phase to the recovery phase. Additionally, *miR-30a-5p* may be a favorable biomarker for early clinical diagnosis of HIS [71]. Finally, in a Wistar rat model of focal cerebral ischemia, sEV *miR-126* obtained from the peripheral blood appeared to be more sensitive to the effects of cerebral ischemia. However, total serum *miR-126* levels may be a more specific indicator of ischemic severity [72].

### 19.5.2 Intracerebral Hemorrhage (ICH)

ICH is a common cerebrovascular disease. However, there is still a lack of effective treatments for ICH. In clinical practice, exosomal miRNAs can be used as alternative biomarkers for predictive diagnosis of ICH-induced injuries. For example, neurons can secrete sEVs as carriers and deliver *miR-132* to cerebrovascular endothelial cells, which then play a protective role in the cerebrovascular system and reduce ICH-induced blood-brain barrier damage [73].

## 19.6 Traumatic Encephalopathy

Chronic traumatic encephalopathy (CTE) can lead to long-term movement problems, cognitive impairment, and death. Currently, the diagnosis of CTE relies on neuroimaging. However, this diagnostic method has limitations and difficulties, and a more reliable diagnostic metric is needed [74, 75]. The detection of markers in the CSF has yielded promising results. Several biomarkers, including total tau, soluble triggering receptor expressed on myeloid cells 2, CCL11, neurofilament light chain, GFAP, and ubiquitin-terminal hydrolase L1, can be used to predict clinical outcomes and monitor disease progress [76, 77]. However, these circulating biomarkers appear to be very unstable and not useful for the diagnosis of mild CTE (mCTE) [78].

Owing to their two-layer membrane structure, sEVs are more effective at initially protecting cells from hydrolysis and can better characterize the timeline and severity of CTE [79]. By analyzing changes in sEV (i.e., those sourced from neurons and glial cells) proteomics and genomic profiles, the extent of damage, as well as pathological changes, in CTE can be assessed more quickly and accurately [80]. This noninvasive diagnostic approach is novel and exciting for several reasons. For example, saliva-derived sEVs can harbor gene expression products that can be used as biomarkers for diagnosing mCTE. In addition, tau in EV cargo is not only a circulating biomarker for AD but also a biomarker for CTE [81]. After acute head injury, the levels of prion cell protein ($PrP^C$), coagulation factor XIII (XIIIA), synaptogyrin-3, IL-6, and aquaporin in neuron-derived sEVs continue to increase for several months, and this may be helpful for predicting CTE [75]. Prolonged expression of aquaporin and IL-6 may be responsible for the persistent central nervous system edema and inflammation characteristic of CTE. $PrP^C$, coagulation factor XIIIA, and synaptogyrin-3 bind to and concentrate the neurotoxic form of oligoamyloid peptide or phosphorylated tau (p-tau) and deliver it to neurons at or far from the trauma site. High-throughput sequencing has also been used to analyze changes in miRNA expression in sEVs isolated from the peripheral blood plasma of rats with traumatic brain injuries. In total, 50 significantly differentially expressed miRNAs were identified, including 31 upregulated and 19 downregulated miRNAs. The mitogen-activated protein kinase signaling pathway, regulation of actin cytoskeleton regulation pathway, Rap1 signaling pathway, and Ras signaling pathway were most relevant pathways [82]. However, further validation of these EV biomarkers is warranted.

## 19.7 Psychiatric Disorders

The metabolomics of serum derived-sEVs in schizophrenia (SCZ) was studied using ultra-performance liquid chromatography-tandem mass spectrometry. Twenty-five metabolites, including phospholipids, amino acids, organic acids, and lipid oxidation products, were found to be significantly different between patients with SCZ and controls. These same metabolites also performed well in differentiating between patients and controls in three other psychiatric patient groups, thus providing new insights into the diagnosis of psychiatric disorders [83]. Post-traumatic stress disorder (PTSD) is a mental disorder. miRNA analysis was performed in whole plasma, plasma EVs, and EV delete (EVD) plasma samples from male military veterans. *miR-203a-3p* in EVs and *miR-339-5p* changes in EVD plasma samples were significantly different between patients and controls. The target genes of these two miRNAs are involved in signaling pathways associated with PTSD (e.g., neurotransmitter systems incorporating dopamine and serotonin signaling, inflammatory responses, and cardiovascular disease) [84]. In addition to miRNAs, the coding transcripts of a few proteins in EVs, including *NCAM1* (low expression in PTSD) and *GRIN2C* (high expression in PTSD), have been shown to be altered in patients experiencing PTSD. Kenny et al. found elevated levels of amyloid beta, total tau, and p-tau in plasma and plasma-derived sEVs in military veterans (some of whom also had chronic neuropsychological symptoms). These potential markers are associated with mCTE, which is characterized by chronic neuropsychological symptoms [85].

## 19.8 Other CNS Diseases

EVs can also serve as a potential biomarker for other CNS diseases. For example, ionizing radiation has been shown to cause CNS damage in irradiated mice. Metabolomic and lipidomic analyses of plasma and plasma-derived EVs 2 days after irradiation exposure revealed EV enrichment of many molecules involved in inflammatory responses, including C16 sphinganine, carnitine, and platelet activators [86]. Focal cortical dysplasia (FCD) is a cortical developmental abnormality that is also the most common cause of refractory epilepsy. Compared with that in healthy volunteers, four upregulated miRNAs (*miR-194-2-5p*, *miR-15a-5p*, *miR-132-3p*, and *miR-145-5p*) were identified in the serum sEVs of patients with FCD. KEGG pathway analysis and target miRNA classification determined that the mTOR, PI3K/Akt, p53, cell cycle regulation, and TGF-β signaling pathways were all high-risk pathways that may play key roles in the pathogenesis of FCD and refractory epilepsy [87].

## 19.9 Problems and Perspectives

Studies of EV biomarkers in the CNS are still relatively recent. Therefore, careful experimental design is crucial. To achieve higher clinical utility, researchers must fully understand the clinical problems and needs of these diseases. For example, control groups should include both healthy controls and patients with other diseases with similar clinical symptoms and/or biomarker expression profiles to improve the specificity of disease diagnosis. In addition, the appropriate methods should be determined. In one case, separate CSF EV kits were analyzed and compared to obtain reproducible results [88]. However, extracting the required purity and volume of EVs from circulating bodily fluids is challenging. For samples with a limited number of EVs (i.e., amounts that are insufficient for high-throughput detection), a mixed-sample approach can be adopted. In addition, methods targeting the neuron cell adhesion molecule L1CAM have been introduced to isolate brain-derived EVs from peripheral blood [89]. However, further studies, more specific surface markers, and new methods are needed to isolate EVs from different brain cell types (i.e., astrocytes, microglia, and neurons) and brain regions. The main techniques currently available for detecting EV nucleic acids include next-generation sequencing and gene chip technology, whereas the multi-germplasm spectrum method is mainly used for detecting EV proteins and lipids. Once the testing is complete and expression differences are obtained, the data quality must also be evaluated. Selected molecules can also be further screened using bioinformatics methods, in combination with basic experiments and clinical studies. Validation experiments, including quantitative polymerase chain reaction and western blotting, are essential to determine whether the target biomarkers meet clinical needs. Although analyzing the results of many clinical EVs pilot studies with small sample sizes may be possible, clinical confirmation with larger populations is needed to elucidate the sensitivity and specificity of the candidate biomarkers. With the continued development of basic EV research and improvements in detection technology, EVs, as important components of liquid biopsies, are expected to become new circulating biomarkers for the auxiliary diagnosis and treatment of CNS diseases.

**Acknowledgments** This research was supported by grants from the Hebei Provincial Central Leading Local Science and Technology Development Fund Project (grant no. 216Z7711G), the National Natural Science Foundation of China (NSFC, grant no. 81772667), and Hebei Natural Science Foundation Precision Medicine Joint Project (grant no. H2020201206).

## References

1. Woith E, Fuhrmann G, Melzig MF. Extracellular vesicles-connecting kingdoms. Int J Mol Sci. 2019;20(22):5695.
2. Zheng J, Tan J, Miao YY, et al. Extracellular vesicles degradation pathway based autophagy lysosome pathway. Am J Transl Res. 2019;11(3):1170–83.
3. Yamamoto T, Kosaka N, Ochiya T. Latest advances in extracellular vesicles: from bench to bedside. Sci Technol Adv Mater. 2019;20(1):746–57.
4. Paulaitis M, Agarwal K, Nana-Sinkam P. Dynamic scaling of exosome sizes. Langmuir. 2018;34(32):9387–93.
5. Hirshman BR, Kras RT, Akers JC, et al. Extracellular vesicles in molecular diagnostics: an overview with a focus on CNS diseases. Adv Clin Chem. 2016;76:37–53.
6. Dutta S. a-Synuclein in brain-derived blood exosomes distinguishes multiple system atrophy from Parkinson's disease. In: 6th international multiple system atrophy congress, 2018; 2018.
7. Goetzl EJ, Schwartz JB, Abner EL, et al. High complement levels in astrocyte-derived exosomes of Alzheimer disease. Ann Neurol. 2018;83(3):544–52.
8. Frühbeis C, Kuo-Elsner WP, Müller C, et al. Oligodendrocytes support axonal transport and maintenance via exosome secretion. PLoS Biol. 2020;18(12):e3000621.
9. Thompson AG, Gray E, Heman-Ackah SM, et al. Extracellular vesicles in neurodegenerative disease—pathogenesis to biomarkers. Nat Rev Neurol. 2016;12(6):346–57.
10. Kowal J, Arras G, Colombo M, et al. Proteomic comparison defines novel markers to characterize heterogeneous populations of extracellular vesicle subtypes. Proc Natl Acad Sci U S A. 2016;113(8):E968–77.
11. Akers JC, Gonda D, Kim R, et al. Biogenesis of extracellular vesicles (EV): exosomes, microvesicles, retrovirus-like vesicles, and apoptotic bodies. J Neurooncol. 2013;113(1):1–11.
12. Muralidharan-Chari V, Clancy J, Plou C, et al. ARF6-regulated shedding of tumor cell-derived plasma membrane microvesicles. Curr Biol. 2009;19(22):1875–85.

13. Fauré J, Lachenal G, Court M, et al. Exosomes are released by cultured cortical neurones. Mol Cell Neurosci. 2006;31(4):642–8.
14. Anastasi F, Masciandaro SM, Carratore RD, et al. Proteomics profiling of neuron-derived small extracellular vesicles from human plasma: enabling single-subject analysis. Int J Mol Sci. 2021;22(6):2951.
15. Basu B, Ghosh MK. Extracellular vesicles in glioma: from diagnosis to therapy. Bioessays. 2019;41(7):e1800245.
16. Louis DN, Perry A, Reifenberger G, et al. The 2016 World Health Organization classification of tumors of the central nervous system: a summary. Acta Neuropathol. 2016;131(6):803–20.
17. Lianidou E, Pantel K. Liquid biopsies. Genes Chromosomes Cancer. 2019;58(4):219–32.
18. Wang H, Jiang D, Li W, et al. Evaluation of serum extracellular vesicles as noninvasive diagnostic markers of glioma. Theranostics. 2019;9(18):5347–58.
19. Ciesielski MJ, Galbo P, Figel S, et al. Abstract 2725: circulating CD9-GFAP-survivin exosomes during active specific immunotherapy, a potential biomarker for glioma. Cancer Res. 2017;77(13 Supplement):2725.
20. Kai H, Chuan F, Kaikai Y, et al. The role of PTRF/Cavin1 as a biomarker in both glioma and serum exosomes. Theranostics. 2018;8(6):1540–57.
21. Zhao XL, Xiao ZH, Li B, et al. miRNA-21 may serve as a promising noninvasive marker of glioma with a high diagnostic performance: a pooled analysis of 997 patients. Ther Adv Med Oncol. 2021;13:1758835920987650.
22. Sun B, Zhao XD, Ming JG, et al. Stepwise detection and evaluation reveal miR-10b and miR-222 as a remarkable prognostic pair for glioblastoma. Oncogene. 2019;38(33):6142–57.
23. Kopkova A, Sana J, Machackova T, et al. Cerebrospinal fluid MicroRNA signatures as diagnostic biomarkers in brain tumors. Cancers (Basel). 2019;11(10):1546.
24. Figueroa JM, Skog J, Akers J, et al. Detection of wild-type EGFR amplification and EGFRvIII mutation in CSF-derived extracellular vesicles of glioblastoma patients. Neuro Oncol. 2017;19(11):1494–502.
25. Manterola L, Guruceaga E, Gállego Pérez-Larraya J, et al. A small noncoding RNA signature found in exosomes of GBM patient serum as a diagnostic tool. Neuro Oncol. 2014;16(4):520–7.
26. Puigdelloses M, Gonzalez-Huarriz M, Garcia-Moure M, et al. RNU6-1 in circulating exosomes differentiates GBM from non-neoplastic brain lesions and PCNSL but not from brain metastases. Neurooncol Adv. 2020;2(1):vdaa010.
27. Rennert RC, Hochberg FH, Carter BS. ExRNA in biofluids as biomarkers for brain tumors. Cell Mol Neurobiol. 2017;36(3):353–60.
28. Hallal S, Russell BP, Wei H, et al. Extracellular vesicles from neurosurgical aspirates identifies chaperonin containing TCP1 subunit 6A as a potential glioblastoma biomarker with prognostic significance. Proteomics. 2019;19(1–2):e1800157.
29. Dobra G, Bukva M, Szabo Z, et al. Small extracellular vesicles isolated from serum may serve as signal-enhancers for the monitoring of CNS tumors. Int J Mol Sci. 2020;21(15):5359.
30. Akers JC, Ramakrishnan V, Kim R, et al. miRNA contents of cerebrospinal fluid extracellular vesicles in glioblastoma patients. J Neurooncol. 2015;123(2):205–16.
31. Geng L, Xu J, Zhu Y, et al. Targeting miR-9 in glioma stem cell-derived extracellular vesicles: a novel diagnostic and therapeutic biomarker. Transl Oncol. 2022;22:101451.
32. Yang JK, Yang JP, Tong J, et al. Exosomal miR-221 targets DNM3 to induce tumor progression and temozolomide resistance in glioma. J Neurooncol. 2017;131(2):255–65.
33. Yin J, Zeng A, Zhang Z, et al. Exosomal transfer of miR-1238 contributes to temozolomide-resistance in glioblastoma. EBioMedicine. 2019;42:238–51.
34. Yue X, Lan F, Xia T. Hypoxic glioma cell-secreted exosomal miR-301a activates Wnt/β--catenin signaling and promotes radiation resistance by targeting TCEAL7. Mol Ther. 2019;27(11):1939–49.
35. Guix FX, Corbett GT, Cha DJ, et al. Detection of aggregation-competent tau in neuron-derived extracellular vesicles. Int J Mol Sci. 2018;19(3):663.
36. Li F, Xie XY, Sui XF, et al. Profile of pathogenic proteins and microRNAs in plasma-derived extracellular vesicles in Alzheimer's disease: a pilot study. Neuroscience. 2020;432:240–6.

37. Kapogiannis D, Boxer A, Schwartz JB, et al. Dysfunctionally phosphorylated type 1 insulin receptor substrate in neural-derived blood exosomes of preclinical Alzheimer's disease. FASEB J. 2015;29(2):589–96.
38. Mullins RJ, Mustapic M, Goetzl EJ, et al. Exosomal biomarkers of brain insulin resistance associated with regional atrophy in Alzheimer's disease. Hum Brain Mapp. 2017;38(4):1933–40.
39. Agliardi C, Guerini FR, Zanzottera M, et al. SNAP-25 in serum is carried by exosomes of neuronal origin and is a potential biomarker of Alzheimer's disease. Mol Neurobiol. 2019;56(8):5792–8.
40. Ellegaard Nielsen J, Sofie Pedersen K, Vestergard K, et al. Novel blood-derived extracellular vesicle-based biomarkers in Alzheimer's disease identified by proximity extension assay. Biomedicines. 2020;8(7):199.
41. Aharon A, Spector P, Ahmad RS, et al. Extracellular vesicles of Alzheimer's disease patients as a biomarker for disease progression. Mol Neurobiol. 2020;57(10):4156–69.
42. Yang TT, Liu CG, Gao SC, et al. The serum exosome derived microRNA-135a, -193b, and -384 were potential Alzheimer's disease biomarkers. Biomed Environ Sci. 2018;31(2):87–96.
43. Liu CG, Zhao Y, Lu Y, et al. ABCA1-labeled exosomes in serum contain higher microRNA-193b levels in Alzheimer's disease. Biomed Res Int. 2021;2021:5450397.
44. Dong ZW, Gu HJ, Guo Q, et al. Profiling of serum exosome MiRNA reveals the potential of a MiRNA panel as diagnostic biomarker for Alzheimer's disease. Mol Neurobiol. 2021;58:428–45.
45. Martins TS, Trindade D, Vaz M, et al. Diagnostic and therapeutic potential of exosomes in Alzheimer's disease. J Neurochem. 2021;156(2):162–81.
46. Si X, Tian J, Chen Y, et al. Central nervous system-derived exosomal alpha-Synuclein in serum may be a biomarker in Parkinson's disease. Neuroscience. 2019;413:308–16.
47. Shi M, Liu C, Cook TJ, et al. Plasma exosomal α-synuclein is likely CNS-derived and increased in Parkinson's disease. Acta Neuropathol. 2014;128(5):639–50.
48. Cerri S, Ghezzi C, Sampieri M, et al. The exosomal/total α-Synuclein ratio in plasma is associated with glucocerebrosidase activity and correlates with measures of disease severity in PD patients. Front Cell Neurosci. 2018;12:125.
49. Cao Z, Wu Y, Liu G, et al. α-Synuclein in salivary extracellular vesicles as a potential biomarker of Parkinson's disease. Neurosci Lett. 2019;696:114–20.
50. Kitamura Y, Kojima M, Kurosawa T, et al. Proteomic profiling of exosomal proteins for blood-based biomarkers in Parkinson's disease. Neuroscience. 2018;392:121–8.
51. Wang S, Kojima K, Mobley JA, et al. Proteomic analysis of urinary extracellular vesicles reveal biomarkers for neurologic disease. EBioMedicine. 2019;45:351–61.
52. Ardashirova NS, Fedotova EY, Illarioshkin SN. The role of microRNA in the pathogenesis and diagnostics of Parkinson's disease. Neurochem J. 2020;14(2):127–32.
53. Iguchi Y, Eid L, Parent M, et al. Exosome secretion is a key pathway for clearance of pathological TDP-43. Brain. 2016;139(12):3187–201.
54. Otake K, Kamiguchi H, Hirozane Y. Identification of biomarkers for amyotrophic lateral sclerosis by comprehensive analysis of exosomal mRNAs in human cerebrospinal fluid. BMC Med Genet. 2019;12(1):7.
55. Yelick J, Men Y, Jin S, et al. Elevated exosomal secretion of miR-124-3p from spinal neurons positively associates with disease severity in ALS. Exp Neurol. 2020;333:113414.
56. Kumar S, Vijayan M, Bhatti JS, et al. MicroRNAs as peripheral biomarkers in aging and age-related diseases. Prog Mol Biol Transl Sci. 2017;146:47–94.
57. Galazka G, Mycko MP, Selmaj I, et al. Multiple sclerosis: serum-derived exosomes express myelin proteins. Mult Scler. 2018;24(4):449–58.
58. Kimura K, Hohjoh H, Fukuoka M, et al. Circulating exosomes suppress the induction of regulatory T cells via let-7i in multiple sclerosis. Nat Commun. 2018;9(1):17.
59. Chen C, Wu Y, Li M, et al. Different exosomal microRNA profile in Aquaporin-4 antibody positive neuromyelitis optica spectrum disorders. Front Immunol. 2020;11:1064.

60. Azimi M, Ghabaee M, Naser Moghadasi A, et al. Altered expression of miR-326 in T cell-derived exosomes of patients with relapsing-remitting multiple sclerosis. Iran J Allergy Asthma Immunol. 2019;18(1):108–13.
61. Singh J, Deshpande M, Suhail H, et al. Targeted stage-specific inflammatory microRNA profiling in urine during disease progression in experimental autoimmune encephalomyelitis: markers of disease progression and drug response. J Neuroimmune Pharmacol. 2016;11(1):84–97.
62. He J, Ren M, Li H, et al. Exosomal circular RNA as a biomarker platform for the early diagnosis of immune-mediated demyelinating disease. Front Genet. 2019;10:860.
63. Hartmann A, Muth C, Dabrowski O, et al. Exosomes and the prion protein: more than one truth. Front Neurosci. 2017;11:194.
64. Mahajan SD, Ordain NS, Kutscher H, et al. HIV neuroinflammation: the role of exosomes in cell signaling, prognostic and diagnostic biomarkers and drug delivery. Front Cell Dev Biol. 2021;9:637192.
65. Mfunyi CM, Vaillancourt M, Vitry J, et al. Exosome release following activation of the dendritic cell immunoreceptor: a potential role in HIV-1 pathogenesis. Virology. 2015;484:103–12.
66. Yarandi SS, Duggan MR, Sariyer IK, et al. Emerging role of Nef in the development of HIV associated neurological disorders. J Neuroimmune Pharmacol. 2021;16(2):238–50.
67. Jia HL, He CH, Wang ZY, et al. MicroRNA expression profile in exosome discriminates extremely severe infections from mild infections for hand, foot and mouth disease. BMC Infect Dis. 2014;14:506.
68. Liu WY, Bai XD, Zhang A, et al. Role of exosomes in central nervous system diseases. Front Mol Neurosci. 2019;12:240.
69. Kerr N, García-Contreras M, Abbassi S, et al. Inflammasome proteins in serum and serum-derived extracellular vesicles as biomarkers of stroke. Front Mol Neurosci. 2018;11:309.
70. Chen Y, Song Y, Huang J, et al. Increased circulating exosomal miRNA-223 is associated with acute ischemic stroke. Front Neurol. 2017;8:57.
71. Wang W, Li DB, Li RY, et al. Diagnosis of hyperacute and acute ischaemic stroke: the potential utility of exosomal microRNA-21-5p and microRNA-30a-5p. Cerebrovasc Dis. 2018;45(5–6):204–12.
72. Chen J, Cui C, Yang X, et al. MiR-126 affects brain-heart interaction after cerebral ischemic stroke. Transl Stroke Res. 2017;8(4):374–85.
73. Zhang Y, Han B, He Y, et al. MicroRNA-132 attenuates neurobehavioral and neuropathological changes associated with intracerebral hemorrhage in mice. Neurochem Int. 2017;107:182–90.
74. Turk KW, Budson AE. Chronic traumatic encephalopathy. Continuum (Minneap Minn). 2019;25(1):187–207.
75. Ayubcha C, Revheim ME, Newberg A, et al. A critical review of radiotracers in the positron emission tomography imaging of traumatic brain injury: FDG, tau, and amyloid imaging in mild traumatic brain injury and chronic traumatic encephalopathy. Eur J Nucl Med Mol Imaging. 2021;48(2):623–41.
76. Wang KK, Yang Z, Sarkis G, et al. Ubiquitin C-terminal hydrolase-L1 (UCH-L1) as a therapeutic and diagnostic target in neurodegeneration, neurotrauma and neuro-injuries. Expert Opin Ther Targets. 2017;21(6):627–38.
77. Pierre K, Dyson K, Dagra A, et al. Chronic traumatic encephalopathy: update on current clinical diagnosis and management. Biomedicines. 2021;9(4):415.
78. Kawata K, Mitsuhashi M, Aldret R. A preliminary report on brain-derived extracellular vesicle as novel blood biomarkers for sport-related concussions. Front Neurol. 2018;9:239.
79. Agoston DV, Shutes-David A, Peskind ER. Biofluid biomarkers of traumatic brain injury. Brain Inj. 2017;31(9):1195–203.
80. Karnati HK, Garcia JH, Tweedie D, et al. Neuronal enriched extracellular vesicle proteins as biomarkers for traumatic brain injury. J Neurotrauma. 2019;36(7):975–87.
81. Ling H, Morris HR, Neal JW, et al. Mixed pathologies including chronic traumatic encephalopathy account for dementia in retired association football (soccer) players. Acta Neuropathol. 2017;133(3):337–52.

82. Wang P, Ma H, Zhang Y, et al. Plasma exosome-derived microRNAs as novel biomarkers of traumatic brain injury in rats. Int J Med Sci. 2020;17(4):437–48.
83. Du Y, Chen L, Li X-S, et al. Metabolomic identification of exosome-derived biomarkers for schizophrenia: a large multicenter study. Schizophr Bull. 2021;47(3):615–23.
84. Lee MY, Baxter D, Scherler K, et al. Distinct profiles of cell-free MicroRNAs in plasma of veterans with post-traumatic stress disorder. J Clin Med. 2019;8(7):963.
85. Kenney K, Qu BX, Lai C, et al. Higher exosomal phosphorylated tau and total tau among veterans with combat-related repetitive chronic mild traumatic brain injury. Brain Inj. 2018;32(10):1276–84.
86. Hinzman CP, Baulch JE, Mehta KY, et al. Plasma-derived extracellular vesicles yield predictive markers of cranial irradiation exposure in mice. Sci Rep. 2019;9(1):9460.
87. Chen S-D, Pan H-Y, Huang J-B, et al. Circulating microRNAs from serum exosomes may serve as a putative biomarker in the diagnosis and treatment of patients with focal cortical dysplasia. Cells. 2020;9(8):1867.
88. Sjoqvist S, Otake K, Hirozane Y. Analysis of cerebrospinal fluid extracellular vesicles by proximity extension assay: a comparative study of four isolation kits. Int J Mol Sci. 2020;21(24):9425.
89. Song Z, Xu Y, Deng W, et al. Brain derived exosomes are a double-edged sword in Alzheimer's disease. Front Mol Neurosci. 2020;13:79.

# Extracellular Vesicle Biomarkers for Infectious Diseases

# 20

Rui Zhang, Meng Yuan, Bikash R. Giri, Shun Li, Guofeng Cheng, and Zhiwei Wu

## 20.1 Introduction

During microbial infection of a host, both microorganisms and host cells secrete extracellular vesicles (EVs), which play important roles in inflammatory responses, immune regulation, infection, host death, pathogen clearance, and disease diagnosis. Infectious diseases induced by pathogenic microorganisms are causes of mortality in children and adults worldwide. Timely and accurate diagnosis is essential for pathogen identification and drug efficacy testing in various control strategies. EVs

---

Rui Zhang, Meng Yuan, Bikash R. Giri and Shun Li contributed equally with all other contributors.

R. Zhang · M. Yuan
Center for Public Health Research, Medical School, Nanjing University, Nanjing, People's Republic of China

B. R. Giri · G. Cheng (✉)
Shanghai Tenth People's Hospital, Institute for Infectious Diseases and Vaccine Development, Tongji University School of Medicine, Shanghai, People's Republic of China
e-mail: chengguofeng@tongji.edu.cn

S. Li
School of Life Science and Engineering, Foshan University, Foshan, People's Republic of China

Z. Wu (✉)
Center for Public Health Research, Medical School, Nanjing University, Nanjing, People's Republic of China

State Key Lab of Analytical Chemistry for Life Science, Nanjing University, Nanjing, People's Republic of China
e-mail: wzhw@nju.edu.cn



Q. Wang, L. Zheng (eds.), *Extracellular Vesicles*,
https://doi.org/10.1007/978-981-99-8365-0_20

are virus-sized nanoparticles released from cells and are significant mediators of cell-to-cell communication in healthy and pathological environments through the delivery of biologically active molecular cargo, both locally and systemically [1]. Most types of cells release EVs that then enter bodily fluids and package and transfer a subset of proteins, DNA fragments, mRNAs, microRNAs (miRNAs), and lipids through which they regulate neighboring or distant cells [2–4]. Furthermore, owing to their high delivery efficiency, biocompatibility, and multifunctional properties, EVs are expected to become a new means for drug delivery, disease diagnosis, and precision treatment [5, 6].

## 20.2 EV Biogenesis in Viruses

EVs are small membrane-bound carriers of intracellular cargo derived from multivesicular bodies (MVBs) or the plasma membrane. Unlike viral particles, the membranes of EVs do not enclose a structured core, such as a capsid. The specific properties of EVs define and distinguish them from other types of microvesicles [7]. The assembly of EVs is an active, energy-dependent, and regulated process [8, 9] that specifically requires sphingomyelinase and components of the endosomal sorting complex required for transport (ESCRT) machinery [10]. The EV content is determined by the protein and RNA composition of the cells from which they are derived. The composition of EVs frequently reflects the relative abundances of contents within EV-producing cells. Thus, EVs derived from virus-infected cells contain highly expressed viral miRNAs [11, 12]. The budding of EVs into MVBs requires the ESCRT proteins tumor susceptibility gene 101 (TSG101) and ALG-2-interacting protein X (Alix; also known as programmed cell death 6-interacting protein), which are known markers of EVs [13]. TSG101 and Alix are also required for ESCRT-dependent budding of enveloped viruses, such as herpes simplex virus-1, from the plasma membrane [14]; therefore, it is difficult to determine whether these are specific EVs, that is, using these proteins as markers cannot conclusively establish the EV origin as MVBs or the plasma membrane. Other membrane proteins that localize to lipid rafts in the plasma membrane, including tetraspanins, such as CD63 and CD81, are also enriched in EVs. This process involves palmitoylation [15], which is required for the incorporation of the Epstein-Barr virus (EBV) protein latent membrane protein 1 (LMP1). LMP1 is enriched in lipid rafts and extracellular EVs. The Rab family of small GTPases regulates multiple steps in the trafficking of vesicles to distinct endocytic compartments and functions in the docking of MVBs to the plasma membrane [16]. Some viruses modulate this complex process to facilitate entry, trafficking, and egress [17, 18]. Multiple viruses affect the expression of ESCRT and Rab GTPases to promote viral entry and egress and are therefore likely to modulate the content of EVs. These examples demonstrate that viruses co-opt the cellular vesicular transport system during egress; therefore, they are likely to modulate the content and secretion of EVs through similar interactions. Further studies on EV biology may improve our understanding of the complexities of viral redirection in cellular processes and the determination of how viruses affect EV production and contents, thereby contributing to both viral pathogenesis and persistence.

## 20.3 EV Cargoes as Potential Diagnostic Markers for the Detection of Viral Infections

### 20.3.1 EVs as Carriers of Viruses and Host RNA Species

EVs, which are similar in several aspects to enveloped viral particles, concentrate, transport, and deliver molecules that are implicated in important biological processes. Viruses have been reported to hijack the host endosomal machinery for viral assembly and can modify the cargo of EVs for delivery to uninfected cells, thereby enhancing viral dissemination and pathogenesis. EVs released from infected cells carry viral proteins, genetic material, and other host genetic elements to neighboring cells, helping to establish productive infections and modulating the host response [19–22]. Differences in cargo contents reflect the physiological or pathological state of the infected cells and the overall state of infection. For example, exosomes released from hepatitis C virus (HCV)-infected cells contain full-length viral RNA, which can be successfully transferred to dendritic cells to establish a productive infection [23]. Exosomes secreted by enterovirus 71 (EV71)-infected cells selectively packaged high levels of *miR-146a* into recipient cells by suppressing the type I interferon response and facilitating the replication of exosomal EV71 RNA in the new target cell [24]. In viral infections, circulating EVs are emerging as potential diagnostic tools for early detection and predictors of disease progression, treatment response, and overall survival [25]. The identities and abundances of miRNAs in EVs represent a snapshot of their cellular origins. Human immunodeficiency virus (HIV)-1 was the first virus to be studied in detail in the context of modulation of exosome formation. Exosomes derived from HIV-1-infected cells or patients with HIV infection incorporate the viral transactivating response element transcribed from the integrated provirus, which has been suggested to enhance HIV replication in recipient cells via suppression of apoptosis [26]. The incorporation of genomic HIV-1 RNA into exosomes is further increased if the producer cells express HIV-1 defective for viral genome packaging [27]. Exosome-like microvesicles isolated from serum during infection with human Pegivirus (an RNA virus within the *Pegivirus* genus of the *Flaviviridae* family, previously known as hepatitis G virus) have been shown to carry viral RNA and transmit viral RNA to peripheral blood mononuclear cells in vitro, resulting in productive infection [28]. Two subsequent studies have confirmed that exosomes derived from HCV-infected hepatocytes contain the complete HCV genome and can mediate the transfer of replication-competent subgenomic HCV RNA to permissive naïve cells, leading to viral RNA replication and productive infection [29, 30]. Exosomes derived from HCV-infected patients have also been found to contain negative-sense HCV RNA (replication-competent viral RNA) in association with Ago2, heat-shock protein (HSP) 90, and *miR-122* [31]. Moreover, exosomes derived from human T-lymphotropic virus type 1 (HTLV-1)-infected cells contain *Tax*, HTLV-1 bZIP factor, and *Env* mRNA transcripts, suggesting that exosomes can serve as vehicles to deliver functional HTLV-1 mRNA to recipient cells [21]. The virus can also pack intact virus particles into exosomes to evade antibody-mediated neutralization or to promote viral infection. Feng et al. showed that hepatitis A virus released from cells is cloaked in host-derived membranes, thereby protecting the virion from antibody-mediated neutralization. These enveloped viruses resemble exosomes and small vesicles, which are important in intercellular communication [32]. In addition, electron microscopy studies have also revealed that the isolated CD63-immunoprecipitated EVs produced during severe fever with thrombocytopenia syndrome (SFTS) virus infection contain virions. Virions harbored within these structures are efficiently delivered to uninfected cells and can sustain SFTS viral replication [33]. Wang found that purified exosomes derived from porcine reproductive and respiratory syndrome virus (PRRSV)-infected cells contained viral genomic RNA and partial viral proteins (M, N, and GP5). PRRSV appears to exploit the cellular exosomal delivery system to transmit viral components that can establish a productive infection [34]. These studies suggest that EVs can be used as diagnostic indicators for patient prognosis and treatment efficacy. Cervical and vaginal lavage fluids from patients with human papilloma virus-positive cervical cancer show high levels of EV *miR-21* and *miR-146-a* [35], suggesting that EVs purified from lavage fluid can be used as noninvasive diagnostic markers.

### 20.3.2 EVs as Viral Protein Carriers

The protein compositions of EVs vary depending on cell type and disease state. Several RNA viruses have been shown to modulate the host cell protein composition of EVs and recruit their own proteins into EVs. In clinical diagnosis, Welker [36] found that serum from patients with HCV secrete higher levels of CD81 than healthy volunteers and that CD81 secretion is positively correlated with the degree of liver fibrosis; thus, CD81 may have applications in the diagnosis or tracing of chronic HCV infection. Moreover, exosomes derived from HCV-infected cells have been shown to carry HCV core protein as well as the apolipoproteins ApoE and ApoB [30]. In HCV-positive patients, the cellular membrane protein CD81 is associated with the envelope glycoprotein E2. Extracellular release of E2-CD81 within microvesicles is associated with increased virus fusing ability [37]. LMP1 is a viral protein that contributes to the modification of EV contents and remodeling of the tumor microenvironment. LMP1-modified EVs may be diagnostic molecules for EBV-induced cancer cell proliferation and migration, enhancing tumor, migration, and invasion [38]. Exosomes from HIV-1-infected cells have been shown to incorporate both Gag [39] and Nef proteins [40, 41]. The latter is incorporated into exosomes upon anchoring to lipid raft microdomains through its N-terminal myristoylation and a stretch of basic amino acids residing in its alpha-helix-1, as well as upon interaction with the host cell protein Mortalin [42]. Similar to HIV, the HTLV-1 transactivator protein Tax, a critical factor for the proliferation and transformation of T cells, is incorporated into exosomes secreted from virus-infected cells. Several cellular proteins also seem to be recruited to exosomes in a Tax-dependent manner, many of which are involved in protein synthesis and antigen presentation [21]. During chronic or latent viral infection, a single infected cell may actually release EVs at orders of magnitude greater than infectious viral particles; therefore, purified EVs may have applications in the diagnosis of chronic infection. Table 20.1 lists the protein and RNA species of viruses recruited to EVs. Importantly,

**Table 20.1** Summary of viral EV cargo proteins and RNA species with potential diagnostic value

| Virus | Genome | Protein carried by EVs | RNA species carried by EVs | Function | References |
|---|---|---|---|---|---|
| HIV | ***RNA*** | Nef | TRA RNA<br>VmiR-88<br>VmiR-99 | Helps to establish infection, complete assembly, support budding, and escape immune response | [26, 43–46] |
| SFTSV | ***RNA*** | Virions | | Helps in virus transmission | [33] |
| ZIKV | ***RNA*** | Envelope | Viral RNA | Helps in virus spreading and establishes new infection | [47] |
| HAV | ***RNA*** | Virions | | Is involved in membrane fusion, transportation of viral RNA from<br>eHA V into the cytoplasm, virus packing, and releasing | [32] |
| HCV | ***RNA*** | | Viral RNA | Affects virus assembly, budding, spreading, and immune evasion | [31] |
| EV71 | ***RNA*** | | Viral RNA | Is involved in virus infection, resistant to antibody neutralization, and delivers miRNA | [48] |
| EBOV | ***RNA*** | VP40, NP, GP | | | [49] |
| DENV | ***RNA*** | IFITM1, IFITM2, IFITM3 | | Antiviral activities | [22] |
| HTLV | ***RNA*** | Tax protein | tax, HBZ, and Env mRNAs | Delivers host and viral components, activates the immune response, and stimulates proinflammatory cytokines | [21, 50–53] |
| IAV | ***RNA*** | Rab11 | | Participates in viral egress | [54] |
| RSV | ***RNA*** | Nucleocapsid protein N, attachment protein G, fusion protein F | RSV N, M, and NS1 gene | Participates in viral egress | [55] |
| EBV | ***DNA*** | LMP1, LMP2a, Gp350 | | Virus release | [56] |
| HSV | ***DNA*** | ICP0, ICP4, virion host shutoff protein (vhs) | | Helps virus transmission escape immune response | [57] |
| HBV | ***DNA*** | Viral protein | Viral nucleic acids | Helps virus transmission | [58] |
| HAdV | ***DNA*** | | miRNA | Diagnostic biomarkers | [59] |

EVs are released from most cells and accumulate in body fluids, providing major opportunities for diagnostic and therapeutic applications. Serum EVs differ in quality and quantity between healthy individuals and patients with viral infections, and these differences may promote the use of EVs as specific biomarkers.

Many models have hypothesized that viral RNAs and proteins can be specifically and differentially loaded into EVs [60]; however, few studies have evaluated the use of these molecules in diagnostic applications. Recently, more studies have assessed the applications of EVs in various diseases, a few of which have progressed to clinical trials. There are still many problems that need to be overcome in the application of EVs for diagnostic purposes. For example, methods for separating and purifying EVs need to be improved to confirm the accuracy of EV composition and determine whether the identified components (e.g., proteins, small RNAs) are adequately representative. Extracting EVs from limited body fluids is a simple process, and although the approach can show high accuracy, contamination is also possible. Based on current research, EVs exist in all types of biological fluids; however, their physical stability needs to be further explored to facilitate their application in disease diagnosis. If these problems can be overcome, EV components will be rapidly developed as tools for diagnostic applications.

## 20.4 EVs and Parasite Infections

Several standard techniques are used to diagnose parasitic infections, such as identification of malaria parasites in blood; observation of filarial parasites in blood or evaluation of skin snip biopsies for onchocerciasis; assessment of eggs or larvae in urine or stools for schistosomiasis; identification of soil-transmitted helminths by microscopy; and detection of antibodies or pathogen antigens circulating in the blood for onchocerciasis, lymphatic filariasis, and schistosomiasis or in the urine for schistosomiasis. Although these techniques are currently used in clinical practice, they show relatively low sensitivity in mild infections and low throughput, are time-consuming, require skilled personnel for operation, and lack onsite applicability [61]. In addition, the serological methods used for the diagnosis of parasitic diseases, particularly those used for the diagnosis of helminth infection, have a limited ability to distinguish between previous and current infections [62]. Therefore, it is essential to develop rapid and sensitive high-throughput tools for diagnosing infectious diseases.

EVs have attracted attention owing to their roles in host-parasite interactions and regulation of diverse cellular functions in various types of organs and tissues. EVs formed by endocytic pathways contain a wealth of information in the form of nucleic acids, proteins, and lipids as cargo. Under normal conditions and pathogen infections, almost all types of cells secrete EVs for cellular communication; however, the differences in cargo contents reflect their origin from infected cells and the overall state of the infection. Parasites have adopted several strategies to communicate with their hosts in order to access favorable conditions in their niches and facilitate infection [63]. Several studies have shown that the EVs released by either parasites or host cells are sensed by parasites during communication between the parasite and host cells [64]. Parasite-derived EVs can manipulate target cells by delivering pathogenic, immunomodulatory, and genetic materials to provide favorable conditions for the survival and multiplication of parasites [65, 66]. For example, EVs from *Schistosoma japonicum* regulate host immune responses to promote parasitism, and the parasitic EV cargo contains some host miRNAs, which regulates host macrophage-mediated immune responses during infection [65, 67]. Furthermore, in unpleasant environments, host cells may release EVs to recruit immune cells for defense against parasites. Interestingly, parasites can manipulate host cells by delivering signals, and EVs can enhance their transmission, pathogenicity, and parasitism [65, 66]. The identification of EV biomarkers has driven research aimed at understanding pathogen-derived EVs or delivering EV molecules to the host. These pathogen-derived molecules are more stable than soluble factors in the circulation, which can be exposed to circulating hydrolases. Thus, because of this increased stability of EVs in circulation, EVs are expected to have roles as diagnostic biomarkers for various parasitic diseases.

## 20.5 EV Biogenesis in Helminths

EVs are typically defined based on their size; exosomes are defined as having a diameter of less than 150 nm, whereas large vesicles, including ectosomes, microparticles, and oncosomes, have a maximum diameter of 1000 nm [68]. The formation of EVs starts with inward budding of the late endosome membrane, which forms MVBs comprising a number of intraluminal vesicles (ILVs). These ILVs are then secreted from the cell surface as EVs after the fusion of MVBs with the plasma membrane. ESCRT, composed of cytoplasmic protein complexes, is a key mechanism mediating this process. In conjunction with other accessory proteins, these ESCRT complexes facilitate a unique mode of membrane remodeling that results in membrane bending/budding away from the cytoplasm. The ESCRT-0, -I, and -II complexes cluster ubiquitinated proteins to the endosomal membrane, whereas ESCRT-III produces polymeric filaments that invade the membrane, thereby forming ILVs [69]. AAA-ATPase is essential for the final cutoff of budding ILVs in the MVB lumen [70]. However, other pathways for ILV formation have been proposed, including the ESCRT pathway, the lipid pathway (involving the lipid hydrolases neutral sphingomyelinase and phospholipase D2), the tetraspanin pathway (involving CD63), and a hybrid pathway combining elements of all of these mechanisms [71]. The fusion of MVBs with the plasma membrane has also been shown to be reliant on small GTPases, such as Rab11, Rab27a, and Rab27b [16, 72].

Our current understanding of EV biogenesis in parasites is limited. Detailed proteomic analyses of the proteins secreted by schistosomes revealed the presence of EVs (~100 to 200 nm in size) and cargo content molecules [73, 74]. EVs contain different known EV markers, such as HSP70/HSP90, CD63, actin, T-complex protein 1, glyceraldehyde 3-phosphate dehydrogenase, annexin, elongation factor, and Rab GTPases, indicating that schistosomes secret true exosomes [73–75]. Recent studies using proteomic, transcriptomic, and microscopic approaches suggest that helminths may have multiple mechanisms for EV biogenesis [76–79]. Our initial study on EV biogenesis in *S. japonicum* revealed that the presence of several crucial endosomes is associated with various proteins [80], including Alix, syntenin, Rab11, Rab27a, and Vps4. In addition, as reported in the helminth *Fasciola hepatica* [77], the ESCRT pathway is expected to be involved in vesicle secretion. Structurally, EVs from the schistosome showed cup- or round-shaped vesicles. Exosome-like vesicles are associated with the gut lining of *S. mansoni* [81]. Transmission electron microscopy has demonstrated the presence of vesicular structures at the tegmental syncytium of trematodes that resemble the MVBs detected in mammalian cells [82]. These typical structures fuse with the plasma membrane at the surface, from where the vesicles are shed. Schistosome exosomes contain a considerable number of tegument-associated proteins, suggesting that teguments play a role in exosome secretion. Although detailed ultrastructural studies of schistosomes have been conducted, these putative MVBs have not been described in detail in the literature. Proteomic analysis of EV contents from these schistosomes has revealed the presence of proteins with a wide range of functions. In addition, the proteins secreted by *S. japonicum* were compared with the protein datasets from the schistosome membrane and tegument; the findings revealed a total of 50 proteins detected in the schistosome tegument [74], including several proteins involved in membrane trafficking, such as coatomer subunit beta, coatomer subunit gamma, coatomer subunit delta, sorting nexin-6, sorting nexin-18, protein SEC13 homolog, AP-1 complex subunit beta-1, and syntaxin-binding protein 1. Exosomal membranes are rich in certain phospholipids and lipid-binding proteins, such as annexins, which are organized in specialized lipid rafts called tetraspanin-enriched microdomains (TEMs). TEMs may play roles in exosome biogenesis and cargo sorting by promoting the specific compartmentalization of proteins and receptor molecules into budding ILVs [83]. Moreover, the unique composition of EV membranes has evolved to confer a high level of protection of the cargo content [78]. Our previous studies explored the proteome contents of *S. japonicum* EVs and supported the highly sensitive and specific diagnostic potential of SjCD63 and peptides 68 and 71 [84, 85]. The roles of excretory-secretory products and tegumental proteins in host-parasite interactions have been explored, and therapeutic and diagnostic approaches have been developed against these parasites [73, 86]. Based on proteomics data, our current understanding of exosome biogenesis supports the endosomal origin of EVs;

however, the findings are still not conclusive. Therefore, future studies should address all of these important issues.

The functions of EVs depend on their cellular origin. Several studies have explored the roles of EVs in host-parasite interactions [65, 87–91] and their influence on host pathophysiology [66, 92, 93]. The functions of EVs in host-parasite interactions have been well studied in terms of immunoregulation related to modulation or regulation of the host immune response. For example, *S. japonicum* secretes EVs harboring *miR-125b* and *bantam* miRNA, which induce macrophage proliferation and tumor necrosis factor-α production by regulating *Pros1*, *Fam212b*, and *Clmp* genes [65]. During parasite infection, *S. mansoni*-derived EV *miR-10* targets *MAPK7* and suppresses the Th2 immune response by regulating Th2 differentiation [88]. *Brugia malayi*-derived EVs downregulate mammalian target of rapamycin phosphorylation and its downstream regulatory protein 4E-BP1, which modulates host immunity during infections [91]. A recent study showed that *Clonorchis sinensis*-derived EV *let-7a-5p* activates M1-like macrophages by targeting *Socs1* and *Clec7a*, which then contribute to the biliary pro-inflammatory response and biliary injuries [87]. A study on EV cargo, including miRNAs and Y RNAs, secreted from the gastrointestinal nematode *Heligmosomoides polygyrus*, demonstrated suppression of type 2 innate responses via regulation of interleukin (IL)-33r and dual-specificity phosphatase 1 [66]. Furthermore, another study revealed that neutralization of *H. polygyrus* EVs facilitates parasite expulsion and that mice vaccinated with worm EV-alum show protective immunity against larval challenge [90]. Parasite-derived EVs can also regulate host pathophysiology through cross-kingdom communication. For example, *S. japonicum* egg EV *miR-71a* inhibits the transforming growth factor (TGF)-β1/SMAD and IL-13/signal transducer and activator of transcription 6 pathways by targeting semaphorin 4D, thereby suppressing liver fibrosis [92]. Additionally, *S. japonicum* egg EV *miR-71a* also blocks liver fibrosis by regulating the balance between Th1/Th2/Th17 and regulatory T cells. The schistosome egg-secreted EV cargo *sja-miR-2162* promotes liver fibrosis by upregulating collagen and α-smooth muscle actin production by targeting the *Tgfbr3* gene [93]. These results suggest that helminth parasite-secreted EVs play important roles in host-parasite interactions and regulate host pathophysiology; this mechanism should be further explored to facilitate the development of vaccines and diagnostic tools. A schematic representation of helminth EV-mediated functions is shown in Fig. 20.1.

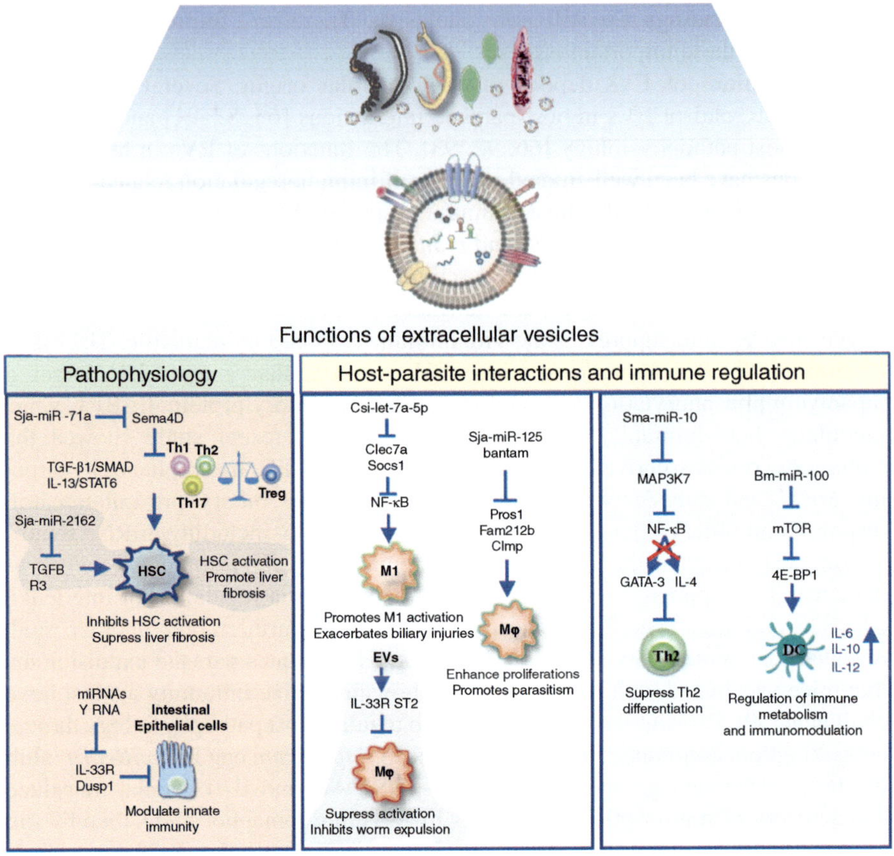


**Fig. 20.1** Schematic representation of helminth EV-mediated functions in host-parasite interactions and pathophysiology regulation. *Sema4D* semaphorin 4D, *Th* T helper cell, *TGFβ1* transforming growth factor beta 1, *STAT6* signal transducer and activator of transcription 6, *TGFBR3* transforming growth factor beta receptor III, *HSC* hepatic stellate cell, *Dusp1* dual specificity phosphatase 1, *Clec7a* C-type lectin domain family 7 member A, *Socs1* suppressor of cytokine signaling 1, *Mφ* macrophage, *Pros1* protein S1, *Fam212b* family with sequence similarity 212, member B, *Clmp* coxsackie- and adenovirus receptor-like membrane protein, *MAP3K7* mitogen-activated protein kinase kinase kinase 7, *mTOR* mammalian target of rapamycin, *4E-BP1* 4E-binding protein 1, *DC* dendritic cells

## 20.6 EV Cargoes as Potential Diagnostic Markers for the Detection of Parasite Infections

Currently available techniques for detection of helminth and protozoan infections in humans and animals lack sufficient sensitivity and specificity for accurate diagnosis of both epidemiological and individual infections; therefore, alternative diagnostic approaches are urgently needed. It is also difficult to precisely distinguish differences according to the stage and intensity of infection. Because EV cargo represents the cellular origin of the EVs, these vesicles are considered new biomarkers for the diagnosis of various diseases, including infectious diseases. For example, EVs carrying CD63, cathepsin B, cysteine proteases, and an uncharacterized protein (B3GUU4) show diagnostic potential in *S. japonicum*-infected hosts [84, 85]. During the infection process, pathogens take advantage of EVs for cross-species and cross-border communication to not only regulate host gene expression for survival but also enhance parasitism through the functions of miRNAs [65, 67]. In infectious diseases, the application of pathogen-specific miRNAs is expected to provide better diagnostic tools because pathogens express unique miRNA sets corresponding to host homologs [94]. Furthermore, signaling molecules, such as miRNAs detected in EVs derived from the host or pathogen, could be exploited for the development of novel diagnostic strategies [95, 96]. *S. mansoni* EVs contain high levels of *sma-miR-71* and *sma-miR-125b*, moderate levels of *sma-bantam* miRNA, and *sma-miR-36-3p* [73]. Further analyses of these schistosome-specific miRNAs in infected host-isolated EVs demonstrated notable differences in the levels of these miRNAs compared with those in control mice. Moreover, microvesicle cargo miRNAs (*bantam* miRNA and *miR-2c-3p*) isolated from *S. mansoni*-infected patients showed a combined sensitivity and specificity of 91% and 82%, respectively, suggesting great potential in clinical diagnosis and follow-up research [97]. Similarly, deep sequencing of infected host plasma showed the presence of *S. japonicum* miRNAs (*bantam* miRNA, *miR-3479*, *miR-10*, *miR-3096*, and novel *miR-0001*), suggesting that EVs carrying parasite-specific miRNAs have potential applications in the development of novel biomarkers [98]. Another study reported a high abundance of *bantam* miRNA in infected individuals [99], suggesting that *bantam* miRNA could be a potential biomarker for the detection of schistosome infection. Analysis of EVs from the filarial worm *Brugia malayi* demonstrated the presence of multiple miRNAs, including *miR-34*, *lin-4*, *miR-71*, and *miR-100c*, which were adequately divergent from the host miRNA sequences [100] and could potentially serve as biomarkers of infection. In another study, miRNAs from *Haemonchus contortus* adults and L4 larvae were found to be secreted into EVs, and high abundances of some miRNAs, including *hco-miR-5352-3p*, were detected in infected sheep tissues [101]. EVs from *Heligmosomoides bakeri* contain specific miRNAs, such as *miR-71*, *miR-49*, *miR-63*, *miR-259*, and *miR-240*, whereas EVs from *Trichuris muris* contain *miR-1*, *miR-1822*, and *miR-252* [102]. Similarly, EVs from *Trichinella spiralis* muscle larvae harbor high abundances of *miR-540-3p*, *miR-8325-3p*, and *miR-1692* [103]. However, further studies are needed to demonstrate the diagnostic potential of these miRNAs in host circulation. The concept of using EV cargo to characterize the diagnosis of infectious diseases is relatively new but highly promising. Several studies have characterized EVs from different helminth parasites; however, very few studies have shed light on the diagnostic potential of EV cargo. In addition, these studies of parasites have evaluated different life stages and various species, such as nematodes (*Ascaris sum* [104], *Brugia malayi* [100, 105], *Haemonchus contortus*, *Heligmosomoides polygyrus* [66], *Heligmosomoides bakeri* [102], *Nippostrongylus brasiliensis* [106], *Teladorsagia circumcincta* [107], *Trichuris muris* [89, 102, 108, 109], *Trichinella spiralis* [103]), trematodes (*S. japonicum* [65, 74], *S. mansoni* [73, 75, 110], *S. haematobium* [111], *Dicrocoelium dendriticum* [82], *Echinostoma caproni* [112], *Fasciola hepatica* [113], *Opisthorchis viverrini* [114]), and cestodes (*Echinococcus multilocularis* [115, 116], *Echinococcus granulosus* [117–121], *Taenia pisiformis* [122], *Taenia crassiceps* [123], *Mesocestoides corti* [123], *Taenia asiatica* [124], and *Hymenolepis diminuta* [125]). Some antigens, such as antigen p29, antigen B, and Eg13, have been detected in EV cargo from *Echinococcus granulosus*. Similarly, *Taenia crassiceps* metacestode EVs also contain antigens p29, fatty acid binding protein, 14-3-3, EM18/H17g, and Ts8B1, which have been shown to have immune diagnostic potential [126–128]. A proteomic analysis of patients with cystic echinococcosis (CE) revealed Src and Lyn as potential biomarkers and suggested that further validation in larger cohorts would be critical for CE diagnosis and management [120]. A summary of helminth EV cargo proteins and miRNAs with diagnostic potential is present in Table 20.2.

In comparison with other parasitic diseases, protozoa infection-associated EVs have not been thoroughly evaluated using high-throughput untargeted proteomics. Exosomes and microvesicles are associated with malaria pathogenesis via modulation of host immunity and cell-to-cell communication. An early report on exosomes from mice infected with *Plasmodium yoelii* revealed the presence of both exosome markers and parasite proteins [130]. This evidence suggests that EVs and their cargo can serve as biomarkers of malaria. Increased EV numbers, particularly for microvesicles derived from platelets and red blood cells in the circulation of human patients with cerebral malaria, can act as potential biomarkers for inflammatory events and malaria infection-related morbidity [131, 132]. However, the roles of EVs as biomarkers of severe malaria are still unclear. A recent study on EVs from the *Plasmodium falciparum*-infected erythrocyte proteome revealed the enrichment of parasitic kinases that could be involved in the phosphorylation of the host membrane and cytoskeletal proteins [133], further supporting their potential diagnostic value. No miRNAs have been reported in malarial parasites; however, host EV miRNAs and circulating miRNAs may have roles in malaria diagnosis [134]. Chagas disease and African trypanosomiasis, neglected tropic diseases caused by *Trypanosoma cruzi* and *Trypanosoma brucei*, result in major economic losses and high morbidity rates. During host-trypanosome interactions, the release of EVs by parasites at different stages involves several key factors that contribute to the transition from migration to internalization [135–138]. A recent clinical study showed that *Trypanosoma cruzi*-infected patients release fewer EVs into circulation than uninfected controls [139]. Similarly, *Trypanosoma cruzi* trypomastigote EVs, isolated from mammalian cell cargo, contain parasite surface glycoproteins, including major surface glycoproteins similar to mucin [135, 137]; these EVs were used for diagnosis and to confirm drug efficacy in chronically infected children and adolescents from an endemic area in a placebo-controlled random trial [140]. Although the presence of small RNA species has been reported in trypanosomes [136], the diagnostic capabilities of these RNAs have not yet been discussed. Studies of leishmaniasis, a disease caused by *Leishmania* spp., have demonstrated the importance of EVs both in the infection process and in disease progression [141]. Although several studies have been conducted on EVs from *Leishmania* spp., very little information is available on the diagnostic prospects of these EVs. *Trichomonas vaginalis* is the causative agent of trichomoniasis, a common sexually transmitted infection. Similar to other pathogens, *Trichomonas vaginalis* releases both exosomes and

**Table 20.2** Summary of helminth EV cargo proteins and miRNAs as potential diagnostic biomarkers

| Species | Stages | Number of proteins associated with EVs | Number of EV cargo miRNAs identified | EV specific proteins | EV specific miRNAs | References |
|---|---|---|---|---|---|---|
| *Nematodes* | | | | | | |
| *Ascaris suum* | Adults, L3, L4 | 268 | 29, 51, 40 | – | – | [104] |
| *Brugia malayi* | Adults male, female, and L3 | 20, 74, 31/31 | 52 for L3 | – | *Lin-4*, *miR-34*, *miR-71*, and *miR-100c* potential biomarker for filarial infection | [100, 105] |
| *Haemonchus contortus* | Adults, L4 | – | 40, 37 | – | *hco-miR-5352-3p* and *hco-miR-50* can be detected in infected sheep tissue | [101] |
| *Heligmosomoides polygyrus* | Adults | 362 (139 enriched in EVs) | 263 (high confidence not limited to EVs) | – | | [66] |
| *H. bakeri* | Adults | – | 94 | – | Unique miRNAs: *miR-71*, *miR-49*, *miR-63*, *miR-259*, and *miR-240* | [102] |
| *Nippostrongylus brasiliensis* | Adults | 81 | 52 | – | – | [106] |
| *Teladorsagia circumcincta* | L4 | 85 | – | – | – | [107] |
| *Trichuris muris* | Adults | 364/125/73/– | 56/14/130/110 | – | Unique miRNAs: *miR-1*, *miR-1822*, and *miR-252*; enriched miRNAs: *miR-59*, *miR-72* and *miR-44* | [89, 102, 108, 109] |
| *Trichinella spiralis* | Muscle, larvae | – | >50 | – | *miR-540-3p*, *miR-8325-3p*, and *miR-1692* are top abundant miRNAs | [103] |

(continued)

**Table 20.2** (continued)

| Species | Stages | Number of proteins associated with EVs | Number of EV cargo miRNAs identified | EV specific proteins | EV specific miRNAs | References |
|---|---|---|---|---|---|---|
| *Trematodes* | | | | | | |
| *S. japonicum* | Adults | 403/– | 35/38 | SjCD63, cathepsin B-like cysteine proteinase, and an uncharacterized protein (B3GUU4) showed diagnostic potential | *Bantam* miRNA and *miR-10* detected in infected host plasma | [65, 74, 84, 85] |
| *S. mansoni* | Schistosomula and adults | 109/83/130 | 167/–/143 | Saposin B domain-containing protein was abundant | *sma-miR-125a*, *sma-miR-125b*, *sma-bantam*, and *sma-miR-71a* detected in infected mice serum | [73, 75, 110] |
| *S. haematobium* | Adults | 387 | – | – | – | [111] |
| *Dicrocoelium dendriticum* | Adults | 84 | 42 | – | – | [82] |
| *Echinostoma caproni* | Adults | 45 | – | – | – | [112] |
| *Fasciola hepatica* | Adults | 180/380/392 (surface proteins)/79 | 55/– | – | – | [77, 79, 112, 129] |
| *Opisthorchis viverrini* | Adults | 108 | – | – | – | [114] |
| *Cestodes* | | | | | | |
| *Echinococcus multilocularis* | Metacestodes | 433/>84 | 18/– | – | – | [115, 116] |

| | | | | | | |
|---|---|---|---|---|---|---|
| *E. granulosus* | Metacestodes and protoscolex | 322/112/663/901/936, 149 | – | Src family kinases, Lyn kinases, antigen B, antigen p29, and antigen Eg13 | – | [117–121] |
| *Taenia pisiformis* | Cysticercus | 87 | 59 | – | – | [122] |
| *T. crassiceps* | Metacestodes | 195 | >7 | Immunodiagnostic antigen p29, FABP, 14-3-3, Em18/H17g, and Ts8B1 | | [123] |
| *T. asiatica* | Adults | 455 | 48 | – | – | [124] |
| *Mesocestoides corti* | Metacestodes | 64 | >1 | – | | [123] |
| *Hymenolepis diminuta* | Adults | 269 | – | – | – | [125] |

microvesicles [142, 143]. Interestingly, the number of vesicles is increased after exposure to the host, suggesting that EVs may be involved in host-parasite interactions. *Giardia* spp., which release EVs, are a common cause of diarrhea in resource-limited countries [144, 145]. Proteomic analysis of *G. intestinalis* and *G. duodenale* EVs revealed the presence of factors important for *Giardia* pathogenesis [144] and regulation of the host immune response [145], respectively. Other protozoan parasites, such as *Toxoplasma*, *Entamoeba*, and *Neospora*, release EVs, and proteomics analyses have identified many proteins that could have potential diagnostic value [146–148]. The above studies confirmed the roles of EVs in host-parasite interactions; however, the differences in cargo composition reported in the host during infection may have diagnostic value, and this should be assessed in further in-depth studies. Additionally, a detailed understanding of the biogenesis of EVs in these parasites and the application of more advanced throughput techniques will help us to precisely understand the differences in cargo during disease progression and pathogenesis. These investigations will provide new avenues for the application of EVs as clinical prognostic and diagnostic tools. Furthermore, such improvements in our understanding of EVs may facilitate the development of vaccines, diagnostic tools, and therapies for different parasitic diseases. A summary of protozoa-secreted EV cargo proteins and small RNAs with diagnostic potential is presented in Table 20.3.

**Table 20.3** Summary of protozoa EV cargo proteins and small RNAs with potential diagnostic value

| Species | Stages | Number of proteins associated with EVs | RNAs | EV specific proteins/RNAs | References |
|---|---|---|---|---|---|
| *Trypanosoma brucei* | | 156 | | | [149] |
| *T. cruzi* | Epimastigotes and trypomastigotes | ~80/–/256 | –/small RNAs/– | Mucins antigen proposed for serodiagnosis | [136–138] |
| *Toxoplasma gondii* | Tachyzoites | 346 | | | [146] |
| *Trichomonas vaginalis* | Trophozoites | 592/215 | –/Small RNAs | | [142, 143] |
| *Leishmania* spp. | Promastigotes | 329 | | | [141] |
| *Entamoeba histolytica* | Trophozoites | 359 | Small RNAs | | [147] |
| *Giardia intestinalis* | Trophozoites | 138 | | | [144] |
| *Giardia duodenalis* | Trophozoites | 154 | | | [145] |
| *Plasmodium falciparum* | Trophozoites | 387 | *Pf*-derived kinases | | [133] |
| *Neospora caninum* | Tachyzoites | 705 | | | [148] |

## 20.7 Conclusion and Perspectives

In recent years, many studies have assessed the applications of EVs in various diseases, and a few clinical trials of EVs are currently underway. Because the interactions between parasites and hosts are mediated by EVs at various stages of infection, exploration of their cargo in patient blood, urine, and saliva may yield powerful tools for diagnostic purposes. Various studies have characterized EV cargo RNA and protein contents in order to identify a reliable panel of components for the diagnosis of parasitic diseases. Although many attempts have been made to diagnose schistosomiasis, echinococcosis, malaria, and leishmania based on EVs, much work is still needed to apply these approaches to the rapid and reliable detection of parasitic infections. To date, only a few species of EVs have been characterized and analyzed; therefore, more studies are needed to explore the diagnostic potential of their cargo.

## References


1. Thery C, Zitvogel L, Amigorena S. Exosomes: composition, biogenesis and function. Nat Rev Immunol. 2002;2:569–79.
2. Valadi H, et al. Exosome-mediated transfer of mRNAs and microRNAs is a novel mechanism of genetic exchange between cells. Nat Cell Biol. 2007;9:654–9.
3. Jeppesen DK, et al. Reassessment of exosome composition. Cell. 2019;177:428–445.e418.
4. Pathan M, et al. Vesiclepedia 2019: a compendium of RNA, proteins, lipids and metabolites in extracellular vesicles. Nucleic Acids Res. 2019;47:D516–9.
5. Kalluri R. The biology and function of exosomes in cancer. J Clin Invest. 2016;126:1208–15.
6. Kahlert C, Kalluri R. Exosomes in tumor microenvironment influence cancer progression and metastasis. J Mol Med (Berl). 2013;91:431–7.
7. Meckes DG Jr, Raab-Traub N. Microvesicles and viral infection. J Virol. 2011;85:12844–54.
8. Hergenreider E, et al. Atheroprotective communication between endothelial cells and smooth muscle cells through miRNAs. Nat Cell Biol. 2012;14:249–56.
9. Gibbings DJ, Ciaudo C, Erhardt M, Voinnet O. Multivesicular bodies associate with components of miRNA effector complexes and modulate miRNA activity. Nat Cell Biol. 2009;11:1143–9.
10. Kosaka N, et al. Secretory mechanisms and intercellular transfer of microRNAs in living cells. J Biol Chem. 2010;285:17442–52.
11. Chugh PE, et al. Systemically circulating viral and tumor-derived microRNAs in KSHV-associated malignancies. PLoS Pathog. 2013;9:e1003484.
12. Meckes DG Jr, et al. Human tumor virus utilizes exosomes for intercellular communication. Proc Natl Acad Sci U S A. 2010;107:20370–5.
13. Kowal J, Tkach M, Thery C. Biogenesis and secretion of exosomes. Curr Opin Cell Biol. 2014;29:116–25.
14. Kharkwal H, Smith CG, Wilson DW. Herpes simplex virus capsid localization to ESCRT-VPS4 complexes in the presence and absence of the large tegument protein UL36p. J Virol. 2016;90:7257–67.
15. Verweij FJ, et al. Exosomal sorting of the viral oncoprotein LMP1 is restrained by TRAF2 association at signalling endosomes. J Extracell Vesicles. 2015;4:26334.
16. Ostrowski M, et al. Rab27a and Rab27b control different steps of the exosome secretion pathway. Nat Cell Biol. 2010;12:19–30; sup pp 11–13.
17. Fotheringham JA, Raab-Traub N. Epstein-Barr virus latent membrane protein 2 induces autophagy to promote abnormal acinus formation. J Virol. 2015;89:6940–4.

18. Inoue J, et al. HBV secretion is regulated through the activation of endocytic and autophagic compartments mediated by Rab7 stimulation. J Cell Sci. 2015;128:1696–706.
19. Madison MN, Okeoma CM. Exosomes: implications in HIV-1 pathogenesis. Viruses. 2015;7:4093–118.
20. Liu Z, Zhang X, Yu Q, He JJ. Exosome-associated hepatitis C virus in cell cultures and patient plasma. Biochem Biophys Res Commun. 2014;455:218–22.
21. Jaworski E, et al. Human T-lymphotropic virus type 1-infected cells secrete exosomes that contain Tax protein. J Biol Chem. 2014;289:22284–305.
22. Zhu X, et al. IFITM3-containing exosome as a novel mediator for anti-viral response in dengue virus infection. Cell Microbiol. 2015;17:105–18.
23. Cosset FL, Dreux M. HCV transmission by hepatic exosomes establishes a productive infection. J Hepatol. 2014;60:674–5.
24. Fu Y, et al. Exosome-mediated miR-146a transfer suppresses type I interferon response and facilitates EV71 infection. PLoS Pathog. 2017;13:e1006611.
25. Vojtech L, et al. Exosomes in human semen carry a distinctive repertoire of small non-coding RNAs with potential regulatory functions. Nucleic Acids Res. 2014;42:7290–304.
26. Narayanan A, et al. Exosomes derived from HIV-1-infected cells contain trans-activation response element RNA. J Biol Chem. 2013;288:20014–33.
27. Columba Cabezas S, Federico M. Sequences within RNA coding for HIV-1 Gag p17 are efficiently targeted to exosomes. Cell Microbiol. 2013;15:412–29.
28. Chivero ET, et al. Human pegivirus RNA is found in multiple blood mononuclear cells in vivo and serum-derived viral RNA-containing particles are infectious in vitro. J Gen Virol. 2014;95:1307–19.
29. Longatti A, Boyd B, Chisari FV. Virion-independent transfer of replication-competent hepatitis C virus RNA between permissive cells. J Virol. 2015;89:2956–61.
30. Ramakrishnaiah V, et al. Exosome-mediated transmission of hepatitis C virus between human hepatoma Huh7.5 cells. Proc Natl Acad Sci U S A. 2013;110:13109–13.
31. Bukong TN, Momen-Heravi F, Kodys K, Bala S, Szabo G. Exosomes from hepatitis C infected patients transmit HCV infection and contain replication competent viral RNA in complex with Ago2-miR122-HSP90. PLoS Pathog. 2014;10:e1004424.
32. Feng Z, et al. A pathogenic picornavirus acquires an envelope by hijacking cellular membranes. Nature. 2013;496:367–71.
33. Silvas JA, Popov VL, Paulucci-Holthauzen A, Aguilar PV. Extracellular vesicles mediate receptor-independent transmission of novel Tick-Borne Bunyavirus. J Virol. 2016;90:873–86.
34. Wang T, Fang L, Zhao F, Wang D, Xiao S. Exosomes mediate intercellular transmission of porcine reproductive and respiratory syndrome virus. J Virol. 2018;92:e01734–17.
35. Liu J, et al. Increased exosomal microRNA-21 and microRNA-146a levels in the cervicovaginal lavage specimens of patients with cervical cancer. Int J Mol Sci. 2014;15:758–73.
36. Welker MW, et al. Soluble serum CD81 is elevated in patients with chronic hepatitis C and correlates with alanine aminotransferase serum activity. PLoS One. 2012;7:e30796.
37. Masciopinto F, et al. Association of hepatitis C virus envelope proteins with exosomes. Eur J Immunol. 2004;34:2834–42.
38. Nkosi D, et al. Epstein-Barr virus LMP1 promotes Syntenin-1- and Hrs-induced extracellular vesicle formation for its own secretion to increase cell proliferation and migration. mBio. 2020;11:e00589–20.
39. Fang Y, et al. Higher-order oligomerization targets plasma membrane proteins and HIV gag to exosomes. PLoS Biol. 2007;5:e158.
40. Lenassi M, et al. HIV Nef is secreted in exosomes and triggers apoptosis in bystander CD4+ T cells. Traffic. 2010;11:110–22.
41. de Carvalho JV, et al. Nef neutralizes the ability of exosomes from $CD4^+$ T cells to act as decoys during HIV-1 infection. PLoS One. 2014;9:e113691.
42. Arenaccio C, Chiozzini C, Columba-Cabezas S, Manfredi F, Federico M. Cell activation and HIV-1 replication in unstimulated $CD4^+$ T lymphocytes ingesting exosomes from cells expressing defective HIV-1. Retrovirology. 2014;11:46.

43. Mfunyi CM, et al. Exosome release following activation of the dendritic cell immunoreceptor: a potential role in HIV-1 pathogenesis. Virology. 2015;484:103–12.
44. Campbell TD, Khan M, Huang MB, Bond VC, Powell MD. HIV-1 Nef protein is secreted into vesicles that can fuse with target cells and virions. Ethn Dis. 2008;18:S2-14-19.
45. Sampey GC, et al. Exosomes from HIV-1-infected cells stimulate production of pro-inflammatory cytokines through trans-activating response (TAR) RNA. J Biol Chem. 2016;291:1251–66.
46. Bernard MA, et al. Novel HIV-1 miRNAs stimulate TNFalpha release in human macrophages via TLR8 signaling pathway. PLoS One. 2014;9:e106006.
47. Zhou W, Woodson M, Sherman MB, Neelakanta G, Sultana H. Exosomes mediate Zika virus transmission through SMPD3 neutral sphingomyelinase in cortical neurons. Emerg Microbes Infect. 2019;8:307–26.
48. Mao L, et al. Enterovirus 71 transmission by exosomes establishes a productive infection in human neuroblastoma cells. Virus Genes. 2016;52:189–94.
49. Pleet ML, et al. Extracellular vesicles and Ebola virus: a new mechanism of immune evasion. Viruses. 2019;11:410.
50. Otaguiri KK, et al. TAX-mRNA-carrying exosomes from human T cell lymphotropic virus type 1-infected cells can induce interferon-gamma production in vitro. AIDS Res Hum Retroviruses. 2018;34:1075–82.
51. Anderson MR, et al. Viral antigens detectable in CSF exosomes from patients with retrovirus associated neurologic disease: functional role of exosomes. Clin Transl Med. 2018;7:24.
52. Gutzeit C, et al. Correction: exosomes derived from Burkitt's lymphoma cell lines induce proliferation, differentiation, and class-switch recombination in B cells. J Immunol. 2019;203:769–70.
53. Rezaie J, et al. The versatile role of exosomes in human retroviral infections: from immunopathogenesis to clinical application. Cell Biosci. 2021;11:19.
54. Bruce EA, Digard P, Stuart AD. The Rab11 pathway is required for influenza A virus budding and filament formation. J Virol. 2010;84:5848–59.
55. Chahar HS, Corsello T, Kudlicki AS, Komaravelli N, Casola A. Respiratory syncytial virus infection changes cargo composition of exosome released from airway epithelial cells. Sci Rep. 2018;8:387.
56. Sato Y, et al. Elimination of LMP1-expressing cells from a monolayer of gastric cancer AGS cells. Oncotarget. 2017;8:39345–55.
57. Heilingloh CS, et al. L particles transmit viral proteins from herpes simplex virus 1-infected mature dendritic cells to uninfected bystander cells, inducing CD83 downmodulation. J Virol. 2015;89:11046–55.
58. Yang Y, et al. Exosomes mediate hepatitis B virus (HBV) transmission and NK-cell dysfunction. Cell Mol Immunol. 2017;14:465–75.
59. Huang F, et al. Identification of potential diagnostic biomarkers for pneumonia caused by adenovirus infection in children by screening serum exosomal microRNAs. Mol Med Rep. 2019;19:4306–14.
60. Raab-Traub N, Dittmer DP. Viral effects on the content and function of extracellular vesicles. Nat Rev Microbiol. 2017;15:559–72.
61. Ullah H, Qadeer A, Rashid M, Rashid MI, Cheng G. Recent advances in nucleic acid-based methods for detection of helminth infections and the perspective of biosensors for future development. Parasitology. 2020;147:383–92.
62. Hinz R, Schwarz NG, Hahn A, Frickmann H. Serological approaches for the diagnosis of schistosomiasis—a review. Mol Cell Probes. 2017;31:2–21.
63. Kochin BF, Bull JJ, Antia R. Parasite evolution and life history theory. PLoS Biol. 2010;8:e1000524.
64. Coakley G, Maizels RM, Buck AH. Exosomes and other extracellular vesicles: the new communicators in parasite infections. Trends Parasitol. 2015;31:477–89.
65. Liu J, et al. *Schistosoma japonicum* extracellular vesicle miRNA cargo regulates host macrophage functions facilitating parasitism. PLoS Pathog. 2019;15:e1007817.

66. Buck AH, et al. Exosomes secreted by nematode parasites transfer small RNAs to mammalian cells and modulate innate immunity. Nat Commun. 2014;5:5488.
67. Giri BR, Cheng G. Host miR-148 regulates a macrophage-mediated immune response during *Schistosoma japonicum* infection. Int J Parasitol. 2019;49:993–7.
68. Shah R, Patel T, Freedman JE. Circulating extracellular vesicles in human disease. N Engl J Med. 2018;379:2180–1.
69. Schorey JS, Cheng Y, Singh PP, Smith VL. Exosomes and other extracellular vesicles in host pathogen interactions. EMBO Rep. 2015;16:24–43.
70. Hasegawa T, et al. The AAA-ATPase VPS4 regulates extracellular secretion and lysosomal targeting of alpha-synuclein. PLoS One. 2011;6:e29460.
71. Colombo M, Raposo G, Thery C. Biogenesis, secretion, and intercellular interactions of exosomes and other extracellular vesicles. Annu Rev Cell Dev Biol. 2014;30:255–89.
72. Savina A, Fader CM, Damiani MT, Colombo MI. Rab11 promotes docking and fusion of multivesicular bodies in a calcium-dependent manner. Traffic. 2005;6:131–43.
73. Samoil V, et al. Vesicle-based secretion in schistosomes: analysis of protein and microRNA (miRNA) content of exosome-like vesicles derived from *Schistosoma mansoni*. Sci Rep. 2018;8:3286.
74. Zhu L, et al. Molecular characterization of *S. japonicum* exosome-like vesicles reveals their regulatory roles in parasite-host interactions. Sci Rep. 2016;6:25885.
75. Sotillo J, et al. Extracellular vesicles secreted by *Schistosoma mansoni* contain protein vaccine candidates. Int J Parasitol. 2016;46:1–5.
76. Bennett APS, de la Torre-Escudero E, Robinson MW. Helminth genome analysis reveals conservation of extracellular vesicle biogenesis pathways but divergence of RNA loading machinery between phyla. Int J Parasitol. 2020;50:655–61.
77. Cwiklinski K, et al. The extracellular vesicles of the helminth pathogen, *Fasciola hepatica*: biogenesis pathways and cargo molecules involved in parasite pathogenesis. Mol Cell Proteomics. 2015;14:3258–73.
78. de la Torre-Escudero E, Bennett APS, Clarke A, Brennan GP, Robinson MW. Extracellular vesicle biogenesis in helminths: more than one route to the surface? Trends Parasitol. 2016;32:921–9.
79. de la Torre-Escudero E, et al. Surface molecules of extracellular vesicles secreted by the helminth pathogen *Fasciola hepatica* direct their internalisation by host cells. PLoS Negl Trop Dis. 2019;13:e0007087.
80. Qadeer A, Giri BR, Ullah H, Cheng G. Transcriptional profiles of genes potentially involved in extracellular vesicle biogenesis in *Schistosoma japonicum*. Acta Trop. 2021;217:105851.
81. Wilson RA. Proteomics at the schistosome-mammalian host interface: any prospects for diagnostics or vaccines? Parasitology. 2012;139:1178–94.
82. Bernal D, et al. Surface analysis of *Dicrocoelium dendriticum*. The molecular characterization of exosomes reveals the presence of miRNAs. J Proteomics. 2014;105:232–41.
83. Andreu Z, Yanez-Mo M. Tetraspanins in extracellular vesicle formation and function. Front Immunol. 2014;5:442.
84. Chen Y, et al. Preliminary evaluation of the diagnostic potential of *Schistosoma japonicum* extracellular vesicle proteins for Schistosomiasis japonica. Acta Trop. 2020;201:105184.
85. Wang L, et al. Molecular characterization, expression profile, and preliminary evaluation of diagnostic potential of CD63 in *Schistosoma japonicum*. Parasitol Res. 2018;117:3625–31.
86. Yoshino TP, et al. Excreted/secreted *Schistosoma mansoni* venom allergen-like 9 (SmVAL9) modulates host extracellular matrix remodelling gene expression. Int J Parasitol. 2014;44:551–63.
87. Yan C, et al. Csi-let-7a-5p delivered by extracellular vesicles from a liver fluke activates M1-like macrophages and exacerbates biliary injuries. Proc Natl Acad Sci U S A. 2021;118:e2102206118.
88. Meningher T, et al. Schistosomal extracellular vesicle-enclosed miRNAs modulate host T helper cell differentiation. EMBO Rep. 2020;21:e47882.

89. Eichenberger RM, et al. Characterization of *Trichuris muris* secreted proteins and extracellular vesicles provides new insights into host-parasite communication. J Extracell Vesicles. 2018;7:1428004.
90. Coakley G, et al. Extracellular vesicles from a helminth parasite suppress macrophage activation and constitute an effective vaccine for protective immunity. Cell Rep. 2017;19:1545–57.
91. Ricciardi A, et al. Extracellular vesicles released from the filarial parasite *Brugia malayi* downregulate the host mTOR pathway. PLoS Negl Trop Dis. 2021;15:e0008884.
92. Wang L, et al. Sja-miR-71a in Schistosome egg-derived extracellular vesicles suppresses liver fibrosis caused by schistosomiasis via targeting semaphorin 4D. J Extracell Vesicles. 2020;9:1785738.
93. He X, et al. A schistosome miRNA promotes host hepatic fibrosis by targeting transforming growth factor beta receptor III. J Hepatol. 2020;72:519–27.
94. Cucher MA, Ancarola ME, Kamenetzky L. The challenging world of extracellular RNAs of helminth parasites. Mol Immunol. 2021;134:150–60.
95. Ketprasit N, et al. The characterization of extracellular vesicles-derived microRNAs in Thai malaria patients. Malar J. 2020;19:285.
96. Cai P, Gobert GN, You H, Duke M, McManus DP. Circulating miRNAs: potential novel biomarkers for hepatopathology progression and diagnosis of Schistosomiasis japonica in two murine models. PLoS Negl Trop Dis. 2015;9:e0003965.
97. Meningher T, et al. Schistosomal MicroRNAs isolated from extracellular vesicles in sera of infected patients: a new tool for diagnosis and follow-up of human schistosomiasis. J Infect Dis. 2017;215:378–86.
98. Cheng G, Luo R, Hu C, Cao J, Jin Y. Deep sequencing-based identification of pathogen-specific microRNAs in the plasma of rabbits infected with *Schistosoma japonicum*. Parasitology. 2013;140:1751–61.
99. Hoy AM, et al. Parasite-derived microRNAs in host serum as novel biomarkers of helminth infection. PLoS Negl Trop Dis. 2014;8:e2701.
100. Zamanian M, et al. Release of small RNA-containing exosome-like vesicles from the human filarial parasite Brugia malayi. PLoS Negl Trop Dis. 2015;9:e0004069.
101. Gu HY, et al. Conservation of a microRNA cluster in parasitic nematodes and profiling of miRNAs in excretory-secretory products and microvesicles of *Haemonchus contortus*. PLoS Negl Trop Dis. 2017;11:e0006056.
102. White R, et al. Extracellular vesicles from *Heligmosomoides bakeri* and *Trichuris muris* contain distinct microRNA families and small RNAs that could underpin different functions in the host. Int J Parasitol. 2020;50:719–29.
103. Yang Y, et al. Extracellular vesicles derived from *Trichinella spiralis* muscle larvae ameliorate TNBS-induced colitis in mice. Front Immunol. 2020;11:1174.
104. Hansen EP, et al. Exploration of extracellular vesicles from *Ascaris suum* provides evidence of parasite-host cross talk. J Extracell Vesicles. 2019;8:1578116.
105. Harischandra H, Yuan W, Loghry HJ, Zamanian M, Kimber MJ. Profiling extracellular vesicle release by the filarial nematode *Brugia malayi* reveals sex-specific differences in cargo and a sensitivity to ivermectin. PLoS Negl Trop Dis. 2018;12:e0006438.
106. Eichenberger RM, et al. Hookworm secreted extracellular vesicles interact with host cells and prevent inducible colitis in mice. Front Immunol. 2018;9:850.
107. Tzelos T, et al. A preliminary proteomic characterisation of extracellular vesicles released by the ovine parasitic nematode, *Teladorsagia circumcincta*. Vet Parasitol. 2016;221:84–92.
108. Shears RK, Bancroft AJ, Hughes GW, Grencis RK, Thornton DJ. Extracellular vesicles induce protective immunity against Trichuris muris. Parasite Immunol. 2018;40:e12536.
109. Tritten L, et al. Excretory/secretory products from the gastrointestinal nematode *Trichuris muris*. Exp Parasitol. 2017;178:30–6.
110. Nowacki FC, et al. Protein and small non-coding RNA-enriched extracellular vesicles are released by the pathogenic blood fluke *Schistosoma mansoni*. J Extracell Vesicles. 2015;4:28665.

111. Mekonnen GG, et al. *Schistosoma haematobium* extracellular vesicle proteins confer protection in a heterologous model of schistosomiasis. Vaccines (Basel). 2020;8:416.
112. Marcilla A, et al. Extracellular vesicles from parasitic helminths contain specific excretory/secretory proteins and are internalized in intestinal host cells. PLoS One. 2012;7:e45974.
113. (!!! INVALID CITATION !!! 80, 114, 115).
114. Chaiyadet S, et al. Carcinogenic liver fluke secretes extracellular vesicles that promote cholangiocytes to adopt a tumorigenic phenotype. J Infect Dis. 2015;212:1636–45.
115. Zheng Y, et al. Regulatory effects of *Echinococcus multilocularis* extracellular vesicles on RAW264.7 macrophages. Vet Parasitol. 2017;235:29–36.
116. Ding J, et al. miRNA-seq of *Echinococcus multilocularis* extracellular vesicles and immunomodulatory effects of miR-4989. Front Microbiol. 2019;10:2707.
117. Siles-Lucas M, et al. Isolation and characterization of exosomes derived from fertile sheep hydatid cysts. Vet Parasitol. 2017;236:22–33.
118. Nicolao MC, Rodriguez Rodrigues C, Cumino AC. Extracellular vesicles from *Echinococcus granulosus* larval stage: isolation, characterization and uptake by dendritic cells. PLoS Negl Trop Dis. 2019;13:e0007032.
119. Zhou X, et al. Extracellular vesicles derived from *Echinococcus granulosus* hydatid cyst fluid from patients: isolation, characterization and evaluation of immunomodulatory functions on T cells. Int J Parasitol. 2019;49:1029–37.
120. Fratini F, et al. Proteomic analysis of plasma exosomes from cystic Echinococcosis patients provides in vivo support for distinct immune response profiles in active vs inactive infection and suggests potential biomarkers. PLoS Negl Trop Dis. 2020;14:e0008586.
121. Wu J, et al. Comparative analysis of different extracellular vesicles secreted by *Echinococcus granulosus protoscoleces*. Acta Trop. 2021;213:105756.
122. Wang LQ, et al. Characterization of exosome-like vesicles derived from *Taenia pisiformis* cysticercus and their immunoregulatory role on macrophages. Parasit Vectors. 2020;13:318.
123. Ancarola ME, et al. Cestode parasites release extracellular vesicles with microRNAs and immunodiagnostic protein cargo. Int J Parasitol. 2017;47:675–86.
124. Liang P, et al. Identification and molecular characterization of exosome-like vesicles derived from the *Taenia asiatica* adult worm. Acta Trop. 2019;198:105036.
125. Mazanec H, Konik P, Gardian Z, Kuchta R. Extracellular vesicles secreted by model tapeworm *Hymenolepis diminuta*: biogenesis, ultrastructure and protein composition. Int J Parasitol. 2021;51:327–32.
126. Yang D, et al. Expression and immunolocalisation of TpFABP as a candidate antigen for the serodiagnosis of rabbit *Taenia pisiformis* cysticercosis. Parasite. 2013;20:53.
127. Deckers N, Dorny P. Immunodiagnosis of *Taenia solium* taeniosis/cysticercosis. Trends Parasitol. 2010;26:137–44.
128. González G, et al. Molecular characterization of P-29, a metacestode-specific component of *Echinococcus granulosus* which is immunologically related to, but distinct from, antigen 5. Mol Biochem Parasitol. 2000;105:177–85.
129. Davis CN, et al. The importance of extracellular vesicle purification for downstream analysis: a comparison of differential centrifugation and size exclusion chromatography for helminth pathogens. PLoS Negl Trop Dis. 2019;13:e0007191.
130. Martin-Jaular L, Nakayasu ES, Ferrer M, Almeida IC, Del Portillo HA. Exosomes from *Plasmodium yoelii*-infected reticulocytes protect mice from lethal infections. PLoS One. 2011;6:e26588.
131. Antwi-Baffour S, et al. Plasma mEV levels in Ghanain malaria patients with low parasitaemia are higher than those of healthy controls, raising the potential for parasite markers in mEVs as diagnostic targets. J Extracell Vesicles. 2019;9:1697124.
132. Campos FM, et al. Augmented plasma microparticles during acute *Plasmodium vivax* infection. Malar J. 2010;9:327.
133. Dekel E, et al. 20S proteasomes secreted by the malaria parasite promote its growth. Nat Commun. 2021;12:1172.

134. Chamnanchanunt S, Fucharoen S, Umemura T. Circulating microRNAs in malaria infection: bench to bedside. Malar J. 2017;16:334.
135. Ribeiro KS, et al. Proteomic analysis reveals different composition of extracellular vesicles released by two *Trypanosoma cruzi* strains associated with their distinct interaction with host cells. J Extracell Vesicles. 2018;7:1463779.
136. Bayer-Santos E, et al. Proteomic analysis of *Trypanosoma cruzi* secretome: characterization of two populations of extracellular vesicles and soluble proteins. J Proteome Res. 2013;12:883–97.
137. Bautista-Lopez NL, et al. Characterization and diagnostic application of *Trypanosoma cruzi* Trypomastigote excreted-secreted antigens shed in extracellular vesicles released from infected mammalian cells. J Clin Microbiol. 2017;55:744–58.
138. Fernandez-Calero T, et al. Profiling of small RNA cargo of extracellular vesicles shed by Trypanosoma cruzi reveals a specific extracellular signature. Mol Biochem Parasitol. 2015;199:19–28.
139. Madeira RP, et al. New biomarker in Chagas disease: extracellular vesicles isolated from peripheral blood in chronic Chagas disease patients modulate the human immune response. J Immunol Res. 2021;2021:6650670.
140. de Andrade AL, et al. Randomised trial of efficacy of benznidazole in treatment of early Trypanosoma cruzi infection. Lancet. 1996;348:1407–13.
141. Silverman JM, et al. An exosome-based secretion pathway is responsible for protein export from Leishmania and communication with macrophages. J Cell Sci. 2010;123:842–52.
142. Nievas YR, et al. Membrane-shed vesicles from the parasite *Trichomonas vaginalis*: characterization and their association with cell interaction. Cell Mol Life Sci. 2018;75:2211–26.
143. Twu O, et al. *Trichomonas vaginalis* exosomes deliver cargo to host cells and mediate hostratioparasite interactions. PLoS Pathog. 2013;9:e1003482.
144. Gavinho B, et al. Peptidylarginine deiminase inhibition abolishes the production of large extracellular vesicles from Giardia intestinalis, affecting host-pathogen interactions by hindering adhesion to host cells. Front Cell Infect Microbiol. 2020;10:417.
145. Zhao P, et al. Extracellular vesicles secreted by *Giardia duodenalis* regulate host cell innate immunity via TLR2 and NLRP3 inflammasome signaling pathways. PLoS Negl Trop Dis. 2021;15:e0009304.
146. Wowk PF, et al. Proteomic profiling of extracellular vesicles secreted from *Toxoplasma gondii*. Proteomics. 2017;17:1600477.
147. Sharma M, et al. Characterization of extracellular vesicles from *Entamoeba histolytica* identifies roles in intercellular communication that regulates parasite growth and development. Infect Immun. 2020;88:e00349–20.
148. Li S, et al. Extracellular vesicles secreted by *Neospora caninum* are recognized by Toll-like receptor 2 and modulate host cell innate immunity through the MAPK signaling pathway. Front Immunol. 2018;9:1633.
149. Szempruch AJ, et al. Extracellular vesicles from *Trypanosoma brucei* mediate virulence factor transfer and cause host anemia. Cell. 2016;164:246–57.

# Extracellular Vesicle Biomarkers for Renal Diseases

# 21

Lin-Li Lv, An-Ran Shen, Tao-Tao Tang, He-qun Zou, and Xiao-lin Li

## 21.1 Introduction

Extracellular vesicles (EVs) are secreted by almost all cell types and carry diverse cargoes under different pathophysiological conditions. EVs may participate in renal development and physiology via the transfer of useful cargo. Importantly, recent studies have revealed the critical roles of EV-mediated intercellular communication in kidney disease, and these disease-associated EVs can be readily detected in urine, reflecting the potential applications of EVs as biomarkers for renal disease. Urinary EVs (uEVs) originate from different fragments of nephrons and thereby represent ideal biomarkers for detection using noninvasive liquid biopsy approaches. Recent studies have identified potential biomarkers from uEVs for kidney diseases, such as acute kidney injury (AKI) and primary and secondary chronic kidney diseases (CKDs). Moreover, omics technology has accelerated the pace of biomarker discovery and provided an in-depth understanding of uEVs. However, optimization and standardization of uEV research are urgently needed for future translational studies.

L.-L. Lv (✉) · A.-R. Shen · T.-T. Tang
Institute of Nephrology, Zhongda Hospital, Southeast University School of Medicine, Nanjing, Jiangsu Province, China
e-mail: lvlinli@seu.edu.cn

H.-q. Zou (✉) · X.-l. Li (✉)
Department of Nephrology, South China Hospital, Health Science Center, Shenzhen University, Shenzhen, People's Republic of China



Q. Wang, L. Zheng (eds.), *Extracellular Vesicles*,
https://doi.org/10.1007/978-981-99-8365-0_21

## 21.2 Roles of EVs in Renal Pathophysiology

### 21.2.1 EVs in Renal Development and Physiology

Despite numerous studies evaluating the roles of EVs in kidney injury and repair, their functions in kidney physiology are still largely unclear. EVs may participate in physiological processes in the kidneys by transferring specific information to target cells or by removing unwanted cellular materials to maintain cell homeostasis. Moreover, emerging evidence suggests that EV-mediated signaling pathways may also contribute to kidney development and tissue repair after kidney injury.

#### 21.2.1.1 Aquaporins (AQPs) in uEVs

AQPs are water channel proteins mainly expressed by kidney tubule epithelial cells. AQPs are a family of transmembrane proteins that are common components of urinary EVs [1]. To date, at least eight aquaporins (AQP1–4, AQP6–8, and AQP11) have been found in renal epithelial cells; however, not all AQPs are detectable in uEVs. AQP1 and AQP2 are common AQPs in uEVs [2, 3]. Thus, it is possible that these AQPs expressed in the apical membrane of epithelial cells can be sorted into EVs and secreted from the apical membrane into the urine space. Urinary exosomal AQPs levels are usually correlated with renal protein levels. However, a reverse correlation between urinary exosomal AQPs and renal levels has been reported in some conditions, which suggests diverse sorting mechanisms of AQPs in renal epithelial cells [4]. Vasopressin and urine alkalinization have also been shown to affect AQP2 levels in uEVs. The roles of EV AQPs in renal physiology are unclear. However, studies have shown that AQPs in uEVs may be promising biomarkers for multiple kidney diseases, including renal transplantation, urinary tract obstruction, and urinary concentration defects [4, 5].

#### 21.2.1.2 EVs in Renal Development

Krause et al. reported the presence of exosomes and vesicles from an embryonic kidney ureteric bud immortalized cell line, and these exosomes were found to carry proteins that are important for the structural organization of the kidney. Furthermore, their cargo was shown to contain different microRNAs (miRNAs) with the potential to influence the Wnt pathway, an important signaling mechanism involved in nephrogenesis. When metanephric mesenchyme (MM) cells were treated with embryonic kidney-derived exosomes in an ex vivo organ culture setting, MM was not fully induced; however, an obviously different cellular organization and fewer apoptotic cells were observed. Despite these findings, the mechanisms through which exosomes participate in nephrogenesis remain unclear [6].

Epithelial and mesenchymal interactions are crucial for the development of different organs. Thus, studying the roles of exosome-mediated crosstalk between the epithelium and mesenchyme may provide important clues for regenerative therapy in kidney diseases [7].

#### 21.2.1.3 EV-Mediated Cellular Crosstalk in Renal Physiology

In the nephron, the glomeruli and different fragments of tubules are closely connected. Tubuloglomerular feedback is an important negative feedback mechanism that controls renal blood flow. Tubuloglomerular crosstalk also occurs during kidney injury and promotes renal damage [8]. EVs participate in nephron communication via the transfer of mRNA, miRNAs, and proteins. EVs may transfer signals from the glomeruli to tubules in the nephron when podocyte-derived EVs are secreted and pass through the urinary tract. Podocyte-specific proteins, such as nephrin, podocin, NEPH1, podocalyxin, and complement receptor type 1 (CR1), can be detected in urinary exosomes from normal human urine, suggesting that exosomes may be secreted by podocytes under normal conditions [9]. Although an in vitro study reported that podocyte microparticles communicate with proximal tubule epithelial cells and induce fibrotic responses via p38 mitogen-activated protein kinase and CD36, the physiological roles of this communication mechanism under normal conditions remain unclear [10].

Tubule epithelial cells are the most populous cell type in the renal structure and perform diverse functions involved in kidney pathophysiology. EVs derived from proximal tubule cells can be taken up by both distal tubule and collecting duct cells. Furthermore, EVs represent a novel mechanism of cellular communication among different tubule fragments [11, 12]. Thus, there are multiple routes of communication mediated by EVs between different parts of the nephron.

### 21.2.2 EV-Mediated Cellular Communication in Renal Inflammation and Fibrosis

Tubulointerstitial inflammation and fibrosis are the main pathological characteristics of various renal diseases. Sustained tubulointerstitial inflammation can lead to renal fibrosis and the development of end-stage renal disease. Recent studies have revealed that external stresses, such as hypoxia, proteinuria, or physical wounding, lead to the secretion of EVs from tubular epithelial cells (TECs), which communicate with neighboring or distant cells [13, 14]. Upon exposure to different insults, TECs secrete EVs and transfer injury signals, including *CCL2* mRNA and miRNAs, to the interstitial space, which activates macrophages [15, 16]. EV-mediated tubule and macrophage crosstalk may be a critical mechanism in tubulointerstitial inflammation. In addition, profibrotic factors, such as *miR-21* and transforming growth factor-β1 (*TGF-β1*) mRNA, can be transferred via tubule epithelial cell-derived EVs to promote tubular phenotype transition and fibroblast activation [17, 18].

Some studies have reported the protective roles of EVs released by intrinsic renal cells, such as TECs. An in vitro study found that exosomes from hypoxic renal proximal tubular cells inhibited apoptosis induced by ATP depletion [13]. An in vivo study showed that the intravenous administration of exosomes from normal human kidney tubular cells prevented damage after hypoxic AKI [19, 20]. Thus, EVs released from resident renal cells may represent an intrinsic recovery mechanism under stress conditions, and the effects of EVs may be determined by the status of parent cells or the severity of injury.

### 21.2.3 EV-Based Therapy in Kidney Disease

The field of EV-based therapy has been growing rapidly in recent years, and its application in the treatment of renal diseases, although still at an early stage, holds great transformative potential. Based on current research, two main therapeutic uses of EVs can be surmised: (1) exploiting EVs as therapeutic agents and (2) employing EVs as therapeutic vectors for drug delivery [14, 21–23].

#### 21.2.3.1 Application of Stem/Progenitor Cell-Derived EVs in Renal Disease

EVs, particularly those derived from stem cells carrying bioactive cargoes, have innate therapeutic potential for tissue regeneration and immunomodulation [24]. In the field of renal disease therapeutics, the application of mesenchymal stromal cell (MSC)-derived EVs has been widely investigated using various kidney injury models [25]. Bruno et al. demonstrated that bone marrow MSC-derived EVs accelerated the recovery of glycerol-induced AKI by promoting TEC proliferation and reducing apoptosis [26]. Profiling studies have revealed that mRNAs and miRNAs involved in cell cycle entry, proliferation, immunoregulation, and antiapoptotic pathways are enriched in MSC-derived EVs [26, 27], which could explain their renoprotective effects. Furthermore, Cao et al. found that placental MSC-derived EVs expressing *miR-200a-3p* activated the Kelch-like ECH-associated protein 1/nuclear factor erythroid 2-related factor 2 signaling pathway in TECs and protected against ischemia-reperfusion injury (IRI)-induced AKI by maintaining mitochondrial structure and function [28]. In other studies, the protective roles of MSC-derived EVs delivering biomolecules, such as hepatocyte growth factor mRNA [29] growth factors and angiogenin [30], and interleukin (IL)-10 [31], have also been confirmed in IRI-induced AKI, diabetic nephropathy (DN), and metabolic syndrome, respectively. Moreover, a randomized, placebo-controlled, phase 2/3 clinical pilot study was conducted to evaluate the feasibility and efficacy of MSC-derived EVs in patients with stage III–IV CKD. Interestingly, two doses of MSC-derived EVs safely and efficiently improved renal function and ameliorated inflammatory immune reactions in patients 1 year after administration [31]. These preclinical and clinical studies support the great therapeutic potential of MSC-derived EVs in renal diseases. However, the underlying mechanisms of action of MSC-derived EVs require further clarification.

#### 21.2.3.2 EVs as Therapeutic Vectors in Kidney Disease

Natural or engineered EVs can be exploited as delivery vectors for various cargoes of interest, including small-molecule drugs, nucleic acids, and proteins. As a therapeutic vector, EVs possess many advantages over synthetic nanoparticles, such as desirable stability and biocompatibility, low immunogenicity and toxicity, and biological barrier permeability [21–23]. Recently, macrophage-derived EVs were employed for the targeted delivery of dexamethasone to inflamed kidneys via integrin $\alpha_L\beta_2$ and $\alpha_4\beta_1$ present on the EV surface. Compared with free dexamethasone treatment, EV-dexamethasone showed a superior capability to suppress renal inflammation and fibrosis in lipopolysaccharide- or adriamycin-induced nephropathy without apparent glucocorticoid-related adverse effects [32]. A subsequent study engineered macrophages to manufacture EVs overexpressing IL-10. Interestingly, EV-mediated delivery enhanced not only the stability of IL-10 but also its targeting to the kidney. Treatment with IL-$10^+$ EVs notably attenuated renal IRI and interrupted the AKI to CKD transition by protecting mitochondrial function in TECs and driving M2 macrophage polarization [33]. These examples indicate that EVs offer a new approach to the treatment of renal diseases. However, in kidney diseases, the glomerular filtration barrier remains a major obstacle for EV-based therapy. Strategies that support the passage of EVs through this barrier are urgently required.

## 21.3 Cellular Origins of uEVs

EVs are found in different biofluids and carry multiple types of molecules, including proteins, lipids, and nucleic acids, providing useful information on parental cells. uEVs represent a unique source of biomarkers for kidney disease owing to the easy, noninvasive nature of sample collection. Since the first proteomic study on uEVs by Pisitkun et al. in 2004 [2], additional studies have highlighted the possible applications of uEVs in the diagnosis of kidney diseases. Subsequent studies have shown that uEVs primarily originate from epithelial cells facing the urinary tract, such as podocytes, different tubule fragments, and circulation. In addition, studies using flow cytometry to characterize specific populations of uEVs have presented direct evidence of the cellular origins of urinary exosomes.

### 21.3.1 Different Fragments of Tubules

EVs released from the apical membrane of TECs are thought to account for a large proportion of uEVs because they face the urinary space directly, whereas EVs released via basolateral secretion are expected to be less dominant. Early studies reported transporter proteins on uEVs originating from different fragments of tubules [34], such as $Na^+$-$Cl^-$ cotransporter (from the distal convoluted tubule), AQP1/2 (from the proximal tubule and collecting duct), $Na^+$-$H^+$ exchanger (NHE3; from the proximal tubule), and $Na^+$-$K^+$-$2Cl^-$ cotransporter (from the thick ascending limb) [35]. Recent in-depth proteomic analyses of urinary exosomes have provided a complete view of cellular sources of EVs. Proteomic and transcriptome data provided evidence that uEVs could be secreted from all fragments of tubules, including the proximal tubule, thick ascending limb of Henle, distal convoluted tubule, and collecting duct [2, 36]. Digital flow cytometry profiling provided direct evidence that EVs in cell-free urine were derived from different fragments of tubules, glomerular juxtaglomerular cells, and podocytes [37].

Glycosylated proteins in uEVs have also provided more specific evidence of the cellular origins in tubules through specific carbohydrate-binding lectins. For example, *Lotus tetragonolobus* lectin labels sodium-glucose cotransporter-2 in proximal tubule cells, *Dolichos Biflorus* agglutinin labels AQP2 in principal collecting duct cells, and peanut agglutinin labels carbonic anhydrase 2 in intercalated collecting duct cells. This may also contribute to the isolation of tubule segment-specific uEVs [35].

### 21.3.2 Podocytes

Podocytes are a crucial component of the glomerular filtration barrier in the kidneys. The interdigitated foot process almost covers the glomerular basement membrane, leaving the filtration slit diaphragm between these two structures. This makes it possible for EVs from podocytes to be detected in the urine. Correspondingly, proteomic studies on uEVs from healthy subjects revealed that uEVs contain podocin and podocalyxin, which are characteristic biomarkers found in podocytes [38]. These proteins can be used to isolate podocyte-derived uEVs through flow cytometry [39]. Moreover, podocyte-derived uEV enrichment requires more specific affinity reagents, such as podocyte-specific CR1-immunocoated beads, which function via immunoadsorption. uEVs isolated using this method share 97% overlap with podocyte-derived EVs assessed by proteomic analysis [9].

### 21.3.3 Circulation

In addition to the urinary system, uEVs can also originate from other systems at relatively low concentrations. The blood contains large amounts of EVs released from diverse organs and tissues, including tumors. Because urine is mainly formed through blood filtration, it is possible that smaller EVs originating from the circulation may pass through the glomerular filtration barrier, particularly when the filtration barrier is damaged in kidney disease. In a recent study, researchers performed tracking analysis of uEVs via RNA sequencing, revealing that uEVs could be derived from multiple organs, including the bladder, lungs, and colon, possibly through the secretion of EVs from those organs into circulation and subsequent filtration into the urine. Moreover, in a study using miRNA sequencing, researchers discovered that urine and serum EV miRNAs are positively correlated in diabetic kidney disease, sharing 84 common miRNA species [40]. These findings have been confirmed in many proteomic analyses, suggesting that uEVs contain an extensive protein pool separate from that characteristic of the urinary system. For example, according bioinformatic analyses, proteins related to various neurological diseases (e.g., Parkinson's disease, Alzheimer's disease, and Huntington disease) are stably expressed in uEVs and are present at levels that exceed kidney markers under conditions without kidney disease [41]. However, it remains unclear when and how EVs pass through the glomerular filtration barrier under different pathophysiological conditions.

## 21.4 Profiling Methods for uEVs

### 21.4.1 uEV Isolation

Currently, there are several methods for the separation of uEVs, including ultracentrifugation, density gradient isolation using sucrose or Percoll reagent, antibody-based affinity capture, ultrafiltration, and polymer-based precipitation [42]. Each of these methods has advantages and disadvantages. Ultracentrifugation [43, 44] can maintain the integrity of EVs, which is necessary for RNA and protein analysis; however, this approach is time-consuming and labor-intensive, and special equipment is required. Density gradient isolation using sucrose or Percoll reagent [45, 46] can remove high-abundance proteins, which is required for protein analysis and can be used to separate EVs into different groups; however, this approach is time-consuming, complicated, and not suitable for large-volume samples. By contrast, ultrafiltration [47–49] is simple, fast, and can only be used for small specimens. However, this method is not suitable for subsequent RNA and protein analyses because of the potential for protein contamination. The polymer-based precipitation method [46] is simple, rapid, and standardized and is conducive to miRNA analysis. However, this method is also not suitable for subsequent protein analyses because of the potential for protein contamination. The antibody-based affinity capture method [50] can be used to obtain specific types of EVs, whereas the microfluidic method [51, 52] is simple, rapid approach that can be used to divide EVs into different groups; however, this method shows low extraction efficiency and is unsuitable for large sample sizes. Hydraulic dialysis [49, 53, 54] can process a large number of specimens without special equipment and maintains EV integrity but requires the subsequent processing of high-abundance proteins.

Other methods have also emerged. For example, in 2019, Campos-Silva et al. developed a new method for analyzing EVs through immune capture isolation and the optimal analysis of flow cells. Using this new method, EVs can be characterized from a very small volume of body fluids, eliminating the need for special treatment and improving the process efficiency and detection sensitivity [55]. In 2021, the research team of Professor Wei-guo Tao at Purdue University described a dual-functional magnetic bead method that achieved effective isolation of uEVs with low contamination, a high recovery rate (> 80%), and a short separation time (<1 h) [56].

In addition to these separation method, other factors, including the urine donor, collection time and void, urine container, preservation time and temperature before processing, collection volume and freezing aliquot, freezing temperature, and storage time, could also affect the integrity and identity of uEVs. These factors are described in detail in the position paper of the International Society for Extracellular

Vesicles [57]. Standard processes should be established to make data from different studies comparable.

In addition, it is difficult to accurately isolate the various EV subtypes and identify the cells from which these EVs originate. These steps will require the development of more reliable isolation methods and the identification of specific markers [58].

### 21.4.2 Transcriptome Analysis of uEVs

Techniques for uEV RNA analysis mainly include RNA-seq, which describes the quantity and sequences of RNAs using next-generation sequencing, RNA array analysis, which describes the quantity of predefined RNA sequences, and quantitative polymerase chain reaction (qPCR), which describes the quantity of predefined RNA sequences [59]. RNA-seq is highly sensitive and can detect different RNA species, which can then be used for various analyses in different studies. However, this approach is costly and must be supported by data analysis, which depends on the completeness and accuracy of reference databases. RNA array analysis is simpler in terms of data analysis, and a large number of reference databases are available. However, this approach requires affinitive probes and is only suitable for highly expressed genes. By contrast, qPCR is inexpensive, but it also requires affinitive probes and lacks normalization parameters.

RNA is more stable in uEVs than in total urine because the lipid membrane of EVs can protect the RNA from RNA-degrading enzymes. The RNA in uEVs mirrors the pathological state of donor cells. Indeed, studies have suggested that EVs selectively envelop specific RNAs, communicate with neighboring receptor cells, and play regulatory roles. RNAs, particularly miRNAs, are major molecules used for researching biomarkers of kidney diseases and are the basis for targeted treatment strategies. Transcriptome analysis has revealed that various types of RNAs, including miRNAs, long noncoding RNAs, circulating RNAs, and mRNAs, can be found in uEVs. To date, transcriptome analysis of uEVs has been used to investigate biomarkers of AKI, glomerular disease, focal segmental glomerulosclerosis, DN, IgA nephropathy, autoimmune nephritis, lupus nephritis, type 1 diabetes mellitus, and type 2 diabetes mellitus [42].

### 21.4.3 Proteomics Analysis of uEVs

uEVs are released from cells of the urogenital tract or organs, such as the kidneys and bladder, prostate, and uterovaginal tract. More than 3000 proteins in uEVs have been identified using mass spectrometry [60]. Proteomic analyses have also identified specific proteins in uEVs that come from the cells of renal glomeruli and different cell types in the urogenital tract, prostate, and bladder, indicating their potential use as biological markers of kidney disease. Several technologies, including aptamers and proximity extension assays [61], immuno-based capture and detection of proteins, and low-density array profiling [62], have been used to analyze the uEV proteome.

To date, online proteomic databases, including Exocarta, Vesiclepedia, and EVpedia, are available. Another commonly used database, the Kidney Systems Biology Project (NHLBI-KSBP) launched by the National Heart, Lung, and Blood Institute (NHLBI), also includes proteomic information. Both Exocarta and NHLBI-KSBP provided urine-specific protein information. Proteomic analysis of uEVs is used to screen molecular markers for early diagnosis and prognosis evaluation of disease; this approach has been applied in prostate tumors and kidney diseases. Quantitative phosphoproteomics of uEV samples from patients with prostate cancer and healthy individuals revealed 121 upregulated phosphoproteins in patients with cancer compared with those in healthy individuals, suggesting that these targets may have applications in noninvasive biomarker screening and early cancer diagnosis [56].

## 21.5 uEVs Biomarkers for Kidney Disease

Because uEV content is protected from degradation by proteases and nucleases in the urine based on the vesicle structure, detection of EV secretion and EV contents in the urine may be used as a type of liquid biopsy analysis to identify renal pathology. Moreover, uEVs can originate from a wide range of sources. Hence, uEVs could be promising noninvasive biomarkers for both acute and chronic kidney injury (Fig. 21.1).

### 21.5.1 uEV Biomarkers for AKI

AKI is characterized by acute deterioration of kidney function with a rapid increase in serum creatinine levels. AKI develops in 10–15% of patients admitted to the hospital, and its incidence in intensive care can reach more than 50% [63]. During the past decade, numerous studies have evaluated the potential applications of several novel biomarkers, such as kidney injury molecule-1 (KIM-1), neutrophil gelatinase-associated lipocalin (NGAL), tissue inhibitor of metalloproteinase 2, and insulin-like growth factor binding protein 7, for early diagnosis and risk stratification in patients with AKI [64, 65]. Therefore, there is an urgent need to identify new biomarkers for AKI to better manage the disease.

Studies have identified several possible proteins, including NHE3, AQP1, activated transcription factor (ATF-3), and fetuin-A, in EVs as potential biomarkers. The urinary NHE3 concentration can predict AKI and differentiate prerenal azotemia, acute tubular necrosis, and other causes of AKI [66]. AQP1 and AQP2 are important renal water regulatory proteins found in uEVs. The concentration of uEV AQP1/2 decreases in ischemia reperfusion-induced AKI [67], which occurs 6–96 h after injury [68]. In sepsis-induced AKI, ATF-3 in uEVs is a biomarker that appears earlier than the serum creatinine elevation. Urinary NGAL is an AKI early

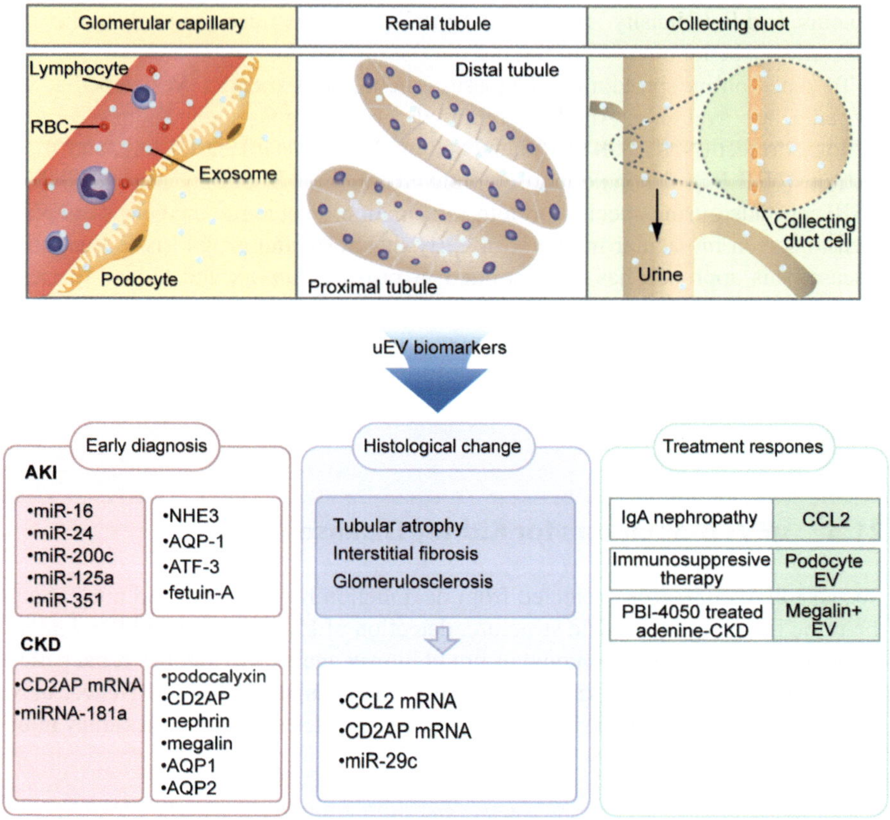


**Fig. 21.1** Urinary extracellular vesicle biomarkers of kidney disease. Urinary EVs commonly originate from glomerular podocytes, epithelial cells lining the urinary tract and circulation. Urinary EV secretion and cargoes could be altered under different pathophysiological conditions, and these EVs could therefore be useful noninvasive biomarkers for kidney diseases. Studies have revealed that urinary EVs may represent novel biomarkers of both acute and chronic kidney disease reflecting renal function, renal histological changes, and treatment responses

biomarker that is upregulated 1 day prior to sepsis-induced AKI, and ATF-3 and uEVs show comparative sensitivity and higher specificity [60]. Unlike these proteins, fetuin-A is synthesized in the liver but is upregulated in uEVs in human patients with AKI and in a rat model of AKI. This increase can be observed 2 days before creatinine levels begin to increase [61].

Urinary miRNAs in uEVs are additional promising biomarkers for AKI. Indeed, urinary miRNAs are more stable than proteins in stored samples, less unstable than circulatory miRNAs, and conserved among different species [62]. A previous study demonstrated that miRNAs arise earlier than changes in blood urea nitrogen and serum creatinine levels, which are almost equal to those in KIM-1 [69]. A study using a rat IRI model reported that *miR-16*, *miR-24*, and *miR-200c* levels were increased in urinary exosomes from animals with experimental AKI. The release of exosomal *miR-125a* and *miR-351* is associated with injury in the renal medulla, which could be regulated by TGF-β signaling [70].

### 21.5.2 uEV Biomarkers for CKD

CKD is a major public health burden worldwide, with an estimated global prevalence of 13.4% (11.7–15.1%) [71]. However, the early diagnosis of CKD remains a challenge because the disease typically progresses silently until an advanced stage is reached. Although measurement of serum creatinine, estimation of the glomerular filtration rate, analysis of the urine albumin/creatinine ratio, and urinalysis are routinely used as screening tests for CKD, it is still difficult to identify early CKD and monitor disease progression. In addition, different pathologies may be present in patients with the same clinical syndromes. Thus, renal biopsy is essential for the diagnosis and management of renal disease [72]; however, it is difficult to monitor the disease owing to the invasive nature of renal biopsy. Hence, noninvasive biomarkers that reflect early disease and histological changes in CKD are urgently needed.

In particular, podocyte-derived EVs have attracted considerable attention as biomarkers of kidney disease. Increased numbers of podocyte-derived EVs in the urine are associated with early glomerular damage in metabolic disease, renovascular hypertension, preeclampsia, and lupus [73, 74]. However, in terms of molecular cargo, *CD2AP* mRNA levels are lower in uEVs from patients with CKD than in those form healthy controls. Additionally, *CD2AP* mRNA levels are negatively correlated with serum creatinine and blood urea nitrogen levels, which decrease with increasing proteinuria severity [75]. This suggests that the numbers and contents of EVs exhibit different diagnostic value.

Tubule-derived EVs are usually detected in the urine after secretion from different tubule segments. Tubule EVs participate in kidney injury by communicating with interstitial cells [15, 76]. Thus, it is reasonable to assume that tubular EVs may contain valuable diagnostic information related to kidney disease. This was confirmed by in vitro cell research mimicking CKD conditions, which demonstrated that uEVs originate from renal proximal tubule cells [77]. A recent RNA-seq study revealed 30 noncoding RNA in uEVs that could be used for the early diagnosis of CKD and identified *miR-181a* as the most abundant and stable biomarker for CKD [77]. Although some evidence suggests that tubular EVs are altered in kidney disease, based on the detection of tubular markers, such as megalin, AQP1, and AQP2 in exosome fractions, studies on purified tubular EVs from the urine are still lacking. In addition, uEVs from extrarenal cell sources could also provide insights into renal pathophysiology. Burbano et al. reported that urinary microparticles containing high mobility group box 1 originating mainly from leukocytes can discriminate between active and inactive lupus nephritis [78].

Renal pathology is essential for the management of kidney disease; however, owing to the invasiveness of renal biopsy, this approach is not practical for the early diagnosis and dynamic monitoring of patients. EVs secreted from renal resident cells are present in the urine and may be analyzed as a noninvasive approach for detecting histological changes in the kidneys. In IgA nephropathy, the levels of urinary exosomal *CCL2* mRNA were investigated in groups of patients stratified by MEST-C scores of renal biopsies. The levels of *CCL2* mRNA expression did not vary between patients with and without endocapillary hypercellularity or segmental glomerulosclerosis. However, *CCL2* shows higher expression levels in patients with more severe tubular atrophy and interstitial fibrosis [3]. Moreover, *CD2AP* mRNA levels in urinary exosomes were negatively correlated with the severity of tubulointerstitial fibrosis and glomerulosclerosis [75]. *MiR-29c* expression in urinary exosomes correlates with the degree of tubulointerstitial fibrosis in patients with CKD [79]. Hence, urinary exosomes may provide a novel approach for detecting histological damage in the kidneys.

Interestingly, proximal tubule-derived EVs (megalin$^+$ EVs) appear to be indicative of a response to therapy in a mouse model of adenine-CKD treated with the small molecule PBI-4050 [80]. Consistent with these findings, in a follow-up study of patients with IgA nephropathy, urinary exosome excretion decreased with a reduction in proteinuria after treatment. Importantly, baseline urinary exosomal *CCL2* mRNA expression was higher in patients whose IgA nephropathy progressed than in patients with stable disease, suggesting that *CCL2* mRNA may be a reliable biomarker for disease remission and for predicting IgA nephropathy progression [3]. In addition, in patients with idiopathic membranous nephropathy, a decrease in urinary podocyte EVs was associated with disease remission after immunosuppressive therapy [81].

### 21.5.3 uEV Biomarkers in Diabetes and DN

Although the gold standard for the diagnosis of DN is renal biopsy, this approach is invasive and not applicable to all patients; thus, uEVs may be promising diagnostic biomarkers that can be collected using a noninvasive method. uEV miRNAs and proteins have potential applications as early diagnostic markers of DN based on EV proteomics and genomic analysis.

MiRNAs are small noncoding RNAs 18–22 nucleotides in length and can be encapsulated in uEVs. In patients with type 1 diabetes and incipient DN, *miR-145* levels are increased, suggesting potential applications as a candidate biomarker [82]. Moreover, in patients with type 2 diabetes, *miR-192* levels were shown to be increased in early DN in the microalbuminuric group but decreased in the normoalbuminuric, control, and macroalbuminuric groups. uEV *miR-192* may be a noninvasive tool for detection of early DN in these patients [83]. A combination of *let-7i-5p*, *miR-15b-5p*, *miR-24-3p*, and *miR-27b-3p* in urinary microvesicles can be used to diagnose patients with type 2 DN [84]. Additionally, *miR-320c* may represent a novel candidate marker for the early progression of type 2 DN [85]. Prabu et al. demonstrated that *miR-30a-5p* may be associated with severe kidney damage in an Indian patient population with type 2 diabetes mellitus; indeed, *miR-30a-5p* levels were higher in the uEVs of patients with microalbuminuria but not in those of patients with normoalbuminuria or macroalbuminuria [84].

EV proteins reflect the pathophysiological state of their original cells and can be used as diagnostic markers of diseases, including DN. Urinary exosomal regucalcin could be explored for early diagnosis and for monitoring disease progression in patients with diabetic kidney disease because its expression in urinary exosomes is reduced in a manner similar to that in DN kidney tissue [86]. A panel of three proteins (alpha-1-microglobulin/bikunin precursor, methyltransferase mixed-lineage leukemia protein 3, and voltage-dependent anion-selective channel 1) that are altered in urinary exosomes from patients with DN could improve disease diagnosis and follow-up [87]. Moreover, the levels of urinary exosomal Wilm's tumor 1 (WT1) protein were significantly higher in patients with proteinuria than in those without proteinuria in urinary exosomes from spot urine samples of patients with type 1 diabetes mellitus; thus, WT1 could be used as an early noninvasive marker of DN [88].

With the development of genomics and proteomics, uEV miRNAs and proteins have been identified as new potential diagnostic markers of DN. However, it is necessary to conduct large-sample, multicenter studies to confirm their value in clinical applications.

## 21.6 Challenges in uEV Biomarker Studies and Future Perspectives

With the continuous technical development of uEV separation, transcriptomic, and proteomic methods, uEVs, as an important component of noninvasively obtained liquid biopsy samples, will play important roles in disease diagnosis and prognosis. In addition, the intersection between proteomics and other high-throughput technologies, such as genomics and bioinformatics, represents a systems biology research model and future focus of life science research.

However, more work is required to translate research on uEVs to the bedside. The following concrete problems remain to be solved: (1) the development of high-throughput and low-cost uEV sample separation technology; (2) how to accurately distinguish different cell sources for kidney EVs; (3) which type of EVs (exosomes, microvesicles, etc.) is most appropriate for kidney disease diagnosis; (4) which EV molecular markers or marker combination is most suitable for detecting related kidney diseases; and (5) the value of single vesicle analysis technology in the diagnosis of kidney disease. In summary, more extensive investigations are necessary to achieve the application of uEVs in clinical practice.

**Acknowledgments** We would like to thank Editage (www.editage.cn) for English language editing.

## References


1. Oshikawa S, Sonoda H, Ikeda M. Aquaporins in urinary extracellular vesicles (exosomes). Int J Mol Sci. 2016;17(6)
2. Pisitkun T, Shen RF, Knepper MA. Identification and proteomic profiling of exosomes in human urine. Proc Natl Acad Sci U S A. 2004;101(36):13,368–73.
3. Feng Y, Lv LL, Wu WJ, Li ZL, Chen J, Ni HF, et al. Urinary exosomes and exosomal CCL2 mRNA as biomarkers of active histologic injury in IgA nephropathy. Am J Pathol. 2018;188(11):2542–52.
4. Sonoda H, Yokota-Ikeda N, Oshikawa S, Kanno Y, Yoshinaga K, Uchida K, et al. Decreased abundance of urinary exosomal aquaporin-1 in renal ischemia-reperfusion injury. Am J Physiol Renal Physiol. 2009;297(4):F1006–16.
5. Li ZZ, Zhao ZZ, Wen JG, Xing L, Zhang H, Zhang Y. Early alteration of urinary exosomal aquaporin 1 and transforming growth factor beta1 after release of unilateral pelviureteral junction obstruction. J Pediatr Surg. 2012;47(8):1581–6.
6. Krause M, Rak-Raszewska A, Naillat F, Saarela U, Schmidt C, Ronkainen VP, et al. Exosomes as secondary inductive signals involved in kidney organogenesis. J Extracell Vesicles. 2018;7(1):1422675.
7. Jiang N, Xiang L, He L, Yang G, Zheng J, Wang C, et al. Exosomes mediate epithelium-mesenchyme crosstalk in organ development. ACS Nano. 2017;11(8):7736–46.
8. Wang J, Zhong J, Yang HC, Fogo AB. Cross talk from tubules to glomeruli. Toxicol Pathol. 2018;46(8):944–8.
9. Prunotto M, Farina A, Lane L, Pernin A, Schifferli J, Hochstrasser DF, et al. Proteomic analysis of podocyte exosome-enriched fraction from normal human urine. J Proteome. 2013;82:193–229.
10. Munkonda MN, Akbari S, Landry C, Sun S, Xiao F, Turner M, et al. Podocyte-derived microparticles promote proximal tubule fibrotic signaling via p38 MAPK and CD36. J Extracell Vesicles. 2018;7(1):1432206.
11. Gildea JJ, Seaton JE, Victor KG, Reyes CM, Bigler Wang D, Pettigrew AC, et al. Exosomal transfer from human renal proximal tubule cells to distal tubule and collecting duct cells. Clin Biochem. 2014;47(15):89–94.
12. Street JM, Birkhoff W, Menzies RI, Webb DJ, Bailey MA, Dear JW. Exosomal transmission of functional aquaporin 2 in kidney cortical collecting duct cells. J Physiol. 2011;589(Pt 24):6119–27.
13. Zhang W, Zhou X, Yao Q, Liu Y, Zhang H, Dong Z. HIF-1-mediated production of exosomes during hypoxia is protective in renal tubular cells. Am J Physiol Renal Physiol. 2017;313(4):F906–F13.
14. Lv LL, Wu WJ, Feng Y, Li ZL, Tang TT, Liu BC. Therapeutic application of extracellular vesicles in kidney disease: promises and challenges. J Cell Mol Med. 2018;22(2):728–37.
15. Lv LL, Feng Y, Wen Y, Wu WJ, Ni HF, Li ZL, et al. Exosomal CCL2 from tubular epithelial cells is critical for albumin-induced tubulointerstitial inflammation. J Am Soc Nephrol. 2018;29(3):919–35.
16. Choo YW, Kang M, Kim HY, Han J, Kang S, Lee JR, et al. M1 macrophage-derived nanovesicles potentiate the anticancer efficacy of immune checkpoint inhibitors. ACS Nano. 2018;12:8977.
17. Zhou Y, Xiong M, Fang L, Jiang L, Wen P, Dai C, et al. miR-21-containing microvesicles from injured tubular epithelial cells promote tubular phenotype transition by targeting PTEN protein. Am J Pathol. 2013;183(4):1183–96.
18. Borges FT, Melo SA, Ozdemir BC, Kato N, Revuelta I, Miller CA, et al. TGF-beta1-containing exosomes from injured epithelial cells activate fibroblasts to initiate tissue regenerative responses and fibrosis. J Am Soc Nephrol. 2013;24(3):385–92.

19. Dominguez JM 2nd, Dominguez JH, Xie D, Kelly KJ. Human extracellular microvesicles from renal tubules reverse kidney ischemia-reperfusion injury in rats. PLoS One. 2018;13(8):e0202550.
20. Dominguez JH, Liu Y, Gao H, Dominguez JM 2nd, Xie D, Kelly KJ. Renal tubular cell-derived extracellular vesicles accelerate the recovery of established renal ischemia reperfusion injury. J Am Soc Nephrol. 2017;28(12):3533–44.
21. Ela S, Mager I, Breakefield XO, Wood MJ. Extracellular vesicles: biology and emerging therapeutic opportunities. Nat Rev Drug Discov. 2013;12(5):347–57.
22. Tang TT, Lv LL, Lan HY, Liu BC. Extracellular vesicles: opportunities and challenges for the treatment of renal diseases. Front Physiol. 2019;10:226.
23. Tang TT, Wang B, Lv LL, Liu BC. Extracellular vesicle-based nanotherapeutics: emerging frontiers in anti-inflammatory therapy. Theranostics. 2020;10(18):8111–29.
24. Riazifar M, Pone EJ, Lotvall J, Zhao W. Stem cell extracellular vesicles: extended messages of regeneration. Annu Rev Pharmacol Toxicol. 2017;57:125–54.
25. Grange C, Skovronova R, Marabese F, Bussolati B. Stem cell-derived extracellular vesicles and kidney regeneration. Cell. 2019;8(10)
26. Bruno S, Grange C, Deregibus MC, Calogero RA, Saviozzi S, Collino F, et al. Mesenchymal stem cell-derived microvesicles protect against acute tubular injury. J Am Soc Nephrol. 2009;20(5):1053–67.
27. Bruno S, Tapparo M, Collino F, Chiabotto G, Deregibus MC, Soares Lindoso R, et al. Renal regenerative potential of different extracellular vesicle populations derived from bone marrow mesenchymal stromal cells. Tissue Eng Part A. 2017;23(21–22):1262–73.
28. Cao H, Cheng Y, Gao H, Zhuang J, Zhang W, Bian Q, et al. In vivo tracking of mesenchymal stem cell-derived extracellular vesicles improving mitochondrial function in renal ischemia-reperfusion injury. ACS Nano. 2020;14(4):4014–26.
29. Ju GQ, Cheng J, Zhong L, Wu S, Zou XY, Zhang GY, et al. Microvesicles derived from human umbilical cord mesenchymal stem cells facilitate tubular epithelial cell dedifferentiation and growth via hepatocyte growth factor induction. PLoS One. 2015;10(3):e0121534.
30. Jiang ZZ, Liu YM, Niu X, Yin JY, Hu B, Guo SC, et al. Exosomes secreted by human urine-derived stem cells could prevent kidney complications from type I diabetes in rats. Stem Cell Res Ther. 2016;7:24.
31. Eirin A, Zhu XY, Puranik AS, Tang H, McGurren KA, van Wijnen AJ, et al. Mesenchymal stem cell-derived extracellular vesicles attenuate kidney inflammation. Kidney Int. 2017;92(1):114–24.
32. Tang TT, Lv LL, Wang B, Cao JY, Feng Y, Li ZL, et al. Employing macrophage-derived microvesicle for kidney-targeted delivery of dexamethasone: an efficient therapeutic strategy against renal inflammation and fibrosis. Theranostics. 2019;9(16):4740–55.
33. Tang TT, Wang B, Wu M, Li ZL, Feng Y, Cao JY, et al. Extracellular vesicle-encapsulated IL-10 as novel nanotherapeutics against ischemic AKI. Sci Adv. 2020;6(33):eaaz0748.
34. McKee JA, Kumar S, Ecelbarger CA, Fernandez-Llama P, Terris J, Knepper MA. Detection of Na+ transporter proteins in urine. J Am Soc Nephrol. 2000;11(11):2128–32.
35. Svenningsen P, Sabaratnam R, Jensen BL. Urinary extracellular vesicles: origin, role as intercellular messengers and biomarkers; efficient sorting and potential treatment options. Acta Physiol (Oxf). 2020;228(1):e13346.
36. Miranda KC, Bond DT, McKee M, Skog J, Paunescu TG, Da Silva N, et al. Nucleic acids within urinary exosomes/microvesicles are potential biomarkers for renal disease. Kidney Int. 2010;78(2):191–9.
37. Turco AE, Lam W, Rule AD, Denic A, Lieske JC, Miller VM, et al. Specific renal parenchymal-derived urinary extracellular vesicles identify age-associated structural changes in living donor kidneys. J Extracell Vesicles. 2016;5:29,642.
38. Pisitkun T, Shen R, Knepper MA. Identification and proteomic profiling of exosomes in human urine. PNAS. 2004;101:13,368–73.
39. Lytvyn Y, Xiao F, Kennedy CR, Perkins BA, Reich HN, Scholey JW, et al. Assessment of urinary microparticles in normotensive patients with type 1 diabetes. Diabetologia. 2017;60(3):581–4.

40. Park S, Lee K, Park IB, Kim NH, Cho S, Rhee WJ, et al. The profiles of microRNAs from urinary extracellular vesicles (EVs) prepared by various isolation methods and their correlation with serum EV microRNAs. Diabetes Res Clin Pract. 2020;160:108010.
41. Wang S, Kojima K, Mobley JA, West AB. Proteomic analysis of urinary extracellular vesicles reveal biomarkers for neurologic disease. EBioMedicine. 2019;45:351–61.
42. Merchant ML, Rood IM, Deegens JKJ, Klein JB. Isolation and characterization of urinary extracellular vesicles: implications for biomarker discovery. Nat Rev Nephrol. 2017;13(12):731–49.
43. Gamez-Valero A, Ines Lozano-Ramos S, Bancu I, Lauzurica-Valdemoros R, Borras FE. Urinary extracellular vesicles as source of biomarkers in kidney diseases. Front Immunol. 2015;6
44. Musante L, Saraswat M, Ravida A, Byrne B, Holthofer H. Recovery of urinary nanovesicles from ultracentrifugation supernatants. Nephrol Dialysis Transpl. 2013;28(6):1425–33.
45. Mitchell PJ, Welton J, Staffurth J, Court J, Mason MD, Tabi Z, et al. Can urinary exosomes act as treatment response markers in prostate cancer? J Transl Med. 2009;7
46. Alvarez ML, Khosroheidari M, Ravi RK, DiStefano JK. Comparison of protein, microRNA, and mRNA yields using different methods of urinary exosome isolation for the discovery of kidney disease biomarkers. Kidney Int. 2012;82(9):1024–32.
47. Rood IM, Deegens JKJ, Merchant ML, Tamboer WPM, Wilkey DW, Wetzels JFM, et al. Comparison of three methods for isolation of urinary microvesicles to identify biomarkers of nephrotic syndrome. Kidney Int. 2010;78(8):810–6.
48. Cheruvanky A, Zhou H, Pisitkun T, Kopp JB, Knepper MA, Yuen PST, et al. Rapid isolation of urinary exosomal biomarkers using a nanomembrane ultrafiltration concentrator. Am J Physiol Renal Physiol. 2007;292(5):F1657–F61.
49. Musante L, Tataruch DE, Holthofer H. Use and isolation of urinary exosomes as biomarkers for diabetic nephropathy. Front Endocrinol. 2014;5
50. Taylor DD, Zacharias W, Gercel-Taylor C. Exosome isolation for proteomic analyses and RNA profiling. Methods Mol Biol (Clifton, NJ). 2011;728:235–46.
51. Santana SM, Antonyak MA, Cerione RA, Kirby BJ. Microfluidic isolation of cancer-cell-derived microvesicles from heterogeneous extracellular shed vesicle populations. Biomed Microdevices. 2014;16(6):869–77.
52. Liga A, Vliegenthart AD, Oosthuyzen W, Dear JW, Kersaudy-Kerhoas M. Exosome isolation: a microfluidic road-map. Lab Chip. 2015;15(11):2388–94.
53. Musante L, Tataruch D, Gu D, Benito-Martin A, Calzaferri G, Aherne S, et al. A simplified method to recover urinary vesicles for clinical applications, and sample banking. Sci Rep. 2014;4:7532.
54. Liu X, Chinello C, Musante L, Cazzaniga M, Tataruch D, Calzaferri G, et al. Intraluminal proteome and peptidome of human urinary extracellular vesicles. Proteomics Clin Appl. 2015;9(5–6):568–73.
55. Campos-Silva C, Suárez H, Jara-Acevedo R, Linares-Espinós E, Martinez-Piñeiro L, Yáñez-Mó M, et al. High sensitivity detection of extracellular vesicles immune-captured from urine by conventional flow cytometry. Sci Rep. 2019;9(1):2042.
56. Sun J, Han S, Ma L, Zhang H, Zhan Z, Aguilar HA, et al. Synergistically bifunctional paramagnetic separation enables efficient isolation of urine extracellular vesicles and downstream phosphoproteomic analysis. ACS Appl Mater Interfaces. 2021;13(3):3622–30.
57. Erdbrügger U, Blijdorp CJ, Bijnsdorp IV, Borràs FE, Burger D, Bussolati B, et al. Urinary extracellular vesicles: a position paper by the urine task force of the International Society for Extracellular Vesicles. J Extracell Vesicles. 2021;10(7):e12093.
58. Tkach M, Théry C. Communication by extracellular vesicles: where we are and where we need to go. Cell. 2016;164(6):1226–32.
59. Erdbrugger U, Blijdorp CJ, Bijnsdorp IV, Borras FE, Burger D, Bussolati B, et al. Urinary extracellular vesicles: a position paper by the urine task force of the International Society for Extracellular Vesicles. J Extracell Vesicles. 2021;10(7):e12093.

60. Panich T, Chancharoenthana W, Somparn P, Issara-Amphorn J, Hirankarn N, Leelahavanichkul A. Urinary exosomal activating transcriptional factor 3 as the early diagnostic biomarker for sepsis-induced acute kidney injury. BMC Nephrol. 2017;18(1):10.
61. Zhou H, Pisitkun T, Aponte A, Yuen PS, Hoffert JD, Yasuda H, et al. Exosomal Fetuin-A identified by proteomics: a novel urinary biomarker for detecting acute kidney injury. Kidney Int. 2006;70(10):1847–57.
62. Harrill AH, Sanders AP. Urinary microRNAs in environmental health: biomarkers of emergent kidney injury and disease. Curr Environ Health Rep. 2020;7(2):101–8.
63. Ronco C, Bellomo R, Kellum JA. Acute kidney injury. Lancet. 2019;394(10212):1949–64.
64. Vijayan A, Faubel S, Askenazi DJ, Cerda J, Fissell WH, Heung M, et al. Clinical use of the urine biomarker [TIMP-2] x [IGFBP7] for acute kidney injury risk assessment. Am J Kidney Dis. 2016;68(1):19–28.
65. Malhotra R, Siew ED. Biomarkers for the early detection and prognosis of acute kidney injury. Clin J Am Soc Nephrol. 2017;12(1):149–73.
66. du Cheyron D, Daubin C, Poggioli J, Ramakers M, Houillier P, Charbonneau P, et al. Urinary measurement of Na+/H+ exchanger isoform 3 (NHE3) protein as new marker of tubule injury in critically ill patients with ARF. Am J Kidney Dis. 2003;42(3):497–506.
67. Asvapromtada S, Sonoda H, Kinouchi M, Oshikawa S, Takahashi S, Hoshino Y, et al. Characterization of urinary exosomal release of aquaporin-1 and -2 after renal ischemia-reperfusion in rats. Am J Physiol Renal Physiol. 2018;314(4):F584–601.
68. Sonoda H, Yokota-Ikeda N, Oshikawa S, Kanno Y, Yoshinaga K, Uchida K, et al. Decreased abundance of urinary exosomal aquaporin-1 in renal ischemia-reperfusion injury. Am J Physiol-Renal Physiol. 2009;297(4):F1006–16.
69. Pavkovic M, Riefke B, Ellinger-Ziegelbauer H. Urinary microRNA profiling for identification of biomarkers after cisplatin-induced kidney injury. Toxicology. 2014;324:147–57.
70. Sonoda H, Lee BR, Park KH, Nihalani D, Yoon JH, Ikeda M, et al. miRNA profiling of urinary exosomes to assess the progression of acute kidney injury. Sci Rep. 2019;9(1):4692.
71. Lv JC, Zhang LX. Prevalence and disease burden of chronic kidney disease. Adv Exp Med Biol. 2019;1165:3–15.
72. Dhaun N, Bellamy CO, Cattran DC, Kluth DC. Utility of renal biopsy in the clinical management of renal disease. Kidney Int. 2014;85(5):1039–48.
73. Medeiros T, Myette RL, Almeida JR, Silva AA, Burger D. Extracellular vesicles: cell-derived biomarkers of glomerular and tubular injury. Cell Physiol Biochem. 2020;54(1):88–109.
74. Zhang LH, Zhu XY, Eirin A, Nargesi AA, Woollard JR, Santelli A, et al. Early podocyte injury and elevated levels of urinary podocyte-derived extracellular vesicles in swine with metabolic syndrome: role of podocyte mitochondria. Am J Physiol Renal Physiol. 2019;317(7):F12–22.
75. Lv LL, Cao YH, Pan MM, Liu H, Tang RN, Ma KL, et al. CD2AP mRNA in urinary exosome as biomarker of kidney disease. Clin Chim Acta. 2014;428:26–31.
76. Lv LL, Feng Y, Wu M, Wang B, Li ZL, Zhong X, et al. Exosomal miRNA-19b-3p of tubular epithelial cells promotes M1 macrophage activation in kidney injury. Cell Death Differ. 2020;27(1):210–26.
77. Khurana R, Ranches G, Schafferer S, Lukasser M, Rudnicki M, Mayer G, et al. Identification of urinary exosomal noncoding RNAs as novel biomarkers in chronic kidney disease. RNA (New York, NY). 2017;23(2):142–52.
78. Burbano C, Gomez-Puerta JA, Munoz-Vahos C, Vanegas-Garcia A, Rojas M, Vasquez G, et al. HMGB1(+) microparticles present in urine are hallmarks of nephritis in patients with systemic lupus erythematosus. Eur J Immunol. 2019;49(2):323–35.
79. Lv LL, Cao YH, Ni HF, Xu M, Liu D, Liu H, et al. MicroRNA-29c in urinary exosome/microvesicle as a biomarker of renal fibrosis. Am J Physiol Renal Physiol. 2013;305(8):F1220–7.
80. Thibodeau JF, Simard JC, Holterman CE, Blais A, Cloutier MP, Medeiros T, et al. PBI-4050 via GPR40 activation improves adenine-induced kidney injury in mice. Clin Sci (Lond). 2019;133(14):1587–602.

81. Lu J, Hu ZB, Chen PP, Lu CC, Zhang JX, Li XQ, et al. Urinary levels of podocyte-derived microparticles are associated with the progression of chronic kidney disease. Ann Transl Med. 2019;7(18):445.
82. Barutta F, Tricarico M, Corbelli A, Annaratone L, Pinach S, Grimaldi S, et al. Urinary exosomal microRNAs in incipient diabetic nephropathy. PLoS One. 2013;8(11):e73798.
83. Jia Y, Guan M, Zheng Z, Zhang Q, Tang C, Xu W, et al. miRNAs in urine extracellular vesicles as predictors of early-stage diabetic nephropathy. J Diabetes Res. 2016;2016:7932765.
84. Prabu P, Rome S, Sathishkumar C, Gastebois C, Meugnier E, Mohan V, et al. MicroRNAs from urinary extracellular vesicles are non-invasive early biomarkers of diabetic nephropathy in type 2 diabetes patients with the 'Asian Indian phenotype'. Diabetes Metab. 2019;45(3):276–85.
85. Delić D, Eisele C, Schmid R, Baum P, Wiech F, Gerl M, et al. Urinary exosomal miRNA signature in type II diabetic nephropathy patients. PLoS One. 2016;11(3):e0150154.
86. Zubiri I, Posada-Ayala M, Benito-Martin A, Maroto AS, Martin-Lorenzo M, Cannata-Ortiz P, et al. Kidney tissue proteomics reveals regucalcin downregulation in response to diabetic nephropathy with reflection in urinary exosomes. Transl Res. 2015;166(5):474–84.e4.
87. Zubiri I, Posada-Ayala M, Sanz-Maroto A, Calvo E, Martin-Lorenzo M, Gonzalez-Calero L, et al. Diabetic nephropathy induces changes in the proteome of human urinary exosomes as revealed by label-free comparative analysis. J Proteome. 2014;96:92–102.
88. Kalani A, Mohan A, Godbole MM, Bhatia E, Gupta A, Sharma RK, et al. Wilm's tumor-1 protein levels in urinary exosomes from diabetic patients with or without proteinuria. PLoS One. 2013;8(3):e60177.

# Extracellular Vesicle Biomarkers for Thrombosis

# 22

Bo Li, Tingting Luo, Shouping Liu, Houmei Feng, Qian Wang, and Lei Zheng

## 22.1 Roles of EVs in Thrombosis

Thrombosis is an important part of physiological hemostasis but may also cause serious diseases, including acute coronary syndrome, systolic stroke, deep vein thrombosis, and other thrombotic diseases, which are all associated with high disability and mortality rates. Recently, researches have demonstrated that EVs participate in thrombosis. The coagulant activity of EVs is mainly derived from tissue factor (TF) and phosphatidylserine (PS) carried on the surface of EVs, which have a larger coagulation surface area, faster migration speed, and better permeability because of their small size compared with cells. Moreover, the population of EVs in thrombotic diseases is obviously higher than that in healthy individuals, suggesting EVs' potential applications as biomarkers for predicting the occurrence, development, and severity of thrombotic diseases.

Bo Li and Tingting Luo have contributed equally to this work.

B. Li · T. Luo · S. Liu · H. Feng · L. Zheng (✉)
Department of Laboratory Medicine, Nanfang Hospital, Southern Medical University, Guangzhou, China

Guangdong Engineering and Technology Research Center for Rapid Diagnostic Biosensors, Nanfang Hospital, Southern Medical University, Guangzhou, China
e-mail: nfyyzhenglei@smu.edu.cn

Q. Wang (✉)
Laboratory Medicine Center, Zhujiang Hospital, Southern Medical University, Guangzhou, Guangdong, China
e-mail: wangqian@smu.edu.cn



Q. Wang, L. Zheng (eds.), *Extracellular Vesicles*,
https://doi.org/10.1007/978-981-99-8365-0_22

### 22.1.1 TF EVs in Thrombosis

TF, also called coagulation factor III, is a key initiator of coagulation. Current models have shown coagulation includes three stages: initiation, amplification, and propagation. Coagulation also has intrinsic and extrinsic pathways. TF can initiate the extrinsic pathway and active Coagulation Factor VII (FVII) into Activated FVII (FVIIa); the complex TF/FVIIa then actives Coagulation Factor IX (FIX) and Coagulation Factor X (FX) to generate Activated FIX (FIXa) and Activated FX (FXa), respectively, and this leads to the production of a prothrombinase complex, which turns Coagulation Factor II (FII) (prothrombin) into thrombin [1]. The above mechanism is the initiation stage in the extrinsic pathway, which is followed by a series of amplification and propagation steps. TF is an important part of the beginning of the extrinsic pathway because the activity of FVIIa is dependent on TF. Accordingly, overexpression of TF can induce pathological thrombosis [1].

TF is a transmembrane protein released following vessel injury. In fact, many TF-positive EVs ($TF^{+}EVs$), including platelet-derived EVs, activated monocyte-derived EVs, cancer cell-derived EVs, and epithelial cell-derived EVs, can be detected in patients with thrombosis-related diseases. Under specific pathological conditions, large numbers of $TF^{+}EVs$ can be detected in circulation. Therefore, $TF^{+}EVs$ can serve as thrombosis biomarkers. In prospective studies, cancer cell-derived $TF^{+}EVs$ were found to act as biomarkers of venous thrombosis and to predict survival time [2]. Moreover, epithelial cell-derived EVs expressing TF and PS are associated with thrombotic events [3]. In addition, monocytes and epithelial cell-derived EVs are TF positive in sickle cell disease, which is a prothrombotic state leading to thrombotic disease [4].

### 22.1.2 PS EVs in Thrombosis

PS contributes to the assembly of the prothrombinase complex by supporting the negatively charged membrane. The membrane can interact with FVa and then bind FXa to assemble the prothrombinase complex on the membrane surface [5]. PS is rich in eukaryotic lipid membranes but is distributed unevenly. PS and phosphatidylethanolamine are mainly concentrated in the cytoplasmic leaflet of the plasma membrane, whereas phosphatidylcholine and sphingomyelin are mostly found in the exoplasmic leaflet; all of these lipids are regulated by three phospholipid translocases. Flippase changes the localization of PS and phosphatidylethanolamine from the outer to inner membrane, whereas floppase changes the localization of phosphatidylcholine from the inner to outer membrane. These two phospholipid translocases are both Adenosine Triphosphate (ATP) dependent, whereas scramblase translocates phospholipids nonselectively and in an ATP-independent manner [6]. This imbalance in phospholipid translocases can expose PS to the membrane surface. Several types of EVs, including platelet EVs [5], mesenchymal stem cell EVs [7], epithelial cell-derived EVs [3], and erythrocyte EVs [8], express PS on the surface, and this affects thrombosis owing to roles in coagulation. Accordingly, these EVs may have potential applications as biomarkers of thrombosis.

### 22.1.3 Other EV-Linked Molecules in Thrombosis

In addition to TF and PS, some membrane proteins on EVs are involved in thrombosis. Epithelial cells and active platelets express P-selectin, which is involved in the interactions of leukocytes with endothelial cells and platelets and promotes thrombosis [9]. EVs exposing P-selectin glycoprotein ligand 1 and integrin interact with platelets, thereby facilitating platelet aggregation [10]. Furthermore, platelet aggregation is also induced by integrins, such as like αIIbβ3, another membrane protein overexpressed in cancer cells [10]. Thus, aggregation of platelets increases the risk of thrombotic diseases.

## 22.2 EV Biomarkers for Arterial Thrombosis

Arterial thrombosis leads to a series of life-threatening diseases, including acute myocardial infarction, cerebral ischemic stroke, and mesenteric arterial embolism. EVs act in arterial thrombosis based on endothelial cell-derived EVs, platelet EVs, immune cell-derived EVs, and EVs from fibroblasts, smooth muscle cells, and mesenchymal stromal cells [11]. Endothelial cell-derived EVs play important roles in vascular homeostasis, mainly in coagulation, inflammation, endothelial function, and angiogenesis [12]. Additionally, platelet EVs contribute to the interactions of platelets with endothelial cells and monocytes in atherosclerosis [13].

### 22.2.1 EVs and Myocardial Infarction

Myocardial infarction involves coronary atherosclerotic stenosis following the rupture of coronary atherosclerotic plaques. Platelets then accumulate on the surface of the ruptured plaque, forming a thrombus, which blocks the lumen of the coronary artery. Some studies have shown increases in the numbers of platelet and endothelial cell EVs in the circulation following myocardial infarction [14]. In animal models, acute myocardial infarction temporarily increases the numbers cardiomyocyte- and endothelial cell-derived EVs, which are taken up by monocytes and regulate inflammatory responses [15, 16]. In prospective studies, endothelial cell EVs have been shown to provide information on endothelial cell status. For example, the numbers of $CD31^+$ endothelial cell EVs in patients experiencing their first myocardial infarction are higher than in those with recurring myocardial infarction [17, 18]. Additionally, CD62P, CD42a, and CD41b levels are increased in EVs from patients with ST segment elevation-related myocardial infarction, and analysis of a set of comprehensive indicators including EV concentrations and CD62P/CD42a levels can discriminate patients with ST segment elevation-related myocardial infarction from controls [19]. In patients with acute myocardial infarction, circulating cardiac EVs contain microRNAs (miRNAs; e.g., *miR-1*, *miR-133a*, *miR-208*, and *miR-499*), indicating the occurrence of cardiac damage; these EVs may have potential applications as biomarkers [20].

### 22.2.2 EVs and Ischemic Stroke

Stroke is associated with high mortality and disability rates worldwide. In recent studies, ischemic stroke was shown to account for more than 70% of all strokes and was found to be more common than hemorrhagic stroke [21]. EVs are involved in the development of ischemic stroke through procoagulation and pro inflammatory effects. Moreover, patients who have experienced ischemic stroke exhibit greater numbers of leukocyte-derived EVs in circulation, and this difference may promote ischemic stroke by enhancing inflammation [22]. Circulating endothelial cell EVs and platelet EVs [23] increase the risk of ischemic stroke, and the numbers of $TF^{+}$EVs also increase after acute ischemic stroke. In a case-control study, patients with acute ischemic stroke were found to have three endothelial cell EV phenotypes (i.e., $Endoglin^{+}$, $PS^{+}$, and intercellular adhesion molecule $[ICAM]\text{-}1^{+}$), which were significantly correlated with brain lesion volume; ICAM-$1^{+}$ EVs showed the strongest correlation [24]. In addition to $CD62E^{+}/CD144^{+}/CD41a^{-}$ endothelial cell-derived EVs, EV *miR-126*, *miR-9*, *miR-124*, and *miR-223* are also increased following brain ischemia [14], and all of these miRNAs have been shown to be related to the levels of ischemia and necrosis [25, 26]. In mice, decreased *miR-126* expression in EVs may contribute to cardiac dysfunction after stroke [27].

## 22.3 EV Biomarkers for Venous Thrombosis

Venous thrombosis is a type of venous reflux disorder involving the pathological clotting of blood in veins, leading to complete or incomplete vascular obstruction. Venous thrombosis includes deep vein thrombosis and pulmonary embolism. In recent years, cancer has also been shown to be associated with venous thrombosis, and venous thrombosis has been identified as a common complication in patients with cancer. Many studies have explored the procoagulation mechanisms of particles from cancer cells and in patients with cancer; however, most of these studies have focused on the roles of $TF^{+}$EVs. The risk of venous thrombosis varies according to the type of cancer, and pancreatic cancer is associated with a particularly high risk of thrombosis [28]. Venous thrombosis occurring in response to cancer cell-derived EVs is primarily related to the expression of TF and PS on the EV surface, and cancer $TF^{+}$EVs can serve as a biomarker of venous thrombosis [29].

### 22.3.1 EVs and Deep Venous Thrombosis (DVT)

DVT is a common thrombosis disease with a high incidence in the clinical setting. Approximately 30% of patients with venous thrombosis after anticoagulant therapy still have a risk of recurrence within 5 years [30]. Monocyte $TF^{+}$EVs are increased in patients with initial and recurrent DVT. In addition, patients with recurrent DVT also exhibit increased levels of endothelial and platelet $TF^{+}$EVs compared with patients with initial DVT, although there are no differences between patients with initial DVT and healthy individuals [31]. DVT increases the activation of leukocytes and the binding of endothelial cell EVs and monocytes [32]. Moreover, one study demonstrated that the combination of soluble P-selectin, total EVs, and D-dimer showed 78% sensitivity, 81% specificity, and 77% accuracy for the diagnosis of DVT [33]. In a prospective study, $TF^+$EVs, D-dimer, and soluble P-selectin were found to act as biomarkers of unprovoked DVT and advanced cancer in patients who did not develop thrombosis [34]. Notably, most studies of EVs in DVT are based on patients with cancer or mouse models of cancer, and the relationships between DVT and EVs should be investigated in greater detail in order to elucidate the true clinical diagnostic efficacy of these EVs as biomarkers.

### 22.3.2 EVs and Pulmonary Embolism

Pulmonary thromboembolism is a blockage of the pulmonary artery or its branches. The numbers of circulating procoagulant EVs, particularly platelet-derived EVs, are increased in patients with pulmonary embolism, whereas the numbers of endothelial-derived EVs are not significantly different [35]. Receiver operating characteristic curve analysis demonstrated that the sensitivity of platelet EVs in the diagnosis of pulmonary embolism was greater than 93% and that a combination of platelet EVs, D-dimer, platelet distribution width, and P-selectin could increase the specificity, accuracy, and sensitivity of pulmonary embolism diagnosis [36].

## 22.4 EV Biomarkers for Microcirculatory Intravascular Thrombosis

Microcirculatory intravascular thrombosis is also related to EVs. In thrombotic thrombocytopenic purpura (TTP), endothelial cell-derived EVs that co-express von Willebrand factor and CD62E can participate in the pathogenesis of TTP. The number of EVs is correlated with the severity of TTP [37] and returns to normal when remission is achieved [38]. Disseminated intravascular coagulation (DIC) is a complication of various diseases. Endothelial cell- and leukocyte-derived EVs in sepsis-induced DIC reflect vascular injury [39], and their generation may be related to the activation of white blood cells and endothelial cells in DIC [40].

## 22.5 Conclusion

EVs, particularly $TF^+$EVs, endothelial cell-derived EVs, and platelet-derived EVs, play important roles in thrombosis. EVs can act as biomarkers for thrombosis; however, their specificity is low. Many studies have demonstrated that combinations of EVs with various indicators can increase the specificity, accuracy, and sensitivity of diagnosis and prognosis (Tables 22.1 and 22.2).

**Table 22.1** Prothrombotic molecules found in EVs [10]

| Molecular | Thrombosis effect | Reference |
|---|---|---|
| TF | Activation of the extrinsic pathway and platelet activation/aggregation in a thrombin-dependent manner | [1] |
| PS | Assembly the prothrombinase complex on the surface | [6] |
| Podoplanin | Platelet aggregation | [41–43] |
| Glycoprotein ligand 1 | Accumulation of the EVs at the site of thrombosis through binding to platelets via P-selection and involved in the interaction between active platelet and endothelial cells and leukocytes | [44] |
| CD41 | Generate a little thrombin in a TF- and FVII-independent way | [45] |
| Integrins | Accumulation of the EVs at the site of thrombosis through binding platelets | |
| Polyphosphate | Activation of the contact pathway, fibrin generation, and platelet activation/aggregation in a thrombin-dependent manner | [46–48] |

**Table 22.2** Summary of studies investigating the roles of EVs in thrombosis [49]

| Type of EVs | Major function in thrombosis | Reference |
|---|---|---|
| Platelet EVs | Increase platelet aggregation and adhesion and promote the generation of thrombin in a factor XII-dependent way | [50, 51] |
| Erythrocyte EVs | Promote thrombosis in a FXI- and FXII-dependent way | [51, 52] |
| Monocyte EVs | Induce endothelial cells, overexpress TF, and involved in fibrin and thrombus generation | [51, 53] |
| Endothelial cell EVs | Stimulate monocytes to express TF and procoagulation activity and bind to platelet | [54, 55] |
| Cancer cell EVs | Increase platelet aggregation and promote TF-dependent coagulation | [56] |

## References

1. Versteeg HH, Heemskerk JWM, Levi M, et al. New fundamentals in hemostasis. Physiol Rev. 2013;93(1):327–58.
2. Hisada Y, Mackman N. Cancer cell-derived tissue factor-positive extracellular vesicles: biomarkers of thrombosis and survival. Curr Opin Hematol. 2019;26(5):349–56.
3. Hromada C, Mühleder S, Grillari J, et al. Endothelial extracellular vesicles—promises and challenges. Front Physiol. 2017;8
4. Shet AS, Aras O, Gupta K, et al. Sickle blood contains tissue factor–positive microparticles derived from endothelial cells and monocytes. Blood. 2003;102(7):2678–83.
5. Lentz BR. Exposure of platelet membrane phosphatidylserine regulates blood coagulation. Prog Lipid Res. 2003;42(5):423–38.
6. Nagata S, Suzuki J, Segawa K, et al. Exposure of phosphatidylserine on the cell surface. Cell Death Differ. 2016;23(6):952–61.
7. Silachev D, Goryunov K, Shpilyuk M, et al. Effect of MSCs and MSC-derived extracellular vesicles on human blood coagulation. Cell. 2019;8(3):258.
8. Rubin O, Delobel J, Prudent M, et al. Red blood cell-derived microparticles isolated from blood units initiate and propagate thrombin generation. Transfusion. 2013;53(8):1744–54.

9. Semenov AV, Romanov YA, Loktionova SA, et al. Production of soluble P-selectin by platelets and endothelial cells. Biochemistry (Mosc). 1999;64(11):1326–35.
10. Almeida VH, Rondon AMR, Gomes T, et al. Novel aspects of extracellular vesicles as mediators of cancer-associated thrombosis. Cell. 2019;8(7):716.
11. Boulanger CM, Loyer X, Rautou P, et al. Extracellular vesicles in coronary artery disease. Nat Rev Cardiol. 2017;14(5):259–72.
12. Dignat-George F, Boulanger CM. The many faces of endothelial microparticles. Arterioscler Thromb Vasc Biol. 2011;31(1):27–33.
13. Badimon L, Suades R, Fuentes E, et al. Role of platelet-derived microvesicles as crosstalk mediators in atherothrombosis and future pharmacology targets: a link between inflammation, atherosclerosis, and thrombosis. Front Pharmacol. 2016;07
14. Wang W, Li Z, Feng J. The potential role of exosomes in the diagnosis and therapy of ischemic diseases. Cytotherapy. 2018;20(10):1204–19.
15. Loyer X, Zlatanova I, Devue C, et al. Intra-cardiac release of extracellular vesicles shapes inflammation following myocardial infarction. Circ Res. 2018;123(1):100–6.
16. Sluijter JPG, Davidson SM, Boulanger CM, et al. Extracellular vesicles in diagnostics and therapy of the ischaemic heart: position paper from the working group on cellular biology of the heart of the European Society of Cardiology. Cardiovasc Res. 2018;114(1):19–34.
17. Bernal-Mizrachi L, Jy W, Jimenez JJ, et al. High levels of circulating endothelial microparticles in patients with acute coronary syndromes. Am Heart J. 2003;145(6):962–70.
18. Min P, Kim J, Chung K, et al. Local increase in microparticles from the aspirate of culprit coronary arteries in patients with ST-segment elevation myocardial infarction. Atherosclerosis. 2013;227(2):323–8.
19. Burrello J, Bolis S, Balbi C, et al. An extracellular vesicle epitope profile is associated with acute myocardial infarction. J Cell Mol Med. 2020;24(17):9945–57.
20. Chistiakov D, Orekhov A, Bobryshev Y. Cardiac extracellular vesicles in normal and infarcted heart. Int J Mol Sci. 2016;17(1):63.
21. Edzie EKM, Gorleku PN, Dzefi-Tettey K, et al. Incidence rate and age of onset of first stroke from CT scan examinations in Cape Coast metropolis`. Heliyon. 2021;7(2):e06214.
22. He Z, Tang Y, Qin C. Increased circulating leukocyte-derived microparticles in ischemic cerebrovascular disease. Thromb Res. 2017;154:19–25.
23. Lee YJ, Jy W, Horstman LL, et al. Elevated platelet microparticles in transient ischemic attacks, lacunar infarcts, and multiinfarct dementias. Thromb Res. 1993;72(4):295–304.
24. Simak J, Gelderman MP, Yu H, et al. Circulating endothelial microparticles in acute ischemic stroke: a link to severity, lesion volume and outcome. J Thromb Haemost. 2006;4(6):1296–302.
25. Jung K, Chu K, Lee S, et al. Circulating endothelial microparticles as a marker of cerebrovascular disease. Ann Neurol. 2009;66(2):191–9.
26. Li P, Qin C. Elevated circulating VE-cadherin+CD144+endothelial microparticles in ischemic cerebrovascular disease. Thromb Res. 2015;135(2):375–81.
27. Chen J, Cui C, Yang X, et al. MiR-126 affects brain-heart interaction after cerebral ischemic stroke. Transl Stroke Res. 2017;8(4):374–85.
28. Hisada Y, Mackman N. Cancer-associated pathways and biomarkers of venous thrombosis. Blood. 2017;130(13):1499–506.
29. Geddings JE, Mackman N. Tumor-derived tissue factor-positive microparticles and venous thrombosis in cancer patients. Blood. 2013;122(11):1873–80.
30. Nakamura M, Yamada N, Ito M. Current management of venous thromboembolism in Japan: current epidemiology and advances in anticoagulant therapy. J Cardiol. 2015;66(6):451–9.
31. Ye R, Ye C, Huang Y, et al. Circulating tissue factor positive microparticles in patients with acute recurrent deep venous thrombosis. Thromb Res. 2012;130(2):253–8.
32. Chirinos JA, Heresi GA, Velasquez H, et al. Elevation of endothelial microparticles, platelets, and leukocyte activation in patients with venous thromboembolism. J Am Coll Cardiol. 2005;45(9):1467–71.
33. Rectenwald JE, Myers DD, Hawley AE, et al. D-dimer, P-selectin, and microparticles: novel markers to predict deep venous thrombosis. A pilot study. Thromb Haemost. 2005;94(6):1312–7.

34. Sánchez-López V, Gao L, Ferrer-Galván M, et al. Differential biomarker profiles between unprovoked venous thromboembolism and cancer. Ann Med (Helsinki). 2020;52(6):310–20.
35. Bal L, Ederhy S, Di Angelantonio E, et al. Factors influencing the level of circulating procoagulant microparticles in acute pulmonary embolism. Arch Cardiovasc Dis. 2010;103(6–7):394–403.
36. Wang M, Fu Y, Xu L, et al. Diagnostic value of platelet-derived microparticles in pulmonary thromboembolism: a population-based study. Exp Ther Med. 2018;16(4):3099–106.
37. Jimenez JJ, Jy W, Mauro LM, et al. Endothelial microparticles released in thrombotic thrombocytopenic purpura express von Willebrand factor and markers of endothelial activation. Br J Haematol. 2003;123(5):896–902.
38. Jimenez JJ, Jy W, Mauro LM, et al. Elevated endothelial microparticles in thrombotic thrombocytopenic purpura: findings from brain and renal microvascular cell culture and patients with active disease. Br J Haematol. 2001;112(1):81–90.
39. Delabranche X, Quenot J, Lavigne T, et al. Early detection of disseminated intravascular coagulation during septic shock. Crit Care Med. 2016;44(10):e930–9.
40. Nomura S. Extracellular vesicles and blood diseases. Int J Hematol. 2017;105(4):392–405.
41. Mir SNP, Riedl J, Pabinger I, et al. The role of podoplanin in cancer-associated thrombosis. Thromb Res. 2018;164(Suppl 1):S34–9.
42. Suzuki-Inoue K, Kato Y, Inoue O, et al. Involvement of the snake toxin receptor CLEC-2, in podoplanin-mediated platelet activation, by cancer cells. J Biol Chem. 2007;282(36):25,993–6,001.
43. Riedl J, Preusser M, Nazari PM, et al. Podoplanin expression in primary brain tumors induces platelet aggregation and increases risk of venous thromboembolism. Blood. 2017;129(13):1831–9.
44. Gasecka A, Böing AN, Filipiak KJ, et al. Platelet extracellular vesicles as biomarkers for arterial thrombosis. Platelets (Edinburgh). 2017;28(3):228–34.
45. Kerris EWJ, Hoptay C, Calderon T, et al. Platelets and platelet extracellular vesicles in hemostasis and sepsis. J Investig Med. 2020;68(4):813–20.
46. Baker CJ, Smith SA, Morrissey JH. Polyphosphate in thrombosis, hemostasis, and inflammation. Res Pract Thromb Haemost. 2019;3(1):18–25.
47. Nickel KF, Ronquist G, Langer F, et al. The polyphosphate-factor XII pathway drives coagulation in prostate cancer-associated thrombosis. Blood. 2015;126(11):1379–89.
48. Nickel KF, Labberton L, Long AT, et al. The polyphosphate/factor XII pathway in cancer-associated thrombosis: novel perspectives for safe anticoagulation in patients with malignancies. Thromb Res. 2016;141(Suppl 2):S4–7.
49. Zarà M, Guidetti GF, Camera M, et al. Biology and role of extracellular vesicles (EVs) in the pathogenesis of thrombosis. Int J Mol Sci. 2019;20(11):2840.
50. Barry OP, Pratico D, Lawson JA, et al. Transcellular activation of platelets and endothelial cells by bioactive lipids in platelet microparticles. J Clin Invest. 1997;99(9):2118–27.
51. Van Der Meijden PE, Van Schilfgaarde M, Van Oerle R, et al. Platelet- and erythrocyte-derived microparticles trigger thrombin generation via factor XIIa. J Thromb Haemost. 2012;10(7):1355–62.
52. van Beers EJ, Schaap MC, Berckmans RJ, et al. Circulating erythrocyte-derived microparticles are associated with coagulation activation in sickle cell disease. Haematologica. 2009;94(11):1513–9.
53. Aharon A, Tamari T, Brenner B. Monocyte-derived microparticles and exosomes induce procoagulant and apoptotic effects on endothelial cells. Thromb Haemost. 2008;100(5):878–85.
54. Sabatier F, Roux V, Anfosso F, et al. Interaction of endothelial microparticles with monocytic cells in vitro induces tissue factor-dependent procoagulant activity. Blood. 2002;99(11):3962–70.
55. Ghosh A, Li W, Febbraio M, et al. Platelet CD36 mediates interactions with endothelial cell-derived microparticles and contributes to thrombosis in mice. J Clin Invest. 2008;118(5):1934–43.
56. Geddings JE, Hisada Y, Boulaftali Y, et al. Tissue factor-positive tumor microvesicles activate platelets and enhance thrombosis in mice. J Thromb Haemost. 2016;14(1):153–66.

# Part IV

# EVs for Disease Therapy and Prevention

# Stem Cell-Derived Extracellular Vesicles (EVs) for Disease Therapy

# 23

Qing Li, Cheng Ji, Yu Chen, Xinyu Niu, Fengtian Sun, Siqi Yin, Hui Shi, Teng Ye, Zhengsheng Chen, Zhifeng Deng, Xianghang Luo, Hui Qian, and Yang Wang

## 23.1 Introduction

In recent years, stem cells have emerged as multipotent cells with the ability to differentiate into several cell types. Thus, stem cells can act as a cell reservoir for regenerative medicine and have been widely applied as therapeutic tools in basic research and clinical trials. Mesenchymal stem cells (MSCs) from different tissues and pluripotent stem cells, including embryonic stem cells (ESCs) and induced pluripotent stem cells (iPSCs), are the main types of stem cells used for tissue regeneration applications. Much of the current interest in the therapeutic applications of stem cells in various disease settings can be linked to the effects of stem cells on promotion of proliferation, inhibition of apoptosis, and suppression of inflammation, and the therapeutic benefits of stem cells are largely dependent on paracrine factors, including extracellular vesicles (EVs). Indeed, some studies have proposed

---

Qing Li and Cheng Ji have contributed equally to this work.

Q. Li · Y. Chen · T. Ye · Z. Chen · Y. Wang (✉)
Institute of Microsurgery on Extremities, Department of Orthopedic Surgery, Shanghai Sixth People's Hospital Affiliated to Shanghai Jiao Tong University School of Medicine, Shanghai, China
e-mail: wangyang63@sjtu.edu.cn

C. Ji · F. Sun · S. Yin · H. Shi · H. Qian (✉)
Zhenjiang Key Laboratory of High Technology Research on Exosomes Foundation and Transformation Application, Jiangsu Key Laboratory of Medical Science and Laboratory Medicine, Jiangsu University, Zhenjiang, Jiangsu, China

X. Niu · Z. Deng
Department of Neurosurgery, Shanghai Sixth People's Hospital Affiliated to Shanghai Jiao Tong University School of Medicine, Shanghai, China

X. Luo (✉)
Xiangya Hospital of Central South University, Changsha, Hunan, China



Q. Wang, L. Zheng (eds.), *Extracellular Vesicles*,
https://doi.org/10.1007/978-981-99-8365-0_23

that the regenerative and immunomodulatory effects of stem cells are mediated mainly by soluble paracrine factors and EVs.

As nanosized bilayer membrane structures containing lipids, microRNAs (miRNAs), mRNAs, DNAs, and proteins, EVs play key roles in cell-to-cell communication. Compared with their parental stem cells, EVs have several advantages for use as therapeutic agents, such as lower immunogenicity, tumorigenicity, and easier management. Moreover, EVs may have a superior safety profile and can be stored without losing function. EVs have emerged as a promising alternative to whole-cell therapies. Stem cell-derived EVs have been widely studied in many animal models, including models of kidney, liver, cardiovascular, immunological, and neurological diseases. Additionally, stem cell-derived EVs can directly target damaged tissues to play reparative roles and can easily cross the blood-brain barrier to repair damaged nervous system tissues. This chapter reviews current preclinical studies on the use of stem cell-derived EVs as therapeutic tools in the treatment of cardiovascular, central nervous system, kidney, endocrine, bone, and aging diseases.

## 23.2 Stem Cell-Derived EVs for Cardiovascular Diseases

Cardiovascular disease is a major economic burden to society and healthcare systems worldwide [1]. The increasing incidence of heart failure has triggered research into adjuvant therapies to further limit excessive tissue loss and enhance heart function. Stem cell-based therapy has emerged as a promising strategy for treating cardiovascular ischemic diseases [2]. However, some important functional deficiencies in stem cell transplantation, such as immune rejection, tumorigenicity, and infusional toxicity, have limited the use of stem cell therapy in clinical applications. Thus, there is a critical need to exploit the powerful regenerative competence of stem cells while avoiding problems associated with cell transplantation.

Accumulating evidence suggests that the discovery of cell-free components that are capable of inducing analogous cell responses in target tissues may provide a promising alternative for cardiac protection and regeneration. Hence, stem cell-derived EVs are considered a novel therapeutic approach for treating cardiovascular diseases. First, EVs play therapeutic roles by serving as vehicles for biological factors to target the cells. Second, EVs can directly affect tissues in cardiac regeneration therapies. With the development of experimental techniques and molecular technologies, EV therapy may become a clinically available and practical treatment for various cardiovascular diseases, including coronary heart disease, heart failure, myocardial infarction, and other related diseases. However, the therapeutic roles of EVs have only been identified in animal experiments and multiple types of cardiac-related cells. Therefore, further experiments are needed in other mammal models and human patients.

### 23.2.1 Stem Cell-Derived EVs for Myocardial Regeneration

The underlying cause of systolic heart failure is the inability of the adult mammalian heart to regenerate the myocardium. Initially, the scar tissue replaces the lost cardiomyocytes and provides strength to the heart to maintain its integrity. However, later progressive matrix deposition by activated myofibroblasts may lead to myocardial stiffening and impaired contraction.

In this field, Mayourian et al. showed that bone marrow-derived MSC (BMSC)-EVs could protect ischemic cardiomyocytes through high *miR-21a-5p* content both in hypoxic-cultured H9c2 cardiomyocytes and in a mouse model of myocardial ischemia reperfusion. They also found that *miR-21-5p* secreted from human BMSC-derived EVs could efficiently enhance cardiac tissue contractility and calcium handling gene expression responses in mature ischemic adult human cardiomyocytes [3]. Furthermore, *miR-21a-5p* exerts cardioprotective effects by reducing MI-mediated cell apoptosis and death in cardiomyocytes via synergistic downregulation of protein levels, including Pellino1, programmed cell death 4, FasL, and phosphatase and tensin homolog, thereby contributing to myocardial salvage [4]. Additionally, Shao et al. demonstrated that BMSC-derived exosomes contain high levels of *miR-29* and *miR-24* and low levels of *miR-21*, *miR-15*, *miR-34*, *miR-130*, and *miR-378*. Compared with BMSCs, EVs can repair injured myocardium by suppressing cardiac fibrosis, inhibiting inflammation, and improving cardiac function, as demonstrated in a rat model of MI [5].

Another study reported that hypoxic treatment of BMSCs could increase *miR-210* expression in EVs. Intramyocardial injection of hypoxic EVs into the infarcted hearts of C57BL/6 mice significantly increased survival, reduced scar size, and improved recovery of cardiac functions. Hypoxic EVs with high levels of *miR-210* enhanced elevated vascular density, decreased fibrosis, and increased the renewal of progenitor cells in the infarcted heart relative to administration of null exosomes [6]. Khan et al. reported that after myocardial infarction, mouse ESC (mESC)-EVs delivery in the heart could stimulate and augment cardiac progenitor cells (CPCs) and cardiomyocyte proliferation-based endogenous myocardial repair, which in part involved the metastasis of ESC-specific *miR-294*. The observed beneficial effects of mESC-EVs in the heart after injury suggest that cardiomyocyte survival and cell cycle entry, enhanced neovascularization, and potentiation of CPC expansion, differentiation, and survival are mediated by *miR-294* delivered via ESC-EVs to the heart [7].

Wang et al. reported that iPSC-EVs transmit cardioprotective miRNAs, including *miR-21* and *miR-210*, which could prevent cardiomyocyte apoptosis in ischemic myocardium. The delivered exosomal *miR-21* and *miR-210* protected recipient H9c2 cardiomyocytes in vitro against oxidative stress-induced apoptosis by suppressing caspase 3/7 activation. This effect confirmed that intramyocardial administration of iPSC-EVs into the mouse ischemic myocardium could protect against myocardial infarction-induced injury [8].

In conclusion, these studies demonstrated that MSC- and ESC-/iPSC-derived EVs could promote cardiac regeneration and modulate both cardiomyocyte- and CPC-based repair programs in the heart. Stem cell-derived EVs represent a novel cell-free system for enhancing endogenous cardiac repair after pathological injury and bypassing the limitations of adoptive cell transplantation.

### 23.2.2 Stem Cell-Derived EVs for Angiogenesis in Cardiovascular Diseases

Many recent preclinical studies and promising results of early clinical trials have supported the potential applications of therapeutic angiogenesis in patients with advanced ischemic heart disease.

Neoangiogenesis can be enhanced by MSC-derived EVs via the transfer of pro-angiogenic miRNAs, including *miR-126*, *miR-210*, *miR-21*, *miR-23a-3p*, and *miR-130a-3p*, which have been suggested to exert cardioprotective effects through the induction of angiogenesis in ischemic hearts after MI [9]. Luo et al. demonstrated that *miR-126*-enriched exosomes isolated from adipose tissue MSCs (AMSCs) significantly induce microvascular generation and migration of endothelial progenitor cells during hypoxic injury. Moreover, exosomal *miR-126* has pro-angiogenic effects and induces blood vessel formation in infarcted areas in a treated rat model of acute myocardial infarction [10]. *MiR-126* can directly inhibit sprouty-related EVH1 domain containing 1 and phosphoinositide-3-kinase (PI3K) regulatory subunit 2 and consequently enhance the vascular endothelial growth factor (VEGF) signaling pathway. Downregulation of *miR-l26* leads to suppression of the mitogen-activated protein kinase and PI3K signaling pathways, which affects angiogenic factor signaling and disrupts angiogenesis [11].

Similarly, as revealed by Xiao et al., exosomal *miR-125b-5p* derived from hypoxia-conditioned MSCs exert cardioprotective effects through modulation of autophagic flux [5]. In summary, hypoxic conditions can increase the levels of cardioprotective miRNAs carried by stem cell-derived EVs, including *miR-126*, *miR-125b-5p*, *miR-26a*, and *miR-210*, which are associated with higher survival and lower apoptosis rates in cardiomyocytes, attenuate infarct size and arrhythmia, increase vascular density, decrease fibrosis, and improve recovery of cardiac function.

### 23.2.3 Stem Cell-Derived EVs for Atherosclerosis (AS)

AS is a complex, multifactorial disease associated with various risk factors, including arterial hypertension, dyslipidemia, diabetes mellitus, and smoking [12]. Although therapeutic medicines for AS have been widely adopted in clinical practice, many patients remain at a high risk of cardiovascular disease despite optimal treatment [13].

Harada-Shiba found that conditioned medium from cultured MSCs (MSC-CM) ameliorates AS in low-density lipoprotein receptor$^{-/-}$ mice and suppresses calmodulin expression and macrophage accumulation in vascular walls. Humoral factors, including hepatocyte growth factor and EVs from MSCs, may be promising therapeutic agents for reducing the residual risk of coronary artery diseases [14]. Additionally, EVs derived from MSCs can inhibit AS development and progression in mice. Furthermore, MSC-derived EVs transfer *miR-221* both in vitro and in vivo to inhibit N-acetyltransferase-1 expression in vascular smooth muscle cells; this results in downregulation of insulin-like growth factor 2/insulin-like growth factor 2 receptor signaling to attenuate atherosclerotic plaque development [15]. Xiao et al. also showed that adipose-derived stem cell-derived EVs can deliver *miR-342-5p* in the context of AS and further demonstrated the potential of *miR-342-5p* as a novel therapeutic target for AS [16].

Stem cell-derived EVs have been shown to have significant effects on AS in disease models. However, studies on this topic remain challenging. Future research should focus on elucidating the intrinsic mechanisms of the therapeutic effects of EVs on AS to establish better treatment strategies.

### 23.2.4 Stem Cell-Derived EVs for Organotypic Vasculature

Blood vessels form one of the body's largest surfaces, exerting gatekeeper functions in tissue homeostasis and adaptation to pathological challenges. Vascular control of the tissue microenvironment is indispensable for the development of hemostasis, inflammation, and metabolism.

Hypertension is a long-term condition that can increase organ susceptibility to insults and lead to severe complications, such as ventricular hypertrophy and chronic kidney disease. Using a deoxycorticosterone acetate-salt-induced hypertensive model, Collin et al. demonstrated that multiple treatments with EVs from adipose-derived MSCs (ASC-EVs) could protect renal tissue by maintaining its filtration capacity. Indeed, ASC-EVs downregulated the pro-inflammatory molecules monocyte chemoattractant protein-1 and plasminogen activating inhibitor-1 and reduced recruitment of macrophages in the kidneys [17]. Sun et al. showed that EVs derived from iPSC-derived MSCs could alleviate aging-associated vascular endothelial dysfunction, arterial stiffening, and hypertension [18]. Moreover, Braun et al. showed that multiple injections of MSC-derived EVs in a model of neonatal chronic bronchopulmonary dysplasia protected the lungs and heart from hypoxia, preserving alveolar growth and inhibiting right heart hypertrophy [19].

EVs derived from BMSCs can also promote elastic matrix regenerative repair in the abdominal aortic aneurysm wall. Intriguingly, these data support the capability of EVs to exert beneficial effects on organotypic vasculature [20].

As discussed above, the cardioprotective capacity of stem cell-derived EVs has been verified in animal models. These studies have demonstrated that stem cell-derived EVs are powerful regulators of cardiovascular regeneration by enhancing cardiac function and angiogenesis after heart injury, providing important translational perspectives for the clinical treatment of patients with cardiovascular diseases. However, further research is needed for testing and fine-tuning in the clinical setting.

## 23.3 Stem Cell-Derived EVs for Central Nervous System Diseases

### 23.3.1 Stem Cell-Derived EVs for Stroke

Stroke is the second leading cause of death globally and the leading cause of disability, with an increasing incidence in developing countries. Ischemic stroke accounts for 80% of all stroke cases and is caused by arterial occlusion. Tissue plasminogen activator and endovascular thrombectomy, the only two effective approaches to treat stroke, are not readily available because of the narrow therapeutic window or limitations associated with contraindications. Therefore, only a small population of patients can benefit from current therapies, and there is an urgent need for improved treatment strategies for stroke.

Angiogenesis and neurogenesis are two compensatory recovery processes that occur after stroke [21] and are limited in adults. Numerous studies have confirmed that MSC-derived EVs play important roles in endogenous angiogenesis and neurogenesis at sites of injury. Xia et al. showed that small EVs (sEVs) derived from iPSC-generated MSCs possess pro-angiogenic effects owing to reduced autophagy caused by the activation of signal transducer and activator of transcription 3 [22]. Additionally, EVs from human urine-derived stem cells (USC-EVs) have pro-neurogenesis functions similar to MSC-derived EVs, offering a simple, noninvasive, and low-cost therapeutic approach. Ling et al. showed that exosomes derived from human USCs (USC-Exos) enhance neurogenesis and alleviate neurological deficits. In an in vitro model of ischemic stroke, USC-Exos exerted pro-neurogenesis functions by transferring *miR-26a* to inhibit histone deacetylase 6 [23].

Immune responses play essential roles in the pathogenesis of ischemic stroke, and modulation of poststroke immune responses ameliorates ischemia-induced brain damage [24]. Specifically, regulatory T cells (Tregs) exert critical immunomodulatory functions after ischemic stroke, through which they ameliorate neurological deficits after brain ischemia [25, 26]. A recent study by Xia et al. showed that sEVs from ESCs modulate neuro-inflammation and protect against ischemic stroke via the expansion of Tregs. Further mechanistic research revealed that this process was partially dependent on the activation of the transforming growth factor (TGF)-β1/Smad signaling pathway [27].

Collectively, these studies demonstrate that stem cell-derived EVs exert notable protective effects against ischemic stroke and are a promising cell-free therapeutic strategy.

### 23.3.2 Stem Cell-Derived EVs for Multiple Sclerosis (MS)

MS, an autoimmune disease of the central nervous system, is characterized by demyelination, chronic inflammation, and loss of axons, even in the early stages of the disease, and is regarded as one of the most common causes of neurological deficits in adults [28]. The clinical course of MS is classified into three phenotypes: relapsing-remitting MS, secondary progressive MS, and primary progressive MS [29].

The mechanism of MS involves the infiltration of circulating inflammatory cells (T cells, B cells, and myeloid cells) through the breakdown of the blood-brain barrier into the central nervous system, which leads to inflammation and demyelination of the central nervous system. Therefore, inhibition of cellular inflammation is crucial to prevent the relapse of MS. Riazifar et al. showed that exosomes derived from human MSCs stimulated by interferon-γ induced remyelination as well as suppressive neuro-inflammation in experimental autoimmune encephalomyelitis (EAE) model mice (a mouse model of MS) by increasing the numbers of $CD4^{+}CD25^{+}FOXP3^{+}$ Tregs and decreasing the levels of pro-inflammation cytokines, such as interleukin (IL)-6, IL-12p70, IL-17A/F, and IL-22 [30]. A recent study generated an engineered murine microglial cell line, and EVs derived from this cell line expressed the "eat me" signal lactadherin (Mfg-e8) on the surface, which facilitate uptake by target cells. These modified EVs resulted in alleviation of clinical symptoms [31]. Therefore, EVs may have therapeutic effects in patients with MS via their anti-inflammatory functions.

Remyelination, another prominent advantage of EVs derived from certain cells, is also worth considering. Oligodendrocyte precursor cells (OPCs) are cells that differentiate into myeline-forming cells and are a key cluster for remyelination in patients with MS. EVs harvested from immunosuppressive stem cells promote OPC recruitment and myelin repair [32]. In a recent study, engineered EVs with elevated *miR-219a-5p* levels were found to promote OPC differentiation and improve clinical outcomes in an EAE animal model [33].

EVs exert their anti-inflammatory and pro-remyelination functions by transferring their contents, including endogenous and/or artificially encapsulated bioactive substances. EVs can easily cross the blood-brain barrier, allowing them to target phagocytes in the cerebral nervous system. As discussed above, EVs are promising bioactive substances against autoimmune diseases of the central nervous system.

### 23.3.3 Stem Cell-Derived EVs for Spinal Cord Injury (SCI)

Traumatic SCI is a devastating disease of the central nervous system that often results in severe limb dysfunction below the injured spinal segment. Initial trauma can physically destroy in situ tissue and trigger a complex delayed inflammatory cascade, which leads to tissue edema, ischemia, and neuronal apoptosis. Therefore, promoting angiogenesis and neurogenesis and reducing secondary inflammation at the injury site are essential for ameliorating SCI. Various studies have demonstrated that stem cell-derived EVs play important roles in SCI therapy and can alleviate sequelae, such as motor dysfunction, suggesting promising therapeutic applications.

Spinal cord microvascular endothelial cells (SCMECs) are essential components of the blood-spinal cord barrier. SCMECs are damaged immediately after SCI, accompanied by angiogenesis disorders, blood supply reduction, and nutrient supply interruption. A recent study revealed that neural stem cell (NSC)-derived EVs (NSC-EVs) have remarkable capacity to facilitate angiogenesis. NSC-EVs improve the migration and proliferation of SCMECs in vitro by delivering VEGF-A. In SCI model mice, NSC-EVs increase microvessel density, reduce the size of the spinal cord cavity, and promote motor function recovery [34].

Severe inflammatory responses activated by SCI often induce secondary damage and consequently result in neuronal apoptosis in the lesion. As one of the resident cell types of the central nervous system, microglia mediate a series of cascade reactions after SCI and exert dual roles in neuro-inflammation and neurogenesis, depending on their polarization. Liu et al. recently reported that MSC-derived EVs mediate microglia shifting from the pro-inflammatory M1 phenotype to the anti-inflammatory M2 phenotype; thus, the expression levels of pro-inflammatory cytokines (e.g., tumor necrosis factor [TNF]-α, IL-1β, and IL-6) were decreased, whereas the expression levels of anti-inflammatory cytokines (e.g., TGF-β, IL-4, and IL-10) were elevated in injured spinal cord tissue. Further mechanistic studies showed that *miR-216a-5p* in MSC-EVs regulates microglia polarization [35].

In summary, the roles of stem cell-derived EVs in SCI have been studied extensively. Given the severity and refractory nature of diseases of the central nervous system, stem cell EVs are expected to have broad clinical applications.

### 23.3.4 Stem Cell-Derived EVs for Neurodegenerative Diseases

Neurodegenerative diseases, including Alzheimer's disease (AD), Parkinson's disease (PD), Huntington disease, vascular dementia (VD), and amyotrophic lateral sclerosis, are age-dependent disorders that are incurable and are rapidly becoming more prevalent. Many combinations of pharmaceutical treatments have been developed; however, the results are not promising. The absence of effective drugs makes clinical treatment difficult.

Aggregation of Aβ plaques is thought to be a cause of AD. Oxidative damage of neurons induced by excessive aggregation of Aβ plaques aggravates cognitive deficits in patients with AD. MSC-derived EVs have neuroprotective and regenerative effects at injury sites and therefore alleviate cognitive deficits in patients with AD. Godoy et al. showed that MSC-derived EVs restore basal neuronal reactive oxygen species (ROS) levels and protect neurons from detrimental oxidative stress induced by Aβ plaques in vitro [36]. Moreover, MSC-derived EVs can modulate the immune response of microglia activated by Aβ peptides. MSC-derived EVs also could promote the shift of microglia from the M1 phenotype (pro-inflammatory phenotype) to the M2 phenotype (anti-inflammatory phenotype).

EVs are also effective in the treatment of other neurodegenerative diseases. PD is characterized by the degeneration of dopaminergic neurons and deficiency of dopamine production. MSC-derived EV-enriched *miR-133b*, which is downregulated in PD, promotes neurite outgrowth [37]. Additionally, Hu et al. showed that hippocampal NSCs become senescent in aging mice and VD model rats. ESC-derived sEVs can reverse senescence-related neurogenesis dysfunction and alleviate cognitive deficits [38, 39]. Increasing evidence has shown that EVs may be effective candidates for the treatment of neurodegenerative diseases in the future.

## 23.4 Stem Cell-Derived EVs for Kidney and Endocrine Diseases

Novel therapies are urgently needed to address the increasing incidence and prevalence of kidney and endocrine diseases. MSC-derived EVs have shown promising results in experimental studies of kidney and endocrine diseases. MSC-derived EVs contain genetic and protein materials that, upon transfer to recipient cells, can activate several repair mechanisms to ameliorate renal injury and endocrine diseases. This section summarizes current knowledge on MSC-derived EVs therapy in kidney and endocrine diseases for consideration of MSC-derived EVs as realistic clinical tools.

### 23.4.1 Stem Cell-Derived EVs for Kidney Diseases

#### 23.4.1.1 Stem Cell-Derived EVs for Acute Kidney Injury (AKI)

AKI is a clinical syndrome that occurs due to the rapid decline of renal function caused by multiple factors and is divided into three types: prerenal, renal, and postrenal [40]. In recent years, stem cell therapy has been widely used in research on kidney diseases. Notably, MSCs derived from various tissues, such as the bone marrow, fat, and umbilical cord, have been shown to repair acute and chronic kidney damage by secreting proteins and RNA, which are carried by EVs. This section discusses the roles of MSC-derived EVs in the treatment of AKI and chronic kidney disease (Fig. 23.1).

**Chemotherapeutic Drug-Induced AKI**

AKI is a relatively common clinical syndrome, and its pathogenesis primarily involves renal tubular epithelial damage, microvascular endothelial damage, and inflammatory reactions. In 2009, Bruno et al. found that BMSC-derived microvesicles play important roles in the repair of glycerol-induced AKI damage in model mice [41]. Furthermore, tail vein injection of BMSC-derived exosomes inhibited the production of the inflammatory factors IL-6 and TNF-$\alpha$, promoted the expression of the anti-inflammatory factor IL-10, and repaired gentamicin-induced AKI in rats. Evidence showed that renal capsule injection of human umbilical cord MSC-derived exosomes could transport 14-3-3$\zeta$ and combine with ATG16, which

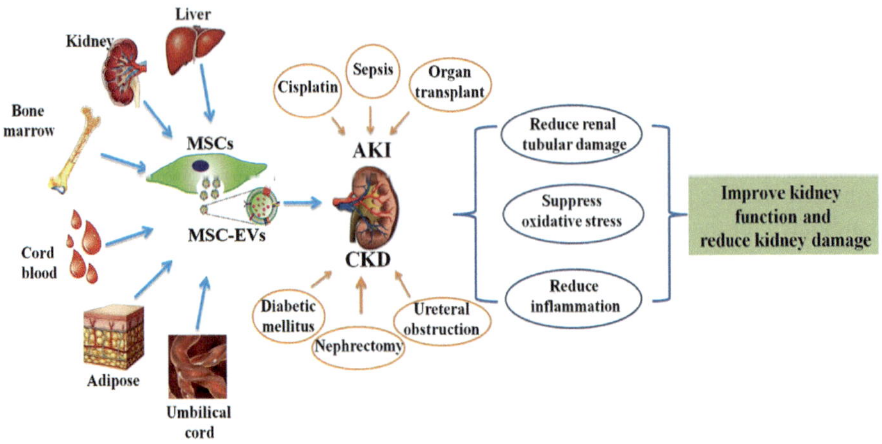


**Fig. 23.1** MSC-derived EVs promote the repair of AKI and chronic kidney disease

subsequently activated autophagy to prevent AKI induced by cisplatin [42]. In summary, EVs, such as stem cell exosomes, may represent a novel therapeutic tool for the treatment of kidney injury.

#### Ischemia-Reperfusion-Induced AKI

Ischemia-reperfusion injury leads to renal cell dysfunction because of extensive apoptosis and inflammation. After the ischemic phase, intravenous injection of MSC-derived exosomes reduces NADPH oxidase 2 and kidney injury [43]. BMSC-derived exosomes were injected into the renal capsules of rats with unilateral renal ischemia to transfer miRNAs that regulate C-X-C motif chemokine ligand 1 (CXCL1) into recipient cells, thereby reducing the number of macrophages and the expression of CXCL1 in macrophages. This effectively suppresses the inflammatory response and regulates inflammation and kidneys [44]. EVs could serve as a promising delivery platform to manipulate IL-10 for effective treatment of ischemic AKI [45]. In summary, these studies have shown that MSC-derived exosomes can repair kidney damage caused by ischemia-reperfusion injury and protect kidney tissues through inhibition of oxidation, regulation of immunity, and promotion of angiogenesis.

### 23.4.1.2 Stem Cell-Derived EVs for Chronic Kidney Injury

Approximately 8–16% of patients will progress to chronic kidney disease and then to end-stage renal disease. The prevalence rate of chronic kidney disease in the adult population in China is 10.8% (approximately 120 million individuals), and the morbidity and fatality rates are increasing every year. Therefore, new treatment methods and early interventions are needed.

**Ischemia-Reperfusion-Induced Chronic Kidney Injury**

This section focuses on the therapeutic roles of MSCs in ischemia-reperfusion-induced chronic kidney injury. Eirin et al. found that multiple injections of BMSC-derived EVs in a 5/6 kidney resection model prevented renal failure [46]. Additionally, multiple injections of purified conditioned medium from human embryonic MSCs intravenously injected twice daily for four consecutive days alleviated the deterioration of kidney function.

**Unilateral Ureteral Obstruction (UUO)-Induced Chronic Kidney Injury**

Renal tubular cell apoptosis and interstitial fibrosis are commonly used models of obstructive renal injury. MSC-derived exosomes were injected intravenously for 7 days to reduce renal tubular damage, inhibit renal interstitial fibrosis, and improve renal function. BMSC-derived exosomes were found to alleviate renal fibrosis and reduce the secretion of α-smooth muscle actin in renal tubular epithelial cells induced by TGF-β1 in vitro. BMSC-derived exosomes also reduce renal fibrosis by transporting endogenous *let-7c*, which downregulates collagen IV, matrix metalloproteinase 9, TGF-β1, and TGF-βR1 [43]. In a UUO rat renal fibrosis model, Ji et al. found that human umbilical cord MSC-derived exosomes injected into the tail vein migrated to the damaged kidney tissue and inhibited renal interstitial fibrosis by delivering the casein kinase 1 delta/β-transducin repeat-containing protein kinase ubiquitin system and mediating YES-associated protein (YAP) ubiquitin degradation [47]. Overall, these findings showed that MSC-derived exosomes regulate the epithelial-mesenchymal transition (EMT), exert antifibrotic effects, and exhibit renal protection in obstructive nephropathy.

### 23.4.2 Stem Cell-Derived EVs for Diabetes and Associated Complications

Characterized by enhanced blood glucose levels, diabetes mellitus is a metabolic disease with an estimated worldwide prevalence of 451 million cases among the adult population, representing a serious threat to human health [48]. Satisfactory therapies for diabetes mellitus are urgently needed. Stem cell-based cell therapy is considered an effective approach for several refractory diseases; however, its potential tumorigenicity limits its application. Increasing evidence indicates that stem cells promote tissue regeneration mainly through paracrine pathways, and EVs are the major components of the paracrine mechanism.

#### 23.4.2.1 Diabetes

Diabetes is a metabolic disease characterized by insufficient insulin secretion and dysfunction and has become a serious threat to human health. Tail vein injection of BMSC-derived EVs inhibits the activation of antigen-presenting cells, ameliorates the T-cell inflammatory response, and increases β-cell mass and insulin production in a type 1 diabetes mellitus murine model. Additionally, Sun et al. found that exosomes derived from human umbilical cord MSCs effectively ameliorate blood

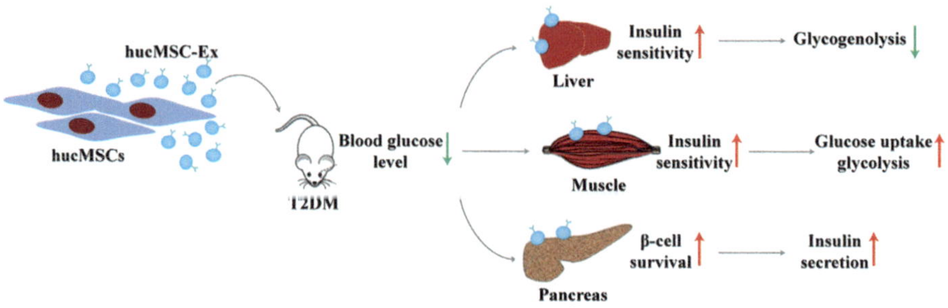


**Fig. 23.2** Schematic diagram of the mechanism through which human umbilical cord MSC-derived exosomes decrease blood glucose levels in a rat model of type 2 diabetes mellitus

glucose levels in streptozotocin-induced type 2 diabetes mellitus rat models. Mechanistically, human umbilical cord MSC-derived exosomes restore glucose homeostasis by promoting the expression and membrane translocation of glucose transporter 4 in the muscle and enhancing glycogen storage. Moreover, human umbilical cord MSC-derived exosomes inhibit the apoptosis of pancreatic β-cells to improve insulin secretion [49]. Enhanced insulin sensitivity and islet remodeling were induced by treatment with human umbilical cord MSC-derived exosomes in a model of type 2 diabetes mellitus (Fig. 23.2).

#### 23.4.2.2 Diabetic Complications

The aggravation of diabetes mellitus results in the emergence of various severe complications, including eye, kidney, heart, skin, and nerve dysfunction. Hyperglycemia duration and diabetes severity are major risk factors for diabetic complications; therefore, strict control of blood glucose is an effective treatment strategy [50]. Accumulating evidence suggests that stem cell-derived EVs exert reparative functions in multiple high glucose-induced tissue injuries (Fig. 23.3).

##### Diabetic Retinopathy

Diabetic retinopathy is a common microvascular complication of diabetes and is the main cause of blindness in adults [51]. As the disease progresses, excessive proliferation of retinal endothelial cells results in the formation of multiple new blood vessels that are unstable and prone to rupture. High glucose levels enhance the degree of oxidative stress, inflammation, and apoptosis in Muller cells, whereas MSC-derived exosomes ameliorate cell damage through *miR-486-3p*-induced regulation of Toll-like receptor 4. Human umbilical cord MSC-derived exosomes deliver *miR-17-3p* to achieve therapeutic effects, including decreased levels of inflammatory factors and VEGF and downregulation of oxidative injury and apoptosis in diabetic retinopathy model mice. High glucose-induced EMT has been recognized as an important risk factor for retinal fibrosis. MSC-derived exosomes suppress the endothelial-mesenchymal transition and tube formation via delivery of the long noncoding RNA (lncRNA) *SNHG7*, which negatively regulates *miR-34a-5p* expression and promotes downstream X-box binding protein 1 expression [52]. These

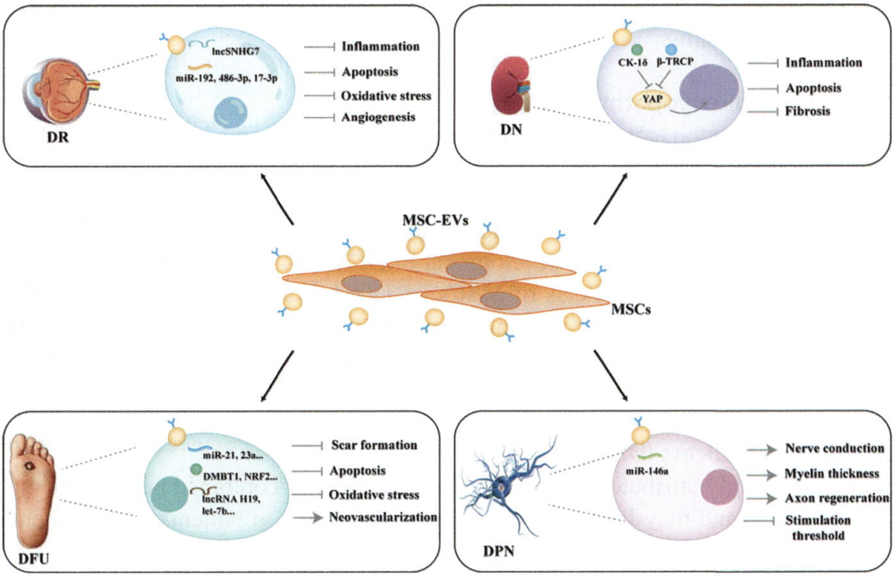


**Fig. 23.3** Therapeutic effects of MSC-derived EV-enriched bioactive molecules in diabetic complications

findings suggest that MSC-derived EVs may be a novel cell-free therapy for diabetic retinopathy.

### Diabetic Nephropathy

Diabetic nephropathy is a metabolic disease characterized by interstitial fibrosis due to persistent hyperglycemia [53]. Recent studies have shown that the Hippo/YAP signaling pathway plays important roles in lower urinary tract embryo development, maintenance of podocyte homeostasis, integrity of the glomerular filtration barrier, regulation of tubular cyst growth, renal epithelial damage in diabetes, and kidney fibrosis. In addition, knockout of YAP in mesangial cells reduces renal interstitial fibrosis; however, the molecular mechanism remains unclear. YAP expression and phosphorylation were increased in animal models of type 1 and type 2 diabetes and in vitro studies. Moreover, expression of epidermal growth factor receptor is increased in a YAP-dependent manner. High YAP activation promotes downstream target gene expression. Therefore, we explored the mechanisms through which human umbilical cord MSC-derived exosome-mediated YAP signaling delays the progression of diabetic nephropathy. Analyses using 10×Genomics single-cell pathway technology have revealed that the mechanism of renal interstitial fibrosis in kidney tissue from patients with diabetic nephropathy may be related to the conversion of macrophages from mesangial cells to myofibroblasts and the inhibition of macrophages by human umbilical cord MSC-derived exosomes. Interaction with mesangial cells can improve renal interstitial fibrosis and delay the progression of diabetic nephropathy [54]; this may represent a novel strategy for the treatment of diabetic nephropathy.

**Diabetic Cardiomyopathy (DCM)**

DCM is a common complication of advanced diabetes and is characterized by abnormal myocardial structure and performance, without other cardiac risk factors. In recent decades, multiple mechanisms, including myocardial cell loss, inflammation, oxidative stress, insulin resistance, and mitochondrial dysfunction, have been proposed to be involved in the development of DCM [50]. Some studies have demonstrated the therapeutic effects of MSCs on DCM. For example, Jin et al. found that ASCs could alleviate myocardial fibrosis and injury in a DCM rat model, partially via the secretion of prostaglandin E2 and the promotion of type 2 macrophage polarization [55].

**Diabetic Foot Ulcers (DFUs)**

DFUs are mainly attributed to ischemic, neuropathic, or combined neuro-ischemic abnormalities and may lead to infection and amputation in patients; thus, patients with DFUs have a 5-year mortality rate that is 2.5 times higher than that in patients without DFU [56]. In the treatment of DFU, which involves increased angiogenesis and removal of oxidation products, existing treatments, including sugar control and debridement therapy, still cannot cure the disease owing to persistent infection and limb ischemia. MSC encapsulated deleted in malignant brain tumors 1, OxOband, nuclear factor erythroid 2-related factor 2, *miR-21*, *miR-23a*, *miR-124a*, *miR-125b*, *miR-126*, lncRNA *H19*, *let-7b*, *mmu_circ_0000250*, *miR-130a*, and *miR-132* contribute to DFU wound healing by mediating proliferation, migration, anti-apoptotic effects, and immune responses and attenuating ROS generation and neovascularization [57]. Therefore, researchers have attempted to modify MSC-derived EVs by engineering them with hydrogels, dressings, and other biomaterials to achieve superior anti-inflammatory properties. Taken together, these findings suggest that MSC-derived EVs may be promising nanovesicles for the treatment of DFUs and that further investigations are required to facilitate clinical applications.

## 23.5 Stem Cell-Derived EVs for Skeletal System Regeneration

### 23.5.1 Stem Cell-Derived EVs in Bone Regeneration

Bone, accounting for approximately 14% of a person's total body weight, is composed of bone cells and extracellular matrix, making it the most important supporting organ in the human body. Bone regeneration is a complex, coordinated process that requires the interactions and cooperation of the bone matrix, bone marrow, and periosteum. The main challenges in bone regeneration treatment are osteoporosis, bone defects, infection, tumor resection, and other complex bone diseases. Stem cell-derived EVs are an important cell-free therapeutic tool and have attracted considerable attention in the field of bone tissue repair and regeneration.

The expression profiles of miRNAs secreted by human BMSCs change during osteogenic differentiation. The enrichment of Wnt signaling molecules is synchronized with the differential distribution of osteogenesis-related EV miRNAs, indicating that BMSC-derived EV miRNAs regulate the osteogenic differentiation function of the cells [58]. ESC-derived sEVs can alleviate age-related bone loss by suppressing BMSCs aging [59].

Better bone regeneration may be achieved after early treatment with EVs or their parental cells. For example, after treating ASCs with TNF-α, the produced EVs can promote the proliferation and osteogenic differentiation of primary osteoblasts [60]. When BMSC-derived EVs are modified with hypoxia inducible factor 1α (HIF1α), they can promote trabecular reconstruction and improve the microvessel density of the necrotic femoral head [61]. Pretreatment of BMSC-derived EVs with low-dose dimethyloxaloacetic acid results in stronger effects on bone regeneration and angiogenesis [62].

In addition to their direct effect on bone cells, EVs derived from stem cells can indirectly promote bone regeneration. Human MSC-derived EVs promote angiogenesis and mineral matrix deposition in bone tissues. Moreover, EVs obtained from human iPSC-derived MSCs can maintain the blood supply to the femoral head, increase capillary density, and prevent bone loss in a hormone-induced femoral head necrosis mouse model [63].

Bone tissue damage and regeneration usually occur in local areas. EVs loaded with novel nanomaterials are an efficient method for bone regeneration. Attaching EVs to β-tricalcium phosphate [64] or hydrogel [65] can also promote bone regeneration by loading stem cell-derived EVs.

### 23.5.2 Stem Cell-Derived EVs in Cartilage Regeneration

Cartilage tissue is responsible for reducing the vibration and friction of the joints during movement. However, there are no blood vessels, nerves, or lymphatic vessels in cartilage tissue, which limits their ability to regenerate. Currently, many studies have confirmed that MSC-derived EVs can effectively promote the repair and regeneration of cartilage tissue.

The direct action of MSC-derived EVs on chondrocytes triggers a series of positive responses. For example, infrapatellar fat pad MSC-derived *miR-100-5p*-enriched EVs can protect cartilage from osteoarthritis (OA) and ameliorate gait abnormalities by inhibiting the mammalian target of rapamycin pathway [66]. Intra-articular injection of ASC-derived EVs significantly protects articular cartilage from degeneration in both the monosodium iodoacetate rat model and the surgical destabilization of the medial meniscus mouse model [67]. Furthermore, CD73 carried by EVs secreted from human ESC-derived MSCs can regulate adenosine-activated AKT and extracellular signal-regulated kinase (ERK) signaling pathways, promote the proliferation and migration of chondrocytes, and inhibit apoptosis [68].

EVs have been proven to be effective in repairing cartilage damage and inhibiting OA; however, they are not suitable for long-term use in the joint cavity. The

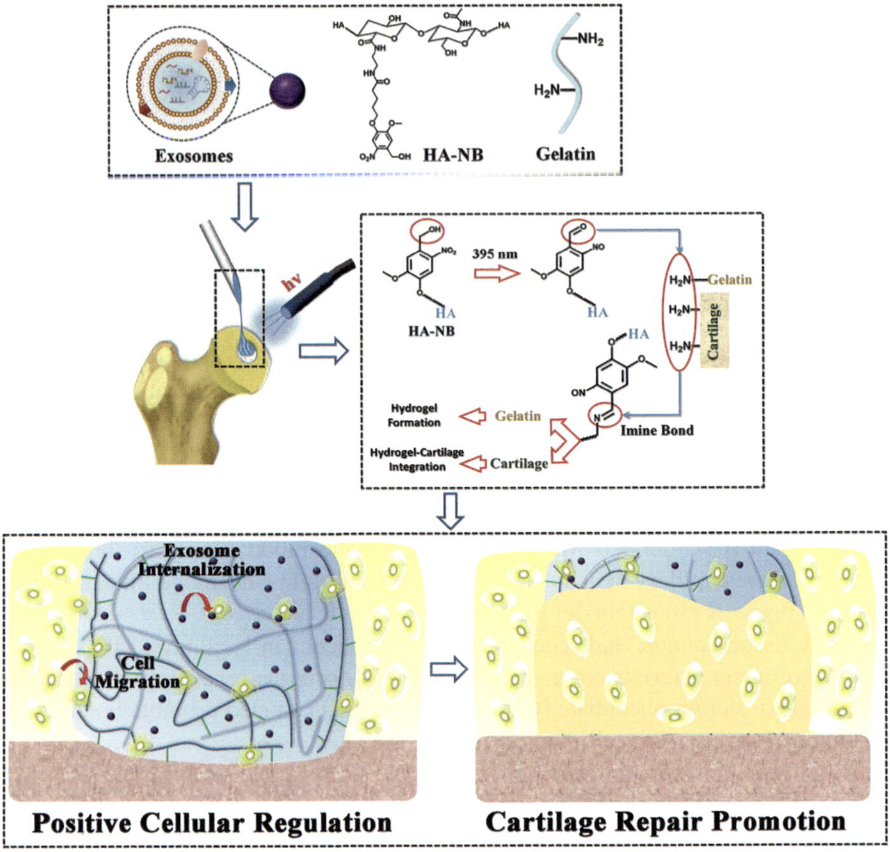


**Fig. 23.4** Construction of a phototriggered-imine crosslinked hydrogel-stem cell extracellular vesicle composite material and schematic diagram for repairing articular cartilage damage. *(Integration of stem cell-derived exosomes with* in situ *hydrogel glue as a promising tissue patch for articular cartilage regeneration. Nanoscale, 2017)*

application of hydrogels as a carrier of EVs provides a more efficient way to repair articular cartilage damage. A tissue-adhesive photocrosslinked hydrogel has been used as a carrier of MSC-derived EVs to repair rabbit articular cartilage damage (Fig. 23.4) [69]. Reversing the surface charge is also an effective strategy to increase intra-articular bioavailability for OA treatment. EVs modified with phosphatidylethanolamine show significantly enhanced cartilage uptake, cartilage penetration, and joint retention capacities compared with unmodified sEVs [70].

### 23.5.3 Stem Cell-Derived EVs Promote Angiogenesis in the Bone

Angiogenesis is the growth of new capillaries from a preexisting vascular network [71]. This important process occurs naturally during organic growth and in response to tissue injury. During tissue repair, angiogenesis is a vital step in restoring the tissue blood supply and supporting cells at the wound site with nutrition and oxygen to promote tissue regeneration and wound healing. Moreover, blood vessels can also supply calcium and phosphate, which are building blocks for mineralization in the bone tissue. Therefore, the vascular network also plays a vital role in bone development, remodeling, and homeostasis. Increasing evidence has demonstrated that enhanced angiogenesis may facilitate bone regeneration.

Recently, stem cell-derived EVs have been reported to play an effective role in tissue regeneration [71], partly due to their ability to promote angiogenesis, particularly in various angiogenesis-related diseases. EVs derived from various stem cell types may deliver pro-angiogenic factors to endothelial cells and induce an endogenous angiogenic response to promote tissue regeneration and repair [72]. As described above, angiogenesis is a critical step in bone regeneration; therefore, researchers have begun to focus on the function of stem cell-derived EVs in orthopedic disorders.

#### 23.5.3.1 Stem Cell-Derived EVs Promote Angiogenesis in Osteonecrosis of the Femoral Head (ONFH)

ONFH is one of the most common orthopedic diseases and the most important cause of total hip arthroplasty in young men and middle-aged people. Although the etiology of ONFH is complicated, local ischemia of the femoral head is the main pathological feature of ONFH. Currently, there is no effective therapy for clinically promoting angiogenesis in the femoral head. Several recent studies have shown that stem cell-derived EVs can promote angiogenesis in ONFH. Liu et al. indicated that EVs secreted from human iPSC-derived MSCs promote angiogenesis in a steroid-induced rat osteonecrosis model, thereby preventing ONFH [63]. The pro-angiogenic effects of these EVs were found to be mediated by activation of the PI3K/AKT signaling pathway in endothelial cells. Li et al. also demonstrated that BMSC-derived EVs can promote angiogenesis in a steroid-induced ONFH model [10]. Moreover, EVs derived from BMSCs with stabilized HIF1$\alpha$ effectively promote angiogenesis. Zuo et al. reported that human $CD34^+$ endothelial progenitor cell (EPC)-derived EVs promote angiogenesis and osteogenesis in a glucocorticoid-induced ONFH model. Upregulation of *miR-26* in EPC-derived EVs could enhance the ability of the EVs to prevent ONFH [73]. These studies indicate that stem cell-derived EVs may effectively prevent the pathogenesis of ONFH by promoting angiogenesis, and this effect can be enhanced by strengthening its pro-angiogenic capacity.

#### 23.5.3.2 Stem Cell-Derived EVs Promote Angiogenesis during Bone Repair

Fracture healing is a complicated physiological process orchestrated by a series of biological events [74]. Despite the remarkable regenerative capacity of bone tissue, fracture healing fails in approximately 5–10% of cases, leading to delayed union or nonunion and the development of large unhealed bone defects. Promoting blood flow to the fracture site has been shown to be essential for successful bone healing.

Zhang et al. reported that EVs derived from human umbilical cord MSCs could significantly enhance angiogenesis and promote bone healing in a rat model of femoral fractures. These pro-angiogenesis effects are mediated by the activation of HIF1α in endothelial cells [75]. Zhang et al. also showed that BMSC-derived EVs could promote angiogenesis and osteogenesis to enhance bone healing in a nonunion model [76]. Jia et al. showed that EPC-derived EVs could downregulate the expression of sprouty-related EVH1 domain containing 1 and activate ERK signaling by transferring *miR-126* to promote angiogenesis in distraction osteogenesis in a large bone defect rat model [77]. Furthermore, preconditioning of stem cells under hypoxic conditions [62, 78] or with bioactive materials [79] to elevate the pro-angiogenic cargo of EVs may enhance their bone healing function.

In conclusion, therapeutic angiogenesis is a promising strategy to enhance bone regeneration. EVs derived from stem cells can accelerate the proliferation, migration, and tube formation of endothelial cells, thereby further promoting angiogenesis and ultimately enhancing bone repair in patients with bone-related diseases. Stem cell-derived EVs have demonstrated great therapeutic potential in orthopedic diseases, and there is an urgent need to develop cell-free therapies for bone regeneration in the future.

## 23.6 Stem Cell-Derived EVs for Delayed Aging

EVs are a general term that encompasses various types of vesicles, including EVs, exosomes, microvesicles, and apoptotic bodies [80, 81]. The potential diagnostic and therapeutic applications of EVs in age-related diseases are being extensively studied [81, 82]. Notably, the EV concentration decreases with age, partly owing to increased EV internalization with aging. EVs contain biologically active molecules that can be delivered to the recipient cells. In order to check whether EV proteins change with age, researchers have analyzed the protein contents of EVs in different cohorts and found that with age, the levels of several apoptotic proteins are decreased, whereas the expression of cancer-related proteins is increased [83]. Additionally, many reports have demonstrated that stem cell-derived EVs are associated with various life processes.

### 23.6.1 Impact of EVs on Aging

Aging is an important risk factor for impaired tissue function and chronic disease. A decline in the availability of systemic nicotinamide adenine dinucleotide ($NAD^+$) plays a crucial role in the regulation of aging in multiple species. In a recent study, researchers found that the level of extracellular nicotinamide phosphoribosyl transferase (eNAMPT) decreased with age. Upregulation of eNAMPT can promote $NAD^+$ production, resist aging, and prolong health. $NAD^+$ levels decline with age at the systemic level, triggering various age-related pathophysiological changes in many model organisms. In mammals, NAMPT is the rate-limiting enzyme in the main $NAD^+$ biosynthetic pathway that converts nicotinamide and 50-phosphoribose pyrophosphate into nicotinamide mononucleotide. Injection of eNAMPT-containing EVs purified from young mice or cultured adipocytes can enhance the mobility of old mice and prolong their life span. These findings prove that this new physiological system mediated by EVs containing eNAMPT plays key roles in maintaining systemic $NAD^+$ biosynthesis and counteracting age-related physiological decline, implying that the eNAMPT contained in EVs may serve as a potential antiaging biological agent in humans [84]. In addition, MSC-derived EVs can slow the process of aging and delay the occurrence of cellular senescence-related diseases [85].

### 23.6.2 Impact of EVs on Cellular Senescence

Cell senescence is characterized by changes in autophagy and metabolic functions, irreversible cell cycle arrest, resistance to apoptosis, activation of DNA damage responses, and abnormal secretion of various pro-inflammatory cytokines, chemokines, growth factors and proteases [86]. Cellular senescence is defined as irreversible cell cycle arrest caused by various biological and pathological conditions [87]. As a strictly regulated process, cellular senescence plays vital roles in tissue homeostasis [88]. Increasing evidence has shown that the continued existence and accumulation of senescent cells are closely related to aging and lead to impaired tissue repair/regeneration potential, resulting in the occurrence of various aging-related diseases [88, 89]. EVs are formed by the fusion of multivesicular bodies in the cytoplasm and plasma membrane and are released from cells into the microenvironment [90]. In the microenvironment, these EVs promote communication between cells through the horizontal transfer of a large number of biologically active proteins and nonprotein molecules [90]. Increasing evidence has shown that EVs derived from BMSCs have similar therapeutic effects in various animal disease models and cell models [91]. MSC-derived exosomes or EVs have been shown to have beneficial effects on senescence or the senescent phenotype in human cells induced by different stress stimuli, suggesting their antiaging potential [92]. Mao et al. found that EVs released from gingiva-derived MSCs (GMSCs) had therapeutic effects on mouse sciatic nerve repair and rat taste bud regeneration [93, 94]. In addition, GMSC-derived EVs can effectively inhibit the senescence of endothelial cells and skin fibroblasts induced by oxidative stress because GMSC-derived EVs have strong inhibitory effects on the senescence of endothelial cells and skin fibroblasts induced by oxidative stress. Furthermore, GMSC-derived EVs may be a potential alternative source of acellular products to alleviate age-related skin and blood vessel dysfunction [95]. In addition, Hu et al. found that in the aging process, hippocampal NSC senescence leads to cell failure, decreased neurogenesis, and cognitive impairment and that ESC-derived sEVs can reverse this phenomenon. Therefore, ESC-derived sEVs may be promising drug candidates for the treatment of age-related diseases [96].

### 23.6.3 Impact of EVs on Mitochondrial Dysfunction

Gradual decreases in cell function and integrity, which clinically manifest as increased vulnerability to adverse consequences and death, are the core of biological aging. Mitochondrial dysfunction, altered intercellular communication, increased oxidative stress, reduced protein homeostasis, genome instability, telomere wear, epigenetic changes, nutrient changes, and stem cell failure are signs of aging. Destruction of the mitochondrial-lysosomal axis combined with the abnormal secretion of EVs may play roles in the pathogenesis of aging and certain disease conditions. In the context of incomplete mitochondrial depolarization, cells can be released into the extracellular cavity through specialized vesicles, thereby removing excess material and damaged organelles [97]. EVs can shuttle between cells, and the transportation of EV cargo has beneficial or harmful effects on neighboring cells, depending on the nature of the material being transported [97]. Therefore, EVs have been promoted as biomarkers to capture the complexity of aging. Cell quality control and EV communication can promote health and extend lifespan [98–103]. Parkinson's disease (PD) is the second most common neurodegenerative disease affecting elderly individuals [104]. The gradual disappearance of dopaminergic neurons in the substantia nigra compact area and the consumption of dopamine in the striatum are pathological features of PD [105], and mitochondrial dysfunction is the main factor in the onset of familial PD [106]. Age-related mitochondrial anomalies and oxidative stress are also conducive to abnormal protein folding and the accumulation of harmful protein aggregates, including α-synuclein [107]. Therefore, mitochondrial dysfunction and protein homeostasis are impaired and trigger PD neuronal dysfunction [108]. Neurodegeneration in PD involves systemic inflammation and mitochondrial dysfunction. EVs are a delivery system through which cells deliver or remove excess substances, and their transport may be associated with inflammation and mitochondrial dysfunction [109]. In addition, mitochondrial dysfunction is an important feature of tissue damage and stem cell aging, and MSC-derived EVs are expected to provide nanotherapeutics for mitochondrial damage-related diseases [110].

### 23.6.4 Impact of EVs on Stem Cell Exhaustion

Aging is an inevitable biological process characterized by the gradual loss of physiological function, leading to various age-related diseases [111]. In mammals, aging of tissue-resident stem cells is one of the main risk factors for aging [111, 112]. Aging tissue-resident stem cells exhibit self-renewal characteristics and reduced multilineage differentiation, which leads to a decline in tissue repair and regeneration [112, 113]. BMSCs are common progenitor cells of osteoblasts that play key roles in maintaining the balance of bone metabolism [114]. The proliferation and osteogenic differentiation of BMSCs are reduced in elderly individuals, and these cells exhibit a variety of aging-related phenotypes, such as cell cycle inhibitor (p16 and p21), aging-related β-galactosidase, and aging-related secretory phenotypes [115]. Similarly, BMSCs extracted from patients with osteoporosis exhibit several abnormalities consistent with aging, including reduced proliferation and mineralization [116]. Therefore, the senescence of BMSCs is considered the main factor driving the pathogenesis of age-related bone loss [117, 118]. In addition, some studies have reported that transplanting young BMSCs into elderly osteoporotic mice significantly slows the aging of BMSCs, suppresses the reduction in bone mass, and prolongs the life span of old mice [119, 120].

EVs derived from adult stem cells can exert effects similar to those of their parent cells in various disease models, including models of myocardial infarction and reperfusion injury [121], liver and kidney damage [41, 122, 123], inflammatory diseases [124], hind limb ischemia [125], and degenerative diseases [126–128]. In addition to adult stem cells, ESCs have also been shown to be a promising source of therapeutic EVs because of their unique ability to proliferate indefinitely and multipotentially [129]. In addition, ESC-derived EVs have no risk of tumorigenesis, and the possibility of immune rejection is very low [130]. Furthermore, EVs derived from ESCs promote regeneration and antiaging. For example, intramyocardial injection of exosomes from mouse ESCs can enhance the healing of myocardial infarction by increasing cardiomyocyte survival and promoting angiogenesis [131]. A study by Gong et al. showed that human ESC-derived sEVs are effective for combatting the aging of BMSCs, thereby preventing age-related bone loss [59].

### 23.6.5 Impact of EVs on Aging-Related Insulin Resistance

Type 2 diabetes [132] plays important roles in chronic metabolic disease. Its incidence and prevalence increase with age, making type 2 diabetes an age-related disease [133]. Type 2 diabetes is a health problem worldwide, affecting more than 4 million people, and complications include cardiovascular disease and damage to the nerves, kidneys, and eyes [133]. A cross-sectional cohort study found that patients with diabetes had higher circulating EV levels than those in the normal blood glucose control group. In addition, in the longitudinal cohort, patients with diabetes had higher circulating EV concentrations over an average of 5 years than those with normal blood glucose levels [134]. These data indicated that the higher circulating EV levels in diabetic individuals may be partly due to insulin resistance-driven EV secretion. Insulin resistance is the main pathological feature of type 2 diabetes, and elderly individuals often develop insulin resistance. EVs released by BMSCs of old mice can be absorbed by fat cells, cardiomyocytes, and liver cells, thereby causing insulin resistance in vivo and in vitro. The amount of *miR-29b-3p* in exosomes released from BMSCs of aging mice is increased significantly. In terms of mechanism, sirtiun 1 has been identified as a downstream target of *miR-29b-3p*, which can regulate insulin resistance. Downregulation of *miR-29b-3p* in exosomes derived from BMSCs can significantly improve insulin resistance in elderly mice. At the same time, the specific overexpression of *miR-29b-3p* in BMSCs induces insulin resistance in young mice. In summary, exosomal *miR-29b-3p* derived from BMSCs can regulate aging-related insulin resistance and can be used as a potential therapeutic target for aging-related insulin resistance [135].

## 23.7 Conclusion and Perspectives

As cell-free stem cell therapy, stem cell-derived EVs have demonstrated great advantages and potential in regenerative medicine. An increasing number of pre-clinical studies have demonstrated the effectiveness of stem cell-derived EVs in the treatment of various diseases. Remarkably, stem cell-derived EVs have also shown powerful antiaging effects and exhibit efficacy in many degenerative disorders. Because the number of individuals suffering from aging-related degenerative diseases continues to increase with aging society, these findings shed light on the health problems of the elderly worldwide. Based on the safety and effectiveness of EVs in preclinical studies, increasing attention has been paid to their clinical relevance. Hundreds of clinical trials for multiple diseases are currently being conducted worldwide. To realize the clinical applications of stem cell-derived EVs, several issues need to be addressed. Further technical advances in automated approaches are required for EV production, including the large-scale production of stem cells and the collection, extraction, and purification of EVs. Additionally, it will be necessary to establish quality control standards and clinical application specifications for EVs. Although there are still some hurdles to this modality, stem cell-derived EV therapy is expected to become widely used in modern medicine in the near future.

## References

1. Arslan F, Lai RC, Smeets MB, Akeroyd L, Choo A, Aguor EN, et al. Mesenchymal stem cell-derived exosomes increase ATP levels, decrease oxidative stress and activate PI3K/Akt pathway to enhance myocardial viability and prevent adverse remodeling after myocardial ischemia/reperfusion injury. Stem Cell Res. 2013;10(3):301–12.
2. Li X, Hacker M. Molecular imaging in stem cell-based therapies of cardiac diseases. Adv Drug Deliv Rev. 2017;120:71–88.
3. Mayourian J, Ceholski DK, Gorski PA, Mathiyalagan P, Murphy JF, Salazar SI, et al. Exosomal microRNA-21-5p mediates mesenchymal stem cell paracrine effects on human cardiac tissue contractility. Circ Res. 2018;122(7):933–44.
4. Luther KM, Haar L, McGuinness M, Wang Y, Lynch Iv TL, Phan A, et al. Exosomal miR-21a-5p mediates cardioprotection by mesenchymal stem cells. J Mol Cell Cardiol. 2018;119:125–37.
5. Xiao C, Wang K, Xu Y, Hu H, Zhang N, Wang Y, et al. Transplanted mesenchymal stem cells reduce autophagic flux in infarcted hearts via the exosomal transfer of miR-125b. Circ Res. 2018;123(5):564–78.
6. Zhu J, Lu K, Zhang N, Zhao Y, Ma Q, Shen J, et al. Myocardial reparative functions of exosomes from mesenchymal stem cells are enhanced by hypoxia treatment of the cells via transferring microRNA-210 in an nSMase2-dependent way. Artif Cells Nanomed Biotechnol. 2018;46(8):1659–70.
7. Khan M, Nickoloff E, Abramova T, Johnson J, Verma SK, Krishnamurthy P, et al. Embryonic stem cell-derived exosomes promote endogenous repair mechanisms and enhance cardiac function following myocardial infarction. Circ Res. 2015;117(1):52–64.
8. Wang Y, Zhang L, Li Y, Chen L, Wang X, Guo W, et al. Exosomes/microvesicles from induced pluripotent stem cells deliver cardioprotective miRNAs and prevent cardiomyocyte apoptosis in the ischemic myocardium. Int J Cardiol. 2015;192:61–9.

9. Esser JS, Saretzki E, Pankratz F, Engert B, Grundmann S, Bode C, et al. Bone morphogenetic protein 4 regulates microRNAs miR-494 and miR-126-5p in control of endothelial cell function in angiogenesis. Thromb Haemost. 2017;117(4):734–49.
10. Luo Q, Guo D, Liu G, Chen G, Hang M, Jin M. Exosomes from MiR-126-overexpressing Adscs are therapeutic in relieving acute myocardial ischaemic injury. Cell Physiol Biochem. 2017;44(6):2105–16.
11. Guo C, Sah JF, Beard L, Willson JK, Markowitz SD, Guda K. The noncoding RNA, miR-126, suppresses the growth of neoplastic cells by targeting phosphatidylinositol 3-kinase signaling and is frequently lost in colon cancers. Genes Chromosomes Cancer. 2008;47(11):939–46.
12. Fava C, Montagnana M. Atherosclerosis is an inflammatory disease which lacks a common anti-inflammatory therapy: how human genetics can help to this issue. A narrative review. Front Pharmacol. 2018;9:55.
13. Tian J, Liu Y, Liu Y, Chen K, Lyu S. Cellular and molecular mechanisms of diabetic atherosclerosis: herbal medicines as a potential therapeutic approach. Oxid Med Cell Longev. 2017;2017:9080869.
14. Takafuji Y, Hori M, Mizuno T, Harada-Shiba M. Humoral factors secreted from adipose tissue-derived mesenchymal stem cells ameliorate atherosclerosis in Ldlr−/− mice. Cardiovasc Res. 2019;115(6):1041–51.
15. Guo Z, Zhao Z, Yang C, Song C. Transfer of microRNA-221 from mesenchymal stem cell-derived extracellular vesicles inhibits atherosclerotic plaque formation. Transl Res. 2020;226:83–95.
16. Xing X, Li Z, Yang X, Li M, Liu C, Pang Y, et al. Adipose-derived mesenchymal stem cells-derived exosome-mediated microRNA-342-5p protects endothelial cells against atherosclerosis. Aging. 2020;12(4):3880–98.
17. Lindoso RS, Lopes JA, Binato R, Abdelhay E, Takiya CM, Miranda KR, et al. Adipose mesenchymal cells-derived EVs alleviate DOCA-salt-induced hypertension by promoting cardio-renal protection. Mol Ther Methods Clin Dev. 2020;16:63–77.
18. Feng R, Ullah M, Chen K, Ali Q, Lin Y, Sun Z. Stem cell-derived extracellular vesicles mitigate ageing-associated arterial stiffness and hypertension. J Extracell Vesicles. 2020;9(1):1783869.
19. Braun RK, Chetty C, Balasubramaniam V, Centanni R, Haraldsdottir K, Hematti P, et al. Intraperitoneal injection of MSC-derived exosomes prevent experimental bronchopulmonary dysplasia. Biochem Biophys Res Commun. 2018;503(4):2653–8.
20. Sajeesh S, Broekelman T, Mecham RP, Ramamurthi A. Stem cell derived extracellular vesicles for vascular elastic matrix regenerative repair. Acta Biomater. 2020;113:267–78.
21. Azizi F, Askari S, Javadpour P, Hadjighassem M, Ghasemi R. Potential role of exosome in post-stroke reorganization and/or neurodegeneration. EXCLI J. 2020;19:1590–606.
22. Xia Y, Ling X, Hu G, Zhu Q, Zhang J, Li Q, et al. Small extracellular vesicles secreted by human iPSC-derived MSC enhance angiogenesis through inhibiting STAT3-dependent autophagy in ischemic stroke. Stem Cell Res Ther. 2020;11(1):313.
23. Ling X, Zhang G, Xia Y, Zhu Q, Zhang J, Li Q, et al. Exosomes from human urine-derived stem cells enhanced neurogenesis via miR-26a/HDAC6 axis after ischaemic stroke. J Cell Mol Med. 2020;24(1):640–54.
24. Gelderblom M, Leypoldt F, Steinbach K, Behrens D, Choe CU, Siler DA, et al. Temporal and spatial dynamics of cerebral immune cell accumulation in stroke. Stroke. 2009;40(5):1849–57.
25. Liesz A, Suri-Payer E, Veltkamp C, Doerr H, Sommer C, Rivest S, et al. Regulatory T cells are key cerebroprotective immunomodulators in acute experimental stroke. Nat Med. 2009;15(2):192–9.
26. Li P, Gan Y, Sun BL, Zhang F, Lu B, Gao Y, et al. Adoptive regulatory T-cell therapy protects against cerebral ischemia. Ann Neurol. 2013;74(3):458–71.
27. Xia Y, Hu G, Chen Y, Yuan J, Zhang J, Wang S, et al. Embryonic stem cell derived small extracellular vesicles modulate regulatory T cells to protect against ischemic stroke. ACS Nano. 2021;15(4):7370–85.

28. Oh J, Vidal-Jordana A, Montalban X. Multiple sclerosis: clinical aspects. Curr Opin Neurol. 2018;31(6):752–9.
29. Vaughn CB, Jakimovski D, Kavak KS, Ramanathan M, Benedict RHB, Zivadinov R, et al. Epidemiology and treatment of multiple sclerosis in elderly populations. Nat Rev Neurol. 2019;15(6):329–42.
30. Riazifar M, Mohammadi MR, Ponc EJ, Ycri A, Lässer C, Segaliny AI, et al. Stem cell-derived exosomes as nanotherapeutics for autoimmune and neurodegenerative disorders. ACS Nano. 2019;13(6):6670–88.
31. Casella G, Colombo F, Finardi A, Descamps H, Ill-Raga G, Spinelli A, et al. Extracellular vesicles containing IL-4 modulate neuroinflammation in a mouse model of multiple sclerosis. Mol Ther. 2018;26(9):2107–18.
32. Lombardi M, Parolisi R, Scaroni F, Bonfanti E, Gualerzi A, Gabrielli M, et al. Detrimental and protective action of microglial extracellular vesicles on myelin lesions: astrocyte involvement in remyelination failure. Acta Neuropathol. 2019;138(6):987–1012.
33. Osorio-Querejeta I, Carregal-Romero S, Ayerdi-Izquierdo A, Mäger I, et al. MiR-219a-5p enriched extracellular vesicles induce OPC differentiation and eae improvement more efficiently than liposomes and polymeric nanoparticles. Pharmaceutics. 2020;12(2)
34. Zhong D, Cao Y, Li CJ, Li M, Rong ZJ, Jiang L, et al. Neural stem cell-derived exosomes facilitate spinal cord functional recovery after injury by promoting angiogenesis. Exp Biol Med (Maywood). 2020;245(1):54–65.
35. Liu W, Rong Y, Wang J, Zhou Z, Ge X, Ji C, et al. Exosome-shuttled miR-216a-5p from hypoxic preconditioned mesenchymal stem cells repair traumatic spinal cord injury by shifting microglial M1/M2 polarization. J Neuroinflammation. 2020;17(1):47.
36. de Godoy MA, Saraiva LM, de Carvalho LRP, Vasconcelos-Dos-Santos A, Beiral HJV, Ramos AB, et al. Mesenchymal stem cells and cell-derived extracellular vesicles protect hippocampal neurons from oxidative stress and synapse damage induced by amyloid-β oligomers. J Biol Chem. 2018;293(6):1957–75.
37. Xin H, Li Y, Buller B, Katakowski M, Zhang Y, Wang X, et al. Exosome-mediated transfer of miR-133b from multipotent mesenchymal stromal cells to neural cells contributes to neurite outgrowth. Stem Cells. 2012;30(7):1556–64.
38. Hu G, Xia Y, Zhang J, Chen Y, Yuan J, Niu X, et al. ESC-sEVs rejuvenate senescent hippocampal NSCs by activating lysosomes to improve cognitive dysfunction in vascular dementia. Adv Sci (Weinh). 2020;7(10):1903330.
39. Hu G, Xia Y, Chen B, Zhang J, Gong L, Chen Y, et al. ESC-sEVs rejuvenate aging hippocampal NSCs by transferring SMADs to regulate the MYT1-Egln3-Sirt1 axis. Mol Ther. 2021;29(1):103–20.
40. Meraz-Munoz A, Langote A, Jhaveri KD, Izzedine H, Gudsoorkar P. Acute kidney injury in the patient with cancer Diagnostics (Basel, Switzerland). 2021;11(4):611.
41. Bruno S, Grange C, Deregibus MC, Calogero RA, Saviozzi S, Collino F, et al. Mesenchymal stem cell-derived microvesicles protect against acute tubular injury. J Am Soc Nephrol JASN. 2009;20(5):1053–67.
42. Wang J, Jia H, Zhang B, Yin L, Mao F, Yu J, et al. HucMSC exosome-transported 14–3-3ζ prevents the injury of cisplatin to HK-2 cells by inducing autophagy in vitro. Cytotherapy. 2018;20(1):29–44.
43. Wang B, Yao K, Huuskes BM, Shen HH, Zhuang J, Godson C, et al. Mesenchymal stem cells deliver exogenous microRNA-let7c via exosomes to attenuate renal fibrosis. Mol Ther. 2016;24(7):1290–301.
44. Shen B, Liu J, Zhang F, Wang Y, Qin Y, Zhou Z, et al. CCR2 positive exosome released by mesenchymal stem cells suppresses macrophage functions and alleviates ischemia/reperfusion-induced renal injury. Stem Cells Int. 2016;2016:1240301.
45. Tang TT, Wang B, Wu M, Li ZL, Feng Y, Cao JY, et al. Extracellular vesicle-encapsulated IL-10 as novel nanotherapeutics against ischemic AKI. Sci Adv. 2020;6(33):eaaz0748.
46. Eirin A, Zhu XY, Puranik AS, Tang H, McGurren KA, van Wijnen AJ, et al. Mesenchymal stem cell-derived extracellular vesicles attenuate kidney inflammation. Kidney Int. 2017;92(1):114–24.

47. Ji C, Zhang J, Zhu Y, Shi H, Yin S, Sun F, et al. Exosomes derived from hucMSC attenuate renal fibrosis through CK1δ/β-TRCP-mediated YAP degradation. Cell Death Dis. 2020;11(5):327.
48. Cho NH, Shaw JE, Karuranga S, Huang Y, da Rocha Fernandes JD, Ohlrogge AW, et al. IDF Diabetes Atlas: Global estimates of diabetes prevalence for 2017 and projections for 2045. Diabetes Res Clin Pract. 2018;138:271–81.
49. Sun Y, Shi H, Yin S, Ji C, Zhang X, Zhang B, et al. Human mesenchymal stem cell derived exosomes alleviate type 2 diabetes mellitus by reversing peripheral insulin resistance and relieving β-cell destruction. ACS Nano. 2018;12(8):7613–28.
50. Tan Y, Zhang Z, Zheng C, Wintergerst KA, Keller BB, Cai L. Mechanisms of diabetic cardiomyopathy and potential therapeutic strategies: preclinical and clinical evidence. Nat Rev Cardiol. 2020;17(9):585–607.
51. Hu J, Dziumbla S, Lin J, Bibli SI, Zukunft S, de Mos J, et al. Inhibition of soluble epoxide hydrolase prevents diabetic retinopathy. Nature. 2017;552(7684):248–52.
52. Cao X, Xue LD, Di Y, Li T, Tian YJ, Song Y. MSC-derived exosomal lncRNA SNHG7 suppresses endothelial-mesenchymal transition and tube formation in diabetic retinopathy via miR-34a-5p/XBP1 axis. Life Sci. 2021;272:119232.
53. Ni W, Fang Y, Xie L, Liu X, Shan W, Zeng R, et al. Adipose-derived mesenchymal stem cells transplantation alleviates renal injury in streptozotocin-induced diabetic nephropathy. J Histochem Cytochem. 2015;63(11):842–53.
54. Kwon S, Kim YC, Park JY, Lee J, An JN, Kim CT, et al. The long-term effects of metformin on patients with type 2 diabetic kidney disease. Diabetes Care. 2020;43(5):948–55.
55. Jin L, Zhang J, Deng Z, Liu J, Han W, Chen G, et al. Mesenchymal stem cells ameliorate myocardial fibrosis in diabetic cardiomyopathy via the secretion of prostaglandin E2. Stem Cell Res Ther. 2020;11(1):122.
56. Hu Y, Rao SS, Wang ZX, Cao J, Tan YJ, Luo J, et al. Exosomes from human umbilical cord blood accelerate cutaneous wound healing through miR-21–3p-mediated promotion of angiogenesis and fibroblast function. Theranostics. 2018;8(1):169–84.
57. Casado-Díaz A, Quesada-Gómez JM, Dorado G. Extracellular vesicles derived from mesenchymal stem cells (MSC) in regenerative medicine: applications in skin wound healing. Front Bioeng Biotechnol. 2020;8:146.
58. Xu JF, Yang GH, Pan XH, Zhang SJ, Zhao C, Qiu BS, et al. Altered microRNA expression profile in exosomes during osteogenic differentiation of human bone marrow-derived mesenchymal stem cells. PloS One. 2014;9(12):e114627.
59. Gong L, Chen B, Zhang J, Sun Y, Yuan J, Niu X, et al. Human ESC-sEVs alleviate age-related bone loss by rejuvenating senescent bone marrow-derived mesenchymal stem cells. J extracell Vesicles. 2020;9(1):1800971.
60. Lu Z, Chen Y, Dunstan C, Roohani-Esfahani S, Zreiqat H. Priming adipose stem cells with tumor necrosis factor-alpha preconditioning potentiates their exosome efficacy for bone regeneration. Tissue Eng Part A. 2017;23(21–22):1212–20.
61. Li H, Liu D, Li C, Zhou S, Tian D, Xiao D, et al. Exosomes secreted from mutant-HIF-1alpha-modified bone-marrow-derived mesenchymal stem cells attenuate early steroid-induced avascular necrosis of femoral head in rabbit. Cell Biol Int. 2017;41(12):1379–90.
62. Liang B, Liang JM, Ding JN, Xu J, Xu JG, Chai YM. Dimethyloxaloylglycine-stimulated human bone marrow mesenchymal stem cell-derived exosomes enhance bone regeneration through angiogenesis by targeting the AKT/mTOR pathway. Stem Cell Res Ther. 2019;10(1):335.
63. Liu X, Li Q, Niu X, Hu B, Chen S, Song W, et al. Exosomes secreted from human-induced pluripotent stem cell-derived mesenchymal stem cells prevent osteonecrosis of the femoral head by promoting angiogenesis. Int J Biol Sci. 2017;13(2):232–44.
64. Zhang J, Liu X, Li H, Chen C, Hu B, Niu X, et al. Exosomes/tricalcium phosphate combination scaffolds can enhance bone regeneration by activating the PI3K/Akt signaling pathway. Stem Cell Res Ther. 2016;7(1):136.

65. Chen S, Tang Y, Liu Y, Zhang P, Lv L, Zhang X, et al. Exosomes derived from miR-375-overexpressing human adipose mesenchymal stem cells promote bone regeneration. Cell Prolif. 2019;52(5):e12669.
66. Wu J, Kuang L, Chen C, Yang J, Zeng WN, Li T, et al. miR-100-5p-abundant exosomes derived from infrapatellar fat pad MSCs protect articular cartilage and ameliorate gait abnormalities via inhibition of mTOR in osteoarthritis. Biomaterials. 2019;206:87–100.
67. Woo CH, Kim HK, Jung GY, Jung YJ, Lee KS, Yun YE, et al. Small extracellular vesicles from human adipose-derived stem cells attenuate cartilage degeneration. J Extracell Vesicles. 2020;9(1):1735249.
68. Li S, Zhao Y, Chen W, Yin L, Zhu J, Zhang H, et al. Exosomal ephrinA2 derived from serum as a potential biomarker for prostate cancer. J Cancer. 2018;9(15):2659–65.
69. Liu X, Yang Y, Li Y, Niu X, Zhao B, Wang Y, et al. Integration of stem cell-derived exosomes with in situ hydrogel glue as a promising tissue patch for articular cartilage regeneration. Nanoscale. 2017;9(13):4430–8.
70. Feng K, Xie X, Yuan J, Gong L, Zhu Z, Zhang J, et al. Reversing the surface charge of MSC-derived small extracellular vesicles by εPL-PEG-DSPE for enhanced osteoarthritis treatment. J Extracell Vesicles. 2021;10(13):e12160.
71. Todorova D, Simoncini S, Lacroix R, Sabatier F, Dignat-George F. Extracellular vesicles in angiogenesis. Circ Res. 2017;120(10):1658–73.
72. Shi Y, Shi H, Nomi A, Lei-Lei Z, Zhang B, Qian H. Mesenchymal stem cell-derived extracellular vesicles: a new impetus of promoting angiogenesis in tissue regeneration. Cytotherapy. 2019;21(5):497–508.
73. Zuo R, Kong L, Wang M, Wang W, Xu J, Chai Y, et al. Exosomes derived from human CD34(+) stem cells transfected with miR-26a prevent glucocorticoid-induced osteonecrosis of the femoral head by promoting angiogenesis and osteogenesis. Stem Cell Res Ther. 2019;10(1):321.
74. Einhorn TA, Gerstenfeld LC. Fracture healing: mechanisms and interventions. Nat Rev Rheumatol. 2015;11(1):45–54.
75. Zhang Y, Hao Z, Wang P, Xia Y, Wu J, Xia D, et al. Exosomes from human umbilical cord mesenchymal stem cells enhance fracture healing through HIF-1alpha-mediated promotion of angiogenesis in a rat model of stabilized fracture. Cell Prolif. 2019;52(2):e12570.
76. Zhang L, Jiao G, Ren S, Zhang X, Li C, Wu W, et al. Exosomes from bone marrow mesenchymal stem cells enhance fracture healing through the promotion of osteogenesis and angiogenesis in a rat model of nonunion. Stem Cell Res Ther. 2020;11(1):38.
77. Jia Y, Zhu Y, Qiu S, Xu J, Chai Y. Exosomes secreted by endothelial progenitor cells accelerate bone regeneration during distraction osteogenesis by stimulating angiogenesis. Stem Cell Res Ther. 2019;10(1):12.
78. Liu W, Li L, Rong Y, Qian D, Chen J, Zhou Z, et al. Hypoxic mesenchymal stem cell-derived exosomes promote bone fracture healing by the transfer of miR-126. Acta Biomater. 2020;103:196–212.
79. Liu L, Yu F, Li L, Zhou L, Zhou T, Xu Y, et al. Bone marrow stromal cells stimulated by strontium-substituted calcium silicate ceramics: release of exosomal miR-146a regulates osteogenesis and angiogenesis. Acta Biomater. 2021;119:444–57.
80. Colombo M, Raposo G, Théry C. Biogenesis, secretion, and intercellular interactions of exosomes and other extracellular vesicles. Annu Rev Cell Dev Biol. 2014;30:255–89.
81. Yáñez-Mó M, Siljander PR, Andreu Z, Zavec AB, Borràs FE, Buzas EI, et al. Biological properties of extracellular vesicles and their physiological functions. J extracell Vesicles. 2015;4:27,066.
82. Kalluri R, LeBleu VS. The biology, function, and biomedical applications of exosomes Science (New York, NY). 2020;367(6478).
83. Qin Y, Zhang C. Endothelial progenitor cell-derived extracellular vesicle-meditated cell-to-cell communication regulates the proliferation and osteoblastic differentiation of bone mesenchymal stromal cells. Mol Med Rep. 2017;16(5):7018–24.
84. Yoshida M, Satoh A, Lin JB, Mills KF, Sasaki Y, Rensing N, et al. Extracellular vesicle-contained eNAMPT delays aging and extends lifespan in mice. Cell Metab. 2019;30(2):329–42.e5.

85. Dorronsoro A, Santiago FE, Grassi D, Zhang T, Lai RC, McGowan SJ, et al. Mesenchymal stem cell-derived extracellular vesicles reduce senescence and extend health span in mouse models of aging. Aging Cell. 2021;20(4):e13337.
86. Gorgoulis V, Adams PD, Alimonti A, Bennett DC, Bischof O, Bishop C, et al. Cellular senescence: defining a path forward. Cell. 2019;179(4):813–27.
87. Childs BG, Durik M, Baker DJ, van Deursen JM. Cellular senescence in aging and age-related disease: from mechanisms to therapy. Nat Med. 2015;21(12):1424–35.
88. Rhinn M, Ritschka B, Keyes WM. Cellular senescence in development, regeneration and disease Development (Cambridge, England). 2019;146(20).
89. Gruber F, Kremslehner C, Eckhart L, Tschachler E. Cell aging and cellular senescence in skin aging - Recent advances in fibroblast and keratinocyte biology. Exp Gerontol. 2020;130:110780.
90. Phinney DGPM. Corrigendum to: concise review: MSC-derived exosomes for cell-free therapy. Stem cells (Dayton, Ohio). 2017;35(9):2103.
91. Furuta T, Miyaki S, Ishitobi H, Ogura T, Kato Y, Kamei N, et al. Mesenchymal stem cell-derived exosomes promote fracture healing in a mouse model. Stem Cells Transl Med. 2016;5(12):1620–30.
92. Liu S, Mahairaki V, Bai H, Ding Z, Li J, Witwer KW, et al. Highly purified human extracellular vesicles produced by stem cells alleviate aging cellular phenotypes of senescent human cells. Stem cells (Dayton, Ohio). 2019;37(6):779–90.
93. Mao Q, Nguyen PD, Shanti RM, Shi S, Shakoori P, Zhang Q, et al. Gingiva-derived mesenchymal stem cell-extracellular vesicles activate schwann cell repair phenotype and promote nerve regeneration. Tissue Eng Part A. 2019;25(11–12):887–900.
94. Zhang Y, Shi S, Xu Q, Zhang Q, Shanti RM, Le AD. SIS-ECM laden with GMSC-derived exosomes promote taste bud regeneration. J Dent Res. 2019;98(2):225–33.
95. Shi HZ, Zeng JC, Shi SH, Giannakopoulos H, Zhang QZ, Le AD. Extracellular vesicles of GMSCs alleviate aging-related cell senescence. J Dent Res. 2021;100:283–92.
96. Hu G, Xia Y, Chen B, Zhang J, Gong L, Chen Y, et al. ESC-sEVs rejuvenate aging hippocampal NSCs by transferring SMADs to regulate the MYT1-Egln3-Sirt1 axis. Mol Ther. 2021;29:103–20.
97. Baixauli F, López-Otín C, Mittelbrunn M. Exosomes and autophagy: coordinated mechanisms for the maintenance of cellular fitness. Front Immunol. 2014;5:403.
98. Burch JB, Augustine AD, Frieden LA, Hadley E, Howcroft TK, Johnson R, et al. Advances in geroscience: impact on healthspan and chronic disease. J Gerontol A Biol Sci Med Sci. 2014;69(Suppl 1):S1–3.
99. Espeland MA, Crimmins EM, Grossardt BR, Crandall JP, Gelfond JA, Harris TB, et al. Clinical trials targeting aging and age-related multimorbidity. J Gerontol A Biol Sci Med Sci. 2017;72(3):355–61.
100. Barzilai N, Crandall JP, Kritchevsky SB, Espeland MA. Metformin as a tool to target aging. Cell Metab. 2016;23(6):1060–5.
101. Sierra F. The emergence of geroscience as an interdisciplinary approach to the enhancement of health span and life span. Cold Spring Harb Perspect Med. 2016;6(4):a025163.
102. Longo VD, Antebi A, Bartke A, Barzilai N, Brown-Borg HM, Caruso C, et al. Interventions to slow aging in humans: are we ready? Aging Cell. 2015;14(4):497–510.
103. Kennedy BK, Berger SL, Brunet A, Campisi J, Cuervo AM, Epel ES, et al. Geroscience: linking aging to chronic disease. Cell. 2014;159(4):709–13.
104. Brooks DJ. The early diagnosis of Parkinson's disease. Ann Neurol. 1998;44(3 Suppl 1):S10–8.
105. Alexander GE. Biology of Parkinson's disease: pathogenesis and pathophysiology of a multisystem neurodegenerative disorder. Dialogues Clin Neurosci. 2004;6(3):259–80.
106. Park JS, Davis RL, Sue CM. Mitochondrial dysfunction in Parkinson's disease: new mechanistic insights and therapeutic perspectives. Curr Neurol Neurosci Rep. 2018;18(5):21.
107. Melo TQ, Copray S, Ferrari MFR. Alpha-synuclein toxicity on protein quality control, mitochondria and endoplasmic reticulum. Neurochem Res. 2018;43(12):2212–23.

108. Cho B, Kim T, Huh YJ, Lee J, Lee YI. Amelioration of mitochondrial quality control and proteostasis by natural compounds in Parkinson's disease models. Int J Mol Sci. 2019;20(20)
109. Picca A, Guerra F, Calvani R, Marini F, Biancolillo A, Landi G, et al. Mitochondrial signatures in circulating extracellular vesicles of older adults with Parkinson's disease: results from the EXosomes in PArkiNson's Disease (EXPAND) study. J Clin Med. 2020;9(2)
110. Zhao M, Liu S, Wang C, Wang Y, Wan M, Liu F, et al. Mesenchymal stem cell-derived extracellular vesicles attenuate mitochondrial damage and inflammation by stabilizing mitochondrial DNA. ACS Nano. 2021;15(1):1519–38.
111. López-Otín C, Blasco MA, Partridge L, Serrano M, Kroemer G. The hallmarks of aging. Cell. 2013;153(6):1194–217.
112. Goodell MA, Rando TA. Stem cells and healthy aging. Science (New York, NY). 2015;350(6265):1199–204.
113. Oh J, Lee YD, Wagers AJ. Stem cell aging: mechanisms, regulators and therapeutic opportunities. Nat Med. 2014;20(8):870–80.
114. Pittenger MF, Mackay AM, Beck SC, Jaiswal RK, Douglas R, Mosca JD, et al. Multilineage potential of adult human mesenchymal stem cells. Science (New York, NY). 1999;284(5411):143–7.
115. Zhou S, Greenberger JS, Epperly MW, Goff JP, Adler C, Leboff MS, et al. Age-related intrinsic changes in human bone-marrow-derived mesenchymal stem cells and their differentiation to osteoblasts. Aging Cell. 2008;7(3):335–43.
116. Yu B, Wang CY. Osteoporosis: the result of an 'Aged' bone microenvironment. Trends Mol Med. 2016;22(8):641–4.
117. Kim M, Kim C, Choi YS, Kim M, Park C, Suh Y. Age-related alterations in mesenchymal stem cells related to shift in differentiation from osteogenic to adipogenic potential: implication to age-associated bone diseases and defects. Mech Ageing Dev. 2012;133(5):215–25.
118. Ganguly P, El-Jawhari JJ, Giannoudis PV, Burska AN, Ponchel F, Jones EA. Age-related changes in bone marrow mesenchymal stromal cells: a potential impact on osteoporosis and osteoarthritis development. Cell Transplant. 2017;26(9):1520–9.
119. Liu S, Liu D, Chen C, Hamamura K, Moshaverinia A, Yang R, et al. MSC transplantation improves osteopenia via epigenetic regulation of notch signaling in lupus. Cell Metab. 2015;22(4):606–18.
120. Kiernan J, Hu S, Grynpas MD, Davies JE, Stanford WL. Systemic mesenchymal stromal cell transplantation prevents functional bone loss in a mouse model of age-related osteoporosis. Stem Cells Transl Med. 2016;5(5):683–93.
121. Lai RC, Arslan F, Lee MM, Sze NS, Choo A, Chen TS, et al. Exosome secreted by MSC reduces myocardial ischemia/reperfusion injury. Stem Cell Res. 2010;4(3):214–22.
122. Gatti S, Bruno S, Deregibus MC, Sordi A, Cantaluppi V, Tetta C, et al. Microvesicles derived from human adult mesenchymal stem cells protect against ischaemia-reperfusion-induced acute and chronic kidney injury. Nephrol Dial Transplant. 2011;26(5):1474–83.
123. Haga H, Yan IK, Takahashi K, Matsuda A, Patel T. Extracellular vesicles from bone marrow-derived mesenchymal stem cells improve survival from lethal hepatic failure in mice. Stem Cells Transl Med. 2017;6(4):1262–72.
124. Cosenza S, Toupet K, Maumus M, Luz-Crawford P, Blanc-Brude O, Jorgensen C, et al. Mesenchymal stem cells-derived exosomes are more immunosuppressive than microparticles in inflammatory arthritis. Theranostics. 2018;8(5):1399–410.
125. Hu GW, Li Q, Niu X, Hu B, Liu J, Zhou SM, et al. Exosomes secreted by human-induced pluripotent stem cell-derived mesenchymal stem cells attenuate limb ischemia by promoting angiogenesis in mice. Stem Cell Res Ther. 2015;6(1):10.
126. Cosenza S, Ruiz M, Toupet K, Jorgensen C, Noël D. Mesenchymal stem cells derived exosomes and microparticles protect cartilage and bone from degradation in osteoarthritis. Sci Rep. 2017;7(1):16,214.
127. Zhang S, Chu WC, Lai RC, Lim SK, Hui JH, Toh WS. Exosomes derived from human embryonic mesenchymal stem cells promote osteochondral regeneration. Osteoarthr Cartil. 2016;24(12):2135–40.

128. Qi X, Zhang J, Yuan H, Xu Z, Li Q, Niu X, et al. Exosomes secreted by human-induced pluripotent stem cell-derived mesenchymal stem cells repair critical-sized bone defects through enhanced angiogenesis and osteogenesis in osteoporotic rats. Int J Biol Sci. 2016;12(7):836–49.
129. Yabut O, Bernstein HS. The promise of human embryonic stem cells in aging-associated diseases. Aging. 2011;3(5):494–508.
130. Basu J, Ludlow JW. Exosomes for repair, regeneration and rejuvenation. Expert Opin Biol Ther. 2016;16(4):489–506.
131. Ratajczak J, Miekus K, Kucia M, Zhang J, Reca R, Dvorak P, et al. Embryonic stem cell-derived microvesicles reprogram hematopoietic progenitors: evidence for horizontal transfer of mRNA and protein delivery. Leukemia. 2006;20(5):847–56.
132. Noren Hooten N, Evans MK. Extracellular vesicles as signaling mediators in type 2 diabetes mellitus. Am J Physiol Cell Physiol. 2020;318(6):C1189–c99.
133. Zheng Y, Ley SH, Hu FB. Global aetiology and epidemiology of type 2 diabetes mellitus and its complications. Nat Rev Endocrinol. 2018;14(2):88–98.
134. Freeman DW, Noren Hooten N, Eitan E, Green J, Mode NA, Bodogai M, et al. Altered extracellular vesicle concentration, cargo, and function in diabetes. Diabetes. 2018;67(11):2377–88.
135. Su T, Xiao Y, Xiao Y, Guo Q, Li C, Huang Y, et al. Bone marrow mesenchymal stem cells-derived exosomal MiR-29b-3p regulates aging-associated insulin resistance. ACS Nano. 2019;13(2):2450–62.

# Extracellular Vesicles in Cancer Immunotherapy

24

Xiurong Ke, Wijnand Helfrich, and Hao Zhang

## 24.1 Introduction

Extracellular vesicles (EVs) are secreted by numerous cell types, including malignant tumor cells. Typically, EV contents reflect those of the cells of origin. Cancer cell-derived EVs are enriched in cancer-associated/cancer-specific antigens that can be taken up by antigen-presenting cells (APCs) and presented to antitumor T cells. Therefore, cancer cell-derived EV-based vaccination represents an attractive prophylactic or therapeutic approach for cancer immunotherapy. Immune cell-derived EVs may modulate various functions of multiple types of immune cells in either an immunostimulatory or immunosuppressive manner, depending mostly on the characteristics and functions of the immune cell type of origin. For example, EVs derived from dendritic cells (DCs) carry surface major histocompatibility complex (MHC) class I and class II molecules that can stimulate $CD8^+$ and $CD4^+$ T cells, respectively. These features make DC-derived EVs ideal candidates for the development of cancer vaccines to induce the T-cell-mediated killing of cancer cells. However, immunosuppressive EVs derived from cancer cells or other immune suppressor cells may impede antitumor immune responses. In this chapter, we focus on the potential

X. Ke
Department of Surgery, Laboratory for Translational Surgical Oncology, University of Groningen, University Medical Center Groningen, Groningen, The Netherlands

Shantou University Medical College, Shantou, Guangdong, China

W. Helfrich
Department of Surgery, Laboratory for Translational Surgical Oncology, University of Groningen, University Medical Center Groningen, Groningen, The Netherlands

H. Zhang (✉)
Institute of Precision Cancer Medicine and Pathology, Jinan University Medical College, Guangzhou, Guangdong, China



Q. Wang, L. Zheng (eds.), *Extracellular Vesicles*,
https://doi.org/10.1007/978-981-99-8365-0_24

applications of EVs in cancer immunotherapy, particularly in cancer vaccination. Moreover, the advantages and challenges of this application are discussed.

## 24.2 Advantages of EVs in Cancer Immunotherapy

Cancer immunotherapy aims to initiate an effective immune response to eliminate malignant cells. In recent decades, many advancements have been made in the field of cancer immunotherapy, particularly with regard to chimeric antigen receptor-modified T cells and the development of immune checkpoint blocking strategies. These advancements support the idea that immunotherapy is coming of age and will lead to durable and long-lasting responses, even in patients with advanced and otherwise incurable malignant disease. However, for most immunotherapeutic approaches, the key challenge is to enhance potency and mitigate cytotoxic side effects. Accordingly, EVs appear to provide a promising platform for combining the high potency of cancer immunotherapy with the benefits of nanomedicine [1].

The possibility of using EVs for cancer immunotherapy was first described in 1998, and many researchers have pursued this concept enthusiastically. Currently, EV-based immunotherapy appears to focus mainly on the use of EVs as cell-free vaccines [2]. The advantages of using EVs in this context are related to the fact that EVs are nonliving nanoscale membrane vesicles. Importantly, for EVs to be therapeutically active, it is not required to preserve "viability." Compared with cell-based immunotherapeutic approaches, the good manufacturing practice handling, scalability, and shelf life of EV-based approaches are relatively simple and cost effective. Moreover, the lipid bilayer of EVs contributes to their stability in serum when introduced into circulation. Additionally, a recent study demonstrated that the surface display of antiphagocytic molecules on EVs, such as CD47, reduces phagocytic removal by circulating myeloid cells, which may significantly increase their plasma half-life [3].

Cellular immunotherapy remains challenging, particularly for the treatment of solid tumors. This is partially due to the physical barriers of solid cancers that appear to curb therapy-harnessed immune cells from infiltrating the tumor site. By contrast, nanosized EVs show an enhanced capacity to overcome these physical barriers. Collectively, these studies support that EVs may have applications in cancer immunotherapy by improving treatment efficacy in solid tumors.

## 24.3 Cancer Vaccines and EV-Based Cancer Vaccination

Vaccination is an effective and safe method to prevent and treat infectious diseases by inducing protective immunity against pathogenic microorganisms. Certain types of cancer are etiologically related to viral pathogens; therefore, considerable efforts have been made to develop prophylactic and therapeutic vaccines against these oncogenic viruses. Currently, prophylactic cancer vaccines directed toward human papillomavirus (HPV) and hepatitis B virus (HBV) are the only two approved by the

U.S. Food and Drug Administration (FDA) [4, 5]. These vaccines aim to protect healthy individuals from infections by these viruses, which are known to be related to cervical cancer and liver cancer, respectively. However, the etiology of most human cancers is not associated with viral infections.

Currently, the cell-based Sipuleucel-T (PROVENGE) vaccine is the only therapeutic vaccine approved by the FDA [6]. Sipuleucel-T has been approved for the treatment of patients with metastatic prostate cancer. The complex Sipuleucel-T procedure includes extraction of APCs from patients with cancer using leukapheresis and in vitro maturation of these APCs into DCs. Subsequently, the DCs are sent to a production facility, where they are incubated with a recombinant fusion protein (PA2024) consisting of prostatic acid phosphatase (PAP) and granulocyte macrophage colony-stimulating factor (GM-CSF). The product (APC8015) is then returned from the production facility to the hospital and reinfused into the patient with prostate cancer. Complete Sipuleucel-T treatment includes three courses at 2-week intervals [6]. Unfortunately, a recent evaluation of Sipuleucel-T in phase III trials yielded rather disappointing results [7, 8].

Importantly, the abovementioned prophylactic and therapeutic vaccines are not authentic anticancer vaccines because they are not designed to induce cancer-selective immunity. In this respect, authentic cancer vaccines are currently under development to mount immune responses toward cancer-restricted neoantigens. Despite significant improvements, the clinical applications of prophylactic and therapeutic cancer vaccines remain limited.

Within the past 20 years, accumulating evidence has confirmed the clinical potential of tumor- and DC-derived EV-based cancer vaccination. Most strategies focus on the use of small EVs (sEVs) with diameters less than 200 nm. These sEVs are referred to as exosomes. Below, we discuss cancer cell-derived sEVs and DC-derived sEVs for cancer vaccination in greater detail.

### 24.3.1 Cancer Cell-Derived sEVs for Cancer Vaccination

Cancer cell-derived sEVs interact extensively with immune cells and have been shown to either suppress or promote anticancer immunity depending on the context of the tumor microenvironment (TME) [9]. Moreover, sEVs secreted by cancer cells may contain and/or expose cancer-specific oncogenic proteins and neoantigens as well as MHC class I molecules [10]. Together, these results suggest that cancer cell-derived sEVs could be exploited as cancer vaccine candidates (Fig. 24.1) [11].

The first study indicating that cancer cell-derived sEVs have cancer vaccine potential was conducted using human melanoma-derived sEVs. In this study, melanoma cell-derived sEVs containing melanoma antigen recognized by T cells 1 (MART-1)/MelanA antigen were shown to transfer tumor antigens to DCs and induce activation of tumor-specific $CD8^+$ T cells [2]. The first in vivo evidence for the effectiveness of cancer cell-derived sEV vaccination came from syngeneic and allogeneic mouse tumor models, in which treatment with cancer cell-derived sEVs induced potent $CD8^+$ T-cell cross-priming and antitumor activity compared with

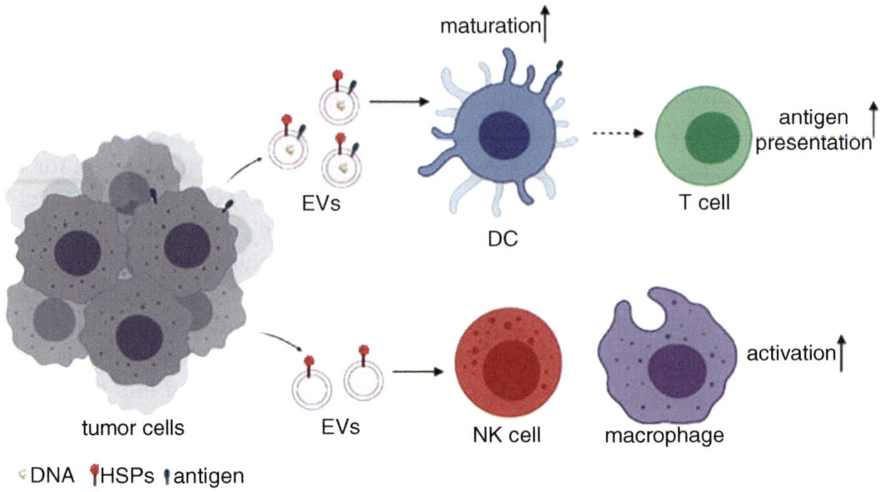


**Fig. 24.1** Cancer cell-derived sEVs interact with different immune cells to promote anticancer immunity (*graph created using* bioRender.com)

that in mice treated with tumor lysates [2]. Thereafter, many studies have demonstrated the capacity of cancer cell-derived sEVs to stimulate immune responses against different types of human cancer. For example, by introducing heat stress to carcinoembryonic antigen (CEA)-positive cancer cells from human colon carcinoma and human lung carcinoma, Dai et al. found that sEVs from these stressed cells had increased levels of heat shock protein (HSP) 70 and MHC class I [12]. These sEVs promoted the maturation of DCs and induced potent activation of CEA-specific T cells derived from human leukocyte antigen-matched CEA$^{pos}$ patients, as evidenced by the increased killing of CEA$^{pos}$ SW480 target cancer cells [12]. Subsequently, HSP70 exposure on the EV membrane was demonstrated to activate macrophages and natural killer (NK) cells [13, 14]. Rao et al. provided evidence that hepatocellular carcinoma (HCC)-derived sEVs elicited a stronger antitumor response than cell lysates in a syngeneic mouse model using the murine HCC cell line Hepa1-6 [15]. Intriguingly, both murine and human HCC-derived sEVs exert tumor-specific cytolytic effects in allogeneic individuals, indicating that shared antigens in sEVs can be cross-presented by DCs independent of MHC haplotypes [15]. In another study on leukemia, DCs pulsed with cancer cell-derived sEVs induced stronger specific cytotoxic T-lymphocyte responses than non-pulsed DCs [16]. More up-to-date studies on the utilization of cancer cell-derived sEVs as cancer vaccines have been reviewed elsewhere [17, 18].

Notably, sEVs isolated from ascites of patients with advanced melanoma contain high concentrations of MART-1 antigen, indicating that they are probably secreted by peritoneal or pleural cancer cells [19]. Furthermore, upon co-culture with DCs from MART-1$^{pos}$ patients, these ascites-derived sEVs induce the activation of autologous cytotoxic T cells [19]. This study highlighted that ascites-derived sEVs could be used as an alternative source of therapeutic sEVs.

Chemotherapy and radiation therapy not only induce DNA damage in cancer cells but also increase the DNA content in cancer cell-derived sEVs [20]. These DNA-containing sEVs activate DCs by inducing their cytokine secretion capacity via the cyclic GMP-AMP synthase-stimulator of interferon genes cytosolic DNA-sensing signaling, resulting in enhanced antitumor immunity in mice [21]. This finding implies that cancer-derived DNA-containing sEVs may be used as a source of therapeutic cancer vaccines.

### 24.3.2 DC-Derived sEVs for Cancer Vaccination

In 1996, sEVs secreted by B cells were shown to stimulate $CD4^+$ T cells in vitro [22]. This finding has shifted the focus of research in this field to DCs. In addition to MHC class I and class II molecules, DC-derived sEVs also carry functional co-stimulatory molecules, such as CD86 and CD80, on their surfaces [2]. In DC-based cancer immunotherapy, tumor antigens are used to pulse DCs, thereby promoting their antitumor effects. Therefore, initial studies on DC-derived sEVs employed sEVs from mouse-derived DCs pulsed with antigenic peptides from tumor antigens [2]. In syngeneic mouse tumor models, DC-derived sEVs induce significant T-cell-dependent tumor rejection [2]. Subsequently, evidence of DC-derived sEVs as a potential vaccine for HCC was provided by applying sEVs derived from α-fetoprotein (AFP)-expressing DCs into a carcinogen-induced HCC mouse model [23]. In this study, mice vaccinated with antigen-loaded DC-derived sEVs showed prolonged survival and a more immunostimulatory TME, with increasing numbers of interferon-γ-expressing $CD8^+$ T lymphocytes and reduced numbers of $CD25^+Foxp3^+$ regulatory T cells [23]. In another study on glioma, sEVs derived from chaperone-rich cell lysate-pulsed DCs inhibited the tumor growth of murine intracranial glioma by modulating Cbl-b and c-Cbl signaling [24]. A comprehensive summary of existing human and murine studies on DC-derived sEVs has been presented elsewhere [17].

In addition to facilitating $CD4^+$ and $CD8^+$ T-cell activation, in vivo studies have shown that DC-derived sEVs can stimulate the proliferation of NK cells via pro-inflammatory cytokines, including tumor necrosis factor and interleukin-15Rα [25, 26]. Moreover, B-cell adaptive immune responses, including specific B-cell activation and IgG secretion, have been reported to be involved in DC-derived sEV vaccinations [27, 28]. These additional effects indicate that DC-derived sEVs can coordinate T-cell and B-cell adaptive immune responses as well as innate immunity (Fig. 24.2).

DC-derived sEVs have shown promising ex vivo antitumor immunity toward human cancer cells and murine tumor models; however, the results from clinical trials remain unsatisfactory. Initial trials of sEVs derived from autologous DCs did not cause serious adverse effects in patients with metastatic melanoma or non-small cell lung cancer (NSCLC) [29, 30]. However, in a subsequent phase II clinical trial testing the efficacy of DC-derived sEVs vaccination in patients with advanced-stage NSCLC, the primary end point (progression-free survival) was not achieved [31].

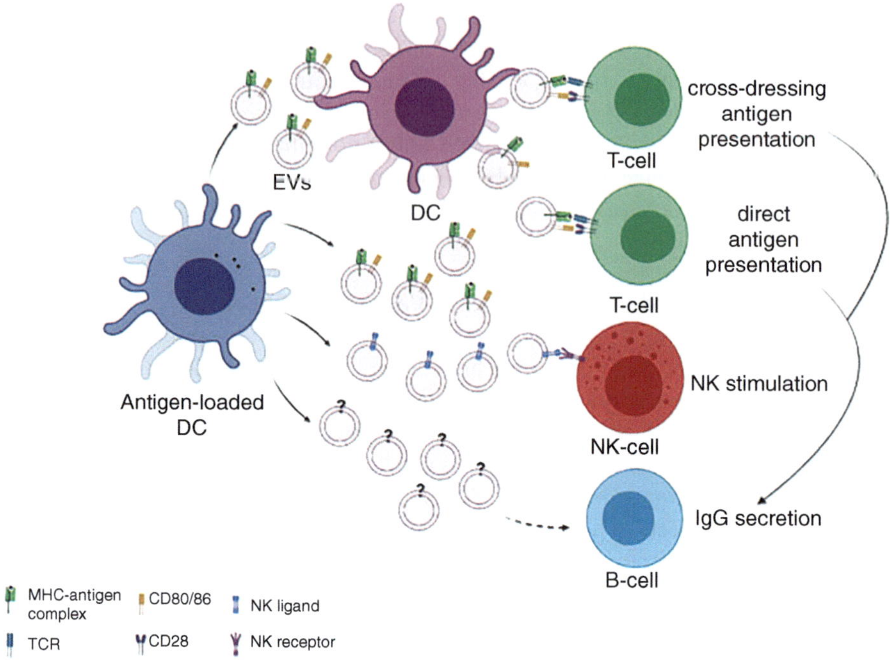


**Fig. 24.2** DC-derived sEVs coordinate adaptive and innate immunity (*graph created using* bioRender.com)

These observations suggest that further improvements are required to convert DC-derived sEVs into clinically useful cancer vaccines.

## 24.4 Challenges in EV-Based Cancer Vaccination

Currently, the successful application of EV-based cancer vaccinations in the clinical setting remains challenging for several reasons.

First, EVs show high heterogeneity. Current isolation methods for EVs rely heavily on the particle size to discriminate various EV populations. However, it is now clear that even EVs of the same size may contain heterogeneous EV subtypes. In this respect, further technological improvements are required to facilitate the isolation of homogeneous EV populations.

Another challenge in EV-based cancer vaccination is the selection of appropriate tumor-specific antigens. Cancer is characterized by highly heterogeneous immunological features. Therefore, it is unlikely that a one-fits-all vaccine can be formulated for cancers of different histological origins. Identification of immunogenic neoantigens using omics tools and bioinformatics may accelerate the development of EV-based cancer vaccines.

Considerable efforts have been made to increase the amount of tumor antigen loaded into sEVs. Current approaches include incubation, transfection, electroporation, sonication, and in situ synthesis. However, these loading strategies either have a rather low cargo loading efficiency or raise concerns due to harmful side effects. Novel approaches are required to achieve optimal antigen-loading efficiency with high biocompatibility and limited side effects.

Finally, it is important to appreciate that EVs can have both pro-cancer and anticancer effects and may therefore alter the TME. Overall, the TME tends to be immunosuppressive in most tumors. The anticancer effects of immunostimulatory EVs may be counteracted by immunosuppressive EVs and/or other immunosuppressive factors secreted by cancer cells. A possible solution to this problem is the use of cancer vaccines in combination with immune checkpoint inhibitors. The efficacy of such combinatorial regimens should be evaluated in the preclinical setting.

In conclusion, EVs represent a promising platform for cancer vaccines. A substantial amount of work is still needed to advance EV-based cancer vaccination from a scientific theory to a clinical therapeutic approach.

## References

1. Hood JL, Wickline SA. A systematic approach to exosome-based translational nanomedicine. Wiley Interdiscip Rev Nanomed Nanobiotechnol. 2012;4(4):458–67.
2. Zitvogel L, Regnault A, Lozier A, Wolfers J, Flament C, Tenza D, et al. Eradication of established murine tumors using a novel cell-free vaccine: dendritic cell-derived exosomes. Nat Med. 1998;4(5):594–600.
3. Koh E, Lee EJ, Nam GH, Hong Y, Cho E, Yang Y, et al. Exosome-SIRPalpha, a CD47 blockade increases cancer cell phagocytosis. Biomaterials. 2017;121:121–9.
4. Wang R, Pan W, Jin L, Huang W, Li Y, Wu D, et al. Human papillomavirus vaccine against cervical cancer: opportunity and challenge. Cancer Lett. 2020;471:88–102.
5. Chen DS. Toward elimination and eradication of hepatitis B. J Gastroenterol Hepatol. 2010;25(1):19–25.
6. Handy CE, Antonarakis ES. Sipuleucel-T for the treatment of prostate cancer: novel insights and future directions. Future Oncol. 2018;14(10):907–17.
7. Huang CC, Lee CC, Chu NS. A case of cerebral nocardiosis successfully treated with trimethoprim-sulfamethoxazole. J Formos Med Assoc. 1991;90(4):407–10.
8. Cantor JC. Expanding health insurance coverage: who will pay? J Health Polit Policy Law. 1990;15(4):755–78.
9. Becker A, Thakur BK, Weiss JM, Kim HS, Peinado H, Lyden D. Extracellular vesicles in cancer: cell-to-cell mediators of metastasis. Cancer Cell. 2016;30(6):836–48.
10. Lobb RJ, Hastie ML, Norris EL, van Amerongen R, Gorman JJ, Moller A. Oncogenic transformation of lung cells results in distinct exosome protein profile similar to the cell of origin. Proteomics. 2017;17(23–24)
11. Lindenbergh MFS, Stoorvogel W. Antigen presentation by extracellular vesicles from professional antigen-presenting cells. Annu Rev Immunol. 2018;36:435–59.
12. Dai S, Wan T, Wang B, Zhou X, Xiu F, Chen T, et al. More efficient induction of HLA-A*0201-restricted and carcinoembryonic antigen (CEA)-specific CTL response by immunization with exosomes prepared from heat-stressed CEA-positive tumor cells. Clin Cancer Res. 2005;11(20):7554–63.
13. Vega VL, Rodriguez-Silva M, Frey T, Gehrmann M, Diaz JC, Steinem C, et al. Hsp70 translocates into the plasma membrane after stress and is released into the extracellular environment in a membrane-associated form that activates macrophages. J Immunol. 2008;180(6):4299–307.

14. Gastpar R, Gehrmann M, Bausero MA, Asea A, Gross C, Schroeder JA, et al. Heat shock protein 70 surface-positive tumor exosomes stimulate migratory and cytolytic activity of natural killer cells. Cancer Res. 2005;65(12):5238–47.
15. Rao Q, Zuo B, Lu Z, Gao X, You A, Wu C, et al. Tumor-derived exosomes elicit tumor suppression in murine hepatocellular carcinoma models and humans in vitro. Hepatology. 2016;64(2):456–72.
16. Yao Y, Wang C, Wei W, Shen C, Deng X, Chen L, et al. Dendritic cells pulsed with leukemia cell-derived exosomes more efficiently induce antileukemic immunities. PLoS One. 2014;9(3):e91463.
17. Markov O, Oshchepkova A, Mironova N. Immunotherapy based on dendritic cell-targeted/-derived extracellular vesicles-a novel strategy for enhancement of the anti-tumor immune response. Front Pharmacol. 2019;10:1152.
18. Naseri M, Bozorgmehr M, Zoller M, Ranaei Pirmardan E, Madjd Z. Tumor-derived exosomes: the next generation of promising cell-free vaccines in cancer immunotherapy. Onco Targets Ther. 2020;9(1):1779991.
19. Andre F, Schartz NE, Movassagh M, Flament C, Pautier P, Morice P, et al. Malignant effusions and immunogenic tumour-derived exosomes. Lancet. 2002;360(9329):295–305.
20. Diamond JM, Vanpouille-Box C, Spada S, Rudqvist NP, Chapman JR, Ueberheide BM, et al. Exosomes shuttle TREX1-sensitive IFN-stimulatory dsDNA from irradiated cancer cells to DCs. Cancer Immunol Res. 2018;6(8):910–20.
21. Kitai Y, Kawasaki T, Sueyoshi T, Kobiyama K, Ishii KJ, Zou J, et al. DNA-containing exosomes derived from cancer cells treated with topotecan activate a STING-dependent pathway and reinforce antitumor immunity. J Immunol. 2017;198(4):1649–59.
22. Raposo G, Nijman HW, Stoorvogel W, Liejendekker R, Harding CV, Melief CJ, et al. B lymphocytes secrete antigen-presenting vesicles. J Exp Med. 1996;183(3):1161–72.
23. Lu Z, Zuo B, Jing R, Gao X, Rao Q, Liu Z, et al. Dendritic cell-derived exosomes elicit tumor regression in autochthonous hepatocellular carcinoma mouse models. J Hepatol. 2017;67(4):739–48.
24. Bu N, Wu H, Zhang G, Zhan S, Zhang R, Sun H, et al. Exosomes from dendritic cells loaded with chaperone-rich cell lysates elicit a potent T cell immune response against intracranial glioma in mice. J Mol Neurosci. 2015;56(3):631–43.
25. Viaud S, Terme M, Flament C, Taieb J, Andre F, Novault S, et al. Dendritic cell-derived exosomes promote natural killer cell activation and proliferation: a role for NKG2D ligands and IL-15Ralpha. PLoS One. 2009;4(3):e4942.
26. Munich S, Sobo-Vujanovic A, Buchser WJ, Beer-Stolz D, Vujanovic NL. Dendritic cell exosomes directly kill tumor cells and activate natural killer cells via TNF superfamily ligands. Onco Targets Ther. 2012;1(7):1074–83.
27. Qazi KR, Gehrmann U, Domange Jordo E, Karlsson MC, Gabrielsson S. Antigen-loaded exosomes alone induce Th1-type memory through a B-cell-dependent mechanism. Blood. 2009;113(12):2673–83.
28. Colino J, Snapper CM. Exosomes from bone marrow dendritic cells pulsed with diphtheria toxoid preferentially induce type 1 antigen-specific IgG responses in naive recipients in the absence of free antigen. J Immunol. 2006;177(6):3757–62.
29. Escudier B, Dorval T, Chaput N, Andre F, Caby MP, Novault S, et al. Vaccination of metastatic melanoma patients with autologous dendritic cell (DC) derived-exosomes: results of the first phase I clinical trial. J Transl Med. 2005;3(1):10.
30. Morse MA, Garst J, Osada T, Khan S, Hobeika A, Clay TM, et al. A phase I study of dexosome immunotherapy in patients with advanced non-small cell lung cancer. J Transl Med. 2005;3(1):9.
31. Besse B, Charrier M, Lapierre V, Dansin E, Lantz O, Planchard D, et al. Dendritic cell-derived exosomes as maintenance immunotherapy after first line chemotherapy in NSCLC. Onco Targets Ther. 2016;5(4):e1071008.

# 25 Platelet-Derived Extracellular Vesicles in Disease Therapy

Caoyi Liu, Yang Gao, Xu Liu, Peipei Xu, and Yu Liu

## 25.1 Introduction

Platelets are important blood cells in the human body and have been shown to influence many physiological and pathophysiological processes, including hemostasis, inflammation, immunity, angiogenesis, and tumor growth. Platelets are vesicle-related cells in the blood. On the one hand, platelets are regarded as vesicles themselves because they do not contain nuclei; on the other hand, platelets secrete extracellular vesicles (EVs), and platelet-derived EVs (PEVs) account for 70–90% of all vesicles in plasma. Because of their important functions, platelets and PEVs may have important clinical applications, including applications as biomarkers of diseases, drug targeting carriers, and separate components or products in the treatment of clinical diseases. In this section, the properties, functions, and applications of platelets and PEVs are reviewed.

Caoyi Liu, Yang Gao, and Xu Liu have contributed equally to this work.

C. Liu · Y. Gao · Y. Liu (✉)
Transfusion Medicine Research Center, Institute of Blood Transfusion, Chinese Academy of Medical Sciences & Peking Union Medical College,
Chengdu, Sichuan Province, People's Republic of China
e-mail: liuyu@ibt.pumc.edu.cn

X. Liu · P. Xu (✉)
Department of Hematology, Nanjing Drum Tower Hospital, Medical School of Nanjing University, Nanjing, Jiangsu Province, People's Republic of China



Q. Wang, L. Zheng (eds.), *Extracellular Vesicles*,
https://doi.org/10.1007/978-981-99-8365-0_25

## 25.2 Platelets as Vesicles

### 25.2.1 Properties of Platelets

Platelets are nonnucleated cells with a short life span and a diameter of 2–4 μm. The circulation cycle in the blood is 7–10 days in humans and even shorter in mice. The platelet count in healthy people is approximately 150,000–450,000/μL, with some differences according to age, sex, race, and genetic background.

As secretory cells, activated platelets release multiple substances from different storage granules, including α-granules, δ-granules, and lysosomes. Secretion products can influence many physiological and pathophysiological processes other than hemostasis, including inflammation, immunity, angiogenesis, and tumor growth. Platelets also produce other bioactive components, such as eicosanoids and EVs. Conversely, platelets can take up plasma and even cellular components, thus providing another mechanism through which the environment influences platelets.

Although platelets are nonnuclear, they contain large amounts of RNA, including mRNAs, tRNAs, rRNAs, small nucleolar RNAs, microRNAs (miRNA), and other RNAs. Among these, miRNAs are the most abundant, accounting for 80% of total platelet RNAs. MiRNAs are endogenous noncoding RNAs that participate in post-transcriptional regulation. MiRNAs are small, stable in molecular structure, and easy to quantify. Moreover, miRNAs play important roles in platelet production and can participate in the regulation of platelet signaling pathways and gene expression. Furthermore, miRNAs are potential biomarkers of platelet dysfunction. Bruchova et al. [1] reported human platelet miRNA profiles for the first time through research on the differential expression of miRNAs in patients with polycythemia vera. Their study found that compared with healthy controls, patients with polycythemia vera showed higher *miR-26* expression in platelets. These findings facilitate further studies of miRNAs as biomarkers for platelet disorders.

### 25.2.2 Functions of Platelets

Many receptors on the surfaces of platelets confer platelets with various functions, including promoting coagulation and thrombosis, regulating inflammation, controlling infection, and modulating the growth cycle of normal cells and tumor cells.

#### 25.2.2.1 Coagulation and Thrombosis

Platelets are important participants in clotting and thrombosis, and these processes can be roughly divided into four steps: platelet aggregation, conglutination, thrombin, and fibrin formation. Various coagulation factors secreted by platelets promote the coagulation pathway, which results in the conversion of fibrinogen into fibrin monomers, which then polymerize and crosslink with each other to form the matrix bones of a thrombus [2].

#### 25.2.2.2 Regulation of Inflammation

Platelets accelerate inflammation via recruitment of leukocytes and interleukin (IL)-1β, secretion/expression of high mobility group box-1, secretion of migration inhibitory factor-related protein-14, and enhancement of IL-1β synthesis and secretion [3]. By contrast, platelets can also inhibit inflammation by maintaining endothelial integrity and reducing extravasation.

#### 25.2.2.3 Control of Infection

Platelets support all steps of leukocyte extravasation during infection and inflammation, such as combining with leukocytes to assist with intravascular rolling, adhesion, and crawling; secreting chemokines to guide leukocytes to the site of infection; and serving as docking platforms to initiate leukocyte migration [3]. Platelets in the blood bind to invasive pathogens and send them to the spleen for clearance and recognition by splenic dendritic cells. Platelets in the liver form aggregates around bacteria bound by Kupffer cells to support their absorption [4]. Platelets express receptors such as integrins and Toll-like receptors (TLRs) to interact with viruses. The viral load and the reaction of receptor-mediated viropexis are controlled by chemokines (such as C-X-C motif chemokine ligand 4 [CXCL4]) secreted by platelets and human immunodeficiency virus (HIV). Moreover, platelets can internalize respiratory syncytial virus (RSV) to eliminate virus particles or interact directly with monocytes to regulate the recognition of RSV via TLRs, promote the production of interferon α/γ, and reduce viral infection in monocytes [5]. Bacterial (*Staphylococcus aureus*) surface proteins can bind to platelets through plasma proteins that interact with bacteria and platelets [6]. Platelets can also interact with Kupffer cells (KCs) through the platelet adhesion receptor glycoprotein Ib (GPIIb). Bacteria such as *Bacillus cereus* and methicillin-resistant *Staphylococcus aureus* are quickly captured by KCs and trigger the transition of platelets from transient adhesion to GPIIb-mediated continuous adhesion, thereby enveloping the bacteria. *Streptococcus* strain M99 widely binds to, activates, and aggregates rabbit platelets and is quickly killed by platelet bactericidal protein [7]. *Toxoplasma gondii* tachyzoites can induce human platelet activation to secrete platelet-derived growth factor (PDGF), which inhibits the intracellular growth of virulent strains of *T. gondii* [8]. Platelets may be an innate defense mechanism against *Schistosoma japonicum* infection that can adhere to and kill mechanically transformed *Schistosoma japonicum* larvae in vitro [9]. Additionally, platelets contact red blood cells parasitized by *Plasmodium* and release platelet factor 4 (also known as CXCL4), which specifically aggregates in infected red blood cells and quickly kills parasites through a mechanism that causes their digestive vacuoles to dissolve [10].

#### 25.2.2.4 Alteration of the Growth Cycle of Normal Cells

Platelets contain multiple growth factors that promote their growth. Platelets can also promote endothelial cell proliferation and angiogenesis. $CD34^+$ bone marrow cells can be recruited to damaged endothelial cells via stromal-derived factor-1α, which is secreted by platelets [11]. Subsequently, $CD34^+$ cells begin to bind to the injured endothelial efflux site and differentiate into mature endothelial cell phenotypes [12]. Platelets regulate the development of lymphatic vessels by directly interacting with lymphatic endotheliocytes and activating Slp-76 signaling [13]. Abundant growth factors, such as vascular endothelial growth factor (VEGF), epidermal growth factor, keratinocyte growth factor, and connective tissue growth factor, are present in the alpha granules of platelets, and all of these growth factors enable platelets to accelerate the pace of damage repair [14]. Another study indicated that platelets and platelet lysates can resist apoptosis induced by mitochondrion-damaging agents [15].

#### 25.2.2.5 Alteration of the Growth Cycle of Tumor Cells

Platelets regulate the growth of tumor cells via various mechanisms and can inhibit tumor growth, promote tumor cell proliferation, help tumor cells resist blood flow shear force and immune damage, and enhance tumor metastasis. Platelets prevent tumor cell metastasis by blocking the ruptured endothelium and releasing vascular inhibitors, such as angiopoietin-1, to support vascular integrity [3]. Platelets have greater effects on tumor cells than on inhibition of tumor metastasis. Tumor cells can activate endotheliocytes by releasing VEGF. Activated vascular endothelial cells promote local platelet recruitment by upregulating von Willebrand factor, P-selectin, $\alpha v\beta 3$ integrin, and tissue factor [3]. There are many adhesion receptors on the surface of platelets, which enable them to quickly bind to inflamed endothelial cells during cancer progression, thereby enhancing platelet activation. Activated platelets release substances containing pro-angiogenic factors to stimulate angiogenesis and regulate vascular permeability and tension [3]. Platelets can adhere to circulating tumor cells (CTCs), covering their surface and acting as a protective barrier against the mechanical force of blood flow and attack by natural killer (NK) T cells [16]. In addition to physically encapsulating CTCs, platelets can downregulate NK group 2D receptors on NK cells, which are the main recognition receptors for detecting and eliminating transformed cells [17]. Platelets can form an ecological environment that assists in the implantation of tumor cells by releasing metastasis-promoting factors and mediating the fibrin reticular structure [3]. Platelets are the main source of PDGF-β in the blood; PDGF-β accelerates migration and promotes the expression of the invasive phenotype in tumor cells [18].

### 25.2.3 Applications of Platelets

Platelets contain several growth factors [19]. To amplify these growth-promoting effects, various platelet concentrates have been designed and used clinically, including platelet-rich plasma (PRP), leukocyte-rich PRP, platelet-rich fibrin (PRF), advanced PRF, and injectable PRF concentrated growth factor.

Platelets are used as drug carriers because they are readily available for chemical ligation with other biomolecules owing to the abundant primary amine or thiol residues on the platelet surface. Similar to other drug carriers, platelets can take up drugs at relatively high loading concentrations and act as a potential strategy to deliver complexes with or without modification. Platelets are used as drug carriers for wound tracking and tumor targeting.

Platelets have been reported to contribute to surgical wound healing by actively deforming and collapsing the fibrin network to contract clots and shorten bleeding time. In the injured vessel intima, delivery of antithrombotic agents by platelets or platelet mimics is crucial in thrombus-targeted treatment. Platelets have been effectively exploited as carriers of antithrombotic agents, particularly arterial clots. Hu et al. developed a platelet membrane-coated tissue plasminogen activator delivery system to dissolve both preexisting and nascent clots in the bone microenvironment [20]. Similarly, Wang et al. created an endothelium-protective anti-restenotic system by covering dendritic polymer-based nanoclusters with platelet membranes to target injured arterial walls [21].

The lethal effects of cytotoxic drugs on tumor cells were increased via promotion of apoptosis when the cytotoxic drugs were carried by platelets, indicating the potential applications of platelets in targeted tumor treatment. When platelets encounter narrow vessels, their self-activation and adhesion trigger drug release. This phenomenon implies that platelets could be used as site-specific drug carriers with minimal side effects while maximizing cytotoxic efficacy. For example, platelets in conjunction with doxorubicin have been shown to improve the antitumor activity of doxorubicin in vitro and in vivo in B-cell lymphoma [22]. Moreover, platelets can also serve as checkpoint inhibitor delivery platforms for immunotherapy by delivering monoclonal antibodies against programmed death 1 and programmed death ligand 1 in primary melanoma, acute myeloid leukemia, and triple-negative breast cancer [23]. Tumor necrosis factor (TNF)-related apoptosis-inducing ligand can be expressed on the surfaces of platelets originating from genetically manipulated megakaryocytes or hepatic stellate cells [24]. The stability of therapeutic genetically engineered platelets may also facilitate the applications of these cells. Additionally, the simultaneous release of pro-inflammatory cytokines during platelet activation can synergistically enhance the efficiency of T-cell activation and residual tumor elimination [25].

## 25.3 PEVs

### 25.3.1 Properties of PEVs

EVs in the blood are mainly released by red blood cells, leukocytes, platelets, megakaryocytes, and endothelial cells [26]. PEVs account for 70–90% of all vesicles in the plasma [27]. The diameter of PEVs is approximately 40–1000 nm. According to their size and source, PEVs can be divided into two categories: exosomes released from multivesicular bodies and α particles released through exocytosis, which measure approximately 40–100 nm in diameter, and microvesicles (MVs) released by budding through the plasma membrane, which measure approximately 100–1000 nm in diameter. In addition, PEVs can be divided into three categories according to their structure: single-layer membrane structure PEVs, multivesicular PEVs (large vesicles with multiple small vesicles inside), and high-electron-density PEVs containing various organelles [28].

PEV surface molecules, PEV cargo, and the relationships of PEVs with pathophysiological processes and diseases are current research hotspots. Multiple molecules, including platelet endothelial adhesion molecules, CD41, CD42a, P-selectin, platelet factor 4, and glycoprotein IIb/IIIa, are expressed on the surfaces of PEVs [8].

PEVs contain lipids, nucleic acids (mRNAs and miRNAs), growth factors, cytokines, chemokines, procoagulant and anticoagulant factors, pro-inflammatory and anti-inflammatory factors, and pro-angiogenic and anti-angiogenic factors. PEV proteomics analyses [29] can provide information on the functions of PEVs in cellular physiological processes or disease conditions. Proteomics and functional analyses of PEVs have shown that the functional effects of PEVs of different sizes on platelets and endothelial cells are different. The RNA components of PEVs have only been reported in a few studies. Studies evaluating the abundances of circular RNAs (circRNAs) in PEVs have identified 1117 and 154 circRNAs in microvesicles and exosomes, respectively [30]. Additionally, PEVs are important transport carriers that transfer miRNAs to cells and regulate gene expression in target cells. PEVs exhibit specific miRNA profiles. One study found 182 miRNAs in PEVs from 5 donors, 48 of which were co-expressed in 5 donors [31].

Different platelet activators induce platelets to produce PEVs through various mechanisms, and the PEVs produced vary in number, size, shape, membrane molecule and specific marker expression, and cargo, conferring them with different functions [32].

There are significant differences in the numbers of EVs produced from platelets in response to various agonists, and the activation pathways may vary in their ability to induce MVs or exosomes. Indeed, one study showed that EV formation increased after thrombin and collagen costimulation (TC) and $Ca^{2+}$ ionophore activation. However, contrary to the vesicles produced by the $Ca^{2+}$ ionophore, TC activation seemed to produce more exosomes than total EVs [32].

Surface receptor expression levels on PEVs vary depending on the mechanisms through which the PEVs are produced from platelets. For example, activated $\alpha$IIb$\beta$3 integrins are expressed on PEVs formed by platelet aging or ultrasonication, whereas CD63 expression is increased on PEVs when platelets are activated by thrombin or $Ca^{2+}$ ionophores. Phosphatidylserine (PS) is expressed on the surfaces of PEVs produced from activated and sonicated platelets, whereas CD40L and CX3CR1 receptors are only abundant in PEVs produced from sonicated platelets. P-selectin is expressed on all PEVs, and PEVs can be internalized into monocytes through via this receptor [33].

In a study comparing the proteomes of PEVs produced from thrombin- or shear-stress platelets, 30 protein features were found to be differentially regulated [34]. In addition, some proteins are common in PEVs induced using different activation methods. Studies have shown that 18 and 19 proteins are shared in MVs and exosomes, respectively, produced by different activation methods. Thirteen of these shared proteins, including common platelet surface proteins and functional proteins, were common in both MVs and exosomes, whereas multimerin-1 was only found in exosomes [32].

The numbers of PEVs in the plasma of healthy individuals of different sexes [35] and ages [36] differ dramatically. Moreover, the number, cargo, and surface markers of PEVs are altered in the context of various diseases, such as thrombosis, inflammation, infection, and cancer.

Notably, the number of PEVs in men is significantly higher than that in women [37]. However, in other studies, the numbers of PEVs in men and women showed the opposite result, that is, women had an elevated number of PEVs compared with men [35]. Additionally, PEV subpopulations, but not the total number of PEVs, were found to increase with age [36].

However, the numbers of PEVs are also altered in different disease states. PEV levels in the blood are increased in patients with several prothrombotic and inflammatory disorders, including cardiovascular diseases, autoimmune and infectious diseases, and cancer.

### 25.3.2 Functions and Mechanisms of PEVs

PEVs have been extensively studied and have been shown to participate in a variety of critical biological and pathological processes, including coagulation and hemostasis, inflammation, immunoregulation, and tumor progression, via the transportation of different RNA and protein cargos. Moreover, PEVs induced under different activation conditions show diverse functions. For example, aging- and activation-induced microparticles show differences in their ability to stimulate the formation of bioactive mediators in THP-1 cells and monocytes. Only the former causes the release of complement factor C5a and stimulates monocyte differentiation into the resident macrophage phenotype, whereas only the latter stimulates the development of a more inflammatory type of monocyte, releasing the reactive cytokine TNFα [33].

PEVs mainly affect the biological functions of target cells through the following mechanisms: (1) formation of signal complexes to stimulate target cells by combining with specific ligands expressed on the surface of target cells; (2) intercellular transfer of surface receptors; and (3) transportation of cargo, such as proteins, biologically active lipids, and mRNAs, which may cause epigenetic reprogramming of cells to target cells. Moreover, PEVs can be used as carriers to deliver infectious particles and complete organelles [38].

#### 25.3.2.1 Coagulation and Hemostasis

PEVs are not only accelerators but also participate in the regulation of coagulation by maintaining the balance of coagulation and anticoagulation through anticoagulant proteins, which are indispensable regulators of the blood coagulation system. Because of the high expression of PS on the surface and the prothrombotic proteins annexin-V, factor X, and prothrombin, PEVs have strong coagulation ability. PS, the procoagulant binding site formed by the coagulation complex of dystonase and prothrombinase, increases the rate and degree of thrombin generation and promotes thrombus formation. Integrin αIIbβ3 (CD41/CD61) on PEVs also supports fibrin clot formation. Platelets can realize a functional autocrine self-amplification mechanism to exert their procoagulant potential by releasing PEVs [39]. By contrast, PEVs express high levels of CD41 on their surfaces, resulting in the production of low levels of thrombin through a pathway independent of transcription factors and factor VII. Thrombin activates protein C and exerts anticoagulant effects [40].

#### 25.3.2.2 Inflammation and Immune Regulation

PEVs may play a dual role in inflammation, instigating either pro- or anti-inflammatory effects, depending on the surface markers, cargo, and microenvironment [27]. Additionally, PEVs can promote inflammation by releasing chemokines to recruit leukocytes and promote the interaction between monocytes and endothelial cells by binding P-selectin and P-selectin glycoprotein ligand-1 [41]. The combination of PEVs and monocytes can promote the expression of IL-1β and TNF-α [42]. PEVs can also activate neutrophils and macrophages through CLEC5A receptors, inducing the formation of neutrophil extracellular traps and the release of pro-inflammatory cytokines [43]. PEVs can promote inflammation by transferring receptors (i.e., CD41, CD61, CD184, CD62P, and PAR-1) expressed on their surfaces and cargo to receptor cells (monocytes, myoblasts, and hematopoietic stem cells) [44]. For example, arachidonic acid can be transferred from PEVs to monocytes and endothelial cells to induce the production of cyclooxygenase-2 and prostanoids and thereby recruit inflammatory cells to the site of infection [38]. CD145 can be transferred from PEVs to B cells to modulate adaptive immunity, stimulate antigen-specific IgG production, and modulate germinal center formation through cooperative activity with $CD4^+$ T cells [45]. In addition, gpIIb/IIIa (CD41/CD61) can be transferred to neutrophils and activate nuclear factor-κB in response to granulocyte-macrophage colony-stimulating factor and enhanced inflammation [46]. By contrast, PEVs can inhibit inflammation. PEVs can transfer arachidonate 12-lipoxygenase to mast cells, which increases the synthesis of a negative regulator of inflammation, lipoxin A4 [44]. Incubation with T cells and PEVs can reduce the release of interferon-γ, TNF-α, and IL-6 to inhibit inflammation [47]. In addition, PEVs can block the release of TNF-α from zymosan-activated macrophages and modify the differentiation of monocytes into DCs to reduce inflammation and phagocytosis [48]. The production of PEVs by platelet activation determines the effect of PEVs on monocytes [49]. PEVs also play roles in viral infections. For example, patients with HIV infection show higher levels of PEVs [50]. C-X-C motif chemokine receptor 4 (CXCR4), a chemokine receptor that mediates HIV entry into target cells, can be transferred from PEVs to CXCR4-negative cells, facilitating infection by the X4 HIV strain [51]. Many viruses have been found to activate platelets and promote the release of PEVs [52], thus affecting the level of PEVs in the blood.

#### 25.3.2.3 PEVs in the Occurrence and Development of Tumors

PEVs also have dual effects on the occurrence and development of tumors. The promotional effect of PEVs on tumors is mainly reflected in two aspects: the promotion of tumor angiogenesis and the enhancement of tumor cell migration, invasion, and metastasis. After co-incubation with PEVs, the transfer of matrix metalloproteinase (MMP) 1 from PEVs to cancer cells and increased expression of *MMP-9*, *VEGF*, and *IL-8* mRNA in cancer cells led to an increase in angiogenesis in the tumor microenvironment (TME) [53]. Moreover, PEV molecules from α granules, such as VEGF, PDGF, and fibroblast growth factor, have pro-angiogenic properties. Because of the high expression of PS on the surface, PEV may be ligands for the PS receptor (PSR) on immune cells, and activation of the PS/PSR pathway may inhibit innate and adaptive immune response in the TME and circulation [54]. PEV-mediated miRNA metastasis, induced by miRNAs such as *miR-126* and *miR-223*, can affect the occurrence and development of tumors [55]. In addition, PEVs can downregulate gene expression by transferring miRNAs to exert antitumor effects. For example, *miR-24* can be delivered to tumor cells to induce apoptosis and suppress tumor growth [55]. Moreover, increased levels of *miR-123* and *miR-233* in newly formed PEVs can decrease cell migration [56]. Aspirin can inhibit the progression of cancer [57] due to the decrease in cargo protein levels in plasma PEVs [58]; however, aspirin has several side effects. Newly designed drugs can inhibit the release of PEVs caused by cancer cells rather than through the functions of platelets, suggesting broad application prospects, and PEVs related to tumor metastasis can avoid side effects related to coagulation function owing to their specific targeting effects.

#### 25.3.2.4 Tissue Repair

PRP is a plasma component in which the platelet level is higher than that of whole blood. Although PRP is widely used in tissue repair, the mechanisms underlying PRP-induced tissue repair remain unclear. With in-depth studies of PEVs, their potential applications in tissue repair have attracted much attention. PEVs may be one of the effectors of PRP and platelet lysates, which can boost angiogenesis and restore endothelial integrity after vascular injury. Furthermore, PEVs promote wound healing in a manner similar to that of PRP, and this effect can induce cell proliferation, migration, and angiogenesis through the phosphoinositide 3-kinase/AKT and mitogen-activated protein kinase/extracellular signal-regulated kinase signaling pathways and crosstalk between transforming growth factor β and YES-associated protein [59]. Some studies have also shown that PEVs can be a feasible transfusion product to stop bleeding, inhibit vascular permeability, and reduce traumatic endothelial lesions in patients with trauma [60]. Bone marrow stromal cells treated with PEVs show significant increases in cell proliferation, migration, and osteogenesis [61]. Additionally, PEVs can significantly increase chondrocyte proliferation and inhibit apoptosis [62].

### 25.3.3 Applications of PEVs

The surface molecules and cargo of PEVs and their roles in physiological and pathological processes may have important clinical applications, including applications as biomarkers of diseases, drug targeting carriers, and separate components or products in the treatment of clinical diseases.

#### 25.3.3.1 Disease Markers

Cancer cells can activate platelets to promote PEV release. PEV levels are increased in patients with various types of cancers, including breast cancer, oral cancer, non-small cell lung cancer, pancreatic cancer, lung cancer, and colorectal cancer [27]. According to the unique characteristics of patients with cancer, PEVs can be used as biomarkers to reflect disease development [63]. PEVs, which express PS on their surface, are also elevated in many cardiovascular and metabolic diseases, including atherosclerosis, acute coronary syndrome, hypertension, heart failure, type 2 diabetes, and obesity [64]. In addition to thrombosis and inflammatory diseases, PEV levels are elevated in some autoimmune and infectious diseases [65]. Plasma PEV levels are decreased in patients with severe bleeding tendency, and PEVs may be used as biomarkers for predicting dengue fever-related bleeding [66]. These studies suggest that PEVs may be biomarkers for the diagnosis and monitoring of cancer, chronic inflammatory diseases, and infectious diseases, among other conditions. Indeed, PEV levels and cargo characteristics may become diagnostic and prognostic markers for different clinical stages of various tumors; however, their sensitivity and specificity need to be further studied and verified.

#### 25.3.3.2 Drug Carriers

Encapsulating drugs in vesicles can prevent them from being destroyed by degradation enzymes or chemicals before reaching the target cells; therefore, PEVs have the potential to act as natural drug carriers. PEVs can be engineered from autologous platelets in large quantities and stored for several days, suggesting that PEVs may be biocompatible and lack immunogenicity. PEVs also can carry a variety of molecules, including macromolecules (DNA, RNAs, and proteins) and small molecules (doxorubicin, curcumin, and paclitaxel) [67]. HL60 and K562 cells (leukemic cell lines) and blast cells can take up doxorubicin-loaded PEVs through P-selectin ligands and integrins; these PEVs have a higher transfer efficiency to leukemic cells, improve the therapeutic effects of the cargo drugs, and minimize the side effects of the drugs [68]. The sensitivity of BT549 cells to cisplatin chemotherapy can be increased by PEVs containing *miR-126* and *miR-223* [56]. Owing to their ability to target tumor cells, PEVs can be used to deliver suicide gene mRNAs and proteins to the target tissue, suggesting potential applications as a new cancer treatment strategy [69]. However, the clinical applications of drug-loaded PEVs are unclear, and an improved understanding of their clinical functions and further development of vesicle operation technologies are urgently needed.

#### 25.3.3.3 Others

Variations have been observed in the amounts, surface molecules, and bioactive media of PEVs induced under different activation conditions [33]. Different activation conditions can be selected according to the relevant clinical treatment in order to improve clinical efficacy.

## References


1. Bruchova H, Merkerova M, Prchal JT. Aberrant expression of microRNA in polycythemia vera. Haematologica. 2008;93(7):1009–16.
2. van der Meijden PEJ, Heemskerk JWM. Platelet biology and functions: new concepts and clinical perspectives. Nat Rev Cardiol. 2019;16(3):166–79.
3. Gaertner F, Massberg S. Patrolling the vascular borders: platelets in immunity to infection and cancer. Nat Rev Immunol. 2019;19(12):747–60.
4. Danese S, Katz JA, Saibeni S, et al. Activated platelets are the source of elevated levels of soluble CD40 ligand in the circulation of inflammatory bowel disease patients. Gut. 2003;52(10):1435–41.
5. Kullaya VI, de Mast Q, van der Ven A, et al. Platelets modulate innate immune response against human respiratory syncytial virus in vitro. Viral Immunol. 2017;30(8):576–81.
6. Shannon O. Platelet interaction with bacterial toxins and secreted products. Platelets. 2015;26(4):302–8.
7. Sullam PM, Frank U, Yeaman MR, et al. Effect of thrombocytopenia on the early course of streptococcal endocarditis. J Infect Dis. 1993;168(4):910–4.
8. Chumpitazi BF, Simon J, Polack B, et al. Human platelet inhibition of toxoplasma gondii growth. Clin Exp Immunol. 1998;111(2):325–33.
9. Stanley RG, Ngaiza JR, Wambayi E, et al. Platelets as an innate defence mechanism against Schistosoma mansoni infections in mice. Parasite Immunol. 2003;25(10):467–73.
10. Love MS, Millholland MG, Mishra S, et al. Platelet factor 4 activity against P. Falciparum and its translation to nonpeptidic mimics as antimalarials. Cell Host Microbe. 2012;12(6):815–23.
11. Stellos K, Langer H, Daub K, et al. Platelet-derived stromal cell-derived factor-1 regulates adhesion and promotes differentiation of human CD34+ cells to endothelial progenitor cells. Circulation. 2008;117(2):206–15.
12. Brill A, Elinav H, Varon D. Differential role of platelet granular mediators in angiogenesis. Cardiovasc Res. 2004;63(2):226–35.
13. Bertozzi CC, Schmaier AA, Mericko P, et al. Platelets regulate lymphatic vascular development through CLEC-2-SLP-76 signaling. Blood. 2010;116(4):661–70.
14. Everts PA, Knape JT, Weibrich G, et al. Platelet-rich plasma and platelet gel: a review. J Extra Corpor Technol. 2006;38(2):174–87.
15. Demers M, Ho-Tin-Noé B, Schatzberg D, et al. Increased efficacy of breast cancer chemotherapy in thrombocytopenic mice. Cancer Res. 2011;71(5):1540–9.
16. Nieswandt B, Hafner M, Echtenacher B, et al. Lysis of tumor cells by natural killer cells in mice is impeded by platelets. Cancer Res. 1999;59(6):1295–300.
17. Kopp HG, Placke T, Salih HR. Platelet-derived transforming growth factor-beta down-regulates NKG2D thereby inhibiting natural killer cell antitumor reactivity. Cancer Res. 2009;69(19):7775–83.
18. Labelle M, Begum S, Hynes RO. Direct signaling between platelets and cancer cells induces an epithelial-mesenchymal-like transition and promotes metastasis. Cancer Cell. 2011;20(5):576–90.
19. Nikolidakis D, Jansen JA. The biology of platelet-rich plasma and its application in oral surgery: literature review. Tissue Eng Part B Rev. 2008;14(3):249–58.
20. Hu Q, Qian C, Sun W, et al. Engineered nanoplatelets for enhanced treatment of multiple myeloma and thrombus. Adv Mater. 2016;28(43):9573–80.
21. Wang B, Chen G, Urabe G, et al. A paradigm of endothelium-protective and stent-free anti-restenotic therapy using biomimetic nanoclusters. Biomaterials. 2018;178:293–301.
22. Xu P, Zuo H, Chen B, et al. Doxorubicin-loaded platelets as a smart drug delivery system: an improved therapy for lymphoma. Sci Rep. 2017;7:42632.
23. Hu Q, Sun W, Wang J, et al. Conjugation of haematopoietic stem cells and platelets decorated with anti-PD-1 antibodies augments anti-leukaemia efficacy. Nat Biomed Eng. 2018;2(11):831–40.

24. Fu Y, Li S, Zhang Z, et al. Phylogeographic analysis of severe fever with thrombocytopenia syndrome virus from Zhoushan Islands, China: implication for transmission across the ocean. Sci Rep. 2016;6:19,563.
25. Lu Y, Hu Q, Jiang C, et al. Platelet for drug delivery. Curr Opin Biotechnol. 2019;58:81–91.
26. Menck K, Bleckmann A, Wachter A, et al. Characterisation of tumour-derived microvesicles in cancer patients' blood and correlation with clinical outcome. J Extracell Vesicles. 2017;6(1):1340745.
27. Żmigrodzka M, Witkowska-Piłaszewicz O, Winnicka A. Platelets extracellular vesicles as regulators of cancer progression-an updated perspective. Int J Mol Sci. 2020;21(15)
28. Ponomareva AA, Nevzorova TA, Mordakhanova ER, et al. Intracellular origin and ultrastructure of platelet-derived microparticles. J Thromb Haemost. 2017;15(8):1655–67.
29. Capriotti AL, Caruso G, Cavaliere C, et al. Proteomic characterization of human platelet-derived microparticles. Anal Chim Acta. 2013;776:57–63.
30. Preußer C, Hung L-H, Schneider T, et al. Selective release of circRNAs in platelet-derived extracellular vesicles. J Extracell Vesicles. 2018;7(1):1424473.
31. Qu M, Zou X, Fang F, et al. Platelet-derived microparticles enhance megakaryocyte differentiation and platelet generation via miR-1915-3p. Nat Commun. 2020;11(1):4964.
32. Aatonen MT, Ohman T, Nyman TA, et al. Isolation and characterization of platelet-derived extracellular vesicles. J Extracell Vesicles. 2014;3:3.
33. Vasina EM, Cauwenberghs S, Staudt M, et al. Aging- and activation-induced platelet microparticles suppress apoptosis in monocytic cells and differentially signal to proinflammatory mediator release. Am J Blood Res. 2013;3(2):107–23.
34. Shai E, Rosa I, Parguiña AF, et al. Comparative analysis of platelet-derived microparticles reveals differences in their amount and proteome depending on the platelet stimulus. J Proteomics. 2012;76:287–96. (Spec No).
35. Robert S, Poncelet P, Lacroix R, et al. Standardization of platelet-derived microparticle counting using calibrated beads and a Cytomics FC500 routine flow cytometer: a first step towards multicenter studies? J Thromb Haemost. 2009;7(1):190–7.
36. van der Zee PM, Biró E, Ko Y, et al. P-selectin- and CD63-exposing platelet microparticles reflect platelet activation in peripheral arterial disease and myocardial infarction. Clin Chem. 2006;52(4):657–64.
37. Ueba T, Haze T, Sugiyama M, et al. Level, distribution and correlates of platelet-derived microparticles in healthy individuals with special reference to the metabolic syndrome. Thromb Haemost. 2008;100(2):280–5.
38. Dovizio M, Bruno A, Contursi A, et al. Platelets and extracellular vesicles in cancer: diagnostic and therapeutic implications. Cancer Metastasis Rev. 2018;37(2–3):455–67.
39. Mause SF. Platelet microparticles: reinforcing the hegemony of platelets in atherothrombosis. Thromb Haemost. 2013;109(1):5–6.
40. Berckmans RJ, Nieuwland R, Böing AN, et al. Cell-derived microparticles circulate in healthy humans and support low grade thrombin generation. Thromb Haemost. 2001;85(4):639–46.
41. Šibíková M, Živný J, Janota J. Cell membrane-derived microvesicles in systemic inflammatory response. Folia Biol. 2018;64(4):113–24.
42. Bei JJ, Liu C, Peng S, et al. Staphylococcal SSL5-induced platelet microparticles provoke proinflammatory responses via the CD40/TRAF6/NFκB signalling pathway in monocytes. Thromb Haemost. 2016;115(3):632–45.
43. Sung PS, Huang TF, Hsieh SL. Extracellular vesicles from CLEC2-activated platelets enhance dengue virus-induced lethality via CLEC5A/TLR2. Nat Commun. 2019;10(1):2402.
44. Edelstein LC. The role of platelet microvesicles in intercellular communication. Platelets. 2017;28(3):222–7.
45. Sprague DL, Elzey BD, Crist SA, et al. Platelet-mediated modulation of adaptive immunity: unique delivery of CD154 signal by platelet-derived membrane vesicles. Blood. 2008;111(10):5028–36.
46. Salanova B, Choi M, Rolle S, et al. Beta2-integrins and acquired glycoprotein IIb/IIIa (GPIIb/IIIa) receptors cooperate in NF-kappaB activation of human neutrophils. J Biol Chem. 2007;282(38):27960–9.

47. Dinkla S, van Cranenbroek B, van der Heijden WA, et al. Platelet microparticles inhibit IL-17 production by regulatory T cells through P-selectin. Blood. 2016;127(16):1976–86.
48. Sadallah S, Eken C, Martin PJ, et al. Microparticles (ectosomes) shed by stored human platelets downregulate macrophages and modify the development of dendritic cells. J Immunol. (Baltimore, Md : 1950). 2011;186(11):6543–52.
49. Vajen T, Mause SF, Koenen RR. Microvesicles from platelets: novel drivers of vascular inflammation. Thromb Haemost. 2015;114(2):228–36.
50. Corrales-Medina VF, Simkins J, Chirinos JA, et al. Increased levels of platelet microparticles in HIV-infected patients with good response to highly active antiretroviral therapy. J Acquir Immune Defic Syndr. 2010;54(2):217–8.
51. Rozmyslowicz T, Majka M, Kijowski J, et al. Platelet- and megakaryocyte-derived microparticles transfer CXCR4 receptor to CXCR4-null cells and make them susceptible to infection by X4-HIV. AIDS (London, England). 2003;17(1):33–42.
52. Boilard E, Paré G, Rousseau M, et al. Influenza virus H1N1 activates platelets through FcγRIIA signaling and thrombin generation. Blood. 2014;123(18):2854–63.
53. Janowska-Wieczorek A, Wysoczynski M, Kijowski J, et al. Microvesicles derived from activated platelets induce metastasis and angiogenesis in lung cancer. Int J Cancer. 2005;113(5): 752–60.
54. Park M, Kang KW. Phosphatidylserine receptor-targeting therapies for the treatment of cancer. Arch Pharm Res. 2019;42(7):617–28.
55. Michael JV, Wurtzel JGT, Mao GF, et al. Platelet microparticles infiltrating solid tumors transfer miRNAs that suppress tumor growth. Blood. 2017;130(5):567–80.
56. Gasperi V, Vangapandu C, Savini I, et al. Polyunsaturated fatty acids modulate the delivery of platelet microvesicle-derived microRNAs into human breast cancer cell lines. J Nutr Biochem. 2019;74:108242.
57. Elwood PC, Morgan G, Pickering JE, et al. Aspirin in the treatment of cancer: reductions in metastatic spread and in mortality: a systematic review and meta-analyses of published studies. PLoS One. 2016;11(4):e0152402.
58. Goetzl EJ, Goetzl L, Karliner JS, et al. Human plasma platelet-derived exosomes: effects of aspirin. FASEB J. 2016;30(5):2058–63.
59. Guo S-C, Tao S-C, Yin W-J, et al. Exosomes derived from platelet-rich plasma promote the re-epithelization of chronic cutaneous wounds via activation of YAP in a diabetic rat model. Theranostics. 2017;7(1):81–96.
60. Miyazawa B, Trivedi A, Togarrati PP, et al. Regulation of endothelial cell permeability by platelet-derived extracellular vesicles. J Trauma Acute Care Surg. 2019;86(6):931–42.
61. Torreggiani E, Perut F, Roncuzzi L, et al. Exosomes: novel effectors of human platelet lysate activity. Eur Cell Mater. 2014;28:137–51; discussion 51.
62. Liu X, Wang L, Ma C, et al. Exosomes derived from platelet-rich plasma present a novel potential in alleviating knee osteoarthritis by promoting proliferation and inhibiting apoptosis of chondrocyte via Wnt/β-catenin signaling pathway. J Orthop Surg Res. 2019;14(1):470.
63. Mege D, Panicot-Dubois L, Ouaissi M, et al. The origin and concentration of circulating microparticles differ according to cancer type and evolution: a prospective single-center study. Int J Cancer. 2016;138(4):939–48.
64. Wei H, Malcor JM, Harper MT. Lipid rafts are essential for release of phosphatidylserine-exposing extracellular vesicles from platelets. Sci Rep. 2018;8(1):9987.
65. Aatonen M, Grönholm M, Siljander PRM. Platelet-derived microvesicles: multitalented participants in intercellular communication. Semin Thromb Hemost. 2012;38(1):102–13.
66. Punyadee N, Mairiang D, Thiemmeca S, et al. Microparticles provide a novel biomarker to predict severe clinical outcomes of dengue virus infection. J Virol. 2015;89(3):1587–607.
67. Fais S, O'Driscoll L, Borras FE, et al. Evidence-based clinical use of nanoscale extracellular vesicles in nanomedicine. ACS Nano. 2016;10(4):3886–99.
68. Kailashiya J, Gupta V, Dash D. Engineered human platelet-derived microparticles as natural vectors for targeted drug delivery. Oncotarget. 2019;10(56):5835–46.
69. Kim SM, Kim HS. Engineering of extracellular vesicles as drug delivery vehicles. Stem Cell Investig. 2017;4:74.

# Plant-Derived Extracellular Nanovesicles for Disease Therapy

# 26

Ying Wang, Kewei Zhao, Xiubo Shi, Qing Zhao, Qianxin Liang, Manlin Xie, and Yanfang Zou

## 26.1 Introduction

Natural plants, traditional Chinese medicines (TCMs), and their derivatives have been widely used throughout human history and are now available in many pharmaceutical markets. Companies have continued to explore natural plants, TCMs, and their derivatives against different diseases, such as microbial infections, inflammatory diseases, and cancer. Plant-derived extracellular nanovesicles (PENVs), which contain microRNAs (miRNAs), bioactive lipids, and proteins, serve as extracellular messengers to mediate cell–cell communication in a manner similar to those secreted by mammalian cells. Given their specific origin and cargo, PENVs can function in interspecies communication and may serve as natural therapeutic agents against various diseases, including inflammatory diseases, liver diseases, and cancer [1–3]. Additionally, TCMs can induce changes in the composition and surface factors of extracellular vesicles (EVs), thereby alleviating disease. Thus, these EVs are closely related to the study of TCM syndromes. EVs derived from Chinese medicines can be used as biomarkers for clinical disease diagnosis, which is closely related to the concept of microscopic syndrome differentiation in TCM syndrome differentiation and treatment.

Ying Wang and Kewei Zhao have contributed equally to this work.

Y. Wang (✉) · X. Shi
Cancer Research Institute, School of Basic Medical Sciences, Southern Medical University, Guangzhou, People's Republic of China

K. Zhao (✉) · Q. Zhao · Q. Liang · M. Xie · Y. Zou
Guangdong Engineering Research Center of Chinese herbal vesicles, The Third Affiliated Hospital of Guangzhou University of Chinese Medicine, Guangzhou, Guangdong, China



Q. Wang, L. Zheng (eds.), *Extracellular Vesicles*,
https://doi.org/10.1007/978-981-99-8365-0_26

The unique design and transport capabilities of PENVs highlight their potential applications in drug delivery [4]. Accordingly, PENVs are emerging as groundbreaking reagents in therapeutics and targeted drug delivery. PENV-based drug delivery systems present many advantages over the synthesized nanoplatforms and mammal-derived ENVs. PENVs exhibit superior biocompatibility, good stability in the gastrointestinal tract, and enhanced cellular uptake. In particular, the lipid bilayer structure of PENVs shrouds their cargo securely and prevents enzymatic decomposition by proteinases and nucleases. PENVs can be used for large-scale production and elimination of potential biohazards to the environment. Furthermore, PENVs have therapeutic capabilities in cross-kingdom communication, such as antitumor, regenerative, and anti-inflammatory activities. Compared with ENVs from mammalian cells, PENVs can overcome problems with unwanted genetic or protein transfer and detrimental immunogenic responses before clinical application [5–9]. These features, coupled with the discovery of their intrinsic therapeutic actions, make PENVs ideal candidates for drug delivery applications.

## 26.2 Physical Properties of PENVs

PENVs exhibit many properties suitable for clinical applications, such as nanosize, stability in the gastrointestinal tract, relatively high internalization rate, prolonged circulation time, and the ability to pass through the blood-brain barrier but not the placental barrier [10]. In general, PENVs maintain stable sizes and zeta potential under physiological conditions, although the sizes of PENVs can also be altered in a pH-dependent manner [3]. In addition, other factors can affect the size distribution, such as the solvents used, buffer concentration, and severe temperature changes.

PENVs can be efficiently taken up, mainly by phagocytosis, endocytosis, or membrane fusion [11–13]. Thus, PENVs have better intracellular uptake ability than synthetic nanoparticles. Indeed, more than 80% of grapefruit ENVs (GFENVs) can be taken up by cells, whereas only 40% of cationic 1,2-dioleoyl-3-trimethylammoniumpropane/dioleoylphosphatidylethanolamine liposomes are internalized. In addition, researchers showed that more than 20% of B cells and 14% of T cells take up the GFENVs within 12 h of treatment; for these cells, such an uptake rate cannot be achieved by any currently commercially available transfection reagent [5]. The uptake efficiency of PENVs is temperature-dependent, and metabolic energy is required for this process.

PENVs have been administered to experimental animals via oral [14], intravenous [15], intramuscular, and intranasal routes [5]. The biodistribution and bioavailability of PENVs are affected by multiple factors, such as selection of administration route, natural characteristics of the source plant, surface charge, and size dimension [14]. 1,1′-Dioctadecyl-3,3,3′,3′-tetramethylindotricarbocyanine iodide (DiR)-labeled GFENVs were injected 72 h after tail vein or intraperitoneal injection. DiR signals were distributed in the liver, lung, kidney, and splenic tissues, whereas intramuscular injections of DiR-labeled GFENVs were mainly localized in the muscle. After intranasal administration of DiR-labeled GFENVs, the majority of GFENVs were located in the lung and brain [5]. Ginger ENVs (GIENVs) were detected in the stomach, colon, and upper ileum, even at 12 h after oral administration. They preferred to accumulate in the intestines after oral administration. Intravenously administered DiR-labeled folic acid (FA)-targeted GIENVs exhibit increased accumulation in the liver, and relatively lower fluorescence is visible in other organs, including the heart, kidney, and spleen. FA-targeted GIENVs have also been detected in blood circulation after 48 h [12]. Accordingly, PENVs can be modified, thereby enhancing their accumulation at the target site for the desired duration with no toxicity.

## 26.3 Biochemical Characterization

PENVs contain RNA, lipids, and proteins, which differ from those found in mammalian ENVs. Plant and mammalian cell-derived ENVs can be characterized using common techniques for chemical component characterization [16]. PENVs deliver many miRNAs to facilitate intercellular communication. Deep sequencing showed that GIENVs contain at least 125 different miRNAs, each containing between 15 and 27 nucleotides [2]. Interestingly, in silico predictions have shown that 124 of these miRNAs can target and regulate the expression of human genes by binding to their 3′-untranslated regions, suggesting that PENVs have the potential to mediate interspecies communication at the molecular level. *Arabidopsis thaliana* ENVs are laden with miRNAs, small interfering RNAs (siRNAs; carrying 18–24 nucleotides), and tiny RNAs (RNAs with 10–17 nucleotides) [17]. In addition, researchers have found that miRNAs in PENVs have the potential to regulate human mRNAs. They isolated PENVs from 11 edible fruits and vegetables and subjected them to corresponding PENV small RNA libraries using Illumina sequencing. In total, 418 miRNAs (32–127 per species) were identified in the 11 PENVs. Target prediction and functional analyses revealed that highly expressed miRNAs are closely associated with inflammatory responses and cancer-related pathways [18]. In addition, PENVs have been shown to contain unusual noncoding RNAs. However, it is unknown how ENVs process these RNAs, and their influence on uptake into recipient cells, and consequently on cell communication, has not yet been reported.

The lipids present in PENVs play important biologically active roles in recipient cells, and lipid profiles of PENVs are different from those of ENVs from mammalian cells. PENV lipids are cholesterol-free because most edible plants and vegetables lack cholesterol [10]. However, mammalian ENVs and synthetic liposomes typically contain high levels of phosphatidylcholine (PC) and a certain percentage of cholesterol (10–20%), which can stabilize the lipid bilayer structure [19, 20]. Two distinct groups of lipids, i.e., phospholipids and glycerol lipids, are vital for the stability, uptake, and other biological functions of PENVs [5]. Phosphatidic acid (PA; the dominant lipid mediator) modulates membrane fission and fusion and can also influence PENV uptake by triggering cytoskeleton rearrangement and protein regulation during vesicular endocytosis [21]. PENVs contain diverse lipids, such as phosphatidylethanolamine (PE), PC, phosphatidylinositol (PI), and diacylglycerol (DG). PC can enhance the translocation of PENVs from the intestines to the liver [2, 22]. Moreover, PENVs from different plants exhibit variations in lipid contents. A deep lipidomic analysis of active GIENVs revealed a unique cholesterol-free profile; that is, the lipids isolated from 30–45% sucrose density gradient ultracentrifugation comprised 42% PA, 27% digalactosyldiacylglycerol, and 19% monogalactosyldiacylglycerol [2]. However, the lipid profiles of grapefruit ENVs (GFENVs) was different from that of the GIENVs, containing 24% PE, 25% PC, 13% PI, and only 10% DG [5]. A thorough understanding of each lipid component of PENVs and how they modulate their biological actions will be beneficial for the development of effective PENV-based therapeutic strategies (Fig. 26.1).

Abundant surface-associated proteins from ENVs derived from mammalian cells play key roles in intercellular communication. PENVs have a relatively low protein content and different protein content compared with mammalian cell-derived ENVs [2]. Plants are less advanced than animals; thus, they do not require complicated ensembles of surface-associated proteins to communicate with different organs (as observed in animals). Thus, it seems logical that PENV proteins are simpler and less abundant than mammalian ENV proteins. Among transmembrane proteins, CD63, CD81, and CD9 have been identified as possible markers of mammalian cell-derived ENVs [19, 23, 24]. However, relatively few reports have characterized PENV proteins, and available results are inconsistent. GIENV proteins tend to be predominantly cytosolic (mainly actins and proteolytic enzymes), and fewer membrane transporter proteins have been identified, among which aquaporins and chloride channels appear to contribute to the substructure of PENVs [2]. Twenty-eight proteins have been detected in GRENVs, whereas approximately 137 proteins have been isolated from GFENVs through mass spectrometry analysis [11, 25].

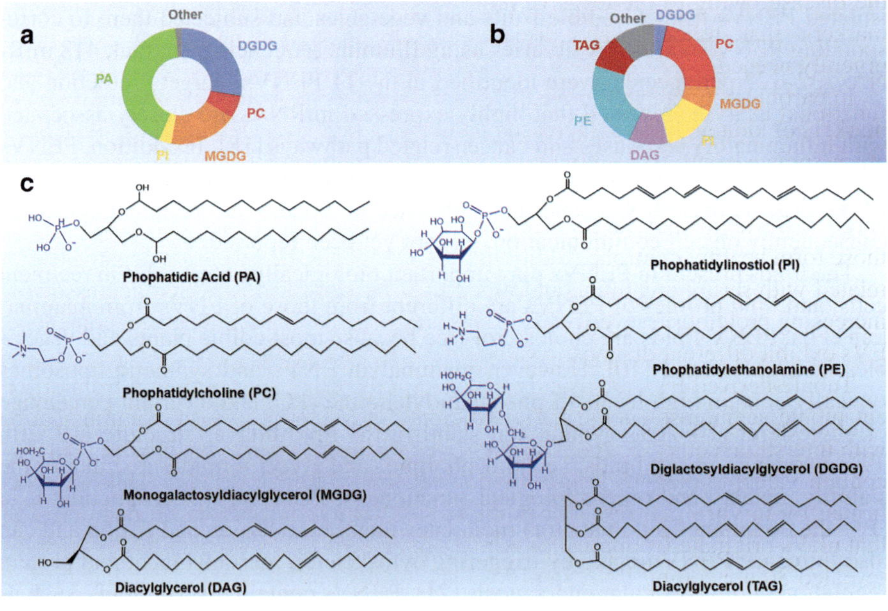


**Fig. 26.1** Lipids from ginger- and grapefruit-derived ENVs. (**a**) Lipid compositions of GIENVs. (**b**) Lipid compositions of GFENVs. (**c**) Representative structures of phospholipids and glycolipids

Raimondo et al. revealed that the ENVs derived from lemon (*Citrus limon L.*) juice have a relatively high protein content and that approximately 57% of the proteins overlap with those found in mammalian cell-derived ENVs, regardless of the cell origin. Another study identified 1018 significantly represented proteins from *Citrus clementina* juice-derived ENVs [26]. In ginseng-derived ENVs (GSENVs), 3129 proteins were analyzed and classified into three categories: biological processes, cellular compartments, and molecular functions [27]. Additionally, grapes, lemons, and *Arabidopsis* have been found to harbor heat shock proteins and aquaporin proteins. It is necessary to identify a broad range of PENV proteins in order to determine their roles in biological and pharmacological activities.

More than 50% of pharmaceutical agents comprise plant secondary metabolites or their derivatives [28–30]. The potential of secondary metabolites as natural resources for curing diseases has been extensively investigated in humans. GIENVs have bioactive secondary metabolites, for example, 6-gingerol and 6-shogaol, which can be used to prevent inflammatory bowel disease and colitis-associated cancer [2], and sulforaphane compounds from broccoli-derived ENVs (BENVs) can inhibit colitis in a mouse model [31].

## 26.4 Applications of PENVs (Fig. 26.2)

### 26.4.1 Therapeutic Effects on Inflammation

Some PENVs have immunological regulatory roles on gastrointestinal hemostasis, positively influence intestinal regeneration, and protect the gut from inflammatory diseases [3, 11, 25, 31]. GFENVs affect intestinal immune homeostasis by fortifying the anti-inflammatory capability of intestinal macrophages, ultimately alleviating dextran sulfate sodium (DSS)-induced colitis in mice with no toxicity. The inflammatory profile was altered by GFENV treatment, as demonstrated by downregulation of interleukin (IL)-6 and Il-1β as well as monocyte chemoattractant

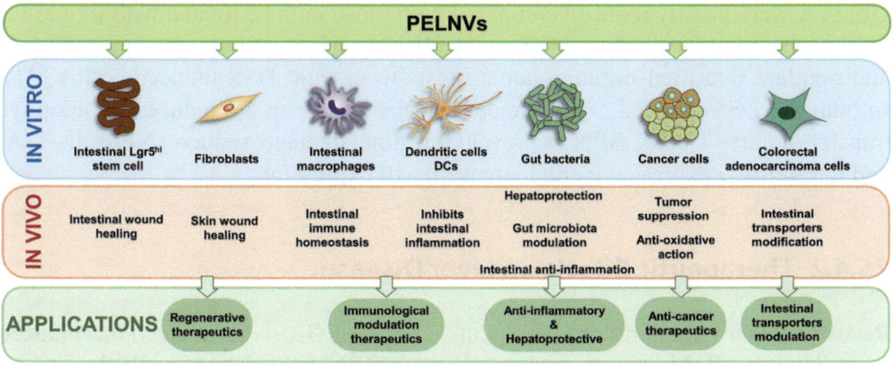


**Fig. 26.2** Summary of the biological functions of PENVs derived from plants

protein-1, C-X-C motif chemokine ligand (CXCL)-9, and CXCL-10, chemokines involved in the recruitment of inflammatory monocytes and T cells [11].

Oral administration of GIENVs can target the colon, reduce acute colitis, prevent chronic colitis, and accelerate intestinal wound healing [2]. The contents of GIENVs, such as the bioactive components gingerols, shogaols, and ginger miRNAs, can downregulate pro-inflammatory cytokines (e.g., tumor necrosis factor [TNF]-α, IL-6, and IL-1β), upregulate anti-inflammatory cytokines (e.g., IL-10 and IL-22), increase the survival and proliferation of intestinal epithelial cells in colitis models, attenuate damaging factors, and promote wound healing [14]. Moreover, PENVs from ginger, carrots, grapes, and grapefruit are taken up by $F4/80^+$ intestinal macrophages and $Lgr5^+$ intestinal stem cells 6 h after oral administration in mice [3]. All four PENVs increase the numbers of Wnt/transcription factor 4-positive intestinal cells when orally administered to mice.

Additionally, PENVs may influence the gut microbiota and have positive effects on colitis [22]. RNAs from ginger PENVs, carrot PENVs, and GRENVs were extracted and encapsulated in PENV-like exosomes made from PENV-extracted lipids. In particular, GIENV RNAs induced several species of *Lactobacillus*, whereas carrot ENV RNAs seemed to have no effect on *Lactobacillus* levels. PA is involved in the uptake of these nanovesicles by *Lactobacillus rhamnosus* and plays roles in maintaining the duration and amount of PENV accumulation in the gut. Mice fed PENVs containing these RNA are taken up by *L. rhamnosus* in vivo and provides superior protection against DSS-induced colitis compared with nanovectors transporting scrambled RNA.

Administration of GRENVs by gavage induces the proliferation of intestinal stem cells, which are crucial for epithelial cell differentiation and required for intestinal tissue homeostasis and repair. GRENV treatment reduces mortality rates in mice treated with DSS, i.e., 100% mortality was observed within 13 days in the control group but 25 days in GRENV-treated mice. These results indicated that GRENVs accelerated mucosal epithelium regeneration and induced rapid restoration of the intestinal architecture in DSS-induced colitis [25]. Methotrexate (MTX) was incorporated into GFENVs (GFMTX) and administered to mice in a DSS-induced colitis model. Colon tissue damage and inflammatory cell infiltration in mice treated with GFMTX were greatly reduced compared with those in mice treated with free MTX or phosphate-buffered saline (PBS) [11]. BENVs specifically target dendritic cells and regulate intestinal immune homeostasis to prevent DSS-induced colitis [31]. In both the DSS-induced colitis model and the colitis model induced by adoptive transfer of naïve T cells, BENVs prevent intestinal damage; reduce TNF-α, IL-17A, and interferon-γ expression; and increase IL-10 expression.

### 26.4.2 Therapeutic Effects on Liver Disease

Researchers have explored the therapeutic effects of GRENVs in alcoholic-induced liver disease [32]. Mice were pretreated with GRENVs by gavage or PBS as a control and then subjected to continuous administration of GRENVs or PBS after treatment with an ethanol-rich diet. Mice had fewer lipid droplets in hepatocytes, lower liver triglyceride levels, and reduced liver weights compared with mice fed alcohol alone. GRENVs suppress reactive oxygen species (ROS) production, which contributes to hepatoprotection. Additionally, GRENVs can induce hepatocyte metabolism and detoxifying and antioxidant defenses, which are vital for hepatoprotection.

Fulminant hepatic failure (FHF) can be induced in mice via administration of D-galactosamine and a low dose of lipopolysaccharide (LPS; GalN/LPS), stimulating many of the clinical manifestations of FHF in humans [33]. To evaluate the protective effects of shiitake mushroom-derived ENVs (SENVs), mice were intraperitoneally injected with PBS or SENVs. Compared with GalN/LPS, pretreatment of mice with SENVs alleviated pathological liver modifications, hemorrhage, and cell death [33].

### 26.4.3 Therapeutic Effects on Cancer

Researchers have demonstrated that citrus lemon-derived ENVs (CLENVs) can exert antiproliferative effects on various cancer cells, including lung cancer, colon cancer, and leukemia cells, but exhibit low toxicity in normal cells [1]. CLENVs can exert important effects in cancer development, including modulation of the balance between pro- and anti-apoptotic signals. When cancer cells are treated with CLENVs, the expression of the pro-apoptotic molecules Bad and Bax is increased, whereas the expression of anti-apoptotic molecules, such as survivin and Bcl-xl, is decreased. CLENVs can upregulate TNF-related apoptosis-inducing ligand (TRAIL) receptor as well as death receptor 5 together with upregulation and release of TRAIL; thus, CLENVs may act through a mechanism in which an autocrine loop induced by CLENVs leads to cancer cell death.

MiRNAs are closely related to tumor proliferation and apoptosis. Plant-derived miRNAs can span species and play regulatory roles in the human body. GIENVs have been shown to curtail colorectal tumorigenesis by downregulating pro-inflammatory cytokines, suppressing intestinal epithelial cell proliferation and apoptosis, and stimulating the overexpression of cyclin D1 mRNA, supporting early cancer onset and tumor progression [2]. Some research groups have investigated whether BENVs can inhibit lung tumor cell proliferation via metabolic activity in preliminary studies [34].

In addition, PENVs can exert antitumor activities in cancer immunotherapy. Cao et al. demonstrated that GSENVs significantly promote the polarization of bone marrow-derived macrophages from the M2 to M1 phenotype in vitro. GSENVs significantly suppress melanoma growth by increasing the presence of M1 macrophages in tumor tissues of mice bearing B16F10 cells. This polarization depends on the interaction between Toll-like receptor-4 and some components of GSENVs, including ceramide, lipids, and proteins [27].

### 26.4.4 Therapeutic Effects on the Pathogenicity of *Porphyromonas gingivalis*

GRENVs are selectively taken up by the periodontal pathogen *P. gingivalis* in a GRENV PA-dependent manner via interactions with hemin-binding protein 35 (HBP35) on the surface of *P. gingivalis* cells [35]. The degree of unsaturation plays a critical role in GRENV-mediated interactions with HBP35. Upon binding to HBP35, the pathogenic mechanisms of *P. gingivalis* are significantly suppressed following interaction with GRENV cargo molecules, including PA and miRNAs. Treatment with GRENVs significantly blocks *P. gingivalis*-induced alveolar bone loss in a mouse model.

### 26.4.5 Protection of the Vascular System Against Various Stressors

Blueberry-derived ENVs (BBENVs) modulate the expression of inflammatory genes in response to TNF-α [36]. Indeed, pretreatment with BBENVs counters TNF-α-induced ROS generation and loss of cell viability and modulates the differential expression of 29 genes (fold change >1.5) induced by TNF-α compared with the control, revealing the involvement of the BBENVs in 340 canonical pathways, 121 Kyoto Encyclopedia of Genes and Genomes pathways, and 121 Gene Ontology biological processes. Analysis of the intersection between differentially expressed genes and miRNAs contained in BBENVs revealed a set of candidate target genes, including prostaglandin I2 synthase, mitogen-activated protein kinase 14, and phosphodiesterase 7A. Thus, BBENVs could be considered candidate therapeutic carriers of bioactive compounds that protect the vascular system against various stressors.

### 26.4.6 Drug Carriers

To avoid the complex interactions of PENV cargo, lipids derived from PENVs have been prepared and used to formulate empty nanocarriers. The lipids obtained from GFENVs have been used to encapsulate curcumin, FAs, and zymosan A, demonstrating good stability at 4 °C for more than 1 month and maintaining the biological functions of the drugs [5]. GFENVs with FAs were administered intravenously to facilitate the transport of the chemotherapeutic drug paclitaxel (PTX) to the tumor location; this inhibited tumor growth in CT26 and SW620 tumor cell xenograft models [12].

Lipids from GFENVs coated with inflammatory-related receptor-enriched membranes of activated leukocytes were used to deliver doxorubicin (DOX) or curcumin to the inflamed tumor site [6]. Compared with blank GFENVs, lipids from GFENVs with receptor-enriched membranes had better targeting effects in various mouse models of inflammation-driven disorders. Moreover, lipids from GFENVs encapsulated with *miR-18a* were taken up by liver Kupffer cells and were visualized by confocal microscopy 1 and 24 h after intravenous injection [37]. GFENV *miR-18a* treatment simulated the activity of M1 macrophages and blocked that of M2 macrophages. Fourteen days after intrasplenic injection of CT26 colon tumor cells, the numbers and sizes of liver metastases were reduced, and the survival of mice was prolonged. GIENV lipids encapsulating siRNAs were prepared with relatively uniform sizes (189.5 nm) [14]. Loading of *CD98* siRNA into GIENV lipids resulted in delivery of a low dose of CD98 siRNA and reduced the expression of colonic *CD98*. Additionally, Zhang et al. reported the activity of GIENVs in colon cancer [12]. The chemotherapeutic drug DOX encapsulated in GIENVs prolonged the release of DOX from GIENVs and induced rapid apoptosis. Collectively, these findings suggested that the combination of reassembled lipids with therapeutic drugs could be a promising drug delivery system with enhanced antitumor activity.

In addition, PENVs can encapsulate or conjugate with a variety of cargo, such as small molecule drugs, siRNAs, DNA expression vectors, proteins, and synthetic nanoparticles, enabling delivery to different tissues [6, 12, 37, 38]. Notably, FA-targeted GFENVs mixed with polyethylenimine were used to deliver RNAs and treat glioma through an intranasal route [39]. Mice bearing GL26 cells intranasally treated with FA-pGEFNV/*miR-17* showed inhibition of tumor growth. Furthermore, aptamer-conjugated GFENVs can be used to deliver DOX and P-glycoprotein siRNA for the treatment of colon cancer [40]. Wang et al. reported a biomimetic delivery system for patching DOX-loaded heparin-based nanoparticles (EV-DNs) onto the surface of natural GFENVs. The patching strategy allows an unprecedented fourfold drug-loading capacity compared with traditional encapsulation of EVs. Biomimetic EV-DNs can bypass the blood–brain barrier (BBB)/blood–brain tumor barrier and penetrate glioma tissues through receptor-mediated transcytosis and membrane fusion. A high abundance of EV-DNs can be detected in glioma tissues, enabling maximal brain tumor uptake of EV-DNs and strong antiglioma efficacy in vivo [38]. In addition, the in vivo tolerogenic effects of sulforaphane (SFN) contained in BENVs was reported in DSS-induced acute colitis in mice [31]. Both knockout and knock-in strategies were used to verify the effects of BENV SFN, and BENV SFN has been shown to enhance the effects of BENVs based on histological scores and reduced colon length in comparison with the SFN-knockout strategy [11]. In addition, encapsulation of MTX into GFENVs showed fewer toxic effects and higher inhibitory effects than free MTX for DSS-induced colitis in mice. Additionally, PENVs have the potential to be exploited as personalized oral delivery tools; indeed, GFENVs have been used to specifically target intestinal cells and macrophages. Therefore, recipient cells can be modulated according to the therapeutic goals of the treatment.

When PENVs are topically applied to the skin, they will penetrate via a "lipid-rich channel" coating on hair follicles [41, 42]. PENVs structurally resemble liposomes and penetrate the skin via a transfollicle route [43]. Due to similarities in the membrane surface between PENVs and mammalian cells, PENVs can efficiently pass through the stratum corneum via both transcellular and intercellular routes through the lipid fusion effect. Additionally, BENVs can penetrate deeply into the skin tissue owing to their high lipophilicity. Enhanced fusion of BENVs with keratinocytes can be enhanced via transport of an encapsulated fluorescent agent into the cells and the presence of broccoli proteins in the plasmalemma of keratinocytes [44].

### 26.4.7 Clinical Trials

Three PENV-based clinical trials were registered by the same sponsor (University of Louisville) [45]. GRENVs are currently in clinical trials and are being tested in humans to avert oral mucositis associated with chemoradiation therapy for head and neck cancer (ClinicalTrials.gov: NCT01668849). Furthermore, aloe- and ginger-derived ENVs are also undergoing human clinical trials for the treatment and mitigation of symptoms (insulin resistance and chronic inflammation) associated with polycystic ovary syndrome (ClinicalTrials.gov: NCT03493984). Clinical trials are underway to investigate PENVs with curcumin for the treatment of normal colon tissue and colon tumors (ClinicalTrials.gov: NCT01294072). In particular, the increasing interest in the antitumor action of PENVs is related to their multiple functionalities, including their intrinsic therapeutic capability and the ability to deliver chemotherapeutic drugs. Moreover, owing to their unique transcytosis, elevated permeability, and retention effects, PENVs have promising potential in the fabrication of drug delivery systems for cancer treatment.

## 26.5 ENVs in Chinese Herbal Medicine

### 26.5.1 Overview of ENVs and TCM Treatment

ENVs are closely related to TCM syndromes [46] because the contents of ENVs can be used as biomarkers for clinical disease diagnosis; this is closely related to the concept of microscopic syndrome differentiation in TCM syndrome differentiation and treatment. Microscopic syndrome differentiation involves syndrome standardization, and its development is inseparable from the scientific thinking of combining disease and syndrome. The presence of disease markers that can be used for diagnosis in ENVs is the basis for this approach. Additionally, TCMs can induce changes through modulation of the composition and surface factors of ENVs, thereby alleviating disease. A series of studies on ENVs stimulated by the specific mechanisms and pathways of TCMs has promoted the development of TCMs, which may have applications in the clinical setting and could provide insights and a theoretical basis for the treatment of various diseases.

TCMs have good clinical diagnosis and treatment effects and can be used to prevent and treat diseases through multiple targets and channels. However, there are problems with TCMs, including the low solubility of active ingredients, nonselective distribution in the body, and reduced bioavailability; these features limit the applications of TCMs. Based on their characteristics and physiological functions,

ENVs may be used to improve the therapeutic efficacy of TCMs to achieve good clinical outcomes owing to the synergistic advantages of TCMs and ENVs [47, 48].

Clinical trials have verified the safety of ENVs in the treatment of many diseases [49], and disease-related studies on ENVs have been carried out in vitro and in vivo. ENVs exhibit efficacy in the treatment of tumors, cardiovascular and cerebrovascular diseases, and other refractory diseases. As traditional medicines, TCMs can effectively regulate relevant organs through multiple channels and targets, thereby exerting clinical efficacy. With the development of big data and precision medicine, it is necessary to analyze, purify, and study the active ingredients of TCMs and to exploit safer and more effective drug carriers, such as EVs.

### 26.5.2 Applications of TCM-Derived ENVs in Tumor Treatment

To maintain the overall balance of the human body, TCM-based treatments for tumors do not act directly on tumors but rather enhance the body's immunity by strengthening the body and regulating the internal organs, yin and yang, qi, and blood; correcting the disordered internal environment; and restoring normal physiological functions, thereby inducing cell proliferation or apoptosis, alleviating clinical symptoms, and improving patient quality of life [50].

#### 26.5.2.1 Turmeric

Turmeric is a dried rhizome of *Curcuma longa* L., a plant in the ginger family. Turmeric breaks the blood, promotes qi, regulates menstruation, and relieves pain. In recent years, two types of turmeric extracts (curcumin and β-elemene) have been studied in the context of ENVs.

Curcumin can inhibit various oncogene-driven cell signaling pathways, thereby preventing or treating many types of cancers, including colorectal cancer, pancreatic cancer, breast cancer, prostate cancer, and lung cancer [51]. Curcumin also affects the expression of miRNA, small noncoding regulatory RNAs that modulate gene expression by targeting the 3′-untranslated region of mRNA, thereby affecting the occurrence and progression of diseases [52]. β-Elemene, a natural compound extracted from turmeric, is a new type of noncytotoxic anticancer drug. Studies have shown that β-elemene may be effective against tumor chemotherapy resistance [53–56].

##### Curcumin Upregulates Transcription Factor 21 (TCF21) to Inhibit ENV-Induced Lung Cancer

Curcumin downregulates the methyltransferase DNMT1, and DNMT1 has inhibitory effects on the lung cancer biomarker TCF21. Downregulation of TCF21 is associated with cancer stem cells and promotes tumor cell migration, invasion, and metastasis related to epithelial transformation; thus, curcumin can inhibit the appearance of cancer stem cells by upregulating the expression of TCF21, thereby inhibiting EV-induced lung cancer [57].

### *miR-21* Plays Key Roles in the Anticancer Effects of Curcumin

Curcumin also affects many miRNAs, which are important regulators of oncogenes and tumor suppressors. Curcumin can reduce *miR-21* levels by increasing the excretion of *miR-21* in ENVs and inhibiting the transcription of *miR-21*, an important tumor suppressor. Inhibition of *miR-21* mediates the anticancer effects of curcumin, including cell proliferation, apoptosis, migration, stem cell maintenance, and drug resistance, through the phosphatase and tensin homolog (PTEN)/phosphatidylinositol 3-kinase (PI3K)/AKT, programmed cell death 4, and nuclear factor (NF)-κB pathways [58].

### Curcumin Reverses the Immunosuppressive Effects of Natural Killer (NK) Cells on Tumor Cytotoxicity Mediated by Breast Cancer ENVs

This suppression is mediated by damage to the ubiquitin-proteasome system [59].

### Curcumin Regulates the Functions of ENVs Derived from Pancreatic Cancer Cells

Pancreatic cancer cells treated with curcumin can secrete curcumin-carrying ENVs and deliver curcumin to recipient pancreatic cancer cells, thereby promoting antitumor effects via ENVs. Indeed, ENVs can deliver curcumin to exert anticancer effects on pancreatic cancer cells, and these anticancer effects can expand to the tumor microenvironment [60].

### Curcumin-Loaded ENVs Targeting Neuropilin-1 (NRP-1) Promote Simultaneous Imaging and Treatment of Gliomas In Vivo and In Vitro

Glioma treatment is limited by two main factors: timely detection at the onset or recurrence and BBB restriction. ENVs have strong cargo-loading capacity and the ability to bypass the BBB. Therefore, ENVs are promising targets for drug delivery. NRP-1 is a transmembrane glycoprotein that is overexpressed in glioma cells and the tumor vascular endothelium, but is downregulated or not expressed in normal nerve cells and other tissues, suggesting that NRP-1 may have applications as a targeting ligand for glioma. Curcumin-loaded ENVs with NRP-1 targeting capacity can bypass the BBB smoothly, thereby providing good efficacy for targeted imaging and treatment of gliomas [61].

### Milk-Derived ENVs Can be Used as Carriers to Improve the Tissue Bioavailability and Efficacy of Curcumin

ENV preparation overcomes the low oral bioavailability of curcumin and increases its tissue accumulation. Compared with free curcumin, milk-derived ENVs as carriers of curcumin show significantly stronger anti-inflammatory activity in human lung cancer and breast cancer cells and show enhanced therapeutic effects in many cancer cell lines (including breast cancer, lung cancer, and cervical cancer cells). Therefore, milk-derived ENVs can be effective nanocarriers for curcumin and other plant bioactive substances [62].

#### Grapefruit-Derived Nanocarriers (GFENVs) Activate Leukocyte Transport Pathways to Deliver Therapeutic Agents to Inflammatory Tumor Sites

Encapsulation of GFENVs with activated leukocyte membranes enriched in inflammation-related receptors can enhance the homing ability of GFENVs to inflammation sites and tumor tissues; injection of GFENVs loaded with DOX or curcumin significantly enhances the inhibitory effects of adriamycin or curcumin on the growth of breast tumors and colon tumors and reduces DSS-induced colitis [6].

#### β-Elemene Reverses Resistance in Breast Cancer by Mediating the Expression of Multidrug Resistance (MDR)-Related Genes in ENVs

Adriamycin and docetaxel are the two most commonly used chemotherapeutic drugs for breast cancer. An important cause of chemotherapy failure in breast cancer is multidrug resistance (MDR). β-Elemene is a novel, noncytotoxic anticancer drug. Previous studies have shown that β-elemene may be an effective MDR reversal agent in tumor chemotherapy; however, the relevant mechanism is not yet known. Importantly, β-elemene has strong anticancer effects and can induce apoptosis in breast cancer BCA cells by mediating the expression of MDR-related miRNAs, *PTEN* genes, and the gene encoding P-glycoprotein in ENVs, thereby reversing the MDR of breast cancer cells and improving treatment efficacy. Elucidation of the molecular mechanisms through which β-elemene reverses tumor resistance and strategic in-depth exploration of the mechanisms for reducing MDR gene expression will help establish alternative methods for improving chemotherapy treatment [63].

Furthermore, β-elemene promotes the release of ENVs by upregulating p53 mRNA and protein in A549 cells, and the released ENVs have inhibitory effects on cell growth, suggesting efficacy in the treatment of human lung cancer [64].

### 26.5.2.2 PTX

PTX is a tricyclic diterpenoid isolated from the bark and tender leaves of *Taxus chinensis* and is the main anticancer and antitumor component of *Taxus chinensis*. PTX has strong anticancer effects, with efficacy rates of up to 75%. PTX is mainly used to treat advanced breast cancer, lung cancer [65], ovarian cancer, head and neck cancer, soft tissue cancer, and gastrointestinal cancers. PTX exerts antitumor effects through three main mechanisms: modulation of microtubule depolymerization and stability [66], stimulation of the immune response [67], and induction of cancer cell apoptosis [65].

#### PTX Enriches Survivin Protein in Breast Cancer Cell-Derived ENVs to Promote Cell Survival and Drug Resistance

PTX is a chemotherapeutic agent that stabilizes microtubules. Treatment of highly aggressive MDA-MB-231 breast cancer cells with PTX causes the cells to produce more ENVs and increases survivin expression in these ENVs. These ENVs can metastasize to other cancer or normal cells within the tumor microenvironment and promote their survival [68]. Therefore, combining PTX with ENV biogenesis inhibitors may overcome PTX resistance and improve the therapeutic efficacy of this chemotherapeutic drug.

**ENVs Derived from Autologous Prostate Cancer Cells Deliver PTX to Parent Cells**

PTX can be introduced into the cells through intracellular pathways to increase its cytotoxicity in autologous prostate cancer [69].

**Matrix-Derived *miR-21*-Induced ENVs Confer PTX Resistance in Ovarian Cancer Cells by Targeting Apoptotic Peptidase Activating Factor 1 (APAF1)**

ENVs harboring *miR-21* can confer chemoresistance and aggressive phenotypes in ovarian cancer cells through release from neighboring stromal cells, indicating that preventing ENV-dependent transport of *miR-21* from stromal cells may be a new strategy to inhibit ovarian cancer growth. In addition, as a direct target of *miR-21*, APAF1 can confer chemoresistance in ovarian cancer, indicating that strategies based on upregulation of APAF1 in ovarian cancer cells can increase PTX sensitivity in ovarian cancer cells [70].

**PTX-Resistant Gastric Cancer Cells Promote the Epithelial-Mesenchymal Transition and Drug Resistance in PTX-Sensitive Cells Through Delivery of *miR-155-5p* by ENVs**

PTX is widely used as a first-line chemotherapy agent for gastric cancer; however, in most cases, tumor resistance to PTX limits the effectiveness of chemotherapy and leads to treatment failure. Therefore, an in-depth understanding of the mechanisms of PTX chemotherapy resistance may help to establish novel treatment methods and more comprehensive means for gastric cancer treatment. The expression of *miR-155-5p* is significantly increased in the PTX-resistant gastric cancer cell line MGC-803R. Additionally, ENVs harboring *miR-155-5p* can inhibit the expression of GATA3 and tumor protein p53-inducible nuclear protein 1 to promote the epithelial-mesenchymal transition and PTX resistance in PTX-sensitive cells. Therefore, targeting *miR-155-5p* may be a new strategy for overcoming PTX resistance in gastric cancer [71].

#### 26.5.2.3 Shikonin

Shikonin is a naphthoquinone compound isolated from TCMs and has been used to treat various types of tumors. Importantly, Shikonin has anti-inflammatory, antioxidant, anticancer, wound healing, and antibacterial therapeutic effects [72, 73].

**ENVs Targeting Pre-adipocytes Can Inhibit Signal Transduction from Pre-adipocytes to Cancer Stem Cells in Early Breast Cancer**

Shikonin inhibits the differentiation of pre-adipocytes through the *miR-140*/SOX9 axis. Because pre-adipocytes can promote the progression of breast cancer, this indicates that shikonin may be an effective chemopreventive compound that can target tumor cells and activate the tumor microenvironment. Moreover, shikonin-treated pre-adipocytes secrete ENVs with high levels of *miR-140*, which can reduce tumor cell migration in nearby ductal carcinoma cells in situ by targeting SOX9 signaling. This strongly suggests that intercellular signals mediated by ENVs play important roles in regulating breast tumorigenesis [74].

**Shikonin Inhibits Human Breast Cancer Cell Proliferation by Reducing the Number of Tumor-Derived ENVs**

*miR-128* in tumor-derived ENVs downregulates Bax in recipient human breast cancer cells (MCF-7) and promotes cell proliferation. Shikonin can also modulate the secretion of ENVs from MCF-7 breast cancer cells, and EV levels are positively correlated with cell proliferation. Therefore, shikonin treatment reduces the secretion of ENVs and inhibits MCF-7 cells [75].

#### 26.5.2.4 Celastrol (CEL)

CEL is a plant-derived triterpenoid isolated from *Tripterygium wilfordii*. This compound inhibits heat shock protein 90 and the NF-κB activation pathway and activates transforming growth factor (TGF) β/Smad signaling. Furthermore, CEL has anti-inflammatory, antiproliferative, and antitumor effects [76–79].

*Tripterygium wilfordii* effectively inhibits the proliferation of human non-small cell lung cancer through inhibition of cell cycle progression and NF-κB activation and induces apoptosis in cancer cells mediated by chronic endoplasmic reticulum stress. In addition, encapsulating CEL in milk-derived ENVs can enhance therapeutic efficacy, reduce dose-related toxicity, and improve the safety and effectiveness of CEL in the treatment of lung cancer [80].

#### 26.5.2.5 Camptothecin

Camptothecin is a quinoline and isoquinoline alkaloid extracted. Because colon cancer is sensitive to this compound, camptothecin can effectively inhibit tumor growth and DNA topoisomerase to prevent DNA replication and RNA synthesis; thus, this compound may have broad applications in clinical practice. However, the hydrophobicity and lactone instability of camptothecin limit its applications in clinical treatment. Therefore, using ENVs as carriers to deliver camptothecin may be a good strategy for solving this problem.

Camptothecin-loaded ENVs have better ability to penetrate cell membranes than camptothecin alone, enabling the drug to exert stronger effects on killing tumor cells and inhibiting tumor growth and thereby resulting in lower cytotoxicity [81].

#### 26.5.2.6 Rotenone

Rotenone is an isoflavone and respiratory chain complex-1 inhibitor that induces the production of ROS and causes oxidative damage to mitochondria. Mitochondria regulate cell necrosis, apoptosis, mitochondrial dysfunction, and antioxidant production through mitochondrial enzyme activity. Rotenone can induce autophagy and mitochondrial damage in several organisms, including insects and fishes.

Rotenone can induce autophagy in retinal pigment epithelial cells by inhibiting mitochondrial activity and releasing additional cell microvesicles and ENVs and can also induce the release of ENVs from prostate cancer and breast cancer stem cells [82].

#### 26.5.2.7 Small Compounds that Replenish Qi

The small Yiqi compound is composed of astragalus and codonopsis. ENVs derived from triple-negative breast cancer may transmit drug resistance information. Drug-resistant cell ENVs can cause drug-sensitive cells to obtain a certain degree of drug resistance, and Yiqi small compounds can block the transmission of these ENVs.

### 26.5.3 Applications of TCM-Derived ENVs in the Treatment of Other Diseases

ENVs carry specific biological information about their cells of origin and mediate the disease process [83]. Therefore, ENVs can be used as specific biomarkers for disease diagnosis and for analysis of the specific mechanisms of various diseases and syndromes [46]. Additionally, ENVs can be used as carriers to deliver drugs to corresponding target organs, thereby exerting important biological effects. ENVs can also play important roles in regulating the immune and inflammatory pathways.

#### 26.5.3.1 Curcumin

##### Curcumin Inhibits Chronic Myeloid Leukemia (CML) Cell Growth Through ENV *miR-21* and Affects the Angiogenic Phenotype

Curcumin can inhibit leukemia cell growth by selective sorting of *miR-21* in CML cells, which leads to a decrease in cellular *miR-21* and subsequent upregulation of PTEN. Simultaneously, curcumin may alter the components of CML ENVs at the protein and miRNA levels, thereby regulating endothelial barrier tissue and reducing the angiogenic phenotype. Accordingly, curcumin may have applications in the treatment of CML via dual effects on cancer cells and the tumor microenvironment [84].

##### Curcumin Promotes Lipid Excretion Through ENVs

Curcumin may promote the excretion of cholesterol and other lipids from cells through ENVs and microvesicles, thereby improving U18666A-induced lysosomal cholesterol accumulation and transport [85].

##### Curcumin Modulates Intracellular Lipid Homeostasis

Curcumin stimulates the removal of cholesterol and other lipids that accumulate in lysosomal chambers and normalizes intracellular lipid homeostasis, which may reduce the adverse metabolic effects of antipsychotic drugs [86].

##### Curcumin-Induced ENVs Affect Endothelial Cell (EC) Function

ENVs produced by curcumin-treated cells can reduce oxidative stress, tight junctions, adhesion junction proteins, and EC layer permeability, thereby reducing EC dysfunction during hyperhomocysteinemia [87].

**Curcumin-Induced ENVs Modulate LPS-Mediated Inflammation**

ENVs triggered by curcumin can reduce LPS-mediated inflammation by reversing LPS-induced expression of pro-inflammatory genes in buffalo granule cells [48].

**ENVs Act as Curcumin Carriers to Improve the Efficacy of Curcumin**

Curcumin-encapsulated ENVs are highly bioavailable, soluble, and safe and can reach high concentrations in the blood. Furthermore, these ENVs show no toxic effects, do not induce immune responses, and exhibit good therapeutic potential [88]. Therefore, curcumin-encapsulated ENVs are safer carriers to facilitate the treatment of specific target organs or cells as alternative to other synthetic nanoparticles and may enhance efficacy in the treatment of oxidation stress, brain disease, cholesterol-related diseases, and endothelial dysfunction [89]. For example, nanoformulations combining curcumin (anti-inflammatory and neuroprotective molecules) and embryonic stem cell ENVs can promote neurovascular recovery in mice after ischemia reperfusion injury [90], and inflammatory drugs encapsulated by ENVs can also be administered through the nasal cavity to the brain to treat brain inflammatory diseases [91]. However, there are still major limitations to overcome, including the standardization of ENV separation, purification, and characterization methods; effective drug loading and surface modification strategies; and optimal administration routes.

#### 26.5.3.2 Ginger

Diet can regulate the composition and function of the intestinal microbiota in the human body and other mammals, thereby preventing or treating various diseases. Plant-derived EV-like nanoparticles (PENVs) can be absorbed by the intestinal flora and contain RNA that changes the composition of the flora and physiology of the host. Ginger ENVs (GENVs) are preferentially taken up by *Lactobacillus* species in a GENV lipid-dependent manner and contain miRNAs that target various genes in *Lactobacillus rhamnosus* (LGG). The targeting effect of LGG monooxygenase ycnE mediated by GELN *mdo-miR-7267-3p* improves the production of indole-3-carboxylic acid (I3A). As ligands for aryl hydrocarbon receptors, GENV RNAs and I3A induce the production of IL-22, which is related to improvement of barrier function, thereby alleviating symptoms of colitis in mice. Accordingly, plant products and their effects on microbial communities can be used to target specific host processes to alleviate disease processes [22].

#### 26.5.3.3 *Astragalus* Polysaccharide (APS)

APS is a clinically safe and effective antitumor polysaccharide and can improve human immune function, promote cell metabolism, increase the activity of macrophages, and activate T cells, B cells, NK cells, and other immune cells.

Moreover, APS upregulates ENV surface factors secreted by dendritic cells, thereby promoting the maturation of bone marrow monocyte-derived dendritic cells in patients with leukemia [92].

#### 26.5.3.4 Catalpol

Catalpol is a low-molecular-weight natural compound extracted from iridoid glycosides in *Rehmannia glutinosa*. This compound promotes nerve repair and remodeling through antioxidative mechanisms, regulates cellular immunity, and increases the expression of neurotrophic factors. Furthermore, catalpol may have applications in the treatment of neurodegenerative diseases [93–95].

Catalpol can promote the secretion of ENVs from SH-SY5Y cells, and catalpol ENVs have neuroprotective effects on the damage caused by low serum in SH-SY5Y cells [96].

#### 26.5.3.5 Berberine (BBR)

BBR is the main component of the rhizome of *Coptis rhizoma*. BBR lowers blood sugar, regulates abnormal lipid metabolism, exerts anti-inflammatory effects, improves insulin resistance, and reduces kidney damage.

ENVs derived from glomerular mesangial cells show elevation of TGFβ1 levels under high-glucose conditions, leading to podocyte damage, whereas BBR protects podocytes by regulating the TGFβ1R/PI3K/AKT pathway [97].

#### 26.5.3.6 Buyang Huanwu Decoction

The Buyang Huanwu decoction is composed of seven Chinese herbal medicines, including astragalus, angelica, red peony root, chuanxiong, safflower, peach kernel, and earthworms. This decoction is commonly used to treat sequelae of cerebrovascular accidents, coronary heart disease, polio sequelae, and other causes of hemiplegia, paraplegia, unilateral upper or lower extremity dysfunction, and blood stasis syndrome.

Exposure to Buyang Huanwu decoction can increase the concentration of angiogenic miRNAs and vascular endothelial growth factor in ENVs secreted by mesenchymal stem cells, which mediates the activation of endothelial cells in vitro and promotes rat brain angiogenesis in ischemic injury [98].

#### 26.5.3.7 Tongxinluo

Tongxinluo is composed of ginseng, leech, whole scorpion, red peony root, cicada slough, soil beetle, centipede, sandalwood, odorifera, frankincense (prepared), jujube seeds (fried), and borneol. It is mainly used for the treatment of coronary heart disease and angina pectoris, which are syndromes of heart qi deficiency, blood stasis, and collateral obstruction.

Tongxinluo-pretreated cardiomyocytes activate the RISK pathway in cocultured human myocardial microvascular ECs (HCMECs) through ENVs to increase the phosphorylation of p70 S6K1 and glycogen synthase kinase-3β, which may reduce hypoxia-reoxygenation damage and protect HCMECs [99].

## References


1. Raimondo S, Naselli F, Fontana S, Monteleone F, Lo Dico A, Saieva L, et al. Citrus Limon-derived nanovesicles inhibit cancer cell proliferation and suppress CML xenograft growth by inducing TRAIL-mediated cell death. Oncotarget. 2015;6(23):19514–27.
2. Zhang M, Viennois E, Prasad M, Zhang Y, Wang L, Zhang Z, et al. Edible ginger-derived nanoparticles: a novel therapeutic approach for the prevention and treatment of inflammatory bowel disease and colitis-associated cancer. Biomaterials. 2016;101:321–40.
3. Mu J, Zhuang X, Wang Q, Jiang H, Deng Z-B, Wang B, et al. Interspecies communication between plant and mouse gut host cells through edible plant derived exosome-like nanoparticles. Mol Nutr Food Res. 2014;58(7):1561–73.
4. Samanta S, Rajasingh S, Drosos N, Zhou Z, Dawn B, Rajasingh J. Exosomes: new molecular targets of diseases. Acta Pharmacol Sin. 2018;39(4):501–13.
5. Wang Q, Zhuang X, Mu J, Deng Z-B, Jiang H, Zhang L, et al. Delivery of therapeutic agents by nanoparticles made of grapefruit-derived lipids. Nat Commun. 2013;4:1867.
6. Wang Q, Ren Y, Mu J, Egilmez NK, Zhuang X, Deng Z, et al. Grapefruit-derived nanovectors use an activated leukocyte trafficking pathway to deliver therapeutic agents to inflammatory tumor sites. Cancer Res. 2015;75(12):2520–9.
7. Schuh CMAP, Cuenca J, Alcayaga-Miranda F, Khoury M. Exosomes on the border of species and kingdom intercommunication. Transl Res. 2019;210:80–98.
8. Fan Y, Zhang Q. Development of liposomal formulations: from concept to clinical investigations. Asian J Pharm Sci. 2013;8(2):81–7.
9. Fujita D, Arai T, Komori H, Shirasaki Y, Wakayama T, Nakanishi T, et al. Apple-derived nanoparticles modulate expression of organic-anion-transporting polypeptide (OATP) 2B1 in Caco-2 cells. Mol Pharm. 2018;15(12):5772–80.
10. Yang C, Zhang M, Merlin D. Advances in plant-derived edible nanoparticle-based lipid nano-drug delivery systems as therapeutic nanomedicines. J Mater Chem B. 2018;6(9):1312–21.
11. Wang B, Zhuang X, Deng Z-B, Jiang H, Mu J, Wang Q, et al. Targeted drug delivery to intestinal macrophages by bioactive nanovesicles released from grapefruit. Mol Ther. 2014;22(3):522–34.
12. Zhang M, Xiao B, Wang H, Han MK, Zhang Z, Viennois E, et al. Edible ginger-derived nano-lipids loaded with doxorubicin as a novel drug-delivery approach for colon cancer therapy. Mol Ther. 2016;24(10):1783–96.
13. Lu M, Xing H, Yang Z, Sun Y, Yang T, Zhao X, et al. Recent advances on extracellular vesicles in therapeutic delivery: challenges, solutions, and opportunities. Eur J Pharm Biopharm. 2017;119:381–95.
14. Zhang M, Wang X, Han MK, Collins JF, Merlin D. Oral administration of ginger-derived nanolipids loaded with siRNA as a novel approach for efficient siRNA drug delivery to treat ulcerative colitis. Nanomedicine (Lond). 2017;12(16):1927–43.
15. Li Z, Wang H, Yin H, Bennett C, Zhang H-G, Guo P. Arrowtail RNA for ligand display on ginger exosome-like nanovesicles to systemic deliver siRNA for cancer suppression. Sci Rep. 2018;8(1):14644.
16. Ha D, Yang N, Nadithe V. Exosomes as therapeutic drug carriers and delivery vehicles across biological membranes: current perspectives and future challenges. Acta Pharm Sin B. 2016;6(4):287–96.
17. Baldrich P, Rutter BD, Karimi HZ, Podicheti R, Meyers BC, Innes RW. Plant extracellular vesicles contain diverse small RNA species and are enriched in 10- to 17-nucleotide "tiny" RNAs. Plant Cell. 2019;31(2):315–24.
18. Xiao J, Feng S, Wang X, Long K, Luo Y, Wang Y, et al. Identification of exosome-like nanoparticle-derived microRNAs from 11 edible fruits and vegetables. PeerJ. 2018;6:e5186.
19. de Gassart A, Geminard C, Fevrier B, Raposo G, Vidal M. Lipid raft-associated protein sorting in exosomes. Blood. 2003;102(13):4336–44.

20. Subra C, Laulagnier K, Perret B, Record M. Exosome lipidomics unravels lipid sorting at the level of multivesicular bodies. Biochimie. 2007;89(2):205–12.
21. Wang X, Devaiah SP, Zhang W, Welti R. Signaling functions of phosphatidic acid. Prog Lipid Res. 2006;45(3):250–78.
22. Teng Y, Ren Y, Sayed M, Hu X, Lei C, Kumar A, et al. Plant-derived exosomal microRNAs shape the gut microbiota. Cell Host Microbe. 2018;24(5):637–52.e8.
23. Théry C, Zitvogel L, Amigorena S. Exosomes: composition, biogenesis and function. Nat Rev Immunol. 2002;2(8):569–79.
24. Kowal J, Arras G, Colombo M, Jouve M, Morath JP, Primdal-Bengtson B, et al. Proteomic comparison defines novel markers to characterize heterogeneous populations of extracellular vesicle subtypes. Proc Natl Acad Sci. 2016;113(8):E968–E77.
25. Ju S, Mu J, Dokland T, Zhuang X, Wang Q, Jiang H, et al. Grape exosome-like nanoparticles induce intestinal stem cells and protect mice from DSS-induced colitis. Mol Ther. 2013;21(7):1345–57.
26. Stanly C, Moubarak M, Fiume I, Turiák L, Pocsfalvi G. Membrane transporters in citrus Clementina fruit juice-derived nanovesicles. Int J Mol Sci. 2019;20(24)
27. Cao M, Yan H, Han X, Weng L, Wei Q, Sun X, et al. Ginseng-derived nanoparticles alter macrophage polarization to inhibit melanoma growth. J Immunother Cancer. 2019;7(1):326.
28. Li JWH, Vederas JC. Drug discovery and natural products: end of an era or an endless frontier? Science (New York, NY). 2009;325(5937):161–5.
29. Butler MS. The role of natural product chemistry in drug discovery. J Nat Prod. 2004;67(12):2141–53.
30. Newman DJ, Cragg GM. Natural products as sources of new drugs from 1981 to 2014. J Nat Prod. 2016;79(3):629–61.
31. Deng Z, Rong Y, Teng Y, Mu J, Zhuang X, Tseng M, et al. Broccoli-derived nanoparticle inhibits mouse colitis by activating dendritic cell AMP-activated protein kinase. Mol Ther. 2017;25(7):1641–54.
32. Zhuang X, Deng Z-B, Mu J, Zhang L, Yan J, Miller D, et al. Ginger-derived nanoparticles protect against alcohol-induced liver damage. J Extracell Vesicles. 2015;4:28713.
33. Maes M, Vinken M, Jaeschke H. Experimental models of hepatotoxicity related to acute liver failure. Toxicol Appl Pharmacol. 2016;290:86–97.
34. Di Gioia S, Hossain MN, Conese M. Biological properties and therapeutic effects of plant-derived nanovesicles. Open Med (Warsaw, Poland). 2020;15(1):1096–122.
35. Sundaram K, Miller DP, Kumar A, Teng Y, Sayed M, Mu J, et al. Plant-derived exosomal nanoparticles inhibit pathogenicity of Porphyromonas gingivalis. iScience. 2020;23(2):100869.
36. De Robertis M, Sarra A, D'Oria V, Mura F, Bordi F, Postorino P, et al. Blueberry-derived exosome-like nanoparticles counter the response to TNF-α-induced change on gene expression in EA.hy926 cells. Biomol Ther. 2020;10(5)
37. Teng Y, Mu J, Hu X, Samykutty A, Zhuang X, Deng Z, et al. Grapefruit-derived nanovectors deliver miR-18a for treatment of liver metastasis of colon cancer by induction of M1 macrophages. Oncotarget. 2016;7(18):25683–97.
38. Niu W, Xiao Q, Wang X, Zhu J, Li J, Liang X, et al. A biomimetic drug delivery system by integrating grapefruit extracellular vesicles and doxorubicin-loaded heparin-based nanoparticles for glioma therapy. Nano Lett. 2021;21(3):1484–92.
39. Zhuang X, Teng Y, Samykutty A, Mu J, Deng Z, Zhang L, et al. Grapefruit-derived nanovectors delivering therapeutic miR17 through an intranasal route inhibit brain tumor progression. Mol Ther. 2016;24(1):96–105.
40. Yan W, Tao M, Jiang B, Yao M, Jun Y, Dai W, et al. Overcoming drug resistance in colon cancer by aptamer-mediated targeted co-delivery of drug and siRNA using grapefruit-derived nanovectors. Cell Physiol Biochem. 2018;50(1):79–91.
41. Gu T-W, Wang M-Z, Niu J, Chu Y, Guo K-R, Peng L-H. Outer membrane vesicles derived from E. coli as novel vehicles for transdermal and tumor targeting delivery. Nanoscale. 2020;12(36):18,965–77.

42. Li N, Peng L-H, Chen X, Nakagawa S, Gao J-Q. Transcutaneous vaccines: novel advances in technology and delivery for overcoming the barriers. Vaccine. 2011;29(37):6179–90.
43. Lu M, Zhao X, Xing H, Xun Z, Zhu S, Lang L, et al. Comparison of exosome-mimicking liposomes with conventional liposomes for intracellular delivery of siRNA. Int J Pharm. 2018;550(1–2):100–13.
44. Yepes-Molina L, Martínez-Ballesta MC, Carvajal M. Plant plasma membrane vesicles interaction with keratinocytes reveals their potential as carriers. J Adv Res. 2020;23:101–11.
45. Dad HA, Gu T-W, Zhu A-Q, Huang L-Q, Peng L-H. Plant exosome-like nanovesicles: emerging therapeutics and drug delivery nanoplatforms. Mol Ther. 2021;29(1):13–31.
46. Haiyan J, Zhiyi Y, Yueyun L, Youming J. The application of exosomes in the study of traditional Chinese medicine syndrome. Chin Med Modern Dist Educ China. 2018;16(01):145–8. (In Chinese).
47. Peng Y, Xiao-chen G, Jun-ping Z, Shi-chao L, Ya-ping Z. Research progress of the exosomes as drug delivery vehicles of Chinese herbal drugs. Acta Pharm Sin. 2017;52(11):1667–72. (In Chinese).
48. Vashisht M, Rani P, Sunita OSK, Singh D. Curcumin primed exosomes reverses LPS-induced pro-inflammatory gene expression in buffalo granulosa cells. J Cell Biochem. 2018;119(2):1488–500.
49. Morse MA, Garst J, Osada T, Khan S, Hobeika A, Clay TM, et al. A phase I study of dexosome immunotherapy in patients with advanced non-small cell lung cancer. J Transl Med. 2005;3(1):9.
50. Chu L, Wang T, You J. The significance of exosomes and the holistic view of Chinese medicine in tumour research. Hebei Chin Med. 2018;40(11):1740–3. (In Chinese).
51. Nagaraju GP, Zhu S, Ko JE, Ashritha N, Kandimalla R, Snyder JP, et al. Antiangiogenic effects of a novel synthetic curcumin analogue in pancreatic cancer. Cancer Lett. 2015;357(2):557–65.
52. Teiten MH, Dicato M, Diederich M. Curcumin as a regulator of epigenetic events. Mol Nutr Food Res. 2013;57(9):1619–29.
53. Zhang X, Zhang Y, Li Y. Beta-elemene decreases cell invasion by upregulating E-cadherin expression in MCF-7 human breast cancer cells. Oncol Rep. 2013;30(2):745–50.
54. Zhang B, Zhang K, Liu Z, Hao F, Wang M, Li X, et al. Secreted clusterin gene silencing enhances chemosensitivity of a549 cells to cisplatin through AKT and ERK1/2 pathways in vitro. Cell Physiol Biochem. 2014;33(4):1162–75.
55. Xu HB, Li L, Fu J, Mao XP, Xu LZ. Reversion of multidrug resistance in a chemoresistant human breast cancer cell line by beta-elemene. Pharmacology. 2012;89(5–6):303–12.
56. Guo HQ, Zhang GN, Wang YJ, Zhang YK, Sodani K, Talele TT, et al. Beta-Elemene, a compound derived from Rhizoma zedoariae, reverses multidrug resistance mediated by the ABCB1 transporter. Oncol Rep. 2014;31(2):858–66.
57. Wu H, Zhou J, Zeng C, Wu D, Mu Z, Chen B, et al. Curcumin increases exosomal TCF21 thus suppressing exosome-induced lung cancer. Oncotarget. 2016;7(52):87081–90.
58. Chen J, Xu T, Chen C. The critical roles of miR-21 in anti-cancer effects of curcumin. Ann Transl Med. 2015;3(21):330.
59. Zhang HG, Kim H, Liu C, Yu S, Wang J, Grizzle WE, et al. Curcumin reverses breast tumor exosomes mediated immune suppression of NK cell tumor cytotoxicity. Biochim Biophys Acta. 2007;1773(7):1116–23.
60. Osterman CJ, Lynch JC, Leaf P, Gonda A, Ferguson BH, Griffiths D, et al. Curcumin modulates pancreatic adenocarcinoma cell-derived exosomal function. PLoS One. 2015;10(7):e0132845.
61. Jia G, Han Y, An Y, Ding Y, He C, Wang X, et al. NRP-1 targeted and cargo-loaded exosomes facilitate simultaneous imaging and therapy of glioma in vitro and in vivo. Biomaterials. 2018;178:302–16.
62. Aqil F, Munagala R, Jeyabalan J, Agrawal AK, Gupta R. Exosomes for the enhanced tissue bioavailability and efficacy of curcumin. AAPS J. 2017;19(6):1691–702.

63. Zhang J, Zhang HD, Yao YF, Zhong SL, Zhao JH, Tang JH. Beta-Elemene reverses chemoresistance of breast cancer cells by reducing resistance transmission via exosomes. Cell Physiol Biochem. 2015;36(6):2274–86.
64. Li J, JunYu LA, Wang Y. Beta-Elemene against human lung cancer via up-regulation of P53 protein expression to promote the release of exosome. Lung Cancer. 2014;86(2):144–50.
65. Jiang Y, Jiang W. Clinical pharmacology of paclitaxel (Taxol). Chin J Clin Pharmacol. 1994;04:234–8. (In Chinese).
66. Yang T, Zongqun W, Jie J. Experiment and clinical research progress of Paclitaxel'santi-tumor function. J Dalian Med Univ. 2007;02:197–9. (In Chinese).
67. Xie B, Pan B. Advances in the pharmacology and clinical research of paclitaxel. China Pharm. 2008;15:76–8. (In Chinese).
68. Kreger BT, Johansen ER, Cerione RA, Antonyak MA. The enrichment of survivin in exosomes from breast cancer cells treated with paclitaxel promotes cell survival and chemoresistance. Cancers (Basel). 2016;8(12)
69. Saari H, Lazaro-Ibanez E, Viitala T, Vuorimaa-Laukkanen E, Siljander P, Yliperttula M. Microvesicle- and exosome-mediated drug delivery enhances the cytotoxicity of paclitaxel in autologous prostate cancer cells. J Control Release. 2015;220(Pt B):727–37.
70. Au YC, Co NN, Tsuruga T, Yeung TL, Kwan SY, Leung CS, et al. Exosomal transfer of stroma-derived miR21 confers paclitaxel resistance in ovarian cancer cells through targeting APAF1. Nat Commun. 2016;7:11150.
71. Wang M, Qiu R, Yu S, Xu X, Li G, Gu R, et al. Paclitaxelresistant gastric cancer MGC803 cells promote epithelialtomesenchymal transition and chemoresistance in paclitaxelsensitive cells via exosomal delivery of miR1555p. Int J Oncol. 2019;54(1):326–38.
72. Papageorgiou VP, Assimopoulou AN, Ballis AC. Alkannins and shikonins: a new class of wound healing agents. Curr Med Chem. 2008;15(30):3248–67.
73. Andujar I, Rios JL, Giner RM, Recio MC. Pharmacological properties of shikonin—a review of literature since 2002. Planta Med. 2013;79(18):1685–97.
74. Gernapudi R, Yao Y, Zhang Y, Wolfson B, Roy S, Duru N, et al. Targeting exosomes from preadipocytes inhibits preadipocyte to cancer stem cell signaling in early-stage breast cancer. Breast Cancer Res Treat. 2015;150(3):685–95.
75. Wei Y, Li M, Cui S, Wang D, Zhang CY, Zen K, et al. Shikonin inhibits the proliferation of human breast cancer cells by reducing tumor-derived exosomes. Molecules. 2016;21(6)
76. Yang H, Chen D, Cui QC, Yuan X, Dou QP. Celastrol, a triterpene extracted from the Chinese "thunder of god vine," is a potent proteasome inhibitor and suppresses human prostate cancer growth in nude mice. Cancer Res. 2006;66(9):4758–65.
77. Jin HZ, Hwang BY, Kim HS, Lee JH, Kim YH, Lee JJ. Antiinflammatory constituents of Celastrus orbiculatus inhibit the NF-kappaB activation and NO production. J Nat Prod. 2002;65(1):89–91.
78. Chang FR, Hayashi K, Chen IH, Liaw CC, Bastow KF, Nakanishi Y, et al. Antitumor agents. 228. Five new agarofurans, Reissantins A-E, and cytotoxic principles from Reissantia buchananii. J Nat Prod. 2003;66(11):1416–20.
79. Allison AC, Cacabelos R, Lombardi VR, Alvarez XA, Vigo C. Celastrol, a potent antioxidant and anti-inflammatory drug, as a possible treatment for Alzheimer's disease. Prog Neuro-Psychopharmacol Biol Psychiatry. 2001;25(7):1341–57.
80. Aqil F, Kausar H, Agrawal AK, Jeyabalan J, Kyakulaga AH, Munagala R, et al. Exosomal formulation enhances therapeutic response of celastrol against lung cancer. Exp Mol Pathol. 2016;101(1):12–21.
81. Ren B, Mao Z, Wang L, Zhang F, Meng X. Anti-colon cancer effects of camptothecin-loaded exosomes. Chin J Pract Diagn Ther. 2016;30(03):231–4. (In Chinese).
82. Kumar D, Gupta D, Shankar S, Srivastava RK. Biomolecular characterization of exosomes released from cancer stem cells: possible implications for biomarker and treatment of cancer. Oncotarget. 2015;6(5):3280–91.

83. Wang Bing, Zhang Yuzhu, Chen Hongfeng. Mechanisms of Yiqi compound recipe on inhibiting exosomes transmitting drug resistance information in triple negative breast cancer cells. Acad J Shanghai Univ Tradit Chin Med 2018;32(06):50–55. (In Chinese).
84. Taverna S, Fontana S, Monteleone F, Pucci M, Saieva L, De Caro V, et al. Curcumin modulates chronic myelogenous leukemia exosomes composition and affects angiogenic phenotype via exosomal miR-21. Oncotarget. 2016;7(21):30420–39.
85. Canfran-Duque A, Pastor O, Quintana-Portillo R, Lerma M, de la Pena G, Martin-Hidalgo A, et al. Curcumin promotes exosomes/microvesicles secretion that attenuates lysosomal cholesterol traffic impairment. Mol Nutr Food Res. 2014;58(4):687–97.
86. Canfran-Duque A, Pastor O, Reina M, Lerma M, Cruz-Jentoft AJ, Lasuncion MA, et al. Curcumin mitigates the intracellular lipid deposit induced by antipsychotics in vitro. PLoS One. 2015;10(10):e0141829.
87. Kalani A, Kamat PK, Chaturvedi P, Tyagi SC, Tyagi N. Curcumin-primed exosomes mitigate endothelial cell dysfunction during hyperhomocysteinemia. Life Sci. 2014;107(1–2):1–7.
88. Sun D, Zhuang X, Xiang X, Liu Y, Zhang S, Liu C, et al. A novel nanoparticle drug delivery system: the anti-inflammatory activity of curcumin is enhanced when encapsulated in exosomes. Mol Ther. 2010;18(9):1606–14.
89. Oskouie MN, Aghili MN, Butler AE, Zamani P, Sahebkar A. Therapeutic use of curcumin-encapsulated and curcumin-primed exosomes. J Cell Physiol. 2019;234(6):8182–91.
90. Kalani A, Chaturvedi P, Kamat PK, Maldonado C, Bauer P, Joshua IG, et al. Curcumin-loaded embryonic stem cell exosomes restored neurovascular unit following ischemia-reperfusion injury. Int J Biochem Cell Biol. 2016;79:360–9.
91. Zhuang X, Xiang X, Grizzle W, Sun D, Zhang S, Axtell RC, et al. Treatment of brain inflammatory diseases by delivering exosome encapsulated anti-inflammatory drugs from the nasal region to the brain. Mol Ther. 2011;19(10):1769–79.
92. Lei M. Pro-maturation effect of Astragalus polysaccharide on dendritic cell exosomes in leukemia patients [D]. Lanzhou University; 2009. (In Chinese).
93. Qi-yan C, Zhong-xiang Y. Advances of Catalpol's roles in the nervous system. Progr Modern Biomed. 2010;10(18):3589–91. +97. (In Chinese).
94. Zhu H, Wan D, Zhang F. Advances in the pharmacological effects and mechanisms of cerebral protection of Diosbulbinol. Chin J Tradit Chin Med. 2009;34(23):2983–6. (In Chinese).
95. Zhang X, Yin A, Liu J. Neuroprotective effect and treatment prospect of catalpol in Rehmannia. Liaoning J Tradit Chin Med. 2013;40(06):1273–6. (In Chinese).
96. Zhang Xiaoyi, Zheng Hong, Wang Yongqiang, et al. Protective effects of catalpol exosomes on damaged SH-SY5Y cells induced by low serum medium. Global Tradit Chin Med 2017;10(02):155–158. (In Chinese).
97. Wang Y. TGFβ1R-PI3K/AKT signaling pathway mediates high glucose-induced thylakoid-derived exosome damage to podocytes and the role of berberine [D]. Anhui Medical University; 2018. (In Chinese).
98. Yang J, Gao F, Zhang Y, Liu Y, Zhang D. Buyang Huanwu decoction (BYHWD) enhances Angiogenic effect of mesenchymal stem cell by upregulating VEGF expression after focal cerebral ischemia. J Mol Neurosci. 2015;56(4):898–906.
99. Chen G, Yang Y, Jiang L, Li Q, Tian X, Qi K, et al. Cardiomyocytes pretreated with Tongxinluo reduce hypoxic reoxygenation damage in endothelial cells co-cultured with them by increasing phosphorylation of p70s6k1 via exosomes. China Cycle Magaz. 2016;31:29–30. (In Chinese).

# 27 Extracellular Vesicle Engineering for Clinical Therapy

Jie Wang, Peng Chen, Han Xie, and Bi-Feng Liu

## 27.1 Introduction

Natural extracellular vesicles (EVs) have inherent targeting abilities with homing characteristics; however, their targeting ability is not sufficient to meet clinical application requirements [1–3]. Researchers have recently proposed various approaches to improve the targeting of EVs through surface modifications [4–6]. The composition of organized lipids, proteins, and polysaccharides in the EV membrane is instrumental for protecting EV-encapsulated materials (such as proteins, mRNAs, and microRNAs) during circulation within the body and facilitates the participation of EVs in communication with the microenvironment and the regulatory effects of EVs with regard to homeostasis with recipient cells [7].

Engineering the EV membrane has been shown to be an interesting approach for endowing EVs with additional biological properties [8–10]. This chapter introduces three typical EV membrane modification methods for EV engineering. We then briefly outline isolation and collection strategies for engineered EVs from the perspectives of traditional and emerging technologies. Finally, we summarize the advantages and challenges of engineered EVs.

Jie Wang and Peng Chen have contributed equally to this work.

J. Wang · P. Chen · H. Xie · B.-F. Liu (✉)
Department of Biomedical Engineering, The Key Laboratory for Biomedical Photonics of MOE at Wuhan National Laboratory for Optoelectronics-Hubei Bioinformatics & Molecular Imaging Key Laboratory, Systems Biology Theme, College of Life Science and Technology, Huazhong University of Science and Technology, Wuhan, China
e-mail: bfliu@mail.hust.edu.cn



Q. Wang, L. Zheng (eds.), *Extracellular Vesicles*,
https://doi.org/10.1007/978-981-99-8365-0_27

## 27.2 Chemistry-Based EV Engineering

Chemistry-based EV engineering strategies for improving tissue targeting ability are mainly based on functional phospholipid insertion and chemical groups (primary thiol residues and amines) for modification of the membrane-associated protein copper-catalyzed azide-alkyne cycloaddition (click chemistry) conjugation of targeted ligands [6, 11, 12]. Chemistry-based EV engineering methods are simple, effective, and fast and can be used for pre-isolation of EVs from culture medium or body fluids. Multiple ligands with different properties can be conjugated simultaneously in a high-throughput manner [13, 14].

Functionalized amphipathic molecules can be easily and directly embedded in the lipid membranes of EVs. Among these, polyethylene glycol (PEG)-grafted 1,2-dioleoyl-sn-glycerol-3-phosphoethanolamine (DSPE-PEG) conjugates (e.g., DSPE-PEG-biotin, DSPE-PEG-Arg-Gly-Asp (RGD), DSPE-PEGmaleimide, DSPE-PEG-folate, DSPE-PEG-carboxylic acid, and DSPE-PEG-amine) are the most widely used amphiphilic linear polymers for exosome membrane surface modification in tumor-targeted drug delivery [15]. The chemical insertion of tumor-specific targeting ligands and DSPE-PEG into the lipids of EVs not only extends the half-life of the encapsulated drug in EVs and confers the drug with more time to reach the target site but can also facilitate delivery of the drug to the desired tumor site with high efficiency by engineering EV membranes with targeted ligands, such as antibodies and polypeptides [16, 17]. For example, Wang et al. inserted biotin and avidin into EV membranes by incubating donor cells with DSPE-PEG2000-biotin and DSPE-PEG2000-avidin. Biotin, also known as vitamin H and coenzyme R, is an essential nutrient for maintaining cell growth and development. Biotin receptors are overexpressed on the surface of cancer cells (Fig. 27.1a) [18]. EVs released from donor cells can inherit the dual ligand membrane modification and show better tumor targeting ability than single ligand membrane engineering. Using the same principle, various ligands, including fatty acids, RGD, and aminoethyl anisamide (AA), can be anchored on the EV surface to confer specific cell-type targeting capability [19, 20]. For example, through a phospholipid insertion strategy, an integrated pipeline for cell vesicle isolation, enrichment, and targeted modification was developed [21]. Using a donor cell-assisted membrane modification strategy, the exosomal membrane was equipped with RGD-modified peptides [22]. The isolated exosomes increased the targeting of blood vessels and synergistically promoted angiogenesis, showing potential as new therapeutic tools for angiogenesis therapy and imaging. AA-engineered EVs showed improved uptake efficiency in lung cells and could deliver the encapsulated chemical drugs to the desired site without unknown side effects [23].

Similar to DSPE, cholesterol and diacyl lipids are also capable of self-assembling into EV membranes owing to their hydrophobicity [24]. For example, EV membranes can be simply functionalized with RNA nanoparticles using cholesterol-modified RNA nanoparticles (PSMA-aptamers) to enhance chemotherapeutic efficiency [25–27]. Diacyl lipid-DNA aptamer (sgc8)-conjugate-modified EV membranes have also been proposed to endow EVs with cancer cell-specific treatment ability [28].

Conjugating functional ligands to EV membranes through click chemistry (biorthogonal copper-free azide-alkyne cycloaddition) is another covalent approach to

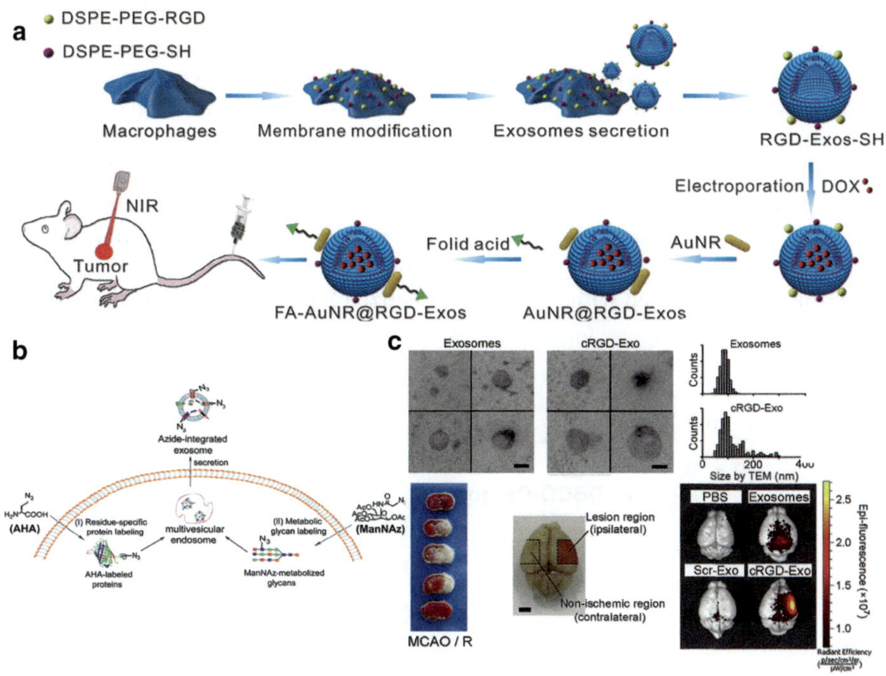


**Fig. 27.1** Chemistry-based EV engineering. (**a**) Surface functionalization of EVs with RGD and thiol-terminated lipids for the addition of diverse surface functionalities for targeted tumor drug delivery. Adapted from Ref. [18]. Reprinted with permission from Advanced Functional Materials. (**b**) Azide functionalization in extracellular vesicles via metabolic engineering. Adapted from Ref. [34]. Reprinted with permission from Royal Society of Chemistry. (**c**) Schematic showing metabolic labeling of cRGD. Adapted from Ref. [35]. Reprinted with permission from Biomaterials

modify EVs [29, 30]. Click chemistry is a highly efficient, simple, and anti-interference chemical reaction [31]. This method has been proven to have no obvious influence on the structural integrity of exosomes and their interactions with recipient cells and can be used for rapid and large-scale production of functionalized exosomes [32]. The representative click chemistry reaction is the Husigen cycloaddition reaction, in which azide and alkyne groups generate triazole rings in a region-selective manner via the catalysis of monovalent copper (Fig. 27.1b) [33, 34]. Click chemistry is particularly suitable for conjugating small and large molecules to the surfaces of EVs. For example, the cyclic (Arg-Gly-Asp-D-Tyr-Lys) peptide [c(RGDyK)] can be conjugated to an EV (cRGD-Exo) membrane through click chemistry (Fig. 27.1c) [35]. cRGD-Exos have been shown to have stronger targeting efficiency for ischemic lesions of the brain than EVs without membrane modification. Subsequently, curcumin was loaded onto cRGD-Exos to obtain cRGD-Exos (cRGD-Exos-cur). Compared with free curcumin, cRGD-Exos-cur could better inhibit inflammation after cerebral ischemia without any obvious toxicity or tissue damage. A copper-free "click chemistry" strategy can also be used to engineer EV membranes with alendronate (Ale), which has a specific high affinity for bone tissue [36]. Bone marrow mesenchymal stem cell-released EV membranes were modified with an alkynyl (DBCO) group using a commercial linker. Then, the Ale molecule with an azide group ($N_3$) conjugate and EV-DBCO were effectively coupled by "click chemistry," and targeted transport of the encapsulated drugs to the bone was achieved for osteoporosis treatment.

Azide functionalization of EVs allows for a wide range of additional functionalities. For example, azides can be linked with cargo-conjugated DBCO using azide-alkyne cycloaddition. Wang et al. reported biotin functionalization in EVs using biotin linked to DBCO by PEG (DBCO-PEG4-biotin; Fig. 27.2) [37]. DBCO-PEG4-biotin was clicked with azide-functionalized EVs. Biotin is a small molecule

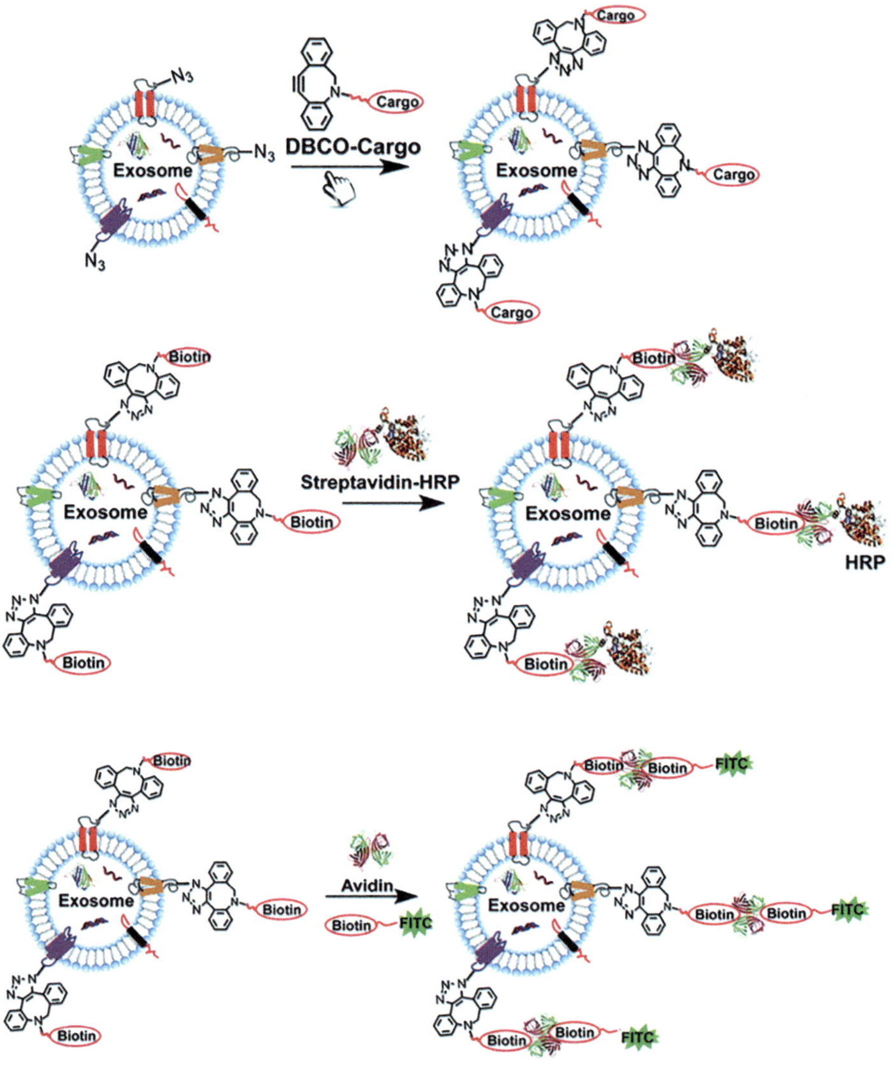


**Fig. 27.2** Click chemistries of azide functionalized EVs. Adapted from Ref. [37]

that shows a strong binding affinity to avidin or streptavidin. The biotinylation of EVs diversifies EV functionalities through various biotin-avidin interaction chemistries that can be widely used in biomedical applications. Biotinylated EVs can be used for protein delivery by integrating proteins of interest with streptavidin, e.g., streptavidin-horseradish peroxidase [34, 38, 39].

Chemistry-based EV surface modification is highly efficient and specific and does not produce any harmful byproducts [40]. However, this approach still suffers a long conjugation process, and the reaction conditions should be carefully controlled to avoid the accumulation of EVs [41].

## 27.3 Biology-Based EV Engineering

Genetic transfection is the most commonly used biology-based EV membrane modification strategy [42]. This method is based on utilization of a gene transfer vector (e.g., lentiviral or retroviral vectors) to insert the gene sequence of the desired ligand into donor cells [43, 44]. The gene sequence can encode the targeted signal peptide or protein in the donor cells such that the EVs released from the donor cells will carry the peptide or protein of interest on the membrane (Fig. 27.3) [34]. The most commonly functionalized transmembrane proteins and homing proteins include lysosome-associated membrane glycoprotein 2b (LAMP-2b), lactadhein (C1C2 domain), tetraspanins (CD63, CD81, CD9), platelet-derived growth factor receptors, and glycosyl-phosphatidylinositol [45]. For example, membrane-bound antigens can be functionalized on EV membranes by taking advantage of the high affinity of the C1C2 domain of lactadherin [46]. Previous reports have demonstrated that EV membranes modified with the C1C2 domain can support immunogenicity by increasing the cytokine levels in EVs [47]. Kooijmans et al. fused the anti-epidermal growth factor receptor (EGFR) protein with the C1C2 domain of lactadherin to obtain a C1C2 fusion protein [48]. The C1C2 domain has high affinity for phosphatidylserine (PS) and is easily enriched on the surface of erythrocytes and EVs secreted in response to Neuro2A. Thus, PS can highly and efficiently interact with and bind to the surface of the EV membrane, and the prepared functionalized EVs can be used as EGFR inhibitors for cancer treatment. SBI Company has also developed the "Xstamp" commercial product, which can efficiently locate a protein of interest on the modified EV surface [49]. This technology is based on the C-terminal fusion of the C1C2 domain of MFG-E8. The protein sequence fused with the XStamp tag effectively displays a protein ligand that can be fused to the secreted EVs [50, 51]. Notably, when there are competitive binding sites, the binding between the molecule and membrane is reversible. Moreover, lectin is a membrane-associated protein, not a transmembrane protein. The use of the lectin C1C2 domain as an anchor point on the EV membrane surface is not as stable as the use of the transmembrane protein of EVs for membrane modification [52–54].

Alvarez-Erviti et al. used immature dendritic cells (imDCs) as a source of engineered EVs by fusing the rabies viral glycoprotein (RVG) gene with the *LAMP2b* gene, which is enriched on the surface of the EV membrane [45]. They then utilized the vector to transfect imDCs, yielding EVs with fusion protein expressed on the

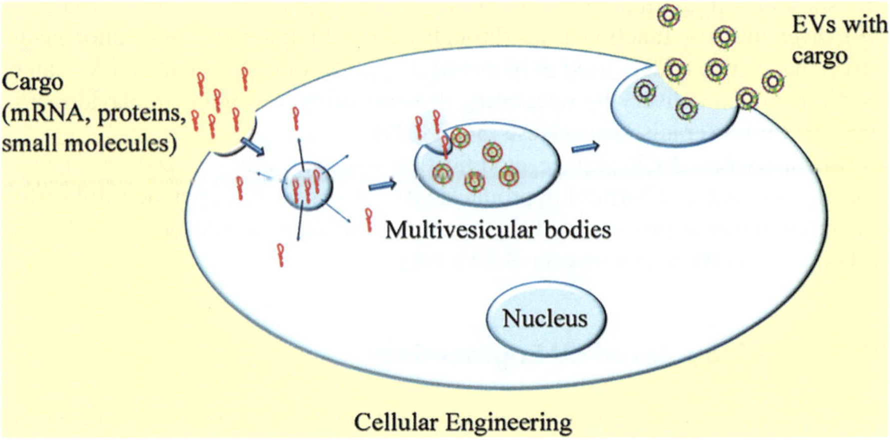


**Fig. 27.3** Biology-based EV engineering. Schematic showing the surface functionalization strategy of EVs using the biological approach. EVs are pre-engineered by cellular engineering of the parent cell. Cargo molecules, such as mRNA, proteins, and small molecules, are incubated with cells that will eventually secrete EVs. Adapted from Ref. [34]. Reprinted with permission from Royal Society of Chemistry

surface and target the acetylcholine receptor in the nervous system. The LAMP2b-RVG protein on the body surface targets nerve cells. Subsequently, they loaded glyceraldehyde 3-phosphate dehydrogenase (GAPDH) small interfering RNA (siRNA) into isolated EVs by electroporation and intravenously injected the functionalized EVs into mice. The results showed that the *Gapdh* gene was knocked out in neurons, microglia, and oligodendrocytes in the mouse brain; RVG-EV pre-exposure did not attenuate this effect, and no nonspecific knockout was found in tissues other than the brain. The use of EVs loaded with siRNA to target the treatment of Alzheimer's has also yielded good results. Compared with wild-type mice, the treatment group showed knockdown of 60% of β-secretase 1 (*BACE1*) mRNA and 62% of BACE1 protein. EVs not only target cargo delivery but also penetrate the blood-brain barrier, suggesting great potential in the treatment of Alzheimer's disease. Hong et al. used liposomes to transfect a plasmid carrying the PH20 hyaluronidase gene into HEK-293T cells [55]. The membrane surface of EVs secreted by modified cells can also be altered using this specific hyaluronidase [56]. The results showed that these hyaluronidase-modified EVs could act on solid tumors and digest the excessive extracellular matrix in the tumor microenvironment to inhibit tumor growth and increase the infiltration of T cells into the tumor.

## 27.4 Physical EV Engineering

Physical modification of EVs is another simple membrane-engineering strategy. Engineering of hybrid EVs can be achieved by fusing the membranes of EVs and liposomes using the freeze-thaw method (Fig. 27.4a) [57]. Using liposomes

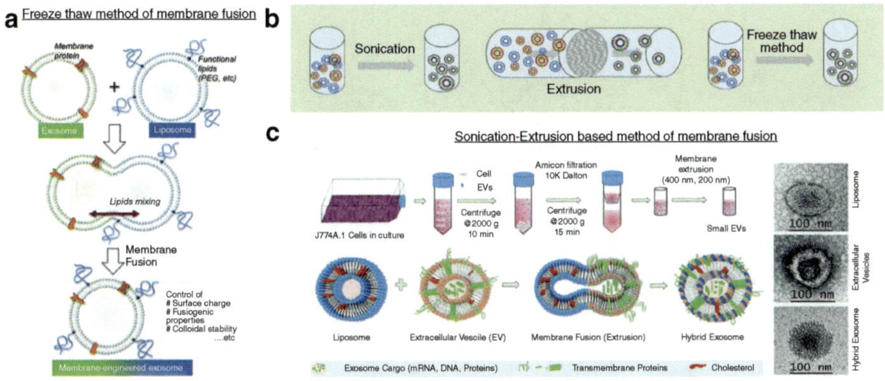


**Fig. 27.4** Surface functionalization strategies using physical methods. (**a**) Schematic showing the freeze-thaw method of membrane fusion. Adapted from Ref. [57]. (**b**) Schematic showing different physical methods for surface functionalization—sonication, extrusion, and freeze-thaw. Adapted from Ref. [34]. (**c**) Schematic showing isolation of extracellular vesicles following the sonication-extrusion method of membrane fusion. Adapted from Ref. [58]. Reprinted with permission from Royal Society of Chemistry

embedded with peptides or antibodies, targeting moieties or PEG can be easily modified on the surface membranes of EVs [11]. First, EVs were isolated from cells expressing the tyrosine kinase receptor human epidermal growth factor receptor 2 (HER2), and these HER2-containing EVs were fused with phospholipid liposomes. Cell uptake studies of hybrid EVs confirmed that the delivery function of EVs could be affected by altering the lipid composition and/or properties of EVs through membrane fusion. When PEG is present on the surface of EVs, the electrostatic repulsion on the surface of anionic EVs is reduced, and the surface of anionic cell membranes promotes EV/cell interactions, thereby enhancing exosome uptake (Fig. 27.4b) [34]. The freeze-thaw method is a relatively simple physical method that allows the destruction of the double-layer membrane by temporarily forming ice crystals and then promoting membrane reconstruction. The freeze-thaw method has been widely used to induce the fusion of EVs and liposome membranes (Fig. 27.4c) [58]. In addition to the simplicity of this method, this physical technique does not require the use of chemical reagents, such as chemistry-based EV modification methods, thus ensuring the purity of EV membranes without contamination.

The surface charge of EVs is another important property that can be exploited for membrane modification [59]. Kim et al. attached cationic lipids to the EV surface through electrostatic interactions and further confirmed that this functionalization could enhance the cell uptake efficiency of EVs (Fig. 27.5) [60]. Although this fusion strategy can endow EV membrane surfaces with transmembrane proteins, and homing proteins can display peptides, homing peptides displayed on the surface of EV membranes are easily degraded by proteases in biofluids, thus losing their targeting ability [61]. Identification of an optimized short-core fragment of the receptor and minimization of the molecular mass of the homing peptide are essential for solving the problem of engineering EVs as targeted therapy carriers.

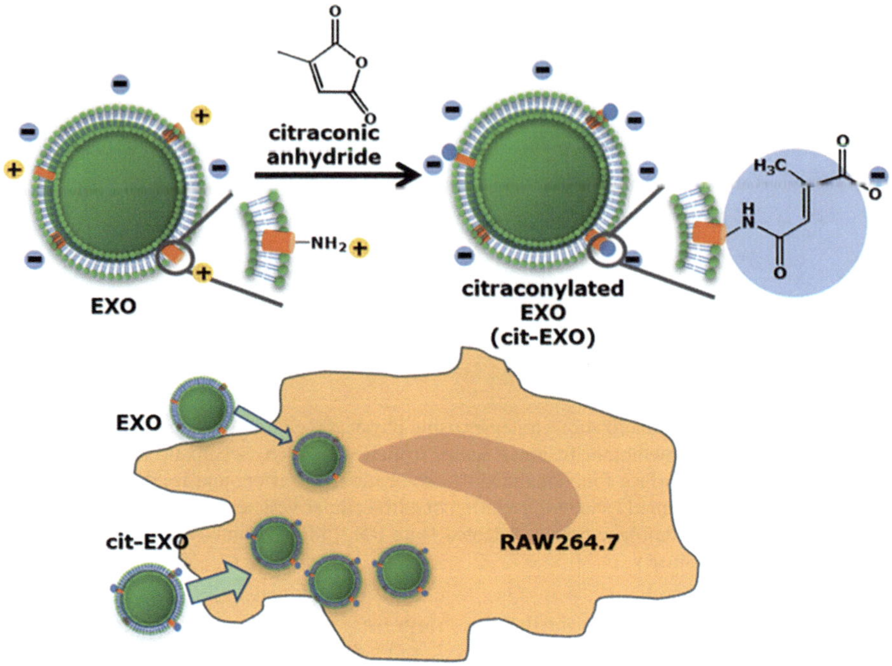


**Fig. 27.5** Surface functionalization through electrostatic interactions. Adapted from Ref. 61

## 27.5 Isolation and Collection Techniques for Engineered EVs

Generally speaking, if the engineering process retains the properties of natural EVs and does not change their size, solubility, density, morphology, or surface markers, traditional EV isolation methods can still be employed. These approaches are not the focus of this chapter, and more detailed information is available in other relevant chapters in this book. Notably, there is no one-size-fits-all method for achieving high-purity and high-quality EV isolation. Ultracentrifugation is the gold standard method and is the most traditional and widely used technique. Burgeoning microfluidics may enable the development of new conceptual isolation strategies with the aid of microscaling effects of laminar flow, unique geometries, and the introduction of external force fields, which can be classified into the following four categories: size-based label-free isolation by membranes or nanowires; dynamic isolation by fluid properties, that is, hydrodynamic properties or external forces (e.g., acoustic or electric fields); immunoaffinity-based isolation by modified microchannels or microbeads with diverse ligands; and orthogonal combinations of the above methodologies [62, 63].

When engineering alters the properties of EVs, the following methods can be applied to isolate the engineered EVs. (1) Low-speed centrifugation isolation was performed when the density of EVs is changed by nanomaterial engineering

(Fig. 27.6a). For example, to obtain gold nanorod-engineered EVs, Wang et al. used a standard benchtop centrifuge to collect the mixture for 5 min at 9000 × *g* [18]. Although this method is convenient and effective, removal of free nanomaterials is difficult using this approach. However, free nanomaterials can be removed by density gradient centrifugation or ultrafiltration. (2) Magnetic isolation can be used when the EV membrane is engineered using magnetic materials (Fig. 27.6b) [64, 65]. Qi et al. designed functional exosomes combined with superparamagnetic nanoparticles to achieve separation, purification, and tumor magnetic targeting of exosomes using an external magnetic field [66]. (3) Affinity reaction isolation can be used for ligand-modified EVs (Fig. 27.6c). Wan et al. reported a rapid EV isolation system utilizing biotin-DSPE-PEG to label the lipid bilayer of EVs and then collected NeutrAvidin-coated magnetic submicrometer particles [67]. (4) Covalent chemical reactions for chemical precursor-modified EVs (Fig. 27.6d). Sun et al. recently proposed EV click chips for early hepatocellular carcinoma detection by synergistically integrating covalent chemistry-mediated EV capture/release, multimarker antibody cocktails, nanostructured substrates, and microfluidic chaotic mixers [68]. Covalent chemistry-mediated EV capture/release was realized using click chemistry-engineered EV capture and disulfide cleavage-driven EV release.

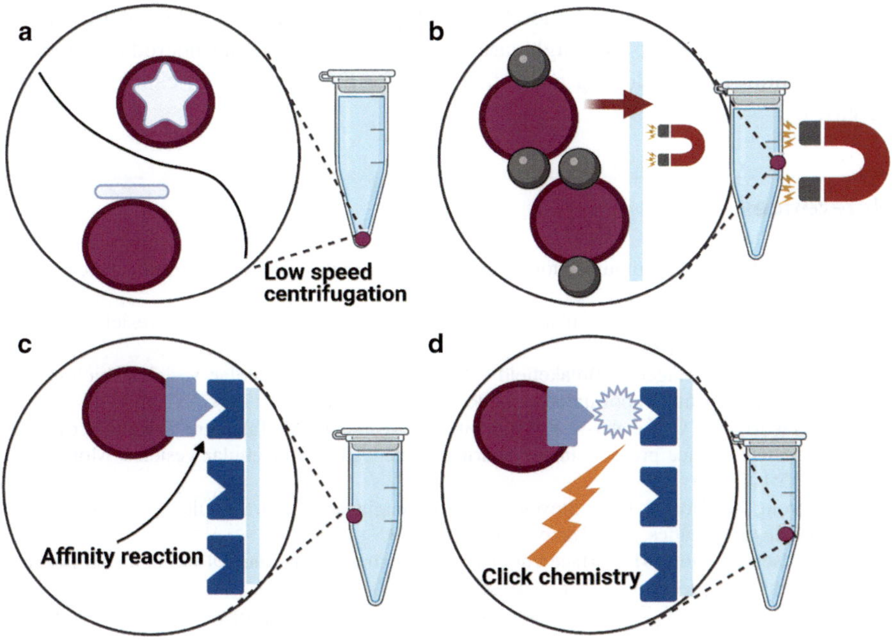


**Fig. 27.6** (**a**) Low-speed centrifugation for the isolation of nanomaterial-engineered EVs. (**b**) Magnetic separation for the isolation of magnetic nanoparticle-engineered EVs. (**c**) Affinity reaction isolation for specific ligand-modified EVs. (**d**) The covalent chemical reaction for specific ligand-modified EVs

## 27.6 Conclusions and Perspectives

In summary, modification of EV surface molecules can be achieved via chemical (e.g., direct insertion and chemical reaction coupling), biological (e.g., genetic transfection), physical (e.g., sonication, extrusion, freeze-thaw, and electrostatic adsorption), and combination engineering. Biological-based EV engineering, usually achieved using pre-engineered parent cells, cannot be used to label or target molecules other than genetically encoded proteins/peptides and is not conducive to mass production. In comparison, chemistry-based/physical EV engineering can potentially increase large-scale production of functionalized EVs. Combination engineering is a future trend that may be used to obtain multifunctional EVs.

Compared with natural EVs, engineered EVs are thought to be an excellent delivery platform for therapy because of their targeted drug delivery and therapy capacity, integration of diagnosis and treatment, multiple cargo loading abilities, and multimodal treatment potential [59]. However, some key points require careful attention during the surface modification of EVs, such as the selection of appropriate targeting ligands or labeling moieties, based on thorough knowledge of the target site and the intended application. Considering their physicochemical stability, unwanted changes on the surface of EVs should be avoided. The reproducibility, stability, safety (e.g., immunogenicity, immunosuppression, and reversion to tumorigenesis), and scalability of engineered EV production should be addressed. Additionally, there is an urgent need to develop efficient and reliable isolation methods for engineered EVs. Although various problems remain, engineered EVs remain a new frontier for developing novel therapies for the treatment of cancer and other diseases.

## References

1. Tkach M, Théry C. Communication by extracellular vesicles: where we are and where we need to go. Cell. 2016;164:1226–32.
2. Robbins PD, Morelli AE. Regulation of immune responses by extracellular vesicles. Nat Rev Immunol. 2014;14:195–208.
3. Andaloussi SE, Mäger I, Breakefield XO, Wood MJ. Extracellular vesicles: biology and emerging therapeutic opportunities. Nat Rev Drug Discov. 2013;12:347–57.
4. Charoenviriyakul C, Takahashi Y, Morishita M, Nishikawa M, Takakura Y. Role of extracellular vesicle surface proteins in the pharmacokinetics of extracellular vesicles. Mol Pharm. 2018;15:1073–80.
5. Armstrong JP, Holme MN, Stevens MM. Re-engineering extracellular vesicles as smart nanoscale therapeutics. ACS Nano. 2017;11:69–83.
6. Richardson JJ, Ejima H. Surface engineering of extracellular vesicles through chemical and biological strategies. Chem Mater. 2019;31:2191–201.
7. Fu S, Wang Y, Xia X, Zheng JC. Exosome engineering: current progress in cargo loading and targeted delivery. NanoImpact. 2020:100261.
8. Li X, et al. Challenges and opportunities in exosome research—perspectives from biology, engineering, and cancer therapy. APL Bioeng. 2019;3:011503.
9. Sawada S-I, et al. Nanogel hybrid assembly for exosome intracellular delivery: effects on endocytosis and fusion by exosome surface polymer engineering. Biomater Sci. 2020;8:619–30.

10. Gilligan KE, Dwyer RM. Engineering exosomes for cancer therapy. Int J Mol Sci. 2017;18:1122.
11. Luan X, et al. Engineering exosomes as refined biological nanoplatforms for drug delivery. Acta Pharmacol Sin. 2017;38:754–63.
12. Liang Y, Duan L, Lu J, Xia J. Engineering exosomes for targeted drug delivery. Theranostics. 2021;11:3183.
13. Zhao Z, McGill J, Gamero-Kubota P, He M. Microfluidic on-demand engineering of exosomes towards cancer immunotherapy. Lab Chip. 2019;19:1877–86.
14. Hwang DW, et al. Chemical modulation of bioengineered exosomes for tissue-specific biodistribution. Adv Ther. 2019;2:1900111.
15. Mukherjee A, et al. Lipid–polymer hybrid nanoparticles as a next-generation drug delivery platform: state of the art, emerging technologies, and perspectives. Int J Nanomedicine. 2019;14:1937.
16. Si Y, et al. Targeted exosomes for drug delivery: biomanufacturing, surface tagging, and validation. Biotechnol J. 2020;15:1900163.
17. Di H, et al. Nanozyme-assisted sensitive profiling of exosomal proteins for rapid cancer diagnosis. Theranostics. 2020;10:9303.
18. Wang J, et al. Designer exosomes for active targeted chemo-photothermal synergistic tumor therapy. Adv Funct Mater. 2018;28:1707360. https://doi.org/10.1002/Adfm.201707360.
19. Lu M, et al. Functionalized extracellular vesicles as advanced therapeutic nanodelivery systems. Eur J Pharm Sci. 2018;121:34–46.
20. Kim H, et al. Engineered extracellular vesicles and their mimetics for clinical translation. Methods. 2019;177:80–94.
21. Wang J, et al. Chemically edited exosomes with dual ligand purified by microfluidic device for active targeted drug delivery to tumor cells. Acs Appl Mater Inter. 2017;9:27,441–52. https://doi.org/10.1021/acsami.7b06464.
22. Chen P, et al. A localized chemical pulse generator for the development of a microfluidic cell-based biosensor. Sensor Actuat B-Chem. 2017;251:112–9. https://doi.org/10.1016/j.snb.2017.05.038.
23. Cao Y, et al. Engineered exosome-mediated near-infrared-II region V2C quantum dot delivery for nucleus-target low-temperature photothermal therapy. ACS Nano. 2019;13:1499–510. https://doi.org/10.1021/acsnano.8b07224.
24. Cheng N, et al. Recent advances in biosensors for detecting cancer-derived exosomes. Trends Biotechnol. 2019;37:1236–54.
25. Zhou J, Rossi JJ. Mechanisms and barriers to RNAi delivery. In: Advanced delivery and therapeutic applications of RNAi. Chichester: John Wiley and Sons, pp. 3–17; 2013.
26. Zhou J, Rossi JJ. Cell-specific aptamer-mediated targeted drug delivery. Oligonucleotides. 2011;21:1–10.
27. Dassie JP, Giangrande PH. Current progress on aptamer-targeted oligonucleotide therapeutics. Ther Deliv. 2013;4:1527–46.
28. Zou J, et al. Aptamer-functionalized exosomes: elucidating the cellular uptake mechanism and the potential for cancer-targeted chemotherapy. Anal Chem. 2019;91:2425–30.
29. Smyth T, et al. Surface functionalization of exosomes using click chemistry. Bioconjug Chem. 2014;25:1777–84.
30. Wang Y, et al. On-board control of wax valve on active centrifugal microfluidic chip and its application for plasmid DNA extraction. Microfluid Nanofluid. 2019;23 https://doi.org/10.1007/s10404-019-2278-y.
31. Song S, et al. In situ one-step fluorescence labeling strategy of exosomes via bioorthogonal click chemistry for real-time exosome tracking in vitro and in vivo. Bioconjug Chem. 2020;31:1562–74.
32. Xu L, et al. Design of experiment (DoE)-driven in vitro and in vivo uptake studies of exosomes for pancreatic cancer delivery enabled by copper-free click chemistry-based labelling. J Extracell Vesicles. 2020;9:1779458.

33. Binder WH, Sachsenhofer R. 'Click'chemistry in polymer and materials science. Macromol Rapid Commun. 2007;28:15–54.
34. Rayamajhi S, Aryal S. Surface functionalization strategies of extracellular vesicles. J Mater Chem B. 2020;8:4552–69.
35. Tian T, et al. Surface functionalized exosomes as targeted drug delivery vehicles for cerebral ischemia therapy. Biomaterials. 2018;150:137–49.
36. Su D, Jing W. Intelligent PID driving and control in the centrifugal microfluidic chip environment. J Environ Prot Ecol. 2020;21:644–53.
37. Wang M, Altinoglu S, Takeda YS, Xu Q. Integrating protein engineering and bioorthogonal click conjugation for extracellular vesicle modulation and intracellular delivery. PLoS One. 2015;10:e0141860.
38. Gu Z, Biswas A, Zhao M, Tang Y. Tailoring nanocarriers for intracellular protein delivery. Chem Soc Rev. 2011;40:3638–55.
39. Ilyas S, Ilyas M, van der Hoorn RA, Mathur S. Selective conjugation of proteins by mining active proteomes through click-functionalized magnetic nanoparticles. ACS Nano. 2013;7:9655–63.
40. Hood JL. Post isolation modification of exosomes for nanomedicine applications. Nanomedicine. 2016;11:1745–56.
41. Li M, et al. Exosomes from different cells: characteristics, modifications, and therapeutic applications. Eur J Med Chem. 2020;207:112784. https://doi.org/10.1016/j.ejmech.2020.112784.
42. Koyama Y, et al. Exosomes derived from tumor cells genetically modified to express Mycobacterium tuberculosis antigen: a novel vaccine for cancer therapy. Biotechnol Lett. 2016;38:1857–66.
43. You B, Xu W, Zhang B. Engineering exosomes: a new direction for anticancer treatment. Am J Cancer Res. 2018;8:1332.
44. O'Loughlin AJ, Woffindale CA, Wood MJA. Exosomes and the emerging field of exosome-based gene therapy. Curr Gene Ther. 2012;12:262–74.
45. Alvarez-Erviti L, et al. Delivery of siRNA to the mouse brain by systemic injection of targeted exosomes. Nat Biotechnol. 2011;29:341–5.
46. Delcayre A, et al. Exosome display technology: applications to the development of new diagnostics and therapeutics. Blood Cell Mol Dis. 2005;35:158–68.
47. Rountree RB, et al. Exosome targeting of tumor antigens expressed by cancer vaccines can improve antigen immunogenicity and therapeutic efficacy. Cancer Res. 2011;71:5235–44.
48. Kooijmans SA, Gitz-Francois JJ, Schiffelers RM, Vader P. Recombinant phosphatidylserine-binding nanobodies for targeting of extracellular vesicles to tumor cells: a plug-and-play approach. Nanoscale. 2018;10:2413–26.
49. Blot XW. XStamp™ exosome targeting technology. Signals. 2015;6:7.
50. Wang L, et al. Exosomes containing miRNAs targeting HER2 synthesis and engineered to adhere to HER2 on tumor cells surface exhibit enhanced antitumor activity. J Nanobiotechnol. 2020;18:153. https://doi.org/10.1186/s12951-020-00711-5.
51. Sabanovic B, Piva F, Cecati M, Giulietti M. Promising extracellular vesicle-based vaccines against viruses, including SARS-CoV-2. Biology. 2021;10:94.
52. Fitzgerald W, et al. A system of cytokines encapsulated in extracellular vesicles. J Extracell Vesicles. 2018;7:191–2.
53. Murphy DE, et al. Extracellular vesicle-based therapeutics: natural versus engineered targeting and trafficking. Exp Mol Med. 2019;51:1–12.
54. Vamvaka E, et al. Cyanovirin-N produced in rice endosperm offers effective pre-exposure prophylaxis against HIV-1 BaL infection in vitro. Plant Cell Rep. 2016;35:1309–19.
55. Hong Y, et al. Exosome as a vehicle for delivery of membrane protein therapeutics, PH20, for enhanced tumor penetration and antitumor efficacy. Adv Funct Mater. 2018;28:1703074.
56. Kim H, et al. Exosomes: cell-derived nanoplatforms for the delivery of cancer therapeutics. Int J Mol Sci. 2021;22:14.
57. Sato YT, et al. Engineering hybrid exosomes by membrane fusion with liposomes. Sci Rep. 2016;6:1–11.

58. Rayamajhi S, Nguyen TDT, Marasini R, Aryal S. Macrophage-derived exosome-mimetic hybrid vesicles for tumor targeted drug delivery. Acta Biomater. 2019;94:482–94.
59. Salunkhe S, Dheeraj, Basak M, Chitkara D, Mittal A. Surface functionalization of exosomes for target-specific delivery and in vivo imaging & tracking: strategies and significance. J Control Release. 2020;326:599–614. https://doi.org/10.1016/j.jconrel.2020.07.042.
60. Kim Y, Mok H. Citraconylated exosomes for improved internalization into macrophages. Appl Biol Chem. 2019;62:1–7.
61. Yang Y, Hong Y, Cho E, Kim GB, Kim I-S. Extracellular vesicles as a platform for membrane-associated therapeutic protein delivery. J Extracell Vesicles. 2018;7:1440131.
62. Lin BQ, et al. Microfluidic-based exosome analysis for liquid biopsy. Small Methods. 2021;5:2001131. https://doi.org/10.1002/smtd.202001131.
63. Jie Wang PM, Kim DH, Liu B-F, Demirci U. Towards microfluidic-based exosome isolation and detection for tumor therapy. Nano Today. 2021;37:101066.
64. Qi H, et al. Blood exosomes endowed with magnetic and targeting properties for cancer therapy. ACS Nano. 2016;10:3323–33. https://doi.org/10.1021/acsnano.5b06939.
65. Zhang W, et al. Magnetic and folate functionalization enables rapid isolation and enhanced tumor-targeting of cell-derived microvesicles. ACS Nano. 2017;11:277–90. https://doi.org/10.1021/acsnano.6b05630.
66. Zhan Q, et al. Engineering blood exosomes for tumor-targeting efficient gene/chemo combination therapy. Theranostics. 2020;10:7889–905. https://doi.org/10.7150/thno.45028.
67. Wan Y, et al. Rapid magnetic isolation of extracellular vesicles via lipid-based nanoprobes. Nat Biomed Eng. 2017;1:0058. https://doi.org/10.1038/s41551-017-0058.
68. Sun N, et al. Purification of HCC-specific extracellular vesicles on nanosubstrates for early HCC detection by digital scoring. Nat Commun. 2020;11:4489. https://doi.org/10.1038/s41467-020-18311-0.

# Other Extracellular Vesicles for Disease Therapy

# 28

Xiubo Shi, Xiumei Hu, Zihao Ou, Bo Situ, Mei Huang, Lei Zheng, and Ying Wang

## 28.1 Introduction

Extracellular vesicles (EVs) are vesicular structures that contain various biomolecules, including proteins, nucleic acids, lipids, and metabolites. Recent research on the use of EVs from different sources to treat related diseases has demonstrated the unique advantages of milk-derived extracellular nanosized vesicles (MENVs) and bacteria-derived EVs (BEVs) in disease therapy.

MENVs carrying lipids, proteins, mRNAs, noncoding RNAs, and DNAs play essential roles in mother-infant communication and regulate the development of the newborn's musculoskeletal and nervous systems as well as the gut microbiome and immunological program. Furthermore, MENVs can ameliorate neonatal diseases, such as necrotizing enterocolitis, and several reports have revealed the potential involvement of MENV nucleic acids in chronic diseases, such as type 2 diabetes, atherogenesis, osteoporosis, Parkinson's disease, breast cancer, and prostate cancer. MENVs have also been proven to be prominent nanocarriers for chemotherapeutic drugs and nucleic acids, which may contribute to their excellent antitumor effects.

BEVs contain various biological components from the parent bacteria and have diverse functions, including roles in virulence factor transport, DNA transfer,

---

Xiubo Shi, Xiumei Hu, and Zihao Ou have contributed equally to this work.

X. Shi · Y. Wang (✉)
Cancer Research Institute, School of Basic Medical Sciences, Southern Medical University, Guangzhou, People's Republic of China

X. Hu · Z. Ou · B. Situ · M. Huang · L. Zheng (✉)
Department of Laboratory Medicine, Nanfang Hospital, Southern Medical University, Guangzhou, People's Republic of China
e-mail: nfyyzhenglei@smu.edu.cn



Q. Wang, L. Zheng (eds.), *Extracellular Vesicles*,
https://doi.org/10.1007/978-981-99-8365-0_28

bacteriophage interception, antibiotics, and eukaryotic host defense factors. These nanosized vesicles can trigger pro-inflammatory signaling cascades that lead to the production of cytokines, chemokines, and antimicrobial peptides and interact with Toll-like receptor 4 (TLR4) to trigger the host immune response, which could provide protection against infection. Moreover, BEVs can be used as carriers for tumor drugs, immune adjuvants, and comprehensive tumor microenvironment regulators to improve the efficacy of antitumor treatment. Therefore, as an approach for cell-free EV therapy, BEVs may have important applications in the clinical setting.

In this chapter, we systematically introduce recent research on MENVs and BEVs in the treatment of various diseases and explore the therapeutic potential and clinical value of these EVs.

## 28.2 MENVs for Disease Therapy

Milk has been extensively studied for hundreds of years, and the first milk-derived exosome-based research was published in recent decades. There is a growing interest in these nanosized vesicles owing to their prominent characteristics as biocarriers and their potential associations with human diseases. Researchers have isolated these vesicles from various milk sources, including humans, cows, goats, horses, pigs, and camels [1]. These MENVs, which have biomolecular lipid bilayer nanostructures, carrying lipids, proteins, mRNAs, noncoding RNAs, and DNAs, play essential roles in mother-infant communication. Furthermore, MENVs regulate the development of the newborn's musculoskeletal and nervous systems, modulate the gut microbiome and immunological program, and ameliorate neonatal diseases, such as necrotizing enterocolitis [2]. Additionally, several reports have revealed the potential involvement of MENV nucleic acids in chronic diseases and cancer [3]. MENVs have also been proven to be prominent nanocarriers for chemotherapeutic drugs and nucleic acids, which may contribute to their excellent antitumor effects [4].

### 28.2.1 Isolation and Characterization

In general, ENVs are isolated by differential ultracentrifugation; however, the isoelectric precipitation method has also been used to isolate MENVs because of their abundant casein protein contents [5, 6]. Briefly, defatted milk was acidified with acetic acid or diluted hydrochloric acid to a pH of 4.6 to deposit casein micelles in whey, followed by differential centrifugation or size-exclusion chromatography to obtain purified MENVs. Notably, during the pretreatment procedure, the untreated milk should be centrifuged as soon as possible to remove the cells and cream layer and avoid contamination by cellular debris [7].

The obtained MENVs have a lipid bilayer, diameter of 30–150 nm [1], density of 1.12–1.21 g/mL, and molecular weight of approximately 10–1200 MDa [8]. MENVs show great stability, enabling them to protect their cargo from digestion, and can be transported to the blood-brain/embryo barrier [9, 10]. Different routes of

MENV administration can result in different pharmacokinetic properties in vivo. Indeed, DiR-labeled MENVs reached a peak at 3 h and accumulated mainly in the liver, spleen, and lungs after administration by intravenous injection. However, these effects peaked at 24 h and were distributed in the heart, lungs, brain, kidneys, liver, spleen, and gastrointestinal tract after oral gavage [9].

### 28.2.2 Components

To date, more than 4000 proteins and peptides have been detected in MENVs [11]. Among these proteins, the most abundant proteins were proteins of the milk fat globule membrane (MFGM), including butyrophilin, xanthine oxidase, adipophilin, and lactadherin [12]. Moreover, MENVs were enriched in Rab proteins, annexins, and exocyst complex proteins EXOC1–8, which are crucial for vesicle fusion and trafficking. Proteins related to immune function, such as major histocompatibility complex (MHC)-I, are also present on the membrane of MENVs, indicating their involvement in mammary health. General Gene Ontology annotations of MENVs are shown in Fig. 28.1.

The lipid composition of MENVs is summarized in Fig. 28.2. Most membranes contained sphingomyelin and phosphatidylethanolamine, whereas few contained phosphatidylethanolamine and phosphatidylserine. The cholesterol content of MENVs was similar to that of MFGM, whereas triacylglycerides were present at low levels [5].

MENVs also contain high abundances of RNAs, including microRNAs, long noncoding RNAs, and circular RNAs [1]. The top 10 most abundant miRNAs regulate various biological pathways in vivo (Table 28.1).

### 28.2.3 Biological Effects and Diseases

Breast and bovine milk are not only essential sources of nutrition but are also crucial for infant maturation [2]. Accumulating evidence suggests that exosome- and RNA-sufficient (ERS) and exosome-depleted (ERD) diets can be used to evaluate the effects of bovine MENV depletion on mice [13]. Compared with mice fed an ERS diet, mice fed an ERD diet displayed significant changes, including loss of spatial learning and memory [14]; accumulation of purine metabolites, including adenosine and ATP [15]; lower birth weight [10]; reduced abundance of the gut microbiome [16]; and hypoplasia of the cecum [17]. Another study reported that MENVs can attenuate intestinal and cecal inflammation through several miRNA-mediated pathways [1].

In studies of western diseases, Melnik et al. reported that MENV miRNAs, particularly *miR-21*, can augment the expression of mammalian target of rapamycin complex 1 [18], which regulates cell growth and is associated with a dozen chronic diseases, including acne, obesity, type 2 diabetes, arterial hypertension, Alzheimer's disease, and cancer (particularly prostate cancer and hepatocellular carcinoma) [3, 19].

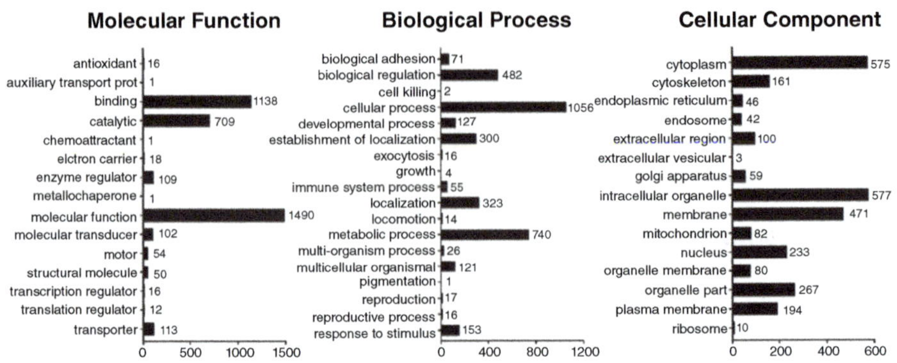


**Fig. 28.1** General Gene Ontology annotations of proteins in MENVs

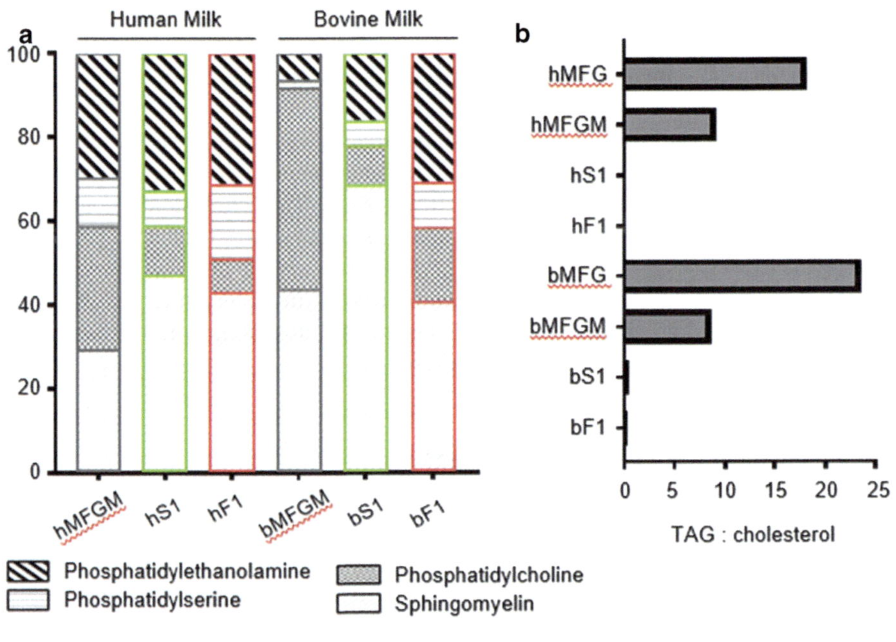


**Fig. 28.2** Lipid profiles of milk membrane fractions. (**a**) Phospholipid profiles of human and bovine MENV and MFGM preparations. (**b**) Triacylglyceride (TAG):cholesterol ratios of MENV, MFGM, and MFG preparations. **hF1* Human fluff MENVs, *hS1* Human milk serum MENVs, *bF1* Bovine fluff MENVs, *bS1* Bovine milk serum MENVs, *bMFGM* Bovine milk fat globule membrane, *bMFG* Bovine milk fat globule, *hMFGM* Human milk fat globule membrane, *hMFG* Human milk fat globule

Recently, Samuel et al. discovered that unpasteurized whole milk and its ENVs had antiproliferative effects against colorectal or breast cancer cells in vitro compared with ENVs from commercial milk and could reduce tumor burden by

**Table 28.1** Top 10 most abundant miRNAs detected in MENVs from human and cow

| | Human (2 mo) | Human (3 mo) | Human (6–8 mo) | Human (3–9 mo) | Cow (not provided) | Cow (not provided) | Cow (3 mo) |
|---|---|---|---|---|---|---|---|
| 1 | miR-148a | miR-148a | miR-22 | miR-30d | miR-2478 | miR-148a | miR-30a |
| 2 | miR-30b | miR-22 | miR-30d | miR-148a | miR-1777b | Let-7a | miR-148a |
| 3 | Let-7f | miR-30d | miR-181a | miR-200a | miR-1777a | Let-7b | miR-141 |
| 4 | miR-146b | Let-7b | miR-148a | miR-200c | Let-7b | miR-21 | miR-22 |
| 5 | miR-29a | miR-200a | miR-30b | Let-7a | miR-1224 | miR-99a | miR-16a |
| 6 | Let-7a | Let-7a | miR-141 | miR-200b | miR-2412 | Let-7f | miR-186 |
| 7 | miR-141 | Let-7f | miR-92a | miR-21 | miR-2305 | Let-7c | miR-182 |
| 8 | miR-182 | miR-146b | miR-26a | Let-7b | Let-7a | miR-200c | miR-181a |
| 9 | miR-200a | miR-24 | miR-375 | Let-7f | miR-200c | miR-26a | miR-191 |
| 10 | miR-378 | miR-21 | miR-30a | miR-30a | miR-141 | miR-30d[a] | miR-27b |

[a]*miR* MicroRNA, *mo* Month

inducing cellular senescence in vivo; by contrast, no effects on HEK-293T embryonic kidney cells were observed [20]. However, these ENVs facilitated pancreatic/breast cancer metastasis to the liver and lungs in a dose-dependent manner owing to the secretion of ENVs by senescent cells, resulting in upregulation of epithelial-mesenchymal transition-associated pathways and enhanced migration. However, when the primary tumor was removed from tumor-bearing mice and the mice were treated with MENVs by gavage, markedly reduced metastasis was observed, suggesting clinical implications in postoperative cancer treatment.

## 28.2.4 Drug Carriers

MENVs are promising tools in drug delivery because of their high biocompatibility, circulation retention, and ability to cross biological barriers, such as the blood-brain barrier. Compared with mammalian cell-derived ENVs, MENVs are easily obtained, have high yields, and are cost efficient. Moreover, daily dietary milk is well tolerated in humans and can induce fewer host immune responses [21]. Thus, MENVs are promising nanocarriers for the delivery of therapeutic agents in various diseases. MENVs have been applied as drug carriers in cancer therapy, as summarized in Table 28.2.

**Table 28.2** Application of MENVs as drug carriers in cancer therapy

| Ref | Cargo | Loading strategy | Surface modification | Cancer type and cell line |
|---|---|---|---|---|
| [22, 23] | Withaferin A, anthocyanidins, curcumin, paclitaxel, docetaxel | Ethanol-assisted incubation | Folate | Breast (T47D & MDA-MB-231)<br>Lung (A549 & H1299)<br>Colon (HCT116)<br>Pancreatic (Panc1 & Mia PaCa2)<br>Prostate (DU145 & PC3)<br>Ovarian (Ovca432) |
| [24] | Doxorubicin (DOX) | Incubation | Hyaluronan (HA) | Breast (MCF-7 & MDA-MB-231)<br>Lung (A549 & HEK293) |
| [25] | DOX with pH-cleavable bond and thermal-activated endoperoxides | | Chlorin e6 (Ce6) | Oral squamous cell carcinomas (HSC-3 & SCC-9 & CAL-27) |
| [26] | DOX | Sonication and extrusion-based active loading (SEAL) | Transferrin | Liver (HepG2) |
| [27] | KRAS inhibiting siRNA | Commercial transfection kits | – | Lung (A549 & H1299) |
| [28] | β-catenin inhibiting siRNA | Three-way junction (3WJ) RNA scaffold | RNA aptamer targeting EpCAM | Liver (Hep3B) |
| [29] | AF-488 siRNA | Membrane protein interaction | – | Colon (Caco2)<br>Liver (HepG2) |

## 28.3 Bacteria-Derived EVs for Disease Therapy

### 28.3.1 The Biogenesis of Bacteria-Derived EVs (BEVs)

Most cells, including eukaryotic, gram-negative, and gram-positive bacterial and archaeal cells, can secrete vesicles into the extracellular environment. Vesicles secreted by gram-negative bacteria are usually derived from the bacterial outer membrane and are therefore called bacterial outer membrane vesicles (OMVs). However, due to the variety and complex formation mechanisms of pathogenic microorganisms, the vesicles secreted by pathogenic microorganisms are not uniformly named. For example, vesicles secreted by gram-positive ($G^+$) bacteria are called cytoplasmic membrane vesicles, whereas vesicles secreted by gram-negative bacteria are called OMVs. To standardize this nomenclature, we refer to the vesicles derived from bacteria as BEVs in this chapter.

Using BEVs, bacteria can deliver various bacterial proteins and lipids into host cells via a secretion mechanism, thereby eliminating the need for bacterial contact with the host cell [30]. Gram-negative OMVs were studied earlier than gram-positive OMVs. EVs from gram-negative bacteria, called OMVs, were first observed using electron microscopy in the 1960s. The average diameter of these EVs was 20–200 nm. Moreover, EVs from gram-negative bacteria have been found under various environmental conditions, such as in vitro experimental conditions, natural environments, tissues, and biological fluids [31].

BEVs contain outer membrane proteins, lipopolysaccharide (LPS), phospholipids, and some periplasmic constituents [30]. Research on secretory vesicles from gram-positive bacteria started more recently, and the mechanisms are still not clear. Notably, it was not until 2009 that Lee et al. first confirmed that *Staphylococcus aureus* can shed BEVs in analyses using mass spectrometry and transmission electron microscopy [32]. In 2010, Tashiro et al. found that the *Pseudomonas* quinolone signaling pathway can promote the secretion of BEVs by *Bacillus subtilis*, whereas excessive stimulation with $MgCl_2$ cannot inhibit the secretion of BEVs, in contrast to the research results for gram-negative bacteria. This suggests that the vesicle secretion mechanisms of gram-positive and gram-negative bacteria may not be exactly the same, and these differences may be related to variations in the peptidoglycan structure [33].

BEVs carry bacterial molecules that can be functionally transported to host cells. For example, *Pseudomonas aeruginosa* uses BEVs to transport virulence factors, such as hemolytic phospholipase C, to host cells, representing the transport function of BEVs [34, 35]. BEVs can also control immune responses. *Lactobacillus plantarum*-derived EVs can regulate the proportions of M1 and M2 macrophages and are used as anti-inflammatory and immunomodulatory substances, thereby altering the phenotypes of inflammatory skin disorders [36, 37].

### 28.3.2 Therapeutic Applications of BEVs in Infectious Diseases

BEVs containing microorganism-associated molecular patterns (MAMPs) can trigger pro-inflammatory signaling cascades that lead to the production of cytokines, chemokines, and antimicrobial peptides and interact with TLR4 [38, 39]. An increasing amount of evidence supports the opinion that BEVs can offer protection from experimental colitis. Shen et al. demonstrated that BEVs derived from *Bacteroides fragilis*-containing polysaccharide can protect animals from experimental colitis by regulating T cells and mucosal tolerance [40]. The regulation of BEVs has been evaluated in other studies. Using BEVs produced by *Neisseria meningitidis* to stimulate human neutrophils triggers the generation of tumor necrosis factor (TNF) and interleukin (IL)-1β and regulates C-X-C motif chemokine ligand 8, C-C motif chemokine ligand (CCL) 3, and CCL437 expression. Interferon (IFN)-γ can further improve these responses, suggesting that in the presence of chronic inflammation and IFN-γ, the immune response triggered by BEVs can be enhanced [41, 42]. Currently, the most popular BEV-based vaccine is the *N. meningitidis* vaccine, which is approved for marketing in the European Union. The vaccine is based on BEVs and is supplemented with three highly immunogenic proteins to produce protective antibodies against meningitis infection [43]. In addition,

BEVs derived from *Escherichia coli* can effectively prevent systemic inflammatory response syndrome by mediating Th1 and Th17 cell responses; this protection can last up to 42 days after the response occurs [44]. BEVs can also promote the development of protective B-cell and T-cell responses in vivo, as demonstrated by the observation that BEVs isolated from *Salmonella* can induce the generation of CD86 and MHC class II molecules from dendritic cells, stimulating the production of TNF-α and IL-12 [45, 46].

Most research on BEVs has demonstrated that their cargo can be used against other bacteria. In studies of bacterial infections, Tashiro et al. found that *P. aeruginosa* can produce OMVs containing various virulence factors to impair host cells or other bacteria, including proteases, alkaline phosphatase, murein hydrolases, phospholipase C, and hemolysin [47–49]. Additionally, BEVs derived from gram-negative bacteria can be used to lyse other bacteria. Li et al. isolated BEVs from 15 strains of gram-negative bacteria (including *Citrobacter*, *Enterobacter*, *Escherichia*, *Klebsiella*, *Morganella*, *Proteus*, *Salmonella*, and *Shigella* strains) and demonstrated that BEVs containing peptidoglycan hydrolases could lyse gram-positive bacteria (including *Mycobacterium*) possessing A1α, A4α, A1γ, A2α, and A4γ peptidoglycan chemotypes but not those containing A3α, A3β, A3γ, A4β, B1α, and B1β chemotypes [50, 51]. In predatory bacteria, BEVs can make it possible to lyse prey cells by packaging various cargoes, such as proteases, hydrolases, and secondary metabolites with antimicrobial activity, in BEVs [52, 53]. This suggests that BEVs can facilitate the use of bacteria as antimicrobial tools.

BEVs also play vital roles in the prevention and treatment of viral infections. In recent research, scholars have investigated the protection of *Escherichia coli* mutant-derived OMV-modified lipid A (fmOMV) against influenza A virus infection and found that administration of fmOMV provided effective protection against virus-induced death and viral pathogenicity for several viruses, processes that were dependent on macrophage recruitment and type I IFN expression [54]. The same outcomes have been observed in animals. The experiment was conducted through intranasal injection of fmOMV into mice, showing significant immunity to complications with a lethal dose of homologous or heterologous influenza virus without adverse effects [55]. Importantly, BEVs were used against the virus after infection and were found to be excellent antigen carriers. Researchers have provided evidence demonstrating that chimeric antigen combined with the receptor binding domain (RBD) of the Middle East respiratory syndrome coronavirus (MERS-CoV; OMVs-H1/RBD) with H1-type hemagglutinin from influenza A virus (H1N1) strain could induce antibodies against both strains, yielding neutralizing antibodies against H1N1pdm09 and sufficient neutralization of MERS-CoV. BEV-based vaccines, which contain various viral antigens, represent a safe and highly efficient method of protection against two different viral infections [56]. Additionally, recombinant BEVs derived from an *Escherichia coli* mutant strain showed an antigen-mapped heterospecies tandem sequence of the M2 protein from the influenza A virus, which demonstrated their capacity to trigger the host immune response and protect against influenza A virus [57].

### 28.3.3 Therapeutic Applications of BEVs in Tumor Diseases

With a preliminary understanding of BEVs, the number of researchers studying BEVs on cancer therapy have increased.

Because BEVs contain large numbers of pathogen-associated molecular patterns, which can activate systemic immune responses, many studies have evaluated the potential applications of BEVs in anticancer therapy. BEVs with size of 30–250 nm, carry proteases, sulfatase, and adhesins, which can be used to avoid phagocytosis, and facilitate cellular entry via different routes, such as lipid rafts, clathrin-dependent endocytosis, and micropinocytosis [30]. Therefore, BEVs can be used as excellent carriers for antitumor drugs. BEVs can strongly activate the innate immune system and can be used as immune adjuvants in tumor treatment. Furthermore, BEVs can effectively drain antigens and transport them to lymph nodes and can stimulate dendritic cell maturation and promote cytokine release [58]. The mechanisms may related to the observations that LPS, lipoprotein, and flagellin in BEVs can effectively activate TLR4 and the inflammasome, TLR2, and TLR5, respectively. At the same time, other studies have shown that BEV surface modification through genetic engineering, such as display of tumor antigens at the Cly A site, can induce a strong antigen-specific antitumor immune response to express tumor antigens [59].

BEVs regulate the tumor immune microenvironment, thereby improving the efficacy of immunotherapy. For example, T cells can be protected from the programmed cell death 1 (PD1)/programmed cell death 1 ligand 1 (PD-L1) immunosuppression axis by binding to tumors using genetically modified BEV PD-1 [60]. BEV PD-1 not only binds to PD-L1 on the tumor cell surface and promotes internalization and reduction but also does not affect their ability to trigger an immune response. The combined effects of immune activation and checkpoint inhibition can lead to the accumulation of T cells in tumor cells, resulting in a greater killing effect on the tumor [61]. In addition, BEVs can improve the efficacy of photothermal therapy for tumors. The effects of LPS on BEVs (Schwarzman response) can lead to the extravasation of red blood cells within tumors. In the near-infrared region, hemoglobin in red blood cells is absorbed and can be used as an endogenous photothermal factor for photothermal therapy [62]. Research has shown that extravasation of red blood cells within the tumor resulted in the accumulation of hemoglobin in the tumor, thereby increasing light absorption in the near-infrared region and ultimately destroying the tumor.

In conclusion, BEVs can theoretically be used as carriers for antitumor drugs, excellent immune adjuvants, comprehensive regulators of the tumor microenvironment, and tools for the photothermal treatment of tumors, thus significantly improving antitumor efficacy. Although few studies have assessed the clinical applications of BEVs, these EVs are still promising tools in the biomedical field.

## References


1. Zeng B, Chen T, Luo J-Y, Zhang L, Xi Q-Y, Jiang Q-Y, et al. Biological characteristics and roles of noncoding RNAs in milk-derived extracellular vesicles. Adv Nutr (Bethesda, Md). 2020;12:1006.
2. Zempleni J, Sukreet S, Zhou F, Wu D, Mutai E. Milk-derived exosomes and metabolic regulation. Ann Rev Animal Biosci. 2019;7:245–62.
3. Melnik BC, Schmitz G. Exosomes of pasteurized milk: potential pathogens of Western diseases. J Transl Med. 2019;17(1):3.
4. Zhong J, Xia B, Shan S, Zheng A, Zhang S, Chen J, et al. High-quality milk exosomes as oral drug delivery system. Biomaterials. 2021;277:121126.
5. Blans K, Hansen MS, Sørensen LV, Hvam ML, Howard KA, Möller A, et al. Pellet-free isolation of human and bovine milk extracellular vesicles by size-exclusion chromatography. J Extracell Vesicles. 2017;6(1):1294340.
6. Yamauchi M, Shimizu K, Rahman M, Ishikawa H, Takase H, Ugawa S, et al. Efficient method for isolation of exosomes from raw bovine milk. Drug Dev Ind Pharm. 2019;45(3):359–64.
7. Zonneveld MI, Brisson AR, van Herwijnen MJC, Tan S, van de Lest CHA, Redegeld FA, et al. Recovery of extracellular vesicles from human breast milk is influenced by sample collection and vesicle isolation procedures. J Extracell Vesicles. 2014;3:3.
8. Brown BA, Zeng X, Todd AR, Barnes LF, Winstone JMA, Trinidad JC, et al. Charge detection mass spectrometry measurements of exosomes and other extracellular particles enriched from bovine milk. Anal Chem. 2020;92(4):3285–92.
9. Manca S, Upadhyaya B, Mutai E, Desaulniers AT, Cederberg RA, White BR, et al. Milk exosomes are bioavailable and distinct microRNA cargos have unique tissue distribution patterns. Sci Rep. 2018;8(1):11321.
10. Sadri M, Shu J, Kachman SD, Cui J, Zempleni J. Milk exosomes and miRNA cross the placenta and promote embryo survival in mice. Reproduction (Cambridge, England). 2020;160(4):501–9.
11. Delosière M, Pires JAA, Bernard L, Cassar-Malek I, Bonnet M. Dataset reporting 4654 cow milk proteins listed according to lactation stages and milk fractions. Data Brief. 2020;29:105105.
12. Reinhardt TA, Lippolis JD, Nonnecke BJ, Sacco RE. Bovine milk exosome proteome. J Proteome. 2012;75(5):1486–92.
13. Parry HA, Mobley CB, Mumford PW, Romero MA, Haun CT, Zhang Y, et al. Bovine milk extracellular vesicles (EVs) modification elicits skeletal muscle growth in rats. Front Physiol. 2019;10:436.
14. Mutai E, Zhou F, Zempleni J. Depletion of dietary bovine Milk exosomes impairs sensorimotor gating and spatial learning in C57BL/6 mice. FASEB J. 2017;2
15. Aguilar-Lozano A, Baier S, Grove R, Shu J, Giraud D, Leiferman A, et al. Concentrations of purine metabolites are elevated in fluids from adults and infants and in livers from mice fed diets depleted of bovine Milk exosomes and their RNA cargos. J Nutr. 2018;148(12):1886–94.
16. Zhou F, Paz HA, Sadri M, Cui J, Kachman SD, Fernando SC, et al. Dietary bovine milk exosomes elicit changes in bacterial communities in C57BL/6 mice. Am J Physiol Gastrointest Liver Physiol. 2019;317(5):G618–G24.
17. Wu D, Kittana H, Shu J, Kachman SD, Cui J, Ramer-Tait AE, et al. Dietary depletion of milk exosomes and their microRNA cargos elicits a depletion of miR-200a-3p and elevated intestinal inflammation and chemokine (C-X-C motif) ligand 9 expression in Mdr1a−/− mice. Curr Dev Nutr. 2019;3(12):nzz122.
18. Melnik BC, John SM, Schmitz G. Milk is not just food but most likely a genetic transfection system activating mTORC1 signaling for postnatal growth. Nutr J. 2013;12:103.
19. Melnik BC. Dairy consumption and hepatocellular carcinoma risk. Ann Transl Med. 2020;9(8):736.

20. Samuel M, Fonseka P, Sanwlani R, Gangoda L, Chee SH, Keerthikumar S, et al. Oral administration of bovine milk-derived extracellular vesicles induces senescence in the primary tumor but accelerates cancer metastasis. Nat Commun. 2021;12(1):3950.
21. Somiya M, Yoshioka Y, Ochiya T. Biocompatibility of highly purified bovine milk-derived extracellular vesicles. J Extracell Vesicles. 2018;7(1):1440132.
22. Munagala R, Aqil F, Jeyabalan J, Gupta RC. Bovine milk-derived exosomes for drug delivery. Cancer Lett. 2016;371(1):48–61.
23. Munagala R, Aqil F, Jeyabalan J, Agrawal AK, Mudd AM, Kyakulaga AH, et al. Exosomal formulation of anthocyanidins against multiple cancer types. Cancer Lett. 2017;393:94–102.
24. Li D, Yao S, Zhou Z, Shi J, Huang Z, Wu Z. Hyaluronan decoration of milk exosomes directs tumor-specific delivery of doxorubicin. Carbohydr Res. 2020;493:108032.
25. Qian Z, Qi X, Honglin Y, Chengwan X, Yumei P, Zhifeng H, et al. Milk-exosome based pH/light sensitive drug system to enhance anticancer activity against oral squamous cell carcinoma. RSC Adv. 2020;10:28314–23.
26. Chen C, Sun M, Wang J, Su L, Lin J, Yan X. Active cargo loading into extracellular vesicles: highlights the heterogeneous encapsulation behaviour. J Extracell Vesicles. 2021;10(13):e12163.
27. Aqil F, Munagala R, Jeyabalan J, Agrawal AK, Kyakulaga A-H, Wilcher SA, et al. Milk exosomes—natural nanoparticles for siRNA delivery. Cancer Lett. 2019;449:186–95.
28. Ishiguro K, Yan IK, Lewis-Tuffin L, Patel T. Targeting liver cancer stem cells using engineered biological nanoparticles for the treatment of hepatocellular cancer. Hepatol Commun. 2020;4(2):298–313.
29. Shandilya S, Rani P, Onteru SK, Singh D. Natural ligand-receptor mediated loading of siRNA in milk derived exosomes. J Biotechnol. 2020;318:1–9.
30. Schwechheimer C, Kuehn MJ. Outer-membrane vesicles from gram-negative bacteria: biogenesis and functions. Nat Rev Microbiol. 2015;13(10):605–19.
31. Kim JH, Lee J, Park J, Gho YS. Gram-negative and Gram-positive bacterial extracellular vesicles. Semin Cell Dev Biol. 2015;40:97–104.
32. Kulp A, Kuehn MJ. Biological functions and biogenesis of secreted bacterial outer membrane vesicles. Annu Rev Microbiol. 2010;64:163–84.
33. Kim MH, Choi SJ, Choi HI, Choi JP, Park HK, Kim EK, et al. Lactobacillus plantarum-derived extracellular vesicles protect atopic dermatitis induced by Staphylococcus aureus-derived extracellular vesicles. Allergy, Asthma Immunol Res. 2018;10(5):516–32.
34. Jan AT. Outer membrane vesicles (OMVs) of gram-negative bacteria: a perspective update. Front Microbiol. 2017;8:1053.
35. Tashiro Y, Ichikawa S, Nakajima-Kambe T, Uchiyama H, Nomura N. Pseudomonas quinolone signal affects membrane vesicle production in not only gram-negative but also gram-positive bacteria. Microbes Environ. 2010;25(2):120–5.
36. Kim W, Lee EJ, Bae IH, Myoung K, Kim ST, Park PJ, et al. Lactobacillus plantarum-derived extracellular vesicles induce anti-inflammatory M2 macrophage polarization in vitro. J Extracell Vesicles. 2020;9(1):1793514.
37. Bomberger JM, Maceachran DP, Coutermarsh BA, Ye S, O'Toole GA, Stanton BA. Long-distance delivery of bacterial virulence factors by Pseudomonas aeruginosa outer membrane vesicles. PLoS Pathog. 2009;5(4):e1000382.
38. Kuehn MJ, Kesty NC. Bacterial outer membrane vesicles and the host-pathogen interaction. Genes Dev. 2005;19(22):2645–55.
39. Kaparakis M, Turnbull L, Carneiro L, Firth S, Coleman HA, Parkington HC, et al. Bacterial membrane vesicles deliver peptidoglycan to NOD1 in epithelial cells. Cell Microbiol. 2010;12(3):372–85.
40. Shen Y, Giardino Torchia ML, Lawson GW, Karp CL, Ashwell JD, Mazmanian SK. Outer membrane vesicles of a human commensal mediate immune regulation and disease protection. Cell Host Microbe. 2012;12(4):509–20.

41. Davis JM, Carvalho HM, Rasmussen SB, O'Brien AD. Cytotoxic necrotizing factor type 1 delivered by outer membrane vesicles of uropathogenic Escherichia coli attenuates polymorphonuclear leukocyte antimicrobial activity and chemotaxis. Infect Immun. 2006;74(8):4401–8.
42. Lapinet JA, Scapini P, Calzetti F, Pérez O, Cassatella MA. Gene expression and production of tumor necrosis factor alpha, interleukin-1beta (IL-1beta), IL-8, macrophage inflammatory protein 1alpha (MIP-1alpha), MIP-1beta, and gamma interferon-inducible protein 10 by human neutrophils stimulated with group B meningococcal outer membrane vesicles. Infect Immun. 2000;68(12):6917–23.
43. Masforrol Y, Gil J, García D, Noda J, Ramos Y, Betancourt L, et al. A deeper mining on the protein composition of VA-MENGOC-BC®: an OMV-based vaccine against N. meningitidis serogroup B and C. Hum Vaccin Immunother. 2017;13(11):2548–60.
44. Alaniz RC, Deatherage BL, Lara JC, Cookson BT. Membrane vesicles are immunogenic facsimiles of salmonella typhimurium that potently activate dendritic cells, prime B and T cell responses, and stimulate protective immunity in vivo. J Immunol (Baltimore, Md : 1950). 2007;179(11):7692–701.
45. Kim OY, Hong BS, Park KS, Yoon YJ, Choi SJ, Lee WH, et al. Immunization with Escherichia coli outer membrane vesicles protects bacteria-induced lethality via Th1 and Th17 cell responses. J Immunol (Baltimore, Md : 1950). 2013;190(8):4092–102.
46. Kim JH, Yoon YJ, Lee J, Choi EJ, Yi N, Park KS, et al. Outer membrane vesicles derived from Escherichia coli up-regulate expression of endothelial cell adhesion molecules in vitro and in vivo. PLoS One. 2013;8(3):e59276.
47. Mashburn LM, Whiteley M. Membrane vesicles traffic signals and facilitate group activities in a prokaryote. Nature. 2005;437(7057):422–5.
48. Li Z, Clarke AJ, Beveridge TJ. A major autolysin of Pseudomonas aeruginosa: subcellular distribution, potential role in cell growth and division and secretion in surface membrane vesicles. J Bacteriol. 1996;178(9):2479–88.
49. Tashiro Y, Yawata Y, Toyofuku M, Uchiyama H, Nomura N. Interspecies interaction between Pseudomonas aeruginosa and other microorganisms. Microbes Environ. 2013;28(1):13–24.
50. Li Z, Clarke AJ, Beveridge TJ. Gram-negative bacteria produce membrane vesicles which are capable of killing other bacteria. J Bacteriol. 1998;180(20):5478–83.
51. Kadurugamuwa JL, Beveridge TJ. Bacteriolytic effect of membrane vesicles from Pseudomonas aeruginosa on other bacteria including pathogens: conceptually new antibiotics. J Bacteriol. 1996;178(10):2767–74.
52. Berleman JE, Allen S, Danielewicz MA, Remis JP, Gorur A, Cunha J, et al. The lethal cargo of Myxococcus xanthus outer membrane vesicles. Front Microbiol. 2014;5:474.
53. Evans AGL, Davey HM, Cookson A, Currinn H, Cooke-Fox G, Stanczyk PJ, et al. Predatory activity of Myxococcus xanthus outer-membrane vesicles and properties of their hydrolase cargo. Microbiology. 2012;158(Pt 11):2742–52.
54. Bae EH, Seo SH, Kim CU, Jang MS, Song MS, Lee TY, et al. Bacterial outer membrane vesicles provide broad-spectrum protection against influenza virus infection via recruitment and activation of macrophages. J Innate Immun. 2019;11(4):316–29.
55. Lee TY, Kim CU, Bae EH, Seo SH, Jeong DG, Yoon SW, et al. Outer membrane vesicles harboring modified lipid A moiety augment the efficacy of an influenza vaccine exhibiting reduced endotoxicity in a mouse model. Vaccine. 2017;35(4):586–95.
56. Shehata MM, Mostafa A, Teubner L, Mahmoud SH, Kandeil A. Bacterial outer membrane vesicles (OMVs)-based dual vaccine for influenza a H1N1 virus and MERS-CoV. Vaccines (Basel). 2019;7(2)
57. Watkins HC, Pagan CL, Childs HR, Posada S, Chau A, Rios J, et al. A single dose and long lasting vaccine against pandemic influenza through the controlled release of a heterospecies tandem M2 sequence embedded within detoxified bacterial outer membrane vesicles. Vaccine. 2017;35(40):5373–80.
58. Kaparakis-Liaskos M, Ferrero RL. Immune modulation by bacterial outer membrane vesicles. Nat Rev Immunol. 2015;15(6):375–87.

59. Cheng K, Zhao R, Li Y, Qi Y, Wang Y, Zhang Y, et al. Bioengineered bacteria-derived outer membrane vesicles as a versatile antigen display platform for tumor vaccination via plug-and-display technology. Nat Commun. 2021;12(1):2041.
60. Pan J, Li X, Shao B, Xu F, Huang X, Guo X, et al. Self-blockade of PD-L1 with bacteria-derived outer-membrane vesicle for enhanced cancer immunotherapy. Adv Mater (Deerfield Beach, Fla). 2022;34:e2106307.
61. Moghimipour E, Abedishirehjin S, Baghbadorani MA, Handali S. Bacteria and archaea: a new era of cancer therapy. J Control Release. 2021;338:1–7.
62. Yang J, Kim EK, Park HJ, McDowell A, Kim YK. The impact of bacteria-derived ultrafine dust particles on pulmonary diseases. Exp Mol Med. 2020;52(3):338–47.

# Extracellular Vesicle-Based Clinical Trials 29

Xingliang Fan, Shu-Bin Fang, and Qing-Ling Fu

## 29.1 From Laboratory to Clinic: Translation of Small Extracellular Vesicle (sEV)-Based Candidates

Differential centrifugation is the most widely used method for isolating sEVs. However, differential centrifugation is a labor-intensive, time-consuming, and non-scalable method that is difficult to apply in the large-scale industrial production of sEVs. The polyethylene glycol (PEG)-based method can enrich EVs rapidly and inexpensively using low-speed centrifugation but has the disadvantages of low purity and nonscalability. In some studies, differential centrifugation or PEG-based precipitation was combined with density gradient ultracentrifugation to obtain higher purity sEVs, resulting in low sEV yields. To purify sEVs, scalable isolation methods are required to obtain large-scale sEVs for clinical applications [1]. Chromatography and filtration-based techniques are attracting increasing interest because they are scalable, relatively simple, robust, and thus readily scaled up to the

X. Fan
Otorhinolaryngology Hospital, The First Affiliated Hospital, Sun Yat-sen University, Guangzhou, China

Institute of Biological and Food Engineering, Guangdong University of Education, Guangzhou, China

S.-B. Fang
Otorhinolaryngology Hospital, The First Affiliated Hospital, Sun Yat-sen University, Guangzhou, China

Q.-L. Fu (✉)
Otorhinolaryngology Hospital, The First Affiliated Hospital, Sun Yat-sen University, Guangzhou, China

Key Laboratory for Stem Cells and Tissue Engineering, Ministry of Education, Sun Yat-sen University, Guangzhou, China
e-mail: fuqingl@mail.sysu.edu.cn



Q. Wang, L. Zheng (eds.), *Extracellular Vesicles*,
https://doi.org/10.1007/978-981-99-8365-0_29

industrial levels for clinical applications. Chromatography-based strategies include size-exclusion chromatography (SEC), immunoaffinity chromatography, and ion-exchange chromatography, which were designed based on the sizes, specific markers, and charges of mesenchymal stromal cell (MSC) sEVs.

In early 2002, Taylor et al. first developed SEC for isolating plasma sEVs [2]. Lim et al. [3] optimized the method for isolating exosomes derived from embryonic stem cell-derived MSCs in 2010. One of the drawbacks of SEC is that the volume of the loading medium is limited. Additionally, the flow-out sEV samples are usually obtained in a very dilute solution. Therefore, concentration of sEVs in loading medium is required and can typically be achieved using ultracentrifugation, ultrafiltration, or precipitation. Notably, the sample concentration can affect vesicle loss, efficiency, and costs.

Ion-exchange chromatography is based on the properties and intensities of charges on the particles. Kim et al. [4] reported that sEVs derived from bone marrow-MSCs were able to bind to anion exchange resin but not cation exchange chromatography; thus, anion exchange chromatography (AEC) was developed for the purification of MSC sEVs (Fig. 29.1). Fang et al. [5–7] further optimized AEC for the purification of functional sEVs derived from induced pluripotent stem cell (iPSC)-derived MSC. Compared with size exclusion- and immunoaffinity-based chromatography, the final samples collected from AEC are usually concentrated and may not need to be further concentrated. However, when using AEC, some proteins with properties and intensities similar to those of sEVs could be released into the supernatants by MSCs. For this reason, Upadhya et al. [8] combined AEC with SEC to obtain higher purity sEVs released by iPSC-derived neural stem cells.

Ultrafiltration-based separation of sEVs can be performed by either direct flow filtration (DFF) or tangential flow filtration (TFF) [9]. The fluid in the TFF runs parallel to the filter rather than being pushed through the membrane. Continuous filtration, reproducibility, and scalability can also be achieved by TFF. However, because TFF simply relies on polymeric ultrafiltration membranes with defined pore sizes for separation of sEVs, the purity and yield rate of sEVs should be of great concern. The final step employing SEC is suitable for the removal of protein aggregates and lipoprotein particles. However, a significant decrease in the total particle number is observed during this step.

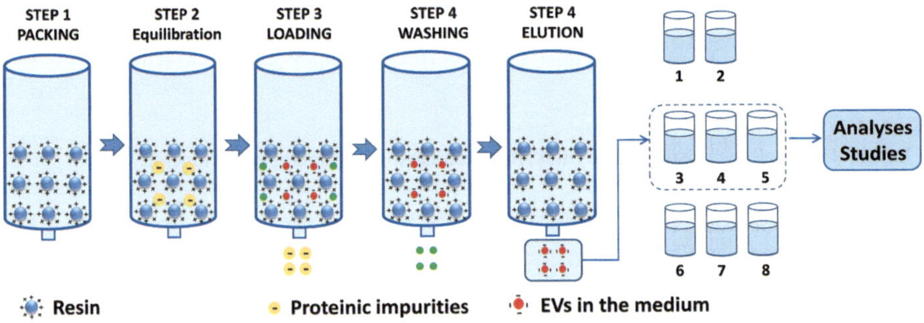


**Fig. 29.1** Schematic diagram of anion exchange chromatography

**Table 29.1** Pros and cons of scalable EV isolation methods

| Method | Pros | Cons |
|---|---|---|
| Size-exclusion chromatography (SEC) | Single-step process<br>Relatively high-purity isolation<br>Isolation of intact vesicles | Low volume of loading samples<br>Contamination of other vesicles with similar size to sEV<br>Diluent final sEV samples |
| Immunoaffinity chromatography | Isolation of subtypes of sEVs<br>High-purity isolation<br>Isolation of intact vesicles | Lower function due to elimination of other sEVs subtypes<br>Diluent final sEV samples |
| Anion exchange chromatography (AEC) | High efficiency<br>Isolation of intact vesicles<br>Concentrated final sEV samples | Contamination of other components with similar charge properties to sEV<br>Need dialysis to remove sodium chloride |
| Tangential-flow filtration (TFF) | High efficiency<br>High yield rate | Contamination of some proteins and further purify may be needed<br>Blockage of membrane pores |

Although the methods described above are theoretically scalable and have great potential for the industrial production of sEVs, each of these methods has its advantages and disadvantages (Table 29.1). Thus, a combination of different isolation methods based on different principles is currently the best way to improve the purity and efficiency of MSC sEVs. For example, in some studies, SEC was combined with ultracentrifugation, a PEG-based method, or DFF for the concentration of sEV samples. However, because ultracentrifugation and DFF are all nonscalable and inefficient, better combination methods are required. In a recent study, Upadhya et al. [8] reported that AEC followed by SEC and DFF were able to isolate high-purity and functional sEVs from iPSC-derived neural stem cells. In fact, DFF could be easily replaced using a scalable TFF in this isolation strategy. Therefore, it appears that the combination of AEC, SEC, and TFF is a promising approach for sEV production in terms of time effectiveness, ease of operation, scalability, and automatization. However, the yield rate of sEVs may be lower and the cost may be higher, although the purity of sEVs can be increased by using a combination of different isolation methods. Studies evaluating the effects of MSC sEVs isolated using AEC alone or in combination with AEC and SEC are ongoing. If the same effects on diseases are obtained, AEC alone can be used for large-scale production of MSC sEVs.

However, because the large-scale clinical manufacturing of sEVs may not be as stringent with regard to segregation of sEVs from copurifying components, there may be an increased percentage of secretome components in the final sEV preparation [1]. Additionally, various components secreted in the final sEV preparation, including some proteins, may influence or promote the therapeutic activity of sEV. Optimistically, the product may resemble more of an sEV-containing preparation or vesicle-enriched secretome fraction (VSF). Therefore, MSC-derived sEV preparations may instead be considered a VSF rather than a pure or at least highly enriched vesicle product. Taken together, these results indicate that as long as we can obtain effective sEV products, we can choose feasible isolation methods.

## 29.2 Current Status of sEVs in Clinical Trials

Many research efforts have focused on developing sEV-based candidates for clinical applications. However, according to www.clinicaltrials.gov reports (reviewed in December 2021), only approximately 13% of registered trials are focused on EVs as therapeutics or drug delivery agents.

Several companies have developed sEV-based products showing potential feasibility and short-term safety. For example, MolecuVax Inc. developed sEV-based technology derived from various immune cells to deliver therapeutic peptides. Jazz Pharmaceutics Inc. is currently working on chemotherapeutics loaded into sEVs to treat cancer. Finally, several companies have manufactured drug-loaded sEVs derived from stem cells and immortalized cell lines based on primary human amniocytes for the treatment of neurodegenerative disorders (Stem Cell Medicine Ltd. and Evox Therapeutics) [10]. As a result, positive feedback from patients has encouraged the development of sEV-based products.

There are two main categories of sEVs being applied in clinical trials, namely, sEVs derived from human specimens and those derived plants, and the results of clinical trials using sEVs from human specimens have been reported. By contrast, plant-derived sEVs are in the beginning stages of development, and patients have been recruited for only one of three planned trials. The use of sEVs as drug carriers has also been investigated in clinical trials because of the vesicle structure of sEVs. Table 29.2 describes the sEVs used in clinical trials with complete clinical results, and Table 29.3 reviews the development of typical sEVs in clinical trials in the recruiting- or nonrecruiting stages.

Three major sources of sEVs, i.e., dendritic cells (DCs), MSCs, and patient-derived tumor cells, have been subjected to clinical trials. The application of human-derived sEVs has mainly focused on cancer therapy and treatments for inflammation-related or chronic diseases. Because sEVs can carry tumor antigens on the surface to induce antitumor immunity in patients or package anticancer drugs as sEV cargo to induce cytotoxicity for the treatment of cancer, sEVs can largely improve outcomes. Moreover, the utilization of MSC sEVs exerting MSC inflammation regulatory effects is a strategy to treat inflammation-related and chronic diseases.

EVs from DCs can be released from immature or mature DCs. DC-derived sEVs harbor tumor peptides that can be injected subcutaneously into patients diagnosed with cancer to induce immune responses. However, only a limited number of clinical trials have used DC sEVs (Table 29.2). Trials employing immature DC-derived sEVs have demonstrated that these sEVs are safe and show comparable efficacy in melanoma and non-small cell lung cancer. However, in non-small cell lung cancer, melanoma antigen gene-specific T-cell responses have also been witnessed [12].

**Table 29.2** Summary of sEVs used in clinical trials with complete results reported

| Indication | Year, phase, patients | Source | Dose | Administration | Purification | Characterization | Bioactivity | EV manipulation | Results |
|---|---|---|---|---|---|---|---|---|---|
| Melanoma [11] | 2000, Phase 1, (*n* = 15) | mDC, autologous | $4 \times 10^{13}$ or $1.3 \times 10^{13}$ MHC Class II molecules | SC (90% of the volume) and ID (10%) injections weekly for 4 weeks | 500-kDa concentration and UC with $D_2O$/ sucrose cushion | CD81 tetraspanin | SEE test of potency | Pulsed with MAGE 3 tumor peptides | No Grade II toxicity; No detected MAGE3-specific $CD4^+$ and $CD8^+$ T cells |
| Non-small cell lung cancer [12] | Not reported. Phase 1, (*n* = 4) | imDCs, autologous | $1.3 \times 10^{13}$MHC Class II molecules | SC (90% of the volume) and ID (10%) injections weekly for 4 weeks | 500-kDa concentration and UC with D2O/ sucrose cushion | Not reported | MHC Class II molecules; ELISPOT for peptide-specific immune response | Pulsed with MAGE-A3, -A4, -A10, and MAGE-3DPO4 tumor peptide | Well-tolerated and only Grade 1–2 adverse events; MAGE-specific T-cell responses in 1/3 patients; increased NK lytic activity in 2/4 patients |
| Non-small cell lung cancer [13] (NCT01159288) | May 2010, Phase 2, (*n* = 22) | Mature-dendritic (mDCs), autologous (induced by rIFN-γ) | $8.5 \times 10^{11}$–$1.0 \times 10^{13}$ MHC Class II molecules | Four ID at 1-week intervals | UF; DF and UC through a 1.21 g/mL sucrose cushion | Exosome marker: Tetraspanin | Activation of LT11 cells Function: MHC class II molecules and CD40, CD86, and ICAM-1/ CD54 | Pulsed with MAGE-A1, -A3, NY-ESO-1, Melan-A/ MART1, MAGE-A3-DP04, EBV tumor peptides | One patient had Grade 3 hepatotoxicity; boosting the NK cell arm of antitumor immunity |

(continued)

**Table 29.2** (continued)

| Indication | Year, phase, patients | Source | Dose | Administration | Purification | Characterization | Bioactivity | EV manipulation | Results |
|---|---|---|---|---|---|---|---|---|---|
| Colon cancer [14] | Not reported, Phase 1, ($n = 40$) | Ascites, autologous | 100–500 μg of protein | Four SC at weekly intervals | Differential centrifugation + sucrose/ $D_2O$ density gradient UC | EV marker: HSPs (including HSC70, HSP70, and HSP90), CD80, ICAM-1, CD71), and LAMP-3; EM | Function: Tumor-associated carcinoembryonic antigen, MHC-I, and MHC-II molecules | ±GM-CSF | Safe, well-tolerated; tumor-specific antitumor CTL response in exosome plus GM-CSF group |
| Chronic kidney diseases [15] | April 2014, Phase 2/3, ($n = 40$) | MSCs, allogeneic | 100 μg/kg/dose | Two doses of MSC-EVs, intraarterial and intravenous injections | UC at 100,000 × g | CD9, CD63; EM | CD45, CD73 | Unmodified | Safe, well-tolerated; improved kidney function; decreased inflammation |

*CTL* Cytotoxic T lymphocyte, *$D_2O$* Heavy water, *DF* Diafiltration, *EM* Electron microscopy, *EV* Extracellular vesicle, *GM-CSF* Granulocyte-macrophage colony-stimulating factor, *HSP* Heat shock protein, *ICAM-1* Intercellular adhesion molecule 1, *ID* Intradermal, *imDCs* Immature dendritic cells, *LAMP-3* Lysosomal-associated membrane protein 3, *MAGE 3* Melanoma-associated antigen 3, *MHC* Major histocompatibility complex, *MSCs* Mesenchymal stem cells, *NK* Natural killer, *rIFN-γ* Recombinant interferon-γ, *SC* Subcutaneous, *UC* Ulceration, *UF* Ultrafiltration

**Table 29.3** Summary of sEVs used in clinical trials by December 2021 (source: clinical trials.gov)

| Indication | Year, phase, patients | Source | Dose | Administration | sEV manipulation | Results/ Status | Link |
|---|---|---|---|---|---|---|---|
| Malignant ascites and pleural effusion (NCT01854866) | May 2013, Phase 2, ($n$ = 30) | Tumor-derived | Not reported | Perfused to the pleural or peritoneal cavity, 4 times/week | Loaded with chemotherapeutic drugs | Unknown status | https://clinicaltrials.gov/ct2/show/NCT01854866?cond=NCT01854866&draw=2&rank=1 |
| Alzheimer Disease (NCT04388982) | July 2020, Phase1/2 | MSCs, allogeneic | 5–20 μg | Nasal drip | Unmodified | Unknown status | https://clinicaltrials.gov/ct2/show/NCT04388982?cond=NCT04388982&draw=2&rank=1 |
| Periodontitis (NCT04270006) | February 2020, early 1 | MSCs | Not reported | Oral | Unmodified | Unknown status | https://clinicaltrials.gov/ct2/show/NCT04270006?cond=NCT04270006&draw=2&rank=1 |
| Malignant pleural effusion (NCT02657460) | January 2016, Phase 2, (n=90) | Malignant pleural effusion | Not reported | Not reported | Loaded with methotrexate | Unknown status | https://clinicaltrials.gov/ct2/show/NCT02657460?cond=NCT02657460&draw=2&rank=1 |
| Metastatic pancreatic cancer (NCT03608631) | March 2020, Phase 1, ($n$ = 28) | MSCs, allogeneic | Not reported | Intravenous on days 1, 4, and 10. Treatment repeated every 14 days for up to 3 courses | KrasG12D siRNA (iExosomes) | Recruiting | https://clinicaltrials.gov/ct2/show/NCT03608631?cond=NCT03608631&draw=2&rank=1 |
| Neuralgia (NCT04202783) | December 2021, n/a | MSCs | 45 mg, 15 mg | Intravenous epineurally | Unmodified | Suspended | https://clinicaltrials.gov/ct2/show/NCT04202783?cond=NCT04202783&draw=2&rank=1 |

(continued)

**Table 29.3** (continued)

| Indication | Year, phase, patients | Source | Dose | Administration | sEV manipulation | Results/ Status | Link |
|---|---|---|---|---|---|---|---|
| Bronchopulmonary dysplasia (NCT03857841) | June 2019, Phase 1, ($n = 18$) | MSCs | −200 pmol phospholipid/kg | Intravenous | Not specified (UNEX-42) | Terminated | https://clinicaltrials.gov/ct2/show/NCT03857841?cond=NCT03857841&draw=2&rank=1 |
| Type 1 diabetes (NCT02138331) | April 2014, Phase 1, ($n = 20$) | MSCs, allogeneic | (1.22–1.51) × $10^6$/kg, Day 0 and 7 | Intravenous | Unmodified | Unknown Status | https://clinicaltrials.gov/ct2/show/NCT02138331?cond=NCT02138331&draw=2&rank=1 |
| Macular holes (NCT03437759) | March 2017, Phase 1, ($n = 44$) | MSCs, allogeneic | 50 μg or 20 μg | Dripped into vitreous cavity | Unmodified | Unknown status | https://clinicaltrials.gov/ct2/show/NCT03437759?cond=NCT03437759&draw=2&rank=1 |
| Acute ischemic stroke (NCT03384433) | April 2019, Phase 1/2, ($n = 5$) | MSCs, allogeneic | 200 μg | Stereotaxic injection | Enriched by miR-124 | Unknown Status | https://clinicaltrials.gov/ct2/show/NCT03384433?cond=NCT03384433&draw=2&rank=1 |
| Chronic Ulcer (NCT04134676) | June 2019, Phase 1 | MSCs | Not reported | Topical | Unmodified | Completed | https://clinicaltrials.gov/ct2/show/NCT04134676?cond=NCT04134676&draw=2&rank=1 |
| Colon cancer (NCT01294072) | January 2011, Phase 1, ($n = 35$) | Plant-derived | Not reported | Tablets taken daily for 7 days | Loaded with curcumin | Recruiting | https://clinicaltrials.gov/ct2/show/NCT01294072?cond=NCT01294072&draw=2&rank=1 |
| Multiple organ failure (NCT04356300) | September 2020, n/a | MSCs | 150 mg once a day for 14 times | Intravenous | Unmodified | Not yet recruiting | https://clinicaltrials.gov/ct2/show/NCT04356300?cond=NCT04356300&draw=2&rank=1 |

| | | | | | | | |
|---|---|---|---|---|---|---|---|
| Dystrophic Epidermolysis Bullosa (NCT04173650) | April 2021, Phase 1/2 | MSCs | Not reported | Topical | Unmodified | Not yet recruiting | https://clinicaltrials.gov/ct2/show/NCT04173650?cond=NCT04173650&draw=2&rank=1 |
| SARS-CoV-2 pneumonia (NCT04276987) | October 2020, Phase 1 | MSCs | $2 \times 10^8$ particles/3 mL | Inhalation | Unmodified | Completed | https://clinicaltrials.gov/ct2/show/NCT04276987?cond=NCT04276987&draw=2&rank=1 |
| SARS-CoV-2 pneumonia (NCT04491240) | July 2020, Phase 1/2 | MSCs | $0.5 \times 10^{10}$–$2 \times 10^{10}$ particles/3 mL | Inhalation | Unmodified | Completed | https://clinicaltrials.gov/ct2/show/NCT04491240?cond=NCT04491240&draw=2&rank=1 |
| SARS-CoV-2 pneumonia (NCT04493242) | September 2020, Phase 2 | Bone marrow | Not reported | Intravenous | Unmodified | Completed | https://clinicaltrials.gov/ct2/show/NCT04493242?cond=NCT04493242&draw=2&rank=1 |
| SARS-CoV-2 pneumonia (NCT04384445) | October 2020, Phase 1/2 | Human amniotic fluid | $0.5 \times 10^{10}$–$2 \times 10^{10}$ particles, 10 μg/drop | Intravenous | Unmodified | Active, not recruiting | https://clinicaltrials.gov/ct2/show/NCT04384445?cond=NCT04384445&draw=2&rank=1 |
| Radiation- and chemotherapy-induced oral mucositis (NCT01668849) | August 2012, Phase 1, ($n = 60$) | Grape derived | Not reported | Oral administration daily for 35 days | Unmodified | Completed | https://clinicaltrials.gov/ct2/show/NCT01668849?term=grape&cond=oral+mucositis&draw=2&rank=1 |
| Insulin resistance and chronic inflammation in polycystic ovary syndrome (NCT03493984) | May 2018, not applicable | Plant-derived (ginger and/or aloe) | Not reported | - | Unmodified | Withdrawn | https://clinicaltrials.gov/ct2/show/NCT03493984?cond=NCT03493984&draw=2&rank=1 |

*sEV* Small extracellular vesicle, *MSCs* Mesenchymal stem cells, *siRNA* Small interfering RNA, *miR* microRNA

Therefore, strategies for DC maturation have been proposed to improve T-cell stimulation in patients with non-small cell lung cancer. Nevertheless, one patient had grade 3 hepatotoxicity, and less than one-third of the patients have experienced progression-free survival for more than 4 months; this was less than the primary endpoint of 50% [13]. Ascites-derived sEVs from patients were collected, and tumor antigens (i.e., carcinoembryonic antigen) were directly derived from a patient with a tumor. Safety and tolerance in phase I trials have been defined for ascites-derived sEVs, which have been utilized as vehicles for anticancer drugs. In addition, a tumor-specific antitumor cytotoxic T-lymphocyte response has also been detected in the ascites-derived sEV plus granulocyte-macrophage colony-stimulating factor group [14].

In another trial, up to $2 \times 10^{12}$ sEVs were administered to patients with advanced colorectal cancer weekly for 4 weeks. The therapies were safe, feasible, and efficient in inducing an antigen-specific T-lymphocyte response [14]. Furthermore, sEVs containing chemotherapeutic drugs or small interfering RNAs (siRNAs) have been used to carry tumor antigens in the management of cancer. There are two registered clinical trials (NCT01854866 and NCT02657460) using chemotherapeutic drugs to treat patients diagnosed with malignant pleural effusion. In the preclinical trial and the trial NCT01854866, methotrexate (MTX) and cisplatin were used as anticancer drugs. The survival rate was higher when EV-loaded MTX was used as an anticancer drug in preclinical trials [16]. In a clinical trial (NCT02657460), both MTX (the encapsulated anticancer medication) and cisplatin (the comparator) were encapsulated into sEVs. $Kras^{G12D}$ siRNA has been promoted as an anticancer drug for the treatment of patients with metastatic pancreatic cancer, and MSC-derived sEVs have been proposed in clinical trials (NCT03608631).

MSC-derived sEVs were used in clinical trials in 2014 (Table 29.2), and extensive reports have been published since 2016 [16]. Most clinical trials using MSC-derived sEVs have focused on chronic diseases, immune diseases, and acute ischemic stroke. One report described a clinical trial using MSC-derived sEVs encapsulated with $Kras^{G12D}$ siRNA. Patients with chronic stroke administered allogeneic MSC-derived sEVs loaded with *miR-124* have been shown to be safe at doses of up to 1.5 million cells/kg body weight, and their behavioral endpoints have also improved in the last 12 months [17]. Two clinical trials reported the roles of MSC-derived sEVs in the treatment of chronic diseases (i.e., chronic kidney disease and bronchopulmonary dysplasia).

Three plant sources have been registered in the clinical trials from the same sponsor (University of Louisville), namely, grapes (NCT01668849) and ginger or aloe (NCT03493984); however, the clinical trial status of plant sEVs has not been disclosed (Table 29.3). Grape source sEVs have been used to treat oral mucositis caused by radiation and chemotherapy. In a previous preclinical study, Ju et al. proved that sEVs derived from grapes can renew intestinal tissues and participate in tissue remodeling when the tissue is pathologically damaged [18]. Another study used ginger or aloe to produce sEVs to treat patients diagnosed with polycystic ovary syndrome, hoping to reduce insulin resistance and chronic inflammation (NCT03493984). In addition, because the molecule is hydrophobic, plant-derived sEVs are used as hydrophobic drug delivery carriers to encapsulate curcumin in the treatment of intestinal diseases (NCT01294072).

## 29.3 Perspectives

The therapeutic potential of sEVs provides exciting new avenues for the treatment of various diseases. However, some problems still need to be resolved. First, the mechanism of action of sEVs is not well understood; however, it is thought that these EVs function through miRNAs. There is no doubt that sEV cargo may also be rich in other ingredients, such as proteins and lipids [19]. However, the contents and functions of sEVs may depend on the metabolic characteristics of the donor and/or conditioned medium, which may make the donor selection and manufacturing process more problematic [19–21]. Furthermore, the short half-life, off-target effects, and insufficient endocytosis remain major limitations of sEVs that need to be overcome during the clinical translation process [22]. Of course, the early phases of biological drug development, including sEV product development, offers the chance for a reasonable and practical approach, referred to as "The process is the product" [23]. Such an approach is common in biopharmaceutical development and suggests that if manufacturing repeatedly follows highly standardized procedures, the resulting batch consistency is probably high. Therefore, the mechanisms of sEV production may not be consistent, at least during the early phases of EV application.

Secondly, there are obvious limitations when considering the isolation, characterization, tracking procedures, and clinical-grade production of sEV-based therapeutics. As described previously, there are many isolation and characterization methods for sEVs. However, other techniques are also widely used for different studies [22, 24, 25], and efforts to standardize these procedures and overcome confusion regarding the nomenclature are ongoing [26]. Similarly, databases built to provide information about the composition and functions of sEVs, such as Evpedia, ExoCarta, and Vesiclepedia, will probably contribute to the clinical translation process [22]. For clinical-grade production, sEVs must be produced and stored under current good manufacturing practice conditions to meet market demands.

## 29.4 Summary

Currently, sEVs are considered novel therapeutic tools and have rapidly progressed into clinical studies. However, these EVs are still in their infancy, and well-designed clinical trials will help to identify their therapeutic activities in humans. sEV-based therapeutics represent a next-generation drug delivery system and could be valuable tools for the treatment of various diseases.

## References


1. Gimona M, Pachler K, Laner-Plamberger S, Schallmoser K, Rohde E. Manufacturing of human extracellular vesicle-based therapeutics for clinical use. Int J Mol Sci. 2017;18(6):1190.
2. Taylor DD, Lyons KS, Gercel-Taylor C. Shed membrane fragment-associated markers for endometrial and ovarian cancers. Gynecol Oncol. 2002;84(3):443–8.
3. Lai RC, Arslan F, Lee MM, Sze NS, Choo A, Chen TS, et al. Exosome secreted by MSC reduces myocardial ischemia/reperfusion injury. Stem Cell Res. 2010;4(3):214–22.
4. Kim DK, Nishida H, An SY, Shetty AK, Bartosh TJ, Prockop DJ. Chromatographically isolated CD63+CD81+ extracellular vesicles from mesenchymal stromal cells rescue cognitive impairments after TBI. Proc Natl Acad Sci U S A. 2016;113(1):170–5.
5. Fang SB, Zhang HY, Meng XC, Wang C, He BX, Peng YQ, et al. Small extracellular vesicles derived from human MSCs prevent allergic airway inflammation via immunomodulation on pulmonary macrophages. Cell Death Dis. 2020;11(6):409.
6. Fang SB, Zhang HY, Wang C, He BX, Liu XQ, Meng XC, et al. Small extracellular vesicles derived from human mesenchymal stromal cells prevent group 2 innate lymphoid cell-dominant allergic airway inflammation through delivery of miR-146a-5p. J Extracell Vesicles. 2020;9(1):1723260.
7. You HJ, Fang SB, Wu TT, Zhang H, Feng YK, Li XJ, et al. Mesenchymal stem cell-derived exosomes improve motor function and attenuate neuropathology in a mouse model of Machado-Joseph disease. Stem Cell Res Ther. 2020;11(1):222.
8. Upadhya R, Madhu LN, Attaluri S, Gitai DLG, Pinson MR, Kodali M, et al. Extracellular vesicles from human iPSC-derived neural stem cells: miRNA and protein signatures, and anti-inflammatory and neurogenic properties. J Extracell Vesicles. 2020;9(1):1809064.
9. Busatto S, Vilanilam G, Ticer T, Lin WL, Dickson DW, Shapiro S, et al. Tangential flow filtration for highly efficient concentration of extracellular vesicles from large volumes of fluid. Cells. 2018;7(12):273.
10. Klyachko NL, Arzt CJ, Li SM, Gololobova OA, Batrakova EV. Extracellular vesicle-based therapeutics: preclinical and clinical investigations. Pharmaceutics. 2020;12(12):1171
11. Escudier B, Dorval T, Chaput N, Andre F, Caby MP, Novault S, et al. Vaccination of metastatic melanoma patients with autologous dendritic cell (DC) derived-exosomes: results of thefirst phase I clinical trial. J Transl Med. 2005;3(1):10.
12. Morse MA, Garst J, Osada T, Khan S, Hobeika A, Clay TM, et al. A phase I study of dexosome immunotherapy in patients with advanced non-small cell lung cancer. J Transl Med. 2005;3(1):9.
13. Besse B, Charrier M, Lapierre V, Dansin E, Lantz O, Planchard D, et al. Dendritic cell-derived exosomes as maintenance immunotherapy after first line chemotherapy in NSCLC. Onco Targets Ther. 2015;5(4):e1071008-e.
14. Dai S, Wei D, Wu Z, Zhou X, Wei X, Huang H, et al. Phase I clinical trial of autologous ascites-derived exosomes combined with GM-CSF for colorectal cancer. Mol Ther. 2008;16(4):782–90.
15. Nassar W, El-Ansary M, Sabry D, Mostafa MA, Fayad T, Kotb E, et al. Umbilical cord mesenchymal stem cells derived extracellular vesicles can safely ameliorate the progression of chronic kidney diseases. Biomater Res. 2016;20:21.
16. Tang K, Zhang Y, Zhang H, Xu P, Liu J, Ma J, et al. Delivery of chemotherapeutic drugs in tumour cell-derived microparticles. Nat Commun. 2012;3:1282.
17. Levy ML, Crawford JR, Dib N, Verkh L, Tankovich N, Cramer SC. Phase I/II study of safety and preliminary efficacy of intravenous allogeneic mesenchymal stem cells in chronic stroke. Stroke. 2019;50(10):2835–41.
18. Ju S, Mu J, Dokland T, Zhuang X, Wang Q, Jiang H, et al. Grape exosome-like nanoparticles induce intestinal stem cells and protect mice from DSS-induced colitis. Mol Ther. 2013;21(7):1345–57.
19. Bogatcheva NV, Coleman ME. Conditioned medium of mesenchymal stromal cells: a new class of therapeutics. Biochemistry (Mosc). 2019;84(11):1375–89.

20. Lener T, Gimona M, Aigner L, Borger V, Buzas E, Camussi G, et al. Applying extracellular vesicles based therapeutics in clinical trials - an ISEV position paper. J Extracell Vesicles. 2015;4:30087.
21. Meng Y, Eirin A, Zhu XY, O'Brien DR, Lerman A, van Wijnen AJ, et al. The metabolic syndrome modifies the mRNA expression profile of extracellular vesicles derived from porcine mesenchymal stem cells. Diabetol Metab Syndr. 2018;10:58.
22. Meng W, He C, Hao Y, Wang L, Li L, Zhu G. Prospects and challenges of extracellular vesicle-based drug delivery system: considering cell source. Drug Deliv. 2020;27(1):585–98.
23. Rohde E, Pachler K, Gimona M. Manufacturing and characterization of extracellular vesicles from umbilical cord-derived mesenchymal stromal cells for clinical testing. Cytotherapy. 2019;21(6):581–92.
24. Gardiner C, Di Vizio D, Sahoo S, Thery C, Witwer KW, Wauben M, et al. Techniques used for the isolation and characterization of extracellular vesicles: results of a worldwide survey. J Extracell Vesicles. 2016;5:5.
25. Konoshenko MY, Lekchnov EA, Vlassov AV, Laktionov PP. Isolation of extracellular vesicles: general methodologies and latest trends. Biomed Res Int. 2018;2018:8545347.
26. Thery C, Witwer KW, Aikawa E, Alcaraz MJ, Anderson JD, Andriantsitohaina R, et al. Minimal information for studies of extracellular vesicles 2018 (MISEV2018): a position statement of the International Society for Extracellular Vesicles and update of the MISEV2014 guidelines. J Extracell Vesicles. 2018;7(1):1535750.

# Part V

# Introduction to the Guidelines and Resources for EV Research

# Introduction to Guidelines for Extracellular Vesicle Research

# 30

Yunyun Hu, Lei Zheng, and Xiaomei Yan

The last decade has witnessed an explosive growth in scientific research on extracellular vesicles (EVs), which play important roles in intercellular communication via bioactive molecules, including proteins, nucleic acids, and lipids [1–3]. Because these biomolecular cargos can reflect the genetic or phenotypic alterations of parental cells and/or tissues, EVs are widely recognized as potential diagnostic and prognostic biomarkers [4–6]. Moreover, given their high abundance, biocompatibility, and ability to pass through biological barriers, EVs are being explored as delivery vehicles for therapeutic agents [7–9]. Nevertheless, the small size and large heterogeneity of EVs, the complexity of EV specimens from different sources, and the diversity of technologies used for EV isolation and characterization make it difficult to reach definitive conclusions in this field. Moreover, various misconceptions and methodological pitfalls have hampered further exploration of the biological functions of EVs. Thus, there is an urgent need to establish criteria and guidance to improve the reproducibility of EV research.

To this end, the International Society for Extracellular Vesicles (ISEV), the International Society on Thrombosis and Haemostasis (ISTH), the International Society on Advancement of Cytometry (ISAC), and other international committees and research groups have made joint efforts to establish guidelines and minimal requirements for studies of EVs [10–15]. In order to facilitate transparent reporting and replication of experiments, the EV-TRACK (Transparent Reporting and

Y. Hu · X. Yan (✉)
Department of Chemical Biology, College of Chemistry and Chemical Engineering, Xiamen University, Xiamen, Fujian, China
e-mail: xmyan@xmu.edu.cn

L. Zheng (✉)
Department of Laboratory Medicine, Nanfang Hospital, Southern Medical University, Guangzhou, Guangdong, China
e-mail: nfyyzhenglei@smu.edu.cn



Q. Wang, L. Zheng (eds.), *Extracellular Vesicles*,
https://doi.org/10.1007/978-981-99-8365-0_30

Centralizing Knowledge in Extracellular Vesicle Research) knowledgebase was created to encourage researchers to upload experimental parameters and provide feedback [16]. Additionally, several public databases have been launched, including ExoCarta: a database of exosomal proteins, RNA, and lipids [17, 18]; EVpedia: an integrated database of high-throughput datasets of vesicular components [19, 20]; Vesiclepedia: a compendium for EVs with continuous community annotation [21]; exoRBase: a database of circular RNAs, long noncoding RNAs, and mRNAs in human blood exosomes; and exRNA Atlas for EV RNA analysis [22]. In this section, ISEV position papers, databases, and other EV-related resources are summarized for researchers to find important guidance and resources for EV research.

## 30.1 General Guidelines

Despite the rapid growth in EV studies, the mechanisms mediating the biogenesis of EV subtypes and the shuttling of EV cargo remain unclear. Since the initial discovery of RNAs in EVs (evRNAs) and the diagnostic and therapeutic potential of these RNAs, the scientific community has recognized the need to minimize the influence of different methodologies for EV isolation and characterization on the results of evRNA profiling. To this end, the first International Society for Extracellular Vesicles (ISEV) Workshop, focusing on evRNA analysis and bioinformatics, was organized and held in New York in October 2012. Subsequently, the first two ISEV position papers were officially released; these papers emphasized the need for standardization of sample collection and processing [10] and RNA analysis [23] approaches. In consideration of the technical challenges encountered when attempting to obtain a completely pure EV isolate free from non-EV components, the ISEV published a position statement titled "Minimal Information for Studies of Extracellular Vesicles" (MISEV2014) [11]. A series of criteria for EV studies were provided, covering EV separation/isolation, characterization, and functional studies. In 2018, based on advances in EV science from 2014 to 2018, MISEV was updated as MISEV2018 [13]. Compared with MISEV2014, the updated version provided clearer explanations of the need for each recommendation. Importantly, a two-page checklist was included as a paradigm to aid researchers in planning studies and interpreting their results in a more transparent way. Additionally, a series of workshops and cosponsored events initiated by ISEV and other research committees gave birth to additional position papers providing guidance on topics including EVs derived from human blood [12], urine [24], and tissues [25]; EVs obtained from pathogens [26] and plants [27]; evRNAs [28]; and EV membranes [29]. Guidelines and recommendations for exploiting EVs as disease biomarkers [30] or therapeutic agents [31, 32] were also summarized. Recently, a position editorial summarized the past contributions of the ISEV committee to improve rigor and standardization in EV research and provided a great deal of guideline papers based on different topics [14]. For ease of reference, the guideline papers for EV research are summarized in Table 30.1.

**Table 30.1** Summary of guideline papers

| Content | Reference(s) |
|---|---|
| *General guidelines* | [14] |
| Minimal information for studies of extracellular vesicles (MISEV) | [11, 13] |
| RNA in EVs (evRNA) | [23, 28] |
| Urine EVs (uEVs) | [24] |
| Tissue-derived EVs | [25] |
| Pathogen EVs | [26] |
| Plant EVs | [27] |
| EV membranes | [29] |
| *Guidelines for sample collection, handling, and storage* | [10, 12, 33–36] |
| *Guidelines for EV isolation and characterization* | |
| Tunable resistive pulse sensing (TRPS) | [37] |
| Nanoparticle tracking analysis (NTA) | [38] |
| Flow cytometry (FCM) | [15, 34, 39–42] |
| Reference materials | [43] |
| Omics profiling | [45–49] |
| *Guidelines for EV clinical research* | [46, 50, 51] |
| Biomarker discovery | [30, 42, 52, 53] |
| Therapeutic agents | [31, 32, 54–57] |

## 30.2 Guidelines for Sample Collection, Handling, and Storage

Standardization of sample collection, handling, and storage is the first step to reduce inconsistency among results from different studies. Among all types of body fluids, human blood represents the most interesting EV source for disease diagnosis. However, there are lots of variabilities and artifacts encountered when analyzing blood EVs. To this end, the International Society on Thrombosis and Haemostasis (ISTH) standardization subcommittee held a series of workshops to determine the major pre-analytical variables and identify the optimal protocols for human blood collection and handling [33, 34]. In 2013, the ISEV published the first position paper on the standardization of sample collection, isolation, and analysis methods in EV or evRNA research [10]. Two years later, another study within the context of the European Metrology Research Program developed standard protocols for handling and storage of plasma EVs. They also expanded the EV sources to other body fluids, such as urine and saliva [35]. On the other hand, owing to the high concentrations of soluble proteins and lipoproteins in the blood and the considerable overlap in both the particle size and density between EVs and lipoproteins, unintentional co-isolation of these two different entities usually occurs. To address this problem, a blood EV working group was set up during the ISEV Biomarker Workshop held in Birmingham (UK) in 2017. Based on the responses to a questionnaire from working group members, the working group published a report providing a roadmap for the collection, handling, and storage of blood needed for EV analysis in 2019 [36].

## 30.3 Guidelines for EV Isolation and Characterization

With advancements in EV science, techniques for the isolation and characterization of EVs are rapidly evolving. However, many of these newly developed approaches are poorly standardized and not widely utilized. To address these problems, the ISEV organized a worldwide survey on "Techniques used for the isolation and characterization of extracellular vesicles" in 2015 and published the results in 2016 [58]. To assess the changing state of the field, a second survey was launched in 2019, and the majority of EV researchers have realized the importance of standardizing methods for EV isolation and characterization [44]. Thus, some standardization studies on EV size and concentration measurements using tunable resistive pulse sensing (TRPS) [37], nanoparticle tracking analysis (NTA) [38], and flow cytometry (FCM) [34, 39–41] have been summarized. Among these approaches, FCM is a key technology capable of detecting, sizing, and phenotyping individual particles. However, owing to limitations of instrument sensitivity and ambiguous data interpretation, controversial results are often obtained in studies from different groups. To address these issues, a working group of EV-FCM researchers from the ISEV, International Society on Advancement of Cytometry (ISAC), and ISTH developed a consensus framework for standardized reporting of EV-FCM experiments, named MIFlowCyt-EV [15]. In addition, some leading groups have also provided an overview of recent standardization initiatives for clinical biomarker exploration of EVs by FCM [42]. Additionally, rapid advancements in characterization techniques also call for the development of reference materials for the standardization of these techniques. An expert view on choosing matched reference materials for different platforms was published in 2020 [43].

Omics profiling (proteomics, transcriptomics, lipidomics, and metabolomics) can provide more comprehensive information about EV molecular composition and has great potential in biomarker discovery. However, variations in EV isolation techniques could yield different -omics results [45, 46]. Sodar et al. summarized the instructions and guidelines for mass spectrometry-based proteomic analysis of EVs to avoid potential pitfalls during sample processing [47]. At the meantime, Barrachina et al. have provided detailed and standardized protocols for proteomic analysis of circulating EVs for cardiovascular disease [48]. Some other technical issues when performing transcriptomics, lipidomics, or metabolomics analysis on EVs have also been discussed [46, 49].

## 30.4 Guidelines for EV Clinical Research

Given that the bioactive molecules in EVs can reflect tumor biology, circulating EVs have been widely accepted as diagnostic and prognostic markers. In addition, EVs are also being actively explored as therapeutic agents owing to their compatibility and stability. Despite the rapid development of EV-based liquid biopsy and therapies [59], different EV subtypes have been enriched and analyzed depending on the specific isolation and characterization methods used, yielding discrepant results among different studies. Thus, it is essential to standardize methods for the isolation and characterization of EVs to facilitate clinical studies and applications of EVs.

In 2017, the ISEV Workshop on EV biomarkers was held in Birmingham, UK to discuss critical issues on biomarker discovery. The meeting report was published in 2018 and called for more efforts focusing on pre-analytical issues and benchmarking of isolation methods [30]. The European Society of Cardiology Working Group published a position paper to provide a set of recommendations for the analysis and translational applications of EVs in ischemic heart disease [50]. To promote the universality of standardization, De Wever and Hendrix wrote a commentary article deciphering eight pillars of a system to translate EV research into clinical applications [51]. In November 2018, the ISEV held a workshop on EV-based clinical theranostics covering the following four key topics: (1) EV isolation methodologies and standardization, (2) EV-omics analysis and biomarker discovery, (3) EV biomarker detection and applications in clinical diagnostics, and (4) EV-based therapies in human disease. Soekmadji et al. summarized this seminar in the form of a meeting report in 2020 [46]. Detailed discussions of each topic will help researchers to conduct studies using best practices.

In addition to these general guidelines, some studies have also focused on EV marker discoveries and EV-based therapeutic applications. Zhao et al. discussed key approaches for EV biomarker discovery and verification and provided recommendations for translating EV-based biomarkers into the clinical setting [52]. Moreover, Barreiro and Holthofer reviewed findings utilizing urinary EVs (uEVs) as highly versatile diagnostic tools and emphasized appropriate standardization for uEVs biomarker discovery [53]. For effective and safe translation of EV-based therapies into clinical trials, a group of members from the ISEV and the European Cooperation in Science and Technology Program of the European Union, namely, the European Network on Microvesicles and Exosomes in Health and Disease, highlighted the safety and regulatory requirements for pharmaceutical manufacturing and clinical application in 2015 [31]. Two years later, a workshop convened by the ISEV and the Society for Clinical Research and Translation of Extracellular Vesicles Singapore (SOCRATES) discussed recent advances in clinical testing and created best-practice models for the therapeutic use of EVs [32]. Specifically, mesenchymal stromal cell (MSC)-derived EVs were frequently discussed as novel therapeutic entities. Strategies and guidelines for manufacturing MSC-derived EVs were provided in some review articles [54–56]. Quantifying the therapeutic efficacy of EV preparations is essential for the development of EV-based therapeutics. Nguyen et al. discussed the challenges and implications of various in vitro and in vivo functional assays for assessing the therapeutic potential of EVs [57]. These guidelines will no doubt facilitate the clinical translation of EVs.

## References


1. Mathieu M, Martin-Jaular L, Lavieu G, et al. Specificities of secretion and uptake of exosomes and other extracellular vesicles for cell-to-cell communication. Nat Cell Biol. 2019;21(1):9–17.
2. Cocozza F, Grisard E, Martin-Jaular L, et al. SnapShot: extracellular vesicles. Cell. 2020;182(1):262.
3. Kalluri R, LeBleu VS. The biology, function, and biomedical applications of exosomes. Science. 2020;367(6478):eaau6977.
4. Melo SA, Luecke LB, Kahlert C, et al. Glypican-1 identifies cancer exosomes and detects early pancreatic cancer. Nature. 2015;523(7559):177–82.
5. Yang KS, Im H, Hong S, et al. Multiparametric plasma EV profiling facilitates diagnosis of pancreatic malignancy. Sci Transl Med. 2017;9(391):eaal3226.
6. Chen G, Huang AC, Zhang W, et al. Exosomal PD-L1 contributes to immunosuppression and is associated with anti-PD-1 response. Nature. 2018;560(7718):382–6.
7. Wiklander OPB, Brennan MA, Lotvall J, et al. Advances in therapeutic applications of extracellular vesicles. Sci Transl Med. 2019;11(492):eaav8521.
8. de Abreu RC, Fernandes H, da Costa Martins PA, et al. Native and bioengineered extracellular vesicles for cardiovascular therapeutics. Nat Rev Cardiol. 2020;17(11):685–97.
9. Reshke R, Taylor JA, Savard A, et al. Reduction of the therapeutic dose of silencing RNA by packaging it in extracellular vesicles via a pre-microRNA backbone. Nat Biomed Eng. 2020;4(1):52–68.
10. Witwer KW, Buzas EI, Bemis LT, et al. Standardization of sample collection, isolation and analysis methods in extracellular vesicle research. J Extracell Vesicles. 2013;2(1):20360.
11. Lotvall J, Hill AF, Hochberg F, et al. Minimal experimental requirements for definition of extracellular vesicles and their functions: a position statement from the International Society for Extracellular Vesicles. J Extracell Vesicles. 2014;3(1):26913.
12. Coumans FAW, Brisson AR, Buzas EI, et al. Methodological guidelines to study extracellular vesicles. Circ Res. 2017;120(10):1632–48.
13. Thery C, Witwer KW, Aikawa E, et al. Minimal information for studies of extracellular vesicles 2018 (MISEV2018): a position statement of the International Society for Extracellular Vesicles and update of the MISEV2014 guidelines. J Extracell Vesicles. 2018;7(1):1535750.
14. Nieuwland R, Falcón-Pérez JM, Théry C, et al. Rigor and standardization of extracellular vesicle research: paving the road towards robustness. J Extracell Vesicles. 2020;10(2):e12037.
15. Welsh JA, Van Der Pol E, Arkesteijn GJA, et al. MIFlowCyt-EV: a framework for standardized reporting of extracellular vesicle flow cytometry experiments. J Extracell Vesicles. 2020;9(1):1713526.
16. EV-TRACK Consortium, Van Deun J, Mestdagh P, et al. EV-TRACK: transparent reporting and centralizing knowledge in extracellular vesicle research. Nat Methods. 2017;14(3):228–32.
17. Mathivanan S, Simpson RJ. ExoCarta: a compendium of exosomal proteins and RNA. Proteomics. 2009;9(21):4997–5000.
18. Mathivanan S, Fahner CJ, Reid GE, et al. ExoCarta 2012: database of exosomal proteins, RNA and lipids. Nucleic Acids Res. 2012;40:1241–4.
19. Kim DK, Kang B, Kim OY, et al. EVpedia: an integrated database of high-throughput data for systemic analyses of extracellular vesicles. J Extracell Vesicles. 2013;2(1):20384.
20. Kim DK, Lee J, Kim SR, et al. EVpedia: a community web portal for extracellular vesicles research. Bioinformatics. 2015;31(6):933–9.
21. Kalra H, Simpson RJ, Ji H, et al. Vesiclepedia: a compendium for extracellular vesicles with continuous community annotation. PLoS Biol. 2012;10(12):e1001450.
22. Subramanian SL, Kitchen RR, Alexander R, et al. Integration of extracellular RNA profiling data using metadata, biomedical ontologies and linked data technologies. J Extracell Vesicles. 2015;4(1):27497.
23. Hill AF, Pegtel DM, Lambertz U, et al. ISEV position paper: extracellular vesicle RNA analysis and bioinformatics. J Extracell Vesicles. 2013;2(1):22859.

24. Erdbrugger U, Blijdorp CJ, Bijnsdorp IV, et al. Urinary extracellular vesicles: a position paper by the urine task force of the International Society for Extracellular Vesicles. J Extracell Vesicles. 2021;10(7):e12093.
25. Li SR, Man QW, Gao X, et al. Tissue-derived extracellular vesicles in cancers and non-cancer diseases: present and future. J Extracell Vesicles. 2021;10(14):e12175.
26. Soares RP, Xander P, Costa AO, et al. Highlights of the Sao Paulo ISEV workshop on extracellular vesicles in cross-kingdom communication. J Extracell Vesicles. 2017;6(1):1407213.
27. Pinedo M, de la Canal L, de Marcos Lousa C. A call for Rigor and standardization in plant extracellular vesicle research. J Extracell Vesicles. 2021;10(6):e12048.
28. Mateescu B, Kowal EJ, van Balkom BW, et al. Obstacles and opportunities in the functional analysis of extracellular vesicle RNA—an ISEV position paper. J Extracell Vesicles. 2017;6(1):1286095.
29. Russell AE, Sneider A, Witwer KW, et al. Biological membranes in EV biogenesis, stability, uptake, and cargo transfer: an ISEV position paper arising from the ISEV membranes and EVs workshop. J Extracell Vesicles. 2019;8(1):1684862.
30. Clayton A, Buschmann D, Byrd JB, et al. Summary of the ISEV workshop on extracellular vesicles as disease biomarkers, held in Birmingham, UK, during December 2017. J Extracell Vesicles. 2018;7(1):1473707.
31. Lener T, Gimona M, Aigner L, et al. Applying extracellular vesicles based therapeutics in clinical trials—an ISEV position paper. J Extracell Vesicles. 2015;4(1):30087.
32. Reiner AT, Witwer KW, van Balkom BWM, et al. Concise review: developing best-practice models for the therapeutic use of extracellular vesicles. Stem Cells Transl Med. 2017;6(8):1730–9.
33. Lacroix R, Judicone C, Poncelet P, et al. Impact of pre-analytical parameters on the measurement of circulating microparticles: towards standardization of protocol. J Thromb Haemost. 2012;10(3):437–46.
34. Lacroix R, Judicone C, Mooberry M, et al. Standardization of pre-analytical variables in plasma microparticle determination: results of the International Society on Thrombosis and Haemostasis SSC collaborative workshop. J Thromb Haemost. 2013;11:1190–3.
35. Yuana Y, Boing AN, Grootemaat AE, et al. Handling and storage of human body fluids for analysis of extracellular vesicles. J Extracell Vesicles. 2015;4(1):29260.
36. Clayton A, Boilard E, Buzas EI, et al. Considerations towards a roadmap for collection, handling and storage of blood extracellular vesicles. J Extracell Vesicles. 2019;8(1):1647027.
37. Vogel R, Coumans FAW, Maltesen RG, et al. A standardized method to determine the concentration of extracellular vesicles using tunable resistive pulse sensing. J Extracell Vesicles. 2016;5(1):31242.
38. Vestad B, Llorente A, Neurauter A, et al. Size and concentration analyses of extracellular vesicles by nanoparticle tracking analysis: a variation study. J Extracell Vesicles. 2017;6(1):1344087.
39. Lacroix R, Robert S, Poncelet P, et al. Standardization of platelet-derived microparticle enumeration by flow cytometry with calibrated beads: results of the International Society on Thrombosis and Haemostasis SSC collaborative workshop. J Thromb Haemost. 2010;8(11):2571–4.
40. Cointe S, Judicone C, Robert S, et al. Standardization of microparticle enumeration across different flow cytometry platforms: results of a multicenter collaborative workshop. J Thromb Haemost. 2017;15(1):187–93.
41. van der Pol E, Sturk A, van Leeuwen T, et al. Standardization of extracellular vesicle measurements by flow cytometry through vesicle diameter approximation. J Thromb Haemost. 2018;16(6):1236–45.
42. Kuiper M, van de Nes A, Nieuwland R, et al. Reliable measurements of extracellular vesicles by clinical flow cytometry. Am J Reprod Immunol. 2021;85(2):e13350.
43. Welsh JA, van der Pol E, Bettin BA, et al. Towards defining reference materials for measuring extracellular vesicle refractive index, epitope abundance, size and concentration. J Extracell Vesicles. 2020;9(1):1816641.

44. Royo F, Thery C, Falcon-Perez JM, et al. Methods for separation and characterization of extracellular vesicles: results of a worldwide survey performed by the ISEV Rigor and standardization subcommittee. Cells. 2020;9(9):1955.
45. Van Deun J, Mestdagh P, Sormunen R, et al. The impact of disparate isolation methods for extracellular vesicles on downstream RNA profiling. J Extracell Vesicles. 2014;3(1):24858.
46. Soekmadji C, Li B, Huang Y, et al. The future of extracellular vesicles as theranostics—an ISEV meeting report. J Extracell Vesicles. 2020;9(1):1809766.
47. Sódar BW, Kovács Á, Visnovitz T, et al. Best practice of identification and proteomic analysis of extracellular vesicles in human health and disease. Expert Rev Proteomics. 2017;14(12):1073–90.
48. Barrachina MN, Calderon-Cruz B, Fernandez-Rocca L, et al. Application of extracellular vesicles proteomics to cardiovascular disease: guidelines, data analysis, and future perspectives. Proteomics. 2019;19(1–2):e1800247.
49. Gandham S, Su X, Wood J, et al. Technologies and standardization in research on extracellular vesicles. Trends Biotechnol. 2020;38(10):1066–98.
50. Sluijter JPG, Davidson SM, Boulanger CM, et al. Extracellular vesicles in diagnostics and therapy of the ischaemic heart: position paper from the working group on cellular biology of the heart of the European Society of Cardiology. Cardiovasc Res. 2018;114(1):19–34.
51. De Wever O, Hendrix A. A supporting ecosystem to mature extracellular vesicles into clinical application. EMBO J. 2019;38(9):e101412.
52. Zhao Z, Fan J, Hsu YS, et al. Extracellular vesicles as cancer liquid biopsies: from discovery, validation, to clinical application. Lab Chip. 2019;19(7):1114–40.
53. Barreiro K, Holthofer H. Urinary extracellular vesicles. A promising shortcut to novel biomarker discoveries. Cell Tissue Res. 2017;369(1):217–27.
54. Rohde E, Pachler K, Gimona M. Manufacturing and characterization of extracellular vesicles from umbilical cord-derived mesenchymal stromal cells for clinical testing. Cytotherapy. 2019;21(6):581–92.
55. Witwer KW, Van Balkom BWM, Bruno S, et al. Defining mesenchymal stromal cell (MSC)-derived small extracellular vesicles for therapeutic applications. J Extracell Vesicles. 2019;8(1):1609206.
56. Borger V, Weiss DJ, Anderson JD, et al. International Society for Extracellular Vesicles and International Society for Cell and Gene Therapy statement on extracellular vesicles from mesenchymal stromal cells and other cells: considerations for potential therapeutic agents to suppress coronavirus disease-19. Cytotherapy. 2020;22(9):482–5.
57. Nguyen VVT, Witwer KW, Verhaar MC, et al. Functional assays to assess the therapeutic potential of extracellular vesicles. J Extracell Vesicles. 2020;10(1):e12033.
58. Gardiner C, Di Vizio D, Sahoo S, et al. Techniques used for the isolation and characterization of extracellular vesicles: results of a worldwide survey. J Extracell Vesicles. 2016;5(1):32945.
59. Min L, Wang B, Bao H, et al. Advanced nanotechnologies for extracellular vesicle-based liquid biopsy. Adv Sci (Weinh). 2021;8(20):e2102789.

# Introduction to Online Nucleic Acid Databases for EV Research

# 31

Jia Hu and Shenglin Huang

## 31.1 Introduction

Extracellular vesicles (EVs) are membranous vesicles that are released by both prokaryotic and eukaryotic cells into the extracellular microenvironment and contain biologically active cargoes, such as nucleic acids, proteins, lipids, and metabolites [1–3]. Based on their biogenesis, EVs can be classified into exosomes, microvesicles (MVs), and apoptotic bodies, among which exosomes and MVs are considered the most common ones [4–6]. Exosomes, smaller than 150 nm in diameter, are released as multivesicular endosomes fused with the cell surface, whereas MVs, measuring 50–1000 nm in size, are formed by the outward budding and fission of the plasma membrane, and are subsequently released into the extracellular space [1, 2, 7, 8]. After secretion, EVs are transported to the proximal environment or distant target cells through a variety of human biological fluids, resulting in the attachment to specific types of target cells [9–11].

Many studies have shown that EVs can eliminate metabolites and participate in intercellular communication by facilitating the exchange of biomolecules between cells [12–14]. Furthermore, EVs are involved in various biological process, including gene regulation, angiogenesis, and immune evasion [2, 15], and are implicated in diverse diseases, such as cancers, cardiometabolic diseases, neurodegenerative diseases, and infectious diseases [9, 16, 17]. Based on these properties and functions, EVs are considered potential diagnostic biomarkers for various diseases [18, 19].

As a subset of EV cargo, RNA species can be categorized as mRNAs, microRNAs (miRNAs), long noncoding RNAs (lncRNAs), circular RNAs (circRNAs), and

J. Hu · S. Huang (✉)
Fudan University Shanghai Cancer Center and Institutes of Biomedical Sciences, Fudan University, Shanghai, China
e-mail: slhuang@fudan.edu.cn



Q. Wang, L. Zheng (eds.), *Extracellular Vesicles*,
https://doi.org/10.1007/978-981-99-8365-0_31

other small RNAs [20]. Recently, an increasing number of studies have evaluated the roles and function of EV RNAs [16, 21–23]. Notably, EV RNAs can reflect the origins and functional state of parent cells based on their highly variable constitution [24, 25]. On the other hand, EV RNAs can be transferred from the original cells to recipient cells, where the proteins encoded by EV mRNAs and noncoding RNAs (ncRNAs) can modulate gene expression or elicit changes in function [26, 27]. In addition, EV RNAs have recently attracted much attention as noninvasive biomarkers of diseases, prognostic indicators, and even therapeutic agents [16, 28].

In this chapter, we summarize several web-accessible databases focusing on EV RNAs, including ExoCarta, miRandola 2017, exoRBase 2.0, EVAtlas, Vesiclepedia 2019, exRNA Atlas, deepBase v3.0, and EVLncRNAs v2.0. These databases can help researchers to query and browse information related to the characteristics and functions of EV RNAs.

## 31.2 Online EV Nucleic Acid Databases

### 31.2.1 ExoCarta

ExoCarta [29] (http://www.exocarta.org/) is a manually curated and freely available web-based database of exosomal proteins, RNA, and lipids identified in exosomes from ten different types of organisms. Currently, ExoCarta contains 41,860 proteins, greater than 7540 RNAs, and 1116 lipids from more than 286 exosomal studies, representing an increase of over twofold in available data compared with the previous release in 2012. Furthermore, ExoCarta has some additional features, as follows: (1) serving as a useful tool to identify candidate exosomal biomarkers and mutant oncogenic proteins; (2) implementing annotation standards proposed by the International Society of Extracellular Vesicles (ISEV) as the minimal requirements for exosomal studies; (3) providing users with dynamic protein-protein interaction networks and biological pathways for exosomal proteins; and (4) allowing users to perform additional functional enrichment and interaction network analyses using Funrich (http://www.funrich.org). Owing to its increasing popularity, ExoCarta may become a widely used resource for exosomal cargo and aid biologists in understanding the biogenesis, secretion, and properties of EVs.

### 31.2.2 miRandola 2017

miRandola 2017 [30] (http://mirandola.iit.cnr.it/) is a manually curated online database of extracellular ncRNAs from 314 EV articles in 14 organisms. The current updated version of miRandola hosts miRNAs, lncRNAs, and circRNAs and the associations of ncRNAs with 25 drugs, 47 sample types, and 197 diseases. The goal of miRandola 2017 is to collect updated information on extracellular ncRNAs from different studies and databases to improve our understanding of the relevance of ncRNAs as noninvasive biomarkers. Additionally, miRandola 2017 includes the following new features: (1) reference to miRNA data from ExoCarta and Vesiclepedia 2019 databases; (2) presentation of user-friendly web interfaces with new graphical visualization and additional functionality to display RNA-disease co-occurrences; (3) classification of extracellular RNAs (exRNAs) into seven types according to their extracellular form; and (4) providing users with the link to the NIH exrna.org Research Portal.

### 31.2.3 exoRBase 2.0

exoRBase 2.0 [31] (http://www.exorbase.org/), an updated version of exoRBase in 2021, is a repository of EV long RNAs (exLRs) derived from RNA-seq data analyses of different human body biofluids. In exoRBase 2.0, there are a total of 19,643 mRNAs, 15,645 lncRNAs, and 79,084 circRNAs, which were obtained from 905 human blood, urine, cerebrospinal fluid (CSF), and bile samples. Compared with the previous version, exoRBase 2.0 has the following new characteristics: (1) addition of 16 urine samples, 5 CSF samples, and 15 bile samples and increasing the number of blood samples from 92 to 869; (2) providing a systematic view of EV transcriptomes in 4 different human biofluids derived from 13 types of cancers; (3) supporting pathway-level functional analysis to interpret EV mRNA expression data and further explore biological mechanisms; and (4) helping researchers to assess the heterogeneity of circulating EV origins in diverse physiological and pathological conditions. exoRBase 2.0 is the only online, free platform that focuses on the identification of novel exLR signatures from human biofluids, and is expected to further aid in the exploration and discovery of new circulating biomarkers for disease diagnosis and treatment.

### 31.2.4 EVAtlas

EVAtlas [32] (Extracellular Vesicle Atlas, http://bioinfo.life.hust.edu.cn/EVAtlas/#/), an updated version of EVmiRNA database, currently contains the expression profiles of seven types of ncRNAs, including miRNAs, Piwi-interacting RNAs (piRNAs), rRNAs, tRNAs, YRNAs, small nuclear RNAs (snRNAs), and small nucleolar RNAs (snoRNAs), in more than 2000 EV samples from 24 human tissues/organs and 40 diseases. Compared with EVmiRNA, EVAtlas has the following new features: (1) in terms of ncRNA types, piRNAs, rRNAs, tRNAs, YRNAs, snRNAs, and snoRNAs, not just miRNAs, are included; (2) users can explore the specific expression profiles of ncRNAs, miRNA targets, miRNA-related drugs, miRNA-regulated pathways, miRNA functions, and related publications; and (3) some improved functionality, allowing users to search for relevant information by inputting the ncRNA symbol directly or browsing the three functional modules (sample, expression, and specific expression modules). EVAtlas is a comprehensive database focusing on ncRNA expression profiles and potential regulatory information in EVs.

### 31.2.5 Vesiclepedia 2019

Vesiclepedia 2019 [33] (http://www.microvesicles.org/) is a free online database of RNAs, proteins, lipids, and metabolites that have been identified in EVs from published and unpublished studies. The current version of Vesiclepedia contains 1254 EV studies, 38,146 RNA entries, 349,988 protein entries, and 639 lipids/metabolites. Additionally, several improvements in Vesiclepedia 2019 have been made to enhance data features and functionality, as follows: (1) supporting users to browse and download detailed gene/molecule information for EVs; (2) enabling users to submit their EV data and participate in data annotation by the biomedical community; (3) providing the biophysical/molecular properties of EVs, experiment details, and the EV-METRIC to help users conduct downstream analysis; and (4) facilitating Vesiclepedia 2019 data analysis and free download through the FunRich Vesiclepedia plugin. This database integrates the top 100 proteins that are most often reported in EV studies and helps researchers to compare their respective EV data with these proteins.

### 31.2.6 exRNA Atlas

The exRNA Atlas [34] (http://exrna-atlas.org/), a free web-accessible database, aims to collect exRNA data and develop an exRNA-mediated mapping of intercellular communication. The current version (4P1) of the exRNA Atlas hosts 2270 exRNA-seq and 3039 exRNA quantitative polymerase chain rection (qPCR) profiles from various human biofluids, collected across 19 different studies. Notably, the additional new features in the exRNA Atlas are as follows: (1) adopting a set of computational deconvolutions and visualization tools, the former being used for variation analysis based on exRNA-seq profiles; (2) revealing six exRNA cargo types (CT1, CT2, CT3A, CT3B, CT3C, and CT4), including both vesicular and nonvesicular carriers, each of which is detectable in different human biofluids (serum, plasma, CSF, saliva, and urine); (3) supporting to download these exRNA data from the NIH Extracellular RNA Communication Consortium (ERCC); and (4) allowing users to analyze exRNA-seq profiles from the Atlas and encouraging them to contribute their data to the Atlas. exRNA Atlas is the first database to develop a map of exRNA cell-cell communication networks in humans.

### 31.2.7 deepBase v3.0

deepBase v3.0 [35] (http://rna.sysu.edu.cn/deepbase3/index.html), a freely accessible web-based database, reveals the expression, evolution, and functions of multiple ncRNAs expressed in tissue, tumor, and exosome samples. The current version of deepBase (version 3.0) integrates 2549 exRNA-seq data sets from the ERCC database, including 81,692 mRNA entries, 81,229 lncRNA entries, 3396 miRNA entries, and 918 circRNA entries. Additionally, deepBase v3.0 has the following features: (1) providing the mRNA/lncRNA expression profiles in reads per kilobase million format and the miRNA/circRNA expression maps in reads per million format; (2) supporting the queries and further to browse expression information based on exRNA types or exRNA IDs; and (3) allowing users to download detailed exRNA expression/variance information in PNG, JPEG, PDF, and SVG formats. The database will aid academic users in exploring and identifying noninvasive ncRNA biomarkers.

### 31.2.8 EVLncRNAs v2.0

EVLncRNAs v2.0 [36] (https://www.sdklab-biophysics-dzu.net/EVLncRNAs2/) is a manually curated and publicly available database that integrates experimentally validated functional lncRNAs (EVlncRNAs). The current version of EVLncRNAs (version 2.0) contains 4010 lncRNAs (33 exosomal lncRNAs), 188 circRNAs, 124 species, 1082 diseases, 6244 interactions, 37 peptide-coding lncRNAs, 8 structural lncRNAs, and 1079 drug/chemoresistance/stress lncRNAs. Additionally, the updates of EVLncRNAs v2.0 include the following features: (1) enabling users to browse detailed exosomal lncRNA annotation information related to structure, function, and disease; (2) allowing users to directly link to four new bioinformatics tools for more relevant knowledge; and (3) providing visualizations of interaction networks, which show the underlying relationships among lncRNAs, biomolecules, and diseases. With its fast expansion, EVLncRNAs v2.0 may serve as a comprehensive lncRNA database to help researchers better understand lncRNAs in various cancers.

### 31.2.9 Comparison of the Above Databases

In this chapter, we introduced online databases that focus on different types of RNAs expressed in EVs. However, the applications and detailed expression information of different databases remain various substantially. Based on RNA subtypes, we can categorize these databases into five subclasses (mRNAs, miRNAs, lncRNAs, circRNAs, and other RNAs). For example, users can search EV mRNAs and obtain related expression information via ExoCarta, exoRBase 2.0, Vesiclepedia 2019, and deepBase v3.0. Moreover, information on miRNAs identified in EVs is included in ExoCarta, miRandola 2017, EVAtlas, Vesiclepedia 2019, exRNA Atlas, and deepBase v3.0, and these databases provide different formats for miRNA expression data. Databases containing lncRNA information include miRandola 2017, exoRBase 2.0, deepBase v3.0, and EVLncRNAs v2.0, whereas miRandola 2017, exoRBase 2.0, and deepBase v3.0 include circRNA-related information. Information on other types of RNAs found in EVs, such as tRNAs, YRNAs, piRNAs, snoRNAs, and snRNAs, is accessible in EVAtlas and exRNA Atlas.

By comparing these databases, it is worth noting that ExoCarta, exoRBase, Vesiclepedia 2019, and EVLncRNAs v2.0 encourage users to upload their data and results, as well as to participate in community annotation. Additionally, we summarized the following features and compared similarities and differences among these databases, including the contents, statistics, program applications, and other important information in Table 31.1.

## 31.3 Applications of Online Nucleic Acid Databases

### 31.3.1 mRNAs

mRNAs, protein-coding RNAs, comprise three basic components: the 5′-untranslated region (UTR), the coding sequence (CDS), and the 3′-UTR [37]. mRNAs carry codes from the DNA in the nucleus and move to the cytoplasm, where they are translated into corresponding proteins via the actions of ribosomes [38]. Many mRNAs residing in EVs can serve as templates and contribute to protein expression in recipient cells [10, 16, 22]. In this section, we summarize the features of several databases that show EV mRNA expression profiles.

ExoCarta currently hosts 4946 mRNAs detected in exosomes isolated from more than ten different species. Users can search mRNAs using the "browse" option and then choose the mRNA of interest (e.g., *PDXP*) from the "Browse Results" feature. On the information page, ExoCarta displays gene information and experiment details related to *PDXP*, the latter of which meets ISEV standards.

exoRBase 2.0 documents 19,643 mRNAs from blood, urine, CSF, and bile samples and provides users more than one way to search/browse mRNAs. For example, users can type the gene symbols of specific mRNAs into the "Search" bar on the "Home" (or Search) page, yielding general expression information in all samples on the "Search result" interface; subsequently, detailed information is available by clicking the gene symbol. Furthermore, exoRBase 2.0 enables users to browse mRNAs by clicking "mRNA" on the "Home" page or "mRNA and lncRNA" on the "Browse" page; and then, detailed information can be obtained by clicking the gene symbol on the "mRNA and lncRNA" interface. On the "Detail" interface, users can acquire more detailed information for relevant mRNAs, including summary information, expression profiles, and comparisons among different datasets.

Vesiclepedia 2019 contains 27,646 mRNAs collected in EVs from 41 organisms and 172 sample sources. Using the "Browse" module, users can scan mRNA evidence, choose the mRNA of interest (e.g., *HSPG2*) from the "Browse Results" feature, and obtain specific mRNA information. In addition, users can download and analyze robust mRNA data by importing the results into FunRich.

deepBase v3.0 comprises 81,692 mRNA entries from various samples. Here, we provide a description of *SAMD5-002* obtained using deepBase v3.0. First, we selected "Exosomal mRNA" on the "Home" page. Subsequently, we found listed mRNA information, including mRNA name, gene symbol, and sample type. Then, we chose the mRNA of interest (e.g., *SAMD5-002)*, yielding detailed expression information, such as the expression/variance values of *SAMD5-002* enriched in

**Table 31.1** Comparison of online nucleic acid databases

| Databases | Websites | Contents | Statistics | Community annotation | Programmatic use | Other information |
|---|---|---|---|---|---|---|
| ExoCarta | http://www.exocarta.org/ | Proteins, RNAs, and lipids | 286 studies, 4946 mRNAs, 2838 miRNAs, 41,860 proteins, 1116 lipids, 10 species | Yes | Python | Dynamic PPI networks and biological pathways |
| miRandola 2017 | http://mirandola.iit.cnr.it/ | miRNA, lncRNA, and circRNA | 7 exRNA forms, 1002 miRNAs, 12 lncRNAs, 8 circRNAs, 25 drugs, 47 sample types, and 197 diseases | No | MySQL | Integration of Vesiclepedia miRNA data |
| exoRBase 2.0 | http://www.exorbase.org/ | mRNA, lncRNA, and circRNA | 905 samples, 19,643 mRNAs, 15,645 lncRNAs, 79,084 circRNAs, 11,536 pathways, and 39 tissue/cell origins | Yes | MySQL | The only online platform for the identification of novel exLR signatures |
| EVAtlas | http://bioinfo.life.hust.edu.cn/EVAtlas/#/ | ncRNA (miRNA, piRNA, rRNA, tRNA, YRNA, snRNA, and snoRNA) | 2030 samples, 24 human tissues, 1506 exosomes and micro-vesicles | No | Python | The first database depositing the most types of ncRNAs |
| Vesiclepedia 2019 | http://www.microvesicles.org/ | RNAs, proteins, lipids, and metabolites | 1254 studies, 27,646mRNA,10,520 miRNA, 349,988 proteins, 639 lipid, 41 species | Yes | Python | FunRich-based Vesiclepedia plugin |
| exRNA Atlas | http://exrna-atlas.org/ | exRNA (miRNA, tRNA, piRNA, snoRNA, snRNA) | 5309 exRNA-seq and exRNA qPCR profiles | No | API | Computational deconvolution |
| deepBase v3.0 | http://rna.sysu.edu.cn/deepbase3/index.html | mRNA, miRNA, lncRNA, and circRNA | 81,692 mRNA entries, 81,229 lncRNA entries, 3396 miRNA entries, and 918 circRNA entries | No | MySQL | The detailed exRNA expression/variance information |

(continued)

**Table 31.1** (continued)

| Databases | Websites | Contents | Statistics | Community annotation | Programmatic use | Other information |
|---|---|---|---|---|---|---|
| EVLncRNAs v2.0 | https://www.sdklab-biophysics-dzu.net/EVLncRNAs2/ | lncRNA | 33 lncRNAs | Yes | MySQL | The visualization of molecules-diseases interaction networks |

different samples. Additionally, the expression information for each mRNA can be downloaded in image or PDF format.

### 31.3.2 MiRNAs

MiRNAs, small ncRNAs 21–23 nt in length, regulate gene expression at the post-transcriptional level by binding to mRNAs and inhibiting their translation into proteins [39]. As post-transcriptional regulators, miRNAs are the most extensively studied molecules in EVs [16, 40, 41]. Thus, comprehensive investigations of miRNA expression profiles in EVs will be helpful to explore their functions and prognostic potential as biomarkers [20, 42]. In this section, we summarize how these databases to present EV miRNA expression profiles.

ExoCarta currently hosts 2838 miRNAs detected in exosomes isolated from more than ten different species. Users can search for detailed miRNA information by entering the miRNA ID (e.g., *let-7a*) in the "Query" module, or clicking "Browse" and "miRNA evidence" respectively. Moreover, users can download miRNA data to import into FunRich for additional enrichment analysis, including analyses of cellular components, molecular functions, biological processes, biological pathways, protein domains, sites of expression, transcriptional factors, clinical phenotypes, context-specific miRNA analysis, and miRNA fold enrichment.

miRandola 2017 includes 1002 miRNAs collected from 314 articles. Users can quickly search for the miRNA of interest (e.g., *hsa-let-7-5p*) by typing its name in the search bar on the "Home" page, and can also browse miRNAs by clicking the "browse" and "miRNA" buttons in sequence. The information for *hsa-let-7-5p*, including the sample source, relevant experiment description, and PubMed ID, will be displayed. Moreover, users can link to Vesiclepedia database for more information on *hsa-let-7-5p*.

EVAtlas provides 3018 miRNA expression profile datasets of EVs from 24 tissues/diseases. In this section, we discuss miRNAs derived from bone marrow exosomes as an example. First, we can search for miRNAs from the bone marrow by clicking "Samples" and "Bone marrow" respectively. EVAtlas then summarizes the expression information for these miRNAs and provides the corresponding Project ID as well as information on related diseases on the "Bone marrow exosome and micro-vesicle" page. Clicking the project ID "SRP046046" can obtain more information on lymphoma, including the mapping distributions of these seven types of RNAs and a heatmap (top 50) of each RNA. Furthermore, the average expression differences between case and control groups will also be displayed in detail. In addition, we can learn how to isolate exosomes from various samples/tissues.

Vesiclepedia 2019 contains 10,520 miRNAs collected in EVs from 41 organisms and 172 sample sources. To view information for specific miRNAs, we take *hsa-let-7a-3p* as an example to perform the following operations. First, users can enter "*hsa-let-7a-3p*" in the "miRNA ID" box in the "Query" module and then click the "Browse" option to get the detailed information. On the "*let-7a*" summary page, users can obtain the related origin information of *hsa-let-7a-3p* that it was derived from CSF and serum, and can also click the link to PubMed to get detailed experimental descriptions. In addition, the page also facilitates user to download the miRNA dataset "Vesiclepedia_miRNA_details.txt," which contains 10,520 miRNAs in the "Download" module. Notably, users can import the miRNA data into FunRich for Venn diagram and enrichment analyses.

exRNA Atlas hosts 5309 exRNA-seq and exRNA qPCR sample profiles, primarily from five human biofluids (serum, plasma, CSF, saliva, and urine), collected across 19 different studies. Users can analyze exRNA-seq profiles from exRNA Atlas, process their own exRNA-seq data, as well as further contribute their data and analysis results to exRNA Atlas, thus creating a cycle of knowledge creation. Moreover, users can utilize the ncRNA search bar in the upper right corner of the "Home" page to search specific miRNAs of interest.

deepBase v3.0 contains 3396 miRNA entries from different biofluids. The expression information of each miRNA is displayed in the chart. For example, the expression information for *hsa-miR-1295b-5p* can be assessed using deepBase v3.0. First, users can click "Exosomes" and "RNA type: miRNA," yielding information on specific miRNAs. The miRNA of interest (e.g., *hsa-miR-1295b-5p*) can then be chosen, and users will be given relevant information, which can be downloaded in PNG, JPEG, PDF, and SVG formats.

### 31.3.3 LncRNAs

LncRNAs, ncRNAs greater than 200 nt in length, are involved in various biological processes, including chromatin organization, gene transcription, mRNA turnover, protein translation, and macromolecular complex assembly [16, 23, 43]. Various lncRNAs have been found in EVs [21, 44]; however, the regulatory mechanisms mediating the abundances of lncRNAs are still unknown under specific physiological and pathological conditions [45]. In this section, we summarize several databases showing the expression profiles of lncRNAs in EVs.

miRandola 2017 includes 12 lncRNAs collected from 314 articles. An example of the information available in this database is *hnf1a-as1*. First, users are allowed to directly search for the lncRNA of interest via the "Search" option and then click on the specific term to browse the results based on the extracellular form of the lncRNA. Next, on the "*hnf1a-as1*" summary page, users can get the related information, including organism and exRNA form information.

exoRBase 2.0 contains 15,645 lncRNAs from 13 types of cancer samples. To obtain information on lncRNAs, users can directly input the gene symbol(s) of one or more lncRNAs into the "Search" box on the "Home/Search" page and then obtain relevant expression information. Owing to the shared browsing modules of mRNAs and lncRNAs, the method for browsing lncRNAs is similar to that of mRNAs. Notably, on the "mRNA and lncRNA" interface, users can choose to browse tissue-specific expression of mRNAs and/or lncRNAs via the "Gene type" and "Tissue specificity" features, browse whether mRNAs/lncRNAs are expressed in single or multiple human biofluids via the "Sample types" button, and select frequently expressed mRNAs and/or lncRNAs in all samples or certain cancer types using the "Detection frequency" and "Different group" boxes. Then, detailed information on the mRNA/lncRNA can be accessed by clicking "Gene symbol."

deepBase v3.0 hosts 81,229 lncRNA entries collected from different samples. We present *RP11-789C17-5-001* as an example to illustrate how to query lncRNAs via deepBase v3.0. First, users can access the "RNA type" page and choose lncRNA by clicking on "Exosomes," where the listed lncRNA information is displayed. Next, users can choose one or multiple lncRNAs (e.g., *RP11-789C17-5-001*) based on their needs and research interests, and the expression information will be displayed in chart form.

EVLncRNAs v2.0 currently contains 33 exosomal lncRNAs, which are used to establish lncRNA/disease interaction networks. Here, we demonstrate how to query a specific lncRNA (e.g., *BCAR4*) and related information. First, we can view the listed lncRNA information using the "Exsomal lncRNAs" module, which includes the names of diseases and interaction targets related to the lncRNA. Then, the lncRNA of interest can be chosen to access detailed information on the lncRNA using the "Detail" option. On the "Detail" page, we can browse basic information, diseases, and interactions associated with *BCAR4*. Additionally, the interaction networks can be downloaded to a local computer.

### 31.3.4 CircRNAs

CircRNAs, a highly heterogeneous class of noncoding regulatory RNAs, form a unique head-to-tail back splicing structure by covalently linking the 3′ end of the RNA to the 5′ end [46–48]. CircRNAs are relatively enriched and stable in cancer EVs, suggesting EV-circRNAs as potential cancer biomarkers [49]. In addition, despite studies on the effects of EV circRNAs on cellular processes and cell fate, the extracellular functions and mechanisms of EV circRNAs are still unclear. In this section, we summarize several databases to show the circRNA expression profiles in EVs.

miRandola 2017 includes 8 circRNAs collected from 314 articles. The circRNA *DNAJC6* can be used as an example of how to use miRandola 2017. First, users can input *DNAJC6* in the search box on the "Home" page and browse to obtain detailed information on *DNAJC6*. From the description page for *DNAJC6*, users can learn that *DNAJC6* is expressed in *Homo sapiens* and may be used as a potential biomarker.

exoRBase 2.0 contains 79,084 circRNAs from 905 blood, urine, CSF, and bile samples of 13 types of cancers. Users can query information on circRNAs by typing a gene symbol or multiple gene symbols into the search bar on the "Home/Search" page, obtain important expression information in all samples, and further click "circID/circBase ID" to obtain detailed information on a specific circRNA. Moreover, users can browse circRNAs using the "exLRs-circRNA" feature on the "Home" page or the "circRNA" button on the "Browse" page. Subsequently, the "circRNA" interface will be displayed, and selection of "circRNA" enables users to browse the circRNA of interest or all circRNAs. Notably, on the "circRNA" page, the "Gene symbol" and "Region" options enable users to screen novel or annotated circRNAs detected in specific genome regions.

deepBase v3.0 contains 918 circRNA entries from different biofluids. Here, we query the expression of a representative circRNA (*hsa_circ_001433*) to demonstrate the application of deepBase v3.0. First, users can access listed information for circRNAs by clicking the "Exosomes" module and circRNA type. On the information page, users can view all circRNAs in tabular form and choose the circRNA of interest (e.g., *hsa_circ_001433*). Then, users can obtain information in the form of a diagram, which allows to make simple comparisons of different expression values as well as further to download corresponding data.

### 31.3.5 Other Nucleic Acids

Other EV ncRNAs, including tRNAs, piRNAs, snoRNAs, and snRNAs, can be found in exRNA Atlas and EVAtlas, and the latter also contains YRNAs. exRNA Atlas is composed of 5309 exRNA-seq and exRNA qPCR profiles collected from 19 different studies on 5 human biofluids (serum, plasma, CSF, saliva, and urine) and enables users to search for specific ncRNAs of interest using the ncRNA search bar, yielding more detailed information. Additionally, as an update to EVmiRNA, EVAtlas contains an additional 1953 snoRNAs, 23,031 piRNAs, 1779 snRNAs, 6594 rRNAs, 432 tRNAs, and 1168 YRNAs in more than 2000 EV samples and provides more basic expression information as well as information on miRNA-related drugs/pathways.

## 31.4 Conclusions and Perspectives

Recent studies have reported major advancements in our understanding of the biological functions and potential applications of EVs. Owing to their capacity to exchange components between cells, EVs participate in various physiological and pathological processes, such as neuronal communication, tumorigenesis, and immune evasion [2, 50]. EVs have also been shown to have potential applications as noninvasive biomarkers in clinical practice [51–53]. Therefore, with the increased interest in EVs, researchers have paid much attention to related EV databases, including ExoCarta, exoRBase, miRandola 2017, EVAtlas, Vesiclepedia 2019, exRNA Atlas, deepBase v3.0, and EVLncRNAs v2.0; these databases have collected and categorized EV expression information from published and unpublished studies. Notably, these databases are web-based and freely accessible without registration, providing users with friendly interfaces and various presentation formats. With the expansion of EV sequencing data, the databases will be continuously updated with additional EV RNA expression profiles. In this chapter, we discussed the main functions and basic features of different databases to aid academic researchers with choosing the appropriate databases to use based on their research interests. We recommend that users browse various database websites for more detailed information. We believe that these databases will be useful resources in for studies of biomarker discovery and the regulatory mechanisms of EVs.

## References


1. van Niel G, D'Angelo G, Raposo G. Shedding light on the cell biology of extracellular vesicles. Nat Rev Mol Cell Biol. 2018;19(4):213–28.
2. Yáñez-Mó M, Siljander PR, Andreu Z, Zavec AB, Borràs FE, Buzas EI, et al. Biological properties of extracellular vesicles and their physiological functions. J Extracell Vesicles. 2015;4:27066.
3. Kalra H, Drummen GP, Mathivanan S. Focus on extracellular vesicles: introducing the next small big thing. Int J Mol Sci. 2016;17(2):170.
4. Théry C, Witwer KW, Aikawa E, Alcaraz MJ, Anderson JD, Andriantsitohaina R, et al. Minimal information for studies of extracellular vesicles 2018 (MISEV2018): a position statement of the International Society for Extracellular Vesicles and update of the MISEV2014 guidelines. J Extracell Vesicles. 2018;7(1):1535750.
5. Raposo G, Stoorvogel W. Extracellular vesicles: exosomes, microvesicles, and friends. J Cell Biol. 2013;200(4):373–83.
6. Xu R, Rai A, Chen M, Suwakulsiri W, Greening DW, Simpson RJ. Extracellular vesicles in cancer—implications for future improvements in cancer care. Nat Rev Clin Oncol. 2018;15(10):617–38.
7. Harding C, Heuser J, Stahl P. Endocytosis and intracellular processing of transferrin and colloidal gold-transferrin in rat reticulocytes: demonstration of a pathway for receptor shedding. Eur J Cell Biol. 1984;35(2):256–63.
8. Tricarico C, Clancy J, D'Souza-Schorey C. Biology and biogenesis of shed microvesicles. Small GTPases. 2017;8(4):220–32.
9. Shah R, Patel T, Freedman JE. Circulating extracellular vesicles in human disease. N Engl J Med. 2018;379(10):958–66.
10. Valadi H, Ekström K, Bossios A, Sjöstrand M, Lee JJ, Lötvall JO. Exosome-mediated transfer of mRNAs and microRNAs is a novel mechanism of genetic exchange between cells. Nat Cell Biol. 2007;9(6):654–9.
11. Wang X, Zhang H, Yang H, Bai M, Ning T, Deng T, et al. Exosome-delivered circRNA promotes glycolysis to induce chemoresistance through the miR-122-PKM2 axis in colorectal cancer. Mol Oncol. 2020;14(3):539–55.
12. Tkach M, Théry C. Communication by extracellular vesicles: where we are and where we need to go. Cell. 2016;164(6):1226–32.
13. Johnstone RM, Adam M, Hammond JR, Orr L, Turbide C. Vesicle formation during reticulocyte maturation. Association of plasma membrane activities with released vesicles (exosomes). J Biol Chem. 1987;262(19):9412–20.
14. Lee Y, El Andaloussi S, Wood MJ. Exosomes and microvesicles: extracellular vesicles for genetic information transfer and gene therapy. Hum Mol Genet. 2012;21(R1):R125–34.
15. Conigliaro A, Costa V, Lo Dico A, Saieva L, Buccheri S, Dieli F, et al. CD90+ liver cancer cells modulate endothelial cell phenotype through the release of exosomes containing H19 lncRNA. Mol Cancer. 2015;14:155.
16. Kim KM, Abdelmohsen K, Mustapic M, Kapogiannis D, Gorospe M. RNA in extracellular vesicles. Wiley Interdiscip Rev RNA. 2017;8(4).
17. Morales-Kastresana A, Musich TA, Welsh JA, Telford W, Demberg T, Wood JCS, et al. High-fidelity detection and sorting of nanoscale vesicles in viral disease and cancer. J Extracell Vesicles. 2019;8(1):1597603.
18. Boukouris S, Mathivanan S. Exosomes in bodily fluids are a highly stable resource of disease biomarkers. Proteomics Clin Appl. 2015;9(3–4):358–67.

19. Mathivanan S, Ji H, Simpson RJ. Exosomes: extracellular organelles important in intercellular communication. J Proteomics. 2010;73(10):1907–20.
20. Skog J, Würdinger T, van Rijn S, Meijer DH, Gainche L, Sena-Esteves M, et al. Glioblastoma microvesicles transport RNA and proteins that promote tumour growth and provide diagnostic biomarkers. Nat Cell Biol. 2008;10(12):1470–6.
21. Ahadi A, Brennan S, Kennedy PJ, Hutvagner G, Tran N. Long non-coding RNAs harboring miRNA seed regions are enriched in prostate cancer exosomes. Sci Rep. 2016;6:24922.
22. Mitsuhashi M, Taub DD, Kapogiannis D, Eitan E, Zukley L, Mattson MP, et al. Aging enhances release of exosomal cytokine mRNAs by Aβ1-42-stimulated macrophages. FASEB J. 2013;27(12):5141–50.
23. Kung JT, Colognori D, Lee JT. Long noncoding RNAs: past, present, and future. Genetics. 2013;193(3):651–69.
24. Wen SW, Lima LG, Lobb RJ, Norris EL, Hastie ML, Krumeich S, et al. Breast cancer-derived exosomes reflect the cell-of-origin phenotype. Proteomics. 2019;19(8):e1800180.
25. Chevillet JR, Kang Q, Ruf IK, Briggs HA, Vojtech LN, Hughes SM, et al. Quantitative and stoichiometric analysis of the microRNA content of exosomes. Proc Natl Acad Sci U S A. 2014;111(41):14888–93.
26. Choi D, Montermini L, Kim DK, Meehan B, Roth FP, Rak J. The impact of oncogenic EGFRvIII on the proteome of extracellular vesicles released from glioblastoma cells. Mol Cell Proteomics. 2018;17(10):1948–64.
27. Cossetti C, Iraci N, Mercer TR, Leonardi T, Alpi E, Drago D, et al. Extracellular vesicles from neural stem cells transfer IFN-γ via Ifngr1 to activate Stat1 signaling in target cells. Mol Cell. 2014;56(2):193–204.
28. Li Z, Zhu X, Huang S. Extracellular vesicle long non-coding RNAs and circular RNAs: biology, functions and applications in cancer. Cancer Lett. 2020;489:111–20.
29. Keerthikumar S, Chisanga D, Ariyaratne D, Al Saffar H, Anand S, Zhao K, et al. ExoCarta: a web-based compendium of exosomal cargo. J Mol Biol. 2016;428(4):688–92.
30. Russo F, Di Bella S, Vannini F, Berti G, Scoyni F, Cook HV, et al. miRandola 2017: a curated knowledge base of non-invasive biomarkers. Nucleic Acids Res. 2018;46(D1):D354–9.
31. Lai H, Li Y, Zhang H, Hu J, Liao J, Su Y, et al. exoRBase 2.0: an atlas of mRNA, lncRNA and circRNA in extracellular vesicles from human biofluids. Nucleic Acids Res. 2022;50:D118. https://doi.org/10.1093/nar/gkab1085.
32. Liu T, Zhang Q, Zhang J, Li C, Miao YR, Lei Q, et al. EVmiRNA: a database of miRNA profiling in extracellular vesicles. Nucleic Acids Res. 2019;47(D1):D89–93.
33. Pathan M, Fonseka P, Chitti SV, Kang T, Sanwlani R, Van Deun J, et al. Vesiclepedia 2019: a compendium of RNA, proteins, lipids and metabolites in extracellular vesicles. Nucleic Acids Res. 2019;47(D1):D516–D9.
34. Murillo OD, Thistlethwaite W, Rozowsky J, Subramanian SL, Lucero R, Shah N, et al. exRNA atlas analysis reveals distinct extracellular RNA cargo types and their carriers present across human biofluids. Cell. 2019;177(2):463–477.e15.
35. Xie F, Liu S, Wang J, Xuan J, Zhang X, Qu L, et al. deepBase v3.0: expression atlas and interactive analysis of ncRNAs from thousands of deep-sequencing data. Nucleic Acids Res. 2021;49(D1):D877–83.
36. Zhou B, Ji B, Liu K, Hu G, Wang F, Chen Q, et al. EVLncRNAs 2.0: an updated database of manually curated functional long non-coding RNAs validated by low-throughput experiments. Nucleic Acids Res. 2020;49(D1):D86–91.
37. Chen S, Zhu X, Huang S. Clinical applications of extracellular vesicle long RNAs. Crit Rev Clin Lab Sci. 2020;57(8):508–21.
38. Black DL. Mechanisms of alternative pre-messenger RNA splicing. Annu Rev Biochem. 2003;72:291–336.
39. Bartel DP. MicroRNAs: target recognition and regulatory functions. Cell. 2009;136(2):215–33.
40. Bronisz A, Godlewski J, Chiocca EA. Extracellular vesicles and MicroRNAs: their role in tumorigenicity and therapy for brain tumors. Cell Mol Neurobiol. 2016;36(3):361–76.

41. Iftikhar H, Carney GE. Evidence and potential in vivo functions for biofluid miRNAs: from expression profiling to functional testing: potential roles of extracellular miRNAs as indicators of physiological change and as agents of intercellular information exchange. Bioessays. 2016;38(4):367–78.
42. Hagiwara S, Kantharidis P, Cooper ME. MicroRNA as biomarkers and regulator of cardiovascular development and disease. Curr Pharm Des. 2014;20(14):2347–70.
43. Ng SY, Johnson R, Stanton LW. Human long non-coding RNAs promote pluripotency and neuronal differentiation by association with chromatin modifiers and transcription factors. EMBO J. 2012;31(3):522–33.
44. Hinger SA, Cha DJ, Franklin JL, Higginbotham JN, Dou Y, Ping J, et al. Diverse long RNAs are differentially sorted into extracellular vesicles secreted by colorectal cancer cells. Cell Rep. 2018;25(3):715–725.e4.
45. Gezer U, Özgür E, Cetinkaya M, Isin M, Dalay N. Long non-coding RNAs with low expression levels in cells are enriched in secreted exosomes. Cell Biol Int. 2014;38(9):1076–9.
46. Salzman J, Chen RE, Olsen MN, Wang PL, Brown PO. Cell-type specific features of circular RNA expression. PLoS Genet. 2013;9(9):e1003777.
47. Rybak-Wolf A, Stottmeister C, Glažar P, Jens M, Pino N, Giusti S, et al. Circular RNAs in the mammalian brain are highly abundant, conserved, and dynamically expressed. Mol Cell. 2015;58(5):870–85.
48. Jeck WR, Sharpless NE. Detecting and characterizing circular RNAs. Nat Biotechnol. 2014;32(5):453–61.
49. Du WW, Yang W, Liu E, Yang Z, Dhaliwal P, Yang BB. Foxo3 circular RNA retards cell cycle progression via forming ternary complexes with p21 and CDK2. Nucleic Acids Res. 2016;44(6):2846–58.
50. Lo Cicero A, Stahl PD, Raposo G. Extracellular vesicles shuffling intercellular messages: for good or for bad. Curr Opin Cell Biol. 2015;35:69–77.
51. Jia Y, Chen Y, Wang Q, Jayasinghe U, Luo X, Wei Q, et al. Exosome: emerging biomarker in breast cancer. Oncotarget. 2017;8(25):41717–33.
52. Szajnik M, Derbis M, Lach M, Patalas P, Michalak M, Drzewiecka H, et al. Exosomes in plasma of patients with ovarian carcinoma: potential biomarkers of tumor progression and response to therapy. Gynecol Obstet (Sunnyvale). 2013;Suppl 4:3.
53. Nair S, Tang KD, Kenny L, Punyadeera C. Salivary exosomes as potential biomarkers in cancer. Oral Oncol. 2018;84:31–40.

# Introduction to Extracellular Vesicles in Systems Biology

# 32

Gong Zhang and Tong Wang

## 32.1 Introduction

Cells exhibit crosstalk with each other, and molecules encapsulated by EVs can act as important messengers. Understanding and deciphering this EV language is critical for improving our understanding of the roles of EVs in biological processes and diseases. In the last 20 years, multiple -omics technologies have been developed, resulting in major advances in EV studies. These advancements have required holistic approaches to interpret the complex information carried by EVs.

Upon publication of the first human genome sequence in 2001 [1], there was renewed interest in systems biology, which was first introduced in the field of cybernetics by Norbert Weiner in 1965 [2]. In 2002, Kitano et al. proposed four principle properties for investigating certain systems in organisms or cells [2], and this framework is still widely accepted [3]. These properties are as follows: (1) systems structures based on interactions, pathways, and the flow of genetic information formed or regulated by genes and gene products; (2) temporal dynamic changes in these systems; (3) potential targets or pathways for modulating or controlling systems shifts, particularly in the context of disease states; and (4) strategies for precise modification or construction of a system. As reviewed by numerous groups,

G. Zhang (✉)
MOE Key Laboratory of Tumor Molecular Biology, and Key Laboratory of Functional Protein Research of Guangdong Higher Education Institutes, Institute of Life and Health Engineering, the First Affiliated Hospital, Jinan University, Guangzhou, Guangdong, China
e-mail: zhanggong@jnu.edu.cn; tongwang@email.jnu.edu.cn

T. Wang (✉)
The First Affiliated Hospital, Institute of Life and Health Engineering, Jinan University, Guangzhou, China

The First Hospital of Changsha, Changsha, China
e-mail: zhanggong@jnu.edu.cn; tongwang@email.jnu.edu.cn



Q. Wang, L. Zheng (eds.), *Extracellular Vesicles*,
https://doi.org/10.1007/978-981-99-8365-0_32

resolution of such systems properties in different biological fields is advancing rapidly [3–9]. In this chapter, we discuss the findings and future directions of EVs from the perspective of systems biology.

## 32.2 EV Proteome Databases

In Chap. 31, databases of EV nucleic acids were introduced. Here, we focus on data on EV proteomes in database resources. To facilitate the identification and characterization of all molecules carried by EVs, various databases have been built to collect information on EV cargo. To date, more than 30 EV databases can be found via search engines.

ExoCarta was the first large-scale database of EV contents. Established in 2009, this database is still one of the most widely used EV databases in the field and is associated with two other databases, Vesiclepedia and exRNA Atlas. Indeed, the homepages of these three databases offer direct links to the other two databases. However, ExoCarta has not been updated since December 2016, whereas Vesiclepedia reproduces the infrastructure of ExoCarta and contains many more studies and entries. Vesiclepedia and ExoCarta provide simple lists of mRNAs, microRNAs (miRNAs), and proteins in various types of exosomes, including those from humans and other species. Although no quantitative information is included, these databases provide external links to Gene Ontology (GO) annotations for RNAs and proteins via FunRich (http://www.funrich.org/), which can be more easily applied than other online GO analysis tools, such as PATHER-DB and AMIGO. In addition, the urinary exosome protein database, which contains only two studies and 1160 proteins, was established in 2009.

Unlike the other generalized databases, such as NCBI, which has been widely used as a reference, the quality and reliability of exosome databases still need to be improved. Therefore, when using such databases, researchers should note the limitations of these databases and consider the following aspects:

1. **Limited number of studies included**. Most users access databases to obtain a global view of EV contents. However, only ExoCarta/Vesiclepedia and EV-TRACK databases contain more than 50 studies. Thus, the other databases may be not sufficiently representative owing to the limited data sources. Such limited data may be still useful if the users need support for the existence or expression of specific molecules in public databases. However, the data may be misleading if global profiling and data mining are major aims.
2. **Existence is not informative**. As the number of datasets is increased, more RNAs and proteins are found in at least one study/dataset. Until its last update in 2016, ExoCarta showed that there were 9769 proteins expressed in exosomes, indicating that half of the human proteome can be found in EVs. Vesiclepedia version 4.1 (last updated August 15, 2018) contains 349,988 protein entries and includes 12,799 human proteins, covering 64.4% of the human proteome. Therefore, one can assume that all proteins and RNAs may be found in exo-

somes. That is, the existence of a protein in EVs is not necessarily relevant. Quantification of expression in various diseases/tissues/cell types may be more important.

3. **Data heterogeneity**. Some of these databases collect data from multiple studies, and each study uses different experimental and computational methods. Therefore, researchers should be cautious when making comparative conclusions and should always keep in mind that direct comparisons of different datasets or across databases should be carried out with appropriate adjustments considering these quantitative deviations.
4. **Data quality control needs to be standardized**. The false-positive rates of many studies are not properly controlled and have not been systematically evaluated or validated. For example, the urinary exosome protein database controls the protein false discovery rate (FDR) to 1.91%, which is higher than the commonly accepted threshold of 1%, especially according to the Human Proteome Project (HPP) Guidelines [10, 11].
5. **Update frequency**. The field of EVs is expanding rapidly, and a considerable number of publications contained experimental -omics data. Unfortunately, the majority of EV databases stopped being updated before 2019. Thus, additional governmental financial support or industrial investment may be required for the construction and maintenance of these valuable EV databases.

## 32.3 Quality Control of EV Proteomics Data

The details of technologies used for EV proteomics analyses were introduced in Chap. 7. Reaching definitive conclusions from high-quality data requires that EV proteome studies overcome two major obstacles. First, the purity of EV isolates can be questionable. In Chap. 6, we presented a discussion of the criteria for determining EV purity and methods for obtaining highly pure EV samples. Overall, ultracentrifugation (UC), particularly density gradient UC, is one of the most accepted methods for the isolation of high-purity EVs. Filtration- or precipitation-based methods suffer from either severe protein contamination or introduction of exogenous chemical materials. By contrast, size-exclusion chromatography (SEC) offers an important additive-free choice to bypass the need for major instruments while isolating EVs with high purity comparable to that obtained with UC [12, 13].

Although absolute purity cannot be achieved using any currently available method, sample type and analytical strategy should be considered when balancing the purity, yield, and laboratory conditions. For example, when carrying out next-generation sequencing (NGS), protein contamination will have less influence on EV-omics than on proteomics, glycomics, or lipodomics. Indeed, Buschmann et al. compared five EV isolation methods in the context of miRNA NGS profiling and found that seemingly nonpure EVs isolated by precipitation and membrane affinity methods were sufficient for miRNA-based biomarker discovery, whereas SEC failed to differentiate healthy individuals from patients [14]. When profiling miRNAs, long noncoding RNA, and mRNAs (RNomics), Yang et al. recently proposed that SEC is superior to precipitation-based methods [15].

However, when dealing with proteins, data quality control becomes far more stringent and complicated in EV studies [16]. Numerous EV proteomics studies have employed shotgun mass spectrometry (MS), which relies on database searches for peptide and protein identification. Such strategies are not sequencing-based and only yield identification lists and probability scores for each item. This may result in major differences in protein identification when the criteria and search parameters vary. For example, Wilhelm et al. reported that 18,097 proteins could be identified in the human proteome [17]. This result, together with a published study by Kim et al. [18], initiated major discussions and debates on MS data quality control [19]. In 2015, Savitski et al. re-analyzed the results from the study by Wilhelm et al. while applying a protein level FDR of less than 0.01, and only 14,714 of the original 18,097 proteins were retained in this re-analysis [20]. In 2016, the HPP released the HPP Mass Spectrometry Data Interpretation Guidelines 2.1, which proposed useful criteria for reporting confident protein identifications, including using a peptide length greater than or equal to 9 amino acids, a protein level FDR less than 1% in database searches, and reporting of greater than or equal to 2 unique peptides for each protein identification [10]. Such guidelines have been widely adopted by proteomics organizations, and in 2019, the HPP Mass Spectrometry Data Interpretation Guidelines 3.0 was released with important updates through years of real-world applications [11]. In the field of EVs, Guo et al. [21] published the first human cell EV proteome using SW620 colorectal cancer (CRC) cells based on the criteria described in the HPP Guideline 2.1 [10]. In this study, UC was used for EV (exosome) isolation, and 1801 total proteins and 313 phosphoproteins were confidently identified in SW620 cell-derived exosomes; among these proteins, 202 new phosphosites were detected (Fig. 32.1a) [21]. Interestingly, 6.4% of phosphosites in SW620 exosomes involved tyrosine phosphorylation (Fig. 32.1b), whereas only 0.6% of phosphosites in whole SW620 cells involved tyrosine phosphorylation

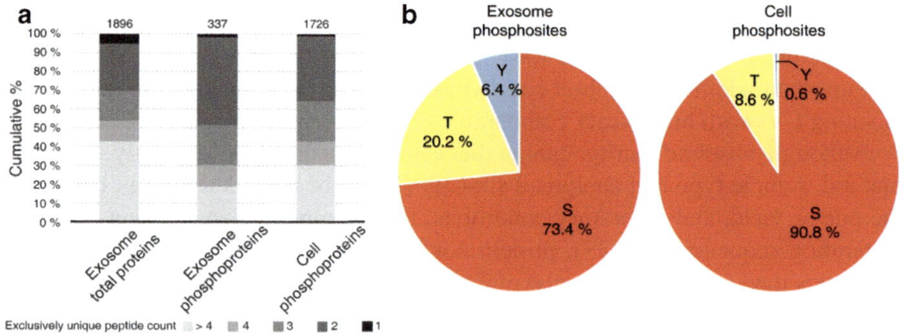


**Fig. 32.1** Proteome and phosphoproteome of SW620 human colorectal cancer cells. (**a**) Total proteins and phosphoproteins identified from exosomes and cells (peptide length ≥ 9 amino acids and protein level FDR < 1%). (**b**, **c**) Phosphosite distributions in exosomes (**b**) and cells (**c**). Y = tyrosine, S = serine, and T = threonine. (Reproduced from ref. [21])

(Fig. 32.1c). These analyses addressed important questions regarding how many proteins and phosphoproteins can be found in human cell-derived EVs under stringent MS criteria.

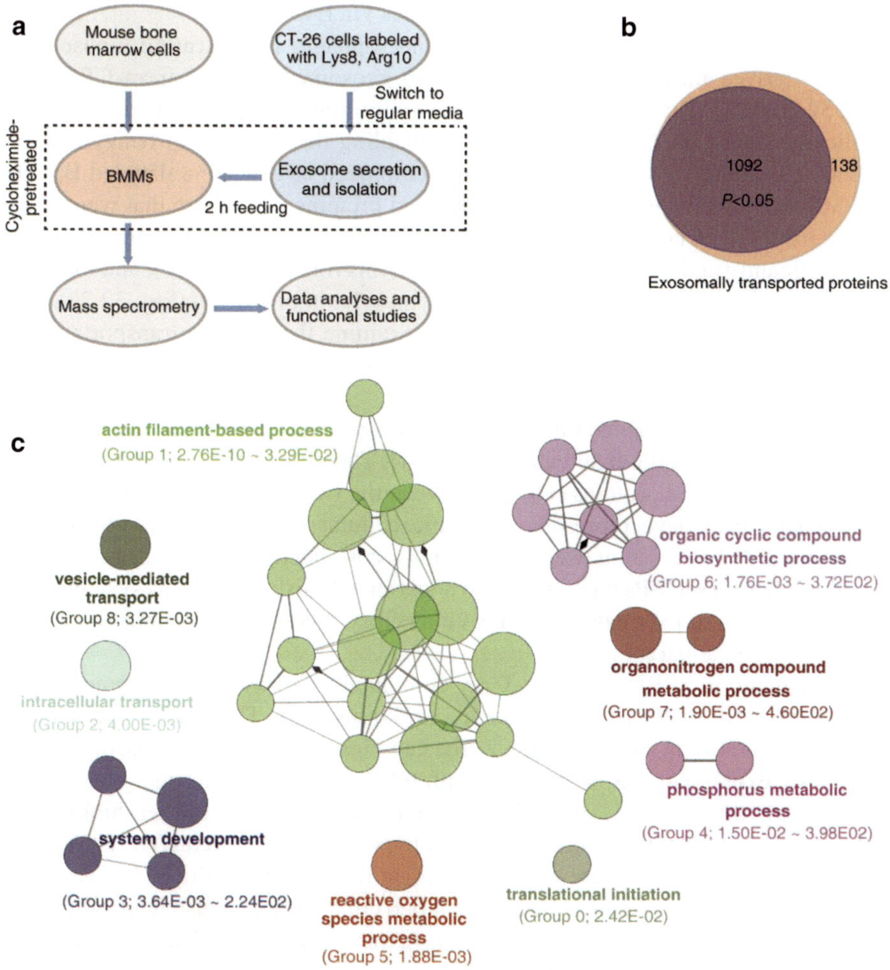


**Fig. 32.2** Tracing the proteome from cancer cells to EVs. (**a**) Metabolic labeling-based proteome tracing with quantification information. CT-26 mouse cells were metabolically labeled with heavy isotopic amino acids, and CT-26 EVs labeled with these heavy isotopic amino acids were harvested and used to educate bone marrow-derived macrophages (BMMs). Mass spectrometry was used to detect isotopic peptides in these educated BMMs from CT-26 EVs. (**b**) Statistical analysis of EV proteins taken up by BMMs. (**c**) Gene Ontology analysis of EV proteins taken up by BMMs. (Reproduced from ref. [23])

## 32.4 Data Integration of EV-Omics

In cells, there is flow of genetic information from DNA to RNA to proteins; this represents the central dogma of molecular biology. However, there is no such flow in EVs. Data integration of EV omics data requires deeper strategic considerations of different scientific questions. For example, in the cancer microenvironment, EVs regulate various cancer phenotypes, including proliferation, angiogenesis, metastasis, and drug resistance, as reviewed in a previous publication [22]. This suggests that cells intentionally export certain molecules via EVs.

Chen et al. modeled tumor-associated macrophages by educating mouse bone marrow-derived macrophages (BMMs) with exosomes secreted from CRC cells (Fig. 32.2a) [23]. They used heavy isotopic amino acids containing medium to metabolically label CT-26 mouse CRC cells, enabling the isolation of heavy isotopic amino acid-labeled EVs. These EVs were employed to educate unlabeled BMMs, and MS analyses were performed. The labeled proteins were those that were transported from CT-26 cells to BMMs via EVs. They found that 1095 EV proteins could be significantly and stably transported through this route (Fig. 32.2b), and these EV proteins were related to regulation of the cytoskeleton of BMMs (Fig. 32.2c). This analysis established a feasible approach to examine the systems transportation of EV proteins from donor to recipient cells, connecting -omics analyses from cells to EVs based on a systems biology perspective.

In another study, Zhang et al. performed MS analysis comparing the proteomes of motile and nonmotile liver cancer cells [24]. They found approximately 450 differentially expressed proteins (DEPs) between the two types of cells. Interestingly, these DEPs may have represented specific cellular subsets because the relative abundances of EV proteins were significantly correlated with cellular proteins. In addition, these EV DEPs were related to sugar metabolism units. This systems analysis suggested that liver cancer cell EVs may carry specific protein units that can be traced back to the parental cells.

With regard to miRNA studies, Gray et al. analyzed EVs secreted from cardiac progenitor cells (CPCs) under hypoxic conditions [25]. Their study employed a subtle computational model to analyze the cue-signal-response relationships of differentially expressed miRNA clusters with tube formation or fibrotic gene expression. Using this strategy, 11 miRNAs were identified and further biologically validated as potential therapeutic targets in hypoxic CPC-derived EVs. They found that EVs from hypoxic CPCs improved cardiac function and reduced fibrosis.

In addition to experimental studies, in silico computation is also helpful for mining the systems rules in EVs. Gezsi et al. reviewed detailed advances in the proteomic, transcriptomic, glycomic, lipidomic, metabolomic, and genomic analyses of EVs [26]. Furthermore, they focused on the correlations between EV biogenesis and disease using three computational models. In Model 1, known gene/phenotype correlations were constructed. Model 2 was used to predict unknown gene/phenotype connections via various pathways. In Model 3, pathways from other species were further included for predicting gene/phenotypes. With these comprehensive analyses, connections between key EV biogenesis-related proteins and numerous diseases were emphasized [26].

## 32.5 Final Remarks

Reductionism in biology is important to build pathways and functional blocks by probing the roles of specific molecules. In both physiological and disease states, EVs tend to function as whole systems rather than single molecules. The systems biology view is relevant to EV research because it can provide a holistic picture of EVs released from different cell types and disease states. Network and spatial-temporal transition rules can help to identify novel targets to precisely interfere with, modulate, or produce EVs in different biological and medical scenarios. Thus, the study of EVs in the context of systems biology is a promising direction that will co-evolve with advances in -omics technologies.

## References

1. Venter JC, Adams MD, Myers EW, Li PW, Mural RJ, Sutton GG, et al. The sequence of the human genome. Science. 2001;291(5507):1304–51.
2. Kitano H. Systems biology: a brief overview. Science. 2002;295(5560):1662–4.
3. Bruggeman FJ, Westerhoff HV. The nature of systems biology. Trends Microbiol. 2007;15(1):45–50.
4. Veenstra TD. Omics in systems biology: current progress and future outlook. Proteomics. 2021;21(3–4):e2000235.
5. Ma W, Chen LS, Ozbek U, Han SW, Lin C, Paulovich AG, et al. Integrative proteo-genomic analysis to construct CNA-protein regulatory map in breast and ovarian tumors. Mol Cell Proteomics. 2019;18(8 suppl 1):S66–81.
6. Ebata K, Yamashiro S, Iida K, Okada M. Building patient-specific models for receptor tyrosine kinase signaling networks. FEBS J. 2022;289(1):90–101.
7. Zimmermann M, Patil KR, Typas A, Maier L. Towards a mechanistic understanding of reciprocal drug-microbiome interactions. Mol Syst Biol. 2021;17(3):e10116.
8. Joshi A, Rienks M, Theofilatos K, Mayr M. Systems biology in cardiovascular disease: a multiomics approach. Nat Rev Cardiol. 2020;18:313.
9. Dhillon BK, Smith M, Baghela A, Lee AHY, Hancock REW. Systems biology approaches to understanding the human immune system. Front Immunol. 2020;11:1683.
10. Deutsch EW, Overall CM, Van Eyk JE, Baker MS, Paik YK, Weintraub ST, et al. Human proteome project mass spectrometry data interpretation guidelines 2.1. J Proteome Res. 2016;15(11):3961–70.
11. Deutsch EW, Lane L, Overall CM, Bandeira N, Baker MS, Pineau C, et al. Human proteome project mass spectrometry data interpretation guidelines 3.0. J Proteome Res. 2019;18(12):4108–16.
12. Thery C, Witwer KW, Aikawa E, Alcaraz MJ, Anderson JD, Andriantsitohaina R, et al. Minimal information for studies of extracellular vesicles 2018 (MISEV2018): a position statement of the International Society for Extracellular Vesicles and update of the MISEV2014 guidelines. J Extracell Vesicles. 2018;7(1):1535750.
13. Xu R, Greening DW, Zhu HJ, Takahashi N, Simpson RJ. Extracellular vesicle isolation and characterization: toward clinical application. J Clin Invest. 2016;126(4):1152–62.

14. Buschmann D, Kirchner B, Hermann S, Marte M, Wurmser C, Brandes F, et al. Evaluation of serum extracellular vesicle isolation methods for profiling miRNAs by next-generation sequencing. J Extracell Vesicles. 2018;7(1):1481321.
15. Yang Y, Wang Y, Wei S, Zhou C, Yu J, Wang G, et al. Extracellular vesicles isolated by size-exclusion chromatography present suitability for RNomics analysis in plasma. J Transl Med. 2021;19(1):104.
16. Rosa-Fernandes L, Rocha VB, Carregari VC, Urbani A, Palmisano G. A perspective on extracellular vesicles proteomics. Front Chem. 2017;5:102.
17. Wilhelm M, Schlegl J, Hahne H, Gholami AM, Lieberenz M, Savitski MM, et al. Mass-spectrometry-based draft of the human proteome. Nature. 2014;509(7502):582–7.
18. Kim MS, Pinto SM, Getnet D, Nirujogi RS, Manda SS, Chaerkady R, et al. A draft map of the human proteome. Nature. 2014;509(7502):575–81.
19. Omenn GS, Lane L, Lundberg EK, Beavis RC, Nesvizhskii AI, Deutsch EW. Metrics for the human proteome project 2015: progress on the human proteome and guidelines for high-confidence protein identification. J Proteome Res. 2015;14(9):3452–60.
20. Savitski MM, Wilhelm M, Hahne H, Kuster B, Bantscheff M. A scalable approach for protein false discovery rate estimation in large proteomic data sets. Mol Cell Proteomics. 2015;14(9):2394–404.
21. Guo J, Cui Y, Yan Z, Luo Y, Zhang W, Deng S, et al. Phosphoproteome characterization of human colorectal cancer SW620 cell-derived exosomes and new phosphosite discovery for C-HPP. J Proteome Res. 2016;15(11):4060–72.
22. Maacha S, Bhat AA, Jimenez L, Raza A, Haris M, Uddin S, et al. Extracellular vesicles-mediated intercellular communication: roles in the tumor microenvironment and anti-cancer drug resistance. Mol Cancer. 2019;18(1):55.
23. Chen Z, Yang L, Cui Y, Zhou Y, Yin X, Guo J, et al. Cytoskeleton-centric protein transportation by exosomes transforms tumor-favorable macrophages. Oncotarget. 2016;7(41):67387–402.
24. Zhang J, Lu S, Zhou Y, Meng K, Chen Z, Cui Y, et al. Motile hepatocellular carcinoma cells preferentially secret sugar metabolism regulatory proteins via exosomes. Proteomics. 2017;17(13–14).
25. Gray WD, French KM, Ghosh-Choudhary S, Maxwell JT, Brown ME, Platt MO, et al. Identification of therapeutic covariant microRNA clusters in hypoxia-treated cardiac progenitor cell exosomes using systems biology. Circ Res. 2015;116(2):255–63.
26. Gezsi A, Kovacs A, Visnovitz T, Buzas EI. Systems biology approaches to investigating the roles of extracellular vesicles in human diseases. Exp Mol Med. 2019;51(3):1–11.

GPSR Compliance

*The European Union's (EU) General Product Safety Regulation (GPSR) is a set of rules that requires consumer products to be safe and our obligations to ensure this.*

*If you have any concerns about our products, you can contact us on ProductSafety@springernature.com*

In case Publisher is established outside the EU, the EU authorized representative is:

Springer Nature Customer Service Center GmbH
Europaplatz 3
69115 Heidelberg, Germany

**Batch number: 10732064**

Printed by Printforce, the Netherlands